ID0984946

PHYSICOCHEMICAL & ENVIRONMENTAL

Plant
Physiology

SECOND EDITION

Park S. Nobel

Department of Organismic Biology, Ecology, and Evolution
University of California, Los Angeles
Los Angeles, California

 ACADEMIC PRESS

San Diego   London   Boston   New York   Sydney   Tokyo   Toronto

This book is printed on acid-free paper. ∞

Copyright © 1999, 1991, 1983, 1974, 1970 by ACADEMIC PRESS

All Rights Reserved.
No part of this publication may be reproduced or transmitted in any form or by any
means, electronic or mechanical, including photocopy, recording, or any information
storage and retrieval system, without permission in writing from the publisher.

Academic Press
a division of Harcourt Brace & Company
525 B Street, Suite 1900, San Diego, California 92101-4495, USA
http://www.apnet.com

Academic Press
24-28 Oval Road, London NW1 7DX, UK
http://www.hbuk.co.uk/ap/

Library of Congress Catalog Card Number: 98-88525

International Standard Book Number: 0-12-520025-0

PRINTED IN THE UNITED STATES OF AMERICA
99 00 01 02 03 04 MM 9 8 7 6 5 4 3 2 1

CONTENTS

Preface   xi
Symbols and Abbreviations   xiii
Representative Principal Equations   xxi

1. Cells and Diffusion   1

I. Cell Structure   1
A. Generalized Plant Cell / B. Leaf Anatomy / C. Vascular Tissue / D. Root Anatomy

II. Diffusion   8
A. Fick's First Law / B. Continuity Equation and Fick's Second Law / C. Time–Distance Relation for Diffusion

III. Membrane Structure   16
A. Membrane Models / B. Organelle Membranes

IV. Membrane Permeability   20
A. Concentration Difference across a Membrane / B. Permeability Coefficient / C. Diffusion and Cellular Concentration

V. Cell Walls   25
A. Chemistry and Morphology / B. Diffusion across Cell Walls / C. Stress–Strain Relations of Cell Walls

Problems   32
References   34

2. Water   36

I. Physical Properties   37
A. Hydrogen Bonding—Thermal Relations / B. Surface Tension / C. Capillary Rise / D. Capillary Rise in the Xylem / E. Tensile Strength, Viscosity / F. Electrical Properties

II. Chemical Potential   45
A. Free Energy and Chemical Potential / B. Analysis of Chemical Potential / C. Standard State / D. Hydrostatic Pressure / E. Water

Activity and Osmotic Pressure / F. The Van't Hoff Relation / G. Matric
Pressure / H. Water Potential

III. Central Vacuole and Chloroplasts   59
A. Water Relations of the Central Vacuole / B. Boyle–Van't Hoff
Relation / C. Osmotic Responses of Chloroplasts

IV. Water Potential and Plant Cells   64
A. Incipient Plasmolysis / B. Höfler Diagram and Pressure–Volume
Curve / C. Chemical Potential and Water Potential of Water Vapor /
D. Plant–Air Interface / E. Pressure in the Cell Wall Water / F. Water
Flux / G. Cell Growth / H. Kinetics of Volume Changes

Problems   78
References   80

3. Solutes   **81**

I. Chemical Potential of Ions   82
A. Electrical Potential / B. Electroneutrality and Membrane Capaci-
tance / C. Activity Coefficients of Ions / D. Nernst Potential /
E. Example of E_{N_K}

II. Fluxes and Diffusion Potentials   90
A. Flux and Mobility / B. Diffusion Potential in a Solution / C. Mem-
brane Fluxes / D. Membrane Diffusion Potential—Goldman Equation / E. Application of the Goldman Equation / F. Donnan Potential

III. Characteristics of Crossing Membranes   104
A. Electrogenicity / B. Boltzmann Energy Distribution and Q_{10}, a
Temperature Coefficient / C. Activation Energy and Arrhenius Plots /
D. Ussing–Teorell Equation / E. Example of Active Transport /
F. Energy for Active Transport / G. Speculation on Active Transport

IV. Mechanisms for Crossing Membranes   116
A. Carriers, Porters, Channels, and Pumps / B. Michaelis–Menten
Formalism / C. Facilitated Diffusion

V. Principles of Irreversible Thermodynamics   123
A. Fluxes, Forces, and Onsager Coefficients / B. Water and Solute
Flow / C. Flux Densities, L_P, and σ / D. Values of Reflection Coeffi-
cients

VI. Solute Movement across Membranes   131
A. The Influence of Reflection Coefficients on Incipient Plasmolysis /
B. Extension of the Boyle–Van't Hoff Relation / C. Reflection Coef-
ficients of Chloroplasts / D. Solute Flux Density

Problems   137
References   139

4. Light   **142**

I. Wavelength and Energy   144
A. Light Waves / B. Energy of Light / C. Illumination, Photon Flux
Density, and Irradiance / D. Sunlight / E. Planck's and Wien's Formulae

II. Absorption of Light by Molecules   153
   A. Role of Electrons in Absorption Event / B. Electron Spin and State
   Multiplicity / C. Molecular Orbitals / D. Photoisomerization / E. Light
   Absorption by Chlorophyll

III. Deexcitation   161
   A. Fluorescence, Radiationless Transition, and Phosphorescence /
   B. Competing Pathways for Deexcitation / C. Lifetimes / D. Quantum
   Yields

IV. Absorption Spectra and Action Spectra   166
   A. Vibrational Sublevels / B. The Franck–Condon Principle / C. Ab-
   sorption Bands and Absorption Coefficients / D. Application of Beer's
   Law / E. Conjugation / F. Action Spectra / G. Absorption and Action
   Spectra of Phytochrome
   Problems   179
   References   181

5. Photochemistry of Photosynthesis   183

I. Chlorophyll—Chemistry and Spectra   185
   A. Types and Structures / B. Absorption and Fluorescence Emission
   Spectra / C. Absorption *in Vivo*—Polarized Light

II. Other Photosynthetic Pigments   191
   A. Carotenoids / B. Phycobilins

III. Excitation Transfers among Photosynthetic Pigments   196
   A. Pigments and the Photochemical Reaction / B. Resonance Trans-
   fer of Excitation / C. Transfers of Excitation between Photosynthetic
   Pigments / D. Excitation Trapping

IV. Groupings of Photosynthetic Pigments   202
   A. Photosynthetic Units / B. Excitation Processing / C. Photosynthetic
   Action Spectra and Enhancement Effects / D. Two Photosystems Plus
   Light-Harvesting Antennae

V. Electron Flow   208
   A. Electron Flow Model / B. Components of the Electron Transfer
   Pathway / C. Types of Electron Flow / D. Photophosphorylation /
   E. Vectorial Aspects of Electron Flow
   Problems   218
   References   220

6. Bioenergetics   222

I. Gibbs Free Energy   223
   A. Chemical Reactions and Equilibrium Constants / B. Interconversion of Chemical and Electrical Energy / C. Redox Potentials

II. Biological Energy Currencies   230
   A. ATP—Structure and Reactions / B. Gibbs Free Energy Change for
   ATP Formation / C. $NADP^+$–NADPH Redox Couple

III. Chloroplast Bioenergetics   236
A. Redox Couples / B. H^+ Chemical Potential Differences Caused by
Electron Flow / C. Evidence for Chemiosmotic Hypothesis /
D. Coupling of Flows

IV. Mitochondrial Bioenergetics   244
A. Electron Flow Components—Redox Potentials / B. Oxidative Phosphorylation

V. Energy Flow in the Biosphere   249
A. Incident Light—Stefan–Boltzmann Law / B. Absorbed Light and
Photosynthetic Efficiency / C. Food Chains and Material Cycles

Problems   253
References   255

7. Temperature—Energy Budgets   **257**

I. Energy Budget—Radiation   258
A. Solar Irradiation / B. Absorbed Infrared Irradiation / C. Emitted
Infrared Radiation / D. Values for a, a_{IR}, and e_{IR} / E. Net Radiation /
F. Examples for Radiation Terms

II. Wind—Heat Conduction and Convection   269
A. Wind—General Comments / B. Air Boundary Layers / C. Boundary Layers for Bluff Bodies / D. Heat Conduction/Convection Equations / E. Dimensionless Numbers / F. Examples of Heat Conduction /Convection

III. Latent Heat—Transpiration   279
A. Heat Flux Density Accompanying Transpiration / B. Heat Flux
Density for Dew or Frost Formation / C. Examples of Frost and Dew
Formation

IV. Further Examples of Energy Budgets   282
A. Leaf Shape and Orientation / B. Shaded Leaves within Plant Communities / C. Heat Storage / D. Time Constants

V. Soil   286
A. Thermal Properties / B. Soil Energy Balance / C. Variations in Soil
Temperature

Problems   290
References   291

8. Leaves and Fluxes   **293**

I. Resistances and Conductances—Transpiration   294
A. Boundary Layer Adjacent to Leaf / B. Stomata / C. Stomatal Con-
ductance and Resistance / D. Cuticle / E. Intercellular Air Spaces /
F. Fick's First Law and Conductances

II. Water Vapor Fluxes Accompanying Transpiration   306
A. Conductance and Resistance Network / B. Values of Conductances / C. Effective Lengths and Resistance / D. Water Vapor

Concentrations and Mole Fractions for Leaves / E. Examples of Water Vapor Levels in a Leaf / F. Water Vapor Fluxes / G. Control of Transpiration

III. CO_2 Conductances and Resistances   315
A. Resistance and Conductance Network / B. Mesophyll Area / C. Resistance Formulation for Cell Components / D. Partition Coefficient for CO_2 / E. Cell Wall Resistance / F. Plasma Membrane Resistance / G. Cytosol Resistance / H. Mesophyll Resistance / I. Chloroplast Resistance

IV. CO_2 Fluxes Accompanying Photosynthesis   324
A. Photosynthesis / B. Respiration and Photorespiration / C. Comprehensive CO_2 Resistance Network / D. Compensation Points / E. Fluxes of CO_2 / F. CO_2 Conductances / G. Range in Photosynthetic Rates / H. Environmental Productivity Index

V. Water-Use Efficiency   337
A. Values of WUE / B. Elevational Effects on WUE / C. Stomatal Control of WUE / D. C_3 versus C_4 Plants

Problems   345
References   347

9. Plants and Fluxes   350

I. Gas Fluxes above the Leaf Canopy   351
A. Wind Speed Profiles / B. Flux Densities / C. Eddy Diffusion Coefficients / D. Resistance of Air above the Canopy / E. Transpiration and Photosynthesis / F. Values for Fluxes and Concentrations / G. Condensation

II. Gas Fluxes within Plant Communities   359
A. Eddy Diffusion Coefficient and Resistance / B. Water Vapor / C. Attenuation of Photosynthetic Photon Flux Density / D. Values of Foliar Absorption Coefficients / E. Light Compensation Point / F. CO_2 Concentrations and Fluxes / G. CO_2 at Night

III. Soil   366
A. Soil Water Potential / B. Darcy's Law / C. Soil Hydraulic Conductivity Coefficient / D. Flux for Cylindrical Symmetry / E. Flux for Spherical Symmetry

IV. Water Movement in the Xylem and the Phloem   373
A. Root Tissues / B. The Xylem / C. Poiseuille's Law / D. Applications of Poiseuille's Law / E. The Phloem / F. Phloem Contents and Speed of Movement / G. Mechanism of Phloem Flow / H. Values for Components of the Phloem Water Potential

V. The Soil–Plant–Atmosphere Continuum   384
A. Values of Water Potential Components / B. Resistances and Areas / C. Specific Resistances and Conductances / D. Capacitance and Time Constants / E. Daily Changes / F. Global Climate Change

Problems   397
References   399

Solutions to Problems   **403**

Appendix I.   Numerical Values of Constants and Coefficients   **439**

Appendix II.   Conversion Factors and Definitions   **445**

Appendix III.   Mathematical Relations   **449**
A. Prefixes (for units of measure) / B. Areas and Volumes / C. Logarithms /
D. Quadratic Equation / E. Trignometric Functions / F. Differential Equa-
tions

Appendix IV.   Gibbs Free Energy and Chemical Potential   **453**
A. Entropy and Equilibrium / B. Gibbs Free Energy / C. Chemical Potential / D. Pressure Dependence of μ_j / E. Concentration Dependence
of μ_j

Index   **461**

PREFACE

Physiology is the study of the function of cells, organs, and organisms—we will specifically consider water relations, solute transport, photosynthesis, transpiration, respiration, and environmental interactions. A physiologist endeavors to understand such topics in physical and chemical terms; accurate models can then be constructed and responses to the internal and the external environment can be predicted. A primary objective of this book is to use elementary chemistry, physics, and mathematics to explain and develop concepts that are key to an understanding of various areas of plant physiology in particular and biology in general. The intent is to provide a rigorous development, not a compendium of facts. References provide further details, although in some cases the enunciated principles carry the reader to the forefront of current research. Calculations are used to indicate the physiological consequences of the various equations, and problems at the end of chapters provide further such exercises (complete solutions are provided in the appendixes, which also have a large list of values for constants and conversion factors).

Chapters 1 through 3 describe water relations and ion transport for plant cells. In Chapter 1, after discussing the concept of diffusion, we consider the physical barriers to diffusion imposed by cellular and organelle membranes. Another physical barrier associated with plant cells is the cell wall, which limits cell size. In the treatment of the movement of water through cells in response to specific forces in Chapter 2, we employ the thermodynamic argument of chemical potential gradients. Chapter 3 considers solute movement into and out of plant cells, leading to an explanation of electrical potential differences across membranes and establishing the formal criteria for distinguishing diffusion from active transport. Based on concepts from irreversible thermodynamics, an important parameter called the reflection coefficient is derived, which permits a precise evaluation of the influence of osmotic pressures on flow. The thermodynamic arguments used to describe ion and water movements are equally applicable to animal cells.

The next three chapters deal primarily with the interconversion of various forms of energy. In Chapter 4 we consider the properties of light and its absorption. After light is absorbed, its radiant energy usually is rapidly converted to heat. However, the arrangement of photosynthetic pigments and their special molecular structures allow some radiant energy from the sun to be converted by plants into chemical energy. In Chapter 5 we discuss the particular features of chlorophyll and the accessory pigments for photosynthesis that allow this energy conversion. Light energy absorbed by chloroplasts leads to the formation of ATP and NADPH. These compounds represent currencies for carrying chemical and electrical (redox potential) energy, respectively. How much energy they actually carry is discussed in Chapter 6.

In the last three chapters we consider the various forms in which energy and matter enter and leave a plant as it interacts with its environment. The physical quantities involved in an energy budget analysis are presented in Chapter 7 so that the relative importance of the various factors affecting the temperature of leaves or other plant parts can be quantitatively evaluated. The resistances (or their reciprocals, conductances) affecting the movement of both water vapor during transpiration and carbon dioxide during photosynthesis are discussed in detail for leaves in Chapter 8. The movement of water from the soil through the plant to the atmosphere is discussed in Chapter 9. Because these and other topics depend on material introduced elsewhere in the book, the text is extensively cross-referenced.

This text is the second edition of *Physicochemical and Environmental Plant Physiology* (Academic Press, 1991), which evolved from *Biophysical Plant Physiology and Ecology* (Freeman, 1983), which evolved from *Introduction to Biophysical Plant Physiology* (Freeman, 1974), which evolved from *Plant Cell Physiology: A Physicochemical Approach* (Freeman, 1970). The text has been updated based on the ever-increasing quality of plant research, such as on cell growth, water relations, membrane channels (including aquaporins), the xanthophyll cycle, the bioenergetics of chloroplasts and mitochondria (especially coupling ratios among electrons, protons, and ATP), CO_2 diffusion within leaves, hydraulic conductances in the soil/air–gap/root regions, and implications of global climate change. References have also been updated (more than 450 changes) but those of historic interest have been retained. Revisions of calculations have been made in every chapter. Sixty-four figures were changed, ranging from a more conventional way of expressing units to entirely new figures that more clearly illustrate the concepts presented.

Many changes have emanated from the best laboratory for any book, the classroom. Moreover, a class on scientific writing critiqued the entire book; comments from Ed Bobich, Paul Chang, Erick De la Barrera, Steve Liao, Vincent Lin, Devang Patel, Timothy Rhee, Jennifer Schaab, Joo Song, Anna Wang, Chi-Man Yue, and especially Matthew Linton are gratefully acknowledged. Manny Alcantara and Linda Mohr painstakingly typed the new material, and Margaret Kowalcyzk competently revised the artwork. The result is a greatly improved written version of my long-term commitment to integrating the physical sciences, engineering, and mathematics to help understand biology, especially for plants.

Park S. Nobel
November 4, 1998

SYMBOLS AND ABBREVIATIONS

Where appropriate, typical units are indicated in parentheses.

Quantity	Description
a	absorptance or absorptivity (dimensionless)
a^{st}	mean area of stomata (m^2)
a_{IR}	absorptance or absorptivity in infrared region (dimensionless)
a_j	activity of species j (same as concentration)[a]
a.t.	subscript indicating active transport
Å	angström (10^{-10} m)
A	electron acceptor
A	area (m^2)
A_λ	absorbance (also called "optical density") at wavelength λ (dimensionless)
ABA	abscisic acid
ADP	adenosine diphosphate
ATP	adenosine triphosphate
b	nonosmotic volume (m^3)
b	optical path length (m)
bl	superscript for boundary layer
c	centi (as a prefix), 10^{-2}
c	superscript for cuticle
c_d	drag coefficient (dimensionless)
c_j	concentration of species j (mol m^{-3})[b]
$\bar{c}_s$	a mean concentration of solute s
cal	calorie
chl	superscript for chloroplast
clm	superscript for chloroplast limiting membranes
cw	superscript for cell wall

Quantity	Description
cyt	superscript for cytosol
C	superscript for conduction
C	capacitance, electrical (F)
C^j	capacitance for water storage in component j (m^3 MPa^{-1})
C'	capacitance/unit area (F m^{-2})
Chl	chlorophyll
Cl	subscript for chloride ion
C_P	volumetric heat capacity (J m^{-3} °C^{-1})
Cyt	cytochrome
d	deci (as a prefix), 10^{-1}
d	depth or distance (m)
d	diameter (m)
dyn	dyne
D	electron donor
D	dielectric constant (dimensionless)
D_j	diffusion coefficient of species j (m^2 s^{-1})
e	electron
e	superscript for water evaporation site
e_{IR}	emissivity or emittance in infrared region (dimensionless)
eV	electron volt
E	light energy (J)
E	kinetic energy (J)
E	electrical potential (mV)
E_j	redox potential of species j (mV)
E_j^{*H}	midpoint redox potential of species j referred to standard hydrogen electrode (mV)
E_M	electrical potential difference across a membrane (mV)
E_{N_j}	Nernst potential of species j (mV)
f	femto (as a prefix), 10^{-15}
F	farad
F	Faraday's constant (coulomb mol^{-1})
F	average cumulative leaf area/ground area (dimensionless)
FAD	flavin adenine dinucleotide (oxidized form)
FADH$_2$	reduced form of flavin adenine dinucleotide
FMN	flavin mononucleotide
g	gram
g_j	conductance of species j (mm s^{-1} with Δc_j and mmol m^{-2} s^{-1} with ΔN_j)
G	giga (as a prefix), 10^9

Quantity	Description
G	Gibbs free energy (J)
Gr	Grashof number (dimensionless)
G/n_j	Gibbs free energy/mole of some product or reactant (J mol^{-1})
h	hour
ha	hectare
h	height (m)
h_c	heat convection coefficient (W m^{-2} °C^{-1})
$h\nu$	a quantum of light energy
H	subscript for heat
i	superscript for inside
i	electrical current (ampere)
ias	superscript for intercellular air spaces
in	superscript for inward
in vitro	in a test tube, beaker, flask (literally, in glass)
in vivo	in a living organism (literally, in the living)
I	electrical current (ampere)
IR	infrared
j	subscript for species j
J	joule
J_j	flux density of species j (mol m^{-2} s^{-1})
J_j^{in}	inward flux density (influx) of species j (mol m^{-2} s^{-1})
J_j^{out}	outward flux density (efflux) of species j (mol m^{-2} s^{-1})
J_{V_j}	volume flux density of species j (m^3 m^{-2} s^{-1}, i.e., m s^{-1})
J_V	total volume flux density (m s^{-1})
k	kilo (as a prefix), 10^3
k	foliar absorption coefficient (dimensionless)
k_j	first-order rate constant for the jth process (s^{-1})
K	temperature on Kelvin scale
K	subscript for potassium ion
K	equilibrium constant (concentration raised to some power)
K_h	hydraulic conductance per unit length (m^4 MPa^{-1} s^{-1})
K^j	thermal conductivity coefficient of region j (W m^{-1} °C^{-1})
K_j	partition coefficient of species j (dimensionless)
K_j	concentration for half-maximal uptake rate of species j (Michaelis constant) (mol m^{-3}, or M)
K_j	eddy diffusion coefficient of gaseous species j (m^2 s^{-1})
K_{pH7}	equilibrium constant at pH 7
l	liter
l	superscript for lower

Quantity	Description
l	length (m), e.g., mean distance across leaf in wind direction
ln	natural or Napierian logarithm (to the base e, where e is $2.71828\cdots$)
log	common or Briggsian logarithm (to the base 10)
L^{soil}	hydraulic conductivity coefficient of the soil ($m^2\ Pa^{-1}\ s^{-1}$)
L_{jk}	Onsager or phenomenological coefficient (flux density per unit force)
L_P	hydraulic conductivity coefficient (in irreversible thermodynamics) ($m\ Pa^{-1}\ s^{-1}$)
L_w	water conductivity coefficient ($m\ Pa^{-1}\ s^{-1}$)
m	milli (as a prefix), 10^{-3}
m	meter
m	molal
m_j	mass per mole of species j (molar mass)($kg\ mol^{-1}$)
max	subscript for maximum
memb	superscript for membrane
mes	superscript for mesophyll
min	subscript for minimum
min	minute
mol	mole, a mass equal to the molecular weight of the species in grams; contains Avogadro's number of molecules
M	mega (as a prefix), 10^6
M	molar ($mol\ liter^{-1}$)
M_j	amount of species j per unit area ($mol\ m^{-2}$)
n	nano (as a prefix), 10^{-9}
n	number of stomata per unit area (m^{-2})
$n(E)$	number of moles with energy of E or greater
n_j	amount of species j (mol)
N	newton
Na	subscript for sodium ion
NAD^+	nicotinamide adenine dinucleotide (oxidized form)
NADH	reduced form of nicotinamide adenine dinucleotide
$NADP^+$	nicotinamide adenine dinucleotide phosphate (oxidized form)
NADPH	reduced form of nicotinamide adenine dinucleotide phosphate
N_j	mole fraction of species j (dimensionless)
Nu	Nusselt number (dimensionless)
o	superscript for outside
0	subscript for initial value (at $t = 0$)
out	superscript for outward
p	pico (as a prefix), 10^{-12}
p	period (s)

Quantity	Description
pH	$-\log(a_{H^+})$
pm	superscript for plasma membrane
ps	superscript for photosynthesis
P	pigment
P	hydrostatic pressure (MPa)
Pa	pascal
P_j	permeability coefficient of species j (m s^{-1})
P_j	partial pressure of gaseous species j (kPa)
PPFD	photosynthetic photon flux density (400–700 nm)
q	number of electrons transferred per molecule (dimensionless)
Q	charge (coulomb)
Q_{10}	temperature coefficient (dimensionless)
r	superscript for respiration
r	radius (m)
r	reflectivity (dimensionless)
r + pr	superscript for respiration plus photorespiration
r_j	resistance for gaseous species j (s m^{-1})
R	electrical resistance (ohm)
R	gas constant (J mol^{-1} K^{-1})
R^j	resistance of component j across which water moves as a liquid (MPa s m^{-3})
Re	Reynolds number (dimensionless)
RH	relative humidity (%)
s	subscript for solute
s	second
s_j	amount of species j (mol)
st	superscript for stoma(ta)
surf	superscript for surface
surr	superscript for surroundings
S	singlet
$S_{(\pi,\pi)}$	singlet ground state
$S_{(\pi,\pi^*)}$	singlet excited state in which a π electron has been promoted to a π^* orbital
S	magnitude of net spin (dimensionless)
S	total flux density of solar irradiation, i.e., global irradiation (W m^{-2})
t	time (s)
ta	superscript for turbulent air
T	superscript for transpiration
T	triplet

Quantity	Description
$T_{(\pi,\pi^*)}$	excited triplet state
T	temperature (K, °C)
u	superscript for upper
u_j	mobility of species j (velocity per unit force)
u_+	mobility of monovalent cation
u_-	mobility of monovalent anion
U	kinetic energy (J mol^{-1})
U_B	minimum kinetic energy to cross barrier (J mol^{-1})
UV	ultraviolet
v	magnitude of velocity (m s^{-1})
v	wind speed (m s^{-1})
v^{wind}	wind speed (m s^{-1})
v_j	magnitude of velocity of species j (m s^{-1})
v_{CO_2}	rate of photosynthesis per unit volume (mol m^{-3} s^{-1})
V	volt
V	subscript for volume
V	volume (m^3)
$\bar{V}_j$	partial molal volume of species j (m^3 mol^{-1})
V_{max}	maximum rate of CO$_2$ fixation (mol m^{-3} s^{-1})
w	subscript for water
wv	subscript for water vapor
W	watt (J s^{-1})
x	distance (m)
z	altitude (m)
z_j	charge number of ionic species j (dimensionless)
α	contact angle (°)
γ_j	activity coefficient of species j (dimensionless, but see a_j)
$\gamma_\pm$	mean activity coefficient of cation–anion pair (dimensionless)
δ	delta, a small quantity of something, e.g., δ^- refers to a small fraction of an electronic charge
δ	distance (m)
δ^{bl}	thickness of air boundary layer (mm)
Δ	delta, the difference or change in the quantity that follows it
ε	volumetric elastic modulus (MPa)
ε_λ	absorption coefficient at wavelength λ (m^2 mol^{-1})
η	viscosity (N s m^{-2}, Pa s)
λ	wavelength of light (nm)
λ_{max}	wavelength position for the maximum absorption coefficient in an absorption band or for the maximum photon (or energy) emission in an emission spectrum

Quantity	Description
μ	micro (as a prefix), 10^{-6}
μ_j	chemical potential of species j (J mol^{-1})
ν	frequency of electromagnetic radiation (s^{-1}, hertz)
ν	kinematic viscosity (m^2 s^{-1})
π	ratio of circumference to diameter of a circle (3.14159)
π	an electron orbital in a molecule or an electron in such an orbital
π^*	an excited or antibonding electron orbital in a molecule or an electron in such an orbital
Π	total osmotic pressure (MPa)
Π_j	osmotic pressure of species j (MPa)
Π_s	osmotic pressure due to solutes (MPa)
ρ	density (kg m^{-3})
ρ	resistivity, electrical (ohm m)
ρ^j	hydraulic resistivity of component j (MPa s m^{-2})
σ	surface tension (N m^{-1})
σ	reflection coefficient (dimensionless)
σ_j	reflection coefficient of species j (dimensionless)
σ_L	longitudinal stress (MPa)
σ_T	tangential stress (MPa)
τ	matric pressure (MPa)
τ	lifetime (s)
τ_j	lifetime for the jth deexcitation process (s)
ϕ_j	osmotic coefficient of species j (dimensionless)
Φ_i	quantum yield or efficiency for ith deexcitation pathway (dimensionless)
Ψ	water potential (MPa)
Ψ_Π	osmotic potential (MPa)
°C	degree Celsius
°	angular degree
*	superscript for a standard or reference state
*	superscript for a molecule in an excited electronic state
*	superscript for saturation of air with water vapor
∞	infinity

[a] The activity, a_j, is often considered to be dimensionless, in which case the activity coefficient, γ_j, has the units of reciprocal concentration ($a_j = \gamma_j c_j$; Eq. 2.5).

[b] We note that mol liter^{-1}, or molarity (M), is a concentration unit of widespread use, although it is not an SI unit.

REPRESENTATIVE PRINCIPAL EQUATIONS

Description	Equation	Page

Chapter 1 Cells and Diffusion

Fick's first law

$$J_j = -D_j \frac{\partial c_j}{\partial x} \quad (1.1)$$

9

Solution to Fick's second law

$$c_j = \frac{M_j}{2(\pi D_j t)^{1/2}} e^{-x^2/4D_j t} \quad (1.5)$$

11

Permeability coefficient

$$P_j = \frac{D_j K_j}{\Delta x} \quad (1.9)$$

22

Elastic property

$$\text{Young's modulus} = \frac{\text{stress}}{\text{strain}} = \frac{\text{force/area}}{\Delta l/l} \quad (1.14)$$

30

Volumetric elastic modulus

$$\varepsilon = \frac{\Delta P}{\Delta V/V} \quad (1.17)$$

32

Chapter 2 Water

Height of capillary rise

$$h = \frac{2\sigma \cos \alpha}{r\rho g} \quad (2.2a)$$

42

Chemical potential of solute species j

$$\mu_j = \mu_j^* + RT \ln a_j + \bar{V}_j P + z_j F E + m_j g h \quad (2.4)$$

49

Van't Hoff relation

$$\Pi_s \cong RT \sum_j c_j \quad (2.10)$$

54

Water potential

$$\Psi = \frac{\mu_w - \mu_w^*}{\bar{V}_w} = P - \Pi + \rho_w g h \quad (2.13a)$$

57

Boyle–Van't Hoff relation

$$\Pi^\circ (V - b) = RT \sum_j \varphi_j n_j = RT n \quad (2.15)$$

60

Water potential of water vapor

$$\Psi_{wv} = \frac{RT}{\bar{V}_w} \ln \frac{\% \text{ relative humidity}}{100} + \rho_w g h \quad (2.22)$$

70

A growth equation

$$\frac{1}{V} \frac{dV}{dt} = \varphi(P^i - P^i_{\text{threshold}}) \quad (2.28)$$

76

Description	Equation		Page

Chapter 3 Solutes

Nernst potential $\quad\quad E_{N_j} = \dfrac{RT}{z_j F} \ln \dfrac{a_j^o}{a_j^i} = 2.303 \dfrac{RT}{z_j F} \log \dfrac{a_j^o}{a_j^i}$ (3.5a) 87

Nernst–Planck equation $\quad\quad J_j = -\dfrac{u_j RT}{\gamma_j} \dfrac{\partial \gamma_j c_j}{\partial x} - u_j c_j z_j F \dfrac{\partial E}{\partial x}$ (3.7) 92

Goldman (constant field) equation

$$E_M = \frac{RT}{F} \ln \frac{\left(P_K c_K^o + P_{Na} c_{Na}^o + P_{Cl} c_{Cl}^i\right)}{\left(P_K c_K^i + P_{Na} c_{Na}^i + P_{Cl} c_{Cl}^o\right)}$$ (3.19) 100

Boltzmann energy distribution $\quad\quad n(E) = n_{total} e^{-E/RT} \quad$ mole basis (3.21b) 107

Arrhenius equation $\quad\quad$ Rate constant $= B e^{-A/RT}$ (3.23) 109

Ussing–Teorell (flux ratio) equation $\quad\quad \dfrac{J_j^{in}}{J_j^{out}} = \dfrac{c_j^o}{c_j^i \, e^{z_j F E_M/RT}}$ (3.24) 111

Influx of solute j in Michaelis–Menten form $\quad\quad J_j^{in} = \dfrac{J_{j\,max}^{in} c_j^o}{K_j + c_j^o}$ (3.27a) 120

Generalized volume flux density $\quad\quad J_V = L_P \left(\Delta P - \sum_j \sigma_j \, \Delta \Pi_j \right)$ (3.39) 128

Solute flux density (irreversible thermodynamics)

$$J_s = \bar{c}_s (1 - \sigma) J_V + \omega \, \Delta \Pi$$ (3.47) 136

Chapter 4 Light

Energy of light $\quad\quad E_\lambda = N h \nu = N h c / \lambda_{vacuum} \quad$ mole basis (4.2b) 146

Wien's displacement law $\quad\quad \lambda_{max} T = 3.67 \times 10^6$ nm K photon basis (4.3a) 152

Lifetimes/rate constants $\quad\quad \dfrac{1}{\tau} = k = \sum_j k_j = \sum_j \dfrac{1}{\tau_j}$ (4.13) 165

Beer's law $\quad\quad A_\lambda = \log \dfrac{J_0}{J_b} = \varepsilon_\lambda c b$ (4.18a) 172

Chapter 6 Bioenergetics

Gibbs free energy change $\quad\quad \Delta G = \Delta G^* + RT \ln \dfrac{(a_C)^c (a_D)^d}{(a_A)^a (a_B)^b}$ (6.5) 225

Redox potential of species j $\quad\quad E_j = E_j^* - \dfrac{RT}{q F} \ln \dfrac{(reduced_j)}{(oxidized_j)}$ (6.9) 229

ΔG for ATP formation

$$\Delta G = -RT \ln (K_{pH\,7}) + RT \ln \frac{[ATP]}{[ADP][phosphate]}$$ (6.13a) 234

Description	Equation	Page

Difference in chemical potential of H^+ across membrane

$$\mu_H^i - \mu_H^o = 5.71\,(pH^o - pH^i) + 96.5\,E_M \quad kJ\,mol^{-1} \qquad (6.16c)$$

241

Stefan–Boltzmann law

$$\text{Maximum radiant energy flux density} = \sigma T^4 \qquad (6.17)$$

250

Chapter 7 Temperature—Energy Budgets

Absorbed minus emitted radiation

$$\text{Net radiation} = \begin{array}{c} a(1+r)S \\ + \\ a_{IR}\sigma[(T^{surr})^4 + (T^{sky})^4] \end{array} - 2e_{IR}\sigma(T^{leaf})^4 \qquad (7.6b)$$

266

Boundary layer for a flat leaf

$$\delta^{bl}_{(mm)} = 4.0\sqrt{\frac{l_{(m)}}{v_{(m\,s^{-1})}}} \qquad (7.8)$$

272

Heat flux density for leaf

$$J_H^C = -2K^{air}\frac{\partial T}{\partial x} = 2K^{air}\frac{(T^{leaf} - T^{ta})}{\delta^{bl}} \qquad (7.11)$$

275

Heat flux density accompanying transpiration

$$J_H^T = J_{wv}H_{vap} = \frac{H_{vap}D_{wv}\,\Delta c_{wv}^{total}}{\Delta x^{total}} \qquad (7.19)$$

279

Heat storage

$$\text{Heat storage rate} = C_P V \frac{\Delta T}{\Delta t} \qquad (7.21)$$

284

Heat conduction, soil

$$J_H^C = -K^{soil}\frac{\partial T}{\partial z} \qquad (7.24)$$

287

Soil temperature

$$T = \bar{T}^{surf} + \Delta T^{surf}e^{-z/d}\cos\left(\frac{2\pi t}{p} - \frac{2\pi t_{max}}{p} - \frac{z}{d}\right) \qquad (7.25)$$

288

Chapter 8 Leaves and Fluxes

Flux density across air boundary layer

$$J_j = D_j\frac{\Delta c_j}{\Delta x} = D_j\frac{\Delta c_j^{bl}}{\delta^{bl}} = g_j^{bl}\Delta c_j^{bl} = \frac{\Delta c_j^{bl}}{r_j^{bl}} \qquad (8.2)$$

296

Conductance along length of stomata

$$g_j^{st} = \frac{J_j}{\Delta c_j^{st}} = \frac{D_j na^{st}}{\delta^{st} + r^{st}} = \frac{1}{r_j^{st}} \qquad (8.5)$$

302

Diffusion coefficient

$$D_j \cong D_{jo}\left(\frac{T}{273}\right)^{1.8}\frac{P_0}{P} \qquad (8.9)$$

306

Water vapor resistance of lower epidermis of a leaf

$$r_{wv}^{total} \cong \frac{1}{D_{wv}}\left(\delta^{ias} + \frac{\delta^{st} + r^{st}}{na^{st}} + \delta^{bl}\right) \qquad (8.14)$$

309

Description	Equation		Page

CO$_2$ resistances

$$r^j_{CO_2} = \frac{1}{A^{mes}/A}\frac{1}{P^j_{CO_2}} = \frac{A}{A^{mes}\,P^j_{CO_2}} = \frac{A\,\Delta x^j}{A^{mes}\,D^j_{CO_2}\,K^j_{CO_2}} = \frac{1}{g^j_{CO_2}}$$ (8.19)   319

Net rate of photosynthesis

$$J_{CO_2} = \frac{c^{ta}_{CO_2} - c^{chl}_{CO_2}}{r^{bl_1}_{CO_2} + r^{leaf_1}_{CO_2} + r^{mes}_{CO_2} + \left(1 + \frac{J^{r+pr}_{CO_2}}{J_{CO_2}}\right)r^{chl}_{CO_2}} = \frac{\Delta c^{total}_{CO_2}}{r^{total}_{CO_2}}$$ (8.26)   332

Water-use efficiency $$WUE = \frac{mol\ CO_2\ fixed}{mol\ H_2O\ transpired}$$ mole basis   (8.31b)   337

Chapter 9 Plants and Fluxes

Wind speed above canopy $$v = \frac{v^*}{k}\ln\frac{z-d}{z_0}$$ (9.1)   352

Attenuation of PPFD $$\ln\frac{J_0}{J} = kF$$ (9.5)   362

Young–Laplace equation $$P = -\sigma\left(\frac{1}{r_1} + \frac{1}{r_2}\right)$$ (9.6)   368

Darcy's law $$J_V = -L^{soil}\frac{\partial P^{soil}}{\partial x}$$ (9.7)   369

Poiseuille's (Hagen–Poiseuille) law

    Volume flow rate per tube $$= -\frac{\pi r^4}{8\eta}\frac{\partial P}{\partial x}, \quad J_V = -\frac{r^2}{8\eta}\frac{\partial P}{\partial x}$$ (9.11)   375

van den Honert relation $$J^j_{V_w}\,A^j = \frac{\Delta\Psi^j}{R^j} \cong constant$$ (9.12)   386

Water capacitance of plant part j

$$C^j = \frac{change\ in\ water\ content\ of\ component\ j}{change\ in\ average\ water\ potential\ along\ component\ j} = \frac{\Delta V^j_w}{\Delta\bar\Psi^j}$$ (9.16)   392

1

Cells and Diffusion

I. **Cell Structure**   **1**
    A. Generalized Plant Cell   1
    B. Leaf Anatomy   3
    C. Vascular Tissue   4
    D. Root Anatomy   6
II. **Diffusion**   **8**
    A. Fick's First Law   9
    B. Continuity Equation and Fick's Second Law   10
    C. Time–Distance Relation for Diffusion   12
III. **Membrane Structure**   **16**
    A. Membrane Models   16
    B. Organelle Membranes   17
IV. **Membrane Permeability**   **20**
    A. Concentration Difference across a Membrane   20
    B. Permeability Coefficient   22
    C. Diffusion and Cellular Concentration   23
V. **Cell Walls**   **25**
    A. Chemistry and Morphology   25
    B. Diffusion across Cell Walls   27
    C. Stress–Strain Relations of Cell Walls   29
    Problems   **32**
    References   **34**

I. Cell Structure

Before formally considering diffusion and related topics, we will outline the structure of certain plant cells and tissues, thus introducing most of the anatomical terms used throughout the book.

A. Generalized Plant Cell

Figure 1-1 depicts a representative leaf cell from a higher plant and illustrates the larger subcellular structures. Surrounding the *protoplast* is the *cell wall*, composed of cellulose

1

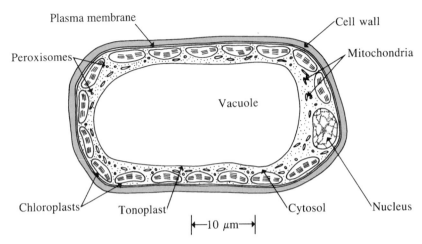

Plasma membrane

Cell wall

Peroxisomes

Mitochondria

Vacuole

Chloroplasts   Tonoplast   Cytosol   Nucleus

|←—10 μm—→|

Figure 1-1. Schematic representation of a mature mesophyll cell from the leaf of a higher plant, suggesting
────── some of the complexity resulting from the presence of many membrane-surrounded subcellular
compartments.

and other polysaccharides, which helps provide rigidity to individual cells as well as to
the whole plant. The cell wall contains numerous relatively large interstices, so it is not
the main permeability barrier to the entry of water or small solutes into plant cells. The
main barrier, a cell membrane known as the *plasma membrane* or *plasmalemma*, occurs
inside the cell wall and surrounds the *cytoplasm*. The permeability of this membrane
varies with the particular solute, so the plasma membrane can regulate what enters and
leaves a plant cell. The cytoplasm contains organelles like *chloroplasts* and *mitochondria*,
which are membrane-surrounded compartments in which energy can be converted from one
form to another. Also in the cytoplasm are microbodies such as peroxisomes, numerous
ribosomes, and proteins, as well as many other macromolecules and structures that influence
the thermodynamic properties of water. The term *cytoplasm* includes the organelles (but
generally not the nucleus), whereas the term *cytosol* refers to the cytoplasmic solution
delimited by the plasma membrane and the tonoplast (to be discussed next) but exterior to
all the organelles.

In mature cells of higher plants and many lower (evolutionarily less advanced) plants,
there is a large central aqueous phase, the *vacuole*, which is surrounded by the *tonoplast*.
The central vacuole is usually quite large and can occupy up to about 90% of the volume of
a mature cell. The aqueous solution in the central vacuole contains mainly inorganic ions
or organic acids as solutes, although considerable amounts of sugars and amino acids may
be present in some species. Water uptake by this vacuole occurs during cell growth.

One immediate impression of plant cells is the great prevalence of membranes. In
addition to surrounding the cytoplasm, membranes also separate various compartments in
the cytoplasm. Diffusion of substances across these membranes is much more difficult than
is diffusion within the compartments. Thus, organelle and vacuolar membranes can control
the contents and consequently the reactions occurring in the particular compartments that
they surround. Diffusion can also impose limitations on the overall size of a cell because
the time for diffusion increases with the square of the distance, as we will quantitatively
consider in the next section.

Although many plant cells share most of the features indicated in Fig. 1-1, they are remarkably diverse in size. The nearly spherical cells of the green alga *Chlorella* are approximately 4×10^{-6} m (4 μm) in diameter. On the other hand, some species of the intertidal green alga *Valonia* have multinucleated cells as large as 20 mm in diameter. The genera *Chara* and *Nitella* include fresh and brackish water green algae having large internodal cells (Fig. 3-9) that may be 100 mm long and 1 mm in diameter. Such large algal cells have proved extremely useful for studying ion fluxes, as we will consider in Chapter 3.

B. Leaf Anatomy

A transverse section of a typical angiosperm leaf can illustrate various cell types and anatomical features that are important for photosynthesis and transpiration. Leaves are generally 4–10 cells thick, which corresponds to a few hundred μm (Fig. 1-2). An *epidermis* occurs on both the upper and the lower sides of a leaf and is usually one cell layer thick. Except for the *guard cells*, the cytoplasm of epidermal cells generally is colorless because very few if any chloroplasts are present (depending on the plant species). Epidermal cells have a relatively thick waterproof *cuticle* on the atmospheric side (Fig. 1-2). The cuticle contains *cutin*, a diverse group of complex polymers composed principally of esters of 16- and 18-carbon monocarboxylic acids that have two or three hydroxyl groups. Cutin is relatively inert and also resists enzymatic degradation by microorganisms; thus, it is often well preserved in fossil material. We will consider its role in preventing water loss from a leaf.

Between the two epidermal layers is the *mesophyll* tissue, which is usually differentiated into chloroplast-containing "palisade" and "spongy" cells. The palisade cells are often

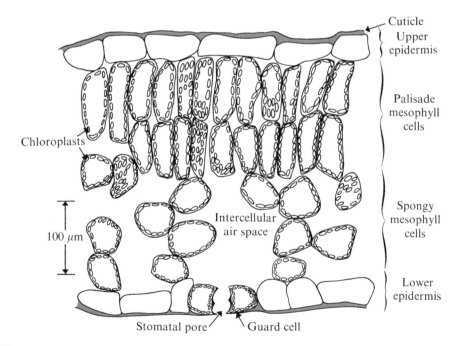

Figure 1-2. Schematic transverse section through a leaf, indicating the arrangement of various cell types. There are often about 30–40 mesophyll cells per stoma.

elongated at right angles to the upper epidermis and are found immediately beneath it (Fig. 1-2). The spongy mesophyll cells, between the palisade mesophyll cells and the lower epidermis, are loosely packed, and intercellular air spaces are conspicuous. In fact, most of the surface area of both spongy and palisade mesophyll cells is exposed to the air in the intercellular spaces. A spongy mesophyll cell is often rather spherical, about 40 μm in diameter, and contains approximately 40 chloroplasts. (As Fig. 1-2 illustrates, the cells are by no means geometrically regular, so dimensions here indicate only approximate size.) A neighboring palisade cell is generally more oblong; it can be 80 μm long, can contain 60 chloroplasts, and might be represented by a cylinder 30 μm in diameter with hemispherical ends. In many leaves about 70% of the chloroplasts are in the palisade cells, which often outnumber the spongy mesophyll cells nearly two to one.

The pathway of least resistance for gases to cross an epidermis—and thus to enter or to exit from a leaf—is through the adjustable space between a pair of guard cells (Fig. 1-2). This pore, and its two surrounding guard cells, is called a *stoma* or *stomate* (plural: stomata and stomates, respectively). When open the stomatal pores allow for the entry of CO_2 into the leaf and for the exit of photosynthetically produced O_2. The inevitable loss of water vapor by transpiration also occurs mainly through the stomatal pores, as we will discuss in Chapter 8. Stomata thus serve as a control, helping to strike a balance between freely admitting the CO_2 needed for photosynthesis and at the same time preventing excessive loss of water vapor from the plant. Air pollutants such as ozone (O_3) and sulfur dioxide (SO_2) also enter plants primarily through the open stomata.

C. Vascular Tissue

The *xylem* and the *phloem* make up the vascular systems found contiguously in the roots, stems (Fig. 1-3), and leaves of plants. In a tree trunk the phloem constitutes a layer of the bark and the xylem constitutes almost all of the wood. The xylem provides structural support for land plants. Water conduction in the xylem of a tree often occurs only in the outermost annual ring,[1] which lies just inside the *vascular cambium* (region of meristematic activity from which xylem and phloem cells differentiate). Outside the functioning phloem are other phloem cells that can be shed as pieces of bark slough off. Phloem external to the xylem, as in a tree, is the general pattern for the stems of plants. As we follow the vascular tissue from the stem along a petiole and into a leaf, we observe that the xylem and the phloem often form a vein, which sometimes conspicuously protrudes from the lower surface of a leaf. Reflecting the orientation in the stem or the trunk, the phloem is found abaxial to the xylem in the vascular tissue of a leaf, i.e., the phloem is located on the side of the lower epidermis. The vascular system branches and rebranches as it crosses a dicotyledonous leaf, becoming smaller (in cross section) at each step. In contrast to the reticulate venation in dicotyledons, monocotyledons characteristically have parallel-veined leaves. Individual mesophyll cells in the leaf are never further than a few cells from the vascular tissue.

The movement of water and nutrients from the soil to the upper portions of a plant occurs primarily in the xylem. The xylem sap generally contains about 10 mol m^{-3} (10 mM)^2

---

[1] The rings in trees are not always annual. In many desert species a ring forms when large xylem cells are produced after a suitable rainy period followed by smaller cells, and this can occur more than once or sometimes not at all in a given year. Moreover, trees from the wet tropics may have no annual rings.

[2] Molarity (moles of solute per liter of solution, symbolized by M) is a useful unit for concentration, but it is not recommended for highly accurate measurements by the international unit convention, Le Système Internationale

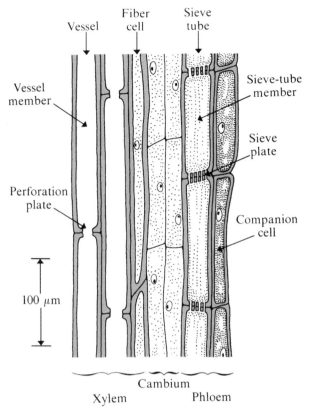

Figure 1-3. Highly idealized longitudinal section through part of a vascular bundle in a stem, illustrating various anatomical aspects of the xylem and the phloem. The new cells forming in the xylem initially contain cytoplasm. Fiber cells, which occur in the xylem, are generally quite tapered and provide structural support; the nucleated companion cells are metabolically involved with the sieve-tube members of the phloem.

inorganic nutrients plus organic forms of nitrogen that are metabolically produced in the root. The xylem is a tissue of various cell types that we will consider in more detail in the final chapter when water movement in plants is quantitatively discussed. The conducting cells in the xylem are the narrow, elongated *tracheids* and the *vessel members* (also called *vessel elements*), which tend to be shorter and wider than the tracheids. Vessel members are joined end-to-end in long linear files, their adjoining end walls or *perforation plates* having from one large to many small holes. The conducting cells lose their protoplasts, and the remaining cell walls thus form a low-resistance channel for the passage of solutions. Xylem sap moves from the root, up the stem, through the petiole, and then to the leaves in these hollow xylem "cells," with motion occurring in the direction of decreasing hydrostatic pressure. Some solutes leave the xylem along the stem on the way to a leaf, and others

d'Unités or Système Internationale (SI). Nevertheless, we will use molarity in addition to the SI unit of mol m^{-3}. We also note that SI as currently practiced allows the American spelling "liter" and "meter" as well as the British spelling "litre" and "metre."

diffuse or are actively transported across the plasma membranes of various leaf cells adjacent to the conducting cells of the xylem.

The movement of most organic compounds throughout the plant takes place in the other vascular tissue, the phloem. A portion of the photosynthetic products made in the mesophyll cells of the leaf diffuses or is actively transported across cellular membranes until it reaches the conducting cells of the leaf phloem. By means of the phloem, the photosynthetic products—which then are often mainly in the form of sucrose—are distributed throughout the plant. The carbohydrates produced by photosynthesis and certain other substances generally move in the phloem toward regions of lower concentration, although diffusion is not the mechanism for the movement, as we will indicate in Chapter 9. The phloem is a tissue consisting of several types of cells. In contrast to the xylem, however, the conducting cells of the phloem contain cytoplasm. They are *sieve cells* and *sieve-tube members* and are joined end-to-end, thus forming a transport system throughout the plant. Although these phloem cells often contain no nuclei at maturity, they remain metabolically active.

D. Root Anatomy

Roots anchor plants in the ground as well as absorb water and nutrients from the soil and then conduct these substances upward to the stem. To help understand uptake into a plant, we will examine the functional zones that occur along the length of a root.

At the extreme tip of a root is the *root cap* (Fig. 1-4a), which consists of relatively undifferentiated cells that are scraped off as the root grows into new regions of the soil. Cell walls in the root cap are often mucilaginous, which can reduce friction with soil particles.

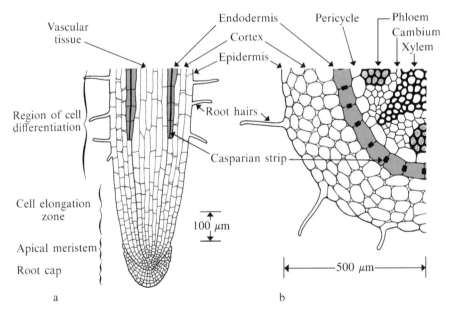

Figure 1-4. Schematic diagrams of a root: (a) longitudinal section, indicating the zones that can occur near the root tip, and (b) cross-sectional view approximately 10 mm back from the tip, indicating the arrangement of the various cell types.

Proximal to the root cap is a meristematic region where the cells rapidly divide. Cells in this *apical meristem* tend to be isodiametric and have thin cell walls. Next is a region of *cell elongation* in the direction of the root axis. Such elongation mechanically pushes the root tip through the soil, causing cells of the root cap to slough off by abrasion with soil particles. Sometimes the region of dividing cells is not spatially distinct from the elongation zone. Also, cell size and the extent of the zones vary with both plant species and physiological status.

The next region indicated in Fig. 1-4a is that of *cell differentiation*, where the cells begin to assume more highly specialized functions. The cell walls become thicker and elongation is greatly diminished. The epidermal cells develop fine projections, radially outward from the root, called *root hairs*. These root hairs greatly increase the surface area across which water and nutrients can enter a plant. As we follow a root toward the stem, the root surface generally becomes less permeable to water and the root interior becomes more involved with conducting water toward the stem. Water movement into the root is discussed in Chapter 9, so the discussion here will be restricted to some of the morphological features.

The region of the root where water absorption most readily occurs generally has little or no waxy cuticle. Figure 1-4b shows a cross section of a root at the level where root hairs are found. Starting from the outside, we observe first the root epidermis and then a number of layers of cells known as the *cortex*. There are abundant intercellular air spaces in the cortex, facilitating the diffusion of O_2 and CO_2 within this tissue (such air spaces generally are lacking in vascular tissue). Inside the cortex is a single layer of cells, the *endodermis*. The radial and transverse walls of the endodermal cells are impregnated with waxy material, including *suberin*, forming a band around the cells known as the *casparian strip*, which prevents passage of water and solutes across that part of the cell wall. Because there are no air spaces between endodermal cells, and the radial walls are blocked by the waterproof casparian strip, water must pass through the lateral walls and enter the cytoplasm of endodermal cells to continue across the root. The endodermal cells can represent the only place in the entire pathway for water movement from the soil, through the plant, to the air where it is mandatory that the water enter a cell's cytoplasm.[3] In the rest of the pathway, water can move in cell walls or in the hollow lumens of xylem vessels, a region referred to as the *apoplast*.

Immediately inside the endodermis is the *pericycle*, which is typically one cell thick in angiosperms. The cells of the pericycle can divide and form a meristematic region that can produce lateral or branch roots in the region just above the root hairs. Radially inside the pericycle is the vascular tissue. The phloem generally occurs in two to eight or more strands located around the root axis. The xylem generally radiates out between the phloem strands, so water does not have to cross the phloem to reach the root xylem. As in stems, the tissue between the xylem and the phloem is the vascular cambium, which through cell division and differentiation produces xylem to the inside and phloem to the outside.

Our rather elementary discussion of leaves, vascular tissues, and roots leads to the following oversimplified but useful picture. The roots take up water from the soil along with nutrients required for growth. These are conducted in the xylem to the leaves. Leaves of a photosynthesizing plant lose the water to the atmosphere along with a release of O_2 and an

---

[3] In the roots of many species a subepidermal layer or layers of *hypodermis* occurs. Radial walls of hypodermal cells can also be blocked with a waxy material analogous to the casparian strip in the endodermis, in which case the layers are often termed an *exodermis*.

uptake of CO_2. Carbon from the latter ends up in photosynthate translocated in the phloem back to the root. Thus the xylem and the phloem serve as the "plumbing" that connects the two types of plant organs that are functionally interacting with the environment. To understand the details of such physiological processes we must turn to fields like calculus, thermodynamics, photochemistry, and physics. Our next step is to bring the abstract ideas of these fields into the realm of cells and plants, which means we need to make calculations using appropriate assumptions and approximations.

We will begin by describing diffusion (Chapter 1). To discuss water (Chapter 2) and solutes (Chapter 3), we will introduce the thermodynamic concept of chemical potential. This leads to a quantitative description of fluxes, electrical potentials across membranes, and the energy requirements for active transport of solutes. Some important energy conversion processes take place in the organelles. For instance, light energy is absorbed (Chapter 4) by photosynthetic pigments located in the internal membranes of chloroplasts (Chapter 5) and then converted into other forms of energy useful to a plant (Chapter 6) or dissipated as heat (Chapter 7). Leaves (Chapter 8) as well as groups of plants (Chapter 9) also interact with the environment through exchanges of water vapor and CO_2. In our problem-solving approach to these topics, we will pay particular attention to dimensions and ranges for the parameters as well as to the insights that can be gained by developing the relevant formulae and then making calculations.

II. Diffusion

Diffusion is a spontaneous process leading to the net movement of a substance from some region to an adjacent one where that substance has a lower concentration. It takes place in both the liquid and the gas phases associated with plants. Diffusion results from the random thermal motions of the molecules either of the solute(s) and the solvent in the case of a solution or of gases in the case of air. The net movement caused by diffusion is a statistical phenomenon—a greater probability exists for molecules to move from the concentrated to the dilute region than vice versa. In other words, more molecules per unit volume are present in the concentrated region than in the dilute one, so more are available for diffusing toward the dilute region than are available for movement in the opposite direction. If isolated from external influences, diffusion of a neutral species tends to even out concentration differences originally present in adjoining regions of a liquid or a gas. In fact, the randomizing tendency of such molecular Brownian movement is a good example of the increase in entropy, or decrease in order, that accompanies all spontaneous processes.

Diffusion is involved in many plant processes, such as gas exchange and the movement of nutrients toward root surfaces. For instance, diffusion is the mechanism for most, if not all, steps by which CO_2 from the air reaches the sites of photosynthesis in chloroplasts. CO_2 diffuses from the atmosphere up to the leaf surface and then diffuses through the stomatal pores. After entering the leaf, CO_2 diffuses within intercellular air spaces (Fig. 1-2). Next, CO_2 diffuses across the cell wall, crosses the plasma membrane of a leaf mesophyll cell, and then diffuses through the cytosol to reach the chloroplasts. Finally, CO_2 enters a chloroplast and diffuses up to the enzymes that are involved in carbohydrate formation. If the enzymes were to fix all the CO_2 in their vicinity, and no other CO_2 were to diffuse in from the atmosphere surrounding the plant, photosynthetic processes would stop (in solution "CO_2" can also occur in the form of bicarbonate, HCO_3^-, and the crossing of membranes does not have to be by diffusion, refinements that we will return to in Chapter 8). In this chapter we will develop the mathematical formulation necessary for understanding both diffusion across a membrane and diffusion in a solution.

A. Fick's First Law

In 1855 Adolph Fick was one of the first to examine diffusion quantitatively. For such an analysis we need to consider the concentration (c_j) of some solute species j in a solution or gaseous species j in air; the subscript j indicates that we are considering only one species out of the many that could be present. We will assume that the concentration of species j in some region is less than in a neighboring one. A net migration of molecules occurs by diffusion from the concentrated to the dilute region. Such a molecular flow down a concentration gradient is analogous to the flow of heat from a warmer to a cooler region. The analogy is actually good (especially for gases) because both processes depend on the random thermal motion of molecules. In fact, the differential equations and their solutions that are used to describe diffusion are those that had previously been developed to describe heat flow.

To express diffusion quantitatively, we will consider a diffusive flux or flow of species j. We will restrict our attention to diffusion involving planar fronts of uniform concentration, a relatively simple situation that fortunately has widespread application to situations of interest in biology. We will let J_j be the amount of species j crossing a certain area per unit time, e.g., moles of particles per meter squared in a second, which is termed the *flux density*.[4] Reasoning by analogy with heat flow, Fick deduced that the "force," or causative agent, leading to the net molecular movement is the concentration gradient. A gradient indicates how a certain parameter changes with distance; the gradient in concentration of species j in the x-direction is represented by $\partial c_j / \partial x$. The partial derivative $\partial c_j / \partial x$ indicates how much c_j changes as we move along the x-axis when other variables, such as time and position along the y-axis, are held constant. In general, *the flux density of some substance is proportional to an appropriate force*, a relation that we will use repeatedly in this text. In the present instance the driving force is the negative of the concentration gradient of species j, which we will represent by $-\partial c_j / \partial x$ for diffusion in one dimension. To help appreciate why a minus sign occurs, recall that the direction of net (positive) diffusion is toward regions of lower concentration. We can now write the following relation showing the dependence of the flux density on the driving force:

$$J_j = -D_j \frac{\partial c_j}{\partial x} \tag{1.1}$$

Equation (1.1) is commonly known as Fick's first law of diffusion, where D_j is the *diffusion coefficient* of species j. For J_j in mol m^{-2} s^{-1} and c_j in mol m^{-3} (hence, $\partial c_j / \partial x$ in mol m^{-4}), D_j has units of m^2 s^{-1}. Because D_j varies with concentration and temperature, it is properly called a coefficient in the general case. In certain applications, however, we can obtain sufficient accuracy by treating D_j as a constant. The partial derivative is used in Eq. (1.1) to indicate the change in concentration in the x-direction of Cartesian coordinates at some moment in time (constant t) and for specified values of y and z. For

---

[4] Although the SI convention recommends the term flux density, much of the diffusion literature refers to J_j as a flux. Moreover, many symbols have been used for flux density (e.g., A, D, E, F, I, J, M, Q, U, and V), some of which (such as A for assimilation and E for evaporation) conflict with those used for other common variables (A for area and E for electric field or potential). We have chosen J because of its lack of conflict and the long precedent for its use (e.g., Lars Onsager used J for the flux densities of heat and mass in the early 1930s).

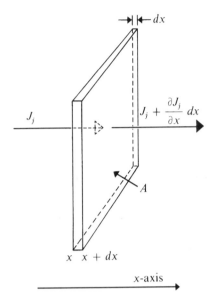

Figure 1-5. Diagram showing the dimensions and the flux densities that form the geometric basis for the continuity equation. The same general figure is used to discuss water flow in Chapter 2 and solute flow in Chapter 3.

most of the cases that we will consider, the flux density in the x-direction has the same magnitude at any value of y and z, meaning that we are dealing with one-dimensional, planar fluxes. By convention, a net flow in the direction of increasing x is positive (from left to right in Fig. 1-5). Because a net flow occurs toward regions of lower concentration, we again note that the minus sign is needed in Eq. (1.1). Fick's first law, which has been amply demonstrated experimentally, is the starting point for our formal discussion of diffusion.

B. Continuity Equation and Fick's Second Law

As we indicated earlier, diffusion in a solution is important for the movement of solutes across plant cells and tissues. How rapid are such processes? For example, if we release a certain amount of material in one location, how long will it take before we can detect that substance at various distances? To discuss such phenomena adequately, we must determine the dependence of the concentration on both time and distance. We can readily derive such a time–distance relationship if we first consider the conservation of mass, which is necessary if we are to transform Eq. (1.1) into an expression that is convenient for describing the actual solute distributions caused by diffusion. In particular, we want to eliminate J_j from Eq. (1.1) so that we can see how c_j depends on x and t.

The amount of solute or gaseous species j per unit time crossing a given area, here considered to be a planar area perpendicular to the x-axis (Fig. 1-5), can change with position along the x-axis. Let us imagine a volume element of thickness dx in the direction of flow and of cross-sectional area A (Fig. 1-5). At x, we will let the flux density across the surface of area A be J_j. At $x + dx$, the flux density has changed to $J_j + (\partial J_j/\partial x)dx$, where $\partial J_j/\partial x$ is the gradient of the flux density of species j in the x-direction; i.e., the rate of change of J_j with position, $\partial J_j/\partial x$, times the distance, dx, gives the overall change in the flux density, $(\partial J_j/\partial x)\,dx$. The change in the amount of species j in the volume $A\,dx$ in unit time for this one-dimensional case is the amount flowing into the volume element per unit time, $J_j A$, minus that flowing out, $[J_j + (\partial J_j/\partial x)\,dx]A$. The change in the amount of species j in the volume element in unit time can also be expressed as the change in the concentration of species j with time, $\partial c_j/\partial t$, multiplied by the volume in which the change in concentration occurs, $A\,dx$. Equating these two different expressions that describe the change in the amount of species j in the volume $A\,dx$, we obtain the following relation:

$$J_j A - \left(J_j + \frac{\partial J_j}{\partial x}\,dx\right)A = \frac{\partial c_j}{\partial t}A\,dx \tag{1.2}$$

The two $J_j A$ terms on the left side of Eq. (1.2) cancel each other. After division through by $A\,dx$, Eq. (1.2) leads to the very useful expression known as the *continuity equation*:

$$-\frac{\partial J_j}{\partial x} = \frac{\partial c_j}{\partial t} \tag{1.3}$$

The continuity equation is a mathematical way of stating that matter cannot be created or destroyed under ordinary conditions. Thus, if the flux density of some species decreases as we move in the x-direction $(\partial J_j/\partial x < 0)$, Eq. (1.3) indicates that its concentration must be increasing with time, as the material is then accumulating locally. If we substitute Fick's first law (Eq. 1.1) into the continuity equation (Eq. 1.3), we obtain Fick's second law. For the important special case of constant D_j, this general equation for diffusion becomes

$$\frac{\partial c_j}{\partial t} = -\frac{\partial}{\partial x}\left(-D_j\frac{\partial c_j}{\partial x}\right) = D_j\frac{\partial^2 c_j}{\partial x^2} \tag{1.4}$$

Solution of Eq. (1.4), which is the most difficult differential equation to be encountered in this book, describes how the concentration of some species changes with position and time as a result of diffusion. To determine the particular function that satisfies this important differential equation, we need to know the specific conditions for the situation under consideration. Nevertheless, a representative solution useful for the consideration of diffusion under simple conditions will be sufficient for the present purpose of describing the characteristics of solute diffusion in general terms. For example, we will assume that there are no obstructions in the x-direction and that species j is initially placed in a plane at the origin $(x = 0)$. In this case, the following expression for the concentration of species j satisfies the differential form of Fick's second law when D_j is constant[5]:

$$c_j = \frac{M_j}{2(\pi D_j t)^{1/2}}e^{-x^2/4D_j t} \tag{1.5}$$

In Eq. (1.5), M_j is the total amount of species j per unit area initially $(t = 0)$ placed in a plane located at the origin of the x-direction (i.e., at $x = 0$, whereas y and z can have any value, which defines the plane considered here), and c_j is its concentration at position x at any

---

[5] To show that Eq. (1.5) is a possible solution of Fick's second law, it can be substituted into Eq. (1.4) and the differentiations performed (M_j and D_j are constant; $\partial ax^n/\partial x = anx^{n-1}$, $\partial e^{ax^n}/\partial x = anx^{n-1}e^{ax^n}$, and $\partial uv/\partial x = u\,\partial v/\partial x + v\,\partial u/\partial x$). The solution of Eq. (1.4) becomes progressively more difficult when more complex conditions or molecular interactions (which cause variations in D_j) are considered.

later time t. For M_j to have this useful meaning, the factor $1/[2(\pi D_j)^{1/2}]$ is necessary in Eq. (1.5) [note that $\int_{-\infty}^{\infty} c_j(x, t)dx = M_j$, where $c_j(x, t)$ is the concentration function that depends on position and time as given by Eq. 1.5 and the probability integral $\int_{-\infty}^{\infty} e^{-a^2u^2} du$, equals $\sqrt{\pi}/a$]. Moreover, the solute can be allowed to diffuse for an unlimited distance in either the plus or the minus x-direction and no additional solute is added at times $t > 0$. Often this idealized situation can be achieved by inserting a radioactive tracer in a plane at the origin of the x-direction. Equation (1.5) is only one of the possible solutions to the second-order partial differential equation representing Fick's second law. The form is relatively simple compared with other solutions, and, more important, the condition of having a finite amount of material released at a particular location is realistic for certain applications to biological problems.

C. Time–Distance Relation for Diffusion

Although the functional form of c_j given by Eq. (1.5) is only one particular solution to Fick's second law (Eq. 1.4) and is restricted to the case of constant D_j, it nevertheless is an extremely useful expression for understanding diffusion. It relates the distance a substance diffuses to the time necessary to reach that distance. The expression involves the diffusion coefficient of species j, D_j, which can be determined experimentally. In fact, Eq. (1.5) is often employed to determine a particular D_j.

Equation (1.5) indicates that the concentration in the plane at the origin of the x-direction ($x = 0$) is $M_j/[2(\pi D_j t)^{1/2}]$, which becomes infinitely large as t is turned back to 0, the starting time. This infinite value for c_j at $x = 0$ corresponds to having all of the solute initially placed in a plane at the origin. For $t > 0$, the solute diffuses away from the origin. The distribution of molecules along the x-axis at two successive times is indicated in Figs. 1-6a and 1-6b, whereas Fig. 1-6c explicitly shows the movement of the concentration profiles along the time axis. Because the total amount of species j does not change (it remains at M_j per unit area of the y–z plane, i.e., in a volume element parallel to the x-axis and extending from x values of $-\infty$ to $+\infty$), the area under each of the concentration profiles is the same.

Comparing Figs. 1-6a and 1-6b, we see that the average distance of the diffusing molecules from the origin increases with time. Also, Fig. 1-6c shows how the concentration profiles flatten out as time increases, as the diffusing solute or gaseous species is then distributed over a greater region of space. In estimating how far molecules diffuse in time t, a useful parameter is the distance x_e at which the concentration has dropped to $1/e$ or 37% of its value in the plane at the origin. Although somewhat arbitrary, this parameter describes the shift of the statistical distribution of the molecules with time. From Eq. (1.5), the concentration at the origin is $M_j/[2(\pi D_j t)^{1/2}]$ (Fig. 1-6). The concentration therefore drops to $1/e\,(= e^{-1})$ of the value at the origin when the exponent of e in Eq. (1.5) is -1. From Eq. (1.5), -1 then equals $-x_e^2/4D_j t_e$, so the distance x_e satisfies

$$x_e^2 = 4D_j t_e \tag{1.6}$$

The distance x_e along the x-axis is also indicated in Figs. 1-6a and 1-6b.

Equation (1.6) is an extremely important relationship that indicates a fundamental characteristic of diffusion processes: The distance a population of molecules of a particular solute or gaseous species diffuses—for the one-dimensional case in which the molecules are released in a plane at the origin—is proportional to the square root of both the diffusion

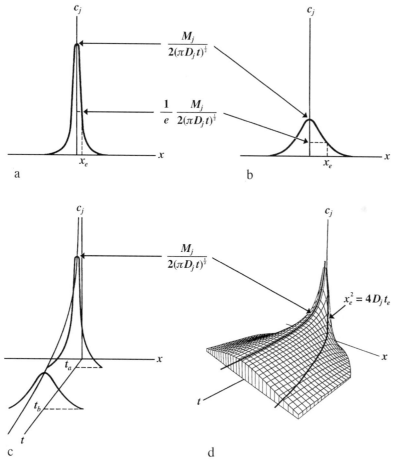

Figure 1-6. Concentration of species j, c_j, as a function of position x for molecules diffusing according to Fick's second law. The molecules were initially placed in a plane at the origin of the x-direction, i.e., at $x = 0$. For a given value of x, c_j is the same throughout a plane in the y- and z-directions. (a) Distribution of concentrations along the x-axis occurring at a time t_a, (b) distribution occurring at a subsequent time t_b, (c) portrayal of the concentration profiles at t_a and t_b, and (d) three-dimensional surface portraying change of concentration with time and position. Note that x_e is the location where the concentration of species j has dropped to $1/e$ of its value at the origin.

coefficient of the species and the time for diffusion. In other words, the time to diffuse a given distance increases with the square of that distance. An individual molecule may diffuse a greater or lesser distance in time t_e than is indicated by Eq. (1.6) (see Fig. 1-6) because the latter refers to the time required for the concentration of species j at position x_e to become $1/e$ of the value at the origin; i.e., we are dealing with the characteristics of a whole population of molecules, not the details of an individual molecule. Furthermore, the factor 4 is rather arbitrary because some criterion other than $1/e$ would cause this numerical factor to be somewhat different, although the basic form of Eq. (1.6) would be preserved. For example, the numerical factor is 2.8 if the criterion is to drop to half of the value at the origin.

Table 1-1. Diffusion Coefficients in Aqueous Solutions and Air[a]

Small solutes in water		Globular proteins in water	
Substance	D_j (m² s⁻¹)	Molecular mass (kDa)	D_j (m² s⁻¹)
Alanine	0.92×10^{-9}	15	1×10^{-10}
Citrate	0.66×10^{-9}	1000	1×10^{-11}
Glucose	0.67×10^{-9}		
Glycine	1.1×10^{-9}	Gases in air	
Sucrose	0.52×10^{-9}	Gas	D_j (m² s⁻¹)
Ca^{2+} (with Cl^-)	1.2×10^{-9}		
K^+ (with Cl^-)	1.9×10^{-9}	CO_2	1.51×10^{-5}
Na^+ (with Cl^-)	1.5×10^{-9}	H_2O	2.42×10^{-5}
CO_2	1.7×10^{-9}	O_2	1.95×10^{-5}

[a] Values are for dilute solutions at 25°C or air under standard atmospheric pressure at 20°C (source: Fasman, 1989; Lide, 1997).

Table 1-1 lists the magnitudes of diffusion coefficients for various solutes in water at 25°C.[6] For ions and other small molecules, D_j's in aqueous solutions are approximately 10^{-9} m² s⁻¹. Because proteins have higher relative molecular masses (i.e., higher molecular weights)[7] than the small solutes, their diffusion coefficients are lower (Table 1-1). Also, because of the greater frictional interaction between water molecules and fibrous proteins than with the more compact globular ones, fibrous proteins often have approximately twofold lower diffusion coefficients than do globular proteins of the same molecular weight.

To illustrate the time–distance consequences of Eq. (1.6), we will quantitatively consider the diffusion of small molecules in an aqueous solution. How long, on average, does it take for a small solute with a D_j of 1×10^{-9} m² s⁻¹ to diffuse 50 μm, the distance across a typical leaf cell? From Eq. (1.6), the time required for the population of molecules to shift so that the concentration at this distance is $1/e$ of the value at the origin is

$$t_e = \frac{(50 \times 10^{-6} \text{ m})^2}{(4)(1 \times 10^{-9} \text{ m}^2 \text{ s}^{-1})} = 0.6 \text{ s}$$

Thus, diffusion is a fairly rapid process over subcellular distances.

Next, let us consider the diffusion of the same substance over a distance of 1 m. The time needed is

$$t_e = \frac{(1 \text{ m})^2}{(4)(1 \times 10^{-9} \text{ m}^2 \text{ s}^{-1})} = 2.5 \times 10^8 \text{ s} \cong 8 \text{ years}$$

Diffusion is indeed not rapid over long distances! Thus, inorganic nutrients in xylary sap do not ascend a tree by diffusion at a rate sufficient to sustain growth. On the other hand,

---

[6] The symbol °C in this text represents degrees on the Celsius temperature scale. By the SI system, the Celsius degree as well as the kelvin unit or a kelvin (abbreviated K) is 1/273.16 of the thermodynamic temperature of the triple point of water (0.01000°C) and absolute zero is at −273.15°C. The term "centigrade" is no longer recommended.

[7] Relative molecular mass, an expression that is preferred over the commonly used "molecular weight," is a dimensionless number indicating the molecular mass of a substance relative to that of a neutral carbon atom with six protons and six neutrons (^{12}C) taken as 12.00000. For proteins, the molecular mass is often expressed in kilodaltons (kDa), where 1 Da is $\frac{1}{12}$ the mass of ^{12}C. For instance, sucrose has a relative molecular mass, or molecular weight, of 342 and a molecular mass of 342 Da.

diffusion is often adequate for the movement of solutes within leaf cells and especially inside organelles such as chloroplasts and mitochondria. In summary, diffusion in a solution is fairly fast over short distances (less than about 100 μm) but extremely slow for very long distances.

In the cytoplasm of living cells *cytoplasmic streaming* causes mechanical mixing, which leads to much more rapid movement than by diffusion. This cytoplasmic streaming, whose cessation is often a good indicator that cellular damage has occurred, requires energy. Energy is usually supplied in the form of adenosine triphosphate (ATP), and movement may involve a similar mechanism to that in muscle.

Diffusion of gases in the air surrounding and within leaves is necessary for both photosynthesis and transpiration. For instance, water vapor evaporating from the cell walls of mesophyll cells diffuses across the intercellular air spaces (Fig. 1-2) to reach the stomata and from there diffuses across an air boundary layer into the atmosphere (considered in detail in Chapter 8). CO_2 diffuses from the atmosphere through the open stomata to the surfaces of mesophyll cells, and the photosynthetically evolved O_2 traverses the same pathway in the reverse direction, also by diffusion. The experimentally determined diffusion coefficients of these three gases in air at sea level (standard atmospheric pressure) and 20°C are about 2×10^{-5} m^2 s^{-1} (Table 1-1). Such diffusion coefficients in air are approximately 10^4 times greater than the D_j describing diffusion of a small solute in a liquid, indicating that diffusion coefficients depend markedly on the medium. In particular, many more intermolecular collisions occur per unit time in a liquid phase than in a less dense gas phase. Thus, a molecule can move further in air than in an aqueous solution before being influenced by other molecules. Most cells in animals are bathed by fluids, so for larger animals circulatory systems are necessary to transport O_2 to their cells and to remove CO_2. Plants, on the other hand, often have conspicuous intercellular air spaces where the large values of D_{O_2} and D_{CO_2} in a gas phase facilitate diffusion.

Although the relation between diffusion coefficients and molecular weight can be complex, especially in solution, molecules with higher molecular weights tend to have lower diffusion coefficients. For instance, in the gaseous phase CO_2 has a lower diffusion coefficient than O_2, which in turn has a lower diffusion coefficient than H_2O (Table 1-1), as would be expected based on their molecular weights (44, 32, and 18, respectively). Diffusion coefficients depend inversely on the viscosity of the medium (discussed in Chapter 3, Section IIA, where the temperature dependence of D_j is also considered). Because diffusion coefficients of gases in air are inversely proportional to ambient pressure (Eq. 8.9), they become larger at higher altitudes.

The pathways for the movement of gas molecules in the intercellular air spaces of a leaf can be quite tortuous. Nevertheless, calculations using Eq. (1.6) can give useful estimates of the diffusion times for the many processes involving gaseous movements in a leaf. An upper limit for the diffusion pathway in such intercellular air spaces of a leaf might be 1000 μm. Using Eq. (1.6) and the diffusion coefficients given in Table 1-1, we can calculate that the times needed for water vapor, CO_2, and O_2 to diffuse 1000 μm in air are from 10 to 16 ms. Diffusion of molecules in a gas is therefore relatively rapid. However, when illumination and temperature are not limiting, the rate of photosynthesis in plants is generally limited by the amount of CO_2 diffusing into the chloroplasts, and the rate of transpiration is determined by diffusion of water vapor from the cell walls within the leaf to the outside air. In the latter case, the limitation posed by diffusion helps prevent excessive water loss from the plant and therefore is physiologically useful.

III. Membrane Structure

The plasma membrane presents a major barrier to the diffusion of solutes into and out of plant cells, the organelle membranes play an analogous role for the various subcellular compartments, and the tonoplast performs this function for the vacuole. For instance, although H_2O and CO_2 readily penetrate the plasma membrane, ATP and metabolic intermediates usually do not diffuse across it easily. Before we mathematically describe the penetration of membranes by solutes, we will briefly review certain features of the structure of membranes.

A. Membrane Models

Evert Gorter and F. Grendel in 1925 estimated that the lipids from erythrocytes (red blood cells), when spread as a monomolecular layer on water, cover an area about twice the surface area of the cells. The amount of lipid present is apparently sufficient to form a double layer in the membrane. Moreover, the penetration of a series of species across membranes often depends primarily on the relative lipid solubility of the molecules. This circumstantial evidence led to the concept of a biological membrane composed primarily of a lipid bilayer.

To help understand the properties of lipid bilayers, we must consider the charge distribution within lipid molecules. The arrangement of atoms in the hydrocarbon (containing only C and H) region of lipid molecules leads to bonding in which no local charge imbalance develops. The hydrocarbon part of the molecule is nonpolar (no substantial, local, intramolecular separation of positive and negative charge), tends to avoid water, and is therefore called *hydrophobic*. Most lipid molecules in membranes also have a phosphate or an amine group, or both, which becomes charged in an aqueous solution. Such charged regions interact electrostatically with the polar parts of other molecules. Because they interact attractively with water, the polar regions are termed *hydrophilic*.

Many of the phospholipids in membranes are glycerol derivatives that have two esterified fatty acids plus a charged side chain joined by a phosphate ester linkage (esterification refers to the chemical joining of an acid and an alcohol with the removal of water). A typical example is phosphatidyl choline (lecithin), a major component of most membranes:

$$
\begin{array}{c}
\quad\quad\quad\quad O \\
\quad\quad\quad\quad \| \\
CH_2-O-C-R \\
| \\
\quad\quad\quad\quad O \\
\quad\quad\quad\quad \| \\
CH-O-C-R' \\
| \\
\quad\quad\quad\quad O \\
\quad\quad\quad\quad \| \\
CH_2-O-P-O-CH_2CH_2\overset{+}{N}(CH_3)_3 \\
| \\
O^-
\end{array}
$$

where R and R′ are the hydrocarbon parts of the fatty acids. Various fatty acids commonly esterified to this phospholipid include palmitic (16C and no double bonds, represented as 16:0), palmitoleic (16:1), stearic (18:0), oleic (18:1), linoleic (18:2), and linolenic (18:3). For example, the major fatty acid in the membranes of higher plant chloroplasts

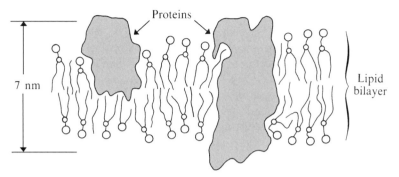

Figure 1-7. Membrane model where globular proteins are interspersed within a lipid bilayer. The ionic "head" of phospholipids is represented by ○—○ and the fatty acid side chains leading to the nonpolar "tail" are indicated by the two wavy lines emanating from the head.

is linolenic. The hydrocarbon side chains of the esterified fatty acids affect the packing of the lipid molecules in a membrane. As the number of double bonds in the fatty acid side chain increases, the side chain tends to be less straight, so the area per lipid molecule in a monolayer increases. This change in intermolecular distances affects the permeability of such lipid layers and, presumably, that of biological membranes.

To form a bilayer in a membrane, the lipid molecules have their nonpolar portions adjacent to each other (Fig. 1-7), facilitating hydrophobic interactions. The polar or hydrophilic regions are then on the outside. These charged regions of the lipids attract water molecules and the charged parts of proteins, both of which are present in membranes. For instance, up to 70% of the dry weight of mitochondrial membranes can be protein, whereas chloroplast membranes are generally <50% protein. Membranes also contain a small amount of steroids and oligosaccharides (composed of 2–10 monosaccharide residues, often bound to proteins). Membranes can be about half water by weight. When the water content of membranes is reduced below about 20%, as can occur during prolonged desiccation, the lipid bilayer configuration and membrane integrity are lost. Thus a minimal water content is necessary to create the hydrophilic environment that stabilizes the lipid bilayer.

Thermodynamic predictions and experimental results indicate that membrane proteins occur as globular forms embedded within the membranes, and indeed some proteins extend all the way across a membrane (Fig. 1-7). Globular proteins have their hydrophobic portions buried within the membrane, and a hydrophilic portion sticks out into the aqueous solution next to the membrane. In such proteins amino acids whose side chains dissociate—aspartate, glutamate, arginine, and lysine—tend to be exposed to water, whereas amino acids with hydrophobic side chains—leucine, isoleucine, and valine—tend to the interior of the membrane where they interact with the fatty acid side chains of the phospholipids, which are also hydrophobic. In this model the lipid occurs as a bilayer, and the hydrophilic portions of the lipids interact directly with water on either side of the membrane (Fig. 1-7). The lipid bilayer arrangement between the globular proteins helps account for the good correlation between lipid solubility and membrane permeation.

B. Organelle Membranes

Both mitochondria and chloroplasts are surrounded by two membranes and have extensive internal membrane systems, and both are highly involved with cellular metabolism.

Specific proteins implementing electron transfer in respiration and in photosynthesis occur in the interior membranes of mitochondria and chloroplasts, respectively. Such membrane subunits move or vibrate thermally because of their own kinetic energy and yet remain in the membrane. Diffusion coefficients of globular proteins within the plane of the membrane are generally about 10^{-14} to 10^{-13} m^2 s^{-1}, compared with nearly 10^{-10} m^2 s^{-1} for the same proteins in solution, suggesting that membranes have rather high viscosities. Such diffusion of membrane proteins allows successive interactions of a bound substrate with various enzymes located in the membrane. The side-by-side location of various components involved with electron transfer in a semisolid part of the membrane system can ensure an orderly, rapid, directed passage of electrons from enzyme to enzyme.

The proteins involved with electron transfer vary in size and shape, and thus the internal membranes of chloroplasts and mitochondria are not uniform and regular. Proteins taking part in ion transport and cell wall synthesis can be embedded in the plasma membrane, and other proteins involved with transport occur in the tonoplast. Many globular proteins in membranes serve a structural role.

Of the two mitochondrial membranes (Fig. 1-8), the outer one is much more permeable than the inner one to sucrose, various small anions and cations (e.g., H$^+$), adenine nucleotides, and many other solutes. The inner membrane invaginates to form the mitochondrial *cristae*, in which are embedded the enzymes responsible for electron transfer and the accompanying ATP formation. For instance, the inner membrane system has various dehydrogenases, an ATPase, and cytochromes (discussed in Chapters 5 and 6). These proteins with enzymatic activity occur in a globular form, and they can be an integral part of the membrane (Fig. 1-7) or loosely bound to its periphery (as for cytochrome c). Electron microscopy has revealed small particles attached by stalks to the cristae; these particles are proteins involved in the phosphorylation accompanying respiration.

Inside the inner membrane of a mitochondrion is a viscous region known as the *matrix* (Fig. 1-8). Enzymes of the tricarboxylic acid (TCA) cycle (also known as the citric acid cycle and the Krebs cycle), as well as others, are located there. For substrates to be catabolized via the TCA cycle, they must cross two membranes to pass from the cytosol to the inside of a mitochondrion. Often the slowest or rate-limiting step in the oxidation of

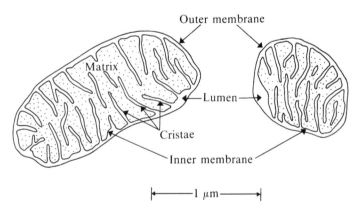

Figure 1-8. Representative mitochondria in cross section, as seen in an electron micrograph of a 50-nm-thick section of plant tissue.

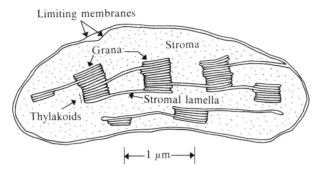

Figure 1-9. Generalized chloroplast in cross section from a leaf mesophyll cell.

such substrates is their entry into the mitochondrial matrix. Because the inner mitochondrial membrane is highly impermeable to most molecules, transport across the membrane using a "carrier" or "transporter" (see Chapter 3) is generally invoked to explain how various substances get into the matrix. These carriers, situated in the inner membrane, might shuttle important substrates from the lumen between the outer and the inner mitochondrial membranes to the matrix. Because of the inner membrane, important ions and substrates in the mitochondrial matrix do not leak out. Such permeability barriers between various subcellular compartments serve to improve the overall efficiency of a cell.

Chloroplasts (Fig. 1-9) are also surrounded by two *limiting* membranes, the outer one of which is more permeable than the inner one to small solutes. These limiting membranes are relatively high in galactolipid and low in protein. The internal *lamellar* membranes of chloroplasts are about half lipid and half protein by dry weight. Chlorophyll and most other photosynthetic pigments are bound to proteins. These proteins, as well as other components involved with photosynthetic electron transfer (see Chapters 5 and 6), are anchored in the lamellar membranes, apparently by hydrophobic forces. Each lamella consists of a pair of closely apposed membranes that are 6–8 nm thick. In many regions of a chloroplast the lamellae form flattened sacs called *thylakoids* (Fig. 1-9). When seen in an electron micrograph, a transverse section of a thylakoid shows a pair of apposed membranes joined at the ends.

The organization of lamellar membranes within chloroplasts varies greatly with environmental conditions during development and among plant groups. The chloroplasts of red algae appear to have the simplest internal structure because the lamellae are in the form of single large thylakoids separated by appreciable distances from each other. For most higher plant chloroplasts, the characteristic feature is stacks of about 10 or more thylakoids known as *grana* (Fig. 1-9); grana are typically 0.4–0.5 μm in diameter, with 10–50 occurring in a single chloroplast. The lamellar extensions between grana are called intergranal or stromal lamellae. The remainder of the chloroplast volume is known as the *stroma*, which contains the enzymes involved with the fixation of CO_2 into the various products of photosynthesis (see Chapter 8). Cyanobacteria and other photosynthetic bacteria do not contain chloroplasts, but their photosynthetic pigments are also generally located in membranes, often in lamellae immediately underlying the cell membrane. In some photosynthetic bacteria the lamellae can pinch off and form discrete subcellular bodies referred to as *chromatophores*, which are about 0.1 μm in diameter.

"Microbodies," which are also quite numerous in plant cells, are generally divided into two classes, *glyoxysomes* and *peroxisomes*, the latter being about three times as prevalent as mitochondria in many leaf cells. Microbodies are usually spherical and 0.4 to 1.5 μm in diameter. In contrast to mitochondria and chloroplasts, they are surrounded by a single membrane. Because both glyoxysomes and peroxisomes carry out only a portion of an overall metabolic pathway, these subcellular compartments depend on reactions in the cytosol and in other organelles. In Chapter 8 we will briefly consider the role of peroxisomes in photorespiration, where O_2 is consumed and CO_2 is released in the light in mesophyll cells. Glyoxysomes contain the enzymes necessary for the breakdown of fatty acids and are prevalent in fatty tissues of germinating seeds. Other membrane-bound structures in plant cells include the endoplasmic reticulum, the Golgi apparatus, and small vesicles associated with them that are involved in the transport of various cellular components.

IV. Membrane Permeability

With this knowledge of the general structure of membranes, we now turn to a quantitative analysis of the interactions between membranes and diffusing solutes. In Chapter 3 we will discuss active transport, which is important for moving specific solutes across membranes, thereby overcoming limitations posed by diffusion.

The rate-limiting step for the movement of many molecules into and out of plant cells is diffusion across the plasma membrane. Because of the close packing and the interactions between the component molecules of a membrane, such diffusion is greatly restricted compared with the relatively free movement in an aqueous phase like the cytosol. In other words, a solute molecule has great difficulty in threading its way between the molecules composing the membrane phase. The average diffusion coefficient of a small solute in a membrane is often about 10^6 times lower than in the adjacent aqueous solutions. A membrane represents a different type of molecular environment than does an aqueous solution, so the relative solubility of a species in the two phases must also be taken into account to describe membrane permeability. As we will show, Fick's first law can be modified to describe the diffusion of molecules across a membrane. Once past the barrier presented by the membrane, the molecules may be distributed throughout the cell relatively rapidly by diffusion as well as by cytoplasmic streaming.

A. Concentration Difference across a Membrane

The driving force for diffusion of uncharged molecules into or out of plant cells can be regarded as the negative concentration gradient of that species across the plasma membrane. Because the actual concentration gradient of some solute species j is not known in the plasma membrane (or in any other membrane for that matter), the driving force is generally approximated by the negative of the average gradient of that species across the membrane:

$$-\frac{\partial c_j}{\partial x} \cong -\frac{\Delta c_j}{\Delta x} = -\frac{c_j^i - c_j^o}{\Delta x} = \frac{c_j^o - c_j^i}{\Delta x} \tag{1.7}$$

where c_j^o is the concentration of species j outside the cell, c_j^i is its concentration in the cytosol, and Δx is the thickness of the plasma membrane that acts as the barrier restricting the penetration of the molecules into the cell (Fig. 1-10). The concentrations c_j^o and c_j^i in

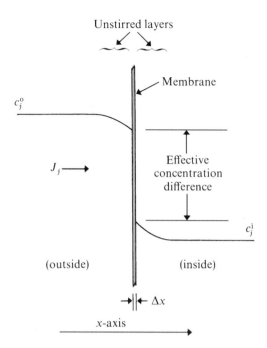

Figure 1-10. The effect of unstirred layers on the concentration of species j near a membrane across which the solute is diffusing.

Eq. (1.7) can also represent values in the phases separated by any other membrane, such as the tonoplast. If $c_j^i < c_j^o$ as we move in the positive x-direction (Fig. 1-10), $\partial c_j / \partial x$ is negative, so $-\partial c_j / \partial x$ is positive; the net flux density then occurs in the positive x-direction (see Eq. 1.1; $J_j = -D_j \, \partial c_j / \partial x$). Alternatively, if we designate a net flow into a cell as positive, which is the usual convention, then the minus sign in $-\partial c_j / \partial x$ is incorporated into the concentration difference $(c_j^o - c_j^i)$ used to describe the flux density, namely, $J_j = D_j(c_j^o - c_j^i)/\Delta x$.

The magnitude of what may be called the "effective" concentration difference across a membrane is made somewhat uncertain by the existence of unstirred boundary layers (Fig. 1-10). Boundary layers, in which turbulent mixing is absent, occur at the interface between any fluid (liquid or gas) and a solid, such as the air boundary layers on either side of a leaf (discussed in Chapter 7). A substance moves across these relatively unstirred layers next to the membrane or other solid by diffusion, indicating that a concentration gradient must also exist in the boundary layers (Fig. 1-10). When the mixing in the solutions adjacent to a membrane is increased—for example, by the turbulence resulting from cytoplasmic streaming on the inside of the cell or by rapid stirring on the outside—the thicknesses of the unstirred layers are reduced. However, they are not eliminated. Under actual experimental conditions with vigorous stirring, the external unstirred layer may be 10–100 μm thick, which is much thicker than the membrane. Because of cytoplasmic streaming in plant cells, the internal unstirred layer is generally thinner than the external one. Although diffusion is more rapid in aqueous solutions than in membranes, an unstirred layer can represent an appreciable distance for molecules to diffuse across. In some cases, diffusion through the unstirred layer can become the rate-limiting factor for the entry into cells or organelles of those molecules that rapidly penetrate the membrane. For convenience, the difference in concentration across a membrane will be represented by $c_j^i - c_j^o$, but this is an overestimate of the effective concentration difference, as is indicated in Fig. 1-10.

The difference in concentration determining the diffusion of molecules across a membrane is the concentration just inside one side of the membrane minus that just within the other side. In Eq. (1.7) the concentrations are those in the aqueous phases on either side of the membrane. Because membranes are quite different as solvents compared with aqueous solutions, the concentrations of a solute just inside the membrane can differ appreciably from those just outside in the aqueous solution. A correction factor must therefore be applied to Eq. (1.7) to give the actual concentration gradient existing in the membrane. This factor is known as the *partition coefficient*, K_j, which is defined as the ratio of the concentration of a solute in the material of the membrane to that in equilibrium outside in the aqueous phase (c_j^o or c_j^i), and so K_j is dimensionless.

As a guide to the situation in a membrane, the partition coefficient is generally determined by measuring the ratio of the equilibrium concentration of some solute in a lipid phase such as olive oil, which mimics the membrane lipids, to the concentration in an adjacent and immiscible aqueous phase (which mimics the solutions on either side of the membrane). This rather simple convention for obtaining partition coeffcients is based on the high lipid content of membranes and the numerous experimental results showing that the relative ease of penetration of a membrane depends mainly on the lipid solubility of the molecules. Partition coefficients vary widely, most being between 10^{-6} and 10 (Wright and Diamond, 1969, in which the effect on K_j of various lipids as solvents is described). For example, an inorganic ion might have a K_j near 10^{-5} for membranes, and a nonpolar hydrophobic substance might have a value near 1. K_j is assumed to be the same coming from either side of the membrane, so $K_j(c_j^o - c_j^i)$ is used for the concentration difference leading to diffusion across a membrane, where K_j is characteristic of solute species j.

B. Permeability Coefficient

We will next use Fick's first law ($J_j = -D_j \, \partial c_j/\partial x$; Eq. 1.1) to obtain an expression for describing the movement of a substance across a membrane. The negative concentration gradient will be replaced by $(c_j^o - c_j^i)/\Delta x$ (Eq. 1.7). Because D_j is the diffusion coefficient of solute species j within the membrane, we must use the actual concentration drop within the membrane, $K_j(c_j^o - c_j^i)$. These various considerations lead us to the following expression describing the diffusion of species j across a membrane or other barrier:

$$J_j = D_j \frac{K_j(c_j^o - c_j^i)}{\Delta x}$$

$$= P_j(c_j^o - c_j^i) \qquad (1.8)$$

where P_j is called the *permeability coefficient* of species j.

Because permeability coefficients have so many applications in physiology, we will define P_j explicitly:

$$P_j = \frac{D_j K_j}{\Delta x} \qquad (1.9)$$

which follows directly from Eq. (1.8). The permeability coefficient conveniently replaces three quantities that describe the diffusion of some solute across a membrane or other barrier.

The partition coefficients of membranes are generally determined by using a lipid phase such as olive oil or ether, not the actual lipids occurring in membranes, and thus K_j will have some uncertainty. Also, measurements of Δx and D_j for membranes tend to be rather indirect. On the other hand, P_j is a single, readily measured quantity characterizing the diffusion of some solute across a membrane or other barrier. The units of P_j are length per time, e.g., $m\,s^{-1}$.

A typical permeability coefficient for a small nonelectrolyte or uncharged molecule (e.g., isopropanol or phenol) is $10^{-6}\,m\,s^{-1}$ for the plasma membrane, whereas P_j for a small ion (e.g., K^+ or Na^+) might be about $10^{-9}\,m\,s^{-1}$. The lower permeability coefficients for charged particles are mainly due to the much lower partition coefficients of these solutes compared with nonelectrolytes (recall that $P_j = D_j K_j/\Delta x$; Eq. 1.9). In other words, because of its charge, an electrolyte tends to be much less soluble in a membrane than is a neutral molecule, so the effective concentration gradient driving the charged species across the membrane is generally smaller for given concentrations in the aqueous phases on either side of the barrier. By comparison, water can rapidly enter or leave cells and can have a permeability coefficient of about $10^{-4}\,m\,s^{-1}$ for Characean algae, although lower values can occur for other plant cells. (Actually, P_{water} is difficult to measure because of the relative importance of the unstirred layers.) Although adequate measurements on plant membranes have not been made for O_2 and CO_2, their P_j's for the plasma membrane are probably quite high. For instance, O_2, which is lipophilic and has a high partition coefficient of 4.4, has a very high permeability coefficient of about $0.3\,m\,s^{-1}$ for erythrocyte membranes at $25°C$ (Fischkoff and Vanderkooi, 1975).

C. Diffusion and Cellular Concentration

Instead of calculating the flux density of some species diffusing into a cell—the amount entering per unit area per unit time—we often focus on the total amount of that species diffusing in over a certain time interval. Let s_j be the amount of solute species j inside the cell, where s_j can be expressed in moles. If that substance is not involved in any other reactions, ds_j/dt represents the rate of entry of species j into the cell. The flux density J_j is the rate of entry of the substance per unit area or $(1/A)\,(ds_j/dt)$, where A is the area of the cellular membrane across which the substance is diffusing. Using this expression for J_j, we can express Eq. (1.8) as follows:

$$ds_j/dt = J_j A = P_j A\left(c_j^o - c_j^i\right) \tag{1.10}$$

When the external concentration of species j (c_j^o) is greater than the internal one (c_j^i), species j will enter the cell and ds_j/dt is positive, as Eq. (1.10) indicates.

A question that often arises in cellular physiology is how the internal concentration of a penetrating solute changes with time. How long, for example, would it take for the internal concentration of an initially absent species to build up to 70% of the external concentration? To solve the general case, we must first express s_j in Eq. (1.10) in a suitable fashion and then integrate. It is useful to introduce the approximation—particularly appropriate to plant cells—that the cell volume does not change during the time interval of interest. In other words, because of the rigid cell wall, we will assume that the plant cell has a nearly constant volume (V) during the entry or the exit of the solute being considered.

The average internal concentration of species j (c_j^i) is equal to the amount of that particular solute inside the cell (s_j) divided by the cellular volume (V), or s_j/V. Therefore,

s_j can be replaced by Vc_j^i; ds_j/dt in Eq. (1.10) thus equals $V\,dc_j^i/dt$ if the cell volume does not change with time. We also assume that species j is not produced or consumed by any reaction within the cell, so the diffusion of species j into or out of the cell is the only process affecting c_j^i. For simplicity, we are treating the concentration of species j as if it were uniform within the cell. (In a more complex case we might need a relation of the form of Eq. 1.10 for each membrane-surrounded compartment.) Moreover, we are presupposing either mechanical mixing or rapid diffusion inside the cell, so the drop in concentration $(c_j^o - c_j^i)$ occurs essentially only across the membrane. Finally, let us assume that P_j is independent of concentration, a condition that is often satisfactorily met over a limited range of concentration. Upon replacement of ds_j/dt by $V\,dc_j^i/dt$, rearrangement to separate variables, and the insertion of integral signs, Eq. (1.10) becomes

$$\int_{c_j^i(0)}^{c_j^i(t)} \frac{dc_j^i}{c_j^o - c_j^i} = \frac{P_j A}{V} \int_0^t dt \qquad (1.11)$$

where $c_j^i(0)$ is the initial internal concentration of species j, i.e., when $t = 0$, and $c_j^i(t)$ is the concentration at a later time t.

The volume outside certain unicellular algae and other membrane-surrounded entities can be large compared with V so that the external concentration (c_j^o) does not change appreciably. Such an approximation can also be appropriate for experiments of short duration or when special arrangements are made to maintain c_j^o at some fixed value. For those cases in which c_j^o is constant, Eq. (1.11) can be integrated to give the following expression (note that $\int dx/(a-x) = -\int d(a-x)/(a-x) = -\ln(a-x) + b$, where ln is the natural logarithm to the base e; see Appendix III for certain logarithmic identities):

$$\frac{P_j A t}{V} = \ln\frac{c_j^o - c_j^i(0)}{c_j^o - c_j^i(t)} \qquad (1.12)$$

Starting with a known c_j^o outside a cell and determining the internal concentration both initially, i.e., for $t = 0$, and at some subsequent time t, we can calculate P_j from Eq. (1.12) if A/V is known. Even when A/V is not known, the relative permeability coefficients for two substances can be determined from the time dependencies of the respective c_j^i's. The previous derivation can easily be extended to the case in which c_j^o is zero. In that case, $P_j A t/V = \ln[c_j^i(0)/c_j^i(t)]$, an expression that can be used to describe the diffusion of a photosynthetic product out of a chloroplast or some substance from a cell into a large external solution initially devoid of that solute. Such studies can be facilitated by using radioactive tracers, which generally are initially present only in one region.

When diffusion occurs from all directions across a membrane surrounding a cell (or an organelle), the time to reach a given internal concentration can be directly proportional to the dimensions of the cell. For convenience in illustrating this, let us consider a spherical cell of radius r. The volume V then equals $(4/3)\pi r^3$ and the surface area A is $4\pi r^2$ (see Appendix III), so V/A is $r/3$. Equation (1.12) then becomes

$$t^{\text{sphere}} = \frac{r}{3P_j}\ln\frac{c_j^o - c_j^i(0)}{c_j^o - c_j^i(t)} \qquad (1.13)$$

For the same P_j, a given level inside a small cell is reached sooner than inside a large cell. When diffusion across the membrane into a spherical cell occurs from all directions, the time to reach a given concentration inside is linearly proportional to the radius (assuming a uniform c_j^i). This contrasts with the time it takes for a planar front to diffuse in one direction, which is proportional to the square of the distance traveled ($x_e^2 = 4D_j t_e$; Eq. 1.6).

Using Eq. (1.13), we can calculate the time required for a substance initially absent from a cell [$c_j^i(0) = 0$] to achieve an internal concentration equal to half the external concentration [$c_j^i(t) = \frac{1}{2}c_j^o$]. These conditions mean that [$c_j^o - c_j^i(0)]/[c_j^o - c_j^i(t)$] equals ($c_j^o - 0)/(c_j^o - \frac{1}{2}c_j^o$), which is 2. Substituting this into Eq. (1.13) indicates that the time needed is $(r/3P_j)\ln 2$, where $\ln 2$ equals 0.693. For a spherical cell 25 μm in radius, which is a reasonable dimension for spongy mesophyll cells in many leaves (see Fig. 1-2), the time for an initially absent nonelectrolyte with a P_j of $1 \times 10^{-6}\,\mathrm{m\,s^{-1}}$ to reach half of the external concentration is

$$t_{\frac{1}{2}}^{\text{sphere}} = \frac{(25 \times 10^{-6}\,\text{m})}{(3)(1 \times 10^{-6}\,\mathrm{m\,s^{-1}})}(0.693) = 6\,\text{s}$$

and the time is 10^3 times longer (about 2 h) for an electrolyte with a P_j of $1 \times 10^{-9}\,\mathrm{m\,s^{-1}}$. Hence, the plasma membrane is an excellent barrier for electrolytes, markedly hindering their entry into or exit from the cell. On the other hand, small nonelectrolytes can fairly readily diffuse in and out of plant cells, depending to a large extent on their relative lipid solubility.

V. Cell Walls

Cell walls play many roles in plants. Their rigidity helps determine the size and the shape of a cell and ultimately the morphology of a plant. This supportive role is performed in conjunction with the internal hydrostatic pressure, which causes a distension of the cell walls. The cell wall is also intrinsically involved in many aspects of the ion and water relations of plant cells. Because it surrounds the plasma membrane of each cell, fluxes of water and solutes into or out of a plant cell protoplast must cross the cell wall, usually by diffusion. The cell wall generally has a large negative charge, so it can interact differently with cations than with anions. Water evaporating from a plant during transpiration comes directly from cell walls (considered in Chapters 2 and 8). The space surrounded by the cell wall in certain specialized cells can act as a channel through which solutions move. In the xylem, for example, the conducting cells have lost their protoplasts, with the resulting pathway for the conduction of solutions in the xylem consisting essentially of hollow pipes, or conduits, made exclusively of cell walls.

Cell walls vary from about 0.1 to 10 μm in thickness, and they are generally divided into three regions: primary wall, secondary wall, and middle lamella. The *primary cell wall* surrounds dividing meristematic cells as well as the elongating cells during the period of cell enlargement. The cell wall often becomes thickened by the elaboration of a *secondary cell wall* inside the primary one (Fig. 1-11), which makes the cell mechanically much less flexible. Hence, cells whose walls have undergone secondary thickening, such as xylem vessel members, are generally incapable of subsequent elongation. The cell wall also includes an amorphous region between contiguous cells called the *middle lamella*. Although it contains some cellulose, the middle lamella may be composed mainly of the calcium salts of pectic acids, which causes adjacent cells to adhere to each other.

A. Chemistry and Morphology

Cellulose is the most abundant organic component in living organisms. It is the characteristic substance of the plant cell wall, constituting from 25 to 50% of the cell wall organic

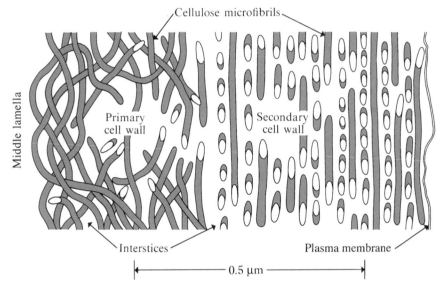

Figure 1-11. Hypothetical thin section through a cell wall, indicating the cellulose microfibrils in the primary and secondary cell walls. The interstices are filled with noncellulosic material including an appreciable amount of water, the solvent for solute diffusion across the cell wall.

material. Cellulose is a linear (unbranched) polysaccharide consisting of 1,4-linked β-D-glucopyranose units:

The polymer is about 0.8 nm in its maximum width, 0.33 nm^2 in cross-sectional area, and can contain about 10,000 glucose residues with their rings in the same plane. In the cell wall these polymers are organized into *microfibrils* that can be 5×9 nm in cross section. These microfibrils apparently consist of an inner core of about 50 parallel chains of cellulose arranged in a crystalline array surrounded by a similar number of cellulose and other polymers in a paracrystalline array. Microfibrils are the basic unit of the cell wall and are readily observed in electron micrographs. Although great variation exists, they tend to be interwoven in the primary cell wall and parallel to each other in the secondary cell wall (Fig. 1-11).

Interstices between the cellulose microfibrils are usually 5–30 nm across. The interstices can contain a matrix of amorphous components occupying a larger volume of the cell wall than do the microfibrils themselves. In fact, by weight, the main constituent of the cell wall is actually water, some consequences of which we will consider here and in Chapter 2.

The cell wall matrix contains noncellulosic polysaccharides such as pectin, lignins (secondary cell wall only), a small amount of protein, bound and free water, appreciable calcium, other cations, and sometimes silicates. Lignins are complex polymers based on phenylpropanoid subunits (a 6-carbon ring, to which is attached a 3-carbon chain as in phenylalanine, a lignin precursor) plus certain other residues. Lignins constitute the second most abundant class of organic molecules in living organisms and are about half as prevalent as cellulose. They are quite resistant to enzymatic degradation, so lignins are important in peat and coal formation. They also make the cell wall more rigid. Because a plant cannot break down the lignin polymers, cells are unable to expand after extensive lignification of their cell walls, as occurs in the secondary walls of xylem vessel members. Pectin consists primarily of 1,4-linked α-D-galacturonic acid residues, the carboxyl groups of which are normally dissociated and have a negative charge (galacturonic acid differs from galactose by having $-COOH$ instead of $-CH_2OH$ in the 6-carbon position). The negative charge of the dissociated carboxyl groups leads to the tremendous cation-binding capacity of cell walls. In particular, much of the divalent cation calcium (Ca^{2+}) is bound, which may help link the various polymers together. Polymers based on 1,4-linked β-D-xylopyranose units (xylans), as well as on many other residues, can also be extracted from cell walls. They are loosely referred to as hemicelluloses, e.g., xylans, mannans, galactans, and glucans. In general, hemicelluloses tend to have lower molecular weights (in the ten thousands) than pectin (about 5–10 times larger) or cellulose. The presence of negatively charged pectins hinders the entry of anions into plant cells. As we will consider next, however, ions and other solutes generally pass through the cell wall much more easily than through the plasma membrane.

B. Diffusion across Cell Walls

How do we calculate the ease with which molecules might diffuse across a cell wall? A good place to start is Eq. (1.9), which indicates that the permeability coefficient of species j equals $D_j K_j / \Delta x$. We must first consider what we mean by D_j and K_j in a cell wall. The diffusion of solutes across a cell wall occurs mainly in the water located in the numerous interstices, which are often about 10 nm across. Thus the movement from the external solution up to the plasma membrane is in aqueous channels through the cell wall. By the definition of a partition coefficient (solubility in barrier/solubility in water), we recognize that K_j is 1 for the water-filled interstices of the cell wall, very low for the solid phases (cellulose, lignin, and the other polymers present in the wall), and at some intermediate value for the cell wall as a whole. Because diffusion takes place primarily in the cell wall water, we will let K_j be 1. If we use this convenient definition for K_j in cell walls, we must define D_j on the basis of the whole cell wall and not on the much smaller area presented by the interstices. The value of D_j in the water of the interstices could be very similar to that in a free solution, whereas its value averaged over the whole cell wall is considerably less because the interstices are not straight channels through the cell wall and because they do not occupy the entire cell wall volume. Such effective diffusion coefficients of small solutes are at least three times lower when averaged over the cell wall than are the D_j's of the same species in an extended aqueous solution.

Next, we calculate the permeability coefficient for a solute that has a diffusion coefficient of 2×10^{-10} m^2 s^{-1} for a cell wall. We will assume a representative value of 1 μm for the cell wall thickness. Using Eq. (1.9) ($P_j = D_j K_j / \Delta x$), we can thus estimate that P_j for the cell wall is

$$P_j = \frac{(2 \times 10^{-10} \text{ m}^2 \text{ s}^{-1})(1)}{(1 \times 10^{-6} \text{ m})} = 2 \times 10^{-4} \text{ m s}^{-1}$$

Most of the permeability coefficients for small solutes crossing the plasma membrane range from 10^{-10} to 10^{-6} m s^{-1}. Hence, a cell wall generally has a higher permeability coefficient than does a membrane, which means that the wall is usually more permeable for small solutes than is the plasma membrane. For comparison, let us consider a permeability coefficient appropriate for an unstirred liquid layer adjacent to a cell wall or membrane. Specifically, D_j for a small solute may be 1×10^{-9} m² s⁻¹ in water, K_j is 1 in the aqueous solution, and let us assume that Δx is 30 μm for the unstirred layer. In such a case we have

$$P_j = \frac{(1 \times 10^{-9}\,\text{m}^2\,\text{s}^{-1})(1)}{(30 \times 10^{-6}\,\text{m})} = 3 \times 10^{-5}\,\text{m}\,\text{s}^{-1}$$

which is less than the value for a cell wall.

Molecules diffuse less readily across a given distance in a plasma membrane than in a cell wall or the adjacent unstirred layer. For the previous numerical values, $D_j K_j$ is 1×10^{-9} m² s⁻¹ in the aqueous solution and 2×10^{-10} m² s⁻¹ in the cell wall, yet for a plasma membrane about 7 nm thick, $D_j K_j$ is 10^{-18}–10^{-14} m² s⁻¹. Membranes do indeed provide very effective barriers for the diffusion of solutes.

Secondary cell walls are often interrupted by localized pits. A pit in the wall of a given cell usually occurs opposite a complementary pit in an adjacent cell, so the cytoplasm of two adjacent cells is brought into close proximity at such a pit pair. The local absence of extensive cell wall substance facilitates the diffusion of molecules from one cell into the other. An easier way for molecules to move between plant cells is by means of the *plasmodesmata* (singular: *plasmodesma*). These are fine, membrane-flanked, cytoplasmic threads that pass from a protoplast, through a pore in the cell wall, directly into the protoplast of a second cell (Fig. 1-12). The pores generally occur in locally thin regions of the primary cell wall, referred to as primary pit-fields, which can contain many plasmodesmata. (If a secondary cell wall is deposited, openings in that wall occur in the regions of the primary

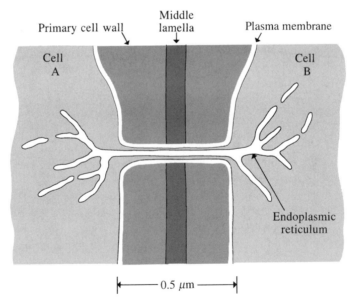

Figure 1-12. Longitudinal section through a plasmodesma, showing that the primary cell walls are locally thinner (which is generally the case) and that the plasma membranes of the two adjacent cells are continuous.

pit-fields.) Plasmodesmata can be 50 nm in diameter (range, 20–200 nm) and typically have a frequency of 2–10 per μm^2 of cell surface. They usually occupy about 0.1–0.5% of the surface area of a cell. The continuum of communicating cytoplasm created by such intercellular connections is commonly referred to as the *symplasm* (*symplast* is defined as the continuum of protoplasts together with the plasmodesmata that interconnect them, and hence means the same as symplasm; compare *apoplast*). The plasmodesmata are not simple aqueous channels between cells because they contain endoplasmic reticulum that is continuous from one cell to the adjacent cell (Fig. 1-12). The plasmodesmata provide a particularly effective pathway for movement of solutes (molecular masses up to about 900 Da) between adjacent cells where permeability barriers in the form of either cell walls or membranes are avoided.

Because solute movement from cell to cell through the symplasm is important physiologically, let us make a rather oversimplified calculation so that we can compare it with the concomitant diffusion across the plasma membrane. We will use Fick's first law presented in Eq. (1.8), namely, $J_j = D_j K_j \, \Delta c_j / \Delta x = P_j \, \Delta c_j$. Let us consider two adjacent cells in which the glucose concentration differs by 1 mol m^{-3} (1 mM). Glucose is rather insoluble in membrane lipids and might have a permeability coefficient of about 1×10^{-9} m s^{-1} for a plasma membrane. For movement from cell to cell across the two plasma membranes in series, the effective P_j is 0.5×10^{-9} m s^{-1} (inclusion of the cell wall as another series barrier for diffusion only slightly reduces the effective P_j in this case; see Problem 1.4 for handling the permeability coefficient of barriers in series). Thus the flux density of glucose across the plasma membranes toward the cell with the lower concentration is

$$J_{glucose}^{plasma\ membrane} = (0.5 \times 10^{-9}\ \mathrm{m\,s^{-1}})(1\ \mathrm{mol\,m^{-3}}) = 5 \times 10^{-10}\ \mathrm{mol\,m^{-2}\,s^{-1}}$$

Let us next consider the flux for the symplasm. In the aqueous part of the plasmodesmata, $D_{glucose}$ should be similar to its value in water, 0.7×10^{-9} m^2 s^{-1} (see Table 1-1), and $K_{glucose}$ is 1. We will let the plasmodesmata be 0.5 μm long. The flux density in the pores thus is

$$J_{glucose}^{pores} = \frac{(0.7 \times 10^{-9}\ \mathrm{m^2\,s^{-1}})(1)}{(0.5 \times 10^{-6}\ \mathrm{m})}(1\ \mathrm{mol\,m^{-3}}) = 1.4 \times 10^{-3}\ \mathrm{mol\,m^{-2}\,s^{-1}}$$

If the aqueous channels or pores occupy 10% of the area of the plasmodesmata, which in turn occupy 0.2% of the surface area of the cells, then the rate of glucose diffusion through the plasmodesmata per unit area of the cells is $(0.1)\,(0.002)\,J_{glucose}^{pores}$, or 3×10^{-7} mol m^{-2} s^{-1}. This is 600-fold greater than the simultaneously occurring diffusion across the plasma membranes.

The permeability coefficients of the plasma membranes for phosphorylated (charged) sugars like ribose-5-phosphate or ribulose-1,5-bisphosphate are less than that for glucose, and the diffusion coefficients in the plasmodesmata are about the same as for glucose. Thus the discrepancy between flux densities of charged species through the plasmodesmata and those across the plasma membranes is even greater than the difference calculated here for glucose. The joining of cytoplasms into a symplasm indeed facilitates the diffusion of solutes from one cell to another.

C. Stress–Strain Relations of Cell Walls

Cell walls of mature plant cells are generally quite resistant to mechanical stretching, especially when appreciable thickening of the secondary cell walls occurs. Nevertheless, cell walls will stretch when a stress is applied, where *stress* equals the force per unit area. Stress can be caused by an external force, such as the pulling on cotton fibers in a shirt, or an internal force, such as that caused by the hydrostatic pressure in a cell. Such applied forces lead to deformation of cell walls, which is quantified as a strain. *Strain* is the length of the stressed material (l) minus the initial unstressed length (l_0) divided by the unstressed length. Hence, strain is the fractional change in length ($\Delta l/l_0$) resulting when stresses occur in a material.

Reversible elastic properties are described by a measure of elasticity known as *Young's modulus*, which is the ratio of applied stress (force per unit area) to the resulting strain (fractional change in length):

$$\text{Young's modulus} = \frac{\text{stress}}{\text{strain}} = \frac{\text{force/area}}{\Delta l/l_0} \tag{1.14}$$

Because $\Delta l/l_0$ is dimensionless, Young's modulus has the dimensions of force per unit area, or pressure. A high value of this modulus of elasticity means that a large stress must be applied to produce an appreciable strain. For instance, Young's modulus for dry cotton fibers, which are nearly pure cellulose, is quite large—about $10^{10}\,\mathrm{N\,m^{-2}}$, or 10,000 MPa, which is 5% of that for steel (Preston, 1974).[8] Because of both the complicated three-dimensional array of microfibrils in the cell wall and the presence of many other components, Young's modulus for a cell wall is considerably less than for pure cellulose. For example, the modulus of elasticity for the cell wall of *Nitella* is about 700 MPa (Kamiya *et al.*, 1963). We will use this value when we indicate the fractional stretching that can occur for plant cells.

The hydrostatic (turgor) pressure (force per unit area in a liquid), P, acts uniformly in all directions in a cell. This internal pressure pushes against the plasma membrane, which itself is closely appressed to the cell wall. This causes the cell to expand and also leads to tensions (stresses) in the cell wall. The magnitude of these stresses varies with the physiological state of the plant as well as with direction, an aspect that we will consider next.

A cylinder, which closely approximates the shape of the large internodal cells of *Chara* or *Nitella* (Fig. 3-9), is a useful geometrical model for evaluating the various cell wall stresses. We will let the radius of the cylinder be r (Fig. 1-13). The force on an end wall is the pressure, P, times the area of the end wall, πr^2, so it equals $P\pi r^2$. This force is balanced by a force arising from the *longitudinal stress*, σ_L, that occurs in the cell wall. The area over which σ_L acts is shown by the cut portion of the cell wall in the left part of Fig. 1-13. The longitudinal stress occurs in an annulus of circumference $2\pi r$ (approximately) and a width equal to the thickness of the cell wall, t. The longitudinal force in the cell wall is thus $(\sigma_L)(2\pi r)(t)$. This force is an equal and opposite reaction to $P\pi r^2$, so $\sigma_L 2\pi rt$ equals $P\pi r^2$, or

$$\sigma_L = \frac{rP}{2t} \tag{1.15}$$

The longitudinal stress acts parallel to the axis of the cylinder and resists the lengthwise deformation of the cell.

A *tangential stress*, σ_T, also exists in the cell wall in response to the internal pressure; it limits the radial expansion of the cell. To determine the magnitude of this stress, we will

---

[8] The newton, abbreviated N, and the pascal, abbreviated Pa, are the SI units for force and pressure, respectively ($1\,\mathrm{N\,m^{-2}} = 1\,\mathrm{Pa}$). Pressures in plant studies have been expressed in bars, where 1 bar $= 10^6$ dynes cm^{-2}, 0.987 atmosphere, or 0.1 MPa. See Appendix II for further conversion factors for pressure.

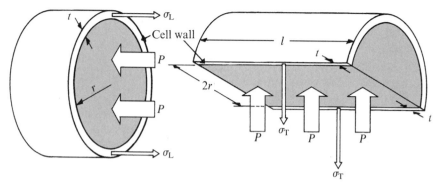

Figure 1-13. Schematic sections of a hypothetical cylindrical cell resembling the internodal cells of *Nitella* or *Chara*, illustrating various dimensions and the stresses (σ_L and σ_T) existing in the cell wall. One way to calculate the stresses is to imagine that the cellular contents are removed, leaving only the cell wall, which has a uniform hydrostatic pressure P acting perpendicular to its inside surface. The projection of this P over the appropriate area gives the force acting in a certain direction, and the reaction to this force is an equal force in the opposite direction in the cell wall. By dividing the force in the cell wall by the area over which it occurs, we can determine the cell wall stress.

consider a cell split in half along its axis (right side of Fig. 1-13). The split part of the cell has an area of $2rl$ in the plane of the cut; this area is acted on by the pressure P, leading to a force of $P2rl$. This force is resisted by the tangential stress in the cell wall. As shown in Fig. 1-13, σ_T acts along two cell wall surfaces, each of width t and length l. The total tangential force in the cut part of the cell wall is thus the area ($2tl$) times the tangential stress. Equating this force ($\sigma_T 2tl$) with that due to the internal pressure in the cell ($P2rl$), we obtain the following relationship for the tangential stress:

$$\sigma_T = \frac{rP}{t} \tag{1.16}$$

The tangential stress in the cell wall given by Eq. (1.16) is twice as large as the longitudinal stress (Eq. 1.15). This simple cylindrical model thus illustrates that the stresses in a cell wall can vary with direction. Young's modulus also varies with direction, reflecting the anisotropic orientation of the microfibrils in the cell wall.

To estimate the magnitudes of the stresses and the resulting strains in the cell wall, let us consider a *Nitella* or *Chara* cell that is 1 mm in diameter with a cell wall that is 5 μm thick. In this case, r/t is

$$\frac{r}{t} = \frac{(0.5 \times 10^{-3}\,\text{m})}{(5 \times 10^{-6}\,\text{m})} = 100$$

A reasonable estimate of P is 0.5 MPa. Using Eq. (1.15), we can calculate the longitudinal stress in the cell wall of such a cell:

$$\sigma_L = \frac{(100)(0.5\,\text{MPa})}{(2)} = 25\,\text{MPa}$$

which is an appreciable tension. By Eq. (1.16), the tangential stress is twice as great, or 50 MPa. Ignoring changes in the radial direction, we can calculate the strain produced by the longitudinal stress using the definition of Young's modulus (Eq. 1.14) and its particular value in the longitudinal direction for the cell wall of *Nitella*, 700 MPa. From Eq. (1.14), the fractional change in length is then

$$\frac{\Delta l}{l_0} = \frac{(25\,\text{MPa})}{(700\,\text{MPa})} = 0.036$$

i.e., about 4%. (If changes in the radial direction were included, the length change would be about 3% in the present case.) Hence, even with an internal pressure of 0.5 MPa, the cell wall (and consequently the whole cell) is not extended very much. The cell wall is indeed rigid and therefore well suited both for delimiting individual cells and for providing the structural support of a plant.

Elastic properties of plant cell walls must be considered when analyzing the cell expansion accompanying growth as well as other features of water movement. For this, we are interested in what volume change (ΔV) is caused by a given pressure change (ΔP); this relation can be quantified using the *volumetric elastic modulus* (ε):

$$\varepsilon = \frac{\Delta P}{\Delta V/V} \tag{1.17}$$

Equation (1.17) indicates that $\Delta V/V = \Delta P/\varepsilon$, so cells with a higher ε will have a smaller fractional change in volume for a given ΔP, i.e., they are more rigid. Values of ε generally range from 1 to 50 MPa, as cells change in volume by 0.2–10% for each 0.1 MPa change in internal hydrostatic pressure. For a typical ε of 10 MPa, a 1% increase in volume due to water influx will be accompanied by a 1% decrease in cellular solute concentration but a 0.1 MPa increase in internal hydrostatic pressure (Eq. 1.17). Also, ε depends on P (it is smaller at lower P), V, and the developmental stage of the cells. We will return to a consideration of the elastic modulus when discussing the water relations of cells in Chapter 2. We note here that both Young's modulus (Eq. 1.14) and the volumetric elastic modulus (Eq. 1.17) represent an applied pressure divided by a fractional change in size.

Before concluding this discussion of cell walls, we note that the case of elasticity or reversible deformability is only one extreme of stress–strain behavior. At the opposite extreme is plastic (irreversible) extension. If the amount of strain is directly proportional to the time that a certain stress is applied, and if the strain persists when the stress is removed, we have viscous flow. The cell wall exhibits intermediate properties and is said to be *viscoelastic*. When a given stress is applied to a viscoelastic material, the resulting strain is approximately proportional to the logarithm of time. Such extension is partly elastic (reversible) and partly plastic (irreversible). Underlying the viscoelastic behavior of the cell wall are the crosslinks between the various polymers. For example, if a bond from one cellulose polymer to another is broken while the cell wall is under tension, a new bond may form in a less strained configuration, leading to an irreversible or plastic extension of the cell wall. The quantity responsible for the tension in the cell wall—which in turn leads to such viscoelastic extension—is the hydrostatic pressure within the cell.

Problems

1.1 A thin layer of some solution is inserted into a long column of water. One hour later the concentration of the solute is 100 mol m^{-3} (0.1 M) at the plane of insertion and 37 mol m^{-3} (0.037 M) at a distance 3 mm away.

(a) What is its diffusion coefficient?

(b) When the concentration 90 mm away is 37% of the value at the plane of insertion, how much time has elapsed?

(c) How many moles of solute per unit area were initially inserted into the column of water?

(d) Suppose that a trace amount of a substance having a diffusion coefficient 100 times smaller than that of the main solute was also initially introduced. For the time in (b), where would its concentration drop to $1/e$ of the value at the plane of insertion?

1.2 Let us suppose that mitochondria with a volume of 0.30 μm^3 each and a density of 1110 kg m^{-3} ($= 1.10$ g cm^{-3}) diffuse like a chemical species.

(a) What is the "molecular weight" of mitochondria?

(b) Suppose that a chemically similar species of molecular weight 200 has a diffusion coefficient of 0.5×10^{-9} m^2 s^{-1}. If diffusion coefficients are inversely proportional to the cube root of molecular weights for this series of similar species, what is $D_{mitochondria}$?

(c) If we assume that Eq. (1.6) can adequately describe such motion, how long would it take on the average for a mitochondrion to diffuse 0.2 μm (a distance just discernible using a light microscope)? How long would it take for the mitochondrion to diffuse 50 μm (the distance across a typical leaf cell)?

(d) If D_{ATP} is 0.3×10^{-9} m^2 s^{-1}, how long would it take ATP to diffuse 50 μm? Is it more expedient for mitochondria or ATP to diffuse across a cell?

1.3 Suppose that an unstirred air layer 1 mm thick is adjacent to a guard cell with a cell wall 2 μm thick.

(a) Assume that an (infinitely) thin layer of $^{14}CO_2$ is introduced at the surface of the guard cell. If D_{CO_2} is 10^6 times larger in air than in the cell wall, what are the relative times for $^{14}CO_2$ to diffuse across the two barriers?

(b) If it takes $^{14}CO_2$ just as long to cross an 8-nm plasma membrane as it does to cross the cell wall, what are the relative sizes of the two diffusion coefficients (assume that the $^{14}CO_2$ was introduced in a plane between the two barriers)?

(c) Assuming that the partition coefficient for CO_2 is 100 times greater in the cell wall than in the plasma membrane, in which barrier is the permeability coefficient larger, and by how much?

1.4 Without correcting for the effect of an unstirred layer 20 μm thick outside a membrane 7.5 nm in thickness, the apparent (total) permeability coefficients were measured to be 1.0×10^{-4} m s^{-1} for D_2O, 2.0×10^{-5} m s^{-1} for methanol, and 3.0×10^{-8} m s^{-1} for L-leucine. For barriers in series, the overall permeability coefficient for species j (P_j^{total}) is related to those of the individual barriers (P_j^i) as follows: $1/P_j^{total} = \sum_i 1/P_j^i$. For purposes of calculation, we will assume that in the present case the unstirred layer on the inner side of the membrane is negligibly thin.

(a) What is P_j for the external unstirred layer for each of the compounds? Assume that D_{D_2O} is 2.6×10^{-9} m^2 s^{-1}, $D_{methanol}$ is 0.80×10^{-9} m^2 s^{-1}, and $D_{leucine}$ is 0.20×10^{-9} m^2 s^{-1} in water at 25°C.

(b) What are the permeability coefficients of the three compounds for the membrane?

(c) From the results in (a) and (b), what are the main barriers for the diffusion of the three compounds in this case?

(d) What are the highest possible values at 25°C of P_j^{total} for each of the three compounds moving across an unstirred layer of 20 μm and an extremely permeable membrane in series?

1.5 Consider a solute having a permeability coefficient of 10^{-6} m s^{-1} for the plasma membrane of a cylindrical *Chara* cell that is 100 mm long and 1 mm in diameter. Assume that its concentration remains essentially uniform within the cell.

(a) How much time would it take for 90% of the solute to diffuse out into a large external solution initially devoid of that substance?

(b) How much time would it take if diffusion occurred only at the two ends of the cell?

(c) How would the times calculated in (a) and (b) change for 99% of the solute to diffuse out?

(d) How would the times change if P_j were 10^{-8} m s^{-1}?

1.6 A cylindrical *Nitella* cell is 100 mm long and 1 mm in diameter, a spherical *Valonia* cell is 10 mm in diameter, and a spherical *Chlorella* cell is 4 μm in diameter.

(a) What is the area/volume in each case?

(b) Which cell has the largest amount of surface area per unit volume?

(c) If it takes 1 s for the internal concentration of ethanol, which is initially absent from the cells, to reach half the external concentration for *Chlorella*, how long would it take for *Nitella* and *Valonia*? Assume that P_{ethanol} is the same for all the cells.

(d) Assume that the cell walls are equal in thickness. For a given internal pressure, which cell would have the highest cell wall stress (consider only the lateral wall for *Nitella*)?

References

Brett, C. T., and Waldron, K. W. (1996). *Physiology and Biochemistry of Plant Cell Walls* (2nd ed.). Chapman & Hall, London.

Crank, J. (1975/1993). *The Mathematics of Diffusion* (2nd ed.). Clarendon/Oxford Univ. Press, Oxford.

Davson, H., and Danielli, J. F. (1952). *The Permeability of Natural Membranes* (2nd ed.). Cambridge Univ. Press, Cambridge, UK.

Diamond, J. M., and Wright, E. M. (1969). Biological membranes: The physical basis of ion and nonelectrolyte selectivity. *Annu. Rev. Physiol.* **31**, 581–646.

Esau, K. (1977). *Anatomy of Seed Plants* (2nd ed.). Wiley, New York.

Fasman, G. B. (1989). *Practical Handbook of Biochemistry and Molecular Biology.* CRC Press, Boca Raton, FL.

Fick, A. (1855). Über diffusion. *Poggendorffs Ann.* **94**, 59–86.

Fischkoff, S., and Vanderkooi, J. M. (1975). Oxygen diffusion in biological and artificial membranes determined by the fluorochrome pyrene. *J. Gen. Physiol.* **65**, 663–676.

Fry, S. C. (1988). *The Growing Plant Cell Wall: Chemical and Metabolic Analysis.* Wiley, New York.

Gebhart, B. (1993). *Heat Conduction and Mass Diffusion.* McGraw-Hill, New York.

Green, K., and Otori, T. (1970). Direct measurements of membrane unstirred layers. *J. Physiol.* **207**, 93–102.

Han, P. (1992). *Tensile Testing.* ASM International, Materials Park, OH.

Jacobson, K., Sheets, E. D., and Simson, R. (1995). Revisiting the fluid mosaic model of membranes. *Science* **268**, 1441–1442.

Jones, M. N., and Chapman, P. (1995). *Micelles, Monolayers and Biomembranes.* Wiley–Liss, New York.

Kamiya, N. (1981). Physical and chemical basis of cytoplasmic streaming. *Annu. Rev. Plant Physiol.* **32**, 205–236.

Kamiya, N., Tazawa, M., and Takata, T. (1963). The relation of turgor pressure to cell volume in *Nitella* with special reference to mechanical properties of the cell wall. *Protoplasma* **57**, 501–521.

Kotyk, A., Janáoek, K., and Koryta, J. (1988). *Biophysical Chemistry of Membrane Functions*. Wiley, Chichester, UK.

Lamb, G. L. (1995). *Introductory Applications of Partial Differential Equations with Emphasis on Wave Propagation and Diffusion*. Wiley, New York.

Lauffer, M. A. (1989). *Motion in Biological Systems*. Liss, New York.

Lide, D. R. (Ed.) (1997). *CRC Handbook of Chemistry and Physics* (78th ed.). CRC Press, Boca Raton, FL.

Leshem, Y. Y. (1992). *Plant Membranes. A Biophysical Approach to Structure, Development and Senescence*. Kluwer, Dordrecht.

Linskens, H. F., and Jackson, J. F. (Eds.) (1996). *Plant Cell Wall Analysis*. Springer-Verlag, Berlin.

Lockhart, J. A. (1965). Cell extension. In *Plant Biochemistry* (J. Bonner and J. E. Varner, Eds.), pp. 826–849. Academic Press, New York.

Lucas, W. J., Ding, B., and van der Schoot, C. (1993). Tansley Review No. 58. Plasmodesmata and the supracellular nature of plants. *New Phytol.* **125**, 435–476.

Merz, K. M., Jr., and Roux, B. (1996). *Biological Membranes: A Molecular Perspective from Computation and Experiment*. Birkhauser, Boston.

Moller, I. M., and Brodelius, P. (Eds.) (1996). *Plant Membrane Biology*. Oxford Univ. Press, Oxford.

Mouritsen, O. G., and Jørgensen, K. (1997). Small-scale lipid–membrane structure: Simulation versus experiment. *Curr. Opin. Struct. Biol.* **7**, 518–527.

Niklas, K. J. (1992). *Plant Biomechanics: An Engineering Approach to Plant Form and Function*. Univ. of Chicago Press, Chicago.

Petty, H. R. (1993). *Molecular Biology of Membranes: Structure and Function*. Plenum, New York.

Preston, R. B. (1974). *The Physical Biology of Plant Cell Walls*. Chapman & Hall, London.

Richter, C., and Dainty, J. (1989). Ion behavior in plant cell walls. I. Characterization of the *Sphagnum russowii* cell wall ion exchanger. *Can. J. Bot.* **67**, 451–459.

Robards, A. W. (1975). Plasmodesmata. *Annu. Rev. Plant Physiol.* **26**, 13–29.

Silver, B. L. (1985). *The Physical Chemistry of Membranes*. Allen & Unwin, Boston.

Singer, S. J., and Nicolson, G. L. (1972). The fluid mosaic model of the structure of cell membranes. *Science* **175**, 720–731.

Starzak, M. E. (1984). *The Physical Chemistry of Membranes*. Academic Press, Orlando, FL.

Stein, W. D. (1986). *Transport and Diffusion across Cell Membranes*. Academic Press, Orlando, FL.

Taylor, B. N. (Ed.) (1991). *The International System of Units (SI)*. U.S. Government Printing Office, Washington, DC.

Taylor, B. N. (Ed.) (1995). *Guide for the International System of Units (SI)*. U.S. Government Printing Office, Washington, DC.

Wright, E. M., and Diamond, J. M. (1969). Patterns of non-electrolyte permeability. *Proc. R. Soc. London Ser. B* **172**, 227–271.

Yeagle, P. L. (1993). *The Membranes of Cells* (2nd ed.). Academic Press, San Diego.

2

Water

I.	**Physical Properties**	**37**
	A. Hydrogen Bonding—Thermal Relations	38
	B. Surface Tension	40
	C. Capillary Rise	40
	D. Capillary Rise in the Xylem	42
	E. Tensile Strength, Viscosity	43
	F. Electrical Properties	44
II.	**Chemical Potential**	**45**
	A. Free Energy and Chemical Potential	46
	B. Analysis of Chemical Potential	49
	C. Standard State	51
	D. Hydrostatic Pressure	52
	E. Water Activity and Osmotic Pressure	52
	F. The Van't Hoff Relation	54
	G. Matric Pressure	56
	H. Water Potential	57
III.	**Central Vacuole and Chloroplasts**	**59**
	A. Water Relations of the Central Vacuole	59
	B. Boyle–Van't Hoff Relation	60
	C. Osmotic Responses of Chloroplasts	62
IV.	**Water Potential and Plant Cells**	**64**
	A. Incipient Plasmolysis	64
	B. Höfler Diagram and Pressure–Volume Curve	66
	C. Chemical Potential and Water Potential of Water Vapor	68
	D. Plant–Air Interface	71
	E. Pressure in the Cell Wall Water	72
	F. Water Flux	74
	G. Cell Growth	75
	H. Kinetics of Volume Changes	77
	Problems	**78**
	References	**80**

WATER IS the main constituent of plant cells, as our discussion of both vacuoles and cell walls in Chapter 1 has suggested. The actual cellular water content varies with cell type and physiological condition. For example, a carrot root is about 85% water by weight, and the young inner leaves of lettuce contain up to 95% water. Water makes up only 5% of certain dry seeds and spores, but when these become metabolically active, an increase in water content is essential for the transformation.

The physical and the chemical properties of water make it suitable for a variety of purposes in plants. It is the medium in which diffusion of solutes takes place in plant cells. Its incompressibility means that water uptake can lead to cell expansion and that intracellular hydrostatic pressures can help support plants. It is well suited to temperature regulation because it has a high heat of vaporization, a high thermal capacity per unit mass, and a high thermal conductivity for a liquid. Water is also an extremely good general solvent, in part owing to the small size of its molecules. Its polar character makes water suitable for dissolving other polar substances. Its high dielectric constant makes it a particularly appropriate solvent for ions. This latter property has far-reaching consequences for life because nearly all biologically important solutes are electrically charged. The mineral nutrients needed for growth and the organic products of photosynthesis are transported throughout the plant in aqueous solutions. In actively growing land plants, a continuous column of water exists from the soil, through the plant, to the evaporation sites in the leaves. Water is relatively transparent to visible irradiation, enabling sunlight to reach chloroplasts within the cells in leaves and to reach plants submerged at appreciable depths in lakes and oceans. Water is also intrinsically involved with metabolism. It is the source of the O_2 evolved in photosynthesis and the hydrogen used for CO_2 reduction. The generation of the important energy currency, ATP, involves the extraction of the components of water from ADP plus phosphate; in other words, such a phosphorylation is a dehydration taking place in an aqueous solution under biological conditions. At the ecological level, the facts that ice floats and that water is densest near 4°C are important for organisms living in streams, rivers, or lakes subject to freezing. Water existing as ice at low temperatures that nevertheless are within the biological realm also has many important consequences for water exchange, metabolism, and ultimately cell survival. Thus an understanding of the physical and chemical properties of water is crucial for an understanding of plant and animal biology.

A number of isotopically different forms of water can be prepared, which greatly facilitates experimental studies. Replacing both of the usual hydrogens with deuterium (^2H) results in "heavy water," or deuterium oxide, with a molecular weight of 20. The role of water in chemical reactions can then be studied by analyzing the deuterium content of substances involved as reactants or products. Tritium (^3H), a radioactive isotope with a half-life of 12.4 years, can also be incorporated into water. Tritiated water has been used to measure the rate of water diffusion in plant tissues. Another alternative for tracing the pathway of water is to replace the usual ^{16}O isotope with ^{18}O. This "labeling" of water with ^{18}O helped determine that the O_2 evolved in photosynthesis comes from H_2O and not from CO_2 (see Chapter 5).

I. Physical Properties

Water differs markedly from substances having closely related electronic structures, namely, CH_4, NH_3, H_2O, HF, and Ne (Fig. 2-1). Each molecule contains 10 protons and 10 electrons,

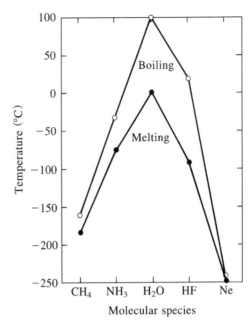

Figure 2-1. Boiling points and melting points for molecules with 10 protons and 10 electrons, showing the high values for water. Most biological processes take place from 0 to 50°C, where water can be in the liquid state but the other substances cannot.

but the number of hydrogen atoms decreases along the series from methane to neon. The relatively high melting point and boiling point for water, compared with substances having a similar electronic structure, are consequences of its strong intermolecular forces. In other words, thermal agitation does not easily disrupt the bonding between water molecules. This strong attraction between molecules is responsible for many characteristic properties of water.

A. Hydrogen Bonding—Thermal Relations

The strong intermolecular forces in water result from the structure of the H_2O molecule (Fig. 2-2). The internuclear distance between the oxygen and each of the two hydrogens is approximately 0.099 nm; the H–O–H bond angle is about 105°. The oxygen atom is strongly electronegative and tends to draw electrons away from the hydrogen atoms. The oxygen atom thus has a partial negative charge (δ^- in Fig. 2-2), and the two hydrogens each have a partial positive charge (δ^+). These positively charged hydrogens are electrostatically attracted to the negatively charged oxygens of two neighboring water molecules. This leads to *hydrogen bonding* between water molecules, with an energy of about 20 kilojoules (kJ) mol^{-1} of hydrogen bonds. Such bonding of water molecules to each other leads to increased order in aqueous solutions. In fact, liquid water becomes nearly crystalline in local regions, which affects the molecular interactions and orientations that occur in aqueous solutions.

Ice is a coordinated crystalline structure in which essentially all the water molecules are joined by hydrogen bonds. When enough energy is added to melt ice, some of these intermolecular hydrogen bonds are broken. The heat of fusion of ice at 0°C is 6.0 kJ mol^{-1}.

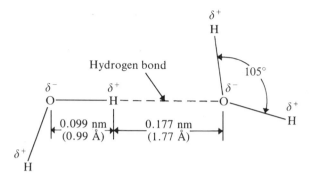

Figure 2-2. Schematic structure of water molecules, indicating the hydrogen bonding resulting from the electrostatic attraction between the net positive charge on a hydrogen (δ^+) in one molecule and the net negative charge on an oxygen (δ^-) in a neighboring water molecule. Depending on the model, δ^+ is about 0.3 of a protonic charge and δ^- is about -0.6.

Total rupture of the intermolecular hydrogen bonds involving both of its hydrogens would require 40 kJ mol^{-1} of water. The heat of fusion thus indicates that (6.0 kJ mol^{-1})/ (40 kJ mol^{-1}), or at most 0.15 (15%), of the hydrogen bonds are broken when ice melts. Some energy is needed to overcome van der Waals attractions,[1] so <15% of the hydrogen bonds are actually broken upon melting. Conversely, over 85% of the hydrogen bonds remain intact for liquid water at 0°C. Because 75.4 J is needed to heat 1 mol of water by 1°C (the thermal capacity of water on a mole basis), (0.0754 kJ mol^{-1} °C^{-1})(25°C), or 1.9 kJ mol^{-1}, is needed to heat water from 0 to 25°C. Even if all this energy were used to break hydrogen bonds (1.9 kJ mol^{-1} could break about 5% of the possible hydrogen bonds in water), over 80% of the bonds would remain intact at 25°C. Such bonding leads to semicrystalline order in aqueous solutions. The extensive intermolecular hydrogen bonding in the liquid state contributes to the unique and biologically important properties of water that we will discuss throughout this chapter.

The energy required to separate molecules from a liquid and move them into an adjacent vapor phase without a change of temperature is called the *heat of vaporization*. For water, the heat of vaporization at 100°C is 2.26 MJ kg^{-1}, or 40.7 kJ mol^{-1}. Per unit mass, this is the highest heat of vaporization of any known liquid and reflects the large amount of energy required to disrupt the extensive hydrogen bonding in aqueous solutions. More pertinent is the heat of vaporization of water at temperatures encountered by plants. At 25°C each mole of water evaporated requires 44.0 kJ (see Appendix I), so a substantial heat loss accompanies the evaporation of water in transpiration. Most of this vaporization energy is needed to break hydrogen bonds so that the water molecules can become separated in the gaseous phase. For example, if 80% of the hydrogen bonds remained at 25°C, then (0.80) (40 kJ mol^{-1}), or 32 kJ mol^{-1}, would be needed to rupture them. Additional energy is needed to overcome the van der Waals forces and for the expansion involved in going from

---

[1] Van der Waals forces are the electrostatic attractions between electrons in one molecule and the nucleus of an adjacent molecule minus the molecules' interelectronic and internuclear repulsive forces. In about 1930, Fritz London showed that these forces are caused by the attraction between an electric dipole in some molecule and the electric dipole induced in an adjacent one. Therefore, van der Waals forces result from random fluctuations of charge and are important only for molecules that are very close together—in particular, for neighboring molecules.

a liquid to a gas. The heat loss accompanying the evaporation of water is one of the principal means of temperature regulation in land plants, dissipating much of the energy gained from the absorption of solar irradiation. (Energy budgets are considered in Chapter 7.)

B. Surface Tension

Water has an extremely high *surface tension*, which is evident at an interface between water and air. Surface tension can be defined as the force per unit length that can pull perpendicular to a line in the plane of the surface. Because of its high surface tension, water can support a steel pin or needle carefully placed on its surface. The surface tension at such an air–water interface is 0.0728 N m^{-1} at $20°C$ (see Appendix I for values at other temperatures). Surface tension is also the amount of energy required to expand a surface by unit area—surface tension has the dimensions of force per unit length and also of energy per unit area.

To see why energy is required to expand the water surface, let us consider water molecules that are brought from the interior of an aqueous phase to an air–water interface. If this involves only an expansion of the surface area—i.e., if there is no accompanying movement of other water molecules from the surface to the interior—then a loss in the water–water attraction from some of the intermolecular hydrogen bonds occurs with no compensating air–water attraction. Energy is thus needed to break the hydrogen bonds that are lost in moving water molecules from the interior of the solution to the air–water interface (0.0728 J to increase the area by 1 m^2). In fact, the term "surface free energy" is more appropriate from a thermodynamic point of view than the conventional term "surface tension."

The surface tension of an aqueous solution usually is only slightly influenced by the composition of an adjacent gas phase, but it can be markedly affected by certain solutes. Molecules are relatively far apart in a gas—dry air at $0°C$ and one standard atmosphere (0.1013 MPa, 1.013 bar, or 760 mm Hg) contains 45 mol m^{-3} compared with $56,000$ mol m^{-3} for liquid water—so the frequency of interactions between molecules in the gas phase and those in the liquid phase is relatively low. Certain solutes, such as sucrose or KCl, do not preferentially collect at the air–liquid interface and consequently have little effect on the surface tension of an aqueous solution. However, because of their hydrophobic regions, fatty acids and certain lipids may become concentrated at interfaces, which can greatly reduce surface tensions. For example, 10 mol m^{-3} (~ 10 mM) caproic acid (a 6-carbon fatty acid) lowers the surface tension by 21% from the value for pure water, and only 0.05 mol m^{-3} capric acid (a 10-C fatty acid) lowers it by 34% (Bull, 1964). Substances such as soaps (salts of fatty acids), other emulsifiers, and denatured proteins with large hydrophobic side chains collect at the interface and can reduce the surface tension of aqueous solutions by over 70%. Such "surfactant" molecules have both polar and nonpolar regions.

C. Capillary Rise

The intermolecular attraction between like molecules in the liquid state, such as the water–water attraction based on hydrogen bonds, is called *cohesion*. The attractive interaction between a liquid and a solid phase, such as between water and the walls of a glass capillary (a cylindrical tube with a small internal diameter), is called *adhesion*. When the water–wall attraction is appreciable compared with the water–water cohesion, the walls are said to be *wettable*, and water then rises in such a vertical capillary. At the opposite extreme, when the intermolecular cohesive forces within the liquid are substantially greater than is the adhesion

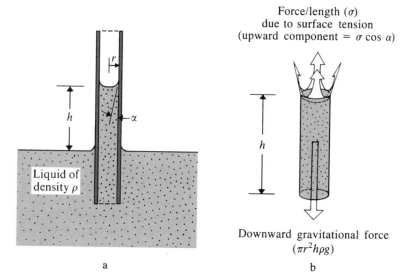

Figure 2-3. Capillary rise of a liquid: (a) variables involved, and (b) force diagram indicating that surface tension projected in the upward direction is balanced by gravity acting downward.

between the liquid and the wall material, the upper level of the liquid in such a capillary is lower than the surface of the solution. Capillary depression occurs for liquid mercury in glass capillaries. For water in glass capillaries or in xylem vessels, the attraction between the water molecules and the walls is great, so the liquid rises. Because capillary rise has important implications in plant physiology, we will discuss its characteristics quantitatively.

As an example appropriate to the evaluation of water ascent in a plant, let us consider a capillary of inner radius r with a wettable wall dipping into some aqueous solution (Fig. 2-3a). The strong adhesion of water molecules to the wettable wall causes the fluid to rise along the inner wall of the capillary. Because a strong water–water cohesion occurs in the bulk solution, water is concomitantly pulled up into the lumen of the capillary as water rises along the wall. In particular, the air–water surface greatly resists being stretched, a property reflected in the high surface tension of water at air–water interfaces. Such resisting minimizes the area of the air–water interface, a condition achieved if water also moves up in the lumen as well as along the inner wall of the capillary. The effect of the rise of water along the wall of the capillary is thus transmitted to a volume of fluid. We will designate the height that the liquid rises in the capillary by h and the contact angle that the liquid makes with the inner capillary wall by α (Fig. 2-3). The extent of the rise depends on α, so the properties of the contact angle will now be examined more closely.

The size of the contact angle depends on the magnitude of the liquid–solid adhesive force compared with that of the liquid–liquid cohesive force. Specifically, Young's equation (also called the Young and Dupré equation) indicates that

$$\text{Adhesion} = \frac{1 + \cos\alpha}{2}\,\text{cohesion} \qquad (2.1)$$

When the adhesive force equals (or exceeds) the cohesive force in the liquid, $\cos\alpha$ is 1; the contact angle α then equals zero ($\cos 0° = 1$). This is the case for water in capillaries made of clear, smooth glass or having walls with polar groups on the exposed surface. When the adhesive force equals half the cohesive force, cos α is zero and the contact angle is then 90° (cos 90° = 0) by Eq. (2.1). In this case, the level of the fluid in the capillary is the same as that in the bulk of the solution. This latter condition is closely approached for water–polyethylene adhesion, where α equals 94°. As the liquid–solid adhesive force becomes relatively less compared with the intermolecular cohesion in the liquid phase, the contact angle (see Fig. 2-3a) increases toward 180°, and capillary depression occurs. For instance, water has an α of about 110° with paraffin, and the contact angle for mercury interacting with a glass surface is about 150°. In such cases, the level of the liquid is lower in the capillary than in the bulk solution (see Fig. 2-3a, but imagine that $\alpha > 90°$, so the liquid curves downward where it intersects the surface of the capillary).

We can calculate the extent of capillary rise by considering the balance of two forces: (1) gravity acting downward and (2) surface tension, which leads to an upward force in the case of a wettable wall in a vertical tube (see Fig. 2-3b). The force pulling upward acts along the inside perimeter of the capillary, a distance of $2\pi r$, with a force per unit length equal to σ, the surface tension. The component of this force acting vertically upward is $2\pi r\sigma \cos \alpha$, where α is the contact angle illustrated in Fig. 2-3a. This upward force is balanced by the gravitational force acting on the liquid of density ρ and volume of approximately $\pi r^2 h$ (some fluid is also held in the "rim" of the meniscus, as indicated in Fig. 2-3a; for narrow capillaries, the fluid in the meniscus increases the effective height of the column by about $r/3$). The gravitational force is the mass involved times the gravitational acceleration g ($F = ma$, Newton's second law of motion). In the present case, gravity acts on a mass of fluid of approximately $\pi r^2 h\rho$, so the gravitational force is $\pi r^2 h\rho g$, acting downward. This force is balanced by $2\pi r\sigma \cos \alpha$ pulling in the opposite direction, so the extent of rise, h, is given by equating the two forces ($\pi r^2 h\rho g = 2\pi r\sigma \cos \alpha$), which leads to

$$h = \frac{2\sigma \cos \alpha}{r\rho g} \tag{2.2a}$$

Equation (2.2a) indicates the readily demonstrated property that the extent of liquid rise in a capillary is inversely proportional to the radius of the tube. For water in glass capillaries as well as in many of the fine channels encountered in plants where the cell walls have a large number of exposed polar groups, the contact angle can be near zero, in which case cos α in Eq. (2.2a) can be set equal to 1. The density of water at 20°C is 998 kg m^{-3} (actually, ρ in Eq. 2.2a is the difference between the liquid density and the density of the displaced air, the latter being about 1 kg m^{-3}), and the acceleration due to gravity, g, is about 9.80 m s^{-2} (see Appendix I). Using a surface tension for water at 20°C of 0.0728 N m^{-1} (see Appendix I), we obtain the following relation between the height of the rise and the capillary radius when the contact angle is zero:

$$h_{(m)} = \frac{1.49 \times 10^{-5} \text{ m}^2}{r_{(m)}} \tag{2.2b}$$

where the subscripts (m) in Eq. (2.2b) mean that the dimensions involved are expressed in meters.

D. Capillary Rise in the Xylem

Although Eq. (2.2) refers to the height of capillary rise only in a static sense, it still has important implications concerning the movement of water in plants. To be specific, let us consider a xylem vessel having a lumen radius of 20 μm. From Eq. (2.2b), we calculate that water will rise in it to the following height:

$$h_{(m)} = \frac{(1.49 \times 10^{-5} \text{ m}^2)}{(20 \times 10^{-6} \text{ m})} = 0.75 \text{ m}$$

Such a capillary rise would account for the extent of the ascent of water in small plants, although it says nothing about the rate of such movement (see Chapter 9, Section IVD). For water to reach the top of a 30-m tree by capillary action, however, the vessel would have to be 0.5 μm in radius. This is much smaller than that observed for xylem vessels, indicating that capillary rise in channels of the size of xylem cells cannot account for the extent of the water rise in tall trees. Furthermore, the lumens of the xylem vessels are not open to the air at the upper end, and thus they are not really analogous to the capillary depicted in Fig. 2-3.

The numerous interstices in the cell walls of xylem vessels form a meshwork of many small, tortuous capillaries, which can lead to an extensive capillary rise of water in a tree. A representative "radius" for these channels in the cell wall might be 5 nm. According to Eq. (2.2b), a capillary of 5 nm radius could support a water column of 3 km—far in excess of the needs of any plant. The cell wall could thus act as a very effective wick for water rise in its numerous small interstices, although the actual rate of such movement up a tree is generally too low to replace the water lost by transpiration.

Because of the appreciable water–wall attraction that can develop both at the top of a xylem vessel and in the numerous interstices of its cell walls, water already present in the lumen of a xylem vessel can be sustained or supported at great heights. The upward force, transmitted to the rest of the solution in the xylem vessel by water–water cohesion, overcomes the gravitational pull downward. The key to sustaining water already present in the xylem vessel against the pull of gravity is the very potent attractive interaction (adhesion) between water and the cell wall surfaces in the vessel. What happens if the lumen of the xylem vessel becomes filled with air? Will water then refill it? The capillary rise of water is not sufficient to refill most air-filled xylem vessels greater than about 1 m in length so that, in general, air-filled vessels are lost for conduction, such as for the inner annual rings of most trees.[2]

E. Tensile Strength, Viscosity

The pulling on water columns that occurs in capillary rise and in the sustaining of water in the xylem requires that water be put under tension, or negative pressure. Water must therefore have substantial *tensile strength*—the maximum tension (force per unit area, or negative pressure) that can be withstood before breaking. The intermolecular hydrogen bonds lead to this tensile strength by resisting the pulling apart of water in a column. According to experiment, water in small glass tubes withstands negative pressures (tensions) of up to approximately 30 MPa at 20°C without breaking, depending on the diameter of the tubes in which the determinations are made and any solutes present in the water. This tensile strength of water is nearly 10% of that of copper or aluminum and is high enough to meet the demands encountered for water movement in plants. The great cohesive forces between water molecules thus allow an appreciable tension to exist in an uninterrupted water column

---

[2] A positive hydrostatic pressure can occur in the root xylem (termed *root pressure*, see Chapter 9), which can help refill vessels in herbaceous species as well as in certain woody species.

in a wettable capillary or tube such as a xylem vessel, which is important for the continuous movement of water from the root through the plant to the surrounding atmosphere during transpiration (discussed in Chapter 9).

In contrast to metals, which are in a stable state, water under tension is actually in a *metastable state* (a situation in which a change is ready to occur but does not without some outside impulse). If gas bubbles form in the water under tension in the xylem vessels, the water column can be ruptured. Thus the introduction of another phase can disrupt the metastable state describing water under tension. Minute air bubbles sometimes spontaneously form in the xylem sap. These usually adhere to the walls of the xylem vessel, and the gas in them slowly redissolves. If, however, they grow large enough or a number of small bubbles coalesce, forming an *embolism*, the continuity of water can be interrupted and that xylem vessel ceases to function, i.e., it cavitates. Freezing of the solution in xylem vessels can lead to bubbles in the ice (the solubility of gases like CO_2, O_2, and N_2 is quite low in ice), and these bubbles can interrupt the water columns upon thawing. Most plants are thus damaged by freezing and thawing of xylem sap and the consequent loss of water continuity in the xylem vessels.

Hydrogen bonding also influences the *viscosity* of water. Viscosity indicates the resistance to flow, reflecting the cohesion within a fluid as well as transfer of molecular momentum between layers of the fluid. It is thus a measure of the difficulty for one layer to slide past an adjacent layer when a shearing force is applied. Because hydrogen bonding can restrict the slipping of adjacent liquid layers, the viscosity of water is relatively high compared with solvents that have little or no hydrogen bonding, e.g., acetone, benzene, chloroform, and other organic solvents with small molecules. The decrease in the viscosity of water as the temperature rises reflects the breaking of hydrogen bonds and also the lessening of other attractive forces (e.g., van der Waals forces) accompanying the greater thermal motion of the molecules.

F. Electrical Properties

Another important physical characteristic of water is its extremely high dielectric constant, which is also a consequence of its molecular structure. A high dielectric constant for a solvent lowers the electrical forces between charged solutes. To quantify the magnitude of electrical effects in a fluid, let us consider two ions having charges Q_1 and Q_2 and separated from each other by a distance r. The electrical force exerted by one ion on the other is expressed by Coulomb's law:

$$\text{Electrical force} = \frac{Q_1 Q_2}{4\pi \varepsilon_0 D r^2} \tag{2.3}$$

where ε_0 is a proportionality constant known as the permittivity of a vacuum and D is a dimensionless quantity called the dielectric constant. D equals unity in a vacuum and is 1.0006 in air at $0°C$ at a pressure of one standard atmosphere.

Any substance composed of highly polar molecules, such as water, generally has a high dielectric constant. Specifically, D for water is 80.2 at $20°C$ and 78.4 at $25°C$—extremely high values for a liquid (the decrease in the dielectric constant of water with increasing temperature mainly reflects increased thermal motion and decreased hydrogen bonding). By contrast, the dielectric constant of the nonpolar liquid hexane at $20°C$ is only 1.87, a low value typical of many organic solvents. Based on these two vastly different dielectric constants and using Eq. (2.3), the attractive electrical force between ions such as Na^+ and Cl^- is (80.2/1.87) or 43 times greater in hexane than in water. The much stronger

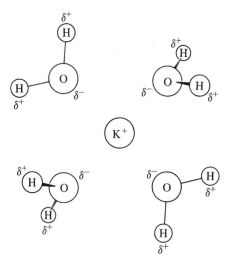

Figure 2-4. Orientation of water molecules around a potassium ion, indicating the charge arrangement that leads to screening of the local electrical field.

attraction between Na^+ and Cl^- in hexane reduces the amount of NaCl that will dissociate, compared with the dissociation of this salt in aqueous solutions. Stated another way, the much weaker electrical forces between ions in aqueous solutions, compared with those in organic solvents, allow a larger concentration of ions to be in solution. Water is thus a good solvent for charged particles.

The electrostatic interaction between ions and water partially cancels or screens out the local electrical fields of the ions (Fig. 2-4). Cations attract the negatively charged oxygen atom of a water molecule, and anions attract its positively charged hydrogen atoms. Water molecules orient around the charged particles and produce local electrical fields opposing the fields of the ions. The resultant screening or shielding diminishes the electrical interaction between the ions and allows more of them to remain in solution, which is the molecular basis for water's high dielectric constant (Fig. 2-4).

The energy of attraction between water and nonpolar molecules is usually less than the energy required to break water–water hydrogen bonds. Nonpolar compounds are therefore not very soluble in water. Certain substances, such as detergents, phospholipids, and proteins, can have a nonpolar and also a polar region in the same molecule. In aqueous solutions these compounds can form aggregates, called *micelles*, in which the nonpolar groups of the molecules are in the center, while the charged or polar groups are external and interact with water. The lack of appreciable attractive electrostatic interaction between polar or charged species and the nonpolar (hydrophobic) regions of membrane lipids underlies the ability of biological membranes to limit the passage of such solutes into and out of cells and organelles.

II. Chemical Potential

The chemical potential of species j indicates the free energy associated with it and available for performing work. Because there are many concepts to be mastered before understanding such a statement, we will first briefly explore the concept of free energy. (Further details are presented later in this chapter, in Chapter 6, and in Appendix IV.) We will introduce specific terms in the chemical potential of species j and then consider the extremely important case of water.

A. Free Energy and Chemical Potential

Plants and animals require a continual input of free energy. If we were to remove the sources of free energy, organisms would drift toward equilibrium with the consequent cessation of life. The ultimate source of free energy is the sun. Photosynthesis converts its plentiful radiant energy into free energy, which is stored first in intermediate energy "currencies" like ATP and then in the altered chemical bonds that result when CO_2 and H_2O react to form a carbohydrate and O_2. When the overall photosynthetic reactions are reversed during respiration, the available free energy is reconverted into suitable energy currencies such as ATP. In turn, this free energy can perform biological work, such as transporting amino acids into cells, pumping blood, powering electrical reactions in the brain, or lifting weights. For every such process on both a molecular and a global scale, the free energy decreases. In fact, the structure of cells, as well as that of ecosystems, is governed by the initial supply of free energy and by the inexorable laws of thermodynamics that describe the expenditure of free energy.

Instead of viewing the whole universe, a thermodynamicist focuses on some small part of it referred to as a "system." Such a system might be K^+ dissolved in water, a plant cell, a leaf, or even a whole plant. Something happens to the system; the K^+ concentration is increased, some protein is made in a cell, a leaf abscises, or water moves up a tree. We say that the system changes from state A to state B. The minimum amount of work needed to cause the change from A to B is the free energy increase associated with it. Alternatively, the free energy decrease in going from state B back to state A represents the maximum amount of work that can be derived from the transition. If there were no frictional or other irreversible losses, we could dispense with the words "minimum" and "maximum." Thus the limits to the work done on or by a system when it undergoes a transition from one state to another involve changes in free energy. Knowledge of the free energy under one condition compared with another allows us to predict the direction of spontaneous change or movement: A spontaneous change in a system at constant temperature and pressure proceeds in the direction of decreasing free energy.

Most systems of interest in biology are subject to a constant pressure (atmospheric) and remain at a constant temperature, at least for short periods. In discussing the energetics of processes for systems at constant temperature and pressure, the appropriate quantity is known as the *Gibbs free energy*. The Gibbs free energy has a very useful property: It decreases for a spontaneous process at constant temperature and pressure. Under such conditions, the decrease in Gibbs free energy equals the maximum amount of energy available for work, whereas if it increases for some transition, the change in Gibbs free energy represents the minimum amount of work required. We will hence restrict our attention to changes that occur when the overall system is at a constant temperature and subjected to a constant pressure. (An equation for the Gibbs free energy is presented in Chapter 6, and Appendix IV deals with a number of formal matters associated with its use.) For present purposes, we note that the Gibbs free energy of an entire system is made up of additive contributions from each of the molecular species present. We will therefore shift our emphasis to the individual components of the system and the manner in which their free energy can be described.

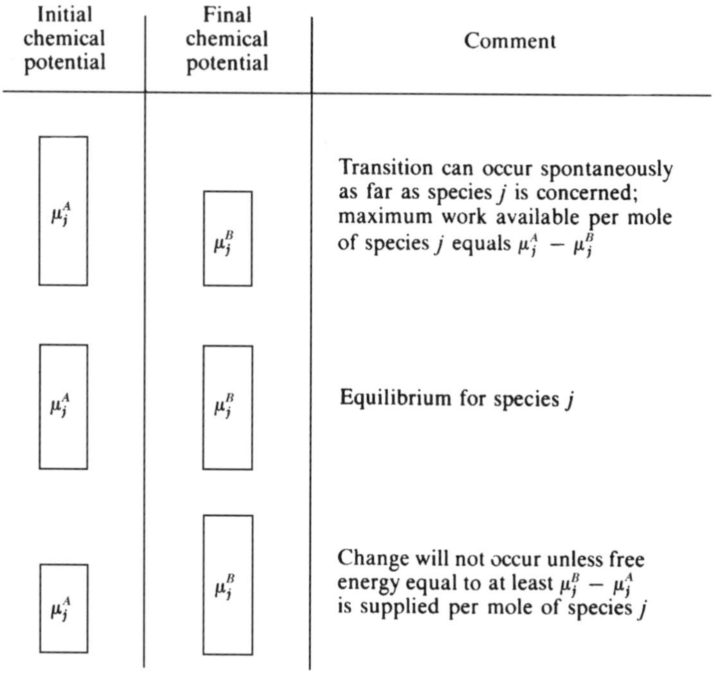

Initial chemical potential	Final chemical potential	Comment
μ_j^A	μ_j^B	Transition can occur spontaneously as far as species j is concerned; maximum work available per mole of species j equals $\mu_j^A - \mu_j^B$
μ_j^A	μ_j^B	Equilibrium for species j
μ_j^A	μ_j^B	Change will not occur unless free energy equal to at least $\mu_j^B - \mu_j^A$ is supplied per mole of species j

Figure 2-5. Possible changes in the chemical potential of species j that can accompany a transition from initial state A to final state B, as might occur in a chemical reaction or in crossing a membrane. The heights of the bars representing μ_j correspond to the relative values of the chemical potential.

To every chemical component in a system we can assign a free energy per mole of that species. This quantity is called the *chemical potential* of species j and is given the symbol μ_j. We can view μ_j as a property of species j, indicating how that species will react to a given change, e.g., a transition of the system from state A to state B. During the transition, the chemical potential of species j changes from μ_j^A to μ_j^B. If $\mu_j^B < \mu_j^A$, then the free energy per mole of species j decreases. Such a process can take place spontaneously as far as species j is concerned—water flowing downhill is an example of such a spontaneous process. In principle we can harness the spontaneous change to do work. The maximum amount of work that can be done per mole of species j is $\mu_j^A - \mu_j^B$ (Fig. 2-5). Suppose that $\mu_j^B = \mu_j^A$. As far as species j is concerned, no work is involved in the transition from state A to state B; species j is then at equilibrium (Fig. 2-5). We will show later that living systems as a whole are far from equilibrium and that many chemical species are not in equilibrium across cellular membranes. Finally, let us consider the case in which $\mu_j^B > \mu_j^A$. A change increasing the chemical potential of a substance can occur only if some other change in the system supplies the free energy required. The pumping of blood along arteries, using ATP to cause contraction of the heart muscles, is an example in which the increase in chemical potential of water in the blood is accompanied by an even larger decrease in the free energy associated with ATP (we will consider ATP in detail in Chapter 6). The minimum amount of work that must be done per mole of species j to cause the energetically uphill transition is $\mu_j^B - \mu_j^A$ (Fig. 2-5).

The condition for a spontaneous change—a decrease in chemical potential—has important implications for discussing fluxes from one region to another. In particular, we can use the chemical potential difference between two locations as a measure of the "driving force" for the movement of that component. The larger the chemical potential difference, $\mu_j^A - \mu_j^B$, the faster the spontaneous change takes place—in this case, the larger is the flux density of species j from region A to region B.

The chemical potential of a substance depends on its chemical composition but is also influenced by other factors. Here we shall discuss in an intuitive way the environmental factors that might influence μ_j. We begin by noting that chemical potential depends on the "randomness" (entropy) of the system and that concentration is a quantitative description of this randomness. For example, the passive process of diffusion discussed in Chapter 1 describes neutral molecules spontaneously moving from some region to another where their concentration is lower. In more formal terms, the net diffusion of species j proceeds in the direction of decreasing μ_j, which in this case is the same as that of decreasing concentration. The concept of chemical potential was thus implicit in the development of Fick's equations in Chapter 1. The importance of concentration gradients (or differences) to the movement of a substance was also established in the discussion of diffusion. Here we will present the effect of concentration on chemical potential in a somewhat more sophisticated manner, using that part of the concentration that is thermodynamically active.

Chemical potential also depends on pressure, which for situations of interest in biology usually means hydrostatic pressure or air pressure (although local pressure effects can be readily incorporated into μ_j, the system as a whole must experience a constant external pressure for the Gibbs free energy to have the useful properties discussed previously). The existence of pressure gradients can cause movements of fluids—the flow of crude oil in long-distance pipelines, blood in arteries, sap in xylem vessels, and air in hurricanes—indicating that pressure differences can affect the direction for spontaneous changes.

Because electrical potential also affects the chemical potential of charged particles, it must be considered when predicting the direction of their movement. By definition, work must be done to move a positively charged particle to a higher electrical potential. Accordingly, if an electrical potential difference is imposed across an electrolyte solution initially containing a uniform concentration of ions, cations will spontaneously move in one direction (toward the cathode, i.e., to regions of lower electrical potential), and anions will move in the opposite direction (toward the anode). To describe the chemical potential of a charged species we must include an electrical term in μ_j.

Another contributor to chemical potential is gravity. We can readily appreciate that position in a gravitational field affects μ_j because work must be done to move a substance vertically upward. Although the gravitational term can be neglected for ion and water movements across plant cells and membranes, it is important for water movement in a tall tree and in the soil.

In summary, the chemical potential of a species depends on its concentration, the pressure, the electrical potential, and gravity. We can compare the chemical potentials of a species on two sides of a barrier to decide whether or not the species is in equilibrium. If μ_j is the same on the two sides, we would not expect a net movement of species j to occur spontaneously across the barrier. We will use a comparison of the relative values of the chemical potential of species j at various locations to predict the directions for spontaneous movement of that chemical substance (toward lower μ_j), just as we can use a comparison of temperatures to predict the direction for heat flow (toward lower T). We will also find that $\Delta\mu_j$ from one region to another gives a convenient measure of the driving force on species j.

B. Analysis of Chemical Potential

Because it has proved experimentally valid, we will represent the chemical potential of any species j by the following sum:

$$\mu_j = \mu_j^* + RT \ln a_j + \bar{V}_j P + z_j FE + m_j gh \qquad (2.4)$$

One measure of the elegance of a mathematical expression is the amount of information that it contains. Based on this criterion, Eq. (2.4) is an extremely elegant relation. After defining and describing the various contributions to μ_j indicated in Eq. (2.4), we will consider the various terms in greater detail for the important special case in which species j is water.

Chemical potential, like electrical or gravitational potential, must be expressed relative to some arbitrary energy level. An unknown additive constant, or reference level, μ_j^*, is therefore included in Eq. (2.4). Because it contains an unknown constant, the actual value of the chemical potential is not determinable. For our applications of chemical potential, however, we are interested in the difference in the chemical potential between two particular locations, so only relative values are important. Specifically, because μ_j^* is added to each of the chemical potentials being compared, it cancels out when the chemical potential in one location is subtracted from that in another to obtain the chemical potential difference between the two locations. Figure 2-5 further illustrates the usefulness of considering differences in chemical potential instead of actual values. Note that the units of μ_j and of μ_j^* are energy per mole of the substance, e.g., J mol^{-1}.

In certain cases we can adequately describe the chemical properties of species j by using the concentration of that solute, c_j. Owing to molecular interactions, however, this usually requires that the total solute concentration be low. Molecules of solute species j interact with each other as well as with other solutes in the solution, and this influences the behavior of species j. Such intermolecular interactions increase as the solution becomes more concentrated. The use of concentrations for describing the thermodynamic properties of some solute thus indicates an approximation, except in the limiting case of infinite dilution in which interactions between solute molecules are negligible. Where exactness is required, *activities*—which may be regarded as "corrected" concentrations—are employed. Consequently, for general thermodynamic considerations, as in Eq. (2.4), the influence of the amount of a particular species j on its chemical potential is handled not by its concentration but by its activity, a_j. The activity of solute j is related to its concentration by means of an *activity coefficient*, γ_j:

$$a_j = \gamma_j c_j \qquad (2.5)$$

The activity coefficient is usually less than 1 because the thermodynamically effective concentration of a species—its activity—is generally less than its actual concentration.

For an ideal solute, γ_j is 1, and the activity of species j equals its concentration. This condition can be approached for real solutes in certain dilute aqueous solutions, especially for neutral species. Activity coefficients for charged species can be appreciably less than 1 because of the importance of their electrical interactions (discussed in Chapter 3).

The activity of a solvent is defined differently from that of a solute. The solvent is the species having the highest mole fraction in a solution; *mole fraction* indicates the fraction of the total number of moles in a system contributed by that species. For a solvent, a_j is $\gamma_j N_j$, where N_j is its mole fraction. An ideal solvent has γ_{solvent} equal to 1, meaning that the interactions of solvent molecules with the surrounding molecules are indistinguishable from their interactions in the pure solvent. An ideal solution has all activity coefficients equal to 1.

In the expression for chemical potential (Eq. 2.4), the term involving the activity is $RT \ln a_j$. Therefore, the greater the activity of species j—or, loosely speaking, the higher its concentration—the larger will be its chemical potential. The logarithmic form can be "justified" in a number of ways, all based on agreement with empirical observations (see Chapter 3 and Appendix IV). The factor RT multiplying $\ln a_j$ in Eq. (2.4), where R is the gas constant (see Appendix I) and T is temperature on the absolute scale, results in units of energy per mole for the activity term.

The term $\bar{V}_j P$ in Eq. (2.4) represents the effect of pressure on chemical potential. Because essentially all measurements in biology are made on systems at atmospheric pressure, P is conveniently defined as the pressure in excess of this, and we will adopt such a convention here. $\bar{V}_j$ is the differential increase in volume of a system when a differential amount of species j is added, with other species, temperature, pressure, electrical potential, and gravitational position remaining constant:

$$\bar{V}_j = \left(\frac{\partial V}{\partial n_j} \right)_{n_i, T, P, E, h} \tag{2.6}$$

The subscript n_i on the partial derivative in Eq. (2.6) means that the number of moles of each species present, other than species j, is held constant when the derivative is taken; the other four subscripts are included to remind us of the additional variables that must be held constant.

$\bar{V}_j$ is called the *partial molal volume* of species j [μ_j is actually the partial molal Gibbs free energy, $(\partial G/\partial n_j)_{n_i, T, P, E, h}$, as discussed in Chapter 6 and Appendix IV]. The partial molal volume of a substance is often nearly equal to the volume of a mole of that species, but because there is in general a slight change in total volume when substances are mixed, the two are not exactly equal. Indeed, the addition of small concentrations of certain salts can cause an aqueous solution to contract—a phenomenon known as electrostriction—which in this special case would result in a negative value for $\bar{V}_j$.

Because work is often expressed as pressure times change in volume, we note that $\bar{V}_j P$ has the correct units for μ_j—energy per mole ($\bar{V}_j$ represents the volume per mole). To help justify the form of the pressure term, let us imagine that the solution containing species j is built up by adding small volumes of species j while the system is maintained at a constant pressure P. The work done to add a mole of species j is then the existing pressure times some volume change of the system characterizing a mole of that species, namely, $\bar{V}_j$. (In Appendix IV the inclusion of the $\bar{V}_j P$ term in the chemical potential of species j is justified more formally.)

The influence of electrical potential on the chemical potential of an ion is expressed by the term $z_j F E$ in Eq. (2.4), where z_j is an integer representing the charge number of species j, F is a constant known as Faraday's constant (Chapter 3), and E is the electrical potential. Because water is uncharged ($z_w = 0$), the electrical term does not contribute to its chemical potential. However, electrical potential is of central importance when discussing ions and the origin of membrane potentials; both of these are examined in detail in Chapter 3, in which we explicitly consider the $z_j F E$ term.

Equation (2.4) also includes a gravitational term $m_j g h$ expressing the amount of work required to raise an object of mass per mole m_j to a vertical height h, where g is the gravitational acceleration (about 9.8 m s^{-2}; see Appendix I for details). In the case of water, which is the primary concern of this chapter, the mass per mole m_w is 0.018016 kg mol^{-1} (18.016 g mol^{-1}). The gravitational term in the chemical potential of water is important for the fall of rain, snow, or hail and also affects the percolation of water downward through porous soil and the upward movement of water in a tree. The gravitational term $m_j gh$ can have units of (kg mol^{-1}) (m s^{-2}) (m), or kg m^2 s^{-2} mol^{-1}, which is J mol^{-1}. (Conversion factors among energy units are given in Appendix II.)

C. Standard State

The additive constant term μ_j^* in Eq. (2.4) is the chemical potential of species j for a specific reference state. From the preceding definitions of the various quantities involved, this reference state is attained when the following conditions hold: The activity of species j is 1 ($RT \ln a_j = 0$); the hydrostatic pressure equals atmospheric pressure ($\bar{V}_j P = 0$); the species is uncharged or the electrical potential is zero ($z_j FE = 0$); we are at the zero level for the gravitational term ($m_j gh = 0$); and the temperature equals the temperature of the system under consideration. Under these conditions, $\mu_j = \mu_j^*$ (Eq. 2.4).

As just indicated, an activity of 1 is defined in different ways for the solute and the solvent. To describe liquid properties, such as the dielectric constant, the heat of vaporization, and the boiling point, the most convenient standard state is that of the pure solvent. For a solvent, $a_{solvent} = \gamma_{solvent} \times N_{solvent}$, so the activity is 1 when the mole fraction $N_{solvent}$ is 1 ($\gamma_{solvent} = 1$ for pure solvent). Specifically, the properties of a solvent are fully expressed when no solute is present. Thus the standard reference state for water is pure water at atmospheric pressure and at the temperature and gravitational level of the system under consideration.

Water has an activity of 1 when N_w (see Eq. 2.8) is 1. The concentration of water on a molality basis (number of moles of a substance per kilogram of water for aqueous solutions) is then 1/(0.018016 kg mol^{-1}) or 55.5 molal (m). The accepted convention for a solute, on the other hand, is that a_j is 1 when $\gamma_j c_j = 1\ m$. For example, if $\gamma_j = 1$, a solution with a 1-m concentration of solute j has an activity of 1 m for that solute. Thus the standard state for an ideal solute is when its concentration is 1 m, in which case $RT \ln a_j$ is zero.[3] A special convention is used for the standard state of a gas such as CO_2 or O_2 in an aqueous solution—namely, the activity is 1 when the solution is in equilibrium with a gas phase containing that gas at a pressure of 1 atm. (At other pressures, the activity is proportional to the partial pressure of that gas in the gas phase.)

The three conventions introduced for the standard state (solvent, solute, and gases) generally do not occur under biological conditions. Essentially never is a solute at a concentration of 1 m, an important gas at a pressure of 1 atm, or a pure solvent present (except sometimes for water). Hence, care must be exercised when accepting chemical data based on standard states and interpreting the consequences in a biological context.

The chemical potential for a solute in the standard state, μ_j^*, depends on the solvent. We will argue that μ_j^* for a polar molecule is smaller in an aqueous solution (in which the species readily dissolves) than in an organic phase such as olive oil (in which it is not as soluble). Consider a two-phase system consisting of water with an overlying layer of olive oil, similar to that discussed in Chapter 1 in relation to measuring partition coefficients. If we put the

---

[3] Because molality involves the moles of a substance per kilogram of solvent, both SI quantitites, it is a legitimate SI unit, whereas moles per liter (M) is not recommended (the analogous SI concentration unit is moles m^{-3} = moles per 1000 liters = mM).

polar solute into our system, shake the water and the olive oil together, and wait for the two phases to separate, the solute concentration will be higher in the bottom or water phase. Let us next analyze this event using symbols, where phase A represents water and phase B is olive oil. Waiting for phase separation is equivalent to waiting for equilibrium, so μ_j^A is then equal to μ_j^B. In the present case, $\mu_j^{*,A} + RT \ln a_j^A = \mu_j^{*,B} + RT \ln a_j^B$ by Eq. (2.4). However, the polar solute is more soluble in water than in olive oil; hence, $RT \ln a_j^A > RT \ln a_j^B$. Because we are at equilibrium, $\mu_j^{*,A}$ must be less than $\mu_j^{*,B}$, as already indicated.

D. Hydrostatic Pressure

Because of rigid cell walls, large hydrostatic pressures can exist in plant cells, whereas hydrostatic pressures in animal cells generally are relatively small. Hydrostatic pressures are involved in plant support and also are important for the movement of water and solutes in the xylem and in the phloem. The term $\bar{V}_w P$ expresses the effect of pressure on the chemical potential of water (see Eq. 2.4), where $\bar{V}_w$ is the partial molal volume of water and P is the hydrostatic pressure in the aqueous solution in excess of the ambient atmospheric pressure. The density of water is about 1000 kg m^{-3}; therefore, when 1 mol or 18.0×10^{-3} kg of water is added to water, the volume increases by 18.0×10^{-6} m^3. Using the definition of $\bar{V}_w$ as a partial derivative (see Eq. 2.6), we need to add only an infinitesimally small amount of water (dn_w) and then observe the infinitesimal change in volume of the system (dV). We thus find that $\bar{V}_w$ for pure water is 18.0×10^{-6} m^3 mol^{-1} (18.0 cm^3 mol^{-1}). Although $\bar{V}_w$ can be influenced by the solutes present, it is generally close to 18.0×10^{-6} m^3 mol^{-1} for a dilute solution, a value that we will use for calculations in this book.

Various units are used for expressing pressures (see Chapter 1, footnote 8). A pressure of one standard atmosphere, or 0.1013 MPa, can support a column of mercury 760 mm high or a column of water 10.35 m high. Research concerning the water relations of plants has used pressures in bars. As indicated in Chapter 1, the SI unit for pressure is the pascal (Pa), which is 1 N m^{-2}; a quantity of convenient size for hydrostatic pressures is often the MPa (1 MPa = 10 bar = 9.87 atm). (An extensive list of conversion factors for pressure units is given in Appendix II, which also includes values for related quantities such as RT.) Pressure is force per unit area and so is dimensionally the same as energy per unit volume, e.g., 1 Pa = 1 N m^{-2} = 1 J m^{-3}. $\bar{V}_w$ has the units of m^3 mol^{-1}, so $\bar{V}_w P$ and hence μ_w can be expressed in J mol^{-1}.

E. Water Activity and Osmotic Pressure

Solutes in an aqueous solution decrease the activity of water (a_w). As a first approximation, the decrease in a_w when increasing amounts of solutes are added is a dilution effect. In other words, the mole fraction of water decreases when solutes are added. As its activity thus decreases, the chemical potential of water is lowered (consider the $RT \ln a_j$ term in μ_j). The presence of solutes also leads to an *osmotic pressure* (Π) in the solution. An increase in the concentration of solutes raises the osmotic pressure, indicating that Π and a_w change in opposite directions. In fact, the osmotic pressure and water activity are related:

$$RT \ln a_w = -\bar{V}_w \Pi \qquad (2.7)$$

where w refers to water. As solutes are added, a_w decreases from its value of 1 for pure water, $\ln a_w$ is therefore negative, and Π (defined by Eq. 2.7) is positive. Using Eq. (2.7), $RT \ln a_w$ in the chemical potential of water (see Eq. 2.4) can be replaced by $-\bar{V}_w \Pi$.

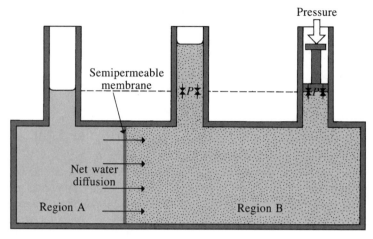

Figure 2-6. Schematic diagram indicating the principle underlying an osmometer in which a semipermeable membrane (permeable to water, but not to solutes) separates pure water (region A) from water containing solutes (region B). Water tends to diffuse toward regions where it has a lower mole fraction, in this case into region B. This causes the solution to rise in the open central column until, at equilibrium, the hydrostatic pressure (P) at the horizontal dashed line is equal to the osmotic pressure (Π) of the solution. Alternatively, we can apply a hydrostatic pressure P to the right-hand column to prevent a net diffusion of water into region B, this P again being equal to Π.

Unfortunately, the use of the terms osmotic pressure and osmotic potential, as well as their algebraic sign, varies. Osmotic pressures were originally measured using an osmometer (Fig. 2-6), a device employing a membrane that ideally is permeable to water but not to the solutes present. When pure water is placed on one side of the membrane and some solution on the other, a net diffusion of water occurs toward the side with the solutes. To counteract this tendency and establish equilibrium, a hydrostatic pressure is necessary on the solution side. This pressure is often called the osmotic pressure.[4] It depends on the presence of the solutes. Does an isolated solution have an osmotic pressure? In the sense of requiring an applied hydrostatic pressure to maintain equilibrium, the answer is no. Yet the same solution when placed in an osmometer can manifest an osmotic pressure. Thus some physical chemistry texts say that a solution has an osmotic potential—i.e., it could show an osmotic pressure if placed in an osmometer—and that osmotic potentials are a property characteristic of solutions. Many plant physiology texts define the negative of the same quantity as the osmotic potential so that the water potential (a quantity that we will introduce later) will depend directly on the osmotic potential, rather than on its negative. We will refer to Π, defined by Eq. (2.7), as the osmotic pressure and to its negative as the osmotic potential. (The osmotic potential will be symbolized by Ψ_Π, a component of the water potential Ψ.)

A direct and convenient way of evaluating osmotic pressure and water activity as de-fined by Eq. (2.7) is to use another *colligative* property. The four colligative properties of a solution—those that depend on the concentration of solutes regardless of their nature,

---

[4] If we represent this applied hydrostatic pressure by Π, then the chemical potential of water on the right side of the semipermeable membrane in Fig. 2-6 is $\mu_w^* + RT \ln a_w + \bar{V}_w \Pi$. This solution is in equilibrium with pure water on the left side of the membrane, where $\mu_w = \mu_w^*$. Hence, $\mu_w^* + RT \ln a_w + \bar{V}_w \Pi = \mu_w^*$, or $RT \ln a_w + \bar{V}_w \Pi = 0$, which is the same as Eq. (2.7).

at least in a dilute enough solution—are the freezing point depression, the boiling point elevation, the lowering of the vapor pressure of the solvent, and the osmotic pressure. Thus if the freezing point of cell sap is measured, $\Pi^{\text{cell sap}}$ can be calculated using the freezing point depression of $1.86°$ C for a 1-m solution together with the Van't Hoff relation.

F. The Van't Hoff Relation

For many purposes in biology, osmotic pressures are directly related to the concentration of solutes instead of expressing Π in terms of the water activity, a_w, as is done in Eq. (2.7). In general, the greater the concentration of solutes, the more negative ln a_w and the larger the osmotic pressure. Thus, some way of expressing a_w in terms of the properties of the solutes is needed. The ensuing derivation not only will show how a_w can be so expressed but also will indicate the many approximations necessary to achieve a rather simple and common expression for Π.

The activity of water equals $\gamma_w N_w$, where γ_w is the activity coefficient of water and N_w is its mole fraction. Thus N_w is given by

$$N_w = \frac{n_w}{n_w + \sum_j n_j} = \frac{n_w + \sum_j n_j - \sum_j n_j}{n_w + \sum_j n_j} = 1 - \frac{\sum_j n_j}{n_w + \sum_j n_j} \tag{2.8}$$

where n_w is the number of moles of water, n_j is the number of moles of solute species j, and the summation $\sum_j$ is over all solutes in the system considered. Equation (2.8) expresses the familiar relation that $N_w = 1-$ the mole fraction of solutes.

For an ideal solution, $\gamma_w = 1$. An activity coefficient of 1 is approached for dilute solutions, in which case $n_w \gg \sum_j n_j$. (The expression $n_w \gg \sum_j n_j$ defines a dilute solution.) Using Eq. (2.8) and assuming a dilute ideal solution, we obtain the following relations for ln a_w:

$$\ln a_w = \ln N_w = \ln\left(1 - \frac{\sum_j n_j}{n_w + \sum_j n_j}\right) \cong -\frac{\sum_j n_j}{n_w + \sum_j n_j} \cong -\frac{\sum_j n_j}{n_w} \tag{2.9}$$

The penultimate step in Eq. (2.9) is based on the series expansion of a logarithm: $\ln(1-x) = -x - x^2/2 - x^3/3 - \cdots$, a series that converges rapidly for $|x| \ll 1$. The last two steps employ the approximation ($n_w \gg \sum_j n_j$) relevant to a dilute solution. Equation (2.9) is thus restricted to dilute ideal solutions; nevertheless, it is a useful expression indicating that, in the absence of solutes ($\sum_j n_j = 0$), ln a_w is zero and a_w is 1, and that solutes decrease the activity of water from the value of one for pure water.

To obtain a familiar form for expressing the osmotic pressure, we can incorporate the approximation for ln a_w given by Eq. (2.9) into the definition of osmotic pressure given by Eq. (2.7) ($RT \ln a_w = -\bar{V}_w \Pi$), which yields

$$\Pi_s \cong -\frac{RT}{\bar{V}_w}\left(-\frac{\sum_j n_j}{n_w}\right) = RT \sum_j \frac{n_j}{\bar{V}_w n_w} = RT \sum_j c_j \tag{2.10}$$

where $\bar{V}_w n_w$ is the total volume of water in the system (essentially the total volume of the system for a dilute aqueous solution), $n_j/\bar{V}_w n_w$ is the number of moles of species j per volume of water and so is the concentration of species j (c_j), and the summations are over all solutes.

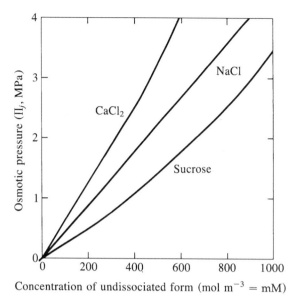

Figure 2-7. Relationship between concentration and osmotic pressure at 20°C for a nonelectrolyte (sucrose) and two readily dissociating salts (NaCl and CaCl$_2$). The different slopes indicate the different degrees of dissociation for the three substances. Data for osmotic pressure are based on the freezing point depression. (Data source: Lide, 1997.)

Osmotic pressure is often expressed by Eq. (2.10), known as the Van't Hoff relation, but this is justified only in the limit of dilute ideal solutions. As we have already indicated, an ideal solution has ideal solutes dissolved in an ideal solvent. The string of equalities in Eq. (2.9) assumes that γ_w is unity, so the subsequently derived expression (Eq. 2.10) strictly applies only when water acts as an ideal solvent ($\gamma_w = 1.00$). To emphasize that we are neglecting any factors that cause γ_w to deviate from 1 and thereby affect the measured osmotic pressure (such as the interaction between water and colloids, which we will discuss later), Π_s instead of Π has been used in Eq. (2.10), and we will follow this convention throughout the book. The effect of solute concentration on osmotic pressure as described by Eq. (2.10) is portrayed in Fig. 2-7 and was first clearly recognized by the botanist Wilhelm Pfeffer in 1877. Both the measurement of osmotic pressure and a recognition of its effects are crucial for an understanding of water relations in biology.

The cellular fluid expressed from young leaves of plants like pea or spinach often contains about 0.3 mol of osmotically active particles per kilogram of water (about 300 mol m^{-3}). This fluid is referred to as 0.3 osmolal by analogy with molality, which refers to the total concentration. For example, 0.1 molal CaCl$_2$ is about 0.26 osmolal because most of the CaCl$_2$ is dissociated (Fig. 2-7). We note that molality (moles of solute/kg solvent) is a concentration unit that is independent of temperature and suitable for colligative properties, e.g., the freezing point depression is 1.86°C for a 1 osmolal aqueous solution. However, mol m^{-3} and molarity (moles of solute/liter of solution, M) are often more convenient units. Below about 0.2 M, molarity and molality are nearly the same for low-molecular weight solutes in an aqueous solution, but at high concentrations the molarity can be considerably less than the molality. In fact, the deviations from linearity in Fig. 2-7 indicate the greater values for osmolality, to which osmotic pressure is proportional, compared with osmolarity. Such deviations are apparent at lower concentrations for sucrose, which has a relatively high molecular weight of 342, than for $CaCl_2$ or NaCl (molecular weights of 111 and 58, respectively). For instance, at 300 mM, sucrose is 321 mm, $CaCl_2$ is 303 mm, and NaCl is 302 mm (at these concentrations, $CaCl_2$ is about 78% dissociated and NaCl about 83% dissociated, leading to the higher osmotic pressures than for sucrose; Fig. 2-7).

Using Eq. (2.10) ($\Pi_s \cong RT \sum_j c_j$), we can calculate the osmotic pressure for the cell sap in a leaf. At 20°C, RT is 0.002437 m^3 MPa mol^{-1} (Appendix II), so Π_s for a 0.3-osmolar solution is

$$\Pi_s = (0.002437 \text{ m}^3 \text{ MPa mol}^{-1})(300 \text{ mol m}^{-3}) = 0.73 \text{ MPa}$$

The osmotic pressure of cell sap pressed out of mature leaves of most plants is 0.6–3 MPa; for comparison, seawater has a Π_s of 2.5 MPa.

G. Matric Pressure

For certain applications in plant physiology, another term is frequently included in the chemical potential of water, namely, $\bar{V}_w \tau$, where τ is the *matric pressure* (also called the *matric potential*). The matric pressure does not represent any new force (all energetic considerations are already fully described by the terms in μ_w for hydrostatic pressure or water activity). However, use of this term is sometimes convenient for dealing explicitly with interactions occurring at interfaces, even though these interfacial forces can also be adequately represented by their contributions to Π or P. In other words, the matric pressure does not represent a new or different force or a new or different contribution to μ_w, but it can be used as a bookkeeping device for handling interfacial interactions. To help make this statement more meaningful, we will briefly consider the influence of liquid–solid interfaces on the chemical potential of water at the surfaces of colloids. "Colloid" is a generic term for solid particles approximately 0.002–1 μm in diameter suspended in a liquid, e.g., proteins, ribosomes, and even some membrane-bounded organelles.

When water molecules are associated with interfaces such as those provided by membranes or colloidal particles, they have less tendency either to react chemically in the bulk solution or to escape into a surrounding vapor phase. Interfaces thus lower the thermodynamic activity of the water (a_w), especially near their surfaces. Solutes also lower the water activity (Eq. 2.10). As a useful first approximation, we can consider that these two effects lowering water activity are additive in a solution containing both ordinary solutes and colloids or other interfaces. Osmotic pressure (Π) depends on the activity of water regardless of the reason for the departure of a_w from 1, i.e., Π still equals $-(RT/\bar{V}_w) \ln a_w$ (Eq. 2.7). Recalling that $a_w = \gamma_w N_w$, we can write the following relations:

$$\Pi = -\frac{RT}{\bar{V}_w} \ln \gamma_w N_w = -\frac{RT}{\bar{V}_w} \ln \gamma_w - \frac{RT}{\bar{V}_w} \ln N_w = \tau + \Pi_s \qquad (2.11)$$

where τ represents a matric pressure resulting from the water–solid interactions at the surfaces of the colloids and other interfaces that lower γ_w from 1. [Note that if we let $\tau = -(RT/\bar{V}_w) \ln \gamma_w$ (Eq. 2.11), then $\gamma_w < 1$ means $\tau > 0$.]

Π_s in Eq. (2.11) is the osmotic pressure of all solutes, including the concentration of the colloids, as represented by Eq. (2.10) ($\Pi_s \cong RT \sum_j c_j$). In other words, in the derivation of Π_s we dealt only with $-(RT/\bar{V}_w) \ln N_w$ because γ_w was set equal to 1. For an exact treatment when many interfaces are present, e.g., in the cytosol of a typical cell, we cannot set γ_w equal to 1 because the activity of water, and hence the osmotic pressure (Π), is affected by proteins, other colloids, and other interfaces. In such a case, Eq. (2.11) suggests a simple way in which a matric pressure may be related to the reduction of the activity coefficient of water caused by the interactions at interfaces. Equation (2.11) should not be viewed as a relation defining matric pressures for all situations but rather for cases in which it might be useful to represent interfacial interactions by a separate term that can be added to Π_s, the effect of the solutes on Π.

Although Π and a_w may be the same throughout some system, both Π_s and τ in Eq. (2.11) may vary. For example, water activity in the bulk of the solution may be predominantly lowered by solutes, whereas at or near the surface of colloids the main factor decreasing a_w from 1 could be the interfacial attraction and binding of water. Such interfacial interactions do not change the mole fraction of water, but they do reduce its activity coefficient, γ_w.

Other areas of plant physiology where matric potentials and pressures have been invoked are descriptions of the chemical potential of water in soil and in cell walls. (Cell walls will be further considered at the end of this chapter, and soil matric potentials will be mentioned in Chapter 9.) Surface tension at the numerous air–water interfaces for the interstices of a cell wall or among soil particles leads to a tension in the water. Such a tension is a negative hydrostatic pressure, i.e., $P < 0$. Although it is not necessary to do so, P is sometimes used to refer only to positive pressures, and a negative P in such pores has been called a positive matric pressure (or potential). Some books define this same negative hydrostatic pressure in the water in cell wall interstices or between soil particles as a negative matric potential. It is more straightforward and consistent to refer to the quantity that reduces μ_w in such pores as a negative P.

H. Water Potential

From the definition of chemical potential (Eq. 2.4) and the formal expression for osmotic pressure (Eq. 2.7), we can express the chemical potential of water (μ_w) as

$$\mu_w = \mu_w^* - \bar{V}_w \Pi + \bar{V}_w P + m_w g h \tag{2.12}$$

where the electrical term ($z_j FE$) is not included because water carries no net charge ($z_w = 0$). The quantity $\mu_w - \mu_w^*$ is important for discussing the water relations of plants. It represents the work involved in moving 1 mol of water from a pool of pure water at atmospheric pressure, at the same temperature as the system under consideration, and at the zero level for the gravitational term to some arbitrary point in a system (at constant pressure and temperature). A difference between two locations in the value of $\mu_w - \mu_w^*$ indicates that water is not in equilibrium—water tends to flow toward the region where $\mu_w - \mu_w^*$ is lower.

A quantity proportional to $\mu_w - \mu_w^*$ that is commonly used in studies of plant water relations is the *water potential*, Ψ, defined as

$$\Psi = \frac{\mu_w - \mu_w^*}{\bar{V}_w} = P - \Pi + \rho_w g h \tag{2.13a}$$

which follows directly from Eq. (2.12) plus the identification of the mass per mole of water (m_w)/volume per mole of water ($\bar{V}_w$), or mass/volume, as the density of water, ρ_w. Equation (2.13a) indicates that an increase in hydrostatic pressure raises the water potential, and an increase in osmotic pressure lowers it.

Because work must be performed to raise an object in the gravitational field of the earth, vertical position also affects the chemical potential. Consequently, the term $\rho_w g h$ is included in the water potential given by Eq. (2.13a), at least when water moves an appreciable distance vertically in the gravitational field. The magnitude of $\rho_w g$ is 0.0098 MPa m^{-1}; if water moves 10 m vertically upward in a tree, the gravitational contribution to the water potential is increased by 0.10 MPa. For certain applications in this text, such as considerations of chemical reactions or the crossing of membranes, little or no change occurs in vertical position, and the gravitational term can then be omitted from μ_j and Ψ.

As we mentioned when introducing osmotic pressure, a number of conventions are used to describe the osmotic and other water potential terms. One such convention is to define Ψ as

$$\Psi = \Psi_P + \Psi_\Pi + \Psi_h \tag{2.13b}$$

where Ψ_p ($= P$) is called the hydrostatic potential or pressure potential, Ψ_Π ($= -\Pi$) is the osmotic potential, and Ψ_h ($= \rho_w g h$) is the gravitational potential. Although uniformity of expression is a cherished ideal, one should not be too alarmed that various conventions are used because persons from many fields have contributed to our understanding of plants. Because Ψ is so important in understanding plant water relations, a method for measuring it is illustrated in Fig. 2-8.

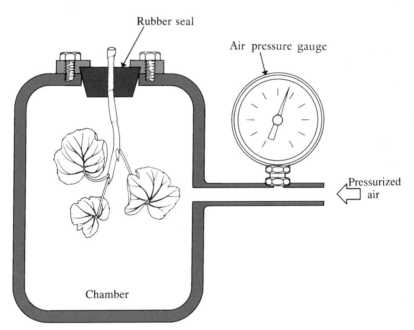

Figure 2-8. Schematic diagram of a "pressure bomb," which can be used to measure the xylem pressure, P^{xylem}, averaged over the material placed in the chamber. To make a measurement, a severed part of a plant is placed in the chamber with its freshly cut end protruding through a rubber seal. The air pressure (P^{air}) in the chamber is then gradually increased until it just causes the exudation of xylem sap at the cut end (generally viewed with a magnifying glass or a dissecting microscope). At this stage, the resulting pressure of the sap, which equals $P^{xylem} + P^{air}$, is zero, so P^{xylem} equals $-P^{air}$. If the xylem osmotic pressure can be ignored, P^{xylem} is approximately equal to Ψ^{xylem}, which can be the same as the water potential of the other tissue in the chamber (if water equilibration has been achieved).

III. Central Vacuole and Chloroplasts

We next consider the water relations of both the large central vacuole and the chloroplasts, using the water potential just defined. Our focus will be on situations in which no net flow of water occurs across the limiting membranes surrounding these subcellular compartments and for the special case of nonpenetrating solutes. Hence, we will be considering the idealization of semipermeable membranes (see Fig. 2-6). The more general and biologically realistic case of penetrating solutes will be discussed in Chapter 3 after a consideration of the properties of solutes and the introduction of concepts from irreversible thermodynamics.

The central vacuole occupies up to 90% of a mature plant cell, so most of the cellular water is in the vacuole. The vacuolar volume is generally about 10^4 μm^3 for the mesophyll cells in a leaf and can be much larger in certain algal cells. Because the vacuole is nearly as large as the cell containing it, special procedures are required to remove the vacuole from a plant cell without rupturing its surrounding membrane, the tonoplast. Chloroplasts are much smaller than the central vacuole, often having volumes near 30 μm^3 *in vivo* (although sizes vary among plant species). When chloroplasts are carefully isolated, suitable precautions will ensure that their limiting membranes remain intact. Such intact chloroplasts can be placed in solutions having various osmotic pressures, and the resulting movement of water into or out of the organelles can be precisely measured.

A. Water Relations of the Central Vacuole

To predict whether and in what direction water will move, we need to know the value of the water potential in the various compartments under consideration. At equilibrium, the water potential is the same in all communicating phases, such as those separated by membranes. For example, when water is in equilibrium across the tonoplast, the water potential is the same in the vacuole as it is in the cytosol. No force then drives water across this membrane, and no net flow of water occurs into or out of the vacuole.

The tonoplast does not have an appreciable difference in hydrostatic pressure across it. A higher internal hydrostatic pressure would cause an otherwise slack (folded) tonoplast to be mechanically pushed outward. The observed lack of such motion indicates that ΔP is close to zero across a typical tonoplast. If the tonoplast were taut, ΔP would cause a stress in the membrane, analogous to the cell wall stresses discussed in Chapter 1 (Section VC), namely, the stress would be $r \, \Delta P/2t$ for a spherical vacuole. However, the tensile strength of biological membranes is low—membranes can rupture when a stress of 0.2–1.0 MPa develops in them. For a tonoplast 7 nm thick with a maximum stress before rupturing of 0.5 MPa surrounding a spherical vacuole 14 μm in radius, the maximum hydrostatic pressure difference across the tonoplast is

$$\Delta P = \frac{2t\sigma}{r} = \frac{(2)(7 \times 10^{-9} \text{ m}) \, (0.5 \text{ MPa})}{(14 \times 10^{-6} \text{ m})} = 0.5 \times 10^{-3} \text{ MPa}$$

which is very small. Thus P is essentially the same in the cytosol as in the vacuole. With this simplifying assumption, and for the equilibrium situation when Ψ is the same in the two phases, Eq. (2.13a) ($\Psi = P - \Pi + \rho_w gh$, where $\Delta h = 0$ across a membrane) gives

$$\Pi^{\text{cytosol}} = \Pi^{\text{vacuole}} \tag{2.14}$$

where Π^{cytosol} is the osmotic pressure in the cytosol and Π^{vacuole} is that in the central vacuole.

The vacuole appears to be a relatively homogeneous aqueous phase, whereas the cytoplasm is a more complex phase containing many colloids and membrane-bounded organelles. Because the vacuole contains few colloidal or other interfaces, any matric pressure in it is negligible compared with the osmotic pressure resulting from the vacuolar solutes. Expressing osmotic pressure by Eq. (2.11) ($\Pi = \tau + \Pi_s$) and assuming that τ^{vacuole} is negligible, we can replace Π^{vacuole} in Eq. (2.14) by Π_s^{vacuole}, i.e., the decrease in vacuolar water activity is due almost solely to the solutes. Water is thus acting like an ideal solvent ($\gamma_w = 1$). On the other hand, water activity in the cytosol presumably is considerably lowered both by solutes and by interfaces. Unlike the case for the vacuole, individual water molecules in the cytosol are never far from proteins, organelles, or other colloids, so interfacial interactions can appreciably affect a_w^{cytosol}. Thus, both Π_s^{cytosol} and τ^{cytosol} contribute to Π^{cytosol}. Hence, Eq. (2.14) indicates that $\tau^{\text{cytosol}} + \Pi_s^{\text{cytosol}} = \Pi_s^{\text{vacuole}}$. Because τ^{cytosol} is positive, we conclude that at equilibrium the osmotic pressure in the vacuole due to solutes, Π_s^{vacuole}, must be larger than Π_s^{cytosol}. Equation (2.14) thus leads to the prediction that the vacuole has a higher concentration of osmotically active solutes than does the cytosol.

Because of the large central vacuole, the cytoplasm occupies a thin layer around the periphery of a plant cell (see Fig. 1-1). Therefore, for its volume, the cytoplasm has a relatively large surface area across which diffusion can occur. Vacuoles also provide large, relatively simple compartments in which hydrostatic pressures lead to the cellular turgidity necessary for the growth and support of a plant. The vacuole can act as a storage reservoir for toxic products or metabolites, e.g., the nocturnal storage of organic acids such as malic acid takes place in the vacuoles of Crassulacean acid metabolism plants (discussed in Chapter 8). Certain secondary chemical products, such as phenolics, alkaloids, tannins, glucosides, and flavonoids (e.g., anthocyanins), can also accumulate in vacuoles.

B. Boyle–Van't Hoff Relation

The volume of a chloroplast or other membrane-bounded body changes in response to variations in the osmotic pressure of the external solution, Π°. This is a consequence of the properties of membranes, which generally allow water to move readily across them but restrict the passage of certain solutes, such as sucrose. For example, if Π° outside a membrane-bounded aqueous compartment were raised, we would expect that water would flow out across the membrane to the region of lower water potential and thereby decrease the compartment volume, whereas in general little movement of solutes occurs in such a case. This differential permeability leads to the "osmometric behavior" characteristic of many cells and organelles. The conventional expression quantifying this volume response to changes in the external osmotic pressure is the Boyle–Van't Hoff relation:

$$\Pi^\circ(V - b) = RT \sum_j \varphi_j n_j = RT n \qquad (2.15)$$

where Π° is the osmotic pressure of the external solution; V is the volume of the cell or organelle; b is the so-called nonosmotic volume (osmotically inactive volume), which is frequently considered to be the volume of a solid phase within volume V that is not penetrated by water; n_j is the number of moles of species j within $V - b$; φ_j is the osmotic coefficient, a correction factor indicating the relative osmotic effect of species j; and $n = \sum_j \varphi_j n_j$ is the apparent number of osmotically active moles in $V - b$.

In this chapter we will derive the Boyle–Van't Hoff relation using the chemical potential of water, and in Chapter 3 we will extend the treatment to penetrating solutes by employing irreversible thermodynamics. Although the Boyle–Van't Hoff expression will be used to interpret the osmotic responses only of chloroplasts, the equations that will be developed are general and can be applied equally well to mitochondria, whole cells, or other membrane-surrounded bodies.

The Boyle–Van't Hoff relation applies to the equilibrium situation for which the water potential is the same on either side of the membranes surrounding chloroplasts. When Ψ^i equals Ψ^o, net water movement across the membrane ceases, and the volume of a chloroplast is constant. (The superscript i refers to the inside of the cell or organelle and the superscript o to the outside.) If we were to measure the chloroplast volume under such conditions, the external solution would generally be at atmospheric pressure ($P^o = 0$). By Eq. (2.13a) ($\Psi = P - \Pi$, when the gravitational term is ignored), the water potential in the external solution is then

$$\Psi^o = -\Pi^o \tag{2.16}$$

Like Ψ^o (Eq. 2.16), the water potential inside the chloroplasts, Ψ^i, also depends on osmotic pressure. However, the internal hydrostatic pressure (P^i) may be different from atmospheric pressure and should be included in the expression for Ψ^i. In addition, macro-molecules and solid–liquid interfaces inside the chloroplasts can lower the activity coefficient of water, which we can represent by a matric pressure τ^i (see Eq. 2.11). To allow for this possibility, the internal osmotic pressure (Π^i) is the sum of the solute and the interfacial contributions, in the manner expressed by Eq. (2.11). Using Eq. (2.13a), Ψ^i is therefore

$$\Psi^i = P^i - \Pi^i = P^i - \Pi^i_s - \tau^i \tag{2.17}$$

We will now reexpress Π^i_s and then equate Ψ^o to Ψ^i. For a dilute solution inside the chloroplasts, $\Pi^i_s = RT \sum_j n^i_j/(\bar{V}_w n^i_w)$ by Eq. (2.10). Equating the water potential outside the chloroplast (Eq. 2.16) to that inside (Eq. 2.17), the condition for water equilibrium across the limiting membranes is

$$\Pi^o = RT\frac{\sum_j n^i_j}{\bar{V}_w n^i_w} + \tau^i - P^i \tag{2.18}$$

Although this expression may look a little frightening, it concisely states that at equilibrium the water potential is the same on both sides of the membranes and that we can explicitly recognize various possible contributors to the two Ψs involved.

To appreciate the refinements that this thermodynamic treatment introduces into the customary expression describing the osmotic responses of cells and organelles, we compare Eq. (2.18) with Eq. (2.15), the conventional Boyle–Van't Hoff relation. The volume of water inside the chloroplast is $\bar{V}_w n^i_w$ because n^i_w is the number of moles of internal water and $\bar{V}_w$ is the volume per mole of water. This factor in Eq. (2.18) can be identified with $V - b$ in Eq. (2.15). Instead of being designated the "nonosmotic volume," b is more appropriately called the "nonwater volume" for it includes the volume of the internal solutes, colloids, and membranes. In other words, the total volume (V) minus the nonwater volume (b) equals the volume of internal water ($\bar{V}_w n^i_w$). We also note that the possible hydrostatic and matric contributions included in Eq. (2.18) are neglected in the usual Boyle–Van't Hoff relation. In summary, although certain approximations and assumptions are incorporated into Eq. (2.18) (e.g., that solutes do not cross the limiting membranes and that the solutions are dilute), it still more satisfactorily describes osmotic responses of cells and organelles than does the conventional Boyle–Van't Hoff relation.

C. Osmotic Responses of Chloroplasts

To illustrate the use of Eq. (2.18) in interpreting osmotic data, we will consider osmotic responses of pea chloroplasts suspended in external solutions of various osmotic pressures. It is customary to plot the volume V versus the reciprocal of the external osmotic pressure, $1/\Pi^\circ$, so certain algebraic manipulations are needed to express Eq. (2.18) in a more convenient form. By transferring $\tau^i - P^i$ to the left side of Eq. (2.18) and then multiplying both sides by $\bar{V}_w n_w^i/(\Pi^\circ - \tau^i + P^i)$, $\bar{V}_w n_w^i$ can be shown to equal $RT \sum_j n_j^i/(\Pi^\circ - \tau^i + P^i)$. The measured chloroplast volume V can be represented by $\bar{V}_w n_w^i + b$, i.e., as the sum of the aqueous and nonaqueous contributions. We thus obtain

$$V = RT \sum_j n_j^i/(\Pi^\circ - \tau^i + P^i) + b \qquad (2.19)$$

for the modified form of the Boyle–Van't Hoff relation.

Figure 2-9 indicates that the volume of pea chloroplasts varies linearly with $1/\Pi^\circ$ over a considerable range of external osmotic pressures. Therefore, $-\tau^i + P^i$ in Eq. (2.19) must be either negligibly small for pea chloroplasts compared with Π° or perhaps proportional to Π°. For simplicity, we will consider only the observed proportionality between V and $1/\Pi^\circ$. In other words, we will assume that $V - b$ (or $\bar{V}_w n_w^i) = RT \sum_j n_j^i/\Pi^\circ$ for pea chloroplasts. We will return to such considerations in Chapter 3, in which we will further

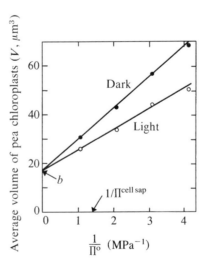

Figure 2-9. Volumes of chloroplasts at various external osmotic pressures, Π°. Pea chloroplasts were isolated from plants in the light or the dark as indicated. [Source: Nobel (1969); used by permission.]

refine the Boyle–Van't Hoff relation to include the more realistic case in which solutes can cross the surrounding membranes.

The relatively simple measurement of the volumes of pea chloroplasts for various external osmotic pressures can yield a considerable amount of information about the organelles. If we measure the volume of the isolated chloroplasts at the same osmotic pressure as in the cytosol, we can determine the chloroplast volume in the plant cell. Cell "sap" expressed from young pea leaves can have an osmotic pressure of 0.70 MPa; such sap comes mainly from the central vacuole, but because we expect Π^{cytosol} to be essentially equal to Π^{vacuole} (Eq. 2.14), $\Pi^{\text{cell sap}}$ is about the same as Π^{cytosol} (some uncertainty exists because during extraction the cell sap can come into contact with water in the cell walls). At an external osmotic pressure of 0.70 MPa (indicated by an arrow in Fig. 2-9), pea chloroplasts have a volume of 29 μm^3 when isolated from illuminated plants and 35 μm^3 when isolated from plants in the dark. Because these volumes occur at approximately the same osmotic pressure as found in the cell, they are presumably reliable estimates of pea chloroplast volumes *in vivo*.

The data in Fig. 2-9 indicate that light affects the chloroplast volume *in vivo*. Although the basis for the size changes is not fully understood, chloroplasts in many plants do have a larger volume in the dark than in the light. The decrease in volume upon illumination of the plants is observed by both phase contrast and electron microscopy as a flattening or decrease in thickness of the chloroplasts. This flattening, which amounts to about 20% of the thickness for pea chloroplasts *in vivo*, is in the vertical direction for the chloroplast depicted in Fig. 1-9. We also note that the slopes of the osmotic response curves in Fig. 2-9 are equal to $RT \sum_j n_j^i$. The slope is 50% greater for pea chloroplasts from plants in the dark than from those in the light, suggesting that chloroplasts in the dark contain more osmotically active particles. In fact, illumination causes an efflux of K^+ and Cl^- from the chloroplasts. Also, the 20% decrease in chloroplast volume in the light *in vivo* reflects an exodus of nearly 40% of the internal water, leading to a concentrating effect on the remaining solutes such as Mg^{2+}, an ion that is important for enzymes involved in photosynthesis.

The intercept on the ordinate in Fig. 2-9 is the chloroplast volume theoretically attained in an external solution of infinite osmotic pressure—a $1/\Pi^\circ$ of zero is the same as a Π° of infinity. For such an infinite Π°, all of the internal water would be removed ($n_w^i = 0$), and the volume, which is obtained by extrapolation, would be that of the nonaqueous components of the chloroplasts. (Some water is tightly bound to proteins and other substances and presumably remains bound even at the hypothetical infinite osmotic pressure; such water is not part of the internal water, $\bar{V}_w n_w^i$.) Thus the intercept on the ordinate of a V-versus-$1/\Pi^\circ$ plot corresponds to b in the conventional Boyle–Van't Hoff relation (Eq. 2.15). This intercept (indicated by an arrow in Fig. 2-9) equals 17 μm^3 for chloroplasts both in the light and in the dark. (The extra solutes in a chloroplast in the dark correspond to <0.1 μm^3 of solids.) Using the chloroplast volumes obtained for a Π° of 0.70 MPa and 17 μm^3 for the nonwater volume b, we find that the fractional water content of pea chloroplasts in the dark is

$$\frac{V - b}{V} = \frac{(35\ \mu\text{m}^3 - 17\ \mu\text{m}^3)}{(35\ \mu\text{m}^3)} = 0.51$$

or 51%. In the light it is 41%. These relatively low water contents in the organelles are consistent with the extensive internal lamellar system (see Fig. 1-9) and the abundance of CO_2 fixation enzymes in chloroplasts. Thus osmometric responses of cells and organelles can be used to provide information on their fractional water content.

IV. Water Potential and Plant Cells

In this section we will shift our emphasis from a consideration of the water relations of subcellular bodies to those of whole cells and extend the development to include the case of water fluxes. Whether water enters or leaves a plant cell, how much, and the rate of movement all depend on the water potential outside compared with that inside. The external water potential Ψ^o can often be varied experimentally, and the direction as well as the magnitude of the resulting water movement will give information about Ψ^i. Moreover, the equilibrium value of Ψ^o can be used to estimate the internal osmotic pressure Π^i. We will also consider various ways of examining the relationships between Ψ^i, Π^i, and P^i.

A loss of water from plant shoots—indeed, sometimes even an uptake—occurs at cell–air interfaces. As we would expect, the chemical potential of water in cells compared with that in the adjacent air determines the direction for net water movement at such locations. Thus we must obtain an expression for the water potential in a vapor phase and then relate this Ψ_{wv} to Ψ for the liquid phases in a cell. We will specifically consider the factors influencing the water potential at the plant cell–air interface, namely, in the cell wall. We will find that $\psi^{cell\ wall}$ is dominated by a negative hydrostatic pressure resulting from surface tension effects in the cell wall pores.

A. Incipient Plasmolysis

For usual physiological conditions, a positive hydrostatic pressure exists inside a plant cell, i.e., it is under turgor. By suitably adjusting the solutes in an external solution bathing such a turgid cell, P^i can be reduced to zero, thereby enabling an estimate of Π^i, as the following argument will indicate (Fig. 2-10).

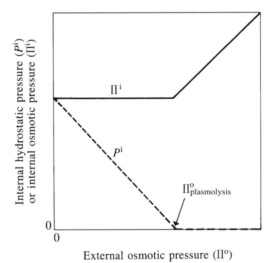

Figure 2-10. Responses of an initially turgid cell ($P^i > 0$) placed in pure water ($\Pi^o = 0$) to changes in the external osmotic pressure (Π^o). At the point of incipient plasmolysis (plasma membrane just beginning to pull away from the cell wall), P^i is reduced to zero, which is hence also called the turgor loss point. Constancy of cell volume and water equilibrium are assumed at each step as Π^o is raised (actually, Π^i increases a few percent as Π^o is raised from 0 to $\Pi^o_{plasmolysis}$ because the cell shrinks slightly accompanying the decrease of P^i to zero).

Let us place the cell in pure water ($\Pi^o = 0$) at atmospheric pressure ($P^o = 0$). Ψ^o is then zero (ignoring the gravitational term, $\Psi = P - \Pi$; Eq. 2.13a), and Ψ^i is $P^i - \Pi^i$ (the inner phase can be the cytosol). If the cell is in equilibrium with this external solution ($\Psi^o = \Psi^i$), P^i must then equal Π^i (Fig. 2-10). Suppose that Π^o is now gradually raised from its initial zero value—for example, by adding solute to the external solution. If the cell remains in equilibrium, then Π^o equals $\Pi^i - P^i$ ($P^o = 0$ because the external solution is at atmospheric pressure). As Π^o is increased, P^i will tend to decrease, whereas Π^i usually does not change very much. More precisely, because the cell wall is quite rigid, the cell will not appreciably change its volume in response to small changes in P^i. [Because the cell wall has elastic properties (see Chapter 1) some water flows out as P^i decreases and the cell shrinks, a matter to which we will return shortly.] If the cell volume does not change in response to changes in P^i and no internal solutes leak out, Π^i will remain constant. As the external osmotic pressure is raised, Π^o will eventually become equal to Π^i. In such a plant cell at equilibrium P^i is 0, so no internal hydrostatic pressure is exerted against the cell wall when $\Pi^o = \Pi^i$ (Fig. 2-10). The cell will thus lose its turgidity. If Π^o is increased further, water will flow out of the cell, plasmolysis will occur as the plasma membrane pulls away from the rigid cell wall, and the internal osmotic pressure will increase (the same amount of solutes in less water; see Fig. 3-13). Ignoring for the moment any overall volume changes of the cell, we find that the condition under which water just begins to move out of the cell—referred to as the point of *incipient plasmolysis* and illustrated in Fig. 2-10—is

$$\Pi^o_{plasmolysis} = \Pi^i \qquad (2.20)$$

Equation (2.20) suggests that only a relatively simple measurement ($\Pi^o_{plasmolysis}$) is needed to estimate the osmotic pressure (Π^i) occurring inside an individual plant cell.

The existence of an internal hydrostatic pressure within a plant cell leads to stresses in its cell wall and a resulting elastic deformation or strain, as indicated in Chapter 1. The decrease of P^i to 0, known as the *turgor loss point*—which occurs in taking a turgid plant cell to the point of incipient plasmolysis—must therefore be accompanied by a contraction of the cell as the wall stresses are relieved. This decrease in volume means that some water will actually flow out of the cell before the point of incipient plasmolysis is reached. If no internal solutes enter or leave as the cell shrinks, then the osmotic pressure inside will increase (the same amount of solutes in a smaller volume). As a useful first approximation, we can assume that the change in osmotic pressure reciprocally follows the change in volume, i.e., the product of Π^i and the cellular volume is approximately constant. With this assumption, the osmotic pressure in the cell—described by Eq. (2.20) and determined by using the technique of measuring the point of incipient plasmolysis—can be corrected to its original value by using the ratio of the initial volume to the final volume of the cell. The change in volume is only a few percent for most plant cells, in which case fairly accurate estimates of Π^i can be obtained from plasmolytic studies alone.

Measurements of $\Pi^o_{plasmolysis}$ often give values near 0.7 MPa for cells in storage tissues like onion bulbs or carrot roots and in young leaves of pea or spinach. These values of the external osmotic pressure provide information on various contributors to the water potential inside the cell, as Eq. (2.20) indicates. When water is in equilibrium within the plant cell at the point of incipient plasmolysis, Π^i in Eq. (2.20) is the osmotic pressure both in the cytosol and in the vacuole. Π^i can be replaced by $\Pi^i_s + \tau^i$ (Eq. 2.11), where Π^i_s is the osmotic pressure contributed by the internal solutes and τ^i is the matric pressure. This relation was invoked when the implications of Eq. (2.14) ($\Pi^{cytosol} = \Pi^{vacuole}$) were discussed, and the various arguments presented at that time extend to the present case. Specifically, because the possible matric pressure in the vacuole is probably negligible, $\Pi^o_{plasmolysis}$ should be a good estimate of $\Pi^{vacuole}_s$. Moreover, $\tau^{cytosol}$ is most probably significant because the water activity in the cytosol can be lowered by the many interfaces present there. Thus $\Pi^o_{plasmolysis}$ is an upper limit for $\Pi^{cytosol}_s$. We see that determination of the external osmotic pressure at the point of incipient plasmolysis provides information on the osmotic pressure existing in different compartments within the plant cell.

B. Höfler Diagram and Pressure–Volume Curve

We next examine the relationship between cellular water potential components and proto-plast volume in a way introduced by Karl Höfler in 1920 and now referred to as a *Höfler diagram* (Fig. 2-11). The protoplast volume is maximal when the cell is in equilibrium with pure water, in which case the internal water potential Ψ^i is zero and P^i equals Π^i (right-hand side of Fig. 2-11, left-hand side of Fig. 2-10). As Ψ^i decreases, P^i decreases until the point of incipient plasmolysis and zero turgor is reached. The decrease in P^i causes a slight shrinkage of the cell and hence a slight concentrating of the solutes, so Π^i increases slightly as P^i decreases. Indeed, the product of protoplast volume and Π^i is essentially constant if the amount of solutes in the cell does not change, as indicated previously. Figure 2-11 also shows that Ψ^i is equal to $-\Pi^i$ when the protoplast volume decreases below the value for incipient plasmolysis, as P^i is then zero (based on Eq. 2.13, $\Psi^i = P^i - \Pi^i = \Psi^i_P + \Psi^i_\Pi$ when the gravitational term is ignored). As the cell goes from incipient plasmolysis to full

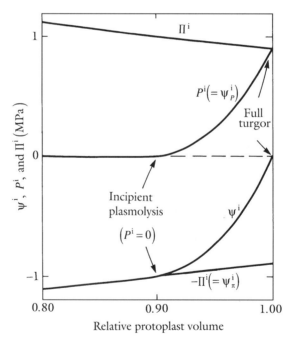

Figure 2-11. A Höfler diagram showing the relationship between the water potential (Ψ^i), hydrostatic pressure (P^i), and osmotic pressure (Π^i) in a plant cell for various protoplast volumes. For a group of cells, relative water content is often used in such diagrams instead of relative proto-plast volume. The nearly 10% decrease in volume from full turgor to incipient plasmolysis is characteristic of plant cells.

turgor, Fig. 2-11 indicates that P^i increases slowly at first as the cell expands and then more rapidly as the cell wall approaches the limit of its expansion. Thus a Höfler diagram directly demonstrates how Ψ^i, P^i, and Π^i change as cell water content and hence volume change, such as during the drying and wetting cycles experienced by plants.

We can readily extend our discussion to include a *pressure–volume*, or *P–V, curve*, which has proved useful for analyzing the water relations of plant organs such as leaves. To obtain such a curve, we can place an excised leaf in a pressure chamber (Fig. 2-8) and increase the air pressure in the chamber until liquid just becomes visible at the cut end of the xylem, which is viewed with a dissecting microscope or a hand-held magnifying lens so that water in individual conducting cells in the xylem can be observed. When the leaf is excised, the tension in the xylem causes the water contained therein to recede into the leaf, so this applied air pressure is needed to force water back to the cut surface. We next increase the air pressure in the pressure chamber to force out some liquid, which is carefully collected and its volume determined, and the new balancing air pressure is recorded. In this way, we obtain data step by step for a plot of the reciprocal of the balancing air pressure in the chamber, which corresponds approximately to the negative of the reciprocal of the tissue water potential, versus the volume of extruded xylem sap, a pressure–volume curve (Fig. 2-12). After forcing out considerable liquid (the maximum air pressure used in pressure chambers rarely exceeds 10 MPa), we can extrapolate the straight-line portion of the curve to the right to obtain the extrudable liquid at an infinite

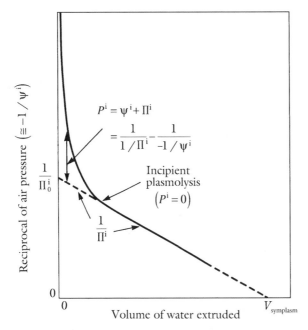

Figure 2-12. Relation between the reciprocal of leaf water potential determined with a pressure chamber (Fig. 2-8) and the volume of xylem sap extruded as the air pressure in the chamber is increased. The solid line indicates a typical range for data points. The reciprocal of the internal osmotic pressure ($1/\Pi^i$) including the value at full turgor ($1/\Pi_0^i$), the internal hydrostatic pressure (P^i), the point of incipient plasmolysis and turgor loss, and the volume of symplastic water (V_{symplasm}) can all be determined from such a P–V curve.

applied air pressure (1/pressure = 0 means that pressure = ∞). This intercept, which occurs when $-1/\Psi^i = 0$, is generally considered to represent the water volume of the symplasm. The apoplastic water volume is obtained as the difference between the total water volume of the tissue (determined from the fresh weight of the fully turgid tissue minus the dry weight obtained after drying the tissue in an oven at about 80°C until no further weight change occurs) minus this symplastic water volume, although some movement of water from the apoplasm to the symplasm may occur as the air pressure is increased in the pressure chamber.

The straight-line portion on the right-hand side of the pressure–volume curve in Fig. 2-12 corresponds to the inverse relationship between internal osmotic pressure Π^i and protoplasm volume V^i discussed previously [$\Pi^i V^i = $ constant $= \Pi^i (V_{symplasm} - V_{extruded})$, where $V_{symplasm}$ is the symplastic water volume at full turgor for all the cells in the chamber and $V_{extruded}$ is the volume of water extruded]. When the cells are at full turgor ($\Psi^i = 0$) and no sap has been extruded, the internal osmotic pressure has its minimum value, Π_0^i, which can be obtained by extrapolating the straight-line portion to the left in Fig. 2-12. The hydrostatic pressure inside the cell (P^i) can be deduced from such a pressure–volume curve using the difference between $-1/\Psi^i$ and the extrapolated line for $1/\Pi^i$. Moreover, because the change in volume (ΔV) for a specific change in pressure (ΔP) can be determined from the left-hand part of the curve in Fig. 2-12, we can also determine the volumetric elastic modulus ε averaged over the cells in the pressure chamber [$\varepsilon = \Delta P/(\Delta V/V)$; Eq. 1.17]. Because pressure–volume curves can be used to estimate symplastic water volume, apoplastic water volume, internal osmotic pressure, internal hydrostatic pressure, the point of incipient plasmolysis, and the volumetric elastic modulus, they have played a prominent role in the study of plant water relations.

C. Chemical Potential and Water Potential of Water Vapor

Water molecules in an aqueous solution continually escape into a surrounding gas phase. At the same time, water molecules also condense back into the liquid. The rates of escape and condensation depend on the chemical activities of water in the gas phase and in the liquid phase, the two rates becoming equal at equilibrium. The gas phase adjacent to the solution then contains as much water as it can hold at that temperature and still remain in equilibrium with the liquid. It is thus saturated with water vapor for the particular solution under consideration. The partial pressure exerted by the water vapor in equilibrium with pure water is known as the *saturation vapor pressure*.

The vapor pressure at equilibrium depends on the temperature and the solution, but it is independent of the relative or absolute amounts of liquid and vapor. When air adjacent to pure water is saturated with water vapor (100% relative humidity), the gas phase has the maximum water vapor pressure possible at that temperature—unless it is supersaturated, a metastable, nonequilibrium situation. The saturation vapor pressure in equilibrium with pure water increases markedly with temperature (Fig. 2-13), e.g., it increases from 0.61 kPa at 0°C to 2.34 kPa at 20°C to 7.38 kPa at 40°C (see Appendix I). Thus heating air at constant pressure causes the relative humidity, which represents the water vapor partial pressure as well as the water content per unit volume of air at a particular temperature relative to their maximum or saturation values at that temperature (Fig. 2-13), to drop dramatically; this can lead to dry environments inside buildings that during cold weather draw air in from the outside and then heat it, leading to dessication of our nasal passages.

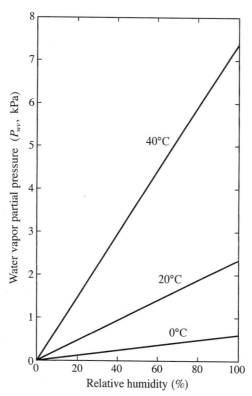

Figure 2-13. Relation between relative humidity and the partial pressure of water vapor at the indicated temperatures.

As solutes are added to the liquid phase and the mole fraction of water is thereby lowered, water molecules have less tendency to leave the solution. Hence, the water vapor pressure in the gas phase at equilibrium becomes less—this is one of the colligative properties of solutions that we mentioned earlier. In fact, for dilute solutions the partial pressure of water vapor (P_{wv}) at equilibrium depends linearly on the mole fraction of water (N_w) in the liquid phase. This is Raoult's law (also mentioned in Appendix IV). For pure water, $N_w = 1$ and P_{wv} has its maximum value, namely P_{wv}^*, the saturation vapor pressure.

The chemical potential of water vapor μ_{wv}, which depends on the partial pressure of water vapor P_{wv}, can be represented as follows:

$$\mu_{wv} = \mu_{wv}^* + RT \ln \frac{P_{wv}}{P_{wv}^*} + m_{wv}gh \tag{2.21}$$

where P_{wv}^* is the saturation vapor pressure in equilibrium with pure liquid water at atmospheric pressure and at the same temperature as the system under consideration, and m_{wv} is the mass per mole of water vapor, which is the same as the mass per mole of water, m_w. To handle deviations from ideality, P_{wv} is replaced by $\gamma_{wv}P_{wv}$, where γ_{wv} is the activity coefficient (more properly, the "fugacity" coefficient) of water vapor. If water vapor obeyed the ideal gas law—$P_j V = n_j RT$, where P_j is the partial pressure of species j in a volume V containing n_j moles of that gas—γ_{wv} would equal unity. This is a good approximation for situations of interest in biology.

We next consider the various parameters at the surface of pure water ($a_w = 1$) at atmospheric pressure ($P = 0$) in equilibrium with its vapor. In such a case, $P_{wv} = P_{wv}^*$, and we can let the zero level of the gravitational term be the surface of the liquid. Hence, $\mu_w = \mu_w^*$ and $\mu_{wv} = \mu_{wv}^*$. Because there is no net gain or loss of water molecules from the liquid phase at equilibrium ($\mu_w = \mu_{wv}$), we conclude that the constant μ_w^* must equal the constant μ_{wv}^* when the two phases are at the same temperature. The arguments are unchanged at other elevations, so the same gravitational term must be included in μ_w and μ_{wv} ($m_w g h = m_{wv} g h$ because $m_w = m_{wv}$). The form for the pressure effects in the chemical potential for water vapor is more subtle and is discussed next and in Appendix IV.

Equations (2.4) and (2.12) indicate that μ_w contains a term dependent on the pressure P applied to the system ($\mu_w = \mu_w^* + RT \ln a_w + \bar{V}_w P + m_w g h$), whereas μ_{wv} has no such term apparent ($\mu_{wv} = \mu_{wv}^* + RT \ln P_{wv}/P_{wv}^* + m_w g h$; Eq. 2.20). Because $\mu_w = \mu_{wv}$ at equilibrium and a pressure can be applied to such a system, P_{wv} must depend on P. This dependency is embodied in the Gibbs equation (see Appendix IV), one form of which is $RT \ln P_{wv}/P_{wv}^0 = \bar{V}_w P$, where P_{wv}^0 is the partial pressure of water vapor at atmospheric pressure. By adding and subtracting $RT \ln P_{wv}^0$ from the right side of Eq. (2.21) and substituting $\bar{V}_w P$ for $RT \ln P_{wv}/P_{wv}^0$, we obtain $\mu_{wv} = \mu_{wv}^* + RT \ln P_{wv}^0/P_{wv}^* + RT \ln P_{wv}/P_{wv}^0 + m_w g h = \mu_{wv}^* + RT \ln P_{wv}^0/P_{wv}^* + \bar{V}_w P + m_w g h$, the latter equality containing the requisite $\bar{V}_w P$ term. In practice the $\bar{V}_w P$ term is generally ignored because most air phases considered for plants are at atmospheric pressure, so $P = 0$ (P is conventionally defined as the pressure in excess of atmospheric). In any case, the difference between P_{wv} and P_{wv}^0 is generally very small, as an increase of 0.01 MPa (0.1 atm) in the gas phase surrounding a plant increases P_{wv} by only 0.007%. To appreciate the motion of water vapor that can be induced by changes in P, we need only consider wind (recall that changes in any term of a chemical potential, which represent changes in energy, can lead to motion). We also note that different air pressures can occur within certain plants—ranging from water lilies to rice—which can lead to internal "winds" causing mass flow of gases (briefly considered in Chapter 8).

We now consider the water potential of water vapor in a gas phase such as air, Ψ_{wv}. Using a definition of Ψ analogous to that in Eq. (2.13a) [$\Psi = (\mu_w - \mu_w^*)/\bar{V}_w$] and defining μ_{wv} by Eq. (2.21), we have

$$
\begin{aligned}
\Psi_{wv} &= \frac{\mu_{wv} - \mu_w^*}{\bar{V}_w} = \frac{\mu_{wv} - \mu_{wv}^*}{\bar{V}_w} \\
&= \frac{RT}{\bar{V}_w} \ln \frac{P_{wv}}{P_{wv}^*} + \rho_w g h \\
&= \frac{RT}{\bar{V}_w} \ln \frac{\% \text{ relative humidity}}{100} + \rho_w g h
\end{aligned}
\tag{2.22}
$$

where $m_{wv}/\bar{V}_w$ (the same as $m_w/\bar{V}_w$) has been replaced by the density of water, ρ_w. We note that $\bar{V}_w$, not $\bar{V}_{wv}$, is used in the definition of Ψ_{wv} in Eq. (2.22). This is necessary because the fundamental term representing free energy per mole is the chemical potential—we want to compare $\mu_j - \mu_j^*$ for the two phases when predicting changes at an interface—and thus the proportionality factor between $\mu_w - \mu_w^*$ or $\mu_{wv} - \mu_{wv}^*$ ($= \mu_{wv} - \mu_w^*$) and the more convenient terms, Ψ or Ψ_{wv}, must be the same in each case, namely, $\bar{V}_w$. Equation (2.22) also indicates that P_{wv}/P_{wv}^* equals the % relative humidity/100 (also see Fig. 2-13b); relative humidity is a readily measured quantity and has been extensively used in studying the water relations of plants and animals.

What happens to P_{wv} as we move vertically upward from pure water at atmospheric pressure in equilibrium with water vapor in the air? If we let h be zero at the surface of the water, $\Psi = 0$ ($\Psi = P - \Pi + \rho_w gh$; Eq. 2.13a), and because the water vapor is by supposition in equilibrium with the liquid phase, Ψ_{wv} is also zero. Equation (2.22) indicates that, as we go vertically upward in the gas phase, $\rho_w gh$ makes an increasingly positive contribution to Ψ_{wv}. Because we are at equilibrium, the other term in Ψ_{wv}, $(RT/\bar{V}_w)\ln(P_{wv}/P^*_{wv})$, must make a compensating negative contribution. At equilibrium P_{wv} must therefore decrease with altitude. We can further appreciate this conclusion simply by noting that gravity attracts the molecules of water vapor and other gases toward the earth, so air pressure is higher at sea level than it is on the top of a mountain.

Great heights are needed before the partial pressure of water vapor is reduced appreciably. In particular, we consider a 7% reduction in P_{wv} from P^*_{wv} (its equilibrium value at the water surface at atmosperic pressure) to $0.93 P^*_{wv}$. Using Eq. (2.21) and values at 20°C in Appendix I, we find that the height necessary is

$$h = -\frac{1}{\rho_w g}\frac{RT}{\bar{V}_w}\ln\frac{P_{wv}}{P^*_{wv}}$$

$$= -\frac{1}{(0.0098 \text{ MPa m}^{-1})}(135 \text{ MPa})\ln\left(\frac{0.93 P^*_{wv}}{P^*_{wv}}\right) = 1000 \text{ m}$$

In other words, not until a height of 1 km does the partial pressure of water vapor at equilibrium decrease by 7% from its value at sea level, indicating that the gravitational term generally has relatively little influence on Ψ_{wv} over the distances involved for an individual plant.

D. Plant–Air Interface

Water equilibrium across the plant–air interface occurs when the water potential in the leaf cells equals that of the surrounding atmosphere. (This presupposes that the leaf and the air are at the same temperature, an aspect that we will reconsider in Chapter 8). To measure Ψ^{leaf}, the leaf can be placed in a closed chamber and the relative humidity adjusted until the leaf does not gain or lose water. Such a determination is experimentally difficult because small changes in relative humidity have large effects on Ψ_{wv}, as well will show next.

Extremely large negative values are possible for Ψ_{wv}. In particular, $RT/\bar{V}_w$ at 20°C is 135 MPa. In the expression for Ψ_{wv} (Eq. 2.22), this factor multiplies $\ln$ (% relative humidity/100), and a wide range of relative humidities can occur in the air. By Eq. (2.22) with $\rho_w gh$ ignored, a relative humidity of 100% corresponds to a water potential in the vapor phase of 0 ($\ln 1 = 0$). This Ψ_{wv} is in equilibrium with pure water at atmospheric pressure, which also has a water potential of 0. For a relative humidity of 99%, Ψ_{wv} given by Eq. (2.22) is

$$\Psi_{wv} = (135 \text{ MPa})\ln\left(\frac{99}{100}\right) = -1.36 \text{ MPa}$$

Going from 100 to 99% relative humidity thus corresponds to a decrease in water potential of 1.36 MPa (Table 2-1). Small changes in relative humidity do indeed reflect large differences in the water potential of air! Finally, we note that a relative humidity of 50% at 20°C leads to a Ψ_{wv} of -94 MPa (Table 2-1).

Table 2-1. Magnitudes of Certain Water Vapor Parameters Useful for Understanding Water Movement at the Plant–Air Interface.[a]

Relative humidity (%)	P_{wv} (kPa)	Ψ_{wv} (MPa)
100.0	2.34	0.00
99.6	2.33	−0.54
99.0	2.32	−1.36
96.0	2.25	−5.51
90.0	2.11	−14.2
50.0	1.17	−93.6
0.0	0.00	$-\infty$

[a]Data are for 20°C. Ψ_{wv} is calculated from Eq. (2.22) ignoring the gravitational term. We note that $P_{wv}/P_{wv}^* \times 100 =$ relative humidity in %.

Because of the large negative values of Ψ_{wv} in air, water tends to diffuse out of leaves. The actual value of Ψ^{leaf} depends on the ambient conditions as well as the plant type and its physiological status. The range of Ψ^{leaf} for most mesophytes is −0.3 to −3 MPa, with −0.5 MPa being typical under conditions of wet soil for leaves of a garden vegetable such as lettuce. From Eq. (2.22), the relative humidity corresponding to a water potential of −0.5 MPa is 99.6%. Such an extremely high value for the relative humidity in equilibrium with a Ψ^{leaf} of −0.5 MPa indicates why it is difficult to determine Ψ^{leaf} by measuring the Ψ^{air} for which no water is gained or lost by the leaf. Even during a rainstorm, the relative humidity of the air rarely exceeds 99%. Because relative humidity is lower than 99.6% under most natural conditions, water is continually lost from a leaf having a water potential of −0.5 MPa. A few desert plants have a Ψ^{leaf} as low as −5.5 MPa. Even in this case of adaptation to arid climates, water still tends to leave the plant unless the relative humidity is about 96% (Table 2-1). Structural modifications and physiological responses are generally more important than a low Ψ^{leaf} for adapting to xerophytic conditions (discussed in Chapters 7 and 8).

E. Pressure in the Cell Wall Water

Plant cells come into contact with air at the cell walls bounding the intercellular air spaces (see Fig. 1-2). Thus the water potential in the cell walls must be considered with respect to Ψ_{wv} in the adjacent gas phase. The main contributing term for Ψ in cell wall water is generally the negative hydrostatic pressure arising from surface tension effects at the numerous air–liquid interfaces of the cell wall interstices near the cell surface. In turn, $P^{cell\ wall}$ can be related to the geometry of the cell wall pores and the contact angles.

The magnitude of the negative hydrostatic pressure that develops in cell wall water can be modeled by considering the pressure that occurs in a liquid within a cylindrical pore—the argument is basically the same as the one presented earlier in this chapter in discussing capillary rise. One of the forces acting on the fluid in a narrow pore of radius r is the result of surface tension, σ. This force equals $2\pi r\sigma \cos\alpha$ (see Fig. 2-3), where α is the contact angle, a quantity that can be essentially zero for wettable walls. Because the adhesive forces at the wall are transmitted to the rest of the fluid by means of cohesion, a tension or negative hydrostatic pressure develops in the fluid. The total force resisting the surface adhesion can be regarded as this tension times the area over which it acts, πr^2, and hence the force is

$\pi r^2 \times (-P)$. Equating the two forces gives the following expression for the pressure that can develop in fluid contained within a cylindrical pore:

$$P = \frac{2\pi r\sigma \cos \alpha}{-\pi r^2} = -\frac{2\sigma \cos \alpha}{r} \tag{2.23}$$

Adhesion of water at interfaces generally creates negative hydrostatic pressures in the rest of the fluid (Eq. 2.23 describes this P near the air–water interface, where the gravitational term can be ignored). Such negative hydrostatic pressures arising from interfacial interactions have sometimes been treated in plant physiology as positive matric pressures, a convention that we mentioned earlier.

The strong water–wall adhesive forces, which are transmitted throughout the cell wall interstices by water–water hydrogen bonding, can greatly reduce the water potential in the cell wall. At 20°C the surface tension of water is 7.28×10^{-8} MPa m (Appendix I), the voids between the fibers in the cell wall are often about 10 nm across ($r = 5$ nm), and $\cos \alpha$ can equal 1 for wettable walls. For water in such cylindrical pores, Eq. (2.23) indicates that P could be

$$P = -\frac{(2)(7.28 \times 10^{-8} \text{ MPa m})(1)}{(5 \times 10^{-9} \text{ m})} = -29 \text{ MPa}$$

(Fig. 2-14a). This is an estimate of the negative hydrostatic pressure or tension that could develop in the aqueous solution within cell wall interstices of typical dimensions, supporting the contention that $\Psi^{\text{cell wall}}$ can be markedly less than zero. Moreover, in such fine pores with hydrophilic surfaces, the hydrogen bonding in water can withstand tensions exceeding 100 MPa. We will next consider the actual pressures that develop in the cell wall water.

Let us begin by considering a Ψ_{wv} of -5.5 MPa, which corresponds to a relative humidity of 96% (Table 2-1). According to Eq. (2.23), water in cylindrical cell wall pores with radii of 26 nm and a contact angle of 0° has a P of -5.5 MPa (Fig. 2-14b), so such water is in equilibrium with air of 96% relative humidity in the intercellular spaces of the leaf. If the relative humidity near the cell wall surface were decreased, water in such interstices could be lost, but it would remain in the finer pores, where a more negative P or larger tension can be created (Figs. 2-14a and 2-14c). Hence, if the plant material is exposed to air of moderate to low relative humidity—as can happen for herbarium specimens—then larger and larger tensions resulting from interfacial interactions develop in the cell walls as the plant material dries. Water is then retained only in finer and finer pores of the wall, as is consistent with Eq. (2.23).

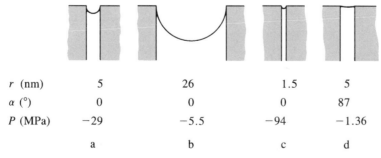

r (nm)	5	26	1.5	5
α (°)	0	0	0	87
P (MPa)	-29	-5.5	-94	-1.36
	a	b	c	d

Figure 2-14. Influence of radius r and contact angle α on the hydrostatic pressure P near the surface in cylindrical pores open to air.

We will now show that the availability of water adjacent to that in wettable cell walls affects $P^{\text{cell wall}}$ and the contact angle in the interstices. Suppose that pure water in mesophyll cell wall interstices 10 nm across is in equilibrium with water vapor in the intercellular air spaces where the relative humidity is 99%. As we calculated previously, Ψ_{wv} for this relative humidity is -1.36 MPa at 20°C. Hence, at equilibrium the (pure) water in the cell wall interstices has a hydrostatic pressure of -1.36 MPa ($\Psi = P - \Pi + \rho_w gh$; Eq. 2.13a). From Eq. (2.23) we can calculate the contact angle for which P can be -1.36 MPa for pores 5 nm in radius:

$$\cos \alpha = -\frac{rP}{2\sigma} = -\frac{(5 \times 10^{-9} \text{ m})(-1.36 \text{ MPa})}{(2)(7.28 \times 10^{-8} \text{ MPa m})} = 0.0467$$

and so α is 87°. For the wettable cell walls, α can be 0°, in which case $\cos \alpha$ is 1 and the maximum negative pressure develops, namely, -29 MPa for the pores 5 nm in radius. On the other hand, when water is available, it can be pulled into the interstices by such possible tensions and thereby cause α for wettable walls to increase toward 90°, which leads to a decrease in $\cos \alpha$ and consequently a decrease in the tension developed. In the present example, α is 87°, P is -1.36 MPa, and the pore 5 nm in radius is nearly filled (Fig. 2-14d). Thus the water in the interstices can be nearly flush with the cell wall surface. In fact, the large tensions that could be present in the cell wall generally do not occur in living cells because water is usually available within the plant and is "pulled" into the interstices, thus nearly filling them, depending of course on their dimensions (see Eq. 2.23).

The only contribution to the water potential of the cell wall that we have been considering is the negative hydrostatic pressure resulting from air–liquid interfaces, as can be the case for pure water. When the gravitational term in Eq. (2.13a) is ignored, $\Psi^{\text{cell wall}} = P^{\text{cell wall}} - \Pi^{\text{cell wall}}$. The bulk solution in the cell wall pores generally has an average osmotic pressure of 0.3–1.5 MPa. The local $\Pi^{\text{cell wall}}$ next to the solid Donnan phase (Chapter 3, Section IIF) is considerably greater because of the many ions present there; to help compensate, $P^{\text{cell wall}}$ is also higher near the Donnan phase than it is in the bulk of the cell wall water (activity coefficients of both ions and water are also lower near the Donnan phase).

F. Water Flux

When the water potential inside a cell differs from that outside, water is no longer in equilibrium and we can expect a net water movement toward the region of lower water potential. This volume flux density of water, J_{V_w}, is generally proportional to the difference in water potential ($\Delta\Psi$) across the membrane or membranes restricting the flow. The proportionality factor indicating the permeability to water flow at the cellular level is expressed by a *water conductivity coefficient*, L_w:

$$J_{V_w} = L_w \Delta\Psi = L_w(\Psi^o - \Psi^i) \tag{2.24}$$

In Eq. (2.24), J_{V_w} is the volume flow of water per unit area of the barrier per unit time. It can have units of $m^3 \ m^{-2} \ s^{-1}$, or $m \ s^{-1}$, which are the units of velocity. In fact, J_{V_w} is the average velocity of water moving across the barrier being considered. To help see why this is so, consider a volume element of cross-sectional area A, extending back from the barrier for a length equal to the average water velocity $\bar{v}_w$ multiplied by dt (e.g., let dx in Fig. 1-5 = $\bar{v}_w \, dt$). In time dt this volume element, $A\bar{v}_w \, dt$, crosses area A of the barrier, and for such a period the volume flow of water per unit area and per unit time, J_{V_w}, equals $(A\bar{v}_w \, dt)/(A \, dt)$, which is $\bar{v}_w$ (also see Fig. 3-3). L_w can have units of $m \ s^{-1} \ Pa^{-1}$, in which case the water potentials would be expressed in Pa. (Experimentally, L_w is usually the same as L_P, a coefficient describing water conductivity that we will introduce in Chapter 3 in a rather different manner.)

When Eq. (2.24) is applied to cells, Ψ^o is the water potential in the external solution, and Ψ^i usually refers to the water potential in the vacuole. L_w then indicates the conductivity for water flow across the cell wall, the plasma membrane and the tonoplast, all in series. For a group of barriers in series the overall water conductivity coefficient of the pathway, L_w, is related to the conductivities of the individual barriers by

$$1/L_w = \sum_j 1/L_{w_j} \qquad (2.25)$$

where L_{w_j} is the water conductivity coefficient of barrier j (see Problem 1.4 for the analogous case of the permeability coefficient for barriers in series). To see why this is so, let us go back to Eq. (2.24), $J_{V_w} = L_w \Delta\Psi$. When the barriers are in series, J_{V_w} is the same across each one, so $J_{V_w} = L_{w_j} \Delta\Psi_j$ where $\Delta\Psi_j$ is the drop in water potential across barrier j, and $\sum_j \Delta\Psi_j = \Delta\Psi$. We therefore obtain the following equalities: $J_{V_w}/L_w = \Delta\Psi = \sum_j \Delta\Psi_j = \sum_j J_{V_w}/L_{w_j} = J_{V_w} \sum_j 1/L_{w_j}$; hence, $1/L_w = \sum_j 1/L_{w_j}$, as indicated in Eq. (2.25). We also note that $1/L_w$ corresponds to a resistance, and the resistance of a group of resistors in series is the sum of the resistances, $\sum_j 1/L_{w_j}$.

We shall now consider a membrane separating two solutions that differ only in osmotic pressure. This will help us to relate L_w to the permeability coefficient of water, P_w, and also to view Fick's first law [$J_j = P_j(c_j^o - c_j^i)$; Eq. 1.8] in a slightly different way. The appropriate form for Fick's first law when describing the diffusion of the solvent water is $J_{V_w} = P_w(N_w^o - N_w^i)$, where N_w is the mole fraction of water. By Eq. (2.24), J_{V_w} also equals L_w ($\Psi^o - \Psi^i$), which becomes L_w ($\Pi^i - \Pi^o$) in our case ($\Psi = P - \Pi + \rho_w gh$; Eq. 2.13a). Thus a volume flux density of water toward a region of higher osmotic pressure (Eq. 2.24) is equivalent to J_{V_w} toward a region where water has a lower mole fraction (Fick's first law applied to water). Let us pursue this one step further. Using Eqs. (2.8)–(2.10), we can obtain the following: $N_w \cong 1 - \sum_j n_j/n_w \cong 1 - \bar{V}_w \Pi_s/RT$. Therefore, $N_w^o - N_w^i$ approximately equals $(1 - \bar{V}_w \Pi_s^o/RT) - (1 - \bar{V}_w \Pi_s^i/RT)$, or $\bar{V}_w(\Pi_s^i - \Pi_s^o)/RT$. When we incorporate this last relation into our expression for the volume flux density of water, we obtain $J_{V_w} = L_w(\Pi^i - \Pi^o) = P_w(N_w^o - N_w^i) = P_w \bar{V}_w(\Pi_s^i - \Pi_s^o)/RT$ (Π^i is here the same as Π_s^i). Hence, we derive the following relationship:

$$L_w = \frac{P_w \bar{V}_w}{RT} \qquad (2.26)$$

which states that the water conductivity coefficient (L_w) is proportional to the water permeability coefficient (P_w).

Next, we will estimate a possible value for L_w. P_w can be 1×10^{-4} m s^{-1} for plasma membranes (Chapter 1), RT is 2.437×10^3 m^3 Pa mol^{-1} at 20°C (Appendix I), and $\bar{V}_w$ is 1.8×10^{-5} m^3 mol^{-1}. Thus L_w can be

$$L_w = \frac{(1 \times 10^{-4}\ \text{m s}^{-1})(1.8 \times 10^{-5}\ \text{m}^3\ \text{mol}^{-1})}{(2.437 \times 10^3\ \text{m}^3\ \text{Pa mol}^{-1})} = 7 \times 10^{-13}\ \text{m s}^{-1}\ \text{Pa}^{-1}$$

L_w for plant cells ranges from about 1×10^{-13} to 2×10^{-12} m s^{-1} Pa^{-1}.

G. Cell Growth

When the value of the water conductivity coefficient is known, the water potential difference necessary to give an observed water flux can be calculated using Eq. (2.24). For the internodal cells of *Chara* and *Nitella*, L_w for water entry is about 7×10^{-13} m s^{-1} Pa^{-1}. For convenience of calculation, we will consider cylindrical cells 100 mm long and 1 mm in diameter as an approximate model for such algal cells (see Fig. 3-9). The surface area across which the water flux occurs is $(2\pi r)(l)$, where r is the cell radius and l is the cell length. Thus the area $= (2\pi)(0.5 \times 10^{-3}$ m$)(100 \times 10^{-3}$ m$)$, or 3.14×10^{-4} m^2. [The area of each end of the cylinder (πr^2) is much less than $2\pi rl$; in any case, a water flux from the external solution across them would not be expected because they are in contact with other cells, not the bathing solution; see Fig. 3-9.] The volume of a cylinder is $\pi r^2 l$, which equals $(\pi)(0.5 \times 10^{-3}$ m$)^2(100 \times 10^{-3}$ m$)$, or 7.9×10^{-8} m^3, in the present case. Internodal cells of *Chara* and *Nitella* grow relatively slowly—a change in volume of about 1% per day is a possible growth rate for the fairly mature cells used in determining L_w. This growth rate means that the water content increases by about 1% of the volume per day (1 day $= 8.64 \times 10^4$ s). The volume flux density for water is the rate of volume increase divided by the surface area across which water enters. Using the indicated value of L_w and Eq. (2.24), we can then calculate the drop in water potential involved:

$$\Delta\Psi = \frac{J_{V_w}}{L_w} = \frac{\frac{1}{A}\frac{dV}{dt}}{L_w} = \frac{\frac{1}{(3.14 \times 10^{-4} \text{ m}^2)} \frac{(0.01)(7.9 \times 10^{-8} \text{ m}^3)}{(8.64 \times 10^4 \text{ s})}}{(7 \times 10^{-13} \text{ m s}^{-1} \text{ Pa}^{-1})} = 42 \text{ Pa}$$

In other words, the internal water potential (Ψ^i) needed to sustain the water influx accompanying a growth of about 1% per day is only 42 Pa less than the outside water potential (Ψ^o). *Chara* and *Nitella* can grow in pond water, which is a dilute aqueous solution often having a water potential near -7 kPa. Thus Ψ^i need be only slightly more negative than -7 kPa to account for the influx of water accompanying a growth of 1% per day.

The enlargement of cells during growth depends on two simultaneously occurring processes—the uptake of water and the increase in area of the cell wall, as was clearly recognized by James Lockhart in 1965. Because water uptake involves spontaneous water movement toward lower water potential, the irreversible aspect of growth depends on the yielding of the cell wall material. To describe growth of cells, we often use the relative rate of volume increase, $(1/V) dV/dt$. For water entering a single cell, $(1/V) dV/dt$ can be represented by $(1/V)J_{V_w} A$, which using Eq. (2.24) we can express as follows:

$$\frac{1}{V}\frac{dV}{dt} = \left(\frac{A}{V}\right) L_w(\Psi^o - \Psi^i) \tag{2.27}$$

Equation (2.27) is often called a *growth equation* (to put this into the form generally found in the literature, L_w should be replaced by L_p and Ψ by $P - \sigma\Pi$, changes based on irreversible thermodynamics that we will introduce in Chapter 3). If water also evaporates from the cell, such as for certain cells in leaves, we should include such water movement in Eq. (2.27).

At the end of Chapter 1 we indicated that cell enlargement requires yielding of the cell wall, an irreversible or plastic process that occurs when the internal hydrostatic pressure exceeds some critical or threshold value, $P^i_{\text{threshold}}$. This leads to another growth equation of the following form:

$$\frac{1}{V}\frac{dV}{dt} = \varphi(P^i - P^i_{\text{threshold}}) \tag{2.28}$$

where φ is a coefficient describing irreversible cell wall yielding $(P^i \geq P^i_{\text{threshold}})$. Our present calculation for *Chara* and *Nitella* growing at a rate of 1% per day indicates that Ψ^o is nearly the same as Ψ^i. In this case, cell extension is controlled primarily by yielding of the cell wall, without which P^i and Ψ^i would increase slightly and thereby decrease the rate of water entry.

For other cases, the uptake or internal production of osmotically active solutes can exert greater control on cell growth, including effects that involve the reversible cell extension caused by changes in internal hydrostatic pressure and are quantitatively described by the volumetric elastic modulus ε. For instance, when reversible or elastic changes in volume exist, then $(1/\varepsilon)dP/dt$ should be added to the right side of Eq. (2.28) [using Eq. 1.17, $(1/\varepsilon)dP/dt = (\Delta V/V\Delta P)dP/dt = (1/V)(dV/dP)dP/dt = (1/V)dV/dt$, which is the relative rate of elastic volume increase].

H. Kinetics of Volume Changes

We next indicate how cell wall and membrane properties influence the kinetics of reversible swelling or shrinking for plant cells. When the water potential outside a cell or group of cells is changed, water movement will be induced. A useful expression describing the time constant for the resulting volume change is

$$t_e = \frac{V}{AL_w(\varepsilon + \Pi^i)} \tag{2.29}$$

where V is the initial volume of a cell, A is the area across which water enters or leaves, ε is the volumetric elastic modulus (Eq. 1.17), and Π^i is the initial internal osmotic pressure. The time constant t_e represents the time to complete all but $1/e$ or 37% of the volume shift. (The half-time, which is the time required to shift halfway from the initial to the final volume, equals $\ln 2$ times t_e.) Equation (2.29) indicates that t_e for swelling or shrinking is small for a cell with a low V/A (a large available surface area per unit volume), a high L_w (which usually means a high permeability of the plasma membrane to water), or a high ε (a rigid cell wall). Figure 2-15 shows a device that can be used to measure ΔV and

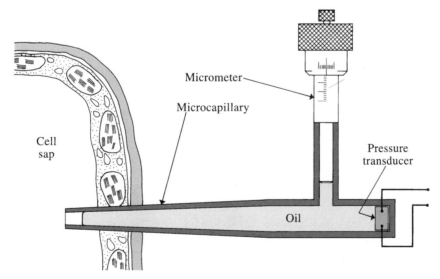

Figure 2-15. Schematic diagram of a "pressure probe," an apparatus that can be used to measure P, L_w (or L_p, see Chapter 3), and ε for individual plant cells. The intracellular hydrostatic pressure is transmitted to the pressure transducer via an oil-filled microcapillary introduced into the cell. Volume can be changed by adjusting the micrometer and observing the motion of the interface between the cell sap and the oil (in the drawing the cell region is greatly enlarged relative to the rest of the apparatus).

ΔP directly, thus allowing a determination of ε [consider Eq. 1.17, $\varepsilon = \Delta P/(\Delta V/V)$] as well as other water-relations parameters [a P–V curve (Fig. 2-12) gives an indirect way of determining P and ε].

We next estimate the time constant for volume changes that might occur for a meso-phyll cell (Fig. 1-2). We will assume that it has a volume of 10^4 μm^3 and that water enters from a neighboring cell over an area of 20 μm^2, so that V/A is 500 μm. We will let L_w be 7×10^{-13} m s^{-1} Pa^{-1}, ε be 20 MPa, and Π^i be 1 MPa (usually Π^i is much less than ε, so Π^i is often ignored when calculating t_e). From Eq. (2.29) the time constant is

$$t_e = \frac{(500 \times 10^{-6} \text{ m})}{(7 \times 10^{-13} \text{ m s}^{-1} \text{ Pa}^{-1})(21 \times 10^6 \text{ Pa})} = 34 \text{ s}$$

The volume changes caused by water transport between adjacent cells would be more rapid if their contacting area were larger and if L_w and ε were higher. Time constants for volume changes have important implications for the water exchanges underlying stomatal opening (discussed in Chapter 8) and capacitance effects (discussed in Chapter 9).

Problems

2.1 The rise of water in a long vertical capillary 2 mm in diameter with a wettable inner wall is 15 mm.

(a) If the capillary is tilted 45° from the horizontal, what is the vertical height of the rise? Assume that the contact angle does not change upon tilting.

(b) If sucrose is added to the solution so that the density becomes 1200 kg m^{-3} ($= 1.2$ g cm^{-3}), what is the height of the rise?

(c) If the capillary extends only 7.5 mm above the main surface of the water, what is the contact angle in the capillary?

(d) If the wall of the capillary is so treated that the contact angle becomes 60°, what is the height of the rise?

(e) What is the rise of water in a capillary similar to the original one, but with a 1 μm radius?

(f) In which of the five cases is the greatest weight supported by capillary (surface tension) forces?

2.2 A neutral solute at equilibrium has a concentration of 0.1 m on one side of a barrier permeable only to water and a C_{solute} of 1 m on the other side. Let the partial molal volume of the solute be 40×10^{-6} m^3 mol^{-1}, the temperature be 20°C, and the activity coefficients of the solute and the water be 1, except in (c).

(a) What is the hydrostatic pressure difference across the barrier?

(b) What would it be if the barrier were permeable to solute only?

(c) If γ_{solute} is 0.5 on the more concentrated side and if other conditions are unchanged, what is ΔP across the barrier permeable to the solute only?

(d) What would happen at equilibrium if the barrier were permeable to both water and solute?

(e) If P on the 0.1-m side is the same as atmospheric pressure, what is the chemical potential of the solute there?

2.3 A solution assumed to be ideal contains 80 g of sorbitol (molar mass of 0.182 kg mol^{-1}) in 1 kg of water at 20°C.

(a) What are N_w and a_w in the solution?

(b) What is the osmotic pressure of the sorbitol solution?

(c) By what percent is the activity of water reduced for an osmotic pressure of 1 MPa at 20°C compared with the value of a_w for pure water?

(d) If we assume that activity coefficients are 1, what concentration of a solute corresponds to a Π of 1 MPa at 20°C?

(e) A 0.25 mol m^{-3} solution of a particular polymer has a measured osmotic pressure at 20°C of 0.01 MPa. What is the osmotic pressure predicted by the Van't Hoff relation (Eq. 2.10)? Explain any discrepancies.

(f) In the vacuole of a certain cell, the mole fraction of water is 0.98, the hydrostatic pressure is 0.8 MPa, and the temperature is 20°C. Assuming activity coefficients are 1, what is the water potential in the vacuole?

2.4 A tank 10 m tall and open at the top is filled with an ideal solution at 20°C. We will assume that the system is in equilibrium and that the zero level for the gravitational term is at the top of the tank, where Ψ is −0.600 MPa.

(a) What is the water potential 0.1 m below the surface and at the bottom of the tank?

(b) What is Π at the upper surface, 0.1 m below the surface, and at the bottom of the tank?

(c) What are P and the gravitational term at the three levels in (b)?

(d) What relative humidity would be in equilibrium with the water in the tank?

2.5 Chloroplasts are isolated from a plant cell whose cytosol has an osmotic pressure of 0.4 MPa at 20°C. When the chloroplasts are placed in solutions at 20°C containing an impermeant solute, the volumes are 36 μm^3 at an external osmotic pressure of 0.33 MPa, 28 μm^3 at 0.5 MPa, and 20 μm^3 at 1.0 MPa. Assume that the activity coefficients are 1.

(a) What is the volume of the chloroplasts in the plant cell?

(b) What is the nonaqueous volume per chloroplast?

(c) What volume fraction of the chloroplast is occupied by water *in vivo*?

(d) What is the amount of osmotically active particles per chloroplast?

2.6 A spherical algal cell 1 mm in diameter has a water conductivity coefficient L_w equal to 10^{-12} m s^{-1} Pa^{-1}. Let the internal osmotic pressure be 1.0 MPa, the internal hydrostatic pressure be 0.6 MPa at 20°C, and the volumetric elastic modulus be 5 MPa.

(a) What is the initial net volume flux density of water into or out of the cell when it is placed in pure water at atmospheric pressure?

(b) What is the time constant for the change in (a)?

(c) What is the water flux density at the point of incipient plasmolysis?

(d) What is the water flux density when the external solution is in equilibrium with a gas phase at 97% relative humidity?

(e) Assume that the water in the cell walls is in equilibrium with the internal cellular water. What are cos α and the contact angle at the air–water interface for cylindrical cell wall pores 20 nm in diameter? Assume that $\Pi^{\text{cell wall}}$ is negligible.

References

Adamson, A. W. (1997). *Physical Chemistry of Surfaces* (6th ed.). Wiley, New York.

Barkla, B. J., and Pantoja, O. (1996). Physiology of ion transport across the tonoplast of higher plants. *Annu. Rev. Plant Physiol. Plant Mol. Biol.* **47**, 159–184.

Boller, T., and Wiemken, A. (1986). Dynamics of vacuolar compartmentation. *Annu. Rev. Plant Physiol.* **37**, 137–164.

Boyer, J. S. (1995). *Measuring the Water Status of Plants and Soils.* Academic Press, San Diego.

Bull, H. B. (1964). *An Introduction to Physical Biochemistry.* Davis, Philadelphia.

Castellan, G. W. (1983). *Physical Chemistry* (3rd ed.). Addison-Wesley, Reading, MA.

Cosgrove, D. (1986). Biophysical control of plant growth. *Annu. Rev. Plant Physiol.* **37**, 377–405.

Dainty, J. (1976). Water relations of plant cells. *Adv. Bot. Res.* **1**, 279–326.

Haines, T. H. (1994). Water transport across biological membranes. *FEBS Lett.* **346**, 115–122.

Hüsken, D., Steudle, E., and Zimmermann, U. (1978). Pressure probe technique for measuring water relations of cells in higher plants. *Plant Physiol.* **61**, 158–163.

Israelachvili, J. N. (1985). *Intermolecular and Surface Forces with Applications to Colloidal and Biological Systems.* Academic Press, New York.

Kotyk, A., Janáoek, K., and Koryta, J. (1988). *Biophysical Chemistry of Membrane Functions.* Wiley, Chichester, UK.

Kramer, P. J., and Boyer, J. S. (1995). *Water Relations of Plants and Soils.* Academic Press, San Diego.

Lide, D. R. (Ed.) (1997). *CRC Handbook of Chemistry and Physics* (78th ed.). CRC Press, Boca Raton, FL.

Lockhart, J. A. (1965). An analysis of irreversible plant cell elongation. *J. Theor. Biol.* **8**, 264–275.

Marin, B. (Ed.) (1987). *Plant Vacuoles: Their Importance in Solute Compartmentation in Cells and Their Applications in Plant Biotechnology.* Plenum, New York.

Maurel, C. (1997). Aquaporins and water permeability of plant membranes. *Annu. Rev. Plant Physiol. Plant Mol. Biol.* **48**, 399–429.

Neumann, A. W., and Spelt, J. K. (Eds.) (1996). *Applied Surface Dynamics*, Surfactant Science Series, Vol. 63. Dekker, New York.

Nobel, P. S. (1969a). The Boyle–Van't Hoff relation. *J. Theor. Biol.* **23**, 375–379.

Nobel, P. S. (1969b). Light-induced changes in the ionic content of chloroplasts in *Pisum sativum. Biochim. Biophys. Acta* **172**, 134–143.

Noggle, J. H. (1996). *Physical Chemistry* (3rd ed). HarperCollins, New York.

Ortega, J. K. E. (1990). Governing equations for plant growth. *Physiol. Plant.* **79**, 116–121.

Pearcy, R. W., Ehleringer, J., Mooney, H. A., and Rundel, P. W. (1989). *Plant Physiological Ecology: Field Methods and Instrumentation.* Chapman & Hall, New York.

Pfeffer, W. F. P. (1877). *Osmotische Untersuchungen. Studien zur Zell Mechanik.* W. Engelmann, Leipzig.

Raven, J. A. (1987). The role of vacuoles. *New Phytologist* **106**, 357–422.

Richter, C., and Dainty, J. (1989). Ion behavior in plant cell walls. I. Characterization of the *Sphagnum russowii* cell wall ion exchanger. *Can. J. Bot.* **67**, 451–459.

Rosen, M. J. (1989). *Surfactants and Interfacial Phenomena* (2nd ed.). Wiley, New York.

Salisbury, F. B., and Ross, C. W. (1991). *Plant Physiology* (4th ed.). Wadsworth, Belmont, CA.

Scholander, P. F., Hammel, H. T., Hemmingsen, E. A., and Bradstreet, E. D. (1964). Hydrostatic pressure and osmotic potential in leaves of mangroves and some other plants. *Proc. Natl. Acad. Sci. USA* **52**, 119–125.

Smith, J. A. C., and Nobel, P. S. (1986). Water movement and storage in a desert succulent: Anatomy and rehydration kinetics for leaves of *Agave deserti. J. Exp. Bot.* **37**, 1044–1053.

Tomos, A. D. (1988). Cellular water relations of plants. In *Water Science Review 3* (F. Franks, Ed.), pp. 186–277. Cambridge Univ. Press, New York.

Tyree, M. T., and Sperry, J. S. (1989). Vulnerability of xylem to cavitation and embolism. *Annu. Rev. Plant Physiol. Plant Mol. Biol.* **40**, 19–38.

Van Holde, K. E. (1985). *Physical Biochemistry* (2nd ed.) Prentice Hall, Englewood Cliffs, NJ.

Wolfe, J., and Steponkus, P. L. (1981). The stress–strain relation of the plasma membrane of isolated plant protoplasts. *Biochim. Biophys. Acta* **643**, 663–668.

3

Solutes

I. **Chemical Potential of Ions**   **82**
    A. Electrical Potential   83
    B. Electroneutrality and Membrane Capacitance   83
    C. Activity Coefficients of Ions   86
    D. Nernst Potential   87
    E. Example of E_{N_K} 89
II. **Fluxes and Diffusion Potentials**   **90**
    A. Flux and Mobility   91
    B. Diffusion Potential in a Solution   93
    C. Membrane Fluxes   96
    D. Membrane Diffusion Potential—Goldman Equation   98
    E. Application of the Goldman Equation   101
    F. Donnan Potential   102
III. **Characteristics of Crossing Membranes**   **104**
    A. Electrogenicity   104
    B. Boltzmann Energy Distribution and Q_{10},
      a Temperature Coefficient   106
    C. Activation Energy and Arrhenius Plots   108
    D. Ussing–Teorell Equation   110
    E. Example of Active Transport   113
    F. Energy for Active Transport   115
    G. Speculation on Active Transport   116
IV. **Mechanisms for Crossing Membranes**   **116**
    A. Carriers, Porters, Channels, and Pumps   117
    B. Michaelis–Menten Formalism   119
    C. Facilitated Diffusion   121
V. **Principles of Irreversible Thermodynamics**   **123**
    A. Fluxes, Forces, and Onsager Coefficients   124
    B. Water and Solute Flow   125
    C. Flux Densities, L_P, and σ 126
    D. Values of Reflection Coefficients   129
VI. **Solute Movement across Membranes**   **131**
    A. The Influence of Reflection Coefficients
      on Incipient Plasmolysis   132
    B. Extension of the Boyle–Van't Hoff Relation   134

    C. Reflection Coefficients of Chloroplasts   135
    D. Solute Flux Density   136
    Problems   **137**
    References   **139**

IN THIS chapter we turn our attention to the properties of solutes. We will compare chemical potentials in the aqueous phases on the two sides of a membrane or across some other region to predict the direction of passive solute fluxes as well as the driving forces leading to such motion. We will also show how the fluxes of charged species can account for the electrical potential differences across biological membranes.

Many solute properties are intertwined with those of the ubiquitous solvent, water. For example, the osmotic pressure term in the chemical potential of water is due mainly to the decrease of the water activity caused by solutes ($RT \ln a_w = -\bar{V}_w \Pi$; Eq. 2.7). The movement of water through the soil to a root and then to its xylem can influence the entry of dissolved nutrients, and the subsequent distribution of these nutrients throughout the plant depends on water movement in the xylem (and the phloem in some cases). In contrast with water, however, solute molecules can carry a net positive or negative electrical charge. For such charged particles, the electrical term must be included in their chemical potential, which leads to a consideration of electrical phenomena in general and an interpretation of the electrical potential differences across membranes in particular. Whether an observed ionic flux of some species into or out of a cell can be accounted for by the passive process of diffusion depends on the differences in both the concentration of that species and the electrical potential between the inside and the outside. Ions can also be actively transported across membranes, in which case metabolic energy is involved.

When both solutes and water traverse the same barrier, we should replace the classical thermodynamic approach with one based on irreversible thermodynamics. The various forces and fluxes are then viewed as interacting with each other, so the movement of water across a membrane influences the movement of solutes, and vice versa. Using this more general approach, we will show that the osmotic pressure difference effective in causing a volume flux across a membrane permeable to both water and solutes is less than the actual osmotic pressure difference across that membrane.

I. Chemical Potential of Ions

We introduced the concept of chemical potential in Chapter 2. Because a substance spontaneously tends to move toward regions where its chemical potential is lower, this quantity is useful for analyzing passive movements of solutes. Using a linear combination of the various contributors, we expressed the chemical potential of any species j by Eq. (2.4): $\mu_j = \mu_j^* + RT \ln a_j + \bar{V}_j P + z_j FE + m_j gh$. Because water is uncharged ($z_w = 0$), the electrical term does not enter into its chemical potential. For ions, however, $z_j FE$ becomes important. In fact, for charged solutes under most conditions of biological interest, differences in the electrical term are usually far larger than are changes in the terms for hydrostatic pressure or gravity. In particular, for movements across membranes, Δh is effectively zero, so we can omit the gravitational term in this chapter. Changes in the $\bar{V}_j P$ term can also usually be ignored because they are relatively small, as we will indicate later. Therefore, the chemical potential generally used when dealing with ions is $\mu_j^* + RT \ln a_j + z_j FE$, commonly referred to as the *electrochemical* potential. This emphasizes the role played by electrical potentials, but the expression "chemical potential" represents the sum of all the different contributors affecting a particular species, so the special term electrochemical potential when dealing with ions is not really necessary.

A. Electrical Potential

The difference in electrical potential E between two locations is a measure of the amount of electrical work involved in moving a charge from one location to the other. Specifically, the work in joules (J) equals the charge in coulombs (C) times the electrical potential difference in volts (V). The zero level for electrical potential is arbitrary. Because both the initial electrical potential and the final electrical potential must be expressed relative to the same zero level, the arbitrariness is of no consequence when the electrical potential difference (final potential − initial potential) is determined. (In keeping with most biological literature, E in electrical terms throughout this book refers to electrical potentials, not electric field intensities, as it does in many physics texts.)

The charge carried by an ion of species j is a positive or negative integer z_j (the *charge number*) × the charge of a proton. For instance, z_j is +1 for potassium (K⁺) and −2 for sulfate (SO₄²⁻). The electrical charge carried by a proton is commonly called the *electronic charge* because it is the same in magnitude as the charge on an electron, although opposite in sign. A single proton has a charge of 1.602×10^{-19} C; thus a mole (Avogadro's number) of protons has a charge equal to (6.022×10^{23} protons mol⁻¹) (1.602×10^{-19} C proton⁻¹), or 9.65×10^4 C mol⁻¹. Such a unit, consisting of Avogadro's number of electronic charges (i.e., 1 mol of single, positive charges), is called Faraday's constant, F. This quantity, which appears in the electrical term of the chemical potential (Eq. 2.4), equals 9.65×10^4 C mol⁻¹ or 9.65×10^4 J mol⁻¹ V⁻¹.

To illustrate the rather small contribution of the pressure term $\bar{V}_j P$ to differences in the chemical potential of a charged substance, we will compare it with the contribution to the electrical term $z_j FE$. We will use a fairly typical electrical potential difference across a biological membrane of 100 mV and a hydrostatic pressure difference (ΔP) of 0.5 MPa and consider a small monovalent cation [$z_j = +1$, $\bar{V}_j = 4.0 \times 10^{-5}$ m³ mol⁻¹ (40 cm³ mol⁻¹)]. Here $\bar{V}_j \Delta P = (4.0 \times 10^{-5}$ m³ mol⁻¹) $(0.5 \times 10^6$ Pa) or 20 J mol⁻¹ (1 Pa = 1 N m⁻² = 1 J m⁻³) and $z_j F \Delta E = (1)(9.65 \times 10^4$ J mol⁻¹ V⁻¹) (0.100 V) or 9650 J mol⁻¹, which is nearly 500 times larger. Hence, contributions of the pressure term to the chemical potential differences of ions across biological membranes are usually negligible compared with electrical contributions and hence can generally be ignored.

B. Electroneutrality and Membrane Capacitance

Another consequence of the relatively large magnitude of electrical effects is the general occurrence of *electroneutrality*. Thus in most aqueous regions that are large compared with atomic dimensions, the total electrical charge carried by the cations is essentially equal in magnitude to that carried by the anions. What is the situation for typical plant cells? Do we ever have an excess of negative or positive charges in cells or organelles? If there were a net charge in some region, such as near a membrane, an electrical potential difference

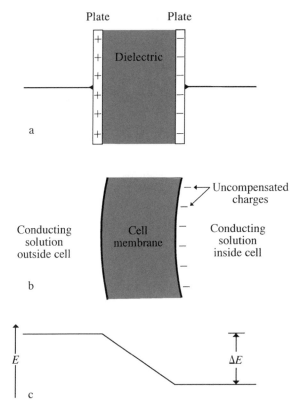

Figure 3-1. (a) A parallel plate capacitor. Each plate represents a conductor in which charges can freely move, thus each plate has a particular electrical potential, reflecting its uncompensated positive or negative charges; between the plates is a region (often called the dielectric) that charges cannot cross, so an electrical potential difference ΔE occurs across this region. (b) Cell membrane as a capacitor. Membranes act as dielectrics (the dielectric constant D, introduced in Chapter 2, is about 3 for the lipid phase) separating the aqueous conducting phases on either side; uncompensated negative charges accumulate on the inner side of a typical cell membrane. (c) Electrical potential difference across dielectric. The higher the capacitance the greater the charge on the plates (a) or the more uncompensated charges adjacent to the membrane (b) for a given ΔE across the dielectric; capacitance is proportional to D.

would exist from one part of the region to another. Can we relate the size of the electrical potential difference to the net charge?

To relate charges and ΔE's, we need to introduce a new term, *capacitance* (see Fig. 3-1). Electrical capacitance is the coefficient of proportionality between a net charge and the resulting electrical potential difference. A high capacitance means that the region has the capacity to have many uncompensated charges separated across it without developing a large electrical potential difference across that region. The magnitude of such an electrical potential difference, ΔE, is related to the capacitance of the region (C) as follows:

$$Q = C\,\Delta E \tag{3.1}$$

where Q is the net charge. The unit for capacitance is the farad (F), which equals $1\ \mathrm{C\ V^{-1}}$. The capacitances of most biological membranes are approximately the same per unit area—about 10 mF m^{-2} (1 μF cm^{-2}). The lipid bilayer of the membrane represents the dielectric phase (Fig. 3-1).

For convenience in estimating the net charge inside a cell that would lead to a typical membrane potential, we will consider a spherical cell of radius r. Suppose that the uncanceled or net charges have a concentration c. For a conductor—a body in which electrical charges can freely move, such as an aqueous solution—the uncanceled or net charges do not remain uniformly distributed. Rather, they repel each other and hence collect at the inner surface of the sphere (see Fig. 3-1b). The quantity c is therefore the hypothetical concentration of the net charge if it were uniformly distributed through the interior of the sphere. The amount of charge Q within the sphere of radius r is then $(4/3)\pi r^3 cF$, where $(4/3)\pi r^3$ is the volume of a sphere. (The concentration c can be expressed in mol m^{-3} and here refers to the average net concentration of uncompensated singly charged particles; F, which is expressed in C mol^{-1}, converts the amount in moles to the electrical charge.) The capacitance C of the sphere is $4\pi r^2 C'$, where $4\pi r^2$ is the surface area of a sphere and C' is the capacitance per unit area. Substituting these values of Q and C into Eq. (3.1) yields $(4/3)\pi r^3 cF = 4\pi r^2 C' \Delta E$, which leads to the following expression for the electrical potential difference in the case of a spherical capacitor:

$$\Delta E = \frac{rcF}{3C'} \tag{3.2}$$

Equation (3.2) gives the electrical potential difference from the center of the sphere to just outside its surface. For a conductor, the internal uncompensated charges are found near the surface. In that case, ΔE actually occurs close to the bounding surface (such as a membrane) surrounding the spherical body under consideration. Equation (3.2) indicates that the electrical potential difference is directly proportional to the average concentration of net charge enclosed and inversely proportional to the capacitance per unit area of the sphere.

To apply Eq. (3.2) to a specific situation, let us consider a spherical cell with a radius of 30 μm and an electrical potential difference across the membrane (inside negative) of -100 mV, a value close to that occurring for many cells. If the membrane capacitance per unit area has a typical value of 10 mF m^{-2} (10^{-2} C V^{-1} m^{-2}), to what net charge concentration in the cell does this electrical potential difference correspond? Using Eq. (3.2), we obtain

$$c = \frac{3C' \Delta E}{rF} = \frac{(3)(10^{-2} \text{ C V}^{-1} \text{ m}^{-2})(-0.100 \text{ V})}{(30 \times 10^{-6} \text{ m})(9.65 \times 10^4 \text{ C mol}^{-1})}$$
$$= -1.0 \times 10^{-3} \text{ mol m}^{-3}$$

The sign of the net charge concentration is negative, indicating more internal anions than cations. The average concentration of the net (uncanceled) charge leading to a considerable electrical potential difference is rather small, an important point that we will consider next.

It is instructive to compare the net charge concentration averaged over the volume of a cell (c in Eq. 3.2) with the total concentration of anions and cations in the cell. Specifically, because the positive and the negative ions in plant cells can each have a total concentration near 100 mol m^{-3} (0.1 M), the previously calculated excess of 10^{-3} mol m^{-3} (1 μM) is only about one extra negative charge per 10^5 anions. Expressed another way, the total charge of the cations inside such a cell equals or compensates that of the anions to within about one part in 100,000. When cations are taken up by a cell to any appreciable extent, anions must accompany them or cations must be released from inside the cell or both.

Otherwise, marked departures from electrical neutrality would occur in some region, and sizable electrical potential differences would develop.

C. Activity Coefficients of Ions

We will now turn our attention to the activity term in the chemical potential—specifically, to the chemical activity itself. As indicated in Chapter 2, the activity of species j, a_j, is its thermodynamically effective concentration. For charged particles in an aqueous solution, this activity can differ appreciably from the actual concentration c_j—a fact that has not always been adequately recognized in dealing with ions. By Eq. (2.5), $a_j = \gamma_j c_j$, where γ_j is the activity coefficient of species j.

A quantitative description of the dependence of the activity coefficients of ions on the concentration of the various species in a solution was developed by Peter Debye and Erich Hückel in the 1920s. In a local region around a particular ion, the electrostatic forces, which are describable by relations such as Eq. (2.3) [Electrical force $= Q_1 Q_2/(4\pi \varepsilon_0 D r^2)$], constrain the movement of other ions. As the concentration increases, the average distance between the ions decreases, thereby facilitating ion–ion interactions. Equation (2.3) indicates, for example, that the electrostatic interaction between two charged particles varies inversely as the square of the distance between them, so electrical forces greatly increase as the ions get closer together. When ions of opposite sign attract each other, the various other interactions of both ions are restricted, thus lowering their thermodynamically effective concentration or activity.

A simple approximate form of the Debye–Hückel equation appropriate for estimating the values of activity coefficients of ions in relatively dilute aqueous solutions at 25°C is

$$\ln \gamma_{\pm} \cong \frac{1.17 \, z_+ z_- \sqrt{\frac{1}{2}\sum_j c_j z_j^2}}{32 + \sqrt{\frac{1}{2}\sum_j c_j z_j^2}} \tag{3.3}$$

where z_+ is the charge number of the cation and z_- is that of the anion, concentrations are expressed in mol m^{-3} (numerically equal to mM), and the summations are over all charged species.[1] Because we cannot have a solution of one type of ion by itself in which to measure or to calculate γ_+ or γ_-, activity coefficients of ions occur as the products of those of cations and anions. Hence, $\gamma_{\pm}$ in Eq. (3.3) represents the mean activity coefficient of some cation–anion pair with charge numbers z_+ and z_-. Because z_- is negative, Eq. (3.3) indicates that $\ln \gamma_{\pm}$ is also negative, and $\gamma_{\pm} < 1$. Thus the activities of ions in aqueous solutions are less than their concentrations, as expected.

To estimate $\gamma_{\pm}$ under conditions approximating those that might occur in a plant or animal cell, let us consider an aqueous solution containing 100 mol m^{-3} (100 mM) of both monovalent cations and anions and 25 mol m^{-3} of both divalent cations and anions. For this solution, $\frac{1}{2}\sum_j c_j z_j^2$, known as the *ionic strength*, is

$$\begin{aligned}
\frac{1}{2}\sum_j c_j z_j^2 &= \tfrac{1}{2}[(100 \text{ mol m}^{-3})(1)^2 + (100 \text{ mol m}^{-3})(-1)^2 \\
&\quad + (25 \text{ mol m}^{-3})(2)^2 + (25 \text{ mol m}^{-3})(-2)^2] \\
&= 200 \text{ mol m}^{-3}
\end{aligned}$$

---

[1] For very dilute aqueous solutions (much more dilute than typically occur in plant or animal cells), $\sqrt{\frac{1}{2}\sum_j c_j z_j^2}$ is neglected compared with 32 in the denominator of Eq. (3.3), leading to $\ln \gamma_{\pm} \cong 0.037 \, z_+ z_- \sqrt{\frac{1}{2}\sum_j c_j z_j^2}$, which is known as the Debye–Hückel limiting law.

From Eq. (3.3), we can calculate $\ln \gamma_\pm$ for the monovalent ions as

$$\ln \gamma_\pm \cong \frac{(1.17)(1)(-1)\sqrt{200}}{32 + \sqrt{200}} = -0.359$$

This corresponds to a mean activity coefficient of only 0.70, a value considerably less than 1.

The activity coefficient of a particular ionic species depends on all the ions in the solution, as indicated by the ionic strength terms in Eq. (3.3). Therefore, even when some particular ionic species is itself dilute, its activity coefficient can nevertheless be appreciably less than 1 because of the many electrostatic interactions with other ions. The departure from 1 for activity coefficients is even greater for divalent and trivalent ions than for monovalent ions, as the $z_+ z_-$ factor in Eq. (3.3) indicates. Although activity coefficients of ions are often set equal to 1 for convenience, this is not always justified. A practical difficulty arising under most experimental situations is that c_j is much easier to determine than is a_j, especially for compartments such as the cytosol or the interior of a chloroplast. For circumstances in which a_j has been replaced by c_j, caution must be exercised in the interpretations or conclusions. The activity coefficients for nonelectrolytes and water are generally much closer to 1 than are those for ions; hence the assumption involved in replacing a_j by c_j for such neutral species is not as severe as it is for the charged substances. For instance, γ_{sucrose} is about 0.96 for 300 mol m^{-3} (0.3 M) sucrose. Equation (3.3) illustrates that activity coefficients of ions can be appreciably less than 1.00 under conditions in cells. An all-inclusive theory for activity coefficients is quite complicated and beyond the scope of this text.

D. Nernst Potential

Having considered the electrical and the activity terms in some detail, we now turn to the role of these quantities in the chemical potential of ions. Specifically, we will consider the deceptively simple yet extremely important relationship between the electrical potential difference across a membrane and the accompanying distribution of ions across it at equilibrium.

When ions of some species j are in equilibrium across a membrane, its chemical potential outside (o) is the same as that inside (i), i.e., $\mu_j^o = \mu_j^i$. A difference in the hydrostatic pressure term generally makes a negligible contribution to the chemical potential differences of ions across membranes, so $\bar{V}_j P$ can be omitted from μ_j in the present case. With this approximation and the definition of chemical potential (Eq. 2.4 without the pressure and the gravitational terms), the condition for equilibrium of ionic species j across the membrane is

$$\mu_j^* + RT \ln a_j^o + z_j FE^o = \mu_j^* + RT \ln a_j^i + z_j FE^i \tag{3.4}$$

The term μ_j^* in Eq. (3.4) is a constant referring to the same standard state of species j in the aqueous solutions on both sides of the membrane, so it can be canceled from the two sides of the equation.

Upon solving Eq. (3.4) for the electrical potential difference $E^i - E^o$ across the membrane at equilibrium, we obtain the following important relationship:

$$E_{N_j} = E^i - E^o = \frac{RT}{z_j F} \ln \frac{a_j^o}{a_j^i} = 2.303 \frac{RT}{z_j F} \log \frac{a_j^o}{a_j^i} \tag{3.5a}$$

which at 25°C becomes

$$E_{N_j} = \frac{25.7}{z_j} \ln \frac{a_j^o}{a_j^i} \quad \text{mV} = \frac{59.2}{z_j} \log \frac{a_j^o}{a_j^i} \quad \text{mV} \tag{3.5b}$$

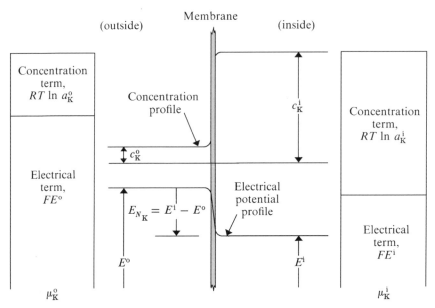

Figure 3-2. Equilibrium of K^+ across a membrane. When K^+ is in equilibrium across some membrane, the side with the higher concentration must be at a lower electrical potential for its chemical potential to be unchanged ($\mu_K^o = \mu_K^i$) in crossing the membrane (μ_K^* is the same on the two sides of the membrane). The electrical potential difference across the membrane is then the Nernst potential for K^+, E_{N_K}. Note that E^o and E^i must be expressed relative to some arbitrary baseline for electrical potentials.

The electrical potential difference E_{N_j} in Eq. (3.5) is called the *Nernst potential* of species j, i.e., $E^i - E^o = \Delta E = E_{N_j}$ in this case, where the subscript N stands for Nernst, who first derived this relation in about 1900.[2] We derived it by assuming equality of the chemical potentials of some charged species on two sides of a membrane. The natural logarithm (ln) is often replaced by 2.303 log, where log is the common logarithm to the base 10; and 2.303 RT/F is replaced by 58.2 mV at 20°C, 59.2 mV at 25°C (as above), and 60.2 mV at 30°C (see Appendices I–III).

Equation (3.5), the Nernst equation, is an equilibrium statement showing how the internal and the external activities of ionic species j are related to the electrical potential difference across a membrane (Fig. 3-2). At equilibrium, a 10-fold difference in the activity of a monovalent ion across some membrane is energetically equivalent to a 59 mV difference in electrical potential (at 25°C). Hence, a relatively small electrical potential difference can balance a large difference in activity across a membrane. For some calculations, γ_j^o/γ_j^i is set equal to 1 (a less stringent assumption than setting both γ_j^o and γ_j^i equal to 1). Under this condition, a_j^o/a_j^i in Eq. (3.5) becomes the ratio of the concentrations, c_j^o/c_j^i ($a_j = \gamma_j c_j$; Eq. 2.5). Such a substitution may be justified when the ionic strengths on the two sides of a membrane are approximately the same, but it can lead to errors when the outside solution is much more dilute than the internal one, as occurs for *Chara* or *Nitella* in pond water.

---

[2] Walter Nernst made many important contributions to the understanding of the physical chemistry of solutions; he was awarded the Nobel prize in chemistry in 1920.

Throughout the rest of this book we will represent the actual electrical potential difference existing across a membrane, $E^i - E^o$, by E_M, where M refers to membrane. Hence, for both equilibrium and nonequilibrium situations, we have $E^i - E^o = \Delta E = E_M$. When a particular ionic species j is in equilibrium across some membrane, $E_M = E_{N_j}$, the Nernst potential for that species. However, regardless of the actual electrical potential difference existing across a membrane (E_M), a Nernst potential for each individual ionic species j can always be calculated from Eq. (3.5) by using the ratio of the outside to the inside activity (a_j^o/a_j^i) or concentration (c_j^o/c_j^i), e.g., $E_{N_j} = (RT/z_j F) \ln (c_j^o/c_j^i)$.

If some ionic species cannot penetrate a membrane or is actively transported across it, E_{N_j} can differ markedly from E_M. In fact, the minimum amount of energy needed to transport ions across a membrane is proportional to the difference between E_{N_j} and E_M. The aqueous phases designated as inside and outside can have more than one membrane intervening between them. For instance, the vacuolar sap and an external solution—the regions often considered experimentally—have two membranes separating them, the plasma membrane and the tonoplast. The thermodynamic arguments remain the same in the case of multiple membranes, with E_M and E_{N_j} referring to the electrical potential differences between the two regions under consideration, regardless of how many membranes occur between them.

E. Example of E_{N_K}

Data obtained on K^+ for the large internodal cells of the green freshwater alga *Chara corallina* are convenient for illustrating the use of the Nernst equation (Vorobiev, 1967). The K^+ activity in the medium bathing the cells, a_K^o (where the element symbol K as a subscript is the conventional notation for K^+), is 0.096 mol m^{-3} (0.096 mM), and a_K^i in the vacuole is 48 mol m^{-3}. By Eq. (3.5) (with a factor of 58.2 because the measurement was made at 20°C), the Nernst potential for K^+ is

$$E_{N_K} = \frac{(58.2 \text{ mV})}{(1)} \log \left(\frac{0.096 \text{ mol m}^{-3}}{48 \text{ mol m}^{-3}} \right) = -157 \text{ mV}$$

The measured electrical potential of the vacuole relative to the external solution, E_M, is -155 mV, which is very close to the calculated Nernst potential for K^+. Thus K^+ in this case may be in equilibrium between the external solution and the vacuole. The K^+ concentration in the vacuole of these *Chara* cells is 60 mol m^{-3} (60 mM). The activity coefficient for K^+ in the vacuole (γ_K^i) = a_K^i/c_K^i ($a_j = \gamma_j c_j$; Eq. 2.5), so γ_K^i is (48 mol m^{-3})/(60 mol m^{-3}), or 0.80, and the K^+ activity coefficient in the bathing solution (γ_K^o) is about 0.96. For this example, in which there are large differences in the internal and the external concentrations, the ratio γ_K^o/γ_K^i is (0.96)/(0.80), or 1.20, which differs appreciably from 1.00. If concentrations instead of activities had been used in Eq. (3.5), E_{N_K} is -162 mV, which is somewhat lower than the measured potential of -155 mV. Calculating from the concentration ratio, the suggestion that K^+ is in equilibrium from the bathing solution to the vacuole could not be made with much confidence, if at all.

Equilibrium does not require that the various forces acting on a substance are zero; rather, it requires that they cancel each other. In the example for *C. corallina*, the factors that tend to cause K^+ to move are the differences in both its activity (or concentration) and the electrical potential across the membranes. The activity of K^+ is much higher in the vacuole than in the external solution (see Fig. 3-2). The activity term in the chemical potential therefore represents a driving force on K^+ directed from inside the cell to the bathing solution. The electrical potential is lower inside the cell, as is indicated in Fig. 3-2. Hence the electrical driving force on the positively charged K^+ tends to cause its entry into the cell. At equilibrium these two tendencies for movement in opposite directions are balanced, and no net K^+ flux occurs. As indicated previously, the electrical potential difference existing across a membrane when K^+ is in equilibrium is the Nernst potential for K^+, E_{N_K}. Generally, c_K^i and a_K^i for both plant and animal cells are much higher than c_K^o and a_K^o, and K^+ is often close to equilibrium across the cellular membranes. From these observations, we expect that the interiors of cells are usually at negative electrical potentials compared with the outside solutions, as is indeed the case. The chemical potentials of ions are generally not equal in all regions of interest, so passive movements toward lower μ_j's occur.

II. Fluxes and Diffusion Potentials

Fluxes of many different solutes occur across biological membranes. Inward fluxes move mineral nutrients into cells, and certain products of metabolism flow out of cells. The primary concern of this section is the passive fluxes of ions toward lower chemical potentials. First, we will indicate that the passive flux density of a solute is directly proportional to the driving force causing the movement. Next, the driving force is expressed in terms of the relevant components of the chemical potential. We then examine the consequences of electroneutrality when there are simultaneous passive fluxes of more than one type of ion. This leads to an expression describing the electrical potential difference across a membrane in terms of the properties of the ions penetrating it.

Before discussing the relation between fluxes and chemical potentials, we will briefly consider fluxes already mentioned or which may be familiar from other contexts. In Chapter 1 we discussed Fick's first law of diffusion (Eq. 1.1), which says that the flux density of (neutral) solute species $j = -D_j \, \partial c_j / \partial x$, where we can consider that the driving force is the negative gradient of the concentration. In Chapter 9 we will use Darcy's law (Eq. 9.7) and Poiseuille's law (Eq. 9.11), both of which indicate that the volume flux density of a solution is proportional to $-\partial P / \partial x$. Ohm's law, which describes electrical effects, can be written as $\Delta E = IR$, where I is the current (charge moving per unit time) and R is the resistance across which the electrical potential difference is ΔE. The current per unit area A is the charge flux density, J_c, which equals $-(1/\rho) \, \partial E / \partial x$, where ρ is the electrical resistivity and the negative gradient of the electrical potential represents the driving force. We usually replace $-\partial E / \partial x$ by $\Delta E / \Delta x$, which leads to $I/A = (1/\rho) \, \Delta E / \Delta x$. For this to conform with the usual expression of Ohm's law ($\Delta E = IR$), $\rho \, \Delta x / A$ should be replaced by R. The gravitational force, $-m_j g$, is the negative gradient of the potential energy in a gravitational field; i.e., $-\partial m_j g h / \partial h = -m_j g$, where the minus sign indicates that the force is directed toward decreasing altitudes, namely, toward the center of the earth. The gravitational force leads to the various forms of precipitation as well as to the percolation of water down through the soil.

We have just considered examples of fluxes depending on each of the four variable terms in the chemical potential ($\mu_j = \mu_j^* + RT \ln a_j + \bar{V}_j P + z_j FE + m_j g h$; Eq. 2.4). Specifically, the activity term ($RT \ln a_j$) leads to Fick's first law, the pressure term ($\bar{V}_j P$) accounts for Darcy's law and Poiseuille's law, the electrical term ($z_j FE$) yields Ohm's law, and the gravitation term ($m_j g h$) is responsible for fluxes caused by gravity. In each case, experiments show that the flux density is directly proportional to an appropriate driving force. We can generalize such relationships because nearly all transport phenomena are represented by the following statement: Flux density is proportional to an appropriate force.

Force is represented by the negative gradient of a suitable potential, which for convenience is often taken as the change in potential over some finite distance. Moreover, we have already shown that the chemical potential is an elegant way of summarizing all the factors that can contribute to the motion of a substance. It should not be surprising, therefore, that in general the flux density of species $j (J_j)$ is proportional to the negative gradient of its chemical potential, $-\partial \mu_j / \partial x$.

A. Flux and Mobility

We now consider a charged substance capable of crossing a particular membrane. When its μ_j depends on position, a net passive movement or flux of that species tends to occur toward the region where its chemical potential is lower. The negative gradient of the chemical potential of solute species j, $-\partial \mu_j / \partial x$, acts as the driving force for this flux. (As in Chapter 1, our discussion will apply to the one-dimensional case, so we will let $\partial \mu_j / \partial y = \partial \mu_j / \partial z = 0$.) The greater $-\partial \mu_j / \partial x$, the larger the flux of species j in the x-direction. As a useful approximation, we will assume that the flux density is proportional to $-\partial \mu_j / \partial x$, where the minus sign means that a net positive flux density occurs in the direction of decreasing chemical potential. The magnitude of a flux density across some plane is also proportional to the local concentration of that species, c_j. That is, for a given driving force on species j, the amount of the substance that moves is proportional to how much of it is actually present— the greater the concentration, the greater the flux density. Thus for the one-dimensional case of crossing a plane perpendicular to the x-axis, the flux density J_j can be expressed as

$$J_j = u_j c_j \left(-\frac{\partial \mu_j}{\partial x} \right)$$

$$= \bar{v}_j c_j \tag{3.6}$$

where u_j is a coefficient called the *mobility* of species j. A mobility is generally the ratio of some velocity to the force causing the motion.

As we have already indicated, the top line of Eq. (3.6) is a representative example from the large class of expressions relating various flows to their causative forces. In this particular case, J_j is the rate of flow of moles of species j across unit area of a plane and can be expressed in mol m^{-2} s^{-1}. Such a molar flux density of a species j divided by its local concentration c_j gives the mean velocity $\bar{v}_j$ with which this species moves across the plane; when c_j is in mol m^{-3}, J_j/c_j can have units of (mol m^{-2} s^{-1})/(mol m^{-3}), or m s^{-1}. Perhaps this important point can be better appreciated by considering it in the following way. Suppose that the average velocity of species j moving perpendicularly toward area A of the plane of interest is $\bar{v}_j$, and consider a volume element of cross-sectional area A extending back from the plane for a distance $\bar{v}_j \, dt$ (Fig. 3-3). In time dt, all molecules in the volume element $(\bar{v}_j \, dt) \times (A)$ will cross area A, so the number of moles of species j crossing the plane in this interval is $[(\bar{v}_j \, dt) A] \times (c_j)$, where c_j is the number of moles per unit volume. Hence, the molar flux density J_j (which is the number of moles crossing unit area per unit time) is $(\bar{v}_j \, dt \, A \, c_j)/(A \, dt)$, or $\bar{v}_j c_j$ (Fig. 3-3). In other words, the mean velocity of species j moving across the plane, $\bar{v}_j$, times the number of those molecules per unit volume that can move, c_j, equals the flux density of that species, J_j, as the bottom line of Eq. (3.6) indicates. The top line of Eq. (3.6) states that this average velocity, J_j/c_j, equals the mobility of species j, u_j, times $-\partial \mu_j / \partial x$, the latter being the force on j that causes it to move. Thus mobility is the proportionality factor between the mean velocity of

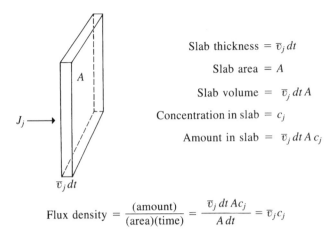

$$\text{Flux density} = \frac{(\text{amount})}{(\text{area})(\text{time})} = \frac{\bar{v}_j \, dt \, A c_j}{A \, dt} = \bar{v}_j c_j$$

Figure 3-3. Geometrical argument showing that the flux density J_j across surface area A equals $\bar{v}_j c_j$, where $\bar{v}_j$ is the mean velocity of species j. Note that $\bar{v}_j \, dt$ is an infinitesimal distance representing the slab thickness. By the definition of $\bar{v}_j$, all the molecules of species j in the slab cross A in time dt.

motion ($\bar{v}_j = J_j / c_j$) and the causative force ($-\partial \mu_j / \partial x$). The greater the mobility of some species, the larger its velocity in response to a given force.

The particular form of the chemical potential to be substituted into Eq. (3.6) depends on the application. For charged particles moving across biological membranes, the appropriate μ_j is $\mu_j^* + RT \ln a_j + z_j FE$. (As we mentioned before, the $\bar{V}_j P$ term makes only a relatively small contribution to the $\Delta \mu_j$ of an ion, so it is not included, and the $m_j g h$ term in Eq. 2.4 can also be ignored here.) For treating the one-dimensional case described by Eq. (3.6), μ_j must be differentiated with respect to x, $\partial \mu_j / \partial x$, which equals $RT \partial \ln a_j / \partial x + z_j F \, \partial E / \partial x$ when T is constant. [Actually, we must assume isothermal conditions for $-\partial \mu_j / \partial x$ to represent the force precisely; i.e., T is constant, and hence $\partial (RT \ln a_j) / \partial x = RT \partial \ln a_j / \partial x$.] The quantity $\partial \ln a_j / \partial x = (1/a_j) \, \partial a_j / \partial x$, which is $(1/\gamma_j c_j) \, \partial \gamma_j c_j / \partial x$ ($a_j = \gamma_j c_j$; Eq. 2.5). Using the previous form of μ_j appropriate for charged solutes and this expansion of $\partial \ln a_j / \partial x$, the net flux density of species j in Eq. (3.6) can be written

$$J_j = -\frac{u_j RT}{\gamma_j} \frac{\partial \gamma_j c_j}{\partial x} - u_j c_j z_j F \frac{\partial E}{\partial x} \tag{3.7}$$

Equation (3.7), which is often called the Nernst–Planck equation, is a general expression for the one-dimensional flux density of species j either across a membrane or in a solution in terms of two components of the driving force—the gradients in activity and in electrical potential.

Before proceeding, we examine the first term on the right-hand side of Eq. (3.7). When γ_j varies across a membrane, $\partial \gamma_j / \partial x$ can be considered to represent a driving force on species j. In keeping with common practice, we will ignore this possible force. That is, we will assume that γ_j is constant (a less severe restriction than setting it equal to 1); in any case $\partial \gamma_j / \partial x$ would be extremely difficult to measure. For constant γ_j, the first term on the right-hand side of Eq. (3.7) becomes $-u_j RT (\partial c_j / \partial x)$, i.e., it is proportional to the concentration gradient. In the absence of electrical potential gradients ($\partial E / \partial x = 0$), or for neutral solutes ($z_j = 0$), Eq. (3.7) indicates that $J_j = -u_j RT (\partial c_j / \partial x)$. This is the flux described by Fick's first law ($J_j = -D_j \partial c_j / \partial x$; Eq. 1.1), with $u_j RT$ taking the place of the diffusion coefficient, D_j.

Because $D_j = u_j RT$, we can appreciate that diffusion coefficients depend on temperature. Moreover, u_j generally is inversely proportional to the viscosity, which decreases as T increases. Thus the dependence of D_j on T can be substantial. For gases in air, D_j depends approximately on $T^{1.8}$. Consequently, the temperature should be specified when the value of a diffusion coefficient is given. In the absence of electrical effects and for constant γ_j, the net flux of species j given by Eq. (3.7) equals $-D_j \partial c_j / \partial x$ when $u_j RT$ is replaced by D_j, as we have just indicated. Fick's first law ($J_j = -D_j \partial c_j / \partial x$; Eq. 1.1) is hence a special case of our general flux relation, the Nernst–Planck equation (Eq. 3.7), for which we first ignored the pressure and gravitational effects and then omitted the electrical effects. This eventual reduction of Eq. (3.7) to Fick's first law is what we should expect because the only driving force considered when we presented Fick's first law (Chapter 1) was the concentration gradient. Such agreement between our present thermodynamic approach and the seemingly more empirical Fick's first law is quite important. It serves to justify the logarithmic term for activity in the chemical potential ($\mu_j = \mu_j^* + RT \ln a_j + \bar{V}_j P + z_j FE + m_j gh$; Eq. 2.4). In particular, if the activity of species j appeared in Eq. (2.4) in a form other than $\ln a_j$, Eq. (3.7) would not reduce to Fick's first law under the appropriate conditions. Because Fick's first law has been amply demonstrated experimentally, such a disagreement between theory and practice would necessitate some modification in the expression used to define chemical potential.[3]

In contrast to the case for a neutral solute, the flux of an ion also depends on an electrical driving force, represented in Eq. (3.7) by $-\partial E / \partial x$. A charged solute spontaneously tends to move in an electrical potential gradient, a cation moving in the direction of lower or decreasing electrical potential. Of course, the concentration gradient also affects charged particles. If a certain type of ion were present in some region but absent in an adjacent one, the ions would tend to diffuse into the latter region. As such charged particles diffuse toward regions of lower concentration, an electrical potential difference is created, which is referred to as a "diffusion potential." Most membrane potentials can be treated as diffusion potentials resulting from different rates of movement of the various ions across a membrane.

B. Diffusion Potential in a Solution

We will now use Eq. (3.7) to derive the electrical potential difference created by ions diffusing down a concentration gradient in a solution containing one type of cation and its accompanying anion. This case is simpler than the biologically more realistic one that follows and thus more clearly illustrates the relationship between concentration gradients and the accompanying electrical potential differences. We will assume that the cations and

---

[3] As indicated in Chapter 2, the terms in the chemical potential can be justified or "derived" by various methods. The forms of some terms in μ_j can be readily appreciated because they follow from familiar definitions of work, e.g., the gravitational term and the electrical term. The comparison with Fick's first law indicates that $RT \ln a_j$ is the appropriate form for the activity term. Another derivation of the $RT \ln a_j$ term is in Appendix IV, together with a discussion of the pressure term for both liquids and gases. Some of these derivations incorporate conclusions from empirical observations. Moreover, the fact that the chemical potential can be expressed as a series of terms that can be added together agrees with experiment. Thus a thermodynamic expression for the chemical potential such as Eq. (2.4) (1) summarizes the results of previous observations, (2) stands the test of experiments, and (3) leads to new and useful predictions.

the anions are initially placed at one side of the solution. In time, they will diffuse across the solution toward regions of lower concentration. In general, one ionic species will have a higher mobility, u_j, than the other. The more mobile ions will tend to diffuse faster than their oppositely charged partners, resulting in a microscopic charge separation. This slight charge separation sets up an electrical potential gradient leading to a *diffusion potential*. Using certain simplifying assumptions, we will calculate the magnitude of the electrical potential difference so created.

For convenience of analysis, let us consider a solution containing only a monovalent cation $(+)$ and its accompanying monovalent anion $(-)$. We will assume that their activity coefficients are constant. As our previous calculations of electrical effects have indicated, solutions are essentially neutral in regions that are large compared with atomic dimensions. Thus, $c_+ = c_-$, so the concentration of either species can be designated c. Furthermore, no charge imbalance develops in time, which means that the flux density of the cation across some plane in the solution, J_+, equals that of the anion across the same plane, J_-. (A very small charge imbalance develops initially, which sets up the electrical potential gradient, but this uncompensated flux density is transitory and in any case is negligible compared with J_+ or J_-.) Both flux densities, J_+ and J_-, can be expressed by Eq. (3.7) and then equated to each other, which gives

$$J_+ = -u_+ RT \frac{\partial c}{\partial x} - u_+ cF \frac{\partial E}{\partial x} = J_- = -u_- RT \frac{\partial c}{\partial x} + u_- cF \frac{\partial E}{\partial x} \qquad (3.8)$$

where the plus sign in the last term occurs because the monovalent anion carries a negative charge $(z_- = -1)$. Rearrangement of Eq. (3.8) yields the following expression for the electrical potential gradient:

$$\frac{\partial E}{\partial x} = \frac{u_- - u_+}{u_- + u_+} \frac{RT}{Fc} \frac{\partial c}{\partial x} \qquad (3.9)$$

Equation (3.9) indicates that a nonzero $\partial E/\partial x$ occurs when the mobility of the cation differs from that of the anion and a concentration gradient exists. If u_- is greater than u_+, the anions move (diffuse) faster than the cations toward regions of lower concentration. As some individual anion moves ahead of its "partner" cation, an electric field is set up in such a direction as to speed up the cation and slow down the anion until they both move at the same speed, thus preserving electrical neutrality.

To obtain the difference in electrical potential produced by diffusion between planes of differing concentration, we must integrate Eq. (3.9). We will restrict our consideration to the steady-state case in which neither E nor c changes with time. At equilibrium, μ_j does not change with time or position for communicating regions and no net flux of that solute occurs $(J_j = 0)$. In a steady state μ_j changes with position but not with time, a condition often used to approximate problems of biological interest; J_j is then constant. A total differential such as dE equals $(\partial E/\partial x)\,dx + (\partial E/\partial t)\,dt$, and dc is $(\partial c/\partial x)\,dx + (\partial c/\partial t)\,dt$. For a steady-state condition, both $\partial E/\partial t$ and $\partial c/\partial t$ are zero. Consequently, $dE = (\partial E/\partial x)\,dx$ and $dc = (\partial c/\partial x)\,dx$. Actually, it is a matter of judgment whether μ_j is constant enough in time to warrant describing the system as being in a steady state. Similarly, constancy of μ_j for appropriate time and distance intervals is necessary before indicating that a system is in equilibrium.

In going along the x-axis from region I to region II, the change in the electrical potential term in Eq. (3.9) is the definite integral $\int_I^{II} (\partial E/\partial x)\,dx$, which becomes $\int_I^{II} dE$ for the steady-state condition and so equals $E^{II} - E^{I}$. The integral of the concentration term is $\int_{I}^{II}(1/c)(\partial c/\partial x)\,dx$, which becomes $\int_{I}^{II} dc/c$, or $\ln c^{II} - \ln c^{I}$, which equals $\ln(c^{II}/c^{I})$. Using these two relations, integration of Eq. (3.9) leads to

$$E^{II} - E^{I} = \frac{u_{-} - u_{+}}{u_{-} + u_{+}} \frac{RT}{F} \ln \frac{c^{II}}{c^{I}} \tag{3.10a}$$

or, at 25°C,

$$E^{II} - E^{I} = 59.2 \frac{u_{-} - u_{+}}{u_{-} + u_{+}} \log \frac{c^{II}}{c^{I}} \quad \text{mV} \tag{3.10b}$$

where ln has been replaced by 2.303 log, and the value 59.2 mV has been substituted for $2.303\,RT/F$ at 25°C (see Appendix I). In the general case, the anions have different mobilities than do the cations. As the ions diffuse to regions of lower chemical potential, an electrical potential difference—given by Eq. (3.10) and called a diffusion potential—is set up across the section where the concentration changes from c^{I} to c^{II}.

An example of a diffusion potential describable by Eq. (3.10) occurs at the open end of the special micropipettes used for measuring electrical potential differences across membranes (Fig. 3-4). The fine tip of the micropipette provides an electrically conducting pathway

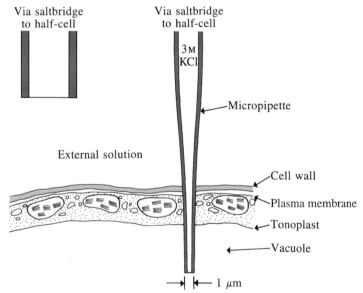

Figure 3-4. Glass micropipette filled with 3 M KCl (a highly conducting solution) and inserted into the vacuole of a plant cell so that the electrical potential difference across the plasma membrane and the tonoplast in series can be measured. The micropipette is carefully inserted into the cell with a micromanipulator while observing with a light microscope. The micropipette tip must be strong enough to penetrate the cell wall and yet fine enough not to disturb the cell substantially either mechanically or electrically. The finer the tip, the higher its resistance—a 1-μm-diameter tip usually has a resistance of about 10^{6} ohms, which is generally acceptable. The micropipette is connected via a saltbridge containing an electrically conducting solution to a half-cell (discussed in Chapter 6). This half-cell plus another one in electrical contact with the external solution are connected to a voltmeter with a high input resistance (generally at least 10^{10} ohms) so that the membrane potentials can be measured without drawing much current.

into the cell or tissue. Ions diffusing through this fine tip give rise to a diffusion potential between the interior of the micropipette and the aqueous compartment into which the tip is inserted. To estimate the magnitude of this potential for KCl as the electrolyte, we will assume that there is 3000 mol m^{-3} (3 M) KCl in the micropipette (Fig. 3-4) and 100 mol m^{-3} in the cell. The chloride mobility u_{Cl} is about 1.04 times u_K, so the diffusion potential calculated from Eq. (3.10b) at 25°C is

$$E^{cell} - E^{micropipette} = (59 \text{ mV})\left(\frac{1.04u_K - u_K}{1.04u_K + u_K}\right) \log\left(\frac{100 \text{ mol m}^{-3}}{3000 \text{ mol m}^{-3}}\right)$$

$$= -2 \text{ mV}$$

For NaCl as the electrolyte, u_{Cl} is 1.52 u_{Na}, and the diffusion potential for the same concentration ratio is -18 mV. To minimize the diffusion potential across the fine tip, KCl is thus a much more suitable electrolyte for micropipettes than is NaCl. In fact, KCl is employed in nearly all micropipettes used for measuring membrane potentials (generally, c_K inside cells is also higher than c_{Na}, and it does not vary as much from cell to cell as does c_{Na}). In any case, the closer u_- is to u_+, the smaller will be the diffusion potential for a given concentration ratio from one region to another.

C. Membrane Fluxes

As is the case for diffusion potentials in a solution, membrane potentials also depend on the different mobilities of the various ions and on their concentration gradients. In this case, however, the "solution" in which the diffusion takes place toward regions of lower chemical potential is the membrane itself. We noted in Chapter 1 that a membrane is often the main barrier, and thus the rate-limiting step, for the diffusion of molecules into and out of cells or organelles. We would therefore expect it to be the phase across which the diffusion potential is expressed. Under biological conditions, a number of different types of ions are present, so the situation is more complex than for the single cation–anion pair analyzed previously. Furthermore, the quantities of interest, such as $\partial E/\partial x$ and γ_j, are those within the membrane, where they are not readily measurable.

To calculate membrane diffusion potentials, we must make certain assumptions. As a start, we will assume that the electrical potential (E) varies linearly with distance across a membrane. This means that $\partial E/\partial x$ is a constant equal to $E_M/\Delta x$, where E_M is the electrical potential difference across the membrane—i.e., $E^i - E^o = \Delta E = E_M$—and Δx is the membrane thickness. This assumption of a constant electric field across the membrane—the electric field equals $-\partial E/\partial x$ in our one-dimensional case—was originally suggested by David Goldman in 1943. It appreciably simplifies the integration conditions leading to an expression describing the electrical potential difference across membranes. As another useful approximation, we will assume that the activity coefficient of species j, γ_j, is constant across the membrane. As noted in Chapter 1, a partition coefficient is needed to describe concentrations within a membrane because the solvent properties of a membrane differ from those of the adjoining aqueous solutions where the concentrations are actually determined. Thus c_j in Eq. (3.7) should be replaced by $K_j c_j$, where K_j is the partition coefficient for species j. Incorporating these various simplifications and conditions, we can rewrite Eq. (3.7) as follows:

$$J_j = -u_j RTK_j \frac{\partial c_j}{\partial x} - u_j K_j c_j z_j F \frac{E_M}{\Delta x} \tag{3.11}$$

Because the two most convenient variables are c_j and x, their differentials should appear on opposite sides of the equation. We therefore transfer the electrical term to the left-hand side of Eq. (3.11), factor out $u_j z_j FE_M / \Delta x$, and divide both sides by $K_j c_j + J_j \Delta x / (u_j z_j FE_M)$, which puts all terms containing concentration on the same side of the equation. After multiplying each side by $dx / (u_j RT)$, we transform Eq. (3.11) to the following:

$$\frac{z_j FE_M}{RT \Delta x} dx = - \frac{K_j \, dc_j}{K_j c_j + \frac{J_j \Delta x}{u_j z_j FE_M}} \tag{3.12}$$

where $(\partial c_j / \partial x) \, dx$ has been replaced by dc_j in anticipation of the restriction to steady-state conditions.

When $J_j \Delta x / (u_j z_j FE_M)$ is constant, Eq. (3.12) can be readily integrated from one side of the membrane to the other. The factors Δx, z_j, F, and E_M are all constants. For convenience, the mobility (u_j) of each species is also assumed to be constant within the membrane. When the flux of species j does not change with time or position, J_j across any plane parallel to and within the membrane is the same, and species j is neither accumulating nor being depleted in any of the regions of interest. Consequently, $\partial c_j / \partial t$ is zero, whereas $\partial c_j / \partial x$ is nonzero—which is the steady-state condition. Our restriction to a steady state therefore means that J_j is constant and $(\partial c_j / \partial x) \, dx = dc_j$, a relation already incorporated into Eq. (3.12). With these restrictions, the quantity $J_j \Delta x / (u_j z_j FE_M)$ is constant and we can integrate Eq. (3.12) from one side of the membrane (the outside, o) to the other (the inside, i). Because $\int K_j \, dc_j / (K_j c_j + b) = \ln(K_j c_j + b)$ and $\int_{x^o}^{x^i} dx = \Delta x$, integration of Eq. (3.12) gives

$$\frac{z_j FE_M}{RT} = \ln \frac{\left(K_j c_j^o + \frac{J_j \Delta x}{u_j z_j FE_M} \right)}{\left(K_j c_j^i + \frac{J_j \Delta x}{u_j z_j FE_M} \right)} \tag{3.13}$$

After taking exponentials of both sides of Eq. (3.13) to put it into a more convenient form, and then multiplying by $K_j c_j^i + J_j \Delta x / (u_j z_j FE_M)$, it becomes

$$K_j c_j^i e^{z_j FE_M / RT} + \frac{J_j \Delta x}{u_j z_j FE_M} e^{z_j FE_M / RT} = K_j c_j^o + \frac{J_j \Delta x}{u_j z_j FE_M} \tag{3.14}$$

A quantity of considerable interest in Eq. (3.14) is J_j, the net flux density of species j. This equation can be solved for J_j, giving

$$J_j = J_j^{in} - J_j^{out} \tag{3.15}$$

$$= \left(\frac{K_j u_j z_j FE_M}{\Delta x} \right) \left(\frac{1}{e^{z_j FE_M / RT} - 1} \right) \left(c_j^o - c_j^i e^{z_j FE_M / RT} \right)$$

where J_j^{in} is the influx or inward flux density of species j, J_j^{out} is its efflux, and their difference is the net flux density. (The net flux density can represent either a net influx or a net efflux, depending on which of the unidirectional components, J_j^{in} or J_j^{out}, is larger.) We will use Eq. (3.15) to derive the Goldman equation, describing the diffusion potential across membranes, and later to derive the Ussing–Teorell equation, a relation obeyed by certain passive fluxes.

Equation (3.15) shows how the passive flux of some charged species j depends on its internal and its external concentrations and on the electrical potential difference across the membrane. For most cell membranes, E_M is negative, i.e., the inside of the cell is at a lower electrical potential than is the outside. For a cation (z_j a positive integer) and a negative E_M, the terms in the first two parentheses on the right-hand side of Eq. (3.15) are both negative; hence their product is positive. For an anion (z_j a negative integer) and a negative E_M, both parentheses are positive; hence their product is also positive. For $E_M > 0$, the product of the first two parentheses is again positive for both anions and cations. Thus the sign of J_j depends on the value of c_j^o relative to that of $c_j^i e^{z_j FE_M/RT}$. When $c_j^o > c_j^i e^{z_j FE_M/RT}$, the expression in the last parentheses of Eq. (3.15) is positive, and a net inward flux density of species j occurs ($J_j > 0$). Such a condition should be contrasted with Eq. (1.8) [$J_j = P_j(c_j^o - c_j^i)$], where the net flux density in the absence of electrical effects is inward when c_j^o is larger than c_j^i, as adequately describes the situation for neutral solutes. However, knowledge of the concentration difference alone is not sufficient to predict the magnitude or even the direction of the flux of ions—we must also consider the electrical potential difference between the two regions.[4]

Although the mathematical manipulations necessary to get from Eq. (3.11) to Eq. (3.15) are lengthy and cumbersome, the resulting expression is extremely important for our understanding of both membrane potentials and passive fluxes of ions. Moreover, throughout this text we have generally presented the actual steps involved in a particular derivation to avoid statements such as "it can easily be shown" that such and such follows from so and so—expressions that can be frustrating and often are untrue. In the present case of the derivation of the Goldman equation, the steps and the equations involved are as follows:

Step	Eq. No.
Basic flux equation (Nernst–Planck equation)	3.7, 3.11
Transformation for integration	3.12
Integration and rearrangement	3.13, 3.14
Net flux density for a single ionic species	3.15
Restriction to three major ionic fluxes	3.16
Insertion of flux densities and rearrangment	3.17, 3.18
Goldman equation	3.19

D. Membrane Diffusion Potential—Goldman Equation

Passive fluxes of ions, which can be described by Eq. (3.15) and are caused by gradients in the chemical potentials of the various solute species, lead to an electrical potential difference (diffusion potential) across a membrane. We can determine the magnitude of this electrical potential difference by considering the contributions from all ionic fluxes across the membrane and the condition of electroneutrality. Certain assumptions are needed, however, to keep the equations manageable. Under usual biological situations not all anions and cations can easily move through membranes. Many divalent cations do not readily enter

---

[4] We might ask what happens to the flux density expressed by Eq. (3.15) as E_M approaches zero. In particular, the quantity within the first parentheses goes to zero and that within the second parentheses goes to infinity. To re-solve this limit situation, we can use the series expansion for an exponential ($e^x = 1 + x + \frac{x^2}{2!} + \cdots$; Appendix III) and keep only the first two terms. The denominator within the second parentheses then becomes $1 + z_j FE_M/RT - 1$, or $z_j FE_M/RT$. Hence, the product of the quantities within the first two parentheses becomes $K_j u_j RT / \Delta x$, which is P_j. Now setting E_M to zero in the last term, Eq. (3.15) becomes $J_j = P_j(c_j^o - c_j^i)$, which is Eq. (1.8), the appropriate form for the flux density in the absence of electrical effects ($E_M = 0$ and/or $z_j = 0$).

or leave plant cells passively, meaning that their mobility in membranes is small. Such ions usually do not make a large enough contribution to the fluxes into or out of plant cells to influence markedly the diffusion potentials across the membranes. Thus we will omit them in the present analysis, which nevertheless is rather complicated.

For many plant cells, the total ionic flux consists mainly of movements of K^+, Na^+, and Cl^-. These three ions generally have fairly high concentrations in and near plant cells and therefore are expected to make substantial contributions to the total ionic flux density. More specifically, a flux density of species j depends on the product of its concentration and its mobility ($J_j = -u_j c_j \partial \mu_j / \partial x$; Eq. 3.6); ions having relatively high local concentrations or moving in the membrane fairly easily (high u_j) will therefore tend to be the major contributors to the total ionic flux density. In some cases there may be a sizable flux density of H^+ or OH^- (which can have high u_j's) as well as of other ions, the restriction here to three ions being partially for algebraic simplicity. However, the real justification for considering only K^+, Na^+, and Cl^- is that the diffusion potentials calculated for the passive fluxes of these three ions across certain membranes are in good agreement with the measured electrical potential differences.

Our previous electrical calculations indicated that aqueous solutions are essentially neutral. In other words, because of the large effects on electrical potentials resulting from small amounts of uncanceled charge, the net charge needed to cause the electrical potential difference across a membrane (E_M) is negligible compared with the ambient concentrations of ions. Furthermore, the steady-state fluxes of the ions across the membrane do not change this condition of electrical neutrality because no net charge is transported by the algebraic sum of the various charge movements across the membrane, i.e., $\sum_j z_j J_j = 0$. When the bulk of the ionic flux density consists of K^+, Na^+, and Cl^- movements, this important condition of electroneutrality can be described by equating the cationic flux densities ($J_K + J_{Na}$) to the anionic one (J_{Cl}), which leads to the following relation (as mentioned previously, the various ions are indicated by using only their element symbols as subscripts):

$$J_K + J_{Na} - J_{Cl} = 0 \tag{3.16}$$

Equation (3.16) describes the net ionic flux densities leading to the electrical potential differences across a membrane. After substituting the expressions for the various J_j's into Eq. (3.16), we will solve the resulting equation for the diffusion potential across a membrane, E_M.

To obtain a useful expression for E_M in terms of measurable parameters, it is convenient to introduce the permeability coefficient for species j, P_j. In Chapter 1, such a permeability coefficient was defined as $D_j K_j / \Delta x$, where D_j is the diffusion coefficient of species j, K_j is its partition coefficient, and Δx is the membrane thickness (Eq. 1.9). Upon comparing Eq. (3.7) with Eq. (1.1) ($J_j = -D_j \partial c_j / \partial x$), we see that $u_j RT$ takes the place of the diffusion coefficient of species j, D_j. The quantity $K_j u_j RT / \Delta x$ can thus be replaced by the permeability coefficient P_j. In this way, the unknown mobility of species j in a given membrane, the unknown thickness of the membrane, and the unknown partition coefficient for the solute can all be replaced by one parameter describing the permeability of that solute crossing a particular membrane.

With all the preliminaries out of the way, let us now derive the expression for the diffusion potential across a membrane for the case in which most of the ionic flux density is due to K^+, Na^+, and Cl^- movements. Using the permeability coefficients of the three ions and substituting in the net flux density of each species as defined by Eq. (3.15), Eq. (3.16) becomes

$$
P_K \left(\frac{1}{e^{FE_M/RT} - 1} \right) \left(c_K^o - c_K^i e^{FE_M/RT} \right)
$$

$$
+ P_{Na} \left(\frac{1}{e^{FE_M/RT} - 1} \right) \left(c_{Na}^o - c_{Na}^i e^{FE_M/RT} \right) \tag{3.17}
$$

$$
+ P_{Cl} \left(\frac{1}{e^{-FE_M/RT} - 1} \right) \left(c_{Cl}^o - c_{Cl}^i e^{-FE_M/RT} \right) = 0
$$

where z_K and z_{Na} have been replaced by 1, z_{Cl} has been replaced by -1, and FE_M/RT has been canceled from each of the terms for the three net flux densities. To simplify this unwieldy expression, the quantity $1/(e^{FE_M/RT} - 1)$ can be canceled from each of the three terms in Eq. (3.17)—note that $1/(e^{-FE_M/RT} - 1)$ in the last term is the same as $-e^{FE_M/RT}/(e^{FE_M/RT} - 1)$. Equation (3.17) then assumes a more manageable form:

$$
P_K c_K^o - P_K c_K^i e^{FE_M/RT} + P_{Na} c_{Na}^o - P_{Na} c_{Na}^i e^{FE_M/RT}
$$

$$
- P_{Cl} c_{Cl}^o e^{FE_M/RT} + P_{Cl} c_{Cl}^i = 0 \tag{3.18}
$$

After solving Eq. (3.18) for $e^{FE_M/RT}$ and taking logarithms, we obtain the following expression for the electrical potential difference across a membrane:

$$
E_M = \frac{RT}{F} \ln \frac{\left(P_K c_K^o + P_{Na} c_{Na}^o + P_{Cl} c_{Cl}^i \right)}{\left(P_K c_K^i + P_{Na} c_{Na}^i + P_{Cl} c_{Cl}^o \right)} \tag{3.19}
$$

Equation (3.19) is generally known as the Goldman, or constant field, equation. As mentioned previously, the electrical field equals $-\partial E/\partial x$, which Goldman in 1943 set equal to a constant (here $-E_M/\Delta x$) to facilitate the integration across a membrane.[5] In 1949 Alan Hodgkin and Bernhard Katz applied the general equation derived by Goldman to the specific case of K^+, Na^+, and Cl^- diffusing across a membrane, so Eq. (3.19) is sometimes referred to as the Goldman–Hodgkin–Katz equation.

Equation (3.19) gives the diffusion potential across a membrane. We derived it by assuming independent passive movements of K^+, Na^+, and Cl^- across a membrane in which $\partial E/\partial x$, γ_j, J_j, and u_j are all constant. We used the negative gradient of its chemical potential as the driving force for the net flux density of each ion. Thus Eq. (3.19) gives the electrical potential difference arising from the different tendencies of K^+, Na^+, and Cl^- to diffuse across a membrane to regions of lower chemical potential. When other ions cross a membrane in appreciable amounts, they will also make a contribution to its membrane potential. However, the inclusion of divalent and trivalent ions in the derivation of an expression for E_M complicates the algebra considerably (e.g., if Ca^{2+} is also considered, Eq. 3.18 has 14 terms on the left-hand side instead of 6, and the equation becomes a quadratic in powers of $e^{FE_M/RT}$). However, the flux densities of such ions are often small, in which case Eq. (3.19) can be adequate for describing the membrane potential.

---

[5] The assumption of a constant electric field in the membrane is actually not essential for obtaining Eq. (3.19); we could invoke Gauss's law and perform a more difficult integration. See Goldman (1943) for a consideration of the constant field situation in a general case.

Table 3-1. Concentrations, Potentials, and Fluxes of Various Ions for *Nitella translucens* in the Light and the Dark.[a]

					Light		Dark	
	c_j^o	c_j^i	E_{N_j} (mV)	$\dfrac{c_j^o}{c_j^i e^{z_j FE_M/RT}}$	J_j^{in}	J_j^{out}	J_j^{in}	J_j^{out}
Ion	(mol m^{-3} = mM)				(nmol m^{-2} s^{-1})			
Na$^+$	1.0	14	-67	17	5.5	5.5	5.5	1.0
K$^+$	0.1	119	-179	0.20	8.5	8.5	2.0	8.5
Cl$^-$	1.3	65	99	0.000085	8.5	8.5	0.5	—

[a] The superscript o refers to concentrations in the external bathing solution and superscript i refers to the cytosol. The Nernst potentials (E_{N_j}) were calculated from Eq. (3.5) using concentration ratios and a numerical factor of 58.2 mV because the temperature was 20°C. The potential across the plasma membrane (E_M) was -138 mV. The fluxes indicated for the dark refer to values soon after cessation of illumination. (Source: MacRobbie, 1962; Spanswick and Williams, 1964.)

E. Application of the Goldman Equation

In certain cases, the quantities in Eq. (3.19)—namely, the permeabilities and the internal and the external concentrations of K$^+$, Na$^+$, and Cl$^-$—have all been measured. The validity of the Goldman equation can then be checked by comparing the predicted diffusion potential with the actual electrical potential difference measured across the membrane.

As a specific example, we will use the Goldman equation to evaluate the membrane potential across the plasma membrane of *Nitella translucens*. The concentrations of K$^+$, Na$^+$, and Cl$^-$ in the external bathing solution and in its cytosol are given in Table 3-1. The ratio of the permeability of Na$^+$ to that of K$^+$, P_{Na}/P_K, is about 0.18 for *N. translucens*. Its plasma membrane is much less permeable to Cl$^-$ than to K$^+$, probably only 0.1 to 1% as much. Thus, for purposes of calculation, we will let P_{Cl}/P_K be 0.003. Using these relative permeability coefficients and the concentrations given in Table 3-1, the Goldman equation (Eq. 3.19) predicts the following membrane potential ($RT/F = 25.3$ mV at 20°C; Appendix I):

$$E_M = (25.3 \text{ mV}) \ln \frac{(P_K)(0.1 \text{ mM}) + (0.18\ P_K)(1.0 \text{ mM}) + (0.003\ P_K)(65 \text{ mM})}{(P_K)(119 \text{ mM}) + (0.18\ P_K)(14 \text{ mM}) + (0.003\ P_K)(1.3 \text{ mM})}$$
$$= -140 \text{ mV}$$

Thus we expect the cytosol to be electrically negative with respect to the external bathing solution, as is indeed the case. In fact, the measured value of the electrical potential difference across the plasma membrane of *N. translucens* is -138 mV at 20°C (Table 3-1). This close agreement between the observed electrical potential difference and that calculated from the Goldman equation supports the contention that the membrane potential is a diffusion potential. This can be checked by varying the external concentration of K$^+$, Na$^+$, and/or Cl$^-$ and seeing whether the membrane potential changes in accordance with Eq. (3.19).

As discussed previously, the different ionic concentrations on the two sides of a membrane help set up the passive ionic fluxes creating the diffusion potential. However, the actual contribution of a particular ionic species to E_M also depends on the ease with which that ion crosses the membrane. Based on the relative permeabilities and concentrations, the major contribution to the electrical potential difference across the plasma membrane of *N. translucens* comes from the K$^+$ flux, with Na$^+$ and Cl$^-$ playing secondary roles. If the Cl$^-$ terms are omitted from Eq. (3.19) (i.e., if P_{Cl} is set equal to zero), the calculated membrane potential is -153 mV, compared with -140 mV when Cl^- is included. This relatively small difference between the two potentials is a reflection of the relatively low permeability coefficient for chloride crossing the plasma membrane of *N. translucens*, so the Cl^- flux has less effect on E_M than does the K^+ flux. The relatively high permeability and the high concentration of K^+ ensure that it will have a major influence on the membrane potential. However, Cl^- must be included in the Goldman equation to predict accurately the membrane potential for many plant cells (Cl^- is less important for E_M of most animal cells).

Restriction to the three ions indicated has proved adequate for treating the diffusion potential across certain membranes, and the Goldman or constant field equation in the form of Eq. (3.19) has found widespread application. However, changes in the amount of Ca^{2+} or H^+ in the external medium cause some deviations from the predictions of Eq. (3.19) for the electrical potential differences across various plasma membranes. Thus the diffusion potential is influenced by the particular membrane being considered, and ions other than K^+, Na^+, and Cl^- may have to be included in specific cases. To allow for the influence of the passive flux of H^+ on E_M, for example, we could include $P_H c_H^o$ in the numerator of the logarithm in the Goldman equation (Eq. 3.19) and $P_H c_H^i$ in the denominator (note that a movement of H^+ in one direction has the same effect on E_M as a movement of OH^- in the other). In fact, H^+ can be the most important ion influencing the electrical potential difference across certain membranes, although its main effect on E_M may be via an electrogenic pump (discussed later), not the passive diffusion described by the Goldman equation. Such a pump actively transports H^+ from the cytosol out across the plasma membrane.

F. Donnan Potential

Another type of electrical potential difference is associated with immobile or fixed charges in some region adjacent to an aqueous phase containing small mobile ions—a *Donnan potential*. When a plant cell is placed in a KCl solution, for example, a Donnan potential arises between the interior of the cell wall and the bulk of the bathing fluid. The electrical potential difference arising from electrostatic interactions at such a solid–liquid interface can be regarded as a special type of diffusion potential. In particular, pectin and other macromolecules in the cell wall have a large number of carboxyl groups (—COOH) from which hydrogen ions dissociate. This gives the cell wall a net negative charge, as indicated in Chapter 1. Cations such as Ca^{2+} are electrostatically attracted to the negatively charged cell wall, and the overall effect is an exchange of H^+ for Ca^{2+} and other cations. Such an attraction of positively charged species to the cell wall can increase the local concentration of solutes to about 600 mol m^{-3} (0.6 M), so a greater osmotic pressure can exist in the cell wall than in surrounding aqueous solutions.

The region containing the immobile charges—such as dissociated carboxyl groups in the case of the cell wall—is generally referred to as the *Donnan phase* (Fig. 3-5). At equilibrium, a distribution of oppositely charged ions electrostatically attracted to these immobile charges occurs between the Donnan phase and the adjacent aqueous one. This sets up an ion concentration gradient, so a Donnan potential is created between the center of the Donnan phase and the bulk of the solution next to it. The sign of the electrical potential in the Donnan phase relative to the surrounding electrolyte solution is the same as the sign of the charge of the immobile ions. For example, because of the dissociated carboxyl groups, the electrical potential in the cell wall is negative with respect to an external solution. Membranes also generally act as charged Donnan phases. In addition, Donnan phases occur

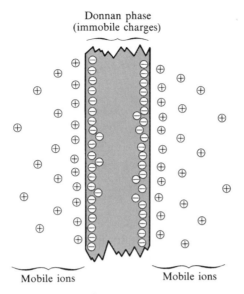

Figure 3-5. Spatial distribution of positively charged mobile ions ($\oplus$) occurring on either side of a Donnan phase in which are embedded immobile negative charges ($\ominus$).

in the cytoplasm, where the immobile charges are due to proteins and other large polymers (RNA and DNA) that have many carboxyl and phosphate groups from which protons can dissociate, leaving the macromolecules with a net negative charge. Because the net electrical charge attracts counterions, locally higher osmotic pressures can occur. Actually, all small ions in the immediate vicinity of a Donnan phase are affected, including H^+, i.e., the local pH.

Figure 3-5 illustrates a negatively charged, immobile Donnan phase with mobile, positively charged ions on either side. For example, layers containing cations are often formed in the aqueous solutions on each side of biological membranes, which generally act as Donnan phases with a net negative charge at physiological pH's. The electrical potential differences or Donnan potentials on either side of a membrane are in opposite directions and are assumed to cancel each other when a diffusion potential across the membrane is calculated. At equilibrium no net movement of the ions occurs, so the chemical potentials of each of the mobile ions (e.g., K^+, Na^+, Cl^-, and Ca^{2+}) have the same values up close to the Donnan phase as they do in the surrounding aqueous phase. The electrical potential difference (Donnan potential) can then be calculated by assuming constancy of the chemical potential. However, this is exactly the same principle that we used in deriving the Nernst potential (Eq. 3.5), $E_{Nj} = E^{II} - E^{I} = (RT/z_j F) \ln(a_j^I/a_j^{II})$, between two aqueous compartments. In fact, because the argument again depends on the constancy of the chemical potential, the equilibrium distribution of any ion from the Donnan phase to the aqueous phase extending away from the barrier must satisfy the Nernst potential (E_{Nj}) for that ion.

The Donnan potential can also be regarded as a special case of a diffusion potential. We can assume that the mobile ions are initially in the same region as the immobile ones. In time, some of the mobile ions will tend to diffuse away. This tendency, based on thermal motion, causes a slight charge separation, thus setting up an electrical potential difference between the Donnan phase and the bulk of the adjacent solution. For the case of a single species of mobile cations with the anions fixed in the membrane (both assumed to be monovalent), the diffusion potential across that part of the aqueous phase next to the membrane can be described by Eq. (3.10), $E^{II} - E^{I} = [(u_- - u_+)/(u_- + u_+)](RT/F) \ln (c^{II}/c^{I})$ that we derived for diffusion toward regions of lower chemical potential in a solution. Fixed anions have zero mobility ($u_- = 0$); hence $(u_- - u_+)/(u_- + u_+)$ here is $-u_+/u_+$, or -1. Equation (3.10) then becomes $E^{II} - E^{I} = -(RT/F) \ln (c^{II}/c^{I})$, which is the Nernst potential (Eq. 3.5) for monovalent cations $[-\ln(c^{II}/c^{I}) = \ln(c^{I}/c^{II})]$. Thus the Donnan potential can also be regarded as a diffusion potential occurring as the mobile ions tend to diffuse away from the immobile charges of opposite sign, which remain fixed in the Donnan phase.

III. Characteristics of Crossing Membranes

Entry of solutes into cells is crucial for many aspects of plant functioning, including nutrient uptake by roots, cell elongation during growth, and the opening of stomatal pores that allows gas exchange with the environment. Solutes cross membranes either passively, by diffusing toward regions of lower chemical potential, or actively, in which energy is needed for the movement. Specifically, *active transport* implies that energy derived from metabolic processes is used to move a solute across a membrane toward a region of higher chemical potential. There are three different aspects to this description of active transport: a supply of energy, movement, and an increase in chemical potential. Although it is not absolutely necessary, the expression "active transport" has conventionally been restricted to the case of movement in the energetically uphill direction, and we will follow such a restriction here.

A difference in chemical potential of a certain species across a membrane does not necessarily imply that active transport of that species is occurring. For example, if the solute cannot penetrate the membrane, the solute is unable to attain equilibrium across it, and μ_j would not be expected to be the same on the two sides. For ions moving across some membrane, the ratio of the influx to the efflux of that ion provides information on whether or not active transport is taking place. A simple but often effective approach for determining whether fluxes are active or passive is to remove possible energy sources. For photosynthesizing plant tissue, this can mean comparing the fluxes in the light with those in the dark. (In addition, we should check whether the permeability of the membrane changes as illumination changes.) Compounds or treatments that disrupt metabolism can also be useful for ascertaining whether metabolic energy is being used for the active transport of various solutes.

We will begin by showing how active transport can directly affect membrane potentials. We then compare the temperature dependencies of metabolic reactions with those for diffusion processes across a barrier to show that a marked enhancement of solute influx caused by increasing the temperature does not necessarily indicate that active transport is taking place. Next we will consider a much more reliable criterion for deciding whether fluxes are passive or not—namely, the Ussing–Teorell, or flux ratio, equation. We then examine a specific case in which active transport is involved, calculate the energy required, and finally speculate on why K^+ and Cl^- are actively transported into plant cells and Na^+ is transported out.

A. Electrogenicity

One of the possible consequences of actively transporting a certain ionic species into a cell or organelle is the development of an excess of electrical charge inside. If the active transport involves an accompanying ion of opposite charge or an equal release of a similarly charged ion, the total charge in the cell or organelle is unaffected. However, if the charge of the actively transported ion is not directly compensated for, the process is *electrogenic*, i.e., it tends to generate an electrical potential difference across the membrane. An electrogenic uptake of an ion that produces a net transport of charge into some cell or organelle will thus affect its membrane potential. We can appreciate this effect by referring to Eq. (3.1) ($Q = C\Delta E$), which indicates that the difference in electrical potential across a membrane, ΔE, equals Q/C, where Q is the net charge enclosed within the cell or organelle and C is the membrane capacitance. The initial movement of net charge across a membrane by active transport leads to a fairly rapid change in the potential difference across the membrane.

To be specific, we will consider a spherical cell of radius r into which an electrogenic flux of chloride occurs by active transport, $J_{a.t.Cl}^{in}$. The amount of charge transported in time t across the surface of the sphere (area $= 4\pi r^2$) is $J_{a.t.Cl}^{in} 4\pi r^2 t$. This active uptake of Cl^- increases the internal concentration of negative charge by the amount moved in divided by the cellular volume, or $J_{a.t.Cl}^{in} 4\pi r^2 t / (4\pi r^3/3)$, which is $3 J_{a.t.Cl}^{in} t/r$. Let us suppose that $J_{a.t.Cl}^{in}$ is 10 nmol m^{-2} s^{-1} and that the cell has a radius of 30 μm. In 1 s the concentration of Cl^- actively transported in is

$$c_{a.t.Cl} = \frac{3 J_{a.t.Cl}^{in} t}{r} = \frac{(3)(10 \times 10^{-9} \text{ mol m}^{-2} \text{ s}^{-1})(1 \text{ s})}{(30 \times 10^{-6} \text{ m})}$$

$$= 1.0 \times 10^{-3} \text{ mol m}^{-3}$$

Assuming a membrane capacitance of 10 mF m^{-2}, we calculated (Section IB) that such a cell has 1.0×10^{-3} mol m^{-3}(1 μM) uncompensated negative charge when the interior is 100 mV negative with respect to the external solution. If no change were to take place in the other ionic fluxes, the electrogenic uptake of Cl^- into this cell would cause its interior to become more negative at the rate of 100 mV s^{-1}. The non-steady-state charging of the membrane capacitance is indeed a rapid process.

The initiation of an electrogenic process causes an adjustment of the passive ionic fluxes across the membrane. In particular, the net charge actively brought in is soon electrically compensated by appropriate passive movements of that ion and other ions into or out of the cell. The actual electrical potential difference across the membrane then results from the diffusion potential caused by these new passive fluxes plus a steady-state contribution from the electrogenic process involving the active transport of some charged species. We can represent the electrical potential difference generated by the active transport of species j, $E_{a.t.j}$, as

$$E_{a.t.j} = z_j F J_{a.t.j} R_j^{\text{memb}} \tag{3.20}$$

where z_j is the charge number of species j, F is Faraday's constant, $J_{a.t.j}$ is the flux density of species j caused by active transport, and R_j^{memb} is the membrane resistance for the specific pathway along which species j is actively transported. $F J_{a.t.j}$ in Eq. (3.20)—which is a form of Ohm's law—is the charge flux density and can have units of (C mol^{-1}) (mol m^{-2} s^{-1}), or C m^{-2} s^{-1}, which is ampere m^{-2}, or current per unit area. R_j^{memb} can be expressed in ohm m^2, so that $z_j F J_{a.t.j} R_j^{\text{memb}}$ in Eq. (3.20) can have units of ampere ohm, i.e., V, the proper unit for electrical potentials.

For many plant cells, R_H^{memb} is 2–20 ohm m^2 for the active transport of H$^+$ ions (protons) out across the plasma membrane (Spanswick, 1981). How large is $E_{a.t.H}$ given by Eq. (3.20) for such a proton "pump"? The electrical potential difference created by the active transport of 20 nmol m^{-2} s^{-1} of H$^+$ out across a typical membrane resistance for H$^+$ of 10 ohm m^2 is

$$E_{a.t._H} = (1)(9.65 \times 10^4 \text{ C mol}^{-1})(-20 \times 10^{-9} \text{ mol m}^{-2}\text{s}^{-1})(10 \text{ ohm m}^2)$$
$$= -1.9 \times 1.0^{-2} \text{ V } (-19 \text{ mV})$$

Because the active transport of H$^+$ is out of the cell in the case considered, $J_{a.t._H}$ is negative, which leads to a decrease in the membrane potential ($E_{a.t._H} < 0$). Measurements in the presence and the absence of metabolic inhibitors have indicated that an electrogenic efflux of H$^+$ can decrease E_M of plant cells by 50 mV or more. On the other hand, R_j^{memb} for ions such as K$^+$, Na$^+$, and Cl$^-$ generally is only 0.1–1 ohm m^2 (Hope, 1971), so active transport of these ions does not generate much of an electrical potential difference. In any case, the actual electrical potential difference across a membrane can be obtained by adding the potential difference caused by active transport of uncompensated charge (Eq. 3.20) to that caused by passive fluxes (predicted by the Goldman equation, Eq. 3.19).

B. Boltzmann Energy Distribution and Q_{10}, a Temperature Coefficient

Most metabolic reactions are markedly influenced by temperature, whereas processes such as light absorption are essentially insensitive to temperature. What temperature dependence should we expect for diffusion? Can we decide whether the movement of some solute into a cell is by active transport or by passive diffusion once we know how the fluxes depend on temperature? To answer such questions, we need an expression describing the distribution of energy among molecules as a function of temperature to determine what fraction of the molecules has the requisite energy for a particular process. In aqueous solutions the relevant energy is generally the kinetic energy of motion of the molecules involved. Hence, we will begin by relating the distribution of kinetic energy among molecules to temperature. The topics introduced here are important for a basic understanding of many aspects of biology—from biochemical reactions to the consequences of light absorption.

Very few molecules possess extremely large kinetic energies. In fact, the probability that a molecule has the kinetic energy E decreases exponentially as E increases. The precise statement of this is the *Boltzmann energy distribution*, which describes the frequency with which specific kinetic energies are possessed by molecules at equilibrium at temperature T:

$$n(E) = n_{\text{total}}e^{-E/kT} \qquad \text{molecule basis} \qquad (3.21a)$$

where $n(E)$ is the number of molecules possessing an energy of E or more out of the total number of molecules, n_{total}, and k is Boltzmann's constant. The quantity $e^{-E/kT}$, which by Eq. (3.21a) equals $n(E)/n_{\text{total}}$, is the *Boltzmann factor*. Equation (3.21a) and Fig. 3-6 indicate that the number of molecules with an energy of zero or greater, $n(0)$, equals $n_{\text{total}}e^{-0/kT}$ or $n_{\text{total}}e^{-0}$, which is n_{total}, the total number of molecules present.

Because of collisions based on thermal motion, energy is continually being gained or lost by individual molecules in a random fashion. Hence, a wide range of kinetic energies is possible, although very high energies are less probable (see Eq. 3.21a). As the temperature is raised, not only does the average energy per molecule become higher but also the relative number of molecules in the "high-energy tail" of the exponential Boltzmann distribution increases substantially (Fig. 3-6).

In this text we will use two analogous sets of expressions: (1) molecule, mass of molecule, photon, electronic charge, k, and kT; and (2) mole, molar mass, mole of photons, Faraday's constant, R, and RT (see Appendix I for numerical values of k, kT, R, and RT).

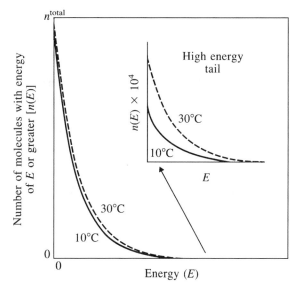

Figure 3-6. Boltzmann energy distributions at 10°C (solid line) and 30°C (dashed line). The inset is a continuation of the right-hand portion of the graph with the scale of the abscissa (E) unchanged and that of the ordinate [$n(E)$] expanded by 10^4. The difference between the two curves is extremely small, except at high energies; although very few molecules are in this "high-energy tail," there are many more such molecules at the higher temperature.

A quantity in the second set, which is more appropriate for most of our applications, is Avogadro's number $N(6.022 \times 10^{23})$ times the corresponding quantity in the first set. In particular, to change the Boltzmann distribution from a molecule to a mole basis, we multiply Boltzmann's constant k (energy molecule^{-1} K^{-1}) by Avogadro's number N (molecules mole^{-1}), which gives the gas constant R (energy mole^{-1}K^{-1}), i.e., $R = kN$. If $n(E)$ and n_{total} are numbers of moles and E is energy per mole, we simply replace k in the Boltzmann energy distribution (Eq. 3.21a) by R:

$$n(E) = n_{total}e^{-E/RT} \qquad \text{mole basis} \qquad (3.21b)$$

For diffusion across a membrane, the appropriate Boltzmann energy distribution indicates that the number of molecules with a kinetic energy per mole of U or greater resulting from velocities in some particular direction is proportional to $\sqrt{T}e^{-U/RT}$ (Davson and Danielli, 1952). A minimum kinetic energy (U_{min}) is often necessary to diffuse past some barrier or to cause a specific reaction. In such circumstances, any molecule with a kinetic energy of U_{min} or greater has sufficient energy for the particular process. For the Boltzmann energy distribution appropriate to this case, the number of such molecules is proportional to $\sqrt{T}e^{-U_{min}/RT}$. (These expressions having the factor $\sqrt{T}$ apply to diffusion in one dimension, e.g., for molecules diffusing across a membrane.) At a temperature 10°C higher, the number is proportional to $\sqrt{(T+10)}e^{-U_{min}/[R(T+10)]}$. The ratio of these two quantities is the Q_{10}, or *temperature coefficient*, of the process:

$$Q_{10} = \frac{\text{rate of process at } T + 10°C}{\text{rate of process at } T} = \sqrt{\frac{T+10}{T}}e^{10U_{min}/[RT(T+10)]} \qquad (3.22)$$

To obtain the form in the exponent of e in Eq. (3.22), we note that

$$\frac{-U_{min}}{R(T+10)} + \frac{U_{min}}{RT} = \frac{-U_{min}T + U_{min}(T+10)}{RT(T+10)} = \frac{10U_{min}}{RT(T+10)}$$

A Q_{10} near 1 is characteristic of passive processes with no energy barrier to surmount, i.e., where $U_{min} = 0$. On the other hand, most enzymatic reactions take place only when the reactants have a considerable kinetic energy, so such processes tend to be quite sensitive to temperature. A value of 2 or greater for Q_{10} is often considered to indicate the involvement of metabolism, as occurs for active transport of a solute into a cell or organelle. However, Eq. (3.22) indicates that any process having an appreciable energy barrier can have a large temperature coefficient.

The Q_{10} for a particular process indicates the minimum kinetic energy required (U_{min}), and vice versa. A membrane often represents an appreciable energy barrier for the diffusion of charged solutes—U_{min} for ions crossing passively can be 50 kJ mol^{-1} (12 kcal mol^{-1} or 0.52 eV molecule^{-1}; Stein, 1986). By Eq. (3.22), this leads to the following temperature coefficient at 20°C:

$$Q_{10} = \sqrt{\frac{(303 \text{ K})}{(293 \text{ K})}} e^{(10 \text{ K})(50 \times 10^3 \text{ J mol}^{-1})/[(8.3143 \text{ J mol}^{-1} \text{ K}^{-1})(293 \text{ K})(303 \text{ K})]}$$

$$= 1.02e^{0.68} = 2.01$$

Therefore, the passive uptake of this ion doubles with only a 10°C increase in temperature, so a passive process can have a rather high Q_{10} if there is an appreciable energy barrier. Thus a large Q_{10} for ion uptake does not necessarily indicate active transport.

A kinetic energy of 50 kJ mol^{-1} or greater is possessed by only a small fraction of the molecules (Eq. 3.21b). For instance, at 20°C the Boltzmann factor then is

$$e^{-E/RT} = e^{-(50 \times 10^3 \text{ J mol}^{-1})/[(8.3143 \text{ J mol}^{-1} \text{ K}^{-1})(293 \text{ K})]} = 1.2 \times 10^{-9}$$

As T is raised, the fraction of molecules in the high-energy part of the Boltzmann distribution increases greatly (Fig. 3-6). Many more molecules then have the requisite kinetic energy, U_{min}, and consequently can take part in the process considered. In particular, at 30°C the Boltzmann factor here becomes 2.4×10^{-9}. Thus the Boltzmann factor for a U_{min} of 50 kJ mol^{-1} essentially doubles for a 10°C rise in temperature, consistent with our Q_{10} calculation for this case.

C. Activation Energy and Arrhenius Plots

An energy barrier requiring a minimum energy U_{min} to cross is related to the concept of *activation energy*, which refers to the minimum amount of energy necessary for some reaction to take place. In the case of a membrane, the required kinetic energy (U_{min}) corresponds to the activation energy for crossing the barrier. We can evaluate U_{min} by determining how the number of molecules diffusing across the membrane varies with temperature, e.g., by invoking Eq. (3.22). For a chemical reaction, we can also experimentally determine how the process is influenced by temperature. If we represent the activation energy per mole by A, the rate constant (see Appendix III, Eq. III.1) for such a reaction varies with temperature

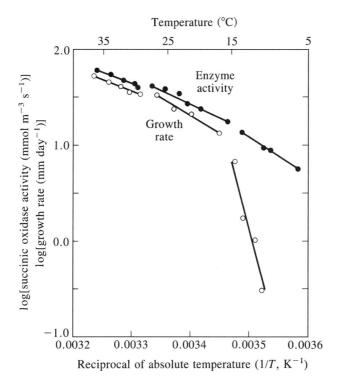

Figure 3-7. Arrhenius plots of mitochondrial succinate oxidase activity (●) and growth rate (○) of the hypocotyl plus radicle of *Vigna radiata*. [Data are from Raison and Chapman (1976); used by permission.]

as follows:

$$\text{Rate constant} = Be^{-A/RT} \tag{3.23}$$

where B is often a constant.[6]

Equation (3.23) is the Arrhenius equation. It was originally proposed on experimental grounds by Svante Arrhenius at the end of the 19th century and subsequently interpreted theoretically. A plot of the logarithm of the rate constant (or rate of some reaction) versus $1/T$ is commonly known as an *Arrhenius plot* (Fig. 3-7); by Eq. (3.23), ln (rate constant) = $\ln B - A/RT$. The slope of an Arrhenius plot $(-A/R)$ can hence be used to determine the activation energy. An enzyme greatly increases the rate of the reaction that it catalyzes by reducing the value of the activation energy needed (see Eq. 3.23), i.e., many more molecules then have enough energy to get over the energy barrier separating the reactants from the products. For instance, the activation energy A for the hydrolysis of sucrose to glucose plus fructose at 37°C is about 107 kJ mol^{-1} in the absence of the enzyme invertase but only 34 kJ mol^{-1} when the enzyme catalyzes the reaction. Such a lowering of A by invertase thus increases the rate constant (see Eq. 3.23) by $e^{\Delta A/RT}$ or $e^{(107-34 \text{ kJ mol}^{-1})/[(8.314 \text{ J mol}^{-1}\text{K}^{-1})(310 \text{ K})]}$, which equals 1.9×10^{12}. Thus by lowering the

---

[6] B can depend on temperature. For diffusion in one dimension, the number of molecules with an energy of at least U_{min} per mole is proportional to $\sqrt{T}e^{-U_{min}/RT}$; B is then proportional to $\sqrt{T}$.

activation energy, the enzyme greatly speeds up the reaction. Many reactions of importance in biochemistry have large values for A, even when catalyzed by enzymes, and are therefore extremely sensitive to temperature.

Let us next consider the activation energies for certain cases of diffusion. For diffusion in water at $20°C$, U_{min} is 17–21 kJ mol^{-1} for solutes with molecular weights from 20 to over 1000 (Stein, 1986). Using Eq. (3.22) we calculate that the Q_{10} for such diffusion is about 1.3. This is a substantial Q_{10}, which is a consequence of the appreciable thermal energy required for a solute to move through the semicrystalline order in aqueous solutions resulting from hydrogen bonds (see Chapter 2). U_{min} for the passive efflux of K$^+$ from many cells is about 60 kJ mol^{-1} near $20°C$, which corresponds to a Q_{10} of 2.3. We again conclude that a purely passive process, such as diffusion across membranes, can have a substantial temperature dependence.

Arrhenius plots have been used to identify processes that change markedly over the range of temperatures that injure chilling-sensitive plants such as corn, cotton, cucumber, rice, soybean, and tomato. Such plants are often severely injured by exposure to temperatures that are low but above freezing, e.g., near $10°C$; prolonged exposure can even result in death. Visible symptoms include wilting, surface pitting of the leaves, and loss of chlorophyll. Damage at the cellular level caused by chilling can be manifested by loss of cytoplasmic streaming, metabolic dysfunction leading to the accumulation of toxic products, and enhanced membrane permeability. Figure 3-7 suggests that the temperature dependence of the growth rate for the chilling-sensitive *Vigna radiata* (mung bean) may change near 15 and $28°C$. The activation energy for a mitochondrial enzyme (succinate oxidase) also may change at these two temperatures (Fig. 3-7), as apparently does the organization of membrane lipids in both mitochondria and chloroplasts. In fact, transitions in the physical properties of membranes (such as a change to a more fluid state) may underlie the changes in slope seen in Fig. 3-7. Changes in the membranes can affect the catalytic properties of enzymes located in them, the permeability of solutes, and the general regulation of cellular metabolism, which in turn affects plant growth.[7]

D. Ussing–Teorell Equation

We now consider ways of distinguishing between active and passive fluxes that are more reliable than determining Q_{10}'s. One of the most useful physicochemical criteria for deciding whether a particular ionic movement across a membrane is active or passive is the application of the Ussing–Teorell, or flux ratio, equation. For ions moving passively, this expression shows how the ratio of the influx to the efflux depends on the internal and the external concentrations of that species and on the electrical potential difference across the membrane. If the Ussing–Teorell equation is satisfied, passive movements can account for the observed flux ratio, so active transport of the ions need not be invoked. We can readily derive this expression by considering how the influx and the efflux could each be determined experimentally, as the following arguments indicate.

---

[7] Apparent breaks in Arrhenius plots for enzyme activity (e.g., Fig. 3-7) most directly reflect changes in protein activation energies, not membrane phase transitions, which are best identified using x-ray diffraction or microcalorimetry. No change in the slope of an Arrhenius plot can occur at a phase transition (e.g., the diffusion of a small solute may be unaffected by the rearrangement of membrane lipids), and sharp breaks may occur in the absence of membrane phase transitions when temperature affects the conformation of a protein. Complicated processes such as growth involve many enzyme-catalyzed reactions and usually lead to continuous changes with temperature that cannot be analyzed with a single rate constant.

For measuring the unidirectional inward component, or influx, of a certain ion (J_j^{in}), the plant cell or tissue can be placed in a solution containing a radioactive isotope of species j. Initially, none of the radioisotope is inside the cells, so the internal specific activity for this isotope equals zero at the beginning of the experiment. (As for any radioisotope study, only some of the molecules of species j are radioactive. This particular fraction is known as the *specific activity*, and it must be determined for both c_j^o and c_j^i in the current experiment.) Because originally none of the radioisotope is inside, its initial unidirectional outward component, or efflux (J_j^{out}), is zero, and the initial net flux density (J_j) of the isotope indicates J_j^{in}. From Eq. (3.15), this influx of the radioisotope can be represented by $(K_j u_j z_j FE_M / \Delta x)[1/(e^{z_j FE_M / RT} - 1)]c_j^o$. After the isotope has entered the cell, some of it will begin coming out. Therefore, only the initial net flux will give an accurate measure of the influx of the radioisotope of species j.

Once the radioactivity has built up inside to a substantial level, we may remove the radioisotope from the outside solution. The flux of the isotope is then from inside the cells to the external solution. In this case, the specific activity for c_j^o equals zero, and c_j^i determines the net flux of the radioisotope. By Eq. (3.15), this efflux of the isotope of species j differs in magnitude from the initial influx only by having the factor c_j^o replaced by $c_j^i e^{z_j FE_M / RT}$, the quantities in the first two parentheses remaining the same. The ratio of these two flux densities—each of which can be separately measured—takes the following relatively simple form:

$$\frac{J_j^{in}}{J_j^{out}} = \frac{c_j^o}{c_j^i \, e^{z_j FE_M / RT}} \tag{3.24}$$

Equation (3.24) was independently derived by both Hans Ussing and Torston Teorell in 1949 and is known as the Ussing–Teorell equation or the flux ratio equation. It is strictly valid only for ions moving passively without interacting with other substances that may also be moving across a membrane. Our derivation uses Eq. (3.15), which gives the passive flux of some charged species across a membrane in response to differences in its chemical potential. Equation (3.15) considers only one species at a time, so possible interactions between the fluxes of different species are not included in Eq. (3.24). The Ussing–Teorell equation can thus be used to determine whether the observed influxes and effluxes are passive (i.e., responses to the chemical potentials of the ions on the two sides of a membrane), or whether additional factors such as interactions between species or active transport are involved. For example, when active transport of species j into a cell is taking place, J_j^{in} is the passive unidirectional flux density (i.e., the one predicted by Eq. 3.24) plus the influx due to active transport.

The ratio of the influx of species j to its efflux, as given by Eq. (3.24), can be related to the difference in its chemical potential across a membrane. This difference causes the passive flux ratio to differ from 1, and we will use it to estimate the minimum amount of energy needed for active transport of that ionic species across the membrane. After taking logarithms of both sides of the Ussing–Teorell equation (Eq. 3.24) and multiplying by RT, we obtain the following equalities:

$$\begin{aligned} RT \ln \frac{J_j^{in}}{J_j^{out}} &= RT \ln \frac{c_j^o}{c_j^i} - z_j FE_M \\ &= RT \ln a_j^o + z_j FE^o - RT \ln a_j^i - z_j FE^i \\ &= \mu_j^o - \mu_j^i \end{aligned} \tag{3.25}$$

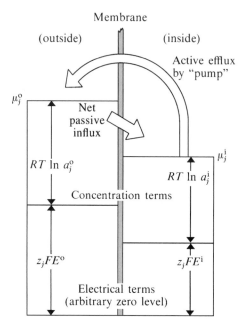

Figure 3-8. Diagram illustrating the situation for an ion not in equilibrium across a membrane. Because $\mu_j^o > \mu_j^i$, there is a net passive flux density into the cell. In the steady state, this net influx is balanced by an equal efflux caused by active transport of species j out of the cell.

where the membrane potential E_M has been replaced by $E^i - E^o$, in keeping with our previous convention. The derivation is restricted to the case of constant $\gamma_j (\gamma_j^o = \gamma_j^i)$, so $\ln (c_j^o/c_j^i) = \ln(\gamma_j^o c_j^o/\gamma_j^i c_j^i)$, or $\ln (a_j^o/a_j^i)$, which is $\ln a_j^o - \ln a_j^i$. Finally, the $\bar{V}_j P$ term in the chemical potential is ignored for these charged species (actually, we need only assume that $\bar{V}_j P^o = \bar{V}_j P^i$) so that μ_j then equals $\mu_j^* + RT \ln a_j + z_j FE$, where μ_j^* is the same on the two sides of the membrane. Thus $\mu_j^o - \mu_j^i$ is $RT \ln a_j^o + z_j FE^o - RT \ln a_j^i - z_j FE^i$, as indicated in Eq. (3.25).

A difference in chemical potential of species j across a membrane causes the ratio of the passive flux densities to differ from 1 (Fig. 3-8), a conclusion that follows directly from Eq. (3.25). When $\mu_j^o = \mu_j^i$, the influx balances the efflux, so no net passive flux density of species j occurs across the membrane ($J_j = J_j^{in} - J_j^{out}$ by Eq. 3.15). This condition ($\mu_j^o = \mu_j^i$) is also described by Eq. (3.4), which was used to derive the Nernst equation. In fact, the electrical potential difference across a membrane when the chemical potentials are equal across it is the Nernst potential, as given by Eq. (3.5): $E_{N_j} = (RT/z_j F) \ln (a_j^o/a_j^i)$. Thus when $E_M = E_{N_j}$ for some species, $J_j^{in} = J_j^{out}$, and no net passive flux density of that ion is expected across the membrane, nor is any energy expended in moving the ion from one side of the membrane to the other. When Eqs. (3.24) and (3.25) are not satisfied for some species, such ions are not moving across the membrane passively, or perhaps not moving independently from other fluxes. One way that this may occur is for the various fluxes to be interdependent, a condition describable by irreversible thermodynamics. Another way is through active transport of the ions, whereby energy derived from metabolism is used to move solutes to regions of higher chemical potential.

E. Example of Active Transport

The previous criteria for deciding whether active transport of certain ions is taking place can be illustrated by using data obtained with the internodal cells of *Nitella translucens* for which all the parameters in the Ussing–Teorell equation have been measured for Na^+, K^+, and Cl^- (Table 3-1). For experimental purposes, this freshwater alga is often placed in a dilute aqueous solution containing 1 mM NaCl, 0.1 mM KCl, plus 0.1 mM $CaCl_2$ (1 mM = 1 mol m^{-3}), which establishes the values for all three c_j^o's (this solution is not unlike the pond water in which *N. translucens* grows and is referred to as "artificial pond water"). The concentrations of Na^+, K^+, and Cl^- measured in the cytosol, the c_j^i's, are given in the third column of Table 3-1. Assuming that activities can be replaced by concentrations, we can calculate the Nernst potentials across the plasma membrane from these concentrations using Eq. (3.5), $E_{N_j} = (58.2/z_j)\log(c_j^o/c_j^i)$ in mV (the numerical factor is 58.2 because the measurements were at 20°C). We thus get

$$E_{N_{Na}} = \frac{(58.2\ \text{mV})}{(1)} \log \frac{(1.0\ \text{mM})}{(14\ \text{mM})} = -67\ \text{mV}$$

Similarly, E_{N_K} is -179 mV and $E_{N_{Cl}}$ is 99 mV (Table 3-1). Direct measurement of the electrical potential difference across the plasma membrane (E_M) gives -138 mV, as indicated earlier in discussing the Goldman equation. Because E_{N_j} differs from E_M in all three cases, none of these ions is in equilibrium across the plasma membrane of *N. translucens*.

A difference between E_M and E_{N_j} for a particular ion indicates departure from equilibrium for that ionic species; it also tells us in which compartment μ_j is higher. Specifically, if the membrane potential is algebraically more negative than the calculated Nernst potential, the chemical potential in the inner aqueous phase (here the cytosol) is lower for a cation (Fig. 3-8) but higher for an anion, compared with the values in the external solution (consider the effect of z_j in the electrical term of the chemical potential, $z_j FE$). Because E_M (-138 mV) is more negative than $E_{N_{Na}}$ (-67 mV), Na^+ is at a lower chemical potential in the cytosol than outside in the external solution. Analogously, we find that K^+ (with $E_{N_K} = -179$ mV) has a higher chemical potential inside, and Cl^- (with $E_{N_{Cl}} = 99$ mV) has a much higher chemical potential inside. If these ions can move across the plasma membrane, this suggests an active transport of K^+ and Cl^- into the cell and an active extrusion of Na^+ from the cell, as is schematically indicated in Fig. 3-9.

We can also consider movement of specific ions into and out of *N. translucens* in terms of the Ussing–Teorell equation to help determine whether active transport needs to be invoked to explain the fluxes. The Ussing–Teorell equation predicts that the quantity on the right-hand side of Eq. (3.24) $(c_j^o/c_j^i e^{z_j FE_M/RT})$ equals the ratio of influx to efflux of the various ions, if the ions are moving passively in response to gradients in their chemical potential. Using the values given in Table 3-1 and noting that RT/F is 25.3 mV at 20°C (Appendix I), we find that this ratio for Na^+ is

$$\frac{J_{Na}^{in}}{J_{Na}^{out}} = \frac{(1.0\ \text{mM})}{(14\ \text{mM})e^{(1)(-138\ \text{mV})/(25.3\ \text{mV})}} = 17$$

Similarly, the expected flux ratio is 0.20 for K^+ and 0.000085 for Cl^- (values given in Table 3-1, column 5). However, the observed influxes in the light equal the effluxes for each of these three ions (Table 3-1, columns 6 and 7). Equal influxes and effluxes are quite reasonable for mature cells of *N. translucens*, which are in a steady-state condition. On the

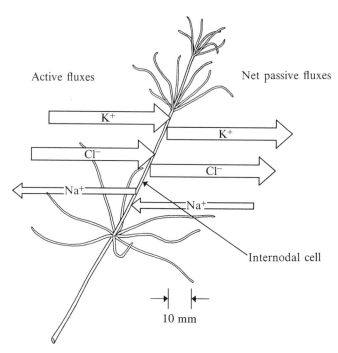

Figure 3-9. For the steady-state condition in the light the three active fluxes across the plasma membrane of the large internodal cell of *Nitella translucens* are balanced by net passive K^+ and Cl^- effluxes and a net passive Na^+ influx.

other hand, if $J_j^{in} = J_j^{out}$, the flux ratios given by Eq. (3.24) are not satisfied for Na^+, K^+, or Cl^-. In fact, active transport of K^+ and Cl^- in and Na^+ out accounts for the marked deviations from the Ussing–Teorell equation for *N. translucens*, as summarized in Fig. 3-9.

As mentioned earlier, another approach for studying active transport is to remove the supply of energy. In the case of *N. translucens*, cessation of illumination causes a large decrease in the Na^+ efflux, the K^+ influx, and the Cl^- influx (last two columns of Table 3-1). However, these are the three fluxes that are toward regions of higher chemical potential for the particular ions involved, so we may reasonably expect all three to be active. On the other hand, some fluxes remain essentially unchanged upon placing the cells in the dark (values in the last two columns of Table 3-1 refer to the fluxes soon after extinguishing the light, not the steady-state fluxes). For instance, the Na^+ influx and the K^+ efflux are initially unchanged when the *Nitella* cells are transferred from the light to the dark, i.e., these unidirectional fluxes toward lower chemical potentials do not depend on energy derived from photosynthesis.

The passive diffusion of ions toward regions of lower chemical potential helps create the electrical potential difference across a membrane, and active transport is needed to maintain the asymmetrical ionic distributions that sustain the passive fluxes. Thus the passive and the active fluxes are interdependent in the ionic relations of cells, and both are crucial for the generation of the observed diffusion potentials. Moreover, active and passive fluxes can occur simultaneously in the same direction. For example, we calculated that J_{Na}^{in} should equal 17 times J_{Na}^{out}, if both flux densities were passive ones obeying the Ussing–Teorell equation. Because J_{Na}^{in} is passive and equal to 5.5 nmol m^{-2} s^{-1}, we expect a passive efflux of Na^+ equaling $(5.5)/(17)$, or 0.3 nmol m^{-2} s^{-1}. The active component of the Na^+ efflux in the light may be $5.5 - 0.3$, or 5.2 nmol m^{-2} s^{-1}. At cessation of illumination, J_{Na}^{out} decreases from 5.5 to 1.0 nmol m^{-2} s^{-1} (Table 3-1). Extinguishing the light removes photosynthesis as a possible energy source for active transport, but respiration could still supply energy in the dark. This can explain why J_{Na}^{out} in the dark does not decrease all the way to 0.3 nmol m^{-2} s^{-1}, the value predicted for the passive efflux.

F. Energy for Active Transport

Suppose that the chemical potential of some species is higher outside than inside a cell, as is illustrated in Fig. 3-8. The minimum amount of energy needed to transport a mole of that species from the internal aqueous phase on one side of some membrane to the external solution on the other side is the difference in chemical potential of that solute across the membrane, $\mu_j^o - \mu_j^i$ (here $\mu_j^o > \mu_j^i$). As we noted in considering Eq. (3.25), the quantity $\mu_j^o - \mu_j^i$ for ions is $RT \ln (a_j^o/a_j^i) - z_j FE_M$. Because the Nernst potential E_{N_j} is $(RT/z_j F) \ln (a_j^o/a_j^i)$ (Eq. 3.5), we can express the difference in chemical potential across the membrane as

$$\mu_j^o - \mu_j^i = z_j FE_{N_j} - z_j FE_M \qquad (3.26a)$$
$$= z_j F(E_{N_j} - E_M)$$

or

$$\mu_j^i - \mu_j^o = z_j F(E_M - E_{N_j}) \qquad (3.26b)$$

Using E_M and the Nernst potentials of Na^+, K^+, and Cl^- for *N. translucens* (Table 3-1), we can calculate $z_j(E_M - E_{N_j})$ for transporting these ions across the plasma membrane of this alga. Such a quantity is $(+1)[(-138$ mV$) - (-67$ mV$)]$ or -71 mV for Na^+, 41 mV for K^+, and 237 mV for Cl^-. By Eq. (3.26), these values for $z_j(E_M - E_{N_j})$ mean that Na^+ is at a higher chemical potential in the external bathing solution, whereas K^+ and Cl^- are at higher chemical potentials inside the cell, as we previously concluded (see also Figs. 3-8 and 3-9). By Eq. (3.26a), the minimum energy required for actively transporting or "pumping" Na^+ out across the plasma membrane of the *Nitella* cell is

$$\mu_{Na}^o - \mu_{Na}^i = (1)(9.65 \times 10^{-2} \text{ kJ mol}^{-1} \text{ mV}^{-1})[-67 \text{ mV} - (-138 \text{ mV})]$$
$$= 6.9 \text{ kJ mol}^{-1}$$

Similarly, to pump K^+ inward requires 4.0 kJ mol^{-1}. The active extrusion of Na^+ from certain algal cells is linked to the active uptake of K^+, with ATP being implicated as the energy source for this coupled exchange process. As we will discuss in Chapter 6, the hydrolysis of ATP under biological conditions usually releases at least 40 kJ mol^{-1} (10 kcal mol^{-1}). For a *Nitella* cell this is more than sufficient energy per mole of ATP hydrolyzed to pump 1 mol of Na^+ out and 1 mol of K^+ in. The transport of Cl^- inward takes a minimum of 23 kJ mol^{-1} according to Eq. (3.26), which is a large amount of energy. Although the mechanism for actively transporting Cl^- into *Nitella* or other plant cells is not fully understood at the molecular level, exchanges with OH^- or cotransport with H^+ apparently are involved. The involvement of proton chemical potential differences across membranes in chloroplast and mitochondrial bioenergetics will be discussed in Chapter 6.

G. Speculation on Active Transport

The active uptake of K^+ and Cl^- together with an active extrusion of Na^+, as for *Nitella*, occurs for many plant cells. We might ask, Why does a cell actively transport K^+ and Cl^- in and Na^+ out? Although no definitive answer can be given to such a question, we shall speculate on possible reasons, based on the principles that we have been considering.

Let us imagine that a membrane-bounded cell containing negatively charged proteins is placed in an NaCl solution, possibly reflecting primeval conditions when life on earth originated. When Na^+ and Cl^- are both in equilibrium across a membrane, $E_M = E_{N_{Cl}}$ and $E_{N_{Na}}$. Using concentrations (instead of activities) in Eq. (3.5), $\log (c_{Na}^o/c_{Na}^i)$ then equals $-\log (c_{Cl}^o/c_{Cl}^i)$, or $c_{Na}^o/c_{Na}^i = c_{Cl}^i/c_{Cl}^o$, and hence $c_{Na}^o c_{Cl}^o = c_{Na}^i c_{Cl}^i$. For electroneutrality in an external solution containing only NaCl, $c_{Na}^o = c_{Cl}^o$. Because $a^2 = bc$ implies that $2a \leq b + c$,[8] we conclude that $c_{Na}^o + c_{Cl}^o \leq c_{Na}^i + c_{Cl}^i$. However, the proteins, which cannot diffuse across the membrane, also make a contribution to the internal osmotic pressure (Π^i), so $\Pi^i > \Pi^o$. When placed in an NaCl solution, water therefore tends to enter such a membrane-bounded cell containing proteins, causing it to swell without limit. An outwardly directed active transport of Na^+ would lower c_{Na}^i and thus prevent excessive osmotic swelling of primitive cells.

An energy-dependent uptake of solutes into a plant cell tends to increase Π^i, which can raise P^i. This higher internal hydrostatic pressure favors cell enlargement and consequently cell growth. For a plant cell surrounded by a cell wall, we might therefore expect an active transport of some species into the cell, e.g., Cl^-. (Animal cells do not have to push against a cell wall to enlarge and do not generally have an active uptake of Cl^-.) Enzymes have evolved that operate efficiently when exposed to relatively high concentrations of K^+ and Cl^-. In fact, many require K^+ for their activity, so an inwardly directed K^+ pump is probably necessary for metabolism as we know it. The presence of a substantial concentration of such ions ensures that electrostatic effects adjacent to a Donnan phase can largely be screened out—otherwise, a negatively charged substrate, e.g., an organic acid or a phosphorylated sugar, might be electrostatically repelled from the catalytic site on an enzyme (proteins are generally negatively charged at cytosolic pH's). Once active transport has set up certain concentration differences across a membrane, the membrane potential is an inevitable consequence of the tendency of such ions to diffuse passively toward regions of lower chemical potential. The roles that such diffusion potentials play in the physiology of plant cells are open to question. It is known, however, that they are essential for the transmission of electrical impulses in excitable cells of animals and certain plants.

IV. Mechanisms for Crossing Membranes

The possible involvement of a "carrier" molecule in the active transport of solutes across plant cell membranes was first suggested by Winthrop Osterhout in the 1930s. A carrier can selectively bind certain molecules and then carry them across a membrane. Carriers provide a cell with the specificity or selectivity needed to control the entry and exit of various solutes. Thus certain metabolites can be specifically taken into a cell, and photosynthetic and metabolic waste products can be selectively moved out across the membranes. At the organelle level, such selectivity plays a key role in maintaining cellular compartmentation. At the organ level, active transport of certain inorganic nutrients into epidermal cells in a

---

[8] To show this, consider that $0 \leq (\sqrt{b} - \sqrt{c})^2 = b + c - 2\sqrt{bc}$, or $2\sqrt{bc} \leq b + c$; hence, if $a^2 = bc$, then $2a \leq b + c$. Here, we let $c_{Na}^o = c_{Cl}^o = a$, $c_{Na}^i = b$, and $c_{Cl}^i = c$.

root allows a plant to obtain and accumulate these solutes from the soil. Even though all the details for binding and moving solutes through membranes are not known, the carrier concept has found widespread application in the interpretation of experimental observations.

A great proliferation in terminology describing how solutes cross membranes has occurred, and new conceptual insights have greatly stimulated research efforts. Carriers are often referred to as transporters, or simply porters. A *symporter* is a porter that causes two different substances to move in the same direction across a membrane, and an *antiporter* causes them to move in opposite directions. Proton (H^+) fluxes appear to be involved with most symporters and antiporters in plant membranes. Also, transmembrane channels with complex and specific properties are increasingly implicated in the movement of ions across plant membranes.

A. Carriers, Porters, Channels, and Pumps

Most transporters are proteins. Small proteins can bind some substance on one side of a membrane, thermally diffuse across the membrane, and then release the substance on the other side. Such mobile carriers may bind a single substance, or they may bind two different substances, like the proton–solute symporter portrayed in Fig. 3-10a. Candidates for transport by a proton symport in plants include inorganic ions such as Na^+ and metabolites

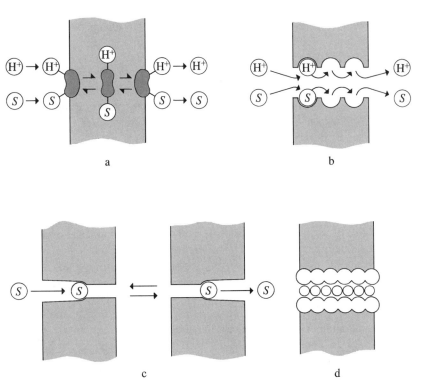

Figure 3-10. Hypothetical structures indicating possible mechanisms for transporters and channels: (a) mobile carrier or porter acting as a symporter for protons (H^+) and some transported solute (S); (b) series of binding sites in a channel across a membrane, acting as a symporter for H^+ and S; (c) sequential conformations of a channel, leading to unidirectional movement of solute; and (d) a protein-lined pore with multiple solute or water molecules in single file, the most accepted version of ion or water (aquaporin) channels.

such as sugars and amino acids. Many substances apparently move in pores or channels, which can be membrane-spanning proteins. Some channels can have a series of binding sites, where the molecule or molecules transported go from site to site through the membrane (Fig. 3-10b). As another alternative, the substance to be transported can first bind to a site accessible from one side of the membrane. Following a conformational change of the protein involved, the substance can subsequently be exposed to the solution on the other side of the membrane (Fig. 3-10c). For mechanisms as speculative as these, there are a number of ways that metabolic energy can be involved—e.g., using ATP to cause protein conformational changes or for active transport of H^+ to maintain an H^+ chemical potential difference across the membrane so that solute transport could be coupled to the passive, energetically downhill flow of protons.

Channels (Figs. 3-10b and 3-10d) can allow more solutes to cross membranes per unit time than can carriers. Individual carrier proteins have maximal processing rates of 10^3–10^5 transported solutes per second, whereas an open channel can allow 10^6–10^8 or more ions to cross a membrane per second. Even though many solute molecules can cross the membrane in a channel, flowing passively in the energetically downhill direction, the channel is still selective for a particular type of ion, possibly because of specific binding sites that depend on ion size and/or charge. Moreover, a "patch-clamp" technique has allowed the recording of electrophysical responses of single ion channels.[9] Specifically, a glass micropipette with a tip diameter of about 1 μm is pressed against some membrane (not through the membrane as in Fig. 3-4), a slight suction is applied to the micropipette such that a small piece of the membrane seals to its tip, and the voltage is then fixed or clamped across this patch of membrane. Opening of a channel or channels for such a patch-clamp preparation leads to ion movements that can be detected as a small current by a sensitive electronic amplifier. Indeed, the opening and closing of even a single channel mediated by the conformational changes of a single protein can be detected. Such studies have indicated that channel conformation can rapidly change from the open (transporting) to the closed (nontransporting) configuration, with such "gating" responses controlling what enters or leaves a plant cell.

Water can also move across membranes in channels, which are often called *aquaporins*. These water channels are individual proteins with a molecular mass of about 30 kDa that span a membrane six times. The pore in the aquaporin is 0.3–0.4 nm in diameter, which allows the passage of water molecules in single file (a water molecule occupies a space about 0.28 nm in diameter; Fig. 2-2). About 3×10^9 water molecules s^{-1} can cross the plasma membrane or the tonoplast through a water channel. Although such channels are specific for water, some small solutes, such as methanol or ethanol, can also cross them.

Potassium conducting channels, which are the predominant ion channels in plant membranes, are apparently involved in the pulvinus motor cells controlling leaf movement for *Samanea saman*, whose leaves tend to move diurnally, and in the opening and the closing of stomatal pores mediated by K^+ uptake and release across the plasma membrane of guard cells, which is a universal phenomenon among plants. The opening of channels is visualized to occur by the opening of "gates," which can be regulated by the membrane potential. In particular, as E_M rises above -40 mV, gates for K^+ channels open so that K^+ can readily cross the membranes of pulvinus motor cells or guard cells. In such cases μ_K is higher in the cells, so K^+ is released from the cells. When E_M for the plasma membrane of guard cells becomes more negative than -100 mV, K^+ channels also open, but in this case μ_K is

---

[9] The Nobel prize in physiology was awarded to Erwin Neher and Bert Sakman in 1991 for their invention of the patch-clamp technique.

lower in the cells and so K^+ enters. The ensuing entry of an accompanying anion and then water (diffusing toward regions of higher Π, a process called *osmosis*) apparently leads to the swelling of guard cells and the opening of stomatal pores (discussed in Chapter 8). Potassium channels can have a frequency of one or more channels per μm^2 of membrane surface area. Cellular control can be exerted on the opening of such K^+ channels because concentrations of cytosolic Ca^{2+} above 3×10^{-4} mol m^{-3} (0.3 μM) can inhibit channel opening. Other ion channels in plant membranes are specific for Ca^{2+} or Cl^-. Besides being sensitive to the electrical potential difference across a membrane, some channels apparently open upon stretching of a membrane.

Channels are involved in plant processes in which large amounts of charged solutes rapidly cross membranes. Thus channels have been implicated in the functioning of chloroplast thylakoids (discussed in Chapter 5) and guard cells (see Chapter 8), in the motion of leaflets in plants sensitive to mechanical stimuli, and in the solar tracking of leaves (see Chapter 7). Also, many plant cells are excitable and can transmit action potentials, a process in which ion channels are undoubtedly involved. In addition, ion channels are involved in the long-term maintenance of specific ion concentrations in plant cells.

Because of electroneutrality and electrogenicity discussed previously, electrically uncompensated transport of charged solutes can have major impacts on the membrane potential, which in turn can control many of the properties of ion channels. An H^+-extruding ATPase (a porter protein that is linked to the hydrolysis of ATP), which is referred to as an H^+ or *proton pump*, can rapidly influence E_M for the plasma membrane, controlling the opening of K^+ and other channels (a pump couples transmembrane solute movement with chemical energy, such as is provided by ATP hydrolysis). A blue-light-activated H^+ pump has been identified in the plasma membrane of guard cells that is apparently involved in stomatal opening. The hyperpolarization (negative shift of E_M) caused by outwardly directed H^+ pumps in the plasma membrane can also subsequently affect symporters using cotransport of H^+ and various solutes (Fig. 3-10a). Also, an inwardly directed H^+ pump in the tonoplast tends to lower the pH and to raise the electrical potential in the vacuole relative to values in the cytosol. Indeed, the most prevalent active transport processes in higher plant cells may be electrogenic proton pumps taking H^+ out of the cytosol across the plasma membrane and across the tonoplast into the vacuole, leading to cytosolic pH's near 7 and vacuolar pH's below 6. The tonoplast H^+–ATPase, which may transport two H^+'s per ATP hydrolyzed, can lead to passive ion uptake into the vacuole, an important process for plant growth. Thus H^+ pumps depending on ATP to move protons to regions of higher chemical potential can generate changes in the electrical potential across membranes that in turn affect many cellular processes. Various actions of plant hormones also most likely involve H^+ pumps.

B. Michaelis–Menten Formalism

When an ion is attached to a particular transporter, a similar ion (of the same or a different solute) competing for the same binding site cannot also be bound. For example, the similar monovalent cations K^+ and Rb^+ appear to bind in a competitive fashion to the same site on a transporter. For some cells, the same carrier might transport Na^+ out of the cell and K^+ in—the sodium–potassium pump alluded to previously. Ca^{2+} and Sr^{2+} may compete with each other for binding sites on another common carrier. Two other divalent cations, Mg^{2+} and Mn^{2+}, are apparently transported by a single carrier that is different from the one for Ca^{2+} and Sr^{2+}. The halides (Cl^-, I^-, and Br^-) may also be transported by a single carrier.

One of the most important variables in the study of carrier-mediated uptake is the external concentration. As the external concentration of a transported solute increases, the rate of uptake eventually reaches an upper limit. We may then presume that all binding sites on the carriers for that particular solute have become filled or saturated. In particular, the rate of active uptake of species j, J_j^{in}, is often proportional to the external concentration of that solute, c_j^o, over the lower range of concentrations, but as c_j^o is raised a maximum rate, $J_{j\,max}^{in}$, is eventually reached. We can describe this kind of behavior by

$$J_j^{in} = \frac{J_{j\,max}^{in}\ c_j^o}{K_j + c_j^o}$$
(3.27a)

where the constant K_j characterizes the affinity of the carrier used for solute species j crossing a particular membrane and is expressed in the units of concentration. For the uptake of many ions into roots and other plant tissues, $J_{j\,max}^{in}$ is 30–300 nmol m^{-2} s^{-1}. Often, two different K_j's are observed for the uptake of the same ionic species into a root. The lower K_j is generally between 6 and 100 μM (6 and 100 mmol m^{-3}), which is in the concentration range of many ions in soil water, and the other K_j can be above 10 mM.

Equation (3.27a) is similar in appearance to the Michaelis–Menten equation describing enzyme kinetics in biochemistry, $v = v_{max}s/(K_M + s)$. The substrate concentration s in the latter relation is analogous to c_j^o in Eq. (3.27a), and the enzyme reaction velocity v is analogous to J_j^{in}. The term in the Michaelis–Menten equation equivalent to K_j in Eq. (3.27a) is the substrate concentration for half-maximal velocity of the reaction, K_M (the Michaelis constant). The lower the K_M, the greater the reaction velocity at low substrate concentrations. Likewise, a low value for K_j indicates that the ion or other solute is more readily bound to some carrier and then transported across the membrane. In particular, Eq. (3.27a) describes a rectangular hyperbola (Fig. 3-11a), a class of equations that has been used to describe many adsorption and other binding phenomena. When c_j^o is small relative to K_j, J_j^{in} is approximately equal to $J_{j\,max}^{in}c_j^o/K_j$ by Eq. (3.27a). The influx of species j is then not only proportional to the external concentration, as already indicated,

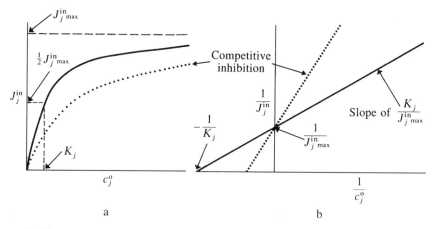

Figure 3-11. Relationship between the external solute concentration (c_j^o) and the rate of influx (J_j^{in}) for active uptake according to Michaelis–Menten kinetics, as given by Eq. (3.27): (a) linear plot and (b) double-reciprocal plot.

but also inversely proportional to K_j. Hence, a low K_j, representing a high affinity for a carrier, indicates that species j is actually favored or selected for active transport into the cell, even when its external concentration is relatively low. We note that Eq. (3.27a) even applies to ions moving in channels where solute binding can occur. Hence, the Michaelis–Menten formalism has found widespread application in describing movement of solutes across membranes.

The two most common ways of graphing data on solute uptake that fit Eq. (3.27a) are illustrated in Fig. 3-11. Figure 3-11a shows that when the external concentration of species j, c_j^o, is equal to K_j, then $J_j^{in} = \frac{1}{2} J_{j\,max}^{in}$, as we can see directly from Eq. (3.27a); i.e., J_j^{in} is then $J_{j\,max}^{in} K_j / (K_j + K_j)$, or $\frac{1}{2} J_{j\,max}^{in}$. Thus K_j is the external concentration at which the rate of active uptake is half-maximal—in fact, the observed values of K_j are convenient parameters for describing the uptake of various solutes. If two different solutes compete for the same site on some carrier, J_j^{in} for species j will be decreased by the presence of the second solute, which is known as *competitive inhibition* of species j. In the case of competitive inhibition, the asymptotic value for the active influx, $J_{j\,max}^{in}$, is not affected because in principle we can raise c_j^o high enough to obtain the same maximum rate for the active uptake of species j. However, the half-maximum rate occurs at a higher concentration, so the apparent K_j is raised if a competing solute is present (Fig. 3-11a).

Often the experimental data are plotted such that a linear relationship is obtained when Eq. (3.27a) is satisfied for the active uptake of species j. Taking reciprocals of both sides of Eq. (3.27a), we note that

$$\frac{1}{J_j^{in}} = \frac{K_j + c_j^o}{J_{j\,max}^{in}\, c_j^o} = \frac{K_j}{J_{j\,max}^{in}\, c_j^o} + \frac{1}{J_{j\,max}^{in}} \tag{3.27b}$$

When $1/J_j^{in}$ is plotted against $1/c_j^o$ (Fig. 3-11b), Eq. (3.27b) yields a straight line with a slope of $K_j / J_{j\,max}^{in}$ and an intercept on the ordinate of $1/J_{j\,max}^{in}$. This latter method of treating the experimental results is widely applied to solute uptake by plant tissues, especially roots. Because K_j increases and $J_{j\,max}^{in}$ remains the same for competitive inhibition, the presence of a competing species causes the slope of the line (i.e., $K_j / J_{j\,max}^{in}$) to be greater, whereas the intercept on the y-axis ($1/J_{j\,max}^{in}$) is unchanged (Fig. 3-11b).

C. Facilitated Diffusion

Equation (3.27) describes the competitive binding of solutes to a limited number of specific sites. In other words, active processes involving metabolic energy do not have to be invoked; if a solute were to diffuse across the membrane only when bound to a carrier, the expression for the influx could also be Eq. (3.27). This passive entry of a solute mediated by a carrier is termed *facilitated diffusion*.

Because facilitated diffusion is important in biology and yet is often misunderstood, we will briefly elaborate on it. Certain molecules passively enter cells more readily than expected from consideration of their molecular structure or from observations with analogous substances, so some mechanism is apparently facilitating their entry. The net flux density is still toward lower chemical potentials and hence is still in the same direction as ordinary diffusion. (The term diffusion usually refers to net thermal motion toward regions of lower concentration; it is used here with a broader meaning—namely, net motion toward regions of lower chemical potential.) To help explain facilitated diffusion, transporters are proposed to act as shuttles for a net passive movement of the specific molecules across the membrane toward regions of lower energy. Instead of the usual diffusion across the barrier—based on random thermal motion of the solutes—transporters selectively bind certain molecules and then release them on the other side of the membrane without an input of metabolic energy. Such facilitation of transport may also be regarded as a special means of lowering the activation energy needed for the solute to cross the energy barrier represented by the membrane. Thus transporters facilitate the influx of solutes in the same way that enzymes facilitate biochemical reactions.

Facilitated diffusion has certain general characteristics. As already mentioned, the net flux is toward lower chemical potential. (According to the usual definition, active transport is in the energetically uphill direction; active transport may employ the same carriers as those for facilitated diffusion.) Facilitated diffusion causes fluxes to be larger than those expected for ordinary diffusion. Furthermore, the transporters can exhibit selectivity, i.e., they can be specific for certain molecules while not binding closely related ones. In addition, carriers in facilitated diffusion become saturated when the external concentration of the species transported is raised sufficiently, a behavior consistent with Eq. (3.27). Finally, because carriers can exhibit competition, the flux density of a species entering a cell by facilitated diffusion can be reduced when structurally similar molecules are added to the external solution. Such molecules compete for the same sites on the carrier and thereby reduce the binding and the subsequent transfer of the original species into the cell.

For convenience, we have been discussing facilitated diffusion into a cell, but exactly the same principles apply for exit, as well as for fluxes at the organelle level. Let us assume that a transporter for K^+ exists in the membrane of a certain cell and that it is used as a shuttle for facilitated duffusion. Not only does the carrier lead to an enhanced net flux density toward the side with the lower chemical potential but also both the unidirectional fluxes J_K^{in} and J_K^{out} can be increased over the values predicted for ordinary diffusion. This increase in the unidirectional fluxes by a carrier is often called *exchange diffusion*. In such a case, the molecules are interacting with a membrane component—namely, the carrier: hence the Ussing–Teorell equation [Eq. 3.24; $J_j^{in}/J_j^{out} = c_j^o/(c_j^i e^{z_j FE_M/RT})$]—is not obeyed because it does not consider interactions with other substances. In fact, observation of departures from predictions of the Ussing–Teorell equation is often how cases of exchange diffusion are discovered.[10]

Both active and passive fluxes across the cellular membranes can occur concomitantly, but these movements depend on concentrations in different ways. For passive diffusion, the unidirectional component J_j^{in} is proportional to c_j^o, as indicated by Eq. (1.8) for neutral solutes [$J_j = P_j(c_j^o - c_j^i)$] and by Eq. (3.15) for ions. This proportionality strictly applies only over the range of external concentrations for which the permeability coefficient is essentially independent of concentration, and the membrane potential must not change in the case of charged solutes. Nevertheless, ordinary passive influxes do tend to be proportional to the external concentration, whereas an active influx or the special passive influx known as facilitated diffusion—either of which can be described by a Michaelis–Menten type of formalism—shows saturation effects at the higher concentrations. Moreover, facilitated diffusion and active transport exhibit selectivity and competition, whereas ordinary diffusion does not.

---

[10] The term "exchange diffusion" has another usage in the literature—namely, to describe the carrier-mediated movement of some solute in one direction across a membrane in exchange for a different solute being transported in the opposite direction. Again, the Ussing–Teorell equation is not obeyed.

V. Principles of Irreversible Thermodynamics

So far we have been using classical thermodynamics—though, it may have been noticed, often somewhat illegitimately. For example, let us consider Eq. (2.24) ($J_{V_w} = L_w \Delta \Psi$). Because the chemical potential changes, represented here by a change in water potential, we expect a net (and irreversible) flow of water from one region to another, which is not an equilibrium situation. Strictly speaking, however, classical thermodynamics is concerned solely with equilibria, not with movement. Indeed, classical thermodynamics might have been better named "thermostatics." Thus we have frequently been involved in a hybrid enterprise—appealing to classical thermodynamics for the driving forces but using nonthermodynamic arguments and analogies to discuss fluxes. One of the objectives of irreversible thermodynamics is to help legitimize the arguments. However, as we shall see, legitimizing them brings in new ideas and considerations.

Irreversible thermodynamics uses the same parameters as classical thermodynamics—namely, temperature, pressure, free energy, activity, and so on. However, these quantities are strictly defined for macroscopic amounts of matter only in equilibrium situations. How can we use them to discuss processes not in equilibrium, the domain of irreversible thermodynamics? This dilemma immediately circumscribes the range of validity of the theory of irreversible thermodynamics: It can deal only with "slow" processes or situations not very far from equilibrium, for only in such circumstances can equilibrium-related concepts such as temperature and free energy retain their validity—at least approximately. We have to assume from the outset that we can talk about and use parameters from classical thermodynamics even in nonequilibrium situations. We must also avoid situations of high kinetic energy, for in such cases the chemical potential of species j, μ_j, does not adequately represent its total energy (kinetic plus potential). Kinetic energy equals $\frac{1}{2} m_j v_j^2$, where m_j is the mass per mole of species j and v_j is its velocity; v_j must exceed 1 m s^{-1} before the kinetic energy becomes relatively important, but such speeds do not occur for solutions in plants.

Another refinement is to recognize that the movement of one species may affect the movement of a second species. A particular flux of some solute may interact with another flux by way of collisions, each species flowing under the influence of its own force. For example, water, ions, and other solutes moving through a membrane toward regions of lower chemical potentials can exert frictional drags on each other. The magnitude of the flux of a solute may then depend on whether water is also flowing. In this way, the fluxes of various species across a membrane become interdependent. Stated more formally, the flux of a solute is not only dependent on the negative gradient of its own chemical potential, which is the sole driving force we have recognized up to now, but also may be influenced by the gradient in the chemical potential of water. Again using Eq. (2.24) as an example, we have considered that the flow of water depends only on the difference in its own chemical potential between two locations and have thus far ignored any coupling to concomitant fluxes of solutes.

A quantitative description of interdependent fluxes and forces is given by irreversible thermodynamics, a subject that treats nonequilibrium situations such as those actually occurring under biological conditions. (Nonequilibrium and irreversible are related because a system in a nonequilibrium situation left isolated from external influences will spontaneously and irreversibly move toward equilibrium.) In this brief introduction to irreversible thermodynamics we will emphasize certain underlying principles and then derive the *reflection coefficient*. To simplify the analysis, we will restrict our attention to isothermal conditions, which approximate many biological situations in which fluxes of water and solutes are considered.

A. Fluxes, Forces, and Onsager Coefficients

In our previous discussion of fluxes, the driving force leading to the flux density of species j, J_j, was the negative gradient in its chemical potential, $-\partial \mu_j / \partial x$. Irreversible thermodynamics takes a more general view—namely, the flux of species j depends not only on the chemical potential gradient of species j, but also potentially on any other force occurring in the system, such as the chemical potential gradient of some other species. A particular force, X_k, can likewise influence the flux of any species. Thus the various fluxes become interdependent, or coupled, because they can respond to changes in any of the forces. Another premise of irreversible thermodynamics is that J_j is linearly dependent on the various forces, so we can treat only those cases that are not too far from equilibrium. Even with this simplification, the algebra often becomes cumbersome, owing to the coupling of the various forces and fluxes.

Using a linear combination of all the forces, we represent the flux density of species j by

$$J_j = \sum_{k}^{n} L_{jk} X_k = L_{j1} X_1 + L_{j2} X_2 + \cdots + L_{jj} X_j + \cdots + L_{jn} X_n \qquad (3.28)$$

where the summation $\sum_{k}^{n}$ is over all forces (all $n X_k$'s), and the L_{jk}'s are referred to as the *Onsager coefficients*, or the phenomenological coefficients, in this case for conductivity.[11] The first subscript on these coefficients (j on L_{jk}) identifies the flux density J_j that we are considering; the second subscript (k on L_{jk}) designates the force, such as the gradient in chemical potential of species k. Each term, $L_{jk} X_k$, is thus the partial flux density of species j due to the particular force X_k. The individual Onsager coefficients in Eq. (3.28) are therefore the proportionality factors indicating what contribution each force X_k makes to the flux density of species j. Equation (3.28) is sometimes referred to as the *phenomenological equation*. Phenomenological equations are used to describe observable phenomena without regard to explanations in terms of atoms or molecules. For instance, Ohm's law and Fick's laws are also phenomenological equations. They also assume linear relations between forces and fluxes.

The phenomenological coefficient $L_{jk} = L_{kj}$, which is known as the *reciprocity relation*. Such an equality of cross-coefficients was derived in 1931 by Onsager from statistical considerations utilizing the principle of "detailed balancing." The argument involves microscopic reversibility, i.e., for local equilibrium, any molecular process and its reverse will be taking place at the same average rate in that region. The Onsager reciprocity relation means that the proportionality coefficient giving the flux density of species k caused by the force on species j equals the proportionality coefficient giving the flux density of j caused by the force on k. (Strictly speaking, conjugate forces and fluxes must be used, as they will later.) The fact that $L_{jk} = L_{kj}$ can be further appreciated by considering Newton's third law—equality of action and reaction. For example, the frictional drag exerted by a moving solvent on a solute is equal to the drag exerted by the moving solute on the solvent. The pairwise equality of cross-coefficients given by the Onsager reciprocity relation reduces the number of coefficients needed to describe the interdependence of forces and fluxes in irreversible thermodynamics and consequently leads to a simplification in solving the sets of simultaneous equations.

---

[11] Lars Onsager received the Nobel prize in chemistry in 1968 for his contributions to irreversible thermodynamics, as did Ilya Prigogine in 1977.

In the next section we will use Eq. (3.28) as the starting point for developing the expression from irreversible thermodynamics that describes the volume flux density. Because the development is lengthy and the details may obscure the final objective, the development of Eq. (3.39), the steps and the equations involved are summarized first:

Step	Eq. No.
Flux equations for a particular species	3.28
Flux equation for water (J_w) and a single solute (J_s) in terms of $\Delta\mu_w$ and $\Delta\mu_s$	3.29, 3.30
Expression of $\Delta\mu_w$ and $\Delta\mu_s$ in terms of ΔP and $\Delta\Pi$	3.31, 3.32
Volume flux density (J_V)	3.33a, b
Diffusional flux density (J_D)	3.34
J_V and J_D in terms of ΔP and $\Delta\Pi$	3.35, 3.36
Reflection coefficient (σ)	3.37
J_V in terms of ΔP, $\Delta\Pi$, and σ	3.38, 3.39

B. Water and Solute Flow

As a specific application of the principles just introduced, we will consider the important coupling of water and solute flow. The driving forces for the fluxes are the negative gradients in chemical potential, which we will assume to be proportional to the differences in chemical potential across some barrier, here considered to be a membrane. In particular, we will represent $-\partial\mu_j/\partial x$ by $\Delta\mu_j/\Delta x$, which in the current case is $(\mu_j^o - \mu_j^i)/\Delta x$. (For convenience, the thickness of the barrier, Δx, will be incorporated into the coefficient multiplying $\Delta\mu_j$ in the flux equations.) To help keep the algebra relatively simple, the development will be carried out for a single nonelectrolyte. The fluxes are across a membrane permeable to both water (w) and the single solute (s), thereby removing the restriction in Chapter 2, in which membranes permeable only to water were considered. Using Eq. (3.28), we can represent the flux densities of water (J_w) and a solute (J_s) by the following linear combination of the differences in chemical potential:

$$J_w = L_{ww}\,\Delta\mu_w + L_{ws}\,\Delta\mu_s \tag{3.29}$$

$$J_s = L_{sw}\,\Delta\mu_w + L_{ss}\,\Delta\mu_s \tag{3.30}$$

Equations (3.29) and (3.30) allow for the possibility that each of the flux densities can depend on the differences in both chemical potentials, $\Delta\mu_w$ and $\Delta\mu_s$. Four phenomenological coefficients are used in these two equations. However, by the Onsager reciprocity relation, $L_{ws} = L_{sw}$. Thus three different coefficients (L_{ww}, L_{ws}, and L_{ss}) are needed to describe the relationship of these two flux densities to the two driving forces. Contrast this with Eq. (3.6), $J_j = u_j\,c_j(-\partial\mu_j/\partial x)$, in which a flux density depends on but one force; accordingly, only two coefficients are then needed to describe J_w and J_s. If the solute were a salt dissociable into two ions, we would have three flux equations (for J_w, J_+, and J_-) and three forces, ($\Delta\mu_w$, $\Delta\mu_+$, and $\Delta\mu_-$); using the approach of irreversible thermodynamics, this leads to nine phenomenological coefficients. Invoking the Onsager reciprocity relations, we would then have six different phenomenological coefficients to describe the movement of water, a cation, and its accompanying anion.

To obtain more convenient formulations for the fluxes of water and the single solute, we generally express $\Delta\mu_w$ and $\Delta\mu_s$ in terms of the differences in the osmotic pressure and the hydrostatic pressure, $\Delta\Pi$ and ΔP, because it is usually easier to measure $\Delta\Pi$ and ΔP than $\Delta\mu_w$ and $\Delta\mu_s$. The expression for $\Delta\mu_w$ is straightforward; the only possible ambiguity is deciding on the algebraic sign. In keeping with the usual conventions for this specific case, $\Delta\mu_w$ is the chemical potential of water on the outside minus that on the inside, $\mu_w^o - \mu_w^i$. From Eq. (2.12) $(\mu_w = \mu_w^* - \bar{V}_w\Pi + \bar{V}_w P + m_w gh)$, $\Delta\mu_w$ is given by

$$\Delta\mu_w = -\bar{V}_w\Delta\Pi + \bar{V}_w\Delta P \tag{3.31}$$

where $\Delta\Pi = \Pi^o - \Pi^i$ and $\Delta P = P^o - P^i$ ($\Delta h = 0$ across a membrane).

To express $\Delta\mu_s$ in terms of $\Delta\Pi$ and ΔP, we will first consider the activity term, $RT \ln a_s$. The differential $RT\, d\,(\ln a_s) = RT\, da_s/a_s$, or $RT\, d\,(\gamma_s c_s)/(\gamma_s c_s)$. When γ_s is constant, this latter quantity becomes $RT\, dc_s/c_s$. By Eq. (2.10) $(\Pi_s = RT \sum_j c_j)$, $RT\, dc_s = d\Pi$ for a dilute solution of a single solute. Hence, $RT\, d\,(\ln a_s)$ can be replaced by $d\Pi/c_s$ as a useful approximation. In expressing the difference in chemical potential across a membrane we are interested in macroscopic changes, not in the infinitesimal changes given by differentials. To go from differentials to differences, $RT\, d(\ln a_s)$ becomes $RT(\Delta \ln a_s)$ and so $d\Pi/c_s$ can be replaced by $\Delta\Pi/\bar{c}_s$, where $\bar{c}_s$ is essentially the mean concentration of solute s, in this case across the membrane. Alternatively, we can simply define $\bar{c}_s$ as that concentration for which $RT(\Delta \ln a_s)$ exactly equals $\Delta\Pi/\bar{c}_s$. In any case, we can replace $RT(\Delta \ln a_s)$ in $\Delta\mu_s$ by $\Delta\Pi/\bar{c}_s$. By Eq. (2.4), $\mu_s = \mu_s^* + RT \ln a_s + \bar{V}_s P$ for a neutral species, and the difference in its chemical potential across a membrane, $\Delta\mu_s$ becomes

$$\Delta\mu_s = \frac{1}{\bar{c}_s}\Delta\Pi + \bar{V}_s\,\Delta P \tag{3.32}$$

The two expressions representing the driving forces, $\Delta\mu_w$ (Eq. 3.31) and $\Delta\mu_s$ (Eq. 3.32), are thus expressed as functions of the same two pressure differences, $\Delta\Pi$ and ΔP, which are experimentally more convenient to measure.

C. Flux Densities, L_P, and σ

Now that we have appropriately expressed the chemical potential differences of water and the solute, we direct our attention to the fluxes. Expressed in our usual units, the flux densities J_w and J_s are the mol of water and solute, respectively, moving across 1 m^2 of membrane surface in a second. A quantity of considerable interest is the volume flux density J_V, which is the rate of movement of the total volume of both water and solute across unit area of the membrane; J_V has the units of volume per unit area per unit time, e.g., m^3 m^{-2} s^{-1}, or m s^{-1}.

The molar flux density of species j (J_j) in mol m^{-2} s^{-1} multiplied by the volume occupied by each mole of species j ($\bar{V}_j$) in m^3 mol^{-1} gives the volume flow for that component (J_{V_j}) in m s^{-1}. Hence, the total volume flux density is

$$J_V = \sum_j J_{V_j} = \sum_j \bar{V}_j J_j \tag{3.33a}$$

For solute and water both moving across a membrane, J_V is the volume flow of water plus that of solute per unit area, which in the case of a single solute can be represented as

$$J_V = \bar{V}_w J_w + \bar{V}_s J_s \tag{3.33b}$$

It is generally simpler and more convenient to measure the total volume flux density (such as that given in Eq. 3.33) than one of the component volume flux densities ($J_{V_j} = \bar{V}_j J_j$). For instance, we can often determine the volumes of cells or organelles under different conditions and relate any changes in volume to J_V. [The volume flux density of water J_{V_w} used in Chapter 2 (e.g., Eq. 2.24) is $\bar{V}_w J_w$.]

Although straightforward, the algebraic substitutions necessary to incorporate the various forces and fluxes (Eqs. 3.29–3.32) into the volume flow (Eq. 3.33b) lead to a rather cumbersome expression for J_V as a linear function of ΔP and $\Delta \Pi$. Hence, the conventional approach, introduced by Ora Kedem and Aharon Katchalsky in 1958 and adopted by essentially all subsequent treatments of irreversible thermodynamics, is to change to a more convenient set of conjugate forces and fluxes. A discussion of the criteria for deciding whether a particular force–flux pair is conjugate would take us into a consideration of the dissipation function and the rate of entropy production, topics outside the scope of this text. The total volume flux density, J_V, is conjugate to ΔP, and the diffusional flux density, J_D, is conjugate to $\Delta \Pi$, where J_D can be represented as

$$J_D = J_s/\bar{c}_s - J_w \bar{V}_w$$
$$\cong J_s/\bar{c}_s - J_w/c_w$$
$$= v_s - v_w \tag{3.34}$$

To obtain the second line, we note that $\bar{V}_w c_w$ is essentially 1 for a dilute aqueous solution, so $\bar{V}_w$ is approximately $1/c_w$ (the volume of a solution V can be represented by the sum of the volumes of individual components, $V = \sum_j V_j$; hence $1 = \sum_j V_j/V = \sum_j n_j \bar{V}_j/V = \sum_j \bar{V}_j c_j$, where n_j is the number of moles of species j and c_j is its concentration; for a solution of a single solute, $\bar{V}_w c_w + \bar{V}_s c_s = 1$, where $\bar{V}_s c_s \ll 1$ or $\bar{V}_w c_w \cong 1$ defines a dilute solution). To obtain the bottom line in Eq. (3.34), we note that the average velocity of species j, v_j, is J_j/c_j (see Eq. 3.6 and Fig. 3-3).

The diffusional flux density (Eq. 3.34) is the difference between the mean velocities of solute and water. In mass flow (such as that described by Poiseuille's law; Eq. 9.10), $v_s = v_w$, so J_D is then zero; such flow is independent of $\Delta \Pi$ and depends only on ΔP. On the other hand, let us consider $\Delta \Pi$ across a membrane that greatly restricts the passage of some solute relative to the movement of water, i.e., a barrier that acts as a differential filter; v_s is then considerably less than v_w, so J_D has a nonzero value in response to its conjugate "force," $\Delta \Pi$. Thus J_D expresses the tendency of the solute relative to water to diffuse in response to a gradient in osmotic pressure.

Next we will use J_V and J_D to express the linear interdependence of conjugate forces and fluxes in irreversible thermodynamics (Eq. 3.28; $J_j = \sum_j L_{jk} X_k$):

$$J_V = L_P \Delta P + L_{PD} \Delta \Pi \tag{3.35}$$
$$J_D = L_{DP} \Delta P + L_D \Delta \Pi \tag{3.36}$$

where the subscripts in both equations are those generally used in the literature. (Although these subscripts—P referring to pressure and D to diffusion—are not consistent with the L_{jk} convention, four coefficients are still needed to describe the dependence of two fluxes on their conjugate forces.) By the Onsager reciprocity relation, which is applicable in the current case of conjugate forces and fluxes, $L_{PD} = L_{DP}$, and again only three coefficients are needed to describe the movement of water and a single solute across a membrane.

L_P is the *hydraulic conductivity coefficient*. It describes the mechanical filtration capacity of a membrane or other barrier, i.e., when $\Delta\Pi$ is zero, L_P relates the total volume flux density, J_V, to the hydrostatic pressure difference, ΔP. When ΔP is zero, Eq. (3.36) indicates that a difference in osmotic pressure leads to a diffusional flow characterized by the coefficient L_D. Membranes also generally exhibit a property called *ultrafiltration*, whereby they offer different resistances to the passage of the solute and water. For instance, in the absence of an osmotic pressure difference ($\Delta\Pi = 0$), Eq. (3.36) indicates a diffusional flux density equal to $L_{DP}\,\Delta P$. Based on Eq. (3.34), v_s is then different from v_w, which results if a membrane restricts the passage of solute more than that of water. Thus the phenomenological coefficient L_{DP} helps describe the relative ease with which solute crosses a membrane compared with water. Such a property of relative selectivity by a barrier is embodied in the *reflection coefficient*, σ, defined by Albert Staverman in 1951 as

$$\sigma = -L_{DP}/L_P = -L_{PD}/L_P \tag{3.37}$$

When a membrane is nonselective, both water and the solute move across it at the same velocity, i.e., $v_s = v_w$ and $\Delta\Pi$ is zero. (Admittedly, the idea of a solution containing a single solute is not realistic from a biological point of view, but it is convenient for illustrating the minimum value for σ.) If $v_s = v_w$, then J_D must be zero by Eq. (3.34). For this to be true for any ΔP and a zero $\Delta\Pi$, Eq. (3.36) indicates that L_{DP} must be zero; hence the reflection coefficient is zero in this case ($\sigma = -L_{DP}/L_P$; Eq. 3.37). Thus σ is zero when the membrane does not select between solute and solvent. At the opposite extreme, the solute does not cross the membrane ($v_s = 0$); hence J_D is $-v_w$ ($J_D = v_s - v_w$; Eq. 3.34). When J_s is zero, $J_V = \bar{V}_w J_w$ (Eq. 3.33b), which is simply v_w (see Eq. 3.34). Therefore, $J_V = -J_D$ when the solute does not cross the membrane, and $L_P\,\Delta P + L_{PD}\,\Delta\Pi = -L_{DP}\,\Delta P - L_D\,\Delta\Pi$ by Eqs. (3.35) and (3.36). Because this is true for any ΔP, L_P must equal $-L_{DP}$, so $-L_{DP}/L_P$ then equals 1. The reflection coefficient is therefore 1 ($\sigma = -L_{DP}/L_P$; Eq. 3.37) when the solute does not cross the membrane, indicating that all solute molecules in that case are reflected by the barrier.

Using the definition of σ (Eq. 3.37), we can rewrite Eq. (3.35) to obtain the following form for the total volume flux density:

$$\begin{aligned} J_V &= L_P\,\Delta P - L_P\sigma\,\Delta\Pi \\ &= L_P(\Delta P - \sigma\,\Delta\Pi) \end{aligned} \tag{3.38}$$

We note that L_P is essentially the same as and hence usually replaces L_w, the water conductivity coefficient that we introduced in Chapter 2 (e.g., in Eqs. 2.24 and 2.29). In the absence of a hydrostatic pressure difference across a membrane, the volume flux density J_V equals $-L_P\sigma\,\Delta\Pi$ by Eq. (3.38). Thus the magnitude of the dimensionless parameter σ determines the volume flux density expected in response to a difference in osmotic pressure across a membrane. It is this use that is most pertinent in biology.

Many different solutes can cross a membrane under usual conditons. Each such species j can be characterized by its own reflection coefficient, σ_j, for that particular membrane. The volume flux density given by Eq. (3.38) can then be generalized to

$$J_V = L_P\left(\Delta P - \sum_j \sigma_j\Delta\Pi_j\right) \tag{3.39}$$

where $\Delta\Pi_j$ is the osmotic pressure difference across the membrane for species j (e.g., $\Delta\Pi_j = RT\,\Delta c_j$ by Eq. 2.10). Although interactions with water are still taken into account, the generalization represented by Eq. (3.39) introduces the assumption that the solutes do not interact with each other as they cross a membrane. Moreover, J_V in Eq. (3.39) refers only to the movement of neutral species—otherwise we would also need a current equation to describe the flow of charge. Nevertheless, Eq. (3.39) is a useful approximation of the actual situation describing the multicomponent solutions encountered by cells, and we will use it as the starting point for our general consideration of solute movement across membranes. Before discussing such movement, however, let us again consider the range of reflection coefficients.

D. Values of Reflection Coefficients

A reflection coefficient characterizes some particular solute interacting with a specific membrane. In addition, σ_j depends on the solvent on either side of the membrane—water is the only solvent that we will consider. Two exteme conditions can describe the passage of solutes: impermeability, which leads to the maximum value of 1 for the reflection coefficient, and nonselectivity, where σ_j is zero. A reflection coefficient of zero may describe the movement of a solute across a very coarse barrier (one with large pores) that cannot distinguish or select between solute and solvent molecules; also, it may refer to the passage through a membrane of a molecule very similar in size and structure to water. Impermeability describes the limiting case in which water can cross some membrane but the solute cannot.

Let us consider the realistic situation of ΔP equaling zero across a membrane bathed on either side by aqueous solutions. By Eq. (3.38) $[J_V = L_P(\Delta P - \sigma \Delta\Pi)]$, the volume flux density (J_V) is then $-L_P\sigma\Delta\Pi$. For a solute having a reflection coefficient equal to zero for that particular membrane, the volume flux density is zero. By Eq. (3.33b) $(J_V = \bar{V}_w J_w + \bar{V}_s J_s)$, a zero J_V implies that $\bar{V}_w J_w = -\bar{V}_s J_s$. In words, the volume flux density of water must be equal and opposite to the volume flux density of the solute to result in no net volume flux density. Conversely, the absence of a net volume flux density across a membrane permeable to both water and the single solute when ΔP is zero but $\Delta\Pi$ is nonzero indicates that the reflection coefficient for that solute is zero. Again, this conditon occurs when the volume of water flowing toward the side with the higher Π is balanced by an equal volume of solute diffusing across the membrane in the opposite direction toward the side where the solute is less concentrated (lower Π). Such a situation of zero volume flux density anticipates the concept of a "stationary state" to be introduced in the next section.

In their interactions with biological membranes, solutes exhibit properties ranging from freely penetrating ($\sigma_j = 0$), indicating nonselectivity by the membrane, to being unable to penetrate ($\sigma_j = 1$), indicating membrane impermeability. Substances retained in or excluded from plant cells have reflection coefficients close to 1 for the cellular membranes. For instance, the σ_j's for sucrose and amino acids are usually near 1.0 for plant cells. Methanol and ethanol enter cells very readily and have reflection coefficients of 0.2–0.4 for the plasma membranes of some *Chara* and *Nitella* internodal cells. On the other hand, σ_j can essentially equal zero for solutes crossing porous barriers, such as those presented by cell walls, or for molecules penetrating very readily across membranes, such as D_2O (^2H_2O). Just as for a permeability coefficient, the reflection coefficient of a species is the same for traversal in either direction across a membrane.

For many small neutral solutes not interacting with carriers in a membrane, the reflection coefficients are correlated with the partition coefficients. For example, when K_j for non-electrolytes is less than about 10^{-4}, σ_j is generally close to 1. Thus compounds that do not

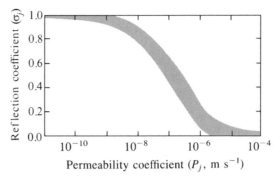

Figure 3-12. Correlation between the reflection coefficients for a series of nonelectrolytes (determined using rabbit gallbladder epithelium) and the permeability coefficients for the same compounds (measured by R. Collander using *Nitella mucronata*). Most of the measurements are in the area indicated. [The curve is adapted from Diamond and Wright (1969); used by permission.]

readily enter the lipid phase of a membrane (low K_j) also do not cross the membrane easily (σ_j near 1). When the partition coefficient is 1 or greater, the solutes can enter the membrane in appreciable amounts, and σ_j is generally close to zero. Considering the intermediate case, the reflection coefficient can be near 0.5 for a small nonelectrolyte having a K_j of about 0.1 for the membrane lipids, although individual molecules differ depending on their molecular weight, branching, and atomic composition. Neglecting frictional effects with other solutes, we see that the intermolecular interactions affecting partition coefficints are similar to those governing the values of reflection coefficients, and therefore the permeability coefficient of a solute ($P_j = D_j K_j / \Delta x$) is correlated with its reflection coefficient for the same membrane. In fact, as Fig. 3-12 illustrates, there is even a correlation between the reflection coefficients of a series of nonelectrolytes determined with animal membranes and the permeability coefficients of the same substances measured for plant membranes (exceptions occur for certain solutes, e.g., those that hydrogen bond to membrane components). Thus as the permeability coefficient goes from small to large values, the reflection coefficient decreases from 1 (describing relative impermeability) to zero (for the opposite extreme of nonselectivity). In the next section we will consider some of the consequences of reflection coefficients' differing from 1 for solutes crossing cellular and organelle membranes.

Osmotic pressures play a key role in plant physiology, so σ_j's are important parameters for quantitatively describing the solute and water relations of plants. In particular, reflection coefficients allow the role of osmotic pressure to be precisely stated. For Poiseuille's law [$J_V = -(r^2/8\eta)\,\partial P/\partial x$; Eq. 9.11], which can adequately describe movement in the xylem and the phloem—as well as in veins, arteries, and household plumbing—the flow is driven by the gradient in hydrostatic pressure. An alternative view of the same situation is that σ equals zero for the solutes. In that case, osmotic pressures have no direct effect on the movement described by Poiseuille's law, indicating nonselectivity. At the opposite extreme of impermeability, σ is 1, so Eq. (3.38) becomes $J_V = L_P(\Delta P - \Delta \Pi) = L_P \Delta \Psi$ (recall that the water potential, Ψ, can be $P - \Pi$; Eq. 2.13a). This is similar to Eq. (2.24), $J_{V_w} = L_w \Delta \Psi$, which we obtained when only water fluxes were considered. This correspondence is expected, because when the solutes are nonpenetrating, $J_V = J_{V_w}$ and $L_w = L_P$. The real usefulness of reflection coefficients comes when σ_j is not at one of its two extremes of zero and 1. For such cases—intermediate between nonselectivity and impermeability— the volume flux density does not depend on the full osmotic pressure difference across

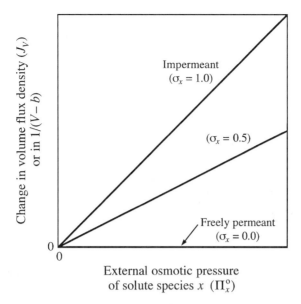

Figure 3-13. Changes in the volume flux density (J_V) across a root or in the reciprocal of the water volume in an organelle $[1/(V - b)]$ as the osmotic pressure of a specific solute in the external bathing solution Π_x^o is increased. Increases in Π_x^o for a nonzero reflection coefficient of solute species $x (\sigma_x > 0)$ decrease the influx or increase the efflux from the root $[J_V = L_P(\Delta P - \sum_j \sigma_j \Delta \Pi_j);$ Eq. 3.39] or decrease the volume for the organelle $[\sigma_x \Pi_x^o + \alpha = \frac{\beta}{V - b};$ Eq. 3.44].

the barrier, but ignoring the osmotic contribution of species j would also be invalid (Fig. 3-13).

VI. Solute Movement across Membranes

We can profitably reexamine certain aspects of the movement of solutes into and out of cells and organelles by using the more general equations from irreversible thermodynamics. One particularly important situation amenable to relatively uncomplicated analysis occurs when the total volume flux density J_V is zero, an example of a *stationary state*. This stationary state, in which the volume of the cell or organelle does not change over the time period of interest, can be brought about by having the net volume flux density of water in one direction across the membrane equal the net volume flux density of solutes in the opposite direction. A stationary state is therefore not the same as a steady state or equilibrium for the cell or organelle; it represents a situation occurring only at a particular time or under some special experimental arrangement. In fact, μ_j can depend on both position and time for a stationary state, but only on position for a steady state and on neither position nor time at equilibrium. When water is moving into a cell, $\Delta\mu_w$ is nonzero (we are not at equilibrium) and, in general, μ_w^i will be increasing with time (we are not even in a steady state). However, we still might have a stationary state, for which the volume is not changing. Our restriction here to cases of zero net volume flux density considerably simplifies the algebra and emphasizes the role played by reflection coefficients. Moreover, a stationary state of no volume change often characterizes the experimental situations under which the Boyle–Van't Hoff relation

(Eqs. 2.15 and 2.18) or the expression describing incipient plasmolysis (Eq. 2.20) is invoked. Thus the derivation of both of these relationships will be reconsidered in terms of irreversible thermodynamics, and we will discuss the role of reflection coeffcients.

What is the relationship between the internal and the external P's and Π's for a stationary state? Our point of departure is Eq. (3.39), $J_V = L_P(\Delta P - \sum_j \sigma_j \Delta \Pi_j)$, where the stationary state condition of zero net volume flux density ($J_V = 0$) leads to the following equalities: $0 = \Delta P - \sum_j \sigma_j \Delta \Pi_j = P^\circ - P^i - \sum_j \sigma_j(\Pi_j^\circ - \Pi_j^i)$. Here, the osmotic pressures indicate the effect of solutes on water activity, whereas in general both osmotic contributions from solutes (Π_s, represented by Π_j for species j) and matric pressures resulting from the presence of interfaces (τ) might occur. Volume measurements for osmotic studies involving incipient plasmolysis or the Boyle–Van't Hoff relation are usually made when the external solution is at atmospheric pressure ($P^\circ = 0$) and when there are no external interfaces ($\tau^\circ = 0$). The stationary state condition of J_V equaling zero then leads to

$$\sum_j \sigma_j \Pi_j^\circ = \sigma^\circ \Pi^\circ = \sum_j \sigma_j \Pi_j^i + \tau^i - P^i \tag{3.40}$$

where the possibilities of interfacial interactions and hydrostatic pressures within the cell or organelle are explicitly recognized by the inclusion of τ^i and P^i. Equation (3.40) applies when the solutes are capable of crossing the barrier, as when molecules interact with real—not idealized—biological membranes. Equation (3.40) also characterizes the external solutes by a mean reflection coefficient, σ°, and the total external osmotic pressure, $\Pi^\circ = \sum_j \Pi_j^\circ$, the latter being relatively easy to measure ($\sigma^\circ = \sum_j \sigma_j \Pi_j^\circ / \sum_j \Pi_j^\circ$, so it represents a weighted mean, with the weights based on the osmotic pressure of each solute species j, Π_j°).

A. The Influence of Reflection Coefficients on Incipient Plasmolysis

In Chapter 2 we used classical thermodynamics to derive the condition for incipient plasmolysis ($\Pi_{\text{plasmolysis}} = \Pi^i$; Eq. 2.20), which occurs when the hydrostatic pressure inside a plant cell P^i just becomes zero. The derivation assumed equilibrium of water, i.e., equal water potential, across a membrane impermeable to solutes. However, the assumptions of water equilibrium and impermeability are often not valid. We can remedy this situation using an approach based on irreversible thermodynamics.

Measurements of incipient plasmolysis can be made for zero volume flux density ($J_V = 0$) and for a simple external solution ($\tau^\circ = 0$) at atmospheric pressure ($P^\circ = 0$). In this case, Eq. (3.40) is the appropriate expression from irreversible thermodynamics, instead of the less realistic condition of water equilibrium used previously. For this stationary state condition, the following expression describes incipient plasmolysis ($P^i = 0$) when the solutes can cross the cell membrane:

$$\sigma^\circ \Pi_{\text{plasmolysis}}^\circ = \sum_j \sigma_j \Pi_j^i + \tau^i \tag{3.41}$$

Because σ° depends on the external solutes present, Eq. (3.41) (a corrected version of Eq. 2.20) indicates that the external osmotic pressure Π° at incipient plasmolysis can vary with the particular solute placed in the solution surrounding the plant cells. Suppose that solute i cannot penetrate the membrane, so $\sigma_i = 1$, a situation often true for sucrose. Suppose that another solute, j, can enter the cells ($\sigma_j < 1$), as is the case for many small nonelectrolytes. If we are at the point of incipient plasmolysis for each of these two solutes as the sole species in the external solution, $\sigma_i \Pi^o_{i\,\text{plasmolysis}}$ must equal $\sigma_j \Pi^o_{j\,\text{plasmolysis}}$ by Eq. (3.41). However, solute i is unable to penetrate the membrane ($\sigma_i = 1$). Hence we obtain the following relationships:

$$\sigma_j = \frac{\sigma_i \Pi^o_{i\,\text{plasmolysis}}}{\Pi^o_{j\,\text{plasmolysis}}} = \frac{\Pi^o_{i\,\text{plasmolysis}}}{\Pi^o_{j\,\text{plasmolysis}}}$$

$$= \frac{\text{effective osmotic pressure of species } j}{\text{actual osmotic pressure of species } j} \tag{3.42}$$

where the effective and the actual osmotic pressures will be discussed next.

Equation (3.42) suggests a straightforward way of describing σ_j. Because by supposition solute species j can cross the cell membrane, $\sigma_j < 1$. Therefore, by Eq. (3.42) $\Pi^o_{j\,\text{plasmolysis}} > \Pi^o_{i\,\text{plasmolysis}}$, where the latter refers to the osmotic pressure of the nonpenetrating solute at the point of incipient plasmolysis. In other words, a higher external osmotic pressure is needed to cause plasmolysis if that solute is able to enter the plant cell. The "actual osmotic pressure" indicated in Eq. (3.42) can be defined by Eq. (2.7) [$\Pi = -(RT/\bar{V}_w)\ln a_w$, where $\Pi = \sum_j \Pi_j$] or by Eq. (2.10) ($\Pi_j = RTc_j$) and also can be measured. When the membrane is impermeable to the solute, $\sigma_j = 1$, and the apparent (effective) osmotic pressure of species j equals the actual Π^o_j. Effective osmotic pressure recognizes that many solutes can cross biological membranes ($\sigma_j < 1$), and hence that the $\Delta \Pi_j$ effective in leading to a net volume flux density is reduced from its actual value [see Eq. 3.39; $J_V = L_P(\Delta P - \sum_j \sigma_j \Delta \Pi_j)$]. In summary, the reflection coefficient of species j indicates how effectively the osmotic pressure of that solute can be exerted across a particular membrane or other barrier.

We can use the condition of incipient plasmolysis to evaluate specific reflection coefficients. Replacing one external solution by another with none of the previous solution adhering to the cells is experimentally difficult. Also, although easy in principle and requiring only a light microscope, determination of when the plasma membrane just begins to pull away from the cell wall is a subjective judgment. Nevertheless, Eq. (3.42) suggests a simple way of considering individual reflection coefficients for various solutes entering plant cells.

To indicate relationships among reflection coefficients, osmotic pressures, and plasmolysis, consider Fig. 3-14. The cell in the upper left of the figure is at the point of incipient plasmolysis ($P^i = 0$) for a nonpenetrating solute ($\sigma^o = 1.0$) in the external solution ($\Pi^o = A$, where A is a constant), i.e., the plasma membrane is just beginning to pull away from the cell wall. By Eq. (3.41) $\sigma^i \Pi^i$ then also equals A. If we place the cell in a second solution containing a penetrating solute ($\sigma^o < 1$), Eq. (3.42) indicates that the external solution must have a higher osmotic pressure for the cell to remain at the point of incipient plasmolysis. For instance, for a second solute with a σ_j of 0.5, the external osmotic pressure at the point of incipient plasmolysis is $2A$ (Fig. 3-14b). Thus when the external solute can enter, Π^o is less effective in balancing the internal osmotic pressure or in leading to a flow of water (Eq. 3.42).

On the other hand, if Π^o were $2A$ for a nonpenetrating external solute, extensive plasmolysis of the cell would occur, as illustrated in Fig. 3-14c. Because $\sigma^o \Pi^o$ is $2A$ in this case, Eq. (3.40) ($\sigma^o \Pi^o = \sigma^i \Pi^i - P^i$, when $\tau^i = 0$) indicates that $\sigma^i \Pi^i$ must also be $2A$, so essentially half of the internal water has left the cell. Finally, if the reflection coefficient were 0.5 and the external osmotic pressure were A, $\sigma^o \Pi^o$ would be $\frac{1}{2}A$, and we would not be at the point of incipient plasmolysis. In fact, the cell would be under turgor with an internal hydrostatic pressure equal to $\frac{1}{2}A$, at least until the concentration of the penetrating solute

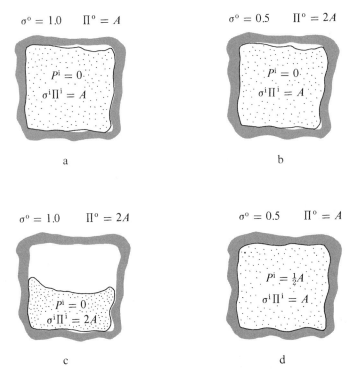

$\sigma^\circ = 1.0 \qquad \Pi^\circ = A$

$P^i = 0$
$\sigma^i \Pi^i = A$

a

$\sigma^\circ = 0.5 \qquad \Pi^\circ = 2A$

$P^i = 0$
$\sigma^i \Pi^i = A$

b

$\sigma^\circ = 1.0 \qquad \Pi^\circ = 2A$

$P^i = 0$
$\sigma^i \Pi^i = 2A$

c

$\sigma^\circ = 0.5 \qquad \Pi^\circ = A$

$P^i = \tfrac{1}{2}A$
$\sigma^i \Pi^i = A$

d

Figure 3-14. Diagrams of sections through a cell showing a cell wall (shaded region) and a plasma membrane (line) for various external osmotic pressures: (a) point of incipient plasmolysis in the presence of a nonpenetrating solute (for clarity of showing the location of the plasma membrane, a slight amount of plasmolysis is indicated), (b) point of incipient plasmolysis with a penetrating solute, (c) extensive plasmolysis, and (d) cell under turgor.

begins to build up inside. We must take into account the reflection coefficients of external and internal solutes to describe conditions at the point of incipient plasmolysis and, by extension, to predict the direction and the magnitude of volume fluxes across membranes.

B. Extension of the Boyle–Van't Hoff Relation

In Chapter 2 we derived the Boyle–Van't Hoff relation assuming that the water potential was the same on both sides of the cellular or organelle membrane under consideration. Not only were equilibrium conditions imposed on water but also we implicitly assumed that the membrane was impermeable to the solutes. However, zero net volume flux density ($J_V = 0$) is a better description of the experimental situations under which the Boyle-Van't Hoff relation is applied. No volume change during the measurement is another example of a stationary state, so the Boyle–Van't Hoff relation will be reexamined from the point of view of irreversible thermodynamics. In this way we can remove two of the previous restrictions—equilibrium for water and impermeability of solutes.

When molecules cross the membranes bounding cells or organelles, the reflection coefficients of both internal and external solutes should be included in the Boyle–Van't Hoff relation. Because $\sigma^\circ < 1$ when the external solutes can penetrate, the effect of the external osmotic pressure on J_V is reduced. Likewise, the reflection coefficients for solutes within the cell or organelle can lessen the contribution of the internal osmotic pressure of each solute. Replacing Π_j^i by $RTn_j^i/(\bar{V}_w n_w^i)$ (Eq. 2.10) in Eq. (3.40) and dividing by σ^o leads to the following Boyle–Van't Hoff relation for the stationary state condition ($J_V = 0$ in Eq. 3.39):

$$\Pi^o = RT \frac{\sum_j \sigma_j n_j^i}{\sigma^o \bar{V}_w n_w^i} + \frac{\tau^i - P^i}{\sigma^o} \tag{3.43}$$

We note that the reflection coefficients of a membrane for both internal and external solutes enter into this extension of the expression relating volume and external osmotic pressure.

As indicated in Chapter 2, $V - b$ in the conventional Boyle–Van't Hoff relation [$\Pi^o(V - b) = RT \sum_j \varphi_j n_j$ (Eq. 2.15)] can be identified with $\bar{V}_w n_w^i$. Comparing Eq. (2.15) with Eq. (3.43), the osmotic coefficient of species j, φ_j, can be equated to σ_j/σ^o as an explicit recognition of the permeation properties of solutes, both internal and external. Indeed, failure to recognize the effect of reflection coefficients on φ_j has led to misunderstandings of osmotic responses.

As discussed in Chapter 2, the volume of pea chloroplasts (as well as other organelles and many cells) responds linearly to $1/\Pi^o$ (Fig. 2-9), indicating that $\tau^i - P^i$ in such organelles may be negligible compared with the external osmotic pressures used. To analyze experimental observations, $\sigma^o \Pi^o$ can be replaced by $\sigma_x \Pi_x^o + \alpha$, where Π_x^o is the external osmotic pressure of solute x whose reflection coefficient (σ_x) is being considered, and α is the sum of $\sigma_j \Pi_j^o$ for all other external solutes. We can replace $RT \sum_j \sigma_j n_j^i$ by β and $\bar{V}_w n_w^i$ by $V - b$. Making these substitutions into Eq. (3.43), we obtain the following relatively simple form for testing osmotic responses in the case of penetrating solutes:

$$\sigma_x \Pi_x^o + \alpha = \frac{\beta}{V - b} \tag{3.44}$$

If we vary Π_x^o and measure V, we can then use Eq. (3.44) to obtain the reflection coefficients for various nonelectrolytes in the external solution.

C. Reflection Coefficients of Chloroplasts

When the refinements introduced by reflection coefficients are taken into account, we can use osmotic responses of cells and organelles to describe quantitatively the permeability properties of their membranes (see Fig. 3-13). As a specific application of Eq. (3.44), we note that the progressive addition of hydroxymethyl groups in a series of polyhydroxy alcohols causes the reflection coefficients to increase steadily from 0.00 to 1.00 for pea chloroplasts (Table 3-2). In this regard, the lipid:water solubility ratio decreases from methanol to ethylene glycol to glycerol to erythritol to adonitol, i.e., the partition coefficient K_x decreases. Because the permeability coefficient $P_x = D_x K_x/\Delta_x$ (Eq. 1.9), we expect a similar decrease in P_x as hydroxymethyl groups are progressively added. Figure 3-12 shows that, as the permeability coefficient decreases, the reflection coefficient generally increases. Consequently, the increase in σ_x of alcohols as —CHOH groups are added can be interpreted as a lowering of K_x. (As we go from the one-C methanol to the five–C adonitol, D_x also decreases, perhaps by a factor of 3, whereas K_x decreases about a 1000-fold, so changes in K_x are the predominant influence on P_x and σ_x in this case.) The reflection coefficients of six-C polyhydroxy alcohols, such as sorbitol and mannitol, are 1 for pea chloroplasts. This indication of relative impermeability suggests that these compounds could serve as suitable

Table 3-2. Reflection Coefficients of Chloroplasts from *Pisum sativum* for Alcohols.[a]

Substance	σ_x	Substance	σ_x
Methanol	0.00	Adonitol	1.00
Ethylene glycol	0.40	Sorbitol	1.00
Glycerol	0.63	Mannitol	1.00
Erythritol	0.90	Sucrose	1.00

[a] The reflection coefficients here apply to the pair of membranes surrounding the organelles, this being the barrier to solute entry or exit encountered in a plant cell (source: Wang and Nobel, 1971).

osmotica in which to suspend chloroplasts, as is indeed the case. From these examples, we see that the esoteric concepts of irreversible thermodynamics can be applied in a relatively simple manner to gain insights into the physiological attributes of membranes.

D. Solute Flux Density

Our final objective in this chapter is to obtain an expression for the solute flux density, J_s, that takes into consideration the coupling of forces and fluxes of irreversible thermodynamics. Using Eqs. (3.33b) and (3.34), we note that

$$J_V + J_D = J_w \bar{V}_w + J_s \bar{V}_s + J_s/\bar{c}_s - J_w \bar{V}_w$$

$$= J_s \left(\bar{V}_s + \frac{1}{\bar{c}_s} \right) \cong J_s/\bar{c}_s \tag{3.45}$$

where the last step applies to a dilute solution ($\bar{V}_s \bar{c}_s \ll 1$, or $\bar{V}_s \ll 1/\bar{c}_s$). After multiplying both sides of Eq. (3.45) by $\bar{c}_s$, and using Eq. (3.36) for J_D, we obtain

$$J_s \cong \bar{c}_s J_V + \bar{c}_s (L_{DP} \Delta P + L_D \Delta \Pi)$$

$$= \bar{c}_s J_V + \bar{c}_s L_{DP} \frac{(J_V - L_{PD} \Delta \Pi)}{L_P} + \bar{c}_s L_D \Delta \Pi$$

$$= \bar{c}_s J_V \left(1 + \frac{L_{DP}}{L_P} \right) + \bar{c}_s \Delta \Pi \left(L_D - \frac{L_{DP} L_{DP}}{L_P} \right) \tag{3.46}$$

where Eq. (3.35) is used to eliminate ΔP. Upon combining the factors multiplying $\Delta \Pi$ into a new coefficient for solute permeability, ω, and using the previous definition of σ (Eq. 3.37), we can rewrite the last line of Eq. (3.46) as

$$J_s = \bar{c}_s (1 - \sigma) J_V + \omega \Delta \Pi \tag{3.47}$$

where σ, ω, and $\Delta \Pi$ all refer to a specific solute. We note that L_P, σ, and ω are the three experimentally convenient parameters introduced by irreversible thermodynamics replacing the two parameters of classical thermodynamics for characterizing the movement of water and a single solute across a membrane.

Equation (3.47) indicates that not only does J_s depend on $\Delta \Pi$, as expected from classical thermodynamics, but also that the solute flux density can be affected by the overall volume flux density, J_V. In particular, the classical expression for J_s for a neutral solute is $P_j \Delta c_j$ (Eq. 1.8), which equals $(P_j/RT) \Delta \Pi_j$ using the Van't Hoff relation (Eq. 2.10; $\Pi_s = RT \sum_j c_j$). Thus, ω is analogous to P_j/RT of the classical thermodynamic description.

The classical treatment indicates that J_s is zero if $\Delta\Pi$ is zero. On the other hand, when $\Delta\Pi$ is zero, Eq. (3.47) indicates that $J_s = \bar{c}_s(1 - \sigma)J_V$; solute molecules are thus dragged across the membrane by the moving solvent, leading to a solute flux density proportional to the local solute concentration and to the deviation of the reflection coefficient from 1. Hence, P_j may not always be an adequate parameter by which to describe the permeability of species j because the interdependence of forces and fluxes introduced by irreversible thermodynamics indicates that water and solute flow can interact with respect to solute movement across membranes.

Problems

3.1 At the beginning of this chapter we calculated that an average concentration of 1 mmol m^{-3}(1 μM) excess monovalent anions can lead to a -100 mV potential change across the surface of a spherical cell 30 μm in radius.

   (a) If the same total amount of charge were concentrated in a layer 3 nm thick at the surface of the sphere, what would be its average concentration there?
   (b) If 10^7 sulfate ions are added inside the sphere, what is the new potential difference across the surface?
   (c) Approximately how much electrical work in joules is required to transport the 10^7 sulfate ions across the surface of the cell?

3.2 For purposes of calculation, let us assume that an external solution is 1 mol m^{-3} (1 mM) KCl, and the solution inside a cell is 160 mol m^{-3} KCl at 20°C.

   (a) If K$^+$ is in equilibrium across the membrane and activity coefficients are 1, what is the electrical potential difference across the membrane?
   (b) If K$^+$ is in equilibrium and the mean activity coefficients are calculated from Eq. (3.3), what would the membrane potential be?
   (c) If 1 mol m^{-3} K$_3$ATP, which fully dissociates to 3K$^+$ and ATP^{-3}, is added inside the cell, and if K$^+$–ATP^{-3} can be considered to act as an ion pair, what are γ_{K-ATP} and a_{ATP}?

3.3 Consider a cell with a membrane potential E_M of -118 mV at 25°C. Suppose that the external solution contains 1 mM KCl, 0.1 mM NaCl, and 0.1 mM MgCl$_2$, and the internal concentration of K$^+$ is 100 mM, that of Ca^{2+} is 1 mM, and that of Mg^{2+} is 10 mM (1 mM = 1 mol m^{-3}). Assume that activity coefficients are 1.

   (a) Are K$^+$ and Mg^{2+} in equilibrium across the membrane?
   (b) If Na$^+$ and Ca^{2+} are in equilibrium, what are their concentrations in the two phases?
   (c) If Cl$^-$ is 177 mV away from equilibrium, such that the passive driving force on it is outward, what is its Nernst potential, and what is the internal concentration of Cl$^-$?
   (d) What is J_{Cl}^{in}/J_{Cl}^{out} for passive fluxes?
   (e) What are $\Delta\mu_{Cl}$ and $\Delta\mu_{Mg}$ across the membrane?

3.4 A 10 mM KCl solution at 25°C is placed outside a cell formerly bathed in 1 mM KCl (1 mM = 1 mol m^{-3}).

   (a) Assuming that some of the original solution adheres to the cell and that the ratio of mobilities (u_{Cl}/u_K) is 1.04, what diffusion potential would be present?

(b) Assume that equilibrium is reached in the bathing solution after a sufficient lapse of time. The membrane may contain many carboxyl groups ($-COOH$) whose H^+'s will dissociate. The ensuing negative charge will attract K^+, and its concentration near the membrane may reach 200 mM. What type of and how large a potential would be associated with this situation?

(c) Suppose that 10 mM NaCl is also in the external solution (with the 10 mM KCl), and that internally there is 100 mM K^+, 10 mM Na^+, and 100 mM Cl^-. Assume that P_{Na}/P_K is 0.20 and P_{Cl}/P_K is 0.01. What diffusion potential would be expected across the membrane?

(d) What would E_M be if P_{Cl}/P_K were 0.00? If P_{Na}/P_K and P_{Cl}/P_K were both 0.00? Assume that other conditions are as in (c).

3.5 Consider an illuminated spherical spongy mesophyll cell 40 μm in diameter containing 50 spherical chloroplasts 4 μm in diameter.

(a) Some monovalent anion produced by photosynthesis has a steady-state net flux density out of the chloroplasts of 10 nmol m^{-2} s^{-1}. If this photosynthetic product is not changed or consumed in any of the cellular compartments, what is the net passive flux density out of the cell in the steady state?

(b) If the passive flux density of the previous substance into the cell at 25°C is 1 nmol m^{-2} s^{-1}, what is the difference in its chemical potential across the cellular membrane?

(c) Suppose that, when the cell is placed in the dark, the influx and the efflux both become 0.1 nmol m^{-2} s^{-1}. If the plasma membrane potential is -118 mV (inside negative) and the same concentration occurs on the two sides of the membrane, what can be said about the energetics of the two fluxes?

(d) If one ATP is required per ion transported, what is the rate of ATP consumption in (c)? Express your answer in μmol s^{-1} per m^3 of cellular contents.

3.6 The energy of activation for crossing biological membranes can represent the energy required to break hydrogen bonds between certain nonelectrolytes and the solvent water, e.g., to enter a cell the solute must first dissolve in the lipid phase of the membrane, and thus the hydrogen bonds with water must be broken.

(a) What will be the Q_{10} for the influx of a solute that forms one H bond per molecule with water if we increase the temperature from 10 to 20°C?

(b) How many hydrogen bonds would have to be ruptured per molecule to account for a Q_{10} of 3.2 under the conditions of (a)?

3.7 Suppose that transporters in the plasma membrane can shuttle K^+ and Na^+ into a cell. We will let the Michaelis constant K_j be 0.010 mM for the K^+ transporter and 1.0 mM for the Na^+ transporter (1 mM $= 1$ mol m^{-3}), and the maximum influx of either ion is 10 nmol m^{-2} s^{-1}.

(a) What is the ratio of influxes, J_K^{in}/J_{Na}^{in}, when the external concentration of each ion is 0.010 mM?

(b) What is J_K^{in}/J_{Na}^{in} when the external concentration of each ion is 100 mM?

(c) If the entry of K^+ is by facilitated diffusion only, what is the rate of K^+ entry when c_K^o is 0.1 mM and ATP is being hydrolyzed at the rate of 10 mmol m^{-2} s^{-1}?

3.8 Consider a cell whose membrane has a hydraulic conductivity coefficient L_P of 10^{-12} m s^{-1} Pa^{-1}. Initially, no net volume flux density occurs when the cell is placed in a solution having the following composition: sucrose ($\Pi_j^o = 0.2$ MPa, $\sigma_j = 1.00$), ethanol (0.1 MPa, 0.30), and glycerol (0.1 MPa, 0.80). The external solution is at atmospheric pressure, and P^i is 0.5 MPa. Inside the cell the osmotic pressure caused by glycerol is 0.2 MPa, sucrose and ethanol are initially absent, and other substances having an osmotic pressure of 1.0 MPa are present.

(a) What is the mean reflection coefficient for the external solution?
(b) What is the mean reflection coefficient for the internal solutes other than glycerol?
(c) Suppose that some treatment makes the membrane nonselective for all solutes present. Is there then a net volume flux density?
(d) If another treatment makes the membrane impermeable to all solutes present, what would be the net volume flux density?

3.9 Consider a cell at the point of incipient plasmolysis in an external solution containing 0.3 m sucrose, a nonpenetrating solute. The concentration of glycine that just causes plasmolysis is 0.4 m. Assume that no water enters or leaves the cell during the plasmolytic experiments.

(a) What is the reflection coefficient of glycine for the cellular membrane?
(b) Suppose that chloroplasts isolated from such a cell have the same osmotic responses as in Problem 2.5. What is the volume of such chloroplasts *in vivo*? Assume that activity coefficients are unity and that the temperature is 20°C.
(c) Suppose that chloroplasts are isolated in 0.3 m sucrose, which has a reflection coefficient of 1.00 for the chloroplasts. If 0.1 mol of glycine is then added per kilogram of water in the isolation medium, and if the chloroplast volume is 23 μm^3, what is the reflection coefficient of glycine for the chloroplast membranes?
(d) What is the external concentration of glycerol ($\sigma_j = 0.60$) in which the chloroplasts have the same initial volume as in 0.3 m sucrose? What is the chloroplast volume after a long time in the glycerol solution?

References

Alberty, R. A., and Silbey, R. J. (1997). *Physical Chemistry* (2nd ed.). Wiley, New York.

Atkins, P. W. (1994). *Physical Chemistry* (5th ed.). Freeman, New York.

Chrispeels, M. J., and Maurel, C. (1994). Aquaporins: The molecular basis of facilitated water movement through living plant cells? *Plant Physiol.* **105**, 9–13.

Davson, H., and Danielli, J. F. (1952). *The Permeability of Natural Membranes* (2nd ed.). Cambridge Univ. Press, Cambridge, UK.

DeFelice, L. J. (1997). *Electrical Properties of Cells: Patch Clamp for Biologists.* Plenum, New York.

Diamond, J. M., and Wright, E. M. (1969). Biological membranes: The physical basis of ion and nonelectrolyte selectivity. *Annu. Rev. Physiol.* **31**, 581–646.

Feher, J. J., and Ford, G. D. (1995). A simple student laboratory on osmotic flow, osmotic pressure, and the reflection coefficient. *Adv. Physiol. Education* **13**, S10–S20.

Goldman, D. E. (1943). Potential, impedance, and rectification in membranes. *J. Gen. Physiol.* **27**, 37–60.

Hodgkin, A. L., and Katz, B. (1949). The effect of sodium ions on the electrical activity of the giant axon of the squid. *J. Physiol.* **108**, 37–77.

Hope, A. B. (1971). *Ion Transport and Membranes.* Butterworths, London.

Hope, A. B., and Walker, N. A. (1975). *The Physiology of Giant Algal Cells.* Cambridge Univ. Press, Cambridge, UK.

Jackson, M. B. (Ed.) (1993). *Thermodynamics of Membrane Receptors and Channels.* CRC Press, Boca Raton, FL.

Jou, D., Casas-Vazquez, J., and Lebon, G. (1996). *Extended Irreversible Thermodynamics* (2nd ed.). Springer-Verlag, Berlin.

Kedem, O., and Katchalsky, A. (1958). Thermodynamic analysis of the permeability of biological membranes to nonelectrolytes. *Biochem. Biophys. Acta.* **27**, 229–246.

Keizer, J. (1987). *Statistical Thermodynamics of Nonequilibrium Processes.* Springer-Verlag, New York.

Kuiken, G. D. C. (1994). *Thermodynamics of Irreversible Processes: Applications to Diffusion and Rheology.* Wiley, New York.

Lehninger, A. L., Nelson, D. L., and Cox, M. M. (1993). *Principles of Biochemistry* (2nd ed.). Worth, New York.

Levine, I. N. (1995). *Physical Chemistry* (4th ed.). McGraw-Hill, New York.

Logan, H., Basset, M., Véry, A.-A., and Sentenac, H. (1997). Plasma membrane transport systems in higher plants: From black boxes to molecular physiology. *Physiol. Plant.* **100**, 1–15.

Maathuis, F. J. M., Ichida, A. M., Sanders, D., and Schroeder, J. L. (1997). Roles of higher plant K^+ channels. *Plant Physiol.* **114**, 1141–1149.

MacRobbie, E. A. C. (1962). Ionic relations of *Nitella translucens. J. Gen. Physiol.* **45**, 861–878.

McGill, P., and Schumaker, M. F. (1996). Boundary conditions for single-ion diffusion. *Biophys. J.* **71**, 1723–1742.

Murphy, R., and Smith, J. A. C. (1994). Derivation of a weighted-average reflection coefficient for mesophyll cell membranes of *Kalanchoë diagremontiana. Planta* **193**, 145–147.

Neher, E., and Sakmann, B. (1976). Single-channel currents recorded from membrane of denervated frog muscle fibers. *Nature* **260**, 799–802.

Nobel, P. S. (1969). The Boyle–Van't Hoff relation. *J. Theor. Biol.* **23**, 375–379.

Noggle, J. H. (1985). *Physical Chemistry.* Little, Brown, Boston.

Onsager, L. (1931). Reciprocal relations in irreversible processes. II. *Phys. Rev.* **38**, 2265–2279.

Osterhout, W. J. V. (1935). How do electrolytes enter the cell? *Proc. Natl. Acad. Sci. USA* **21**, 125–132.

Peracchia, C. (Ed.) (1994). *Handbook of Membrane Channels: Molecular and Cellular Physiology.* Academic Press, San Diego.

Prigogine, I. (1967). *Thermodynamics of Irreversible Processes* (3rd ed.). Wiley (Interscience), New York.

Pytkowicz, R. M. (1979). *Activity Coefficients in Electrolyte Solutions.* CRC Press, Cleveland, OH.

Raison, J. K., and Chapman, E. A. (1976). Membrane phase changes in chilling-sensitive *Vigna radiata* and their significance to growth. *Aust. J. Plant Physiol.* **3**, 291–299.

Richter, C., and Dainty, J. (1989). Ion behavior in plant cell walls. I. Characterization of the *Sphagnum russowii* cell wall ion exchanger. *Can. J. Bot.* **67**, 451–459.

Sanders, D., and Tester, M. (Eds.) (1997). Ion channels. *J. Exp. Bot.* **48**, 353–631.

Schütz, K., and Tyerman, S. D. (1997). Water channels in *Chara corallina. J. Exp. Bot.* **48**, 1511–1518.

Serrano, R. (1985). *Plasma Membrane ATPase of Plants and Fungi.* CRC Press, Boca Raton, FL.

Smallwood, M., Knox, J. P., and Bowles, D. J. (Eds.) (1996). *Membranes: Specialized Functions in Plants.* BIOS Scientific, Oxford.

Spanswick, R. M. (1981). Electrogenic ion pumps. *Annu. Rev. Plant Physiol.* **32**, 267–289.

Spanswick, R. M., and Williams, E. J. (1964). Electrical potentials and Na, K, and Cl concentrations in the vacuole and cytoplasm of *Nitella translucens. J. Exp. Bot.* **15**, 193–200.

Starzak, M. E. (1984). *The Physical Chemistry of Membranes.* Academic Press, Orlando, FL.

Staverman, A. J. (1951). The theory of measurement of osmotic pressure. *Recueil Travaux Chimiques Pays-Bas* **70**, 344–352.

Stein, W. D. (1986). *Transport and Diffusion across Cell Membranes.* Academic Press, Orlando, FL.

Stein, W. D. (1990). *Channels, Carriers, and Pumps: An Introduction to Membrane Transport.* Academic Press, San Diego.

Steudle, E., and Henzler, T. (1995). Water channels in plants: Do basic concepts of water transport change? *J. Exp. Bot.* **46**, 1067–1076.

Tazawa, M., Shimmen, T., and Mimura, T. (1987). Membrane control in the Characeae. *Annu. Rev. Plant Physiol.* **38**, 95–117.

Teorell, T. (1949). Membrane electrophoresis in relation to bio-electrical polarization effects. *Arch. Sci. Physiol.* **3**, 205–219.

Ussing, H. H. (1949). The distinction by means of tracers between active transport and diffusion. *Acta Physiol. Scand.* **19**, 43–56.

Van Holde, K. E. (1985). *Physical Biochemistry* (2nd ed.). Prentice Hall, Englewood Cliffs, NJ.

Voet, D., and Voet, J. G. (1995). *Biochemistry* (2nd ed.). Wiley, New York.

Volkov, A. G., Deamer, D. W., Tanelian, D. L., and Markin, V. S. (1997). *Liquid Interfaces in Chemistry and Biology*. Wiley, New York.

Vorobiev, V. N. (1967). Potassium ion activity in the cytoplasm and the vacuole of cells of *Chara and Griffithsia*. *Nature* **216**, 1325–1327.

Wang, C.-t., and Nobel, P. S. (1971). Permeability of pea chloroplasts to alcohols and aldoses as measured by reflection coefficients. *Biochim. Biophys. Acta* **241**, 200–212.

Ward, J. M. (1997). Patch-clamping and other molecular approaches for the study of plasma membrane transporters demystified. *Plant Physiol.* **114**, 1151–1159.

Yagi, K., and Pullman, B. (Eds.) (1987). *Ion Transport through Membranes*. Academic Press, Orlando, FL.

Zubay, G. L., Parson, W. W., and Vance, D. E. (1995). *Principles of Biochemistry*. Brown, Dubuque, IA.

4

Light

I. **Wavelength and Energy**   **144**
    A. Light Waves   144
    B. Energy of Light   146
    C. Illumination, Photon Flux Density, and Irradiance   148
    D. Sunlight   150
    E. Planck's and Wien's Formulae   152
II. **Absorption of Light by Molecules**   **153**
    A. Role of Electrons in Absorption Event   153
    B. Electron Spin and State Multiplicity   154
    C. Molecular Orbitals   156
    D. Photoisomerization   158
    E. Light Absorption by Chlorophyll   159
III. **Deexcitation**   **161**
    A. Fluorescence, Radiationless Transition,
        and Phosphorescence   161
    B. Competing Pathways for Deexcitation   162
    C. Lifetimes   164
    D. Quantum Yields   166
IV. **Absorption Spectra and Action Spectra**   **166**
    A. Vibrational Sublevels   168
    B. The Franck–Condon Principle   169
    C. Absorption Bands and Absorption Coefficients   171
    D. Application of Beer's Law   173
    E. Conjugation   174
    F. Action Spectra   175
    G. Absorption and Action Spectra of Phytochrome   176
    Problems   **179**
    References   **181**

THROUGH A series of nuclear reactions taking place within the sun, mass is converted into energy in accordance with Einstein's famous relation, $E = mc^2$. By such conversion of mass to energy, the sun maintains an extremely high surface temperature and thus radiates a great amount of energy into space. Some of this radiant energy is incident on the earth, only a small fraction of which is absorbed by plants (discussed in Chapter 6). This absorption initiates a flow of energy through the biosphere (all living things and that portion of the earth that they inhabit).

The first step in the utilization of sunlight for this energy flow is the conversion of its radiant energy into various forms of chemical energy by the primary processes of photosynthesis. This chemical energy may then be stored in plants, mainly in the form of carbohydrates. The stored energy may later be acquired by animals—directly by herbivores, indirectly by carnivores, or both directly and indirectly by omnivores like us. However the energy is acquired, its ultimate source is the solar radiation trapped by photosynthesis. Without this continuous energy input from the sun, living organisms would drift toward equilibrium and hence death.

Sunlight also regulates certain activities of plants and animals by acting as a trigger. The energy to carry out these activities is supplied by metabolic reactions, not directly by the light itself. Examples of light acting as a trigger include vision, phototaxis, phototropism, and the phytochrome regulation of various plant processes.

In this chapter we will be concerned primarily with the physical nature of light and the mechanism of light absorption by molecules. We will discuss how molecular states excited by light absorption can promote endergonic (energy-requiring) reactions or be dissipated by other deexcitation processes. In Chapter 5 we will consider the photochemistry of photosynthesis and in Chapter 6 the bioenergetics of energy conversion, especially that taking place in organelles. In Chapter 7 we will demonstrate how the net energy input by radiation is dissipated by a leaf.

As an introduction to the topic of light, we consider certain historical developments in the understanding of its nature. In 1666 Sir Isaac Newton showed that a prism could disperse white light into many different colors, suggesting that such radiation is a mixture of many components. Soon thereafter, Christiaan Huygens proposed that the propagation of light through space could be by wave motion. In 1802–1804 Thomas Young attributed interference properties to the wave character of light. However, a wave theory of light was not generally accepted until about 1850, when Jean Foucault demonstrated that light travels more slowly in a dense medium such as water than in a rarefied medium such as air, one of the predictions of the wave theory. In 1887 Heinich Hertz discovered that light striking the surface of a metal could cause the release of electrons from the solid—the "photoelectric" effect. However, he also found that wavelengths above a certain value could not eject any electrons at all, no matter what the total energy in the light beam. This important result was contrary to the then-accepted wave theory of light. In an important departure from wave theory, Max Planck proposed in 1901 that radiation was particle-like, i.e., light was describable as consisting of discrete packets, or quanta, each of a specific energy. In 1905 Albert Einstein explained the photoelectric effect of Hertz as a special example of the particle nature of light, indicating that the absorption of a photon of sufficiently short wavelength by an electron in the metal could supply enough energy to cause the ejection or release of that electron, whereas if the wavelengths were longer, then the individual photons were not energetic enough to eject any electrons. The intriguing wave–particle duality of light has subsequently become described in a consistent manner through the development of quantum mechanics. Both wave and particle attributes of light are necessary for a complete description of radiation, and we will consider both aspects in this chapter. (Although this text does not require a background in quantum mechanics, some knowledge of this field is essential for a comprehensive understanding of light.)

I. Wavelength and Energy

Light is often defined as that electromagnetic radiation perceivable by the human eye. Although such a definition may be technically correct, the word *light* is frequently used to refer to a wider range of electromagnetic radiation. In this section we will discuss the range of electromagnetic radiation important in biology, including the subdivisions into various wavelength intervals. The wavelength of light will be related to its energy. After noting various conventions used to describe radiation, we will briefly consider some of the characteristics of solar radiation reaching the earth.

A. Light Waves

The regular and repetitive changes in the intensity of the minute electric and magnetic fields indicate the passage of a light wave (Fig. 4-1). Light can travel in a solid (e.g., certain plastics), in a liquid (water), in a gas (air), and even in a vacuum (the space between the sun and the earth's atmosphere). One way to characterize light is by its *wavelength*—the distance between successive points of the same phase, such as between two successive peaks of a wave (Fig. 4-1). A wavelength, then, is the distance per cycle of the wave. The preferred unit for wavelengths of light is the nanometer (nm, 10^{-9} m), which we will use in this text (1 nm $= 0.001$ μm $= 10$ Å).

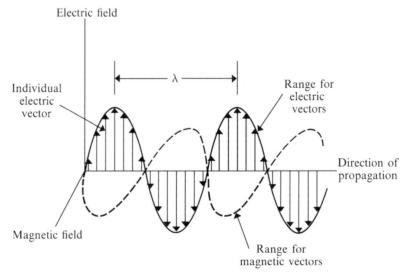

Figure 4-1. Light can be represented by an electromagnetic wave corresponding to oscillations of the local electric and magnetic fields. The oscillating electric vectors at a particular instant in time are indicated by arrows in the diagram. A moment later, the entire pattern of electric and magnetic fields will shift in the direction of propagation of the wave. A wavelength (λ) is the distance between two successive points of the same phase along the wave.

Table 4-1. Definitions and Characteristics of the Various Wavelength Regions of Light[a]

Color	Approximate wavelength range (nm)	Representative wavelength (nm)	Frequency (cycles s^{-1}, or hertz)	Energy (kJ mol^{-1})
Ultraviolet	Below 400	254	11.80×10^{14}	471
Violet	400–425	410	7.31×10^{14}	292
Blue	425–490	460	6.52×10^{14}	260
Green	490–560	520	5.77×10^{14}	230
Yellow	560–585	570	5.26×10^{14}	210
Orange	585–640	620	4.84×10^{14}	193
Red	640–740	680	4.41×10^{14}	176
Infrared	Above 740	1,400	2.14×10^{14}	85

[a]The ranges of wavelengths leading to the sensation of a particular color are arbitrary and vary with individuals. Both frequencies and energies in the table refer to the particular wavelength indicated in column 3 for each wavelength interval. Wavelength magnitudes are those in a vacuum.

The wavelength regions of major interest in biology are the ultraviolet, the visible, and the infrared (Table 4-1). Wavelengths immediately below about 400 nm are referred to as ultraviolet (UV)—meaning on the other side of or beyond violet in the sense of having a shorter wavelength (x-rays and γ-rays occur at even shorter wavelengths). The lower limit for UV is somewhat arbitrary, often near 10 nm, but very little solar radiation occurs at wavelengths <150 nm. The visible region extends from approximately 400 to 740 nm and is subdivided into various bands such as blue, green, and red (Table 4-1). These divisions are based on the subjective color experienced by humans. The infrared (IR) region has wavelengths longer than those of the red end of the visible spectrum, extending up to approximately 100 μm (microwaves and radiowaves occur at even longer wavelengths).

Besides wavelength, we can also characterize a light wave by its frequency of oscillation, v, and by the magnitude of its velocity of propagation, v (i.e., v is the speed of light). These three quantities are related as follows:

$$\lambda v = v \tag{4.1}$$

where λ is the wavelength. In a vacuum the speed of light for all wavelengths is a constant, generally designated as c, which experimentally equals 299,800 km s^{-1}, or about 3.00×10^8 m s^{-1}. Light passing through a medium other than a vacuum has a speed less than c. For example, the speed of light with a wavelength of 589 nm in a vacuum is decreased 0.03% by air, 25% by water, and 40% by dense flint glass; shorter wavelengths are decreased more in speed. In this regard, more electrons are encountered by the light wave per unit path length traversed through an optically dense material, thus slowing the wave more; also, shorter wavelengths have a higher frequency and hence interact more frequently with electrons. The wavelength undergoes a decrease in magnitude equal to the decrease in the speed of propagation in these various media because the unchanging property of a wave propagating through different media is the frequency. We also note that v is the frequency of the oscillations of the local electric and magnetic fields of light (see Fig. 4-1).

For light, $\lambda_{vacuum} v = v_{vacuum}$ by Eq. (4.1), where v_{vacuum} is the constant c. Therefore, if we know the wavelength in a vacuum, we can calculate the frequency. In fact, the wavelengths given for light generally refer to values in a vacuum, as is the case for columns 2 and 3 in Table 4-1 (λ's in air differ only slightly from the magnitudes listed). As a specific example, let us select a wavelength in the blue region of the spectrum, e.g., 460 nm. By Eq. (4.1), the frequency of this blue light is

$$\nu = \frac{(3.00 \times 10^8 \text{ m s}^{-1})}{(460 \times 10^{-9} \text{ m cycle}^{-1})} = 6.52 \times 10^{14} \text{ cycles s}^{-1}$$

Because ν does not change from medium to medium, it is often desirable to describe light by its frequency, as has been done in column 4 of Table 4-1. The wavelength of light is generally expressed in nm, not nm cycle^{-1}; ν is then considered to have units of s^{-1} (not cycles s^{-1}).

B. Energy of Light

In addition to its wave-like properties, light also exhibits particle-like properties, as in the case of the photoelectric effect mentioned previously. Thus light can act as if it were divided (or quantized) into discrete units, which we call *photons*. The light energy (E_λ) carried by a photon is

$$E_\lambda = h\nu = hc/\lambda_{\text{vacuum}} \quad \text{photon basis} \qquad (4.2a)$$

where h is a fundamental physical quantity called *Planck's constant*. By Eq. (4.2a), a photon of light has an energy directly proportional to its frequency and inversely proportional to its wavelength in a vacuum. For most applications in this book, we will describe light by its wavelength λ. To emphasize that the energy of a photon depends on its wavelength, we have used the symbol E_λ in Eq. (4.2a). A *quantum* (plural: quanta) refers to the energy carried by a photon, i.e., $h\nu$ represents a quantum of electromagnetic energy. The terms "photon" and "quantum" are sometimes used interchangeably, a usage that is generally clear but not strictly correct.

The introduction of the constant h by Planck in the early 1900s represented a great departure from the accepted wave theory of light. It substantially modified the classical equations describing radiation and provided a rational basis for determining the energy of photons. Because frequency has the units of time^{-1}, Eq. (4.2a) indicates that the dimensions of Planck's constant are energy × time, e.g., 6.626×10^{-34} J s (see Appendix I). We note that hc is 1240 eV nm (Appendix I), so we can readily calculate the energy per photon of blue light at 460 nm using Eq. (4.2a):

$$E_\lambda = \frac{(1240 \text{ eV nm})}{(460 \text{ nm})} = 2.70 \text{ eV}$$

Instead of energy per photon, we are usually more interested in the energy per Avogadro's number $N(6.022 \times 10^{23})$ of photons (N photons is sometimes called an einstein, but the correct SI unit is mole). On a mole basis, Eq. (4.2a) becomes

$$E_\lambda = N h\nu = Nhc/\lambda_{\text{vacuum}} \quad \text{mole basis} \qquad (4.2b)$$

Let us consider blue light of 460 nm, which has a frequency of 6.52×10^{14} cycles s^{-1} (Table 4-1). Using Eq. (4.2b), we calculate that the energy per mole of 460-nm photons is

$$E_\lambda = (6.022 \times 10^{23} \text{ mol}^{-1})(6.626 \times 10^{-34} \text{ J s})(6.52 \times 10^{14} \text{ s}^{-1})$$

$$= 260 \text{ kJ mol}^{-1}$$

Alternatively, we can calculate the energy by dividing Nhc (119,600 kJ mol^{-1} nm) by the wavelength in a vacuum (see Table 4-1).

Quanta of visible light represent relatively large amounts of energy. The hydrolysis of ATP, the main currency for chemical energy in biology, yields about 40–50 kJ mol^{-1} under physiological conditions (see Chapter 6), whereas, as we just calculated, blue light has five or six times as much energy per mole of photons. Quanta of ultraviolet light represent even higher energies than do those of visible light, e.g., 254-nm photons have 471 kJ of radiant energy mol^{-1} (Table 4-1; 254 nm is the wavelength for a major mercury line in discharge lamps used for their bactericidal action). This is greater than the carbon–carbon bond energy of 348 kJ mol^{-1} or the oxygen–hydrogen bond energy of 463 kJ mol^{-1}. The high quantum energy of UV radiation underlies its mutagenic and bactericidal action because it is energetic enough to cause disruption of various covalent bonds.

The photoelectric effect, for which light leads to the ejection of electrons from the surface of a metal, clearly illustrates the distinction between light energy and the energy of its photons. Robert Millikan found that photons with at least 175 kJ mol^{-1}, representing wavelengths of 683 nm or less, can lead to a photoelectric effect for sodium.[1] For wavelengths beyond 683 nm, however, no matter how much light energy is absorbed, electrons are not ejected from the surface. Hence, a photon of a specific minimum energy may be necessary for a certain reaction. Measurement of the total light energy does not indicate how many photons are involved or what their individual energies are, unless the wavelength distribution is known.

Absorption of radiation by an atom or a molecule leads to a more energetic state of the absorbing species. Such energetic states can also be produced by collisions resulting from the random thermal motion of the molecules. The higher the temperature, the greater the average kinetic energy of the atoms and molecules, and the greater the probability of achieving a relatively energetic state by collision. The number of molecules having a particular kinetic energy can often be approximated by the Boltzmann energy distribution. By Eq. (3.21b) $[n(E) = n_{total}e^{-E/RT}]$, the fraction of atoms or molecules having a kinetic energy of molecular Brownian motion of E or greater at thermal equilibrium equals the Boltzmann factor $e^{-E/RT}$, where we have expressed energy on a mole basis. Based on the kinetic theory of gases, the average kinetic energy of translational motion for molecules in a gas phase is $(3/2)RT$, which equals 3.72 kJ mol^{-1} at 25°C. What fraction of the molecules exceeds this average energy? The Boltzmann factor becomes $e^{-(3RT/2)/(RT)}$, or $e^{-1.5}$, which is 0.22, so 22% of the molecules have kinetic energies that are higher than the average. At what temperature would 44% have such kinetic energies? The Boltzmann factor then equals 0.44, so T would be

$$T = \frac{1}{-\ln(0.44)}\frac{E}{R} = \frac{1}{(0.82)}\frac{(3720 \text{ J mol}^{-1})}{(8.3143 \text{ J mol}^{-1}\text{ K}^{-1})}$$

$$= 546 \text{ K} \quad (273°\text{C})$$

Raising the temperature clearly increases the fraction of molecules with higher energies. However, temperature can be raised to only a limited extent under biological conditions without causing cellular damage.

---

[1] For his experimental work on the photoelectric effect, Millikan was awarded the Nobel prize in physics in 1923. Others whose contributions to the understanding of radiation led to Nobel prizes in physics include Wilhelm Wien in 1911 for heat radiation laws, Planck in 1918 for the quantum concept, Johannes Stark in 1919 for spectral properties, Einstein in 1921 in part for interpreting the photoelectric effect, Niels Bohr in 1922 for atomic radiation, James Franck and Gustav Hertz in 1925 for atom–electron interactions, and Wolfgang Pauli in 1945 for atomic properties.

Let us next consider a kinetic energy E of 260 kJ mol^{-1}, as is possessed by blue light of 460 nm. The fraction of molecules possessing at least this energy is $e^{-E/RT}$, which equals $e^{-(260\ kJ\ mol^{-1})/(2.48\ kJ\ mol^{-1})}$, or only 3×10^{-46} at 25°C! For comparison, the total number of atoms in the entire biomass of all living organisms is about 3×10^{41}. The chance that a particular molecule can gain the equivalent of 260 kJ mol^{-1} by means of thermal collisions is therefore vanishingly small. Hence, the absorption of blue light can lead to energetic states that otherwise simply do not occur at temperatures encountered by plants and animals. Absorption of the relatively high quantum energy of light thus promotes the attainment of very improbable energetic states, a key point in the understanding of photobiology.

C. Illumination, Photon Flux Density, and Irradiance

For many purposes in studying plants, we need to know the amount of incident light. There are three common classes of instruments for measuring such fluxes: (1) photometers or light meters, which measure the available illuminating power, a quantity related to the wavelength sensitivity of the human eye; (2) quantum or photon meters, which measure the number of photons; and (3) radiometers, which measure the total energy of the radiation (Fig. 4-2).

By definition, photometers do not appreciably respond to radiation in the infrared or the ultraviolet (Fig. 4-2a). They are "light" meters in the sense that they mimic human vision; i.e., they respond to photons in the visible region, in the same way as the light meter on a camera. A candle is a unit of luminous intensity, originally based on a standard candle or lamp. The current international unit is called a *candela* (sometimes still referred to as a "candle"), which was previously defined as the total light intensity of 1.67 mm^2 of a *blackbody* radiator (one that radiates maximally) at the melting temperature of pure platinum (2042 K). In 1979 the candela was redefined as the luminous intensity of a monochromatic source with a frequency of 5.40×10^{14} cycles s^{-1} (λ of 555 nm) emitting 0.01840 J s^{-1} or 0.01840 W (1/683 W steradian^{-1}, where W is the abbreviation for watt and steradian is the unit for solid angle).[2] The total light emitted in all directions by a source of 1 candela is 4π lumens; 1 lumen m^{-2} is the photometric illuminance unit, *lux*. Hence, 1 lux is the luminous flux density (illuminance) 1 m from a spherically uniform source of 1 candela, which is pretty dim (similar to the lighting in a movie theater during the picture); indeed, the minimum illumination recommended for reading in libraries and most public conveyances is 300 lux. Measurement in lux is adequate for certain purposes where human vision is involved, but it is not appropriate for studies with plants.

For many purposes in plant studies, it is important to know the photon flux density. For instance, the rate of photosynthesis depends on the rate of absorption of photons (not on the rate of absorption of energy). Some instruments are sensitive only to photons that might be useful photosynthetically (e.g., wavelengths from 400 to 700 nm), the so-called *photosynthetic photon flux* (PPF) or *photosynthetic photon flux density* (PPFD), both of which are expressed in μmol m^{-2} s^{-1} (Fig. 4-2b).

---

[2] A solid angle is the three-dimensional counterpart of the more familiar planar angle. It can be interpreted as a region of space defined by a series of lines radiating from a point, the lines forming a smooth surface (such as a conical surface viewed from its apex). The region delimits a certain surface area A on a sphere of radius r; the solid angle in steradians equals A/r^2. For instance, a hemisphere viewed from its center subtends $2\pi r^2/r^2$ or 2π steradians.

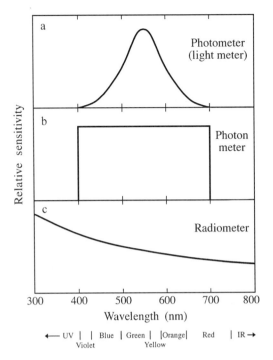

Figure 4-2. Wavelength sensitivities for instruments measuring illumination (a), photon flux density (b), and irradiance (c). Photometers or light meters are used for measuring illumination levels (a), are usually calibrated in lux (or footcandles), and are most sensitive in the middle of the visible spectrum where the human eye is also most sensitive. Photon or quantum meters for determining PPF (PPFD; b) are typically calibrated in μmol photons m^{-2} s^{-1} and often respond essentially uniformly from 400 to 700 nm, the region where photosynthetic pigments absorb maximally. Radiometers ideally respond to radiant energy of all wavelengths (energy is greater per photon at shorter wavelengths; Eq. 4.2) and typically are calibrated in W m^{-2} (J m^{-2} s^{-1}).

Radiometers (e.g., thermocouples, thermopiles, or thermistors that have been treated so as to absorb all wavelengths) respond to radiant energy and thus are sensitive to irradiation in the ultraviolet and infrared as well as in the visible (Fig. 4-2c). Readings are expressed in energy per unit area and time, e.g., J m^{-2} s^{-1}, which is W m^{-2} (conversion factors for radiometric units are given in Appendix II). If the irradiance or radiant energy flux density at a specified wavelength is measured in radiometric units, the value can be converted to a photon flux density by using the energy carried by individual photons (see Eq. 4.2).[3] In general, radiometric units, photometric units, and photon flux densities cannot

---

[3] Various terms are used to describe electromagnetic radiation. *Emittance* is the flux density at the emitting surface, e.g., radiant emittance (expressed in J m^{-2} s^{-1} = W m^{-2}). *Illuminance* is the luminous flux density or "illumination" at some surface (lux). *Intensity* refers to a property of a source; e.g., light intensity designates the rate of light emission for a photometric source (lumens per unit solid angle, or candelas), and radiant intensity is the power emitted per unit solid angle (W steradian^{-1}). *Irradiance* (commonly termed *irradiation*) is the radiant energy flux density received on some surface (W m^{-2}). *Radiance* is the rate of radiant energy emission (power emitted) per unit solid angle per unit area (W steradian^{-1} m^{-2}). *Radiant flux* is the radiant energy emitted or received per unit time (W). *Radiant flux density* is the radiant flux per unit area (W m^{-2}). The term *fluence* is sometimes used for amount per area, so radiant energy per unit area is the energy fluence (J m^{-2}). *Fluence rate* is therefore the fluence per unit time, e.g., the photon fluence rate (mol m^{-2} s^{-1}) is the same as the photon flux density.

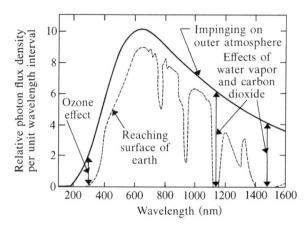

Figure 4-3. Wavelength distributions of the sun's photons incident on the earth's atmosphere and its surface. The curve for the solar irradiation on the atmosphere is an idealized one based on Planck's radiation distribution formula. The spectral distribution and the amount of solar irradiation reaching the earth's surface depend on clouds, other atmospheric conditions, altitude, and the sun's angle in the sky. The pattern indicated by the lower curve is appropriate at sea level on a clear day with the sun overhead.

be unambiguously interconverted (unless the frequency distribution is known), although empirical relations exist among them for certain light sources. For instance, the human eye and photometric devices are far less sensitive at the two extremes of the visible range of wavelengths compared with the center, whereas the photon flux density incident on leaves can be absorbed relatively equally throughout the visible (Fig. 7-4); thus more PPFD (or energy in W m^{-2}) is required at the extremes of the visible spectrum than at 555 nm, where the candela is defined, to give the same response in lux (see Fig. 4-2).

In choosing a lamp for a controlled-environment chamber used for growing plants, both the total energy emitted and the wavelength distribution of the photons should be considered. Tungsten lamps are comparatively poor sources of visible radiation because about 90% of their radiant energy is emitted in the infrared—the actual amount depends on the operating temperature of the filaments—creating appreciable cooling problems when large numbers of such lamps are used. Typical fluorescent lamps emit only about 10% of their energy in the IR, so cooling problems are less severe than with tungsten lamps. On the other hand, the wavelength distribution of the sun's photons in the visible region (Fig. 4-3) is matched far better by tungsten lamps than by fluorescent ones. A close match can be important because the wavelength distribution influences the relative amounts of the two forms of phytochrome discussed at the end of this chapter.

D. Sunlight

Essentially all energy for life originates in the form of electromagnetic radiation from the sun. In radiometric units the radiant flux density of solar irradiation (irradiance) perpendicularly incident on the earth's atmosphere—the "solar constant"—is about 1367 W m^{-2}. The solar constant varies by up to $\pm3.4\%$ from the average due to the eliptical orbit of the earth. The value given is for the mean distance between the earth and the sun (the earth is closest to the sun on January 3, at 1.471×10^8 km, and furthest away on July 4, at 1.521×10^8 km). There are additional variations in solar irradiation based on changes in solar activity, such as occur for sun spots, which lead to the 11-year solar cycle. (In Chapter 6 we will consider the solar constant in terms of the annual photosynthetic yield and in Chapter 7 in terms of the energy balance of a leaf.) Based on the solar constant and averaged for all latitudes, the total daily amount of radiant energy from the sun incident on a horizontal surface just outside the earth's atmosphere averages 29.6 MJ m^{-2} day^{-1}, of which atmospheric conditions such as clouds permit an average of only 58% or 17.0 MJ m^{-2} day^{-1} to reach the earth's surface (at mid-latitudes the amount at the earth's surface on a relatively clear day in the summer can be about 30 MJ m^{-2} day^{-1}). The instantaneous irradiance in the visible region at noon on a cloudless day with the sun approximately overhead can be 420 W m^{-2} (total irradiance of about 850 W m^{-2}; equivalent to about 100,000 lux). If we represent sunlight by yellow light of 570 nm, which by Table 4-1 carries 210 kJ mol^{-1}, then 420 W m^{-2} in the visible region from the sun is a photon flux density of about (420 J m^{-2} s^{-1})/(210,000 J mol^{-1}), or 2.0×10^{-3} mol m^{-2} s^{-1} (2000 μmol m^{-2} s^{-1}) on the earth's surface at sea level.

Figure 4-3 shows the relative number of the sun's photons impinging on the earth's atmosphere and reaching its surface as a function of wavelength. About 5% of the photons incident on the earth's atmosphere are in the ultraviolet (below 400 nm), 28% in the visible, and 67% in the infrared (beyond 740 nm). Most of the ultraviolet component of sunlight incident on the atmosphere is prevented from reaching the surface of the earth by ozone (O_3) present in the stratosphere, 20–30 km above the earth's surface (the absorption of sunlight by this O_3 leads to a pronounced heating of the upper atmosphere). Ozone absorbs some visible radiation (e.g., near 600 nm) and effectively screens out the shorter UV rays by absorbing strongly below 300 nm. Much of the IR from the sun is absorbed by atmospheric water vapor and CO_2 (see Fig. 4-3). Water absorbs strongly near 900 and 1100 nm, and above 1200 nm, having a major IR absorption band at 1400 nm (1.4 μm). Although the amount of water vapor in the air varies with latitude, longitude, and season, the mean vapor concentration is equivalent to a path of liquid water approximately 20 mm thick. The substantial absorption of UV and IR by atmospheric gases causes the solar irradiation at the earth's surface to have a larger fraction in the visible region than that incident on the outer atmosphere. In the example in Fig. 4-3, about 2% of the photons at the earth's surface are in the UV, 45% in the visible, and 53% in the IR.

The radiation environment in water is quite different from that on land. For instance, absorption by water causes most of the IR in sunlight to be removed after penetrating <1 m in lakes or oceans. This, coupled with greater scattering at shorter wavelengths, causes a greater fraction of the photons to be in the visible region at greater depths. However, water attenuates the visible region too, so even at the wavelengths for greatest penetration (approximately 500 nm) in the clearest oceans, only about 1% of the solar photon flux density incident on the water surface penetrates to 200 m. Light is also considerably attenuated with depth in typical freshwater lakes and reservoirs, where substances in the water reduce the flux density of the wavelengths penetrating most readily (near 550 nm) to 1% of the surface values generally by a depth of 10 m.

The wavelength distribution of photons reaching the earth's surface profoundly influences life. For example, the substantial absorption of UV by ozone reduces the potential hazard of mutagenic effects caused by this short-wavelength irradiation. In this regard, prior to the advent of significant amounts of ozone in the upper atmosphere, UV irradiation from the sun would have been a potent factor affecting genetic processes. Even now, exposure to the UV in sunlight can inhibit photosynthesis and decrease leaf expansion. As another example, the peak near 680 nm for photons reaching the earth's surface (Fig. 4-3) coincides with the red absorption band of chlorophyll, thereby favoring photosynthesis. Vision also utilizes the wavelength region where most of the sunlight reaches the earth. Selective pressure must have favored the evolution of photochemical systems capable of using the most abundant wavelengths while avoiding the highly energetic UV and the weak IR.

E. Planck's and Wien's Formulae

The shape of the curve depicting the wavelength distribution of photons incident upon the earth's atmosphere can be closely predicted using Planck's radiation distribution formula. This expression indicates that the relative photon flux density per unit wavelength interval is proportional to $\lambda^{-4}/(e^{hc/\lambda kT} - 1)$, where T is the temperature of the radiation source. Such a formula applies exactly to a perfectly efficient emitter, a so-called blackbody. A blackbody is a convenient idealization describing an object that absorbs all wavelengths— it is uniformly "black" at all wavelengths—and emits in accordance with Planck's radiation distribution formula. This formula is a good approximation for describing radiation from the sun—T is the surface temperature (about 5800 K)—so it was used to obtain the upper curve in Fig. 4-3. Also, the radiation from a tungsten lamp of a few hundred watts can be fairly well described by Planck's radiation distribution formula using a T of 2900 K; i.e., the curve for the relative photon flux density from a tungsten lamp has the same shape as the solid line in Fig. 4-3, although it is shifted toward longer wavelengths because the temperature of a tungsten filament is less than that of the sun's surface. Planck's radiation distribution formula indicates that any object with a temperature greater than 0 K will emit electromagnetic radiation.

If we know the surface temperature of a blackbody, we can predict at what wavelength the radiation from it will be maximal. To derive such an expression, we differentiate Planck's radiation distribution formula with respect to wavelength and set the derivative equal to zero.[4] The relation thus obtained is known as Wien's displacement law:

$$\lambda_{max} T = 3.67 \times 10^6 \text{ nm K} \quad \text{photon basis} \tag{4.3a}$$

where λ_{max} is the wavelength position for maximum photon flux density and T is the surface temperature of the source. Because the surface of the sun is about 5800 K, Wien's displacement law (Eq. 4.3a) predicts that

$$\lambda_{max} = \frac{(3.67 \times 10^6 \text{ nm K})}{(5800 \text{ K})} = 630 \text{ nm}$$

as the upper line in Fig. 4-3 indicates. For a tungsten lamp operating at a temperature of 2900 K (half the temperature of the sun's surface), the position for maximum photon flux density shifts to 1260 nm (a doubling of λ_{max} compared with the sun), consistent with tungsten lamps emitting primarily infrared radiation. From Wien's displacement law, the λ_{max} for a maximum photon flux density from a body at 298 K (25°C) occurs at 1.23×10^4 nm,

---

[4] Using the relations for differentiation in Chapter 1, footnote 5, we obtain: $\delta(\text{Radiation})/\delta\lambda = \delta[\lambda^{-4} \frac{\text{constant}}{(e^{hc/\lambda kT}-1)}]/$ $\delta\lambda = -4\lambda^{-5} \frac{\text{constant}}{(e^{hc/\lambda kT}-1)} - \lambda^{-4} \frac{\text{constant}}{(e^{hc/\lambda kT}-1)^2}(-\frac{hc}{kT}\frac{1}{\lambda^2}e^{hc/\lambda kT}) = 0$. We next multiply everything by $-\lambda^5(e^{hc/\lambda kT} - 1)^2$/constant, leading to $4(e^{hc/\lambda kT} - 1) - \frac{hc}{\lambda kT}e^{hc/\lambda kT} = 0$. Upon setting hc/k equal to 1.439×10^7 nm K (Appendix I), the latter relation can be solved numerically, yielding $\lambda T = 3.67 \times 10^6$ nm K, which is Wien's displacement law (Eq. 4.3a).

which is 12 μm, i.e., far into the IR. Plants emit such infrared radiation, which is a crucial aspect of their overall energy balance, as we will consider in Chapter 7.

When considering the number of photons available for absorption by pigment molecules, as is relevant for discussing photosynthesis, we generally use the spectral distribution of photons per unit wavelength interval (see Fig. 4-3). On the other hand, for applications such as describing the energy gain by leaves exposed to sunlight, we are usually more interested in the spectral distribution of energy per unit wavelength interval. To recast Fig. 4-3 on an energy basis, we need to divide each point on the curves by its wavelength—the ordinate then becomes "Relative energy flux density per unit wavelength interval." Planck's radiation distribution formula indicates that the relative energy flux density per unit wavelength interval is proportional to $\lambda^{-5}/(e^{hc/\lambda kT} - 1)$. Wien's displacement law for maximum energy output then is

$$\lambda_{max}T = 2.90 \times 10^6 \text{ nm K} \quad \text{energy basis} \tag{4.3b}$$

II. Absorption of Light by Molecules

Only light that is absorbed can produce a chemical change, a principle embodied in the Grotthuss–Draper law of photochemistry. This is true whether radiant energy is converted to some other form and then stored or is used as a trigger. Another important principle of photochemistry is the Stark–Einstein law, which specifies that each absorbed photon activates only one molecule. Einstein further postulated that all the energy of the photon is transferred to a single electron during the absorption event, resulting in the movement of this electron to a higher energy state. To help understand light absorption, we will first consider some of the properties of electrons. The fate of the excited electrons will be discussed in the next section.

A. Role of Electrons in Absorption Event

From a classical viewpoint, an electron is a charged particle that can move in some orbit around an atomic nucleus. Its energy depends both on the location of the orbit in space and on how fast the electron moves in its orbit. The increase in energy of an electron upon absorbing a photon could transfer that electron into an orbit at a higher energy than the original orbit or could cause the electron to move more rapidly about the nucleus than it did before the light arrived. The locations of various possible electron orbits and the speeds of electrons in them are both limited to certain discrete, or "allowed," values, a phenomenon that has been interpreted by quantum mechanics. Hence the energy of an electron in an atom or molecule can change only by certain specific amounts. Light of the appropriate wavelength will have the proper energy to cause the electron to move from one possible energetic state to another. For light absorption to occur, therefore, the energy of a photon as given by Eq. (4.2) must equal the difference in energy between some allowed excited state of the atom or molecule and the initial state, the latter usually being the ground (lowest energy) state.

During light absorption the electromagnetic field of the light interacts with some electron. Because electrons are charged particles, they experience a force when they are in an electric field. The oscillating electric field of light (Fig. 4-1) thus represents a periodic driving force acting on the electrons. This electric field—a vector having a specific direction in space, such as along the vertical axis in Fig. 4-1—causes or induces the electrons to move. If the frequency of the electromagnetic radiation causes a large sympathetic oscillation or beating of some electron, that electron is said to be in *resonance* with the light wave. Such a resonating electron leads to an *electric dipole* (local separation of positive and negative charge) in the molecule, as the electron is forced to move first in a certain direction and then in the opposite one in response to the oscillating electric field of light. The displacement of the electron back and forth requires energy—in fact, it may take the entire energy of the photon, in which case the quantum is captured or absorbed. The direction and the magnitude of the induced electric dipole will depend on the resisting, or restoring, forces on the electron provided by the rest of the molecule. These restoring forces depend on the other electrons and the atomic nuclei in the molecule, so they are not the same in different types of molecules. Indeed, the electric dipoles that can be induced in a particular molecule are characteristic of that molecule, which helps explain why each molecular species has its own unique absorption spectrum.

The probability that light will be absorbed depends on both its wavelength and the relative orientation of its electromagnetic field with respect to the possible induced oscillations of electrons in the molecules. Absorption of a photon without ejection of an electron from the absorbing species can take place only if the following two conditions exist: (1) The photon has the proper energy to get to a discrete excited state of the molecule, i.e., has a specific wavelength (see Eq. 4.2), and (2) the electric field vector associated with the light (Fig. 4-1) has a component parallel to the direction of some potential electric dipole in the molecule so that an electron can be induced to oscillate. In other words, the electric field of light must exert a force on some electron. It must exert this force in the direction of a potential electric dipole so that the electron can be induced to move and thus be able to accept the energy of the photon. The probability for absorption is proportional to the square of the cosine of the angle between the electric field vector of light and the direction of the induced electric dipole in the molecule. Within these limits set by the wavelength and the orientation, light energy can be captured by the molecule, placing it in an excited state.

B. Electron Spin and State Multiplicity

Light absorption is affected by the arrangement of electrons in an atom or a molecule, which depends among other things on a property of the individual electrons known as their *spin*. We can view each electron as a charged particle spinning about an axis in much the same way as the earth spins about its axis. Such rotation has an angular momentum, or spin, associated with it. The magnitude of the spin of all electrons is the same, but because spin is a vector quantity, it can have different directions in space. For an electron, only two orientations are found—the spin of the electron is aligned either parallel or antiparallel to the local magnetic field, i.e., either in the same direction or in the opposite direction as the magnetic field. Even in the absence of an externally applied magnetic field, such as that of the earth or some electromagnet, a local internal magnetic field is provided by both the moving charges in the nucleus and the motion of the electrons. A magnetic field therefore always exists with which the electron spin can be aligned.

Angular momentum and hence spin have units of energy × time. It proves convenient to express the spin of electrons in units of $h/(2\pi)$, where h is Planck's constant, which has units of energy × time, e.g., J s. In units of $h/(2\pi)$, the projection along the magnetic field of the spin for a single electron is either $+\frac{1}{2}$ (e.g., when the spin is parallel to the local magnetic field), or $-\frac{1}{2}$ (when the spin is in the opposite direction, or antiparallel). The net spin of an atom or molecule is the vector sum of the spins of all the electrons, each individual electron having a spin of either $+\frac{1}{2}$ or $-\frac{1}{2}$. The magnitude of this net spin is given the symbol S, an extremely important quantity in spectroscopy.

To discuss the spectroscopic properties of various molecules, we introduce the *spin multiplicity*. The spin multiplicity of an electronic state is defined as $2S + 1$, where S is the magnitude of the net spin for the whole atom or molecule. For example, if $S = 0$—so the spin projections of all the electrons taken along the magnetic field cancel each other—then $2S + 1 = 1$, and the state is called a *singlet*. On the other hand, if $S = 1$, $2S + 1 = 3$, and the state is a *triplet*. Singlets and triplets are the two most important spin multiplicities encountered in biology. When referring to an absorbing species, the spin multiplicity is generally indicated by S for singlet and T for triplet. When $S = \frac{1}{2}$, as could occur if there were an odd number of electrons in a molecule, $2S + 1 = 2$; such *doublets* occur for free radicals (molecules with a single unpaired electron—such molecules are generally quite reactive).

Electrons are found only in certain "allowed" regions of space; the particular locus in which some electron can move is referred to as its *orbital*. In the 1920s Pauli noted that, when an electron is in a given atomic orbital, a second electron having its spin in the same direction is excluded from that orbital. This led to the enunciation of the *Pauli exclusion principle* of quantum mechanics: When two electrons are in the same orbital, their spins must be in opposite directions. When a molecule has all of its electrons paired in orbitals with their spins in opposite directions, the total spin of the molecule is zero ($S = 0$), and the molecule is in a singlet state ($2S + 1 = 1$; see Fig. 4-4a). The ground, or unexcited, state of most molecules is a singlet, i.e., all the electrons are in pairs in the lowest energy orbitals. When some electron is excited to an unoccupied orbital, two spin possibilities exist. The spins of the two electrons (which are now in different orbitals) may be in opposite directions,

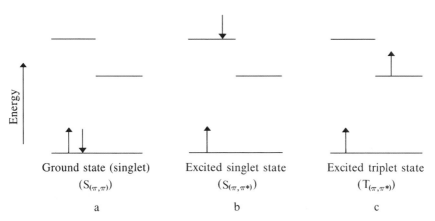

Figure 4-4. Effect of light on a pair of electrons in a molecular orbital. The arrows indicate the directions of the electron spins with respect to the local magnetic field. (a) In the ground state, the two electrons in a filled orbital have their spins in opposite directions. (b) The absorption of a photon can cause the molecule to go to an excited singlet state where the spins of the electrons are still in opposite directions. (c) In an excited triplet state, the spins of the two electrons are in the same direction.

as they were when paired in the ground state (Fig. 4-4b); this electronic configuration is still a singlet state. The two electrons may also have their spins in the same direction—a triplet state (Fig. 4-4c). (Because the electrons are in different orbitals, their spins can be in the same direction without violating the Pauli principle.) An important rule—first enunciated by Friedrich Hund based on empirical observations and later explained using quantum mechanics—is that the level with the greatest spin multiplicity has the lowest energy. Thus an excited triplet state is lower in energy than is the corresponding excited singlet stated, as is illustrated in Fig. 4-4c.

C. Molecular Orbitals

For a discussion of the light absorption event involving the interaction of an electromagnetic wave with some electron, it is easiest to visualize the electron as a small point located at some specific position in the atom or molecule. For the classical description of an electron's energy, we can imagine the electron as moving in some fixed trajectory or orbit about the nucleus. However, to describe the role of electrons in binding atoms together to form a molecule, it is most convenient to imagine the electron as spatially distributed like a cloud of negative charge surrounding the nuclei of adjacent atoms in the molecule. In this last description, involving probability considerations introduced by quantum mechanics, we say that the electrons are located in *molecular orbitals*. Such considerations lead us into a different way of looking at things. Our common experience yields relations such as Newton's laws of motion, which describe events on a scale much larger than atomic or molecular dimensions. In contrast, the main application of quantum mechanics involves molecular, atomic, and subatomic dimensions. At this scale our intuition often fails us. Moreover, many things that otherwise seem to be absolute, such as the position of an object, turn out to be describable only on a relative basis or as a probability. We cannot say that an electron is located at such and such a place. Instead, we must be satisfied with knowing only the probability of finding an electron in some region about a nucleus.

   A logical way to begin our discussion of molecular orbitals is to consider the probabilities of finding electrons in given regions about a single atomic nucleus. The simplest atom is that of hydrogen, which has a single electron. If we were to determine the probability of finding that electron in the various regions of space about its nucleus, we would find that it spends most of its time fairly close to the nucleus (within about 0.1 nm), and that the probability distribution is *spherically symmetric* in space. In other words, the chance of finding the electron is the same in all directions about the nucleus. In atomic theory, this spherically symmetric distribution of the electron about an H nucleus is called an *s* orbital, and the electron is referred to as an *s* electron.

   The next simplest atom is that of helium; He has two *s* electrons moving in the *s* orbital about its nucleus. By the Pauli exclusion principle these two electrons must have their spins in opposite directions. What happens for lithium, which contains three electrons? Two of its electrons are in the same type of orbital as that of He. This orbital is known as the *K* shell and represents the spherically symmetric orbital closest to the nucleus. The third electron is excluded from the *K* shell but occurs in another orbital with spherical symmetry whose probability description indicates that on average its electrons are farther away from the nucleus than is the case for the *K* shell. Electrons in this new orbital are still *s* electrons, but they are in the *L* shell. Beryllium, which has an atomic number of four, can have two *s* electrons in its *K* shell and two more in its *L* shell. What happens when we have five electrons moving about a nucleus, as for an uncharged boron atom? The fifth

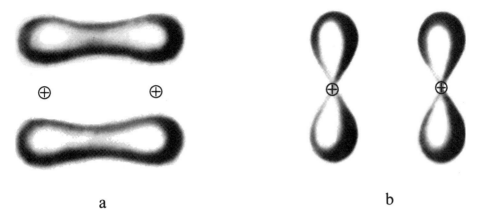

a   b

Figure 4-5. Typical π and π^* orbitals, indicating the spatial distribution about the nuclei ($\oplus$) where the greatest probability of finding the electrons occurs: (a) π orbital (bonding) and (b) π^* orbital (antibonding).

electron is in an orbital that is not spherically symmetric about the nucleus. Instead, the probability distribution has the shape of a dumbbell, although it is still centered about the nucleus (similar in appearance to the distributions depicted in Fig. 4-5b). This new orbital is referred to as a p orbital and electrons in it are called p electrons. The probability of finding a p electron in various regions of space is greatest along some axis through the nucleus and vanishingly small at the nucleus itself.

How do these various atomic orbitals relate to the spatial distribution of electrons in molecules? A molecule contains more than one atom (except for "molecules" like helium or neon), and certain electrons can move between the atoms—this interatomic motion is crucial for holding the molecule together. Fortunately, the spatial localization of electrons in molecules can be described using suitable linear combinations of the spatial distributions represented by various atomic orbitals centered about the nuclei involved. In fact, molecular orbital theory is concerned with giving the correct quantum-mechanical or wave-mechanical description of the probability of finding electrons in various regions of space in molecules by using the already established descriptions for the probability of finding electrons in atomic orbitals.

Some of the electrons in molecules are localized about a single nucleus, and others are delocalized, or shared between nuclei. For the delocalized electrons, the combination of the various atomic orbitals used to describe the spatial positions of the electrons in three dimensions is consistent with a sharing of the electrons between adjacent nuclei. In other words, the molecular orbitals of the delocalized electrons spatially overlap more than one nucleus. This sharing of electrons is responsible for the chemical bonds that prevent the molecule from separating into its constituent atoms, i.e., the negative electrons moving between the positive nuclei hold the molecule together by attracting the nuclei of different atoms. Moreover, these delocalized electrons are usually the electrons involved in light absorption by molecules.

The lowest energy molecular orbital is a σ orbital, which can be constructed by linear combinations of s atomic orbitals. The spatial distribution of the electron cloud for a σ orbital is cylindrically symmetric about the internuclear axis for the pair of atoms involved in the σ bond. Nonbonding or lone-pair electrons contributed by atoms such as oxygen or nitrogen occur in n orbitals and retain their atomic character in the molecule. These n electrons are essentially physically separate from the other electrons in the molecule and do not take part in the bonding between nuclei. We shall devote most of our attention to electrons in π molecular orbitals, which are the molecular equivalent of p electrons in atoms. These π electrons are delocalized in a bond joining two or more atoms. They are of prime importance in light absorption. In fact, most photochemical reactions and spectroscopic properties of biological importance result from the absorption of photons by π electrons. With respect to orbital nomenclature, σ and π are the Greek transliterations for s and p, respectively, and n derives from nonbonding.

The excitation of a π electron by light absorption can lead to an excited state of the molecule in which the electron moves into a π^* orbital, the asterisk referring to an excited, or high-energy, molecular orbital. Figure 4-5 illustrates the probability distributions for electrons in both π and π^* orbitals (the circumscribed regions indicate where the electrons are most likely to be found). In Fig. 4-5a a π orbital is delocalized between two nuclei; the same clouds of negative charge electrostatically attract both nuclei. Such sharing of electrons in the π orbital helps join the atoms together, so we refer to this type of molecular orbital as bonding. As shown in Fig. 4-5b, electrons in a π^* orbital do not help join atoms together; rather, they tend to decrease bonding between atoms in the molecule because the clouds of negative charge in adjacent atoms repel each other. A π^* orbital is therefore referred to as antibonding. The decrease in bonding when going from a π to a π^* orbital results in a less stable (higher energy) electronic state for the molecule, so a π^* orbital is at a higher energy than a π orbital.

The energy required to move an electron from the attractive (bonding) π orbital to the antibonding π^* orbital is obtained by the absorption of a photon of the appropriate wavelength. For molecules such as chlorophylls and carotenoids, the π^* orbitals are often only a few hundred kJ mol^{-1} higher in energy than are the corresponding π orbitals. For such molecules, the absorption of visible light corresponding to photons with energies from 160 to 300 kJ mol^{-1} (Table 4-1) can lead to the excitation of π electrons into the π^* orbitals.

D. Photoisomerization

Light energy can cause molecular changes known as *photoisomerizations*. The three main types of photoisomerization are (1) *cis–trans*[5] isomerization about a double bond; (2) a double bond shift; and (3) a molecular rearrangement involving changes in carbon–carbon bonds, e.g., ring cleavage or formation. *Cis–trans* photoisomerization—which illustrates the consequences of the different spatial distribution of electrons in π and π^* orbitals— makes possible the generally restricted rotation about a double bond. A double bond has two π electrons; light absorption leads to the excitation of one of them to a π^* orbital. The attraction between the two carbon atoms caused by the remaining π electron is mostly canceled by the antibonding, or repulsive, contribution from this π^* electron (see Fig. 4-5). Thus the molecular orbitals of the original two π electrons that prevented rotation about the double bond have been replaced by an electronic configuration permitting relatively easy rotation about the carbon–carbon axis (the most stable configuration in the excited state can be twisted 90° about the C–C axis compared to the orientation in the unexcited state).

---

[5] In the *trans* configuration, the two large groups are on opposite sides of the double bond between them $\left({}^{R}\!\!>\!C\!\!=\!\!C\!<^{}_{R'} \right)$, and in the *cis* form they are on the same side.

Let us consider the absorption of light by a molecule in the *cis* form. When the excited π^* electron drops back to a π orbital, the two large groups can be on the same side (the original *cis* isomer) or on opposite sides (the *trans* isomer) of the double bond. Because the two isomers can have markedly different chemical properties, such use of light to trigger their interconversion can have profound importance in biology. For example, light can cause the photoisomerization of a *cis* isomer of retinal (a carotenoid attached to the lipoprotein opsin in a complex referred to as rhodopsin), yielding a *trans* isomer, which is the basic photochemical event underlying vision.

E. Light Absorption by Chlorophyll

We will use chlorophyll to help illustrate some of the terms introduced previously describing the absorption of light by molecules. (Chlorophyll is discussed in more detail in Chapter 5.) The principal energy levels and electronic transitions of chlorophyll are presented in Fig. 4-6. Chlorophyll is a singlet in the ground state, as are all other pigments of importance in biology. When a photon is absorbed, some π electron is excited to a π^* orbital. If this excited state is a singlet, it is represented by $S_{(\pi,\pi^*)}$ (Figs. 4-4 and 4-6). The first symbol in the subscript is the type of electron (here a π electron) that has been excited to the antibonding orbital indicated by the second symbol (here π^*). We will represent the ground state by $S_{(\pi,\pi)}$, indicating that no electrons are then in excited, or antibonding, orbitals. If the orientation of the spin of the excited π electron became reversed during excitation, it would be in the

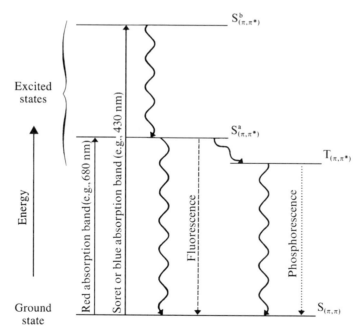

Figure 4-6. Energy level diagram indicating the principal electronic states and some of the transitions of chlorophyll. Straight vertical lines represent the absorption of light; wavy lines indicate radiationless transitions, for which the energy is eventually released as heat; broken lines indicate those deexcitations accompanied by radiation.

same direction as the spin of the electron that remained in the π orbital (see Fig. 4-4c). (Each filled orbital contains two electrons whose spins are in opposite directions.) In this case, the net spin of the molecule in the excited state is 1, i.e., a triplet. The excited triplet state of chlorophyll is represented by $T_{(\pi,\pi^*)}$ in Fig. 4-6.

Chlorophyll has two principal excited singlet states that differ considerably in energy. One of these states, designated $S^a_{(\pi,\pi^*)}$ in Fig. 4-6, can be reached by absorption of red light, e.g., a wavelength of 680 nm. The other state, $S^b_{(\pi,\pi^*)}$, involves a π^* orbital that lies higher in energy and is reached by absorption of blue light, e.g., 430 nm. The electronic transitions caused by the absorption of photons are indicated by solid vertical arrows in Fig. 4-6, and the vertical distances correspond approximately to the differences in energy involved. Excitation of a singlet ground state to an excited triplet state is usually only about 10^{-5} times as probable as going to an excited singlet state, so the transition from $S_{(\pi,\pi)}$ to $T_{(\pi,\pi^*)}$ has not been indicated for chlorophyll in Fig. 4-6. To go from $S_{(\pi,\pi)}$ to $T_{(\pi,\pi^*)}$, the energy of an electron must be substantially increased and the orientation of its spin must be simultaneously reversed. The coincidence of these two events is improbable, so very few chlorophyll molecules are excited directly to $T_{(\pi,\pi^*)}$ from the ground state by the absorption of light.

We will now briefly compare the two most important excitations in photobiology—the transitions of n and π electrons to π^* orbitals. The n electrons have very little spatial overlap with other electrons, and they tend to be higher in energy than the π electrons, which are reduced in energy (stabilized) by being delocalized over a number of nuclei. Therefore, the excitation of an n electron to a π^* orbital generally takes less energy than the transition of a π electron to the same antibonding orbital (Fig. 4-7). Hence, an $S_{(n,\pi^*)}$ state usually occurs at a lower energy—a longer wavelength is required for its excitation—than does the analogous $S_{(\pi,\pi^*)}$. Another difference is that π and π^* orbitals may overlap spatially (see Fig. 4-5), but n and π^* orbitals generally do not. Consequently, the n to π^* transition is not as favored, or probable, as the π to π^* one. Thus transitions to $S_{(\pi,\pi^*)}$ states tend to dominate the absorption properties of a molecular species (as evidenced, for example, by an absorption spectrum) compared with excitations yielding $S_{(n,\pi^*)}$ states.

We next consider the effect of molecular environment on n, π, and π^* orbitals (Fig. 4-7). The n electrons generally interact strongly with water, e.g., by participating in hydrogen bonding, and therefore the energy level of n electrons is considerably lower in water than in an organic solvent. The energy of π^* electrons is also lowered by water, but to a lesser

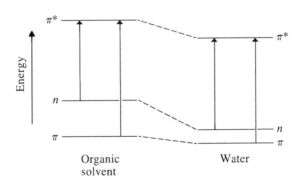

Figure 4-7.   Influence of the solvent on the energies of n, π, and π^* orbitals. The indicated n-to-π^* transition takes more energy in water than in an organic solvent, and the π-to-π^* transition takes less.

extent than for n electrons. The π orbitals are physically deepest within the molecule and hence are the least affected by the solvent. A transition from an n to a π^* orbital therefore takes more energy in water than in an organic solvent (Fig. 4-7). On the other hand, the π to π^* transition takes less energy in an aqueous environment than in an organic one. As an example, let us consider the transition of chlorophyll to an excited singlet state in different solvents. The excitation to its excited singlet state takes less energy in water than in an organic solvent; i.e., when chlorophyll is in water, the required photons have a longer wavelength than when the chlorophyll is dissolved in an organic solvent such as acetone. This lower energy requirement in water is consistent with our statement that the transition for chlorophyll corresponds to the promotion of a π electron to a π^* orbital.

III. Deexcitation

The primary processes of photochemistry involve the light absorption event, which we have already discussed, together with the subsequent deexcitation reactions. We can portray such transitions on an energy level diagram, as in Fig. 4-6 for chlorophyll. In this section we discuss the various deexcitation processes, including a consideration of their rate constants and lifetimes.

One characteristic of the various excitation and deexcitation processes is the time needed for the transitions. A useful estimate of the time for the absorption of a photon is the time required for one cycle of the light wave to pass an electron. This time is the distance per cycle of the wave divided by the speed with which light travels, or λ/v, which equals $1/v$ by Eq. (4.1). Hence the time required for the absorption of a photon is approximately equal to the reciprocal of the frequency of the light, $1/v$, which is the time necessary for one complete oscillation of the electromagnetic field. During one oscillation the electric vector of light can induce an electron to move first in one direction and then in the opposite one, thereby setting up a beating or resonating of the electron. To be specific, let us consider blue light with a wavelength of 460 nm in a vacuum. From its frequency (Table 4-1), we can calculate that the time for one cycle is $1/(6.52 \times 10^{14} \text{ s}^{-1})$, or 1.5×10^{-15} s. Light absorption is indeed extremely rapid!

Times for deexcitation reactions are usually expressed in *lifetimes*. A lifetime is the time required for the number of molecules in a given state to decrease to $1/e$, or 37%, of the initial number. Lifetimes are extremely convenient for describing first-order processes because the initial species in such processes decay (disappear) exponentially with time (see Eq. 4.9 and Appendix III for examples of first-order processes). A *half-time*, the time necessary for the number of species in a given state to decrease by 50%, is also used to describe deexcitation processes; for an exponential decay, one half-time is ln 2, or 0.693, times the duration of a lifetime.

A. Fluorescence, Radiationless Transition, and Phosphorescence

The excess energy of the excited state can be dissipated by various competing pathways. One way deexcitation can occur is by the emission of light known as *fluorescence* (indicated by a dashed line in Fig. 4-6 for the principal transitions of chlorophyll). Fluorescence describes the electromagnetic radiation emitted when a molecule goes from an excited singlet state to a singlet ground state. Fluorescence lifetimes for most organic molecules range from 10^{-9} to 10^{-6} s. We shall see in Chapter 5 that the properties of chlorophyll fluorescence are extremely important for an understanding of the primary events of photosynthesis.

Deexcitation of an excited state often occurs without the emission of any radiation, termed a nonradiative or *radiationless* transition (indicated by wavy lines in Fig. 4-6). For a radiationless transition to a lower energy excited state or back to the ground state, the energy of the absorbed photon is eventually converted to heat, which is passed on by collisions with the surrounding molecules. Radiationless transitions can be extremely rapid from some excited singlet state to a lower energy excited singlet state in the same molecule. For example, the radiationless transition from $S^b_{(\pi,\pi^*)}$ to $S^a_{(\pi,\pi^*)}$ indicated for chlorophyll in Fig. 4-6 takes about 10^{-12} s. This transition is so rapid that hardly any fluorescence is emitted from $S^b_{(\pi,\pi^*)}$, so no fluorescence emission from the upper excited singlet state of chlorophyll is indicated in Fig. 4-6. The excited singlet state lying at the lower energy, $S^a_{(\pi,\pi^*)}$, can decay to the ground state by a radiationless transition. As another alternative, $S^a_{(\pi,\pi^*)}$ can go to $T_{(\pi,\pi^*)}$, also by a radiationless transition (Fig. 4-6). In fact, excited triplet states in molecules are mainly formed by radiationless transitions from excited singlet states at higher energies. The deexcitation of $T_{(\pi,\pi^*)}$ to $S_{(\pi,\pi)}$ can be radiationless, or it can be by radiation known as *phosphorescence* (dotted line in Fig. 4-6), which we consider next.

Phosphorescence is the electromagnetic radiation that can accompany the transition of a molecule from an excited triplet state to a ground state singlet. Because the molecule goes from a triplet to a singlet state, its net spin must change during the emission of this radiation. The lifetimes for phosphorescence are usually from 10^{-3} to 10 s, which are long compared with those for fluorescence (10^{-9}–10^{-6} s). The relatively long times for deexcitation by phosphorescence occur because the molecule goes from one electronic state to another and simultaneously the electron spin is reversed—the coincidence of these two events is rather improbable. In fact, the low probability of forming $T_{(\pi,\pi^*)}$ from $S_{(\pi,\pi)}$ by light absorption has the same physical basis as the long lifetime for deexcitation of the excited triplet state back to the ground state by emitting phosphorescence.

Another, although usually minor, way for an excited molecule to emit radiation is by "delayed fluorescence." Specifically, a molecule in the relatively long-lived excited triplet state $T_{(\pi,\pi^*)}$ can sometimes be supplied enough thermal energy by collisions to put it into the higher energy excited singlet state $S_{(\pi,\pi^*)}$. Subsequent radiation, as the molecule goes from this excited singlet state to the ground state $S_{(\pi,\pi)}$, has the characteristics of fluorescence. However, it is considerably delayed after light absorption compared with normal fluorescence because the excitation spent some time as $T_{(\pi,\pi^*)}$. In any case, the lifetime of an excited state indicates the time course for deexcitation, not the time required for the deexcitation event per se, e.g., the time required for the emission of a photon is essentially the same as for its capture.

B. Competing Pathways for Deexcitation

Each excited state has not only a definite energy but also a specific lifetime, the length of which depends on the particular processes competing for the deexcitation of that state. In addition to fluorescence, phosphorescence, and the radiationless transitions that we have just introduced, the excitation energy can also be transferred to another molecule, putting this second molecule into an excited state while the originally excited molecule returns to its ground state. In another type of deexcitation process, an excited (energetic) electron may leave the molecule that absorbed the photon, as occurs for certain excited chlorophyll molecules. The excited state of pivotal importance in photosynthesis is the lower excited singlet state of chlorophyll, indicated by $S^a_{(\pi,\pi*)}$ in Fig. 4-6. We will use this state to illustrate some of the possible ways for the deexcitation of an excited singlet state, $S_{(\pi,\pi*)}$. The competing pathways and their equations presented in this section are summarized as follows:

Pathway	Eq. No.
Fluorescence	4.4
Radiationless transitions	4.5, 4.6
Excitation transfer	4.7
Electron donation	4.8

The absorbed quantum can be reradiated as electromagnetic energy, $h\nu$, causing the excited molecule $S_{(\pi,\pi*)}$ to drop back to its ground state $S_{(\pi,\pi)}$:

$$S_{(\pi,\pi*)} \xrightarrow{k_1} S_{(\pi,\pi)} + h\nu \tag{4.4}$$

Such fluorescence exponentially decays with time after the exciting light is removed, indicating a first-order process with a rate constant k_1 in Eq. (4.4) (first-order rate constants have units of s^{-1}). In particular, for a first-order process, the rate of disappearance of the excited state, $dS_{(\pi,\pi*)}/dt$, is linearly proportional to the amount of $S_{(\pi,\pi*)}$ present at any time (also see Appendix III). The fluorescence lifetime of $S_{(\pi,\pi*)}$, which is typically about 10^{-8} s, would be the actual lifetime of $S_{(\pi,\pi*)}$ if no other competing deexcitation processes occurred. When the energy of the absorbed photon is dissipated as fluorescence, no photochemical work can be done. Therefore, the fluorescence lifetime is an upper time limit within which any biologically useful reactions can be driven by the lowest excited singlet state of a molecule; i.e., if the reactions take longer than the fluorescence lifetime, most of the absorbed energy will already have been dissipated.

The next deexcitation processes that we will consider are the radiationless transitions by which $S_{(\pi,\pi*)}$ eventually dissipates its excess electronic energy as heat. Two different states can be reached by radiationless transitions from $S_{(\pi,\pi*)}$:

$$S_{(\pi,\pi*)} \xrightarrow{k_2} S_{(\pi,\pi)} + \text{heat} \tag{4.5}$$

$$S_{(\pi,\pi*)} \xrightarrow{k_3} T_{(\pi,\pi*)} + \text{heat} \tag{4.6}$$

Radiationless transitions such as those in Eqs. (4.5) and (4.6) involve deexcitations in which the excess energy is often first passed on to other parts of the same molecule. This causes the excitation of certain vibrational modes for other pairs of atoms within the molecule— vibrations such as those that we will discuss in conjunction with the Franck–Condon principle (see Fig. 4-8). This energy, which has become distributed over the molecule, is subsequently dissipated by collisions with other molecules in the randomizing interchanges that are the basis of temperature. When an excited molecule returns directly to its ground state by a radiationless transition (Eq. 4.5), all of the radiant energy of the absorbed light is eventually converted into the thermal energy of motion of the surrounding molecules. In Eq. (4.6) only some of the excess electronic energy appears as the quantity designated "heat," which in that case represents the difference in energy between $S_{(\pi,\pi*)}$ and $T_{(\pi,\pi*)}$.

As for fluorescence, radiationless transitions also obey first-order kinetics. Although the dissipation of excitation energy as heat in a transition to the ground state (Eq. 4.5) is photochemically wasteful in that no biological work is performed, the transition to $T_{(\pi,\pi^*)}$ (Eq. 4.6) can be quite useful. The lowest excited triplet state usually lasts 10^4–10^8 times longer than does $S_{(\pi,\pi^*)}$, which allows time for many more intermolecular collisions. Because each collision increases the opportunity for a given reaction to occur, $T_{(\pi,\pi^*)}$ can be an important excited state in photobiology.

The absorption of light can lead to a photochemical reaction initiated by a molecule other than the one that absorbed the photon. This phenomenon suggests that electronic excitation can be transferred between molecules, resulting in the excitation of one and the deexcitation of the other. For instance, the excitation energy of $S_{(\pi,\pi^*)}$ might be transferred to a second molecule, represented in the ground state by $S_{2(\pi,\pi)}$:

$$S_{(\pi,\pi^*)} + S_{2(\pi,\pi)} \xrightarrow{k_4} S_{(\pi,\pi)} + S_{2(\pi,\pi^*)} \tag{4.7}$$

This second molecule thereby becomes excited, indicated by $S_{2(\pi,\pi^*)}$, and the molecule that absorbed the photon becomes deexcited and is returned to its ground state. Such transfer of electronic excitation from molecule to molecule underlies the energy migration among the pigments involved in photosynthesis (see Chapter 5). We will assume that Eq. (4.7) represents a first-order reaction, as it does for the excitation exchanges between chlorophyll molecules *in vivo* (in certain cases, Eq. 4.7 can represent a second-order reaction, i.e., $dS_{(\pi,\pi^*)}/dt$ then equals $-k' \, S_{(\pi,\pi^*)}^2$).

As another type of deexcitation process, $S_{(\pi,\pi^*)}$ can take part in a *photochemical* reaction. For example, the excited π^* electron can be donated to a suitable acceptor:

$$S_{(\pi,\pi^*)} \xrightarrow{k_5} D_{(\pi)} + e^* \tag{4.8}$$

where the ejected electron is represented by e^*. The electron removed from $S_{(\pi,\pi^*)}$ is replaced by another one donated from some other compound; $D_{(\pi)}$ in Eq. (4.8), which represents a doublet because one of the π orbitals contains an unpaired electron, then goes back to its original ground state $S_{(\pi,\pi)}$. Photochemical reactions of the form of Eq. (4.8) serve as the crucial link in the conversion of radiant energy into chemical or electrical energy (Chapters 5 and 6). Indeed, Eq. (4.8) can be used to represent the photochemical reaction taking place at the special chlorophyll molecules P_{680} and P_{700} that we will discuss later.

C. Lifetimes

Equations (4.4)–(4.8) represent five competing pathways for the deexcitation of the excited singlet state, $S_{(\pi,\pi^*)}$, and they must all be considered when predicting its lifetime. In this section we will use rate constants and lifetimes of the individual competing pathways to predict the rate constant and lifetime of the excited state. The development and the equations are as follows:

Development	Eq. No.
Disappearance of excited singlet state	4.9
Integration	4.10
Reexpression using lifetime	4.11
Relation between rate constants and lifetimes	4.12, 4.13
Reexpression of general decay equation	4.14

Assuming that each deexcitation process is first order and that no reaction leads to the formation of $S_{(\pi,\pi^*)}$, the disappearance of the excited singlet state then satisfies the following relation:

$$-\frac{dS_{(\pi,\pi^*)}}{dt} = (k_1 + k_2 + k_3 + k_4 + k_5)\, S_{(\pi,\pi^*)} \tag{4.9}$$

where the various k_j's in Eq. (4.9) are the rate constants for the five individual decay reactions (Eqs. 4.4–4.8). After multiplying both sides by $dt/S_{(\pi,\pi^*)}$ and then integrating Eq. (4.9) (see Appendix III), taking exponentials of both sides followed by rearrangement leads to the following expression for the time dependence of the number of molecules in the excited singlet state:

$$S_{(\pi,\pi^*)_t} = S_{(\pi,\pi^*)_0}\, e^{-(k_1+k_2+k_3+k_4+k_5)t} \tag{4.10}$$

where $S_{(\pi,\pi^*)_0}$ represents the number of molecules in the excited singlet state when the illumination ceases ($t = 0$), and $S_{(\pi,\pi^*)_t}$ is the number of excited singlet states remaining at a subsequent time t. Relations such as Eq. (4.10)—showing the amount of some state remaining at various times after illumination or other treatment—are extremely important for describing processes with first-order rate constants.

Because the lifetime of an excited state is the time required for the number of excited molecules to decrease to $1/e$ of the initial value, $S_{(\pi,\pi^*)_t}$ in Eq. (4.10) equals $(1/e)\, S_{(\pi,\pi^*)_0}$ when t equals the lifetime τ, i.e.,

$$S_{(\pi,\pi^*)_\tau} = e^{-1} S_{(\pi,\pi^*)_0} = S_{(\pi,\pi^*)_0}\, e^{-(k_1+k_2+k_3+k_4+k_5)\tau} \tag{4.11}$$

which leads to the following relationship:

$$(k_1 + k_2 + k_3 + k_4 + k_5)\tau = 1 \tag{4.12}$$

Equation (4.12) indicates that the greater the rate constant for any particular deexcitation process, the shorter the lifetime of the excited state.

Equation (4.12) can be generalized to include all competing reactions, leading to the following expression for the lifetime:

$$\frac{1}{\tau} = k = \sum_j k_j = \sum_j \frac{1}{\tau_j} \tag{4.13}$$

where k_j is the first-order rate constant for the jth deexcitation process and τ_j is its lifetime ($\tau_j = 1/k_j$). Also, τ is the lifetime of the excited state, and k in Eq. (4.13) is the overall rate constant for its decay. Using Eq. (4.13) we can reexpress Eq. (4.10) as follows:

$$S_{(\pi,\pi^*)_t} = S_{(\pi,\pi^*)_0}\, e^{-kt} = S_{(\pi,\pi^*)_0}\, e^{-t/\tau} \tag{4.14}$$

Equations (4.12) and (4.13) indicate that, when more than one deexcitation process is possible, τ is less than the lifetime of any individual competing reaction acting alone. In other words, because each deexcitation reaction is independent of the others, the observed rate of decay of an excited state is faster than deactivation by any single competing reaction acting by itself.

If the rate constant for a particular reaction is much larger than for its competitors, the excited state becomes deexcited predominantly by that process. As an example, we will consider an excited triplet state of a molecule that shows delayed fluorescence. Suppose that the lifetime for phosphorescence, τ_P, is 10^{-2} s, in which case k_P is 100 s^{-1}. When sufficient thermal energy is supplied, $T_{(\pi,\pi^*)}$ can be raised in energy to an excited singlet state, which could then emit "delayed" fluorescence if $S_{(\pi,\pi^*)}$ decays to the ground state by emitting electromagnetic radiation. Suppose that the rate constant ($k_{T^* \to S^*}$) for the transition from $T_{(\pi,\pi^*)}$ to $S_{(\pi,\pi^*)}$ is 20 s^{-1}. By Eq. (4.13), k for these two competing pathways is $100 + 20$ s^{-1}, or 120 s^{-1}, which corresponds to a lifetime of $1/(120$ s$^{-1})$, or 0.008 s. Suppose that another molecule, which can readily take on the excitation of $T_{(\pi,\pi^*)}$ of the original species, is introduced into the solution—$k_{transfer}$ might be 10^4 s^{-1}. Because of the relatively large rate constant, such a molecule "quenches" the phosphorescence and delayed fluorescence originating from $T_{(\pi,\pi^*)}$ (i.e., its decay pathway predominates over the other competing processes), so it is generally referred to as a *quencher*. For the three pathways indicated, the overall rate constant is $100 + 20 + 10^4$ s^{-1}, or essentially 10^4 s^{-1}. The deexcitation here is dominated by the quencher because $k_{transfer} \gg k_P + k_{T^* \to S^*}$.

D. Quantum Yields

A *quantum yield* (or *quantum efficiency*), Φ_i, is often used to describe the deexcitation processes following absorption of light. Here Φ_i represents the fraction of molecules in some excited state that will decay by the ith deexcitation reaction out of all the possible competing pathways:

$$\Phi_i = \frac{\text{number of molecules using } i\text{th deexcitation reaction}}{\text{number of excited molecules}}$$
$$= \frac{k_i}{\sum_j k_j} = \frac{\tau}{\tau_i} \qquad (4.15)$$

Equation (4.15) indicates that the rate constant for a particular pathway determines what fraction of the molecules in a given excited state will use that deexcitation process. Hence k_i determines the quantum yield for the ith deexcitation pathway. From Eq. (4.13) and the definition of τ_j given previously, we can also indicate such competition among pathways using lifetimes (see Eq. 4.15). The shorter the lifetime for a particular deexcitation pathway, the larger will be the fraction of the molecules using that pathway, and hence the higher will be its quantum yield. Finally, by Eqs. (4.13) and (4.15), the sum of the quantum yields for all the competing deexcitation pathways, $\sum_i \Phi_i$, equals 1.

To illustrate the use of Eq. (4.15), let us consider the quantum yield for chlorophyll fluorescence, Φ_{Fl}. The fluorescence lifetime τ_{Fl} of the lower excited singlet state of chlorophyll in ether is 1.5×10^{-8} s, and the observed lifetime τ for deexcitation of this excited state in ether is 0.5×10^{-8} s (Clayton, 1965). By Eq. (4.15), the expected quantum yield for fluorescence is $(0.5 \times 10^{-8}$ s$)/(1.5 \times 10^{-8}$ s$)$, or 0.3, which is consistent with the observed Φ_{Fl} of 0.33 for the fluorescence deexcitation of chlorophyll in ether.

IV. Absorption Spectra and Action Spectra

The absorption of radiation causes a molecule to go from its ground state to some excited state in which one of the electrons enters an orbital of higher energy. We have so

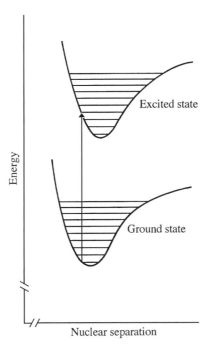

Figure 4-8. Energy curves for the ground state and an excited state showing the range of nuclear separations and the different energies for the various vibrational sublevels. The vertical arrow represents a transition that is caused by the absorption of a photon and that is consistent with the Franck–Condon principle.

far considered that both the ground state and the excited states occur at specific energy levels, as is indicated by the horizontal lines in Fig. 4-6 for chlorophyll. This leads to a consideration of whether only a very limited number of wavelengths are absorbed. For instance, are 430 and 680 nm the only wavelengths absorbed by chlorophyll (Fig. 4-6)? We will find that each electronic energy level is divided or split into various discrete sublevels that differ in energy. The largest splitting is due to *vibrational sublevels*. Vibrational sublevels affect the wavelengths of light involved in photosynthesis and other photochemical processes.

Vibrational sublevels are the result of the vibration of atoms in a molecule, which affects the total energy of the molecule. We can describe this atomic oscillation by the accompanying change in the internuclear distance. Therefore, we will refer to an energy level diagram indicating the range of positions, or trajectories, taken on by the vibrating nuclei (Fig. 4-8). The trajectories of such nuclear vibrations are quantized; i.e., only certain specific vibrations can occur, so only certain energies are possible for the vibrational sublevels of a given state.

For usual plant temperatures, essentially all molecules are in the ground or unexcited state. Moreover, these molecules are nearly all in the lowest vibrational sublevel of the ground state—another consequence of the Boltzmann energy distribution (see Eq. 3.21). The absorption of a photon can cause a transition of the molecule from the lowest vibrational sublevel of the ground state to one of the vibrational sublevels of the excited state. The actual sublevel reached depends on the energy of the absorbed photon. The probability that a photon will be absorbed also depends on its energy. Consideration of this absorption probability as a function of wavelength leads to an absorption spectrum for that particular molecule. The effect or consequence of light absorption on some process, when presented as a function of wavelength, leads to an action spectrum.

A. Vibrational Sublevels

Various vibrational sublevels of both ground and excited states of a molecule are schematically indicated in Fig. 4-8. In principle, such an energy level diagram can be prepared for any pigment, e.g., chlorophyll, carotenoid, or phytochrome. Energy level diagrams are extremely useful for predicting which electronic transitions are most likely to accompany the absorption of light. They can also help explain why certain wavelengths predominate in the absorption process.

The abscissa in Fig. 4-8 represents the distance between a pair of nuclei in the same molecule vibrating back and forth with respect to each other and can be obtained by imagining one nucleus to be situated at the origin of the coordinate system while the position of the other nucleus is plotted relative to this origin. The ordinate represents the total energy of the electrons plus the pair of nuclei. Figure 4-8 shows that the excited state is at a higher energy than the ground state and indicates that the two states are split into many vibrational sublevels differing in energy.

Because internuclear and interelectronic repulsive forces act against the electrostatic attraction between nuclei and shared electrons, only a certain range of internuclear separations can occur for a particular bond in a molecule. As internuclear separation decreases, the nuclei repel each other more and more, and the energy of the molecule increases. Moreover, the clouds of negative charge representing electrons localized on each nucleus have a greater overlap as internuclear separation decreases, resulting in interelectronic repulsion and likewise an increase in molecular energy. These effects account for the steep rise of the energy curves in Fig. 4-8 as nuclear separation becomes less (left-hand side of the figure). At the other extreme, the delocalized (bonding) electrons shared by the two nuclei resist an unlimited increase in nuclear separation, which diminishes the attractive electrostatic interaction between nuclei and electrons; such an increase in internuclear distance corresponds to a stretching of the chemical bond. Thus the energy curves in Fig. 4-8 also rise as the internuclear distance becomes greater (right-hand side of the figure). Because of these two opposing tendencies, the range of possible nuclear separations is confined to an energy trough. It is within these potential energy curves—one for the ground state and another for the excited state—that the trajectories of the nuclear vibrations occur.

A horizontal line in Fig. 4-8 represents the range of nuclear separations corresponding to a specific vibrational sublevel, i.e., nuclei vibrate back and forth along the distance indicated by a horizontal line. As is evident in the figure, both ground and excited states have many vibrational sublevels differing in energy. For the upper vibrational sublevels of a given state, the nuclei vibrate over longer distances (i.e., there is a more extensive range of nuclear separations in Fig. 4-8), which also corresponds to higher vibrational energies. Because the excited state has an electron in an antibonding orbital, it has a greater mean internuclear separation than does the ground state. This increase in bond length is shown in Fig. 4-8 by a slight displacement to greater nuclear separations, i.e., to the right, for the upper curve.

The direction of nuclear motion is reversed at the extremities of the vibrational pathways (horizontal lines in Fig. 4-8), so the velocity of the nuclei must be zero at these turning points. As the turning point at either end of the oscillation range is approached, the nuclei begin to slow down and eventually stop before reversing their direction of motion. Consequently, the nuclei spend most of their time at or near the extreme ends of their trajectory. Therefore, a photon is most likely to arrive at the molecule when the nuclei are at or near the extremes of their vibrational range.[6] Because of our probability consideration, the electronic transition resulting from the absorption of a photon, represented by the vertical arrow in Fig. 4-8, has been initiated from one of the ends of the nuclear oscillation range for the lowest vibrational sublevel of the ground state. This arrow begins from the lowest sublevel because nearly all ground state molecules are in the lowest vibrational sublevel at the temperatures encountered in plants, which, as we noted previously, is a consequence of the Boltzmann energy distribution.

During light absorption, the energy of the photon is transferred to some electron in the molecule. Because the molecular orbital describing the trajectory of an electron in the excited state has a small, but finite, probability of spatially overlapping with the nuclei, an interaction is possible between the excited electron and the nuclei over which it is delocalized. Such interactions generally take place over a rather short time period ($<10^{-13}$ s). An interaction between an energetic electron and the nuclei can cause the excitation of nuclei to higher energy vibrational states. In fact, the transition represented by the arrow in Fig. 4-8 corresponds to both the excitation of an electron leading to an excited state of the molecule and the subsequent excitation of the nuclei to some excited vibrational sublevel. Thus part of the energy of the photon is rapidly passed to nuclear vibrations. The length of the arrow in Fig. 4-8 is proportional to the light energy (or quantum) added to the molecule and therefore represents the energy distributed to the nuclei plus that remaining with the excited electron. Next, we will demonstrate which vibrational sublevel of the excited state has the highest probability of being reached by the excitation process.

B. The Franck–Condon Principle

Franck and Edward Condon in 1926 enunciated a principle, based mainly on classical mechanics, to help rationalize the various bands in absorption spectra and fluorescence emission spectra. We will direct our attention to the nuclei to discuss the effect of the quantized modes of nuclear vibration (see the energy level diagram in Fig. 4-8). The Franck–Condon principle states that the nuclei change neither their separation nor their velocity during the transitions for which the absorption of a photon is most probable. We can use this principle to predict which vibrational sublevels of the excited state are most likely to be involved in the electronic transitions accompanying light absorption.

First we will consider the part of the Franck–Condon principle stating that nuclei do not change their relative position for the most probable electronic transitions caused by light absorption. Because a vertical line in Fig. 4-8 represents no change in nuclear separation, the absorption of a photon has been indicated by a vertical arrow. This condition of constant internuclear distance during light absorption is satisfied most often when the nuclei are moving slowly or have stopped at the extremes of their oscillation range. Thus the origin of the arrow indicating an electronic transition in Fig. 4-8 is at one of the turning points of the lowest vibrational sublevel of the ground state, and the tip is drawn to an extremity of one of the vibrational sublevels in the excited state (the fourth vibrational sublevel for the particular case illustrated).

---

[6] Quantum-mechanical calculations beyond the scope of this text lead to somewhat different conclusions, especially for the lowest vibrational sublevel.

The other condition embodied in the Franck–Condon principle is that a photon has the greatest chance of being absorbed when the velocity (a vector) of the vibrating nuclei does not change. In other words, absorption is maximal when the nuclei are moving in the same direction and at the same speed in both the ground state and the excited state. Again, this condition has the greatest probability of being met when the nuclei are moving slowly or not at all, as occurs at the turning points for a nuclear oscillation, because the nuclei spend the most time at a particular velocity under such conditions. Therefore, the most probable electronic transition represented in a diagram such as that shown in Fig. 4-8 is a vertical line that originates from one of the ends of the horizontal line representing the range of nuclear separations for the lowest vibrational sublevel of the ground state and terminates at the end of the vibrational trajectory for some sublevel of the excited state.

Another way to view the Franck–Condon principle is to consider that the light absorption event is so rapid that the nuclei do not have a chance to move during it. The absorption of a photon requires about 10^{-15} s. In contrast, the period for one nuclear vibration back and forth along an oscillation range (such as those represented by horizontal lines in Fig. 4-8) is generally somewhat more than 10^{-13} s. Thus the nuclei cannot move an appreciable distance during the time necessary for the absorption of a photon, especially when the nuclei are moving relatively slowly near the ends of their vibrational trajectory. Also, nuclear velocity would not change appreciably in a time interval as short as 10^{-15} s.

The time scale for nuclear oscillations has far-reaching consequences for energy dissipation. In particular, as the nuclei oscillate back and forth along their trajectories, they can interact with other nuclei. These encounters make possible the transfer of energy from one nucleus to another (within the same molecule or to adjacent molecules). Thus the time for one cycle of a nuclear vibration, approximately 10^{-13} s, is an estimate of the time in which excess vibrational energy can be dissipated as heat by interactions with other nuclei. As excess energy is exchanged by such processes, the part of the molecule indicated in Fig. 4-8 soon reaches the lowest vibrational sublevel of the excited state; these transitions within the same electronic state, e.g., $S_{(\pi, \pi^*)}$, are usually complete in about 10^{-12} s. Fluorescence lifetimes generally are on the order of 10^{-8} s, so an excited singlet state gets to its lowest vibrational sublevel before appreciable deexcitation can occur by fluorescence. The rapid dissipation of excess vibrational energy causes some of the energy of the absorbed photon to be released as heat. Therefore, fluorescence is generally of lower energy (longer wavelength) than the absorbed light, as we will show for chlorophyll in Chapter 5.

The Franck–Condon principle predicts the most likely transition caused by the absorption of light, but others do occur. These other transitions become statistically less probable the more that the nuclear position or velocity changes during the absorption of the photon. Because transitions from the ground state to either higher or lower vibrational sublevels in the excited state occur with a lower probability than transitions to the optimal sublevel, the absorption of light is not as great at the wavelengths that excite the molecule to such vibrational sublevels of the excited state. Moreover, some transitions begin from an excited vibrational sublevel of the ground state. We can calculate the fraction of the ground state molecules in the various vibrational sublevels from the Boltzmann energy distribution, $n(E) = n_{\text{total}} e^{-E/RT}$ (Eq. 3.21b).

For many molecules the vibrational sublevels of both the ground state and the excited state are 10–20 kJ mol^{-1} apart. Consequently, as the wavelength of incident light is increased or decreased from that for the most intense absorption, transitions involving other vibrational sublevels become important, e.g., at approximately 15 kJ mol^{-1} intervals. An energy difference of 15 kJ mol^{-1} between two photons corresponds to a difference in wavelength of about

40 nm near the middle of the visible region (green or yellow; Table 4-1). Such wavelength spacings can be seen by the various peaks near the major absorption bands of chlorophyll (Fig. 5-3) or the three peaks in the absorption spectra of typical carotenoids (Fig. 5-5). In summary, the amount of light absorbed is maximal at a certain wavelength corresponding to the most probable transition predicted by the Franck–Condon principle. Transitions from other vibrational sublevels of the ground state and to other sublevels of the excited state occur less frequently and help create an absorption spectrum that is characteristic of a particular molecule.[7]

C. Absorption Bands and Absorption Coefficients

Our discussion of light absorption has so far been primarily concerned with transitions from the ground state energy level to those of excited states, which we just expanded to include the occurrence of vibrational sublevels of the states. In addition, vibrational sublevels are subdivided into *rotational* states. In particular, the motion of the atomic nuclei within a molecule can be described by quantized rotational states of specific energy, leading to the subdivision of a given vibrational sublevel into a number of rotational sub-sublevels. The energy increments between rotational states are generally about 1 kJ mol^{-1} (approximately 3 nm in wavelength in the visible region). The further broadening of absorption lines because of a continuum of *translational* energies of the whole molecule is generally much less, often about 0.1 kJ mol^{-1}. Moreover, interactions with the solvent or other neighboring molecules can affect the distribution of electrons in a particular molecule and, consequently, can shift the position of the various energy levels. The magnitude of the shifts caused by intermolecular interactions varies considerably and can be 5 kJ mol^{-1} or more.

The photons absorbed in an electronic transition involving specific vibrational sublevels—including the range of energies due to the various rotational sub-sublevels and other shifts—give rise to an *absorption band*. These wavelengths represent the transition from a vibrational sublevel of the ground state to some vibrational sublevel of an excited state. A plot of the relative efficiency for light absorption as a function of wavelength is an *absorption spectrum*, which generally includes more than one absorption band. Such bands represent transitions to or from different vibrational sublevels and to different excited states. The smoothness of the absorption bands of most pigment molecules indicates that a great range of photon energies can correspond to the transition of an electron from the ground state to some excited state (see absorption spectra in Figs. 4-12, 5-3, 5-5, and 5-7).[8] Because of the large effects that intermolecular interactions can have on electronic energy levels, the solvent should always be specified when presenting an absorption band or spectrum.

Absorption bands and spectra indicate how light absorption varies with wavelength. The absorption at a particular wavelength by a certain species is quantitatively described using

---

[7] Electromagnetic radiation in the far infrared region does not have enough energy to cause an electronic transition, but its absorption can excite a molecule to an excited vibrational sublevel of the ground state. For instance, by Eq. (4.2b) ($E_\lambda = Nhc/\lambda_{\text{vacuum}}$) IR at 4 μm has an energy of (119,600 kJ mol^{-1} nm)/(4000 nm) or 30 kJ mol^{-1}. Absorption of such radiation can excite a molecule to the third vibrational sublevel of the ground state when the vibration sublevels have a typical energy spacing of 15 kJ mol^{-1}.

[8] To "sharpen" an absorption spectrum, the translational broadening of absorption bands can be reduced by substantially decreasing the temperature, e.g., by using liquid nitrogen (boiling point = $-196°$C) to cool the sample. Also, the reduction in temperature decreases the number of the absorbing molecules in excited vibrational sublevels and higher energy rotational sublevels of the ground state—we can predict the relative populations of these states from the Boltzmann factor (see Eq. 3.21).

an *absorption coefficient*, ε_λ (ε_λ is also referred to as an extinction coefficient). Because of its usefulness, we will derive an expression incorporating ε_λ.

Let us consider a monochromatic beam of parallel light of flux density J. Because "monochromatic" refers to light of a single wavelength, J can be expressed as either a photon or an energy flux density. Some of the light may be absorbed in passing through a solution, so the emerging beam will generally have a lower flux density—we will assume that scattering and reflection are negligible (an assumption that must be tested when absorption properties are determined). In a small path length dx along the direction of the beam, J decreases by dJ due to absorption by a substance having a concentration c. Johann Lambert is often credited with recognizing in 1768 that $-dJ/J$ is proportional to dx— actually, Jean Bouguer had expressed this in 1729—and in 1852 August Beer noted that $-dJ/J$ is proportional to c. Upon putting these two observations together, we obtain the following expression:

$$-\frac{dJ}{J} = k_\lambda c \, dx \qquad (4.16)$$

where k_λ is a proportionality coefficient referring to a particular wavelength, and the minus sign indicates that the flux density is decreased by absorption. We can integrate Eq. (4.16) across a solution of a particular concentration, which leads to

$$-\int_{J_0}^{J_b} dJ/J = -\ln\frac{J_b}{J_0} = \int_0^b k_\lambda c \, dx = k_\lambda c b \qquad (4.17)$$

where J_0 is the flux density of the incident beam and J_b is its flux density after traversing a distance b through the solution (see Fig. 4-9).

Equation (4.17) is usually recast into a slightly more convenient form. We can replace the natural logarithm by the common logarithm ($\ln = 2.303 \log$; see Appendices II or III) and $k_\lambda/2.303$ by the absorption coefficient at a specific wavelength, ε_λ. Equation (4.17) then becomes

$$A_\lambda = \log\frac{J_0}{J_b} = \varepsilon_\lambda c b \qquad (4.18a)$$

where A_λ is the absorbance (colloquially, the "optical density") of the solution at a particular wavelength. When more than one absorbing substance is present in a solution, we can

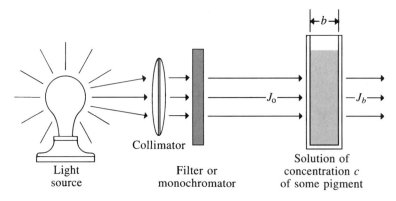

Figure 4-9. Quantities involved in light absorption by a solution as described by Beer's law, $\log(J_0/J_b) = \varepsilon_\lambda c b$ (Eq. 4.18a).

generalize Eq. (4.18a) to give

$$A_\lambda = \log \frac{J_0}{J_b} = \sum_j \varepsilon_{\lambda_j} c_j b \qquad (4.18b)$$

where c_j is the concentration of substance j, and ε_{λ_j} is its absorption coefficient at wavelength λ. Equation (4.18) is generally referred to as Beer's law, although it is also called the Beer–Lambert law, the Lambert–Beer law, and even the Bouguer–Lambert–Beer law.

According to Beer's law, the absorbance at some wavelength is proportional to the concentration of the absorbing substance, to its absorption coefficient at that wavelength, and to the optical path length b (Fig. 4-9). Values of ε_λ for organic compounds can exceed $10^4 \text{ m}^2 \text{ mol}^{-1}$ in the visible region. If ε_λ at some wavelength is known for a particular solute, we can determine its concentration from the measured absorbance at that wavelength by using Beer's law (Eq. 4.18). For laboratory absorption studies, the optical path length b is often 1 cm and c_j is expressed in mol liter^{-1} (i.e., molarity), in which case ε_{λ_j} has units of liter mol^{-1} cm^{-1} and is referred to as the *molar absorption* (or *extinction*) *coefficient* (1 liter mol^{-1} cm^{-1} = 1 M^{-1} cm^{-1} = 10^{-3} mM^{-1}cm^{-1} = 10^3 cm^2 mol^{-1} = 10^{-1} m^2 mol^{-1}). The absorbing solute is usually dissolved in a solvent that does not absorb at the wavelengths under consideration.

D. Application of Beer's Law

As an application of Beer's law, we will estimate the average chlorophyll concentration in leaf cells. The palisade and spongy mesophyll cells in the leaf section portrayed in Fig. 1-2 can correspond to an average thickness of chlorophyll-containing cells of about 200 μm. The maximum molar absorption coefficient ε_λ in the red or the blue bands of chlorophyll (see Figs. 4-6 and 5-3) is about $10^4 \text{ m}^2 \text{ mol}^{-1}$. At the peaks of the absorption bands, about 99% of the incident red or blue light can be absorbed by chlorophyll in a leaf. This corresponds to having an emergent flux density J_b equal to 1% of the incident flux density J_0, so the absorbance A_λ in Eq. (4.18a) equals $\log(100/1)$ or 2—for simplicity, we are ignoring absorption by pigments other than chlorophyll. Using Beer's law (Eq. 4.18a), we find that the average chlorophyll concentration is

$$c = \frac{A_\lambda}{\varepsilon_\lambda b} = \frac{(2)}{(10^4 \text{ m}^2 \text{ mol}^{-1})(200 \times 10^{-6} \text{ m})}$$

$$= 1 \text{ mol m}^{-3} \quad (1 \text{ mM})$$

a value characteristic of the average chlorophyll concentration in the photosynthesizing cells of many leaves.

Chlorophyll is located only in the chloroplasts, which in turn occupy about 3 or 4% of the volume of a mesophyll cell in the leaf of a higher plant. The average concentration of chlorophyll in chloroplasts is thus about 30 times higher than the estimate of chlorophyll concentration in a leaf, or approximately 30 mol m^{-3}. A typical light path across the thickness of a chloroplast (Fig. 1-9) is about 2 μm. Using Beer's law (Eq. 4.18a), we find that the absorbance of a single chloroplast in the red or blue bands is about

$$A_\lambda = (10^4 \text{ m}^2 \text{ mol}^{-1})(30 \text{ mol m}^{-3})(2 \times 10^{-6} \text{ m})$$

$$= 0.6 = \log \frac{J_0}{J_b}$$

Hence, $J_b = J_0/(\text{antilog } 0.6)$, or $0.25J_0$. Therefore, approximately 75% of the incident red or blue light at the peak of absorption bands is absorbed by a single chloroplast, which helps explain why individual chloroplasts appear green under a light microscope.

E. Conjugation

Light absorption by molecules generally involves transitions of π electrons to excited states for which the electrons are in π^* orbitals. These π electrons occur in double bonds; the more double bonds that there are in some molecule, the greater the probability for light absorption by that species. Indeed, both the effectiveness in absorbing electromagnetic radiation and the wavelengths absorbed are affected by the number of double bonds in *conjugation*, where conjugation refers to the alternation of single and double bonds (e.g., C—C=C—C=C—C=C—C) along some part of the molecule. For the molecules that we will consider, this alternation of single and double bonds involves mostly C atoms, but it also includes N atoms and O atoms. The absorption coefficient ε_λ increases with the number of double bonds in conjugation because more delocalized (shared) π electrons can then interact with light (each π bond contains two electrons, so the number of excitable electrons increases proportionally with the number of double bonds in conjugation). Moreover, as the number of double bonds in the conjugated system increases, the absorption bands shift to longer wavelengths.

To help understand why the number of double bonds in conjugation affects the wavelength position for an absorption band, let us consider the shifts in energy for the various orbitals as the number of π electrons in a conjugated system changes. The various π orbitals in a conjugated system occur at different energy levels, with the average energy remaining about the same as for the π orbital in an isolated double bond not part of a conjugated system (Fig. 4-10). The more double bonds in a conjugated system, the more π orbitals there are in the conjugated system, and the greater the energy range from the lowest to the highest energy π orbital. Because the average energy of the π orbitals in a conjugated

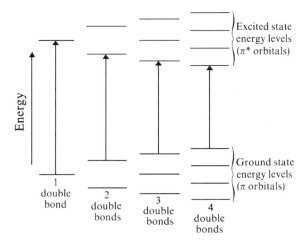

Figure 4-10. Effect on the energy levels of π and π^* orbitals as the number of double bonds in conjugation increases. The vertical arrows represent transitions caused by the absorption of photons—the required energy becomes less as the extent of conjugation increases.

system does not markedly depend on the number of double bonds, the energy of the highest energy π orbital increases as the number of double bonds in conjugation increases. The π^* orbitals are similarly split into various energy levels (also diagrammed in Fig. 4-10). Again, the range of energy levels about the mean for these π^* orbitals increases as the number of double bonds in the conjugated system increases. Consequently, the more π^* orbitals available in the conjugated system, the lower in energy will be the lowest of these.

In Fig. 4-10 we present transitions from the highest energy π orbital to the lowest energy π^* orbital for a series of molecules differing in the number of double bonds in conjugation. We notice that the more delocalized π electrons there are in the conjugated system, the less energy required for a transition. This is illustrated in Fig. 4-10 by a decrease in the length of the vertical arrow (which represents an electronic transition) as the number of double bonds in conjugation increases. Moreover, the most likely or probable electronic transition in this case is the one involving the least amount of energy, i.e., the excitation from the highest energy π orbital to the lowest energy π^* orbital is the transition that predominates (the transition probability depends on the spatial overlap between the quantum-mechanical wave functions describing the trajectories of the electrons in the two states, and such overlap is relatively large between the highest energy π orbital and the lowest energy π^* one).

The decrease in energy separation between the π orbitals and the π^* orbitals as the number of double bonds in conjugation increases (Fig. 4-10) accounts for the accompanying shift of the peaks of the absorption bands toward longer wavelengths. For example, an isolated double bond (C—C=C—C) generally absorbs maximally near 185 nm in the ultraviolet and has a maximum absorption coefficient of nearly 10^3 m^2 mol^{-1}. For two double bonds in conjugation (C—C=C—C=C—C), the maximum absorption coefficient doubles, and the wavelength position for maximum absorption shifts to about 225 nm, i.e., toward longer wavelengths. As the number of double bonds in conjugation in straight-chain hydrocarbons increases from three to five to seven to nine, the center of the absorption band for these hydrocarbons when they are dissolved in hexane shifts from approximately 265 to 325 to 375 to 415 nm, respectively. The maximum absorption coefficient is approximately proportional to the number of double bonds in the conjugated system, so it increases to almost 10^4 m^2 mol^{-1} for the hydrocarbon containing nine double bonds in conjugation. For molecules to absorb strongly in the visible region, an extensive conjugated system of double bonds is necessary, as is the case for pigments such as the chlorophylls and the carotenoids discussed in Chapter 5.

F. Action Spectra

The relative effectiveness of various wavelengths in producing a specified response is of basic importance in photobiology and is presented in an *action spectrum*. An action spectrum is complementary to an absorption spectrum, the latter being the relative probability for the absorption of different wavelengths, e.g., ε_λ versus λ. When many different types of pigments are present, the action spectrum for a particular response can differ greatly from the absorption spectrum of the entire system. However, the Grotthuss–Draper law implies that an action spectrum should resemble the absorption spectrum of the substance that absorbs the light responsible for the specific effect or action being considered.

To obtain an action spectrum for some particular response, we could expose the system to the same photon flux density at each of a series of wavelength intervals and measure the resulting effect or action. The action could be the amount of O_2 evolved, the fraction of seeds germinating, or some other measured change. We could then plot the responses obtained as a function of their respective wavelength intervals to see which wavelengths are most effective in leading to that "action." Another way to obtain an action spectrum is to plot the reciprocal of the number of photons required in the various wavelength intervals to give a particular response. If twice as many photons are needed at one wavelength compared with a second, the action spectrum has half the height at the first wavelength, and thus the effectiveness of various wavelengths can easily be presented (Fig. 4-13). Using the latter approach, the photon flux density is varied until the response is the same for each wavelength interval. This is an important point—if it is to be a true action spectrum, the action or effect measured must be linear with photon flux density for each of the wavelength intervals used, i.e., we must not be approaching light saturation. When we approach light saturation at certain wavelengths, the measured action per photon is less than it should be, compared with the values at other wavelengths. Consequently, an action spectrum is flattened for those wavelengths for which the response approaches light saturation. In the extreme case of light saturation at all wavelengths, the action spectrum is flat because the response is then the same at each wavelength.

We can compare the action spectrum of some response with the absorption spectra of the various pigments suspected of being involved to see which pigment is responsible. If the measured action spectrum closely resembles the known absorption spectrum of some molecule, light absorbed by that molecule may be leading to the action considered. Examples in which the use of action spectra have been important in understanding the photochemical aspects of plant physiology include the study of photosynthesis (Chapter 5) and investigations of the responses mediated by the pigment phytochrome, to which we now turn.

G. Absorption and Action Spectra of Phytochrome

Phytochrome is an important plant pigment that may be present in all eukaryotic photosynthetic organisms. It regulates photomorphogenic aspects of plant growth and development, including seed germination, stem elongation, leaf expansion, formation of certain pigments, chloroplast development, and flowering. Many of these classical phytochrome effects can be caused by very low amounts of light; e.g., 500 μmol m^{-2} of red light (the photons in the visible for $\frac{1}{4}$ s of full sunlight) can saturate most of the processes, and 0.3 μmol m^{-2} can lead to half-saturation for a very sensitive one (Salisbury and Ross, 1992). High irradiance levels can also lead to effects mediated by phytochrome. We will direct our attention first to the structure of phytochrome and then to the absorption spectra for two of its forms. The absorption spectra will subsequently be compared with the action spectra obtained for the promotion as well as for the inhibition of seed germination of lettuce (*Lactuca sativa*).

Phytochrome consists of a protein to which is covalently bound the *chromophore*, or light-absorbing part of the pigment (Fig. 4-11a). The chromophore is a *tetrapyrrole*, as occurs in the chlorophylls and the phycobilins that we will discuss in Chapter 5. Pyrrole refers to a five-membered ring having four carbons, one nitrogen, and two double bonds, $\underset{H}{N}$. Light absorption evidently leads to a photoisomerization about the double bond between rings C and D, which may underlie the conversion of phytochrome to its physiologically active form (Fig. 4-11b).

The chromophore of phytochrome is highly conjugated, as the structure for P$_r$ in Fig. 4-11a indicates (P stands for pigment and the subscript r indicates that it absorbs in the red region). In fact, P$_r$ has eight double bonds in conjugation. Other double bonds are not in the conjugated system because only those alternating with single bonds along the molecule

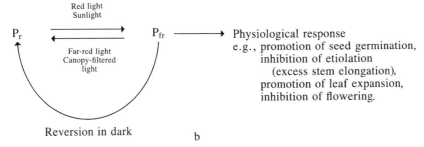

a

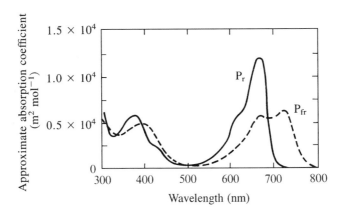

Figure 4-11.   Phytochrome structure, interconversions, and associated physiological responses. (a) Structure
for P_r, indicating the tetrapyrrole forming the chromophore and the convention for lettering the
rings. (b) Light and dark interconversions of phytochrome, indicating some of the reactions
promoted by the physiologically active form, P_{fr}.

are part of the main conjugation. For instance, the double bond at the top of pyrrole ring
C is a branch, or cross-conjugation, to the main conjugation and only slightly affects the
wavelength position for maximum absorption (semiempirical rules exist for predicting the
wavelength position of maximum absorption based on all double bonds that occur, including
those in branches to the main conjugated system). Because of the extensive conjugation,
phytochrome absorbs in the visible region.

The absorption spectra of two forms of phytochrome are presented in Fig. 4-12. P_r has a
major absorption band in the red, with a peak near 667 nm (the exact location varies with

Figure 4-12.   Absorption spectra for the red (P_r) and the far-red (P_{fr}) absorbing forms of phytochrome. Note
that the absorption spectra presented in this text were obtained at or near room temperature.
[Data are replotted from Butler *et al.* (1965); used by permission.]

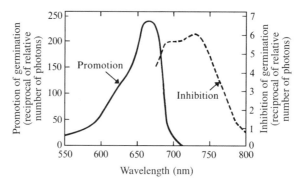

Figure 4-13. Action spectra for the promotion of lettuce seed germination and its reversal, or "inhibition."
[Data are replotted from Hendricks and Borthwick (1965); used by permission.]

the plant species). Upon absorption of red light, P_r can be converted to a form having an absorption band in the far-red, P_{fr} (Figs. 4-11 and 4-12). Although maximum absorption for P_{fr} occurs near 725 nm, appreciable absorption occurs in the near IR up to about 800 nm. In fact, the absorption bands of both pigments are rather broad, and many different wavelengths can be absorbed by each of them (Fig. 4-12). As is indicated in Fig. 4-12, the maximum absorption coefficients are about 10^4 m^2 mol^{-1} for both P_r and P_{fr}. The plant pigments that we will consider in Chapter 5 have similar high values of ε_λ. We also note that the absorption spectra for P_r and P_{fr} overlap considerably, e.g., 660-nm light can be readily absorbed by either pigment.

We next examine the action spectra for seed germination responses with respect to the known absorption properties of phytochrome. Figure 4-13 indicates that a pigment absorbing in the red region promotes the germination of lettuce seeds. If we compare this action spectrum with the absorption spectra given in Fig. 4-12, we see that the pigment absorbing the light that promotes seed germination most likely is the P_r form of phytochrome. This enhancement of seed germination by red light can be reversed or inhibited by subsequent irradiation of the seeds with far-red light (Fig. 4-13). This inhibition induced by far-red light requires fewest photons at wavelengths near 725–730 nm, similar to the position of the peak in the absorption spectrum of P_{fr} (Fig. 4-12). In summary, both the promotion of lettuce seed germination and the reversal of this promotion are apparently controlled by two forms of phytochrome, P_r and P_{fr}, which can be reversibly interconverted by light. (Germination is not enhanced by red light for seeds of all species; enhancement tends to be more common in small seeds rich in fat that come from wild plants.)

To illustrate the ecological consequences of the phytochrome control of seed germination, let us consider a seed present on the surface of the soil under a dense canopy of leaves. Because of their chlorophyll, leaves in the canopy absorb red light preferentially to far-red light; thus very little red light compared with far-red light reaches the seed. The phytochrome in the seed hence occurs predominantly in the P_r form. If this seed were to become exposed to sunlight—as could happen because of a fire or the removal of a shading tree—a larger fraction of the P_r would be converted to the active form, P_{fr} (see Fig. 4-11b). Seed germination would then proceed under light conditions favorable for photosynthesis and thus for growth.

Studies using action spectra have indicated that, on a photon basis, the maximum sensitivity is usually quite different for the opposing responses attributable to the two forms of phytochrome. (Because the expression of phytochrome action involves a multistep process including biochemical reactions, the two responses need not have the same sensitivity.) For lettuce seed germination, about 30 times more far-red photons (e.g., at 730 nm) are required to cause a 50% inhibition than the number of red photons (e.g., at 660 nm) needed to promote seed germination by 50%—compare the different scales used for the ordinates in Fig. 4-13. Thus ordinary sunlight is functionally equivalent to red light because much of the phytochrome is converted to the active form, P_{fr}. Also, P_r has a much larger absorption coefficient in the red region than does P_{fr} (Fig. 4-12).

The P_{fr} form of phytochrome reverts spontaneously to P_r in the dark (Fig. 4-11b), except in most monocots. Reversion can take place in less than 1 h and probably is not used by plants as a timing mechanism per se, although phytochrome may be involved in other aspects of photoperiodism (the influence of day length or night length on plant processes). Also, the interconversions of phytochrome indicated in a simplified fashion in Fig. 4-11b are actually multistep processes involving a number of intermediates. The mechanisms of action of phytochrome are not yet fully understood, although changes in membrane permeability, in the activity of membrane-bound enzymes, and in the transcription or translation of genetic information have all been implicated. Finally, as we might expect, individual plants vary considerably in how they amplify and use the information provided by the P_r–P_{fr} system.

Our discussion of phytochrome leads us to the concept of a *photostationary state*. A photostationary state refers to the relative amounts of interconvertible forms of some pigment that occur in response to a particular steady illumination—such a state should more appropriately be called a "photosteady" state. When illumination is constant, the conversion of P_r to P_{fr} eventually achieves the same rate as the reverse reaction (see Fig. 4-11). The ratio of P_r to P_{fr} for this photostationary state depends on the absorption properties of each form of the pigment for the incident wavelengths (note that the absorption spectra overlap; Fig. 4-12), the number of photons in each wavelength interval, the kinetics of the competing deexcitation reactions, and the kinetics of pigment synthesis or degradation. If the light quality or quantity were to change to another constant condition, we would shift to a new photostationary state for which P_r/P_{fr} is generally different. Because the ratio of P_r to P_{fr} determines the overall effect of the phytochrome system, a description in terms of photostationary states can be useful for discussing the influences of this pigment.

Problems

4.1   Consider electromagnetic radiation with the indicated wavelengths in a vacuum.

    a. If λ is 400 nm, how much energy is carried by 10^{20} photons?

    b. If a mole of 1800-nm photons is absorbed by 10^{-3} m^3 (1 liter) of water at 0°C, what would be the final temperature? Assume that there are no other energy exchanges with the external environment; the volumetric heat capacity of water averages 4.19×10^6 J m^{-3} °C^{-1} over the temperature range involved.

    c. A certain optical filter, which passes all wavelengths below 600 nm and absorbs all those above 600 nm, is placed over a radiometric device. If the meter indicates 1 W m^{-2}, what is the maximum photon flux density in μmol m^{-2} s^{-1}?

    d. What is the illuminance in (c) expressed in lux (lumens m^{-2})?

4.2   Consider electromagnetic radiation having a frequency of 0.9×10^{15} cycles s^{-1}.

    a. The speed of the radiation is 2.0×10^8 m s^{-1} in dense flint glass. What are the wavelengths in a vacuum, in air, and in such glass?

    b. Can such radiation cause an $S_{(\pi,\pi)}$ ground state of a pigment to go directly to $T_{(\pi,\pi^*)}$?

    c. Can such radiation cause the transition of a π electron to a π^* orbital in a molecule having six double bonds in conjugation?

    d. Electromagnetic radiation is often expressed in "wave numbers," which is the frequency divided by the speed of light in a vacuum, i.e., v/c, which is $1/\lambda_{vacuum}$. What is the wave number in m^{-1} in the current case?

4.3   Suppose that the quantum yield for ATP formation—molecules of ATP formed/numbers of excited chlorophyll molecules—is 0.40 at 680 nm, and the rate of ATP formation is 0.20 mol m^{-2} h^{-1}.

    a. What is the minimum photon flux density in μmol m^{-2} s^{-1} at 680 nm?

    b. What is the energy flux density under the conditions of (a)?

    c. If light of 430 nm is used, the ground state of chlorophyll, $S_{(\pi,\pi)}$, is excited to $S^b_{(\pi,\pi^*)}$. Suppose that 95% of $S^b_{(\pi,\pi^*)}$ goes to $S^a_{(\pi,\pi^*)}$ in 10^{-12} s and that the rest of the upper excited singlet state returns to the ground state. What is the energy conversion efficiency of 430 nm light as an energy source for ATP formation compared with 680 nm light?

    d. The hydrolysis of ATP to ADP and phosphate under physiological conditions can yield about 45 kJ of free energy mol^{-1}. What wavelength of light has the same amount of energy mol^{-1}?

4.4   Assume that some excited singlet state can become deexcited by three competing processes: (1) fluorescence (lifetime $= 10^{-8}$ s), (2) a radiationless transition to an excited triplet state (5×10^{-9} s), and (3) a radiationless transition to the ground state (10^{-8} s).

    a. What is the lifetime of the excited singlet state?

    b. What is the maximum quantum yield for all deexcitations leading directly or indirectly to electromagnetic radiation?

    c. Suppose that the molecule is inserted into a membrane, which adds a deexcitation pathway involving intermolecular transfer of energy from the excited singlet state (rate constant $= 10^{12}$ s^{-1}). What is the new lifetime of the excited singlet state?

4.5   The *cis* isomer of some species has an absorption coefficient of 2.0×10^3 m^2 mol^{-1} at 450 nm, where the spectrophotometer has a photon flux density of 10^{17} photons m^{-2} s^{-1}.

    a. What concentration of the *cis* isomer will absorb 65% of the incident 450 nm light for a cuvette (a transparent vessel used in a spectrophotometer) with an optical path length of 10 mm?

    b. What is the absorbance of the solution in (a)? What would be the absorbance if the flux density at 450 nm were halved?

    c. Suppose that the 450-nm light causes a photoisomerization of the *cis* isomer to the *trans* isomer with a quantum yield of 0.50, the other deexcitation pathway being the return to the *cis* form. If the *trans* isomer did not absorb at 450 nm, what would be the initial rate of decrease of the *cis* isomer and rate of change of absorbance at 450 nm under the conditions of (a)?

d. If ε_{450} for the *trans* isomer were 10^3 m^2 mol^{-1}, and 450 nm light led to a photoisomerization of the *trans* isomer with a quantum yield of 0.50 for forming the *cis* isomer, what would be the ratio of *cis* to *trans* after a long time?

4.6 **a.** Suppose that the spacing in wave numbers (see Problem 4.2) between vibrational sublevels for the transition depicted in Fig. 4-8 is 1.2×10^5 m^{-1} and that the most probable absorption predicted by the Franck–Condon principle occurs at 500 nm (the main band). What are the wavelength positions of the satellite bands that occur for transitions to the vibrational sublevels just above and just below the one for the most probable transition?

b. Suppose that the main band has a maximum absorption coefficient of 5×10^3 m^2 mol^{-1}, and each satellite band has an ε_λ one-fifth as large. If 20% of the incident light is absorbed at the wavelengths of either of the satellite bands, what percentage is absorbed at the main-band wavelength?

c. When the pigment is placed in a cuvette with an optical path length of 5 mm, the maximum absorbance is 0.3. What is the concentration?

4.7 A straight-chain hydrocarbon has 11 double bonds in conjugation. Suppose that it has three absorption bands in the visible region, one at 450 nm ($\varepsilon_{450} = 1.0 \times 10^4$ m^2 mol^{-1}), one at 431 nm ($\varepsilon_{431} = 2 \times 10^3$ m^2 mol^{-1}), and a minor band near 470 nm ($\varepsilon_{470} \cong 70$ m^2 mol^{-1}). Upon cooling from 20°C to liquid helium temperatures, the minor band essentially disappears.

a. What is the splitting between vibrational sublevels in the excited state?

b. What transition could account for the minor band? Support your answer by calculation.

c. If the λ_{max} for fluorescence is at 494 nm, what transition is responsible for the 450 nm absorption band?

d. If the double bond in the middle of the conjugated system is reduced (by adding two H's so that it becomes a single bond) and the rest of the molecule remains unchanged, calculate the new λ_{max} for the main absorption band and its absorption coefficient. Assume that for every double bond added to the conjugated system, λ_{max} shifts by 25 kJ mol^{-1}, and that $\varepsilon_{\lambda_{max}}$ is directly proportional to the number of double bonds in conjugation.

References

Alberty, R. A., and Silbey, R. J. (1997). *Physical Chemistry* (2nd ed). Wiley, New York.

Butler, W. L., Hendricks, S. B., and Siegelman, H. W. (1965). In *Chemistry and Biochemistry of Plant Pigments* (T. W. Goodwin, Ed.), pp. 197–210. Academic Press, London.

Clayton, R. K. (1965). *Molecular Physics in Photosynthesis*. Blaisdell, New York.

Clayton, R. K. (1980). *Photosynthesis: Physical Mechanisms and Chemical Patterns*. Cambridge Univ. Press, Cambridge, UK.

Condon, E. (1926). A theory of intensity distribution in band systems. *Phys. Rev.* **28**, 1182–1201.

Cutnell, J. D., and Johnson, K. W. (1995). *Physics* (3rd ed.). Wiley, New York.

Franck, J. (1926). Elementary processes of photochemical reactions. *Trans. Faraday Soc.* **21**, 536–542.

Gates, D. M. (1980). *Biophysical Ecology*. Springer-Verlag, New York.

Halliday, D., and Resnick, R. (1997). *Fundamentals of Physics* (5th ed.). Wiley, New York.

Hart, J. W. (1988). *Light and Plant Growth*. Unwin Hyman, London.

Hendricks, S. B., and Borthwick, H. A. (1965). In *Chemistry and Biochemistry of Plant Pigments* (T. W. Goodwin, Ed.), pp. 405–436. Academic Press, London.

Horspool, W., and Song, P.-S. (Eds.) (1995). *CRC Handbook of Organic Photochemistry and Photobiology*. CRC Press, Boca Raton, FL.

Kagan, J. (1993). *Organic Photochemistry: Principles and Applications*. Academic Press, San Diego.

Kendrick, R. E., and Kronenberg, G. H. M. (1994). *Photomorphogenesis in Plants* (2nd ed.). Kluwer, Dordrecht.

Kirk, J. T. O. (1994). *Light and Photosynthesis in Aquatic Ecosystems* (2nd ed.). Cambridge Univ. Press, Cambridge, UK.

Klessinger, M. (1995). *Excited States and Photochemistry of Excited Molecules*. VCH, New York.

Kohen, E., Santus, R., and Hirschberg, J. G. (1995). *Photobiology*. Academic Press, San Diego.

Kopecký, J. (1992). *Organic Photochemistry: A Visual Approach*. VCH, New York.

Mohr, H., and Schopfer, P. (1995). *Plant Physiology*. Springer-Verlag, Berlin.

Pap, J. M. (1997). Total solar irradiance variability: A review. In *Proceedings of the International School of Physics "Enrico Fermi" Course CXXXIII* (G. Cini Castagnoli and A. Provenzale, Eds.). IOS Press, Amsterdam.

Pap, J. M., and Frölich, C. (1998). Total irradiance variations. *J. Atmosph. Solar–Terrestrial Phys.*, in press.

Pearcy, R. W. (1989). Radiation and light measurements. In *Plant Physiological Ecology: Field Methods and Instrumentation* (R. W. Pearcy, J. Ehleringer, H. A. Mooney, and P. W. Rundel, Eds.), pp. 97–116. Chapman & Hall, London.

Pedrotti, F. L., and Pedrotti, L. S. (1993). *Introduction to Optics* (2nd ed.). Prentice Hall, Englewood Cliffs, NJ.

Ramamurthy, V., and Schanze, K. S. (Eds.) (1997). *Organic Photochemistry*. Dekker, New York.

Rea, M. S. (Ed.) (1993). *Lighting Handbook* (8th ed.). Illuminating Engineering Society of North America, New York.

Sage, L. C. (1992). *Pigment of the Imagination: A History of Phytochrome Research*. Academic Press, San Diego.

Salisbury, F. B., and Ross, C. W. (1992). *Plant Physiology* (4th ed.). Wadsworth, Belmont, CA.

Short, T. W., and Briggs, W. R. (1994). The transduction of blue light signals in higher plants. *Annu. Rev. Plant Physiol. Plant Mol. Biol.* **45**, 143–171.

Smith, H. (1995). Physiological and ecological function within the phytochrome family. *Annu. Rev. Plant Physiol. Plant Mol. Biol.* **46**, 289–315.

Sobel, M. I. (1987). *Light*. Univ. of Chicago Press, Chicago.

Turro, N. J. (1991). *Modern Molecular Photochemistry*. Univ. Science Books, Mill Valley, CA.

Wayne, C. E., and Wayne, R. P. (1996). *Photochemistry*. Oxford Univ. Press, Oxford.

Williams, D. H., and Fleming, I. (1995). *Spectroscopic Methods in Organic Chemistry* (5th ed.). McGraw-Hill, London.

Wilson, R. C., and Hudson, H. S. (1988). Solar luminosity variations in solar cycle 21. *Nature* **332**, 810–812.

Winn, J. S. (1995). *Physical Chemistry*. HarperCollins, New York.

5

Photochemistry of Photosynthesis

I.	**Chlorophyll—Chemistry and Spectra**	**185**
	A. Types and Structures	186
	B. Absorption and Fluorescence Emission Spectra	187
	C. Absorption *in Vivo*—Polarized Light	189
II.	**Other Photosynthetic Pigments**	**191**
	A. Carotenoids	191
	B. Phycobilins	194
III.	**Excitation Transfers among Photosynthetic Pigments**	**196**
	A. Pigments and the Photochemical Reaction	196
	B. Resonance Transfer of Excitation	198
	C. Transfers of Excitation between Photosynthetic Pigments	199
	D. Excitation Trapping	201
IV.	**Groupings of Photosynthetic Pigments**	**202**
	A. Photosynthetic Units	203
	B. Excitation Processing	203
	C. Photosynthetic Action Spectra and Enhancement Effects	205
	D. Two Photosystems Plus Light-Harvesting Antennae	206
V.	**Electron Flow**	**208**
	A. Electron Flow Model	208
	B. Components of the Electron Transfer Pathway	210
	C. Types of Electron Flow	215
	D. Photophosphorylation	216
	E. Vectorial Aspects of Electron Flow	217
	Problems	**218**
	References	**220**

PHOTOSYNTHESIS IS the largest scale synthetic process on earth. About 1.0×10^{14} kg (100 billion tons) of carbon are fixed annually into organic compounds by photosynthetic organisms (often called the *net primary productivity*). This equals about 1% of the world's

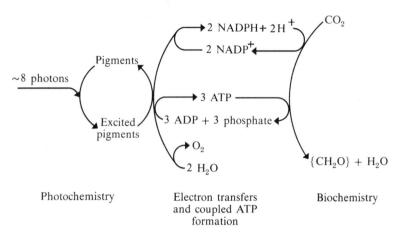

Photochemistry   Electron transfers   Biochemistry
                        and coupled ATP
                        formation

Figure 5-1. Schematic representation of the three stages of photosynthesis. The absorption of light can excite
photosynthetic pigments, leading to the photochemical events in which electrons are donated by
special chlorophylls. The electrons are then transferred along a series of molecules, leading to
the reduction of NADP⁺ to NADPH; ATP formation is coupled to the electron transfer steps.
The biochemistry of photosynthesis can proceed in the dark and requires 3 mol of ATP and 2 mol
of NADPH per mole of CO_2 fixed into a carbohydrate, represented in the figure by {CH_2O}.

known reserves of fossil fuels (coal, gas, and oil), or 10 times the world's current annual
energy consumption. The carbon source used in photosynthesis is the 0.04% CO_2 contained
in the air (about 8×10^{14} kg carbon) and the CO_2 or HCO_3^- dissolved in lakes and oceans
(about 400×10^{14} kg carbon). In addition to the organic compounds, another product of
photosynthesis essential for all respiring organisms is O_2. At the current rate, the entire
atmospheric content of O_2 is replenished by photosynthesis every 2000 years.

Photosynthesis is composed of many individual steps that work together with a remarkably high overall efficiency. We can divide the process into three stages: (1) the photochemical steps, our primary concern in this chapter; (2) electron transfer, to which is coupled ATP
formation, which we consider in both this chapter and Chapter 6; and (3) the biochemical
reactions involving the incorporation of CO_2 into carbohydrates. Figure 5-1 summarizes the
processes involved and introduces the relative amounts of the various reactants and products
taking part in the three stages of photosynthesis. The photochemical reactions, which are
often referred to as the *primary events* of photosynthesis, lead to electron transfer along a
sequence of molecules, resulting in the formation of NADPH and ATP.

In the net chemical reaction for photosynthesis (Fig. 5-1), two H_2Os are indicated as
reactants in the O_2 evolution step, and one H_2O is a product in the biochemical stage.
Hence, the overall net chemical reaction describing photosynthesis is CO_2 plus H_2O yields
carbohydrate plus O_2. Using the energy of each of the chemical bonds in these compounds
leads to the following representation for the net photosynthetic reaction:

$$\overset{463}{H - } O \overset{463}{- H} \; + \; O \overset{800}{=\!=} C \overset{800}{=\!=} O \; \rightarrow$$

$$H \overset{413}{-\!-} \overset{\overset{\frac{1}{2}(348)}{|}}{C} \overset{350}{-\!-} O \overset{463}{-\!-} H \; + \; O \overset{498}{=\!=} O \qquad (5.1)$$

$$\frac{1}{2}(348)$$

where the numbers represent the approximate bond energies in kJ mol^{-1}. A C–C bond, which occurs on two sides of the carbon in {CH$_2$O}, has an energy of 348 kJ mol^{-1}, so $\frac{1}{2}$(348) has been indicated in the appropriate places in Eq. (5.1).

The formulation of photosynthesis in Eq. (5.1) fails to do justice to the complexity of the reactions but does estimate the amount of Gibbs free energy that is stored. The total chemical bond energy is 2526 kJ mol^{-1}(463 + 463 + 800 + 800) for the reactants in Eq. (5.1) and 2072 kJ mol^{-1} for the products (413 + 348 + 350 + 463 + 498). Thus the reactants H$_2$O and CO$_2$ represent the lower energy (i.e., they are more "tightly" bonded) because 454 kJ mol^{-1} is necessary for the bond changes to convert them to the products {CH$_2$O} plus O$_2$. This energy change actually represents the increase in enthalpy required, ΔH, although we are really more concerned here with the change in Gibbs free energy, ΔG (see Chapter 6 and Appendix IV). (For a reaction at constant temperature, $\Delta G = \Delta H - T\Delta S$, where S is the entropy; ΔG is about the same as ΔH for Eq. 5.1.) Although the actual ΔG per mole of C depends somewhat on the particular carbohydrate involved, 454 kJ is approximately the increase in Gibbs free energy per mole of CO$_2$ that reacts according to Eq. (5.1). For instance, the Gibbs free energy released when glucose is oxidized to CO$_2$ and H$_2$O is 479 kJ mol^{-1} of C. In discussing photosynthesis we will frequently use this ΔG, which refers to standard state conditions (25°C, pH 7, 1 molal concentrations, 1 atm pressure).

About eight photons are required in photosynthesis per CO$_2$ fixed and O$_2$ evolved (see Fig. 5-1). Red light at 680 nm corresponds to 176 kJ mol^{-1} (Table 4-1), so 8 mol of such photons have 1408 kJ of radiant energy. Using this as the energy input and 479 kJ as the energy stored per mole of CO$_2$ fixed, the efficiency of energy conversion by photosynthesis is (479 kJ/1408 kJ)(100), or 34%. Actually, slightly more than eight photons may be required per CO$_2$ fixed. Furthermore, the energy for wavelengths < 680 nm, which are also used in photosynthesis, is higher than 176 kJ mol^{-1}. Both of these considerations reduce the efficiency for the utilization of absorbed energy. Nevertheless, photosynthesis is an extremely efficient energy conversion process, considering all the steps involved, each with its inherent energy loss.

Nearly all the enzymes involved in the synthetic reactions of photosynthesis also occur in nonphotosynthetic tissue. Thus the unique feature of photosynthesis is the conversion of radiant energy into chemical energy. This chapter will emphasize the light absorption and the excitation transfer aspects of photosynthesis. We consider the structures and the absorption characteristics of the photosynthetic pigments and the means by which radiant energy is trapped, transferred, and eventually used. Thus the emphasis is on the photo part of photosynthesis.

I. Chlorophyll—Chemistry and Spectra

Chlorophylls represent the principal class of pigments responsible for light absorption in photosynthesis and are found in all photosynthetic organisms. A number of different types of chlorophyll occur, as Mikhail Tswett demonstrated in 1906 using adsorption chromatography. For instance, approximately 1 g of the chlorophylls designated *a* and *b* is present per kilogram fresh weight of green leaves. The empirical formulas were first given by Richard Willstätter; Hans Fischer established the structures of various chlorophylls by 1940. These two investigators, as well as Robert Woodward, who synthesized chlorophyll *in vitro*, all received the Nobel prize in chemistry for their studies on this important plant pigment (in 1915, 1930, and 1965, respectively). We will first consider the structure of chlorophyll *a* (Chl *a*) and then its absorption and fluorescence characteristics.

A. Types and Structures

The various types of chlorophyll are identified by letters or by the taxonomic group of the organisms in which they occur. The most important is Chl a, with a relative molecular mass of 893.5 and the structure in Fig. 5-2. Chl a is found in all photosynthetic organisms except the green and the purple bacteria, i.e., in all species for which O_2 evolution accompanies photosynthesis. It is a tetrapyrrole with a relatively flat porphyrin "head" about 1.5×1.5 nm $(15 \times 15\ \text{Å})$ in the center of which a magnesium atom is coordinately bound. Attached to the head is a long-chain terpene alcohol, phytol, which acts like a "tail" about 2 nm in length containing 20 carbon atoms (Fig. 5-2). This tail provides a nonpolar region that helps bind the chlorophyll molecules to chlorophyll–protein complexes in the lamellar membranes, but it makes no appreciable contribution to the optical properties of chlorophyll in the visible region. The system of rings in the porphyrin head of Chl a is highly conjugated, having nine double bonds in conjugation (plus three other double bonds in branches to the main conjugated system). These alternating single and double bonds of the porphyrin ring provide many delocalized π electrons that can take part in the absorption of light.

    Other chlorophyll forms structurally similar to Chl a occur in nature. For instance, Chl b differs from Chl a by having a formyl group ($-CHO$) in place of a methyl group ($-CH_3$)

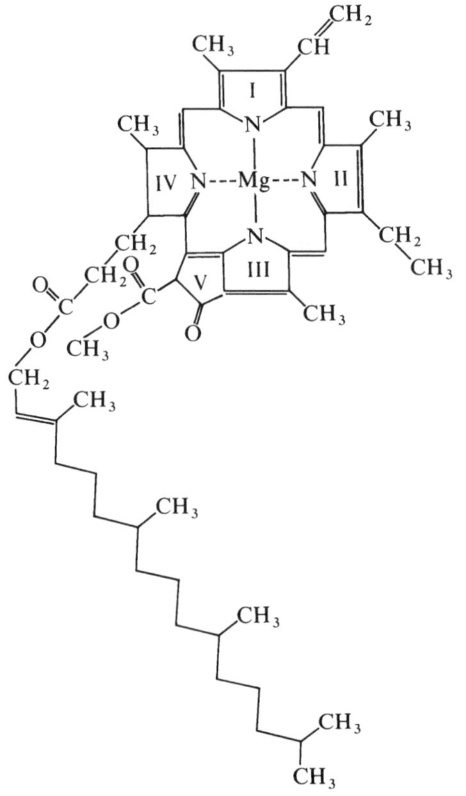

Figure 5-2. Structure of Chl a, illustrating the highly conjugated porphyrin "head" to which is attached a phytol "tail." The convention for numbering the various rings is also indicated. The solid lines to Mg indicate a resonating form with shared electrons in the bonds, and the dashed lines indicate bonds with little electron sharing at that moment (electron sharing varies over time).

on ring II. Chl b is found in virtually all land plants (including ferns and mosses), the green algae, and the Euglenophyta; the ratio of Chl a to Chl b in these organisms is usually about 3:1. Chl b is not essential for photosynthesis because a barley mutant containing only Chl a carries out photosynthesis satisfactorily.

Another type is Chl c, which occurs in the dinoflagellates, diatoms, golden algae, and brown algae. The common chlorophyll of the green and the purple bacteria is bacteriochlorophyll a (BChl a). The purple photosynthetic bacteria contain BChl a (or b in some species), and BChl a plus Chlorobium chlorophyll occur in green photosynthetic bacteria. These bacterial pigments differ from green plant chlorophylls by containing two more hydrogens in the porphyrin ring. They also have different substituents around the periphery of the porphyrin ring. In addition, Chlorobium chlorophyll has the alcohol farnesol (15 C and three double bonds) in place of phytol (20 C and one double bond). The pigment of principal interest in this text is Chl a.

B. Absorption and Fluorescence Emission Spectra

The absorption spectrum of Chl a has a blue band and a red band, so the characteristic color of chlorophyll is green. The band in the blue part of the spectrum has a peak at 430 nm for Chl a in ether (Fig. 5-3). This band is known as the *Soret band*; it occurs in the UV, violet, or blue region for all tetrapyrroles. We will designate the wavelength position for a maximum of the absorption coefficient in an absorption band by λ_{max}. Figure 5-3 indicates that the absorption coefficient at the λ_{max} for the Soret band of Chl a is just over 1.2×10^4 m^2 mol^{-1} (1.2×10^5 M^{-1} cm^{-1}). Such a high value is a consequence of the many double bonds in the conjugated system of the porphyrin ring of chlorophyll. Chl a has a major band in the red region with a λ_{max} at 662 nm when the pigment is dissolved in ether (Fig. 5-3).

Chl a also has minor absorption bands. For instance, Chl a dissolved in ether has a small absorption band at 615 nm, which is 47 nm shorter than the λ_{max} of the main red band (Fig. 5-3). Absorption of light at 615 nm leads to an electronic transition requiring

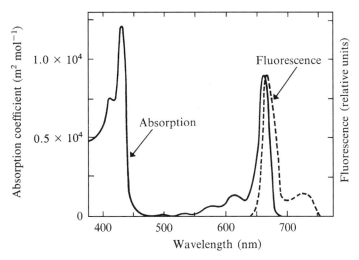

Figure 5-3. Absorption and fluorescence emission spectra of Chl a dissolved in ether. [Data are from Holt and Jacobs (1954); used by permission.]

$14\,kJ\,mol^{-1}$ more energy than the main band at 662 nm. This extra energy is similar to the energy spacing between vibrational sublevels. In fact, this small band on the shorter wavelength (higher energy) side of the red band corresponds to electrons going to the vibrational sublevel in the excited state immediately above the sublevel for the λ_{max} at 662 nm—an aspect to which we will return.

Although chlorophyll absorbs strongly in both the red and the blue, its fluorescence is essentially all in the red region (Fig. 5-3). This is because the upper singlet state of chlorophyll excited by blue light ($S^b_{(\pi,\pi*)}$ in Fig. 4-6) is extremely unstable and goes to the lower excited singlet state $S^a_{(\pi,\pi*)}$ in about 10^{-12} s, i.e., before any appreciable blue fluorescence can take place ($S^b_{(\pi,\pi*)}$ and $S^a_{(\pi,\pi*)}$ are distinct excited electronic states; each has its own energy curve in a diagram such as Fig. 4-8). Because of such rapid energy degradation by a radiationless transition, photons absorbed in the Soret band of chlorophyll are as effective for photosynthesis as the lower energy photons absorbed in the red region. We can observe the red fluorescence of chlorophyll accompanying light absorption by the Soret band if we illuminate a leaf with blue or shorter wavelength light in the dark. With a light microscope we can see the red fluorescence emanating from individual chloroplasts in the leaf's cells when using such exciting light (the red fluorescence is often masked by scattering when using red exciting light, so shorter wavelengths are used in most fluorescence studies).

The transition having a λ_{max} at 662 nm in the absorption spectrum for Chl a dissolved in ether corresponds to the excitation of the molecule from the lowest vibrational sublevel of the ground state to some vibrational sublevel of the lower excited state. We can use the Boltzmann factor [$n(E)/n_{total} = e^{-E/RT}$; Eq. 3.21b] to estimate the fraction of chlorophyll molecules in the first excited vibrational sublevel of the ground state when light arrives. Because RT is $2.48\,kJ\,mol^{-1}$ at 25°C (Appendix II), and the distance between vibrational sublevels is about $14\,kJ\,mol^{-1}$ for chlorophyll, the Boltzmann factor equals $e^{-(14\,kJ\,mol^{-1})/(2.48\,kJ\,mol^{-1})}$, or $e^{-5.65}$, which is 0.0035. Therefore, only about 1 in 300 chlorophyll molecules is in the first excited vibrational sublevel of the ground state when light arrives. Consequently, the absorption of a photon nearly always occurs when chlorophyll is in the lowest vibrational sublevel of the ground state.

In Chapter 4 we argued that fluorescence generally occurs from the lowest vibrational sublevel of the excited singlet state. In other words, any excess vibrational energy is usually dissipated before the rest of the energy of the absorbed photon can be reradiated as fluorescence. However, Fig. 5-3 shows that the wavelength region for most of the fluorescence is nearly coincident with the red band in the chlorophyll absorption spectrum (the difference in energy or wavelength between absorption and fluorescence bands is often called the Stokes shift). In particular, the λ_{max} for fluorescence occurs at 666 nm, which is only $1\,kJ\,mol^{-1}$ lower in energy than the λ_{max} of 662 nm for the red band in the absorption spectrum. The slight shift, which is much less than the distance between vibrational sublevels of $14\,kJ\,mol^{-1}$ for chlorophyll, represents the loss of some rotational energy (rotational subsublevels of a vibrational sublevel are generally about $1\,kJ\,mol^{-1}$ apart). Thus the transition from the lowest vibrational sublevel of the ground state up to the lower excited state (the red absorption band) has essentially the same energy as a transition from the lowest vibrational sublevel of that excited state down to the ground state (the red fluorescence band). The only way for this to occur is to have the lowest vibrational sublevels of both the ground state and the excited state involved in each of the transitions. Hence, the red absorption band corresponds to a transition of the chlorophyll molecule from the lowest vibrational sublevel of the ground state to the lowest vibrational sublevel of the lower excited state, as is depicted in Fig. 5-4.

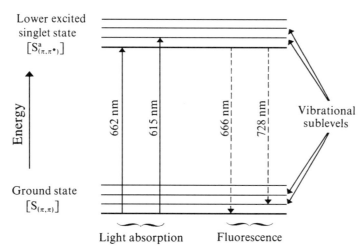

Figure 5-4. Energy level diagram indicating the vibrational sublevels of the ground state $[S_{(\pi,\pi)}]$ and the lower excited singlet state $[S^a_{(\pi,\pi*)}]$ of Chl a. Solid vertical lines indicate the absorption of light by Chl a dissolved in ether; dashed lines represent fluorescence at the specified wavelengths. The lengths of the arrows are proportional to the amounts of energy involved in the various transitions.

The participation of the lowest vibrational sublevels of both the ground state and the lower excited state of Chl a in the major red band can also be appreciated by considering the minor band adjacent to the major red band in both the absorption spectrum and the fluorescence emission spectrum (Figs. 5-3 and 5-4). The shorter wavelength absorption band at 615 nm in ether—14 kJ mol^{-1} higher in energy than the 662 nm band—corresponds to a transition to the first excited vibrational sublevel in the lower excited state. Deexcitations from the lowest vibrational sublevel of the lower excited state to excited vibrational sublevels of the ground state correspond to fluorescence at wavelengths > 700 nm. In fact, a small band near 728 nm in the fluorescence emission spectrum of Chl a (Fig. 5-3) occurs about 62 nm greater in wavelength than the main fluorescence band, indicating an electronic transition with 15 kJ mol^{-1} less energy than the 666-nm band. This far-red band corresponds to fluorescence emitted as the chlorophyll molecule goes from the lowest vibrational sublevel of the lower excited state to the first excited vibrational sublevel of the ground state (Fig. 5-4). In summary, we note that (1) excitations from excited vibrational sublevels of the ground state are uncommon, which is a reflection of the Boltzmann energy distribution; (2) fluorescence from excited vibrational sublevels of an excited state is also uncommon because radiationless transitions to the lowest vibrational sublevel are so rapid; and (3) transitions to excited vibrational sublevels of the ground state can be quite significant (see Fig. 5-4 for Chl a).

C. Absorption *in Vivo*—Polarized Light

The values of λ_{max} for Chl a *in vivo* result from interactions between a chlorophyll molecule and the surrounding molecules, such as the proteins and the lipids in the lamellar membranes as well as adjacent water molecules. In fact, probably all Chl a is associated with proteins in chlorophyll–protein complexes. Hydrophobic interactions among phytol tails of adjacent chlorophylls and with hydrophobic regions in the protein help stabilize these chlorophyll–protein complexes. Because of the interactions of the porphyrin ring with the other molecules in the complex, and especially with the polar amino acids of the protein, the red bands for Chl a *in vivo* are shifted toward longer wavelengths (lower energy) than for Chl a dissolved in ether (λ_{max} at 662 nm), e.g., 670–680 nm. This is an example of the pronounced effect that the solvent or other neighboring molecules can have in determining the electronic energy levels of a pigment. The red absorption band of Chl b in $vivo$ occurs as a "shoulder" on the short wavelength side of the Chl a red band, usually near 650 nm, and its Soret band occurs at slightly longer wavelengths than for Chl a.

A small amount of Chl a occurs in special sites that play a particularly important role in photosynthesis. These Chl a's have λ_{max}'s at approximately 680 and 700 nm and are referred to as P_{680} and P_{700}, respectively (P indicating pigment). P_{700} is a dimer of Chl a molecules (i.e., two Chl a's acting as a unit). P_{680} may also be a dimer.

We can define the $bandwidth$ of an absorption band as the difference in energy between photons on the two sides of the band at wavelengths for which the absorption has dropped to half of that for λ_{max}. Such bandwidths of the red absorption bands of the various Chl a's in $vivo$ are fairly narrow—often about 10 nm at 20°C. At 680 nm a bandwidth of 10 nm is equivalent to 3 kJ mol^{-1}, i.e., a photon having a wavelength of 675 nm has an energy 3 kJ mol^{-1} greater than a photon with a wavelength of 685 nm. An energy of 3 kJ mol^{-1} is smaller than the spacing between vibrational sublevels of 14 kJ mol^{-1} for Chl a. Thus a bandwidth of 3 kJ mol^{-1} results from adjacent molecules and the rotational and the translational broadening of an electronic transition to a single vibrational sublevel of the excited state of Chl a.

The absorption of polarized light by chlorophyll in $vivo$ can provide information on the orientations of individual chlorophyll molecules. [Polarized means that the oscillating electric vector of light (Fig. 4-1) is in some specified direction.] The electronic transition of chlorophyll to the excited singlet state that is responsible for the red absorption band has its electric dipole in the plane of the porphyrin ring—actually, there are two dipoles in the plane in mutually perpendicular directions. Polarized light of the appropriate wavelength with its oscillating electric vector parallel to one of the dipoles is therefore preferentially absorbed by chlorophyll—recall that the probability for absorption is proportional to the square of the cosine of the angle between the induced dipole and the electric field vector of light (see Chapter 4). Absorption of polarized light indicates that the porphyrin rings of a few percent of the Chl a molecules, perhaps including P_{680}, are nearly parallel to the plane of the chloroplast lamellae. However, most of the chlorophyll molecules have their porphyrin heads randomly oriented in the internal membranes of chloroplasts.

The polarization of fluorescence following the absorption of polarized light can tell us whether the excitation has been transferred from one molecule to another. If the same chlorophyll molecules that absorbed polarized light later emit photons when they go back to the ground state, the fluorescence would be polarized to within a few degrees of the direction of the electric vector of the incident light. However, the chlorophyll fluorescence following absorption of polarized light by chloroplasts is not appreciably polarized. This fluorescence depolarization indicates that the excitation energy has been transferred from one chlorophyll molecule to another so many times that the directional aspect becomes randomized, i.e., the chlorophyll molecule emitting fluorescence is randomly aligned relative to the chlorophyll molecule that absorbed the polarized light.

When unpolarized light is incident on chloroplast lamellae that have been oriented in some particular direction, the fluorescence is polarized. The plane of polarization is similar to the plane of the membranes, indicating that the porphyrin rings of the emitting chlorophyll molecules have about the same orientation as the membrane. However, the porphyrin rings of the absorbing chlorophyll molecules are randomly oriented. Again, we conclude that the excitation has been transferred from the absorbing to the emitting molecule.

II. Other Photosynthetic Pigments

Besides chlorophyll, other molecules in photosynthetic organisms also absorb light in the visible region. If these molecules pass their electronic excitations on to Chl a (or to BChl a), they are referred to as auxiliary or *accessory* pigments. In addition to Chl b and Chl c, two groups of accessory pigments important to photosynthesis are the *carotenoids* and the *phycobilins*. These two classes of accessory pigments can absorb yellow or green light, wavelengths for which absorption by chlorophyll is not appreciable.

Fluorescence studies have indicated the sequence of excitation transfer to and from the accessory pigments. For example, light absorbed by carotenoids, phycobilins, and Chl b leads to the fluorescence of Chl a. However, light absorbed by Chl a does not lead to the fluorescence of the accessory pigments, suggesting that excitation energy is not transferred from Chl a to the accessory pigments. Thus accessory pigments can increase the photosynthetic use of white light and sunlight by absorbing at wavelengths where Chl a absorption is low; the excitations are then transferred to Chl a before the photochemical reactions take place.

A. Carotenoids

Carotenoids occur in essentially all green plants, algae, and photosynthetic bacteria. In fact, the dominant pigments for plant leaves are the chlorophylls, which absorb strongly in the red and the blue regions, and the carotenoids, which absorb mostly in the blue and somewhat in the green region of the spectrum (Table 4-1; Fig. 4-2). The predominant colors reflected or transmitted by leaves are therefore green and yellow. In the autumn, chlorophylls in the leaves of deciduous plants can bleach and are usually not replaced, thereby greatly reducing absorption in the red and the blue regions. The remaining carotenoids absorb only in the blue and the green regions, leading to the well-known fall colors of such leaves, namely, yellow, orange, and red. Animals apparently do not synthesize carotenoids. Hence, brightly colored birds such as canaries and flamingoes, as well as many invertebrates, obtain their yellow or reddish colors from the carotenoids in the plants that they eat.

Carotenoids involved in photosynthesis are bound to the chlorophyll–protein complexes, of which various types occur in the lamellar membranes of chloroplasts (see Fig. 1-9). Carotenoids also are found in organelles known as *chromoplasts*, which are about the size of chloroplasts and are often derived from them. Lycopene (red) is in tomato fruit chromoplasts, α- and β-carotenes (orange) occur in carrot root chromoplasts, and the various chromoplasts of flowers contain a great diversity of carotenoids. Such carotenoids are important for attracting pollinators and in seed dispersal.

Carotenoids are 40-carbon terpenoids, also known as isoprenoids. They are composed of eight *isoprene* units, where isoprene is a 5-carbon compound having two double bonds ($CH_2{=}CCH{=}CH_2$). In many carotenoids, the isoprene units at one or both ends of the mole-
$$|$$
$$CH_3$$
cule are part of six-membered rings. Carotenoids are about 3 nm long, and those involved in photosynthesis generally have nine or more double bonds in conjugation.

The wavelength position of the λ_{max} depends on the solvent, on the substitutions on the hydrocarbon backbone, and on the number of double bonds in the conjugated system. We can illustrate this latter point for carotenoids in n-hexane, in which the central maxima of the three observed peaks in the absorption spectra are at 286 nm for 3 double bonds in conjugation, at 347 nm for 5, at 400 nm for 7, at 440 nm for 9, at 472 nm for 11, and at

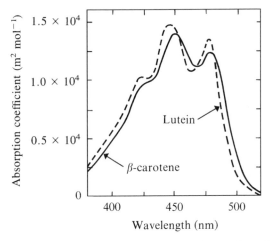

Figure 5-5. Absorption spectra for the two major carotenoids of green plants. [Data for β-carotene (in *n*-hexane) and lutein (in ethanol) are from Zscheile *et al.* (1942); used by permission.]

500 nm for 13 double bonds in conjugation. Thus the greater the degree of conjugation, the longer the wavelength representing λ_{max}, as we discussed in Chapter 4. For the 9–12 double bonds occurring in the conjugated systems of photosynthetically important carotenoids, the maximum absorption coefficient is $>10^4$ m^2 mol^{-1}.

The carotenoids that serve as accessory pigments for photosynthesis absorb strongly in the blue (425–490 nm; Table 4-1) and somewhat in the green (490–560 nm), usually having triple-banded spectra from 400 to 540 nm. For β-carotene in hexane, the three bands are centered at 425, 451, and 483 nm; another major carotenoid in plants, lutein, has peaks at 420, 447, and 477 nm when dissolved in ethanol (absorption spectra in Fig. 5-5). The three absorption bands in each spectrum are about 17 kJ mol^{-1} apart, a reasonable energy spacing between adjacent vibrational sublevels. Hence, the triple-banded spectra characteristic of carotenoids most likely represent transitions to three adjacent vibrational sublevels in the same excited state. The absorption spectra of the carotenoids *in vivo* are shifted about 20–30 nm toward longer wavelengths (lower energy) compared with absorption when the pigments are dissolved in hexane or ethanol.

Carotenoids are subdivided into two groups: the *carotenes*, which are hydrocarbons, and the *xanthophylls*, which are oxygenated. The major carotene in green plants is β-carotene (absorption spectrum in Fig. 5-5; structure in Fig. 5-6); α-carotene is also abundant (α-carotene has the double bond in the right-hand ring shifted one carbon clockwise compared with β-carotene; Fig. 5-6). The xanthophylls exhibit much greater structural diversity than do the carotenes because the oxygen atoms can be in epoxy, hydroxy, keto, or methoxy groups. The most abundant xanthophyll in green plants is lutein (absorption spectrum in Fig. 5-5; structure in Fig. 5-6); antheraxanthin, neoxanthin, violaxanthin, and zeaxanthin are also common. The major carotene of algae is β-carotene, and lutein is the most common xanthophyll, although great variation in the type and the amount of xanthophylls is characteristic of algae. For instance, golden algae, diatoms, and brown algae contain considerable amounts of the xanthophyll fucoxanthin (Fig. 5-6), which functions as the main accessory pigment in these organisms. The distribution and the types of carotenoids in plants have evolutionary implications and taxonomic usefulness.

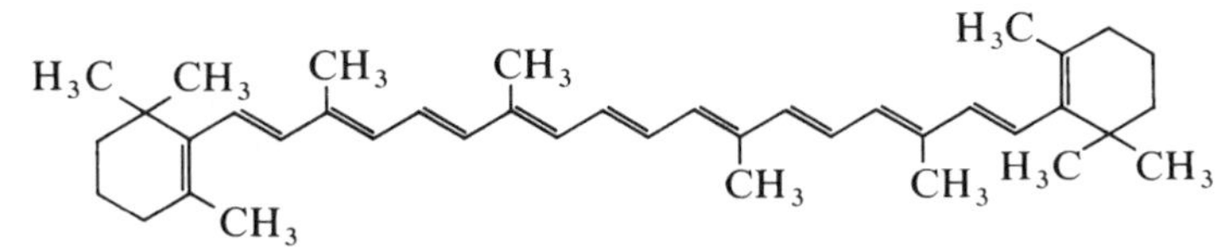

β-carotene

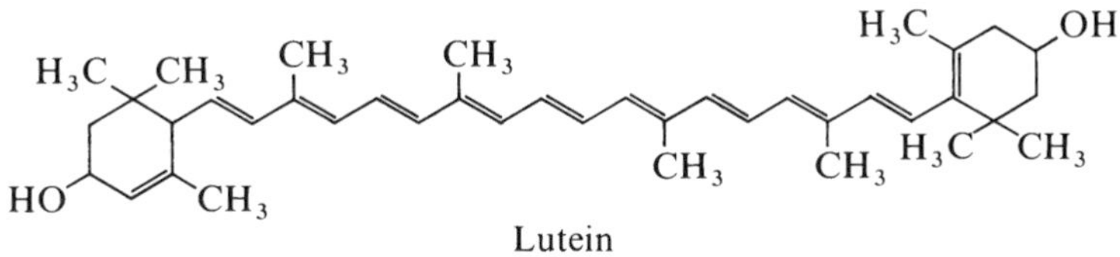

Lutein

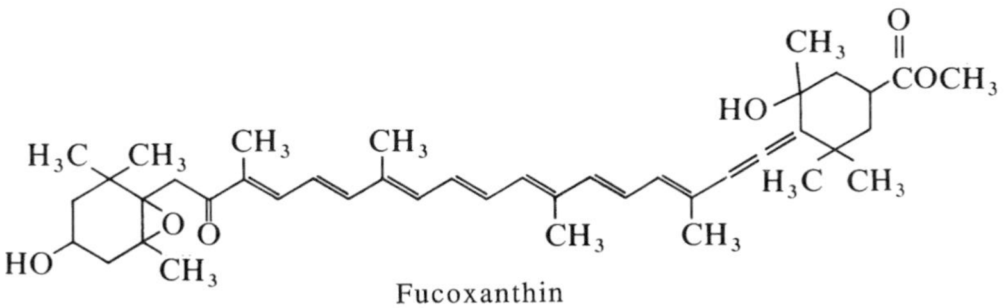

Fucoxanthin

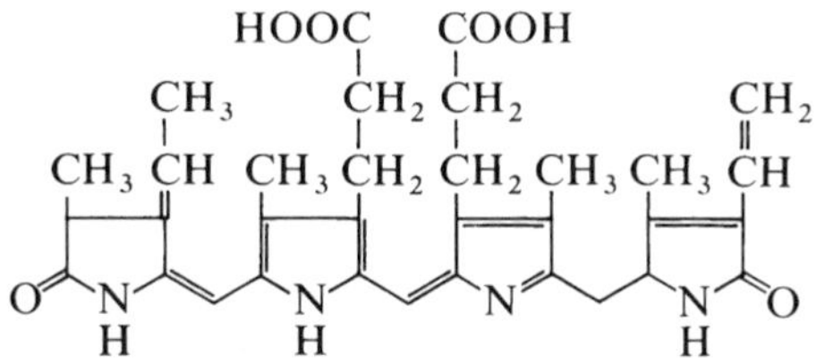

Phycoerythrobilin

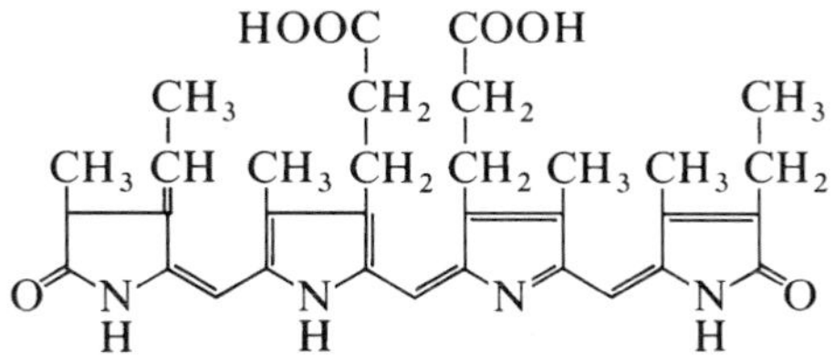

Phycocyanobilin

Figure 5-6. Structure of five important accessory pigments. The phycobilins (lower two structures) occur covalently bound to proteins, i.e., they are the chromophores for phycobiliproteins.

In addition to functioning as accessory pigments for photosynthesis, carotenoids are also important for protecting photosynthetic organisms from the destructive photooxidations that can occur in the presence of light, O_2, and certain pigments. In particular, light absorbed by chlorophyll can lead to excited states of O_2. These highly reactive states can damage chlorophyll, but their interactions with carotenoids prevent harmful effects to the organism. Because photosynthesis in the green and the purple bacteria does not lead to O_2 evolution, it can proceed in the absence of carotenoids. For instance, a mutant of *Rhodopseudomonas spheroides* lacking carotenoids performs photosynthesis in a normal manner in the absence of O_2; when O_2 is introduced in the light, the bacteriochlorophyll becomes photooxidized and the bacteria are killed, a sensitivity not present in related strains containing carotenoids. On the other hand, cyanobacteria, algae, and higher plants produce O_2 as a photosynthetic product, so they must contain carotenoids to survive in the light. Because such oxidations tend to increase with time, the fraction of carotenoids in the form of xanthophylls generally increases in leaves as the growing season progresses. In addition, certain xanthophylls are reduced during the daytime and then oxidized back to the original form at night, especially when the light level is excessive or photosynthesis is inhibited by stress. In particular, plants as well as certain algae (e.g., green algae and brown algae) possess a *xanthophyll cycle*, in which violaxanthin is reduced to antheraxanthin which in turn is reduced to zeaxanthin (both steps are actually deepoxidations) during the daytime and the steps are reversed at night; such a cycle tends to dissipate excess energy or excess reductant when the absorption of photons outpaces their use in photosynthesis.

B. Phycobilins

The other main accessory pigments in photosynthesis are the *phycobilins*. Rudolf Lemberg in the 1920s termed these molecules phycobilins because they occur in algae (red algae and blue-green algae, the latter now referred to as cyanobacteria; *phyco* is derived from the Greek for seaweed) but they structurally resemble bile pigments. Like the chlorophylls, the phycobilins are tetrapyrroles. However, the four pyrroles in the phycobilins occur in an open chain, as is the case for phytochrome, and not in a closed porphyrin ring, as is the case for the chlorophylls. Phycobilins have a relative molecular mass of 586. They occur covalently bound to proteins with molecular masses of 30–35 kDa. These assemblies containing 300–800 phycobilins are organized into phycobilisomes, which are about 40 nm in diameter and are associated with the outer (stromal) surfaces of lamellar membranes in cyanobacteria and red algae, where they function as the main accessory pigments.

Phycobilins generally have their major absorption bands from 500 to 650 nm, with a relatively small Soret band in the UV (Fig. 5-7). These pigments are higher in concentration in many cyanobacteria and red algae than are the chlorophylls and are responsible for the color of certain species. The main phycobilins are phycocyanobilin and phycoerythrobilin (structures in Fig. 5-6; note the great structural similarity between the phycobilins and the chromophore for phytochrome in Fig. 4-11). Phycoerythrobilin plus the protein to which it is covalently attached is called phycoerythrin.[1] Phycoerythrin is soluble in aqueous solutions, so we can obtain absorption spectra for it under conditions similar to those *in vivo*. Phycoerythrin is reddish because it absorbs green and has at least one main band between 530 and

---

[1] The phycobilins are covalently bound to their proteins (referred to as apoproteins) to form phycobiliproteins, whereas chlorophylls and carotenoids are joined to their apoproteins by weaker bonds such as H bonds and hydrophobic interactions.

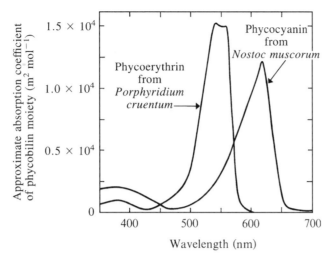

Figure 5-7. Absorption spectra of phycoerythrin from a red alga and phycocyanin from a cyanobacterium. [Data are from Ó hEocha (1965); used by permission.]

570 nm (see absorption spectrum in Fig. 5-7). It occurs throughout the red algae and in some cyanobacteria. Phycocyanin (phycocyanobilin plus protein) appears bluish because it absorbs strongly from 610 to 660 nm (Fig. 5-7). It is the main phycobilin in the cyanobacteria and also is found in the red algae. As is the case for other pigments, the greater the number of double bonds in conjugation in the phycobilins, the longer the wavelengths for λ_{max}. For example, phycoerythrobilin has seven double bonds in the main conjugated system and absorbs maximally in the green region of the spectrum; phycocyanobilin has nine such double bonds and its λ_{max} occurs in the red (see the structures of these compounds in Fig. 5-6). The maximum absorption coefficients of both phycobilins exceed 10^4 m^2 mol^{-1} (see Fig. 5-7).

As we indicated in Chapter 4, both the quantity and the quality of radiation change with depth in water, with wavelengths near 500 nm penetrating the deepest. For instance, only about 10% of the blue and the red parts of the spectrum penetrate to a depth of 50 m in clear water, so chlorophyll would not be a very useful light-harvesting pigment at that depth (see Fig. 5-3 for a chlorophyll absorption spectrum). Although there are many exceptions, changes in the spectral quality (relative amounts of various wavelengths) with depth affect the distribution of photosynthetic organisms according to their pigment types. The predominant accessory pigment in green algae is Chl b, which absorbs mainly in the violet (400–425 nm; see Table 4-1) and the red (640–740 nm). Green algae as well as sea grasses and freshwater plants grow in shallow water, where the visible spectrum is little changed from that of the incident sunlight. Fucoxanthin (Fig. 5-6) is the major accessory pigment in brown algae, such as the kelps, and it absorbs strongly in the blue and the green regions (425–560 nm), helping to extend the range of such plants downward to over 20 m. Marine red algae can occur at even greater depths (e.g., 100 m), and their phycoerythrin absorbs the green light (490–560 nm) that penetrates to such distances. Changes in spectral quality can also induce changes in the synthesis of biliproteins within an organism. For instance, green light induces the synthesis of the green-absorbing phycoerythrin, and red light induces the synthesis of the red-absorbing phycocyanin (see Fig. 5-7) in certain cyanobacteria and red algae.

Interestingly, only two types of pigments appear to be involved in all known photochemical reactions in plants. These are the carotenoids and the tetrapyrroles, the latter class

Table 5-1. Approximate Relative Amounts and Locations of Photosynthetic Pigments[a]

Pigment	Number	Location
Chl a	450	Approximately 40% in the cores of Photosystems I and II, with the remainder in the light-harvesting antennae
Chl b	150	In light-harvesting antennae
P_{680}	~1.6	Trap for Photosystem II
P_{700}	~1.0	Trap for Photosystem I
Carotenoids	120	Most in light-harvesting antennae of Photosystem II
Phycobilins	500	Covalently bound to proteins situated on the outer surface of photosynthetic membranes in cyanobacteria and red algae; serve in light-harvesting antennae of Photosystem II

[a]Data are expressed per 600 chlorophylls and are for representative leaves of green plants (except for the phycobilins) growing at moderate sunlight. Photosystems and the light-harvesting antennae are discussed later in this chapter.

including the chlorophylls, the phycobilins, and phytochrome. The maximum absorption coefficients for the most intense absorption bands are slightly over $10^4 \, \mathrm{m^2 \, mol^{-1}}$ in each case. Cytochromes, which are involved in the electron transport reactions in chloroplasts and mitochondria, are also tetrapyrroles (considered later). Table 5-1 summarizes the relative frequency of the main types of photosynthetic pigments.

III. Excitation Transfers among Photosynthetic Pigments

Chlorophyll is at the very heart of the primary events of photosynthesis. It helps convert the plentiful radiant energy from the sun into chemical free energy that can be stored in various ways. In this section we will represent light absorption, excitation transfer, and the photochemical step as chemical reactions; this will serve as a prelude to a further consideration of certain molecular details of photosynthesis.

A. Pigments and the Photochemical Reaction

The first step in photosynthesis is light absorption by one of the pigments. The absorption event (discussed in Chapter 4) for the various types of photosynthetic pigments described in this chapter can be represented as follows:

$$\left.\begin{array}{c} \text{accessory pigment} \\ \text{or} \\ \text{Chl } a \\ \text{or} \\ \text{trap chl} \end{array}\right\} + h\nu \rightarrow \left\{\begin{array}{c} \text{accessory pigment}^* \\ \text{or} \\ \text{Chl } a^* \\ \text{or} \\ \text{trap chl}^* \end{array}\right. \qquad (5.2)$$

where the asterisk refers to an excited state of the pigment molecule caused by the absorption of a light quantum, $h\nu$. Trap chl indicates a special type of Chl a (e.g., P_{680} or P_{700}) that occurs much less frequently than the other chlorophylls (see Table 5-1); we will consider its important excitation-trapping properties at the end of this section.

Because the photochemical reactions take place only at the trap chl molecules, the excitations resulting from light absorption by either the accessory pigments or the other Chl a's must be transferred to the trap chl before they can be used for photosynthesis. The relative rarity of trap chl compared with the other photosynthetic pigments means that it absorbs only a small fraction of the incident light. In fact, under natural conditions in green plants over 99% of the photons are absorbed by either the accessory pigments or Chl a. The migration of excitations from the initially excited species to the trap chl—the mechanism for which we will discuss later—can be represented as follows:

$$\text{Accessory pigment}^* + \text{Chl } a \rightarrow \text{accessory pigment} + \text{Chl } a^* \qquad (5.3)$$

$$\text{Chl } a^* + \text{trap chl} \rightarrow \text{Chl } a + \text{trap chl}^* \qquad (5.4)$$

In other words, the direction of excitation transfer or migration is from the accessory pigments to Chl a (Eq. 5.3) and from Chl a to the special "trap" chlorophylls (Eq. 5.4) where the actual photochemical reactions take place. Hence, the overall effect of the steps described by Eqs. (5.2)–(5.4) is to funnel the excitations caused by the absorption of light to the trap chl.

A prerequisite for the conversion of radiant energy into a form that can be stored chemically is the formation of reducing and oxidizing species. The *reducing* (electron-donating) and the *oxidizing* (electron-accepting) species that result from light absorption must be fairly stable and located in such a way that they do not interact. (We will discuss the energetics of oxidation and reduction in Chapter 6.) If we denote the molecule that accepts an electron from the excited trap chl by A, this electron transfer step can be represented by

$$\text{Trap chl}^* + A \rightarrow \text{trap chl}^+ + A^- \qquad (5.5)$$

where A^- indicates the reduced state of the acceptor and trap chl$^+$ means that the special chlorophyll has lost an electron. Equation (5.5) represents a photochemical reaction because the absorption of a light quantum (Eq. 5.2) has led to the transfer of an electron away from a special type of chlorophyll, representing a chemical change in that molecule. The electron removed from trap chl* (Eq. 5.5) can be replaced by one coming from a donor, D, which leads to the oxidation of this latter species, D^+, and the return of the trap chl to its unexcited state:

$$\text{Trap chl}^+ + D \rightarrow \text{trap chl} + D^+ \qquad (5.6)$$

The generation of stable reduced (A^-) and oxidized (D^+) intermediates completes the conversion of light energy into chemical potential energy. Combining Eqs. (5.2)–(5.6) gives us the following relation for the net reaction describing the primary events of photosynthesis:

$$A + D + h\nu \rightarrow A^- + D^+ \qquad (5.7)$$

The light-driven change in chemical free energy represented by the conversion of $A + D$ to $A^- + D^+$ (Eq. 5.7) eventually causes chemical reactions leading to the evolution of O_2 from water, the production of a reduced compound (NADPH), and the formation of high-energy phosphates (ADP + phosphate $\rightarrow$ ATP). Such a conversion of light energy into chemical energy represented by Eq. (5.7) is the cornerstone of photosynthesis.

B. Resonance Transfer of Excitation

We have already mentioned examples of excitation transfer among photosynthetic pigments. For instance, light absorbed by the accessory pigments can lead to the fluorescence of Chl a. Studies on the absorption of polarized light by chlorophyll *in vivo*, where the resulting fluorescence is not polarized, provide further evidence that excitations can migrate from molecule to molecule before the energy is emitted as radiation. In this regard, the excitation of the lower excited singlet state of chlorophyll can be passed to a second chlorophyll molecule. This causes the deactivation of the originally excited molecule and the attainment of the lower excited singlet state in the second chlorophyll, a process described by Eq. (4.7), $S_{(\pi,\pi^*)} + S_{2(\pi,\pi)} \rightarrow S_{(\pi,\pi)} + S_{2(\pi,\pi^*)}$. The most widely accepted mechanism for such exchange of electronic excitation between chlorophyll molecules is *resonance transfer* (also called inductive resonance, the Förster mechanism, or weak coupling), which we next consider qualitatively.

On the basis of our discussion in the previous chapter, we might expect that an excited molecule can induce an excited state in a second molecule in close proximity. In particular, the oscillating electric dipole representing the energetic electron in the excited state of the first molecule leads to a varying electric field. This field can cause a similar oscillation or resonance of some electron in a second molecule. A transfer of electronic excitation energy takes place when an electron in the second molecule is induced to oscillate. When excitation transfer is completed, the previously excited electron in the first molecule has ceased oscillating and some electron in the second molecule is now oscillating, leading to an excited state of that molecule. Resonance transfer of excitation between molecules is thus analogous to the process by which light is originally absorbed because an oscillation of some electron in the molecule is induced by a locally varying electric field. Resonance transfer of excitation is most probable when the orientation between the electric dipole in the excited molecule and the potential dipole in the second molecule is proper and the energy of the original dipole is appropriate, an aspect that we will consider next.

For resonance transfer of electronic excitation to occur, the energy available in the excited molecule must match the energy that can be accepted by a second molecule. The wavelengths for fluorescence indicate the energy of the excited singlet state of a molecule (at least after the very rapid radiationless transitions to the lowest vibrational sublevel of that excited state have occurred). Although fluorescence itself is not involved in this type of excitation transfer, the fluorescence emission spectrum usually gives the range of energies available for transfer to a second molecule. The range of wavelengths of light that can sympathetically induce an oscillation of some electron in a second molecule is given by the absorption spectrum of that molecule (see Chapter 4), and therefore the absorption spectrum shows the energies that can be accepted by a molecule. As might be expected from these two considerations, the probability for resonance transfer is high when the overlap in wavelength between the fluorescence band for the excited oscillator (available energy) and the absorption band of an unexcited oscillator (acceptable energy) in a neighboring molecule is large. Because the overlap in the red region between the absorption spectrum and the fluorescence emission spectrum of Chl a is large (Fig. 5-3), excitations can be efficiently exchanged between Chl a molecules by resonance transfer. Figure 5-8 illustrates the various energy considerations involved in resonance transfer of excitation between two dissimilar molecules.

The probability for resonance transfer of electronic excitation decreases as the distance between the two molecules increases. If chlorophyll molecules were uniformly distributed in three dimensions in the lamellar membranes of chloroplasts (Fig. 1-9), they would have

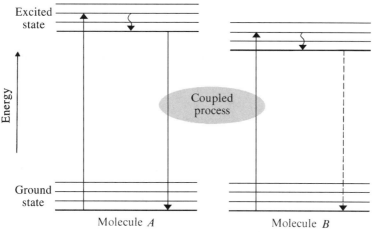

Figure 5-8. Resonance transfer of excitation from molecule A to molecule B. After light absorption by molecule A, a radiationless transition occurs to the lowest vibrational sublevel of its excited state. Next, resonance transfer of the excitation takes place from A to B, causing the second molecule to go to an excited state, while molecule A returns to its ground state. Following a radiationless transition to the lowest vibrational sublevel in the excited state, fluorescence can then be emitted by molecule B as it returns to its ground state. Based on the energy level diagrams (which include the vibrational sublevels for each of these two different pigments), we can conclude that generally the excitation rapidly decreases in energy after each intermolecular transfer between dissimilar molecules.

a center-to-center spacing of approximately 2 nm, an intermolecular distance over which resonance transfer of excitation can readily occur (resonance transfer is effective up to about 10 nm for chlorophyll). Thus both the spectral properties of chlorophyll and its spacing in the lamellar membranes are conducive to an efficient migration of excitation from molecule to molecule by resonance transfer.

C. Transfers of Excitation between Photosynthetic Pigments

In addition to the transfer from one Chl a molecule to another, excitations can also migrate by resonance transfer from the accessory pigments to Chl a. The transfers of excitation among Chl a's can be nearly 100% efficient (i.e., $k_{transfer} \gg$ rate constants for competing pathways; see Eq. 4.15), whereas the fraction of excitation transfers between dissimilar molecules is quite variable. For instance, the transfer of excitation from β-carotene to Chl a is very efficient in certain algae, but the fraction of excitations transferred from most xanthophylls to Chl a is generally low. The transfer of excitation from α-carotene and lutein to chlorophyll usually is intermediate, e.g., about 40% of the excitations of these carotenoids may be transferred to Chl a. Photons absorbed by the carotenoid fucoxanthin—found in the golden algae, diatoms, and brown algae—can approach 100% efficiency of excitation transfer to Chl a. Also, most excitations of phycobilins and Chl b can be transferred to Chl a. In red algae, 90% of the electronic excitations produced by the absorption of photons by phycoerythrin can be passed on to phycocyanin and then to Chl a; these transfers require about 4×10^{-10} s each. We will next consider the direction for excitation transfer between various photosynthetic pigments and then the times involved for intermolecular excitation transfers of chlorophyll.

Some energy is generally lost by each molecule to which the excitation is transferred. Any excess vibrational or rotational energy is usually dissipated rapidly as heat (see Fig. 5-8). Therefore, the λ_{max} for each type of pigment involved in the sequential steps of excitation transfer tends to become longer in the direction in which the excitation migrates. In particular, the fluorescence emission spectrum of some molecule—which must appreciably overlap the absorption spectrum of the receiving molecule for resonance transfer to take place efficiently—occurs at longer wavelengths than the absorption spectrum of that molecule (see Fig. 5-3 for Chl a). Therefore, for a second molecule to become excited by resonance transfer, it should have an absorption band at longer wavelengths (lower energy) than the absorption band for the molecule from which it receives the excitation. Thus the direction for excitation migration by resonance transfer among photosynthetic pigments is usually toward those pigments with longer λ_{max}. We can appreciate this important aspect by considering Fig. 5-8. If excitation to the second excited vibrational sublevel of the excited state is the most probable transition predicted by the Franck–Condon principle for each molecule, then the excitation of molecule A requires more energy than that of B (the pigment to which the excitation is transferred). Hence, λ_{max_A} must be less than λ_{max_B} in the two absorption spectra, consistent with our statement that the excitation migrates toward the pigment with the longer λ_{max}.

As a specific example of the tendency for excitations to migrate toward pigments with longer λ_{max}'s in their absorption spectra, we will consider the transfer of excitations from accessory pigments to Chl a. In red algae and in some cyanobacteria, phycoerythrin has a λ_{max} at about 560 nm and passes excitation energy to phycocyanin, which has an absorption maximum near 620 nm. This excitation can then be transferred to a Chl a with a λ_{max} near 670 nm. The biliprotein allophycocyanin absorbs maximally at 650 nm and apparently intervenes in the transfer of excitation between phycocyanin and Chl a. Because some of the excitation energy is generally dissipated as heat by each molecule (see Fig. 5-8), the excitation represents less energy (longer λ) after each pigment in the sequence. Consequently, the overall direction for excitation migration is essentially irreversible.

As the nuclei vibrate back and forth after the absorption of a photon by some electron, their collisions with other nuclei every 10^{-13} s or so can lead in such short times to the dissipation of any excess energy in the excited vibrational sublevels. In addition, the radiationless transition from the upper excited singlet to the lower excited singlet of Chl a—$S^b_{(\pi,\pi*)}$ to $S^a_{(\pi,\pi*)}$ in Fig. 4-5—is completed within 10^{-12} s. The time for the transfer of the excitation between two Chl a molecules in $vivo$ is somewhat longer—about 1 or 2×10^{-12} s. Thus the originally excited chlorophyll molecule usually attains the lowest vibrational sublevel of the lower excited singlet state before the excitation is transferred to another molecule. The amount of energy resonantly transferred from one Chl a to another therefore generally corresponds to the energy indicated by the fluorescence emission spectrum (Fig. 5-3). An excitation representing this amount of energy can, in principle, be transferred many times by resonance transfer with essentially no further degradation of the energy.

In Chapter 4 we noted that an upper time limit within which processes involving excited singlet states must occur is provided by the kinetics of fluorescence deexcitation. The lifetime for chlorophyll fluorescence from the lower excited singlet state is about 1.5×10^{-8} s. Time is therefore sufficient for approximately 10,000 transfers of excitation among the Chl a molecules—each transfer requiring about 10^{-12} s—before the loss of the excitation by the emission of fluorescence. The number of excitation transfers among Chl a molecules is actually much less than this for reasons that will become clear shortly.

D. Excitation Trapping

The special Chl a's, P_{680} and P_{700}, can absorb at longer wavelengths in the red region than the other types of Chl a, so the excited singlet states in P_{680} and P_{700} are at lower energies. The other Chl a's excite such trap chls by resonance transfer, but P_{680} and P_{700} usually do not pass the excitation back; i.e., they rapidly lose some excitation energy (within 10^{-12} s), so they do not retain enough energy to reexcite the other Chl a's by resonance transfer. The excited singlet states of other Chl a molecules therefore can have their excitations readily passed on to the trap chls, but not vice versa—analogous to the irreversibility of the migration of excitations from the accessory pigments to Chl a. The excitations resulting from the absorption of radiation by the various photosynthetic pigments are thereby funneled into P_{680} or P_{700}. Such collecting of excitations by one species is the net effect of Eqs. (5.2)–(5.4) (with the term trap chl replaced by P_{680} or P_{700}).

Because one of the trap chls is present per approximately 230 chlorophylls (Table 5-1), on average only a few hundred transfers are necessary to get an excitation from Chl a to P_{680} or P_{700}. Thus the 10,000 transfers of excitation from one Chl a to another possible within the fluorescence lifetime do not occur. Because each excitation transfer takes approximately 10^{-12} s, 100 transfers require about 10^{-10} s. In agreement with this, both calculations from mathematical models and ingenious experimentation have shown that over 90% of the excitations of Chl a migrate to P_{680} or P_{700} in $<10^{-9}$ s.

The characteristics of fluorescence provide information on the lifetime of the excited singlet state of chlorophyll *in vivo* and thus on the time available for migration of excitations. Specifically, approximately 1–3% of the light absorbed by Chl a *in vivo* is lost by fluorescence. The amount reradiated depends on the competing deexcitation reactions and hence is higher at higher incident light levels for which the photochemical reactions become saturated. Near full sunlight the quantum yield for fluorescence *in vivo*, Φ_{Fl}, is approximately 0.03. Equation (4.12) ($\Phi_i = \tau/\tau_i$) indicates that this quantum yield is equal to τ/τ_{Fl}, where τ is the lifetime of the excited singlet state and τ_{Fl} is its fluorescence lifetime. A reasonable assumption is that τ_{Fl} *in vivo* is similar to the fluorescence lifetime of Chl a *in vitro*, 1.5×10^{-8} s. Therefore, Eq. (4.12) predicts a lifetime for the excited state of Chl a *in vivo* of $(0.03)\,(1.5 \times 10^{-8}$ s), or 0.5×10^{-9} s. This is another estimate of the average time necessary for the excitation to migrate to the trap chl (not all of the Chl a readily fluoresces *in vivo*, and hence this is only an approximate guide to the lifetime).

P_{680} and P_{700} act as traps for excitations in chloroplast lamellae, and a special type of bacteriochlorophyll with a λ_{max} between 870 and 890 nm (depending on the species) acts in an analogous manner for the green and the purple bacteria. One of the useful features of such excitation traps is to have an excited singlet state lower in energy than the excited singlet states in the other pigment molecules. This lower energy state (longer λ_{max} for absorption) is a consequence of the molecular environment in which the chlorophyll molecules acting as excitation traps are located. Moreover, the longer λ_{max} ensures the directionality for the migration of excitations. Another characteristic of the trap chls is their relative rarity. Thus most of the photosynthetic pigments act as light harvesters, which collect the radiation and channel the excitations toward the trap chl, as illustrated in Fig. 5-9. Processing of the excitation originally caused by light takes place only at the trap chls. This participation in the essentially irreversible, electron-transfer process is the crucial feature of an excitation trap.

When the excitation migrates to a trap such as P_{680} or P_{700}, this special Chl a goes to an excited singlet state, as would any other Chl a. Because the trap chl cannot readily excite other chlorophylls by resonance transfer, it might become deexcited by the emission of

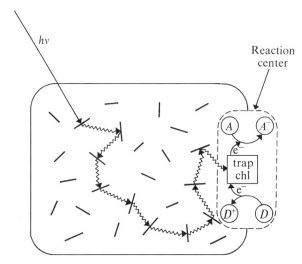

Figure 5-9. Schematic representation of a group of pigments that harvests a light quantum ($h\nu$) and passes the excitation to a special trap chlorophyll. In the reaction center an electron (e^-) is transferred from the trap chl to some acceptor (A^- in the reduced form) and is then replaced by another electron coming from a suitable donor (D^+ in the oxidized form).

fluorescence. However, very little fluorescence from the trap chls is observed *in vivo*. This is explained by the occurrence of a rapid photochemical event (see Eq. 5.5; Trap chl* $+ A \rightarrow$ trap chl$^+ + A^-$), i.e., the donation within 10^{-10} s of an electron to an acceptor prevents the deexcitation of the trap chls by fluorescence, which has a longer lifetime.

As we have just indicated, an excited trap chl can rapidly donate an electron to some acceptor molecule, which is part of the photochemistry of photosynthesis. The donation of the electron initiates the chemical reactions of photosynthesis and the subsequent storage of energy in stable chemical bonds. Moreover, once the trap chl has lost an electron, it can take on another electron from some donor, indicated by D in Eq. (5.6) (Trap chl$^+ + D \rightarrow$ trap chl $+ D^+$) and Fig. 5-9. Thus the photochemical reactions of photosynthesis lead to electron flow. An excitation trap, such as P_{680} or P_{700}, plays a key role in the conversion of radiant energy into forms of energy that are biologically useful. We generally refer to the trap chl plus A and D as a *reaction center* (illustrated in Fig. 5-9), which is the locus for the photochemistry of photosynthesis (Eq. 5.5).[2]

IV. Groupings of Photosynthetic Pigments

We have discussed the absorption of light by photosynthetic pigments and the ensuing transfers of excitation among these molecules, which leads us to a consideration of whether there are discrete units of such pigments acting together in some concerted fashion. Such an ensemble is presented in Fig. 5-9, in which the light-harvesting photosynthetic pigments greatly outnumber the special trap chl molecules, the latter occurring in a one-to-one

---

[2] The Nobel prize in chemistry was awarded to Johan Deisenhofer, Robert Huber, and Hartmut Michel in 1988 for unraveling the structure of the reaction center from the purple photosynthetic bacterium *Rhodopseudomonas viridis* using x-ray crystallography.

relationship with suitable electron acceptors and donors. In this section we will consider whether photosynthetic pigments are organized into functional groups. If they are, how many molecules make up one group? Are various groupings of pigments identical in terms of their function in photosynthesis?

A. Photosynthetic Units

At low light levels, one CO_2 can be fixed and one O_2 evolved for approximately every eight photons absorbed by any of the photosynthetic pigments. Is one O_2 evolved for every eight photons absorbed at high light levels? Data to answer this question were provided in 1932 by Robert Emerson and William Arnold, who exposed the green alga *Chlorella pyrenoidosa* to a series of repetitive intense flashes of light. These flashes excite nearly all chlorophyll molecules and other photosynthetic pigments simultaneously. However, the maximum yield in such experiments is only one evolved O_2 for approximately every 2000 chlorophyll molecules. Assuming that each chlorophyll molecule absorbs one photon, 250 times more photons are needed than the eight needed to produce one O_2 at low light levels. At low light levels sufficient time elapses between the arrival of individual photons for the excitations of the accessory pigments and Chl *a* to be efficiently collected in a trap chl and used for the chemical reactions of photosynthesis. At high light levels, however, many photosynthetic pigments become excited at the same time, and only one excited chlorophyll out of about 250 leads to any photochemical reaction. One possible interpretation is that 250 chlorophylls act together as a *photosynthetic unit*; when any one chlorophyll in this photosynthetic unit becomes excited by light and then has this excitation transferred to the reaction center, the concomitant excitation of other chlorophylls in that unit cannot be used for photosynthesis. However, the calculation indicates only the number of chlorophylls per reaction center and should not be taken to imply structurally definable units.

The previous conclusions can also be considered in terms of Fig. 5-9, in which the trap chl is shown interacting with the electron acceptor *A* and the donor *D*. At high light levels, the rate-limiting step for photosynthesis is not light absorption, excitation transfer, or photochemistry (electron donation by trap chl*), but the subsequent steps leading to O_2 evolution and CO_2 fixation. A brief intense illumination thus leads to more excitations than can be processed by the electron transfer reactions and subsequent biochemical events. In the limit of a very intense flash exciting all photosynthetic pigments simultaneously, one excitation is processed photochemically by each reaction center; all others are dissipated by various nonphotochemical deexcitation processes, such as those discussed in Chapter 4.

B. Excitation Processing

The electron excitation caused by the absorption of a photon can be processed by the chemical reactions leading to CO_2 fixation about once every 5×10^{-3} s (5 ms). This processing time has important consequences for both the efficiency of light use at different photon flux densities and the optimal number of chlorophylls per reaction center.

The highest photon flux density normally encountered by plants occurs when the sun is directly overhead on a cloudless day, in which case the photosynthetic photon flux density (PPFD) for wavelengths from 400 to 700 nm is about 2000 μmol m^{-2} s^{-1} on a horizontal plane. The average chlorophyll concentration in chloroplasts is approximately 30 mol m^{-3}, and in passing through a chloroplast 2 μm thick about 30% of the incident PPFD is absorbed. We can therefore estimate how often an individual chlorophyll molecule absorbs a photon. Specifically, $(0.3) (2000 \times 10^{-6}$ mol photons m^{-2} s^{-1}) or 600×10^{-6} mol photons m^{-2} s^{-1} is absorbed by (30 mol chlorophyll m^{-3}) $(2 \times 10^{-6}$ m) or 60×10^{-6} mol chlorophyll m^{-2}, which is 10 mol photons (mol chlorophyll)$^{-1}$ s^{-1}. Thus 10 photons per second are absorbed on the average by each chlorophyll molecule in a chloroplast exposed to full sunlight.

As we have just calculated, each chlorophyll in an unshaded chloroplast can absorb a photon on the average about once every 0.1 s. When there are 250 chlorophylls per reaction center, 12.5 of these molecules are excited every 5 ms (250 chlorophylls $\times$ 1 excitation per chlorophyll/0.1 s $\times$ 0.005 s). However, because the average processing time per reaction center is about 5 ms, only one of these excitations can be used photochemically—the others are dissipated by nonphotochemical deexcitation reactions. Consequently, although the chemical reactions leading to CO_2 fixation operate at their maximum rate under such conditions of high PPFD, a large fraction of the electronic excitations caused by light absorption are not used for photosynthesis.

Full midday sunlight is seldom incident on a chloroplast under natural conditions because chloroplasts are generally shaded by other chloroplasts in the same cell, by chloroplasts in other cells, and by overlying leaves. Furthermore, the amount of sunlight incident on a plant is much less at sunrise or sunset, during overcast periods, or during the winter than near noon on a clear day in the summer. For the sake of argument, let us consider that the PPFD incident on a chloroplast is one-tenth of that from the direct midday sun, namely 200 μmol m^{-2} s^{-1}. In this case, each chlorophyll in the chloroplast absorbs a photon once every second. When individual chlorophylls are excited every 1 s at this moderate illumination, one chlorophyll out of the 250 per reaction center is excited on average every 4 ms (1000 ms per 250 excitations). This excitation frequency is such that the photons can be efficiently used for photosynthesis with its processing time of 5 ms. In other words, the photons are arriving at a rate such that the excitations produced by nearly all of them can be used. Moreover, the chemical reactions are working at their maximum capacity. Consequently, a reaction center, with its photochemistry and associated enzymatic reactions, will function very effectively at a moderate illumination.

What happens if there were but one chlorophyll molecule per reaction center? This single pigment molecule would be excited about once per second at a PPFD of 200 μmol m^{-2} s^{-1}. If the chemical reactions required 5 ms as used previously, the excitation could easily be processed by the chemical reactions. However, the photochemical step plus the subsequent enzymatic reactions leading to CO_2 fixation would be working at only 0.5% of capacity— $(5 \times 10^{-3}$ s)/(1 s), or 0.005, is the fraction of time they could be used. In other words, although all the absorbed photons would be used for photosynthesis, even the slowest of the chemical steps would be idle more than 99% of the time.

Given the 5-ms processing time for the chemical reactions, approximately 250 chlorophylls per reaction center connected with the appropriate enzyme machinery thus provides a plant with a mechanism for efficiently handling the usual illuminations found in nature— both for harvesting the photons and for using the chemical reactions at a substantial fraction of their capacity. For instance, averaged over the earth's surface and the year, the mean PPFD reaching the ground during the daytime on clear days is about 800 μmol m^{-2} s^{-1}. The total area of all leaves divided by the total land area where vegetation occurs is about 4.3 (this ratio, called the leaf area index, will be discussed in Chapter 9), so the average PPFD on a leaf is somewhat less than 200 μmol m^{-2} s^{-1}. At this PPFD, which can also occur for chloroplasts in cells on the side away from the fully sunlit side of a leaf, the rate of photon absorption is well matched to the rate of excitation processing.

In addition to interspecific variations, the number of chlorophylls per reaction center can depend on the PPFD present during leaf development. Some algae and leaves of land plants developing under low illumination can have over 700 chlorophylls per reaction center, whereas certain leaves developing under full sunlight can have as few as 100. The ratio is fairly low in bacteria, where there are 40–100 bacteriochlorophyll molecules per reaction center.

C. Photosynthetic Action Spectra and Enhancement Effects

The electron excitations resulting from light absorption by any photosynthetic pigment can be transferred to a trap chl and thus lead to photochemical reactions. The absorption spectrum of chloroplasts should therefore match the action spectrum for photosynthesis. However, the action spectrum for CO_2 fixation or O_2 evolution and the overall absorption spectrum for the photosynthetic pigments in the same organism do not always coincide. Among other things, a "red drop" occurs, e.g., the photosynthetic action spectrum drops off much more rapidly in the red region beyond 690 nm than does the absorption spectrum for chlorophylls and other pigments (illustrated in Fig. 5-10 for the case of a green alga).

In 1957 Emerson demonstrated that the relatively low photosynthetic efficiency of *Chlorella* in far-red light—a red drop such as that in Fig. 5-10—can be increased by simultaneously using light of a shorter wavelength along with the far-red light. The photosynthetic rate with the two beams can be 30–40% greater than the sum of the rates of the far-red light and the shorter wavelength light used separately. Also, the quantum yield (Eq. 4.15) for O_2 evolution by *Chlorella* using the 700-nm radiation is higher when the 650-nm light is also present. Such synergism, or enhancement, suggests that photosynthesis involves the cooperation of two distinct photochemical reactions. Light of wavelengths >690 nm mainly powers only one of the two necessary reactions, and thus photosynthesis does not proceed at an appreciable rate. When shorter wavelengths are also used, however, the other necessary reaction takes place, resulting in a marked enhancement of the photosynthetic rate or quantum yield.

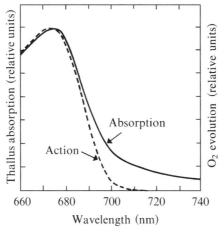

Figure 5-10. Absorption spectrum for an algal thallus and the action spectrum for its O_2 evolution, illustrating the "red drop" in photosynthesis. [Data for *Ulva taeniata* are from Haxo and Blinks (1950). Reproduced from **The Journal of General Physiology** by copyright permission of The Rockefeller University Press.]

D. Two Photosystems Plus Light-Harvesting Antennae

We can use the photosynthetic enhancement effect to study the pigments associated with each of the two photochemical systems involved in photosynthesis. The system containing the pigments absorbing beyond 690 nm was discovered first and is referred to as Photosystem I, a terminology introduced by Louis Duysens, Jan Amesz, and Bert Kamp in 1961. Much of the far-red absorption by Photosystem I is due to a Chl a with a λ_{max} near 680 nm. The special Chl a dimer, P_{700}, is found exclusively in Photosystem I. Therefore, light above 690 nm is absorbed mainly by the long-wavelength-absorbing form of Chl a and P_{700} in Photosystem I.

An action spectrum for the enhancement of photosynthesis in the presence of a constant irradiation with wavelengths longer than 690 nm, which are absorbed by Photosystem I, indicates which pigments are in the other system, Photosystem II. For example, the action spectrum for such photosynthetic enhancement represented in Fig. 5-11 for the red alga *Porphyridium cruentum* resembles the absorption spectrum of phycoerythrin (λ_{max} near 540 nm; see Fig. 5-7). A marked increase in photosynthesis in the cyanobacterium *Anacystis nidulans* occurs when such far-red light is supplemented by light absorbed by phycocyanin (see Fig. 5-11 and the absorption spectrum for phycocyanin in Fig. 5-7). Therefore, in such organisms the phycobilins funnel their excitations mainly into Photosystem II. Studies using photosynthetic enhancement also indicate that the excitations of Chl b preferentially go to

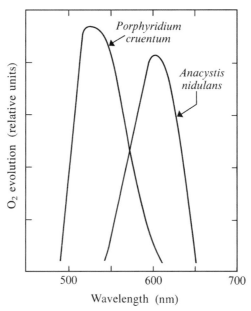

Figure 5-11. Action spectra for the enhancement of O_2 evolution. Cells were exposed to a constant far-red illumination (wavelengths beyond 690 nm that would produce little O_2 evolution) plus a specific photosynthetic photon flux at the various wavelengths indicated on the abscissa. The ordinate represents the rate of O_2 evolution for the two beams acting together minus that produced by the shorter wavelength acting alone. The pigment absorbing the light leading to enhancement of O_2 evolution in the red alga *Porphyridium cruentum* is phycoerythrin, and phycocyanin is responsible in the cyanobacterium *Anacystis nidulans*—see absorption spectra in Fig. 5-7. [Source: Emerson and Rabinowitch (1960); used by permission.]

Table 5-2. Chlorophyll Composition of the Photosystems and Light-Harvesting Antennae[a]

Component examples	Approximate number of chlorophyll molecules
Photosystem I	
Core	100 Chl a
Light-harvesting antenna	70 Chl a, 20 Chl b
Photosystem II	
Core	50 Chl a
Intrinsic light-harvesting antenna	50 Chl a, 35 Chl b
Peripheral light-harvesting antenna	70 Chl a, 50 Chl b

[a]Data are for chloroplasts from leaves of plants growing under moderate sunlight. Component size and chlorophyll numbers vary, especially for the antennae.

Photosystem II, as do those of fucoxanthin in the brown algae. In a related experimental approach, action spectra for the evocation of Chl a fluorescence indicate that light absorbed by the phycobilins and Chl b leads to the fluorescence of Chl a in Photosystem II, perhaps including fluorescence by P_{680}. However, the results are not clear-cut because photons absorbed by Chl b and the phycobilins can lead to photochemical reactions powered by P_{700}.

Steady progress has occurred in the isolation and chemical identification of the photosystems as well as in the recognition that certain photosynthetic pigments in other distinct complexes can transfer excitations to Photosystems I and II. These latter pigments are referred to as being in a *light-harvesting antenna*. For instance, probably all of the Chl b occurs in chlorophyll–protein complexes that are part of light-harvesting antennae (Tables 5-1 and 5-2).

Each photosystem contains various protein subunits that are associated with specific pigments. For instance, the core of Photosystem I consists of polypeptides to which can be bound about 100 molecules of Chl a (Table 5-2). This supramolecular organization contains a reaction center with P_{700} and also interacts with a chlorophyll–protein complex acting as a light-harvesting antenna (Table 5-2). Similarly, Photosystem II is composed of various protein subunits, about 50 Chl a molecules, and a reaction center containing P_{680}. This photosystem interacts with two types of light-harvesting complexes, a tightly bound one and a slightly larger peripheral one (Table 5-2). The peripheral light-harvesting complex can become disconnected from Photosystem II and then transfer excitations to Photosystem I. In particular, when absorption of photons is greatly in excess of what can be processed by Photosystem II, some of the light-harvesting complex associated with Photosystem II becomes detached, apparently after phosphate groups are added that can lead to electrostatic repulsion. Movement of such antenna chlorophylls away from Photosystem II and their ability to pass excitations on to Photosystem I help create a better balance in photon processing by the two photosystems. The phycobiliproteins (organized into phycobilisomes in cyanobacteria and red algae) are also part of a light-harvesting complex, in this case one interacting with Photosystem II. Only a single photosystem occurs in the green and the purple bacteria, which oxidize primarily organic acids or inorganic sulfur-containing compounds instead of water (hence, O_2 evolution does not accompany their photosynthesis); again, a special light-harvesting complex is involved in collecting excitations and directing them toward a trap BChl in the core.

We note that Photosystem II is about 40–90% more abundant than Photosystem I for plants growing in full sunlight, and the relative amounts of various pigments in the complexes can also be influenced by the illumination level during leaf development. In particular, different light-absorbing and light-processing efficiencies between the two photosystems necessitate different stoichiometries. Photosystem I is excluded from most of the appressed membrane regions where lamellae are stacked into grana (see Fig. 1-9), whereas Photosystem II is located mainly in these stacked granal thylakoids. Thus the stromal lamellae and apparently also the exposed end membranes and the margins of the grana are greatly enriched in Photosystem I. Chloroplasts in leaves developing in the shade tend to have more photosynthetic pigments per unit volume, more thylakoids, and a greater relative stacking of thylakoids into grana than do chloroplasts developing under high illumination. Consequently, the Photosystem I-enriched lamellae exposed to the stroma increase from about 40% of the lamellar membranes for leaves developing under low illuminations to 60% under high ones. The increase in the ratio of Photosystem I to Photosystem II with increasing availability of light leads to an increase in the ratio of Chl a to Chl b, as Photosystem I and its associated light-harvesting antenna have a higher fraction of their chlorophyll as Chl a than does Photosystem II with its antennae (Table 5-2). Such relative increases of Photosystem I and Chl a with increasing illumination can be seen between plants, between leaves on a tree, and even between cells in a single leaf.

V. Electron Flow

The photochemical reaction of photosynthesis involves the removal of an electron from an excited state of the special chlorophyll that acts as an excitation trap. The movement of the electron from this trap chl to some acceptor begins a series of electron transfers that can ultimately lead to the reduction of $NADP^+$. The oxidized trap chl, which has lost an elecron, can accept another electron from some donor, as in the steps leading to O_2 evolution. Coupled to the electron transfer reactions in chloroplasts is the formation of ATP, a process known as "photophosphorylation." In this section we will consider some of the components of chloroplasts involved in accepting and donating electrons; a discussion of the energetics of such processes will follow in Chapter 6.

The various steps of photosynthesis vary greatly in the amount of time required. For instance, the absorption of light and the transfer of excitation from both accessory pigments and Chl a molecules to a trap chl take from 10^{-15} to 10^{-9} s after the arrival of a photon. The photochemical event at the reaction center leads to the separation of an electron from the trap chl (Eq. 5.5), causing bleaching—a decrease in the absorption coefficient for wavelengths in the visible region—of this pigment. By observing the kinetics of this bleaching in the far-red region, we can tell when photochemistry occurs. The donation of an electron to the oxidized trap chl (Eq. 5.6) usually occurs 10^{-7} to 10^{-4} s after the arrival of a photon and restores the original spectral properties of the trap chl. The ensuing electron flow to some components can last into the millisecond range. Finally, the overall processing time for a reaction center plus the associated enzymes is about 5 ms per excitation.

A. Electron Flow Model

In 1937 Robert Hill demonstrated that isolated chloroplasts, when placed in an aqueous solution in the presence of a suitable electron acceptor, can evolve oxygen in the light, a process that subsequently became known as the Hill reaction. Oxygen evolution proceeds in the absence of CO_2, suggesting that CO_2 fixation and O_2 evolution are separate processes, contrary to the prevailing belief. Using ^{18}O-labeled H_2O and ^{18}O-labeled CO_2 in differ-ent experiments, Laurens Ruben and Martin Kamen showed in 1941 that the evolved O_2 comes from water and not from CO_2. Subsequent studies have elucidated many of the steps intervening between O_2 evolution and CO_2 fixation in photosynthesis.

We begin our discussion of electron flow in photosynthesis with the water oxidation step:

$$2H_2O \rightarrow O_2 + 4H \qquad (5.8)$$

CO_2 fixation into a carbohydrate involves the reduction of carbon, with four hydrogen atoms being required per carbon atom:

$$CO_2 + 4H \rightarrow \{CH_2O\} + H_2O \qquad (5.9)$$

where $\{CH_2O\}$ represents a general carbohydrate (Fig. 5-1). The movement of the reductant H in Eq. (5.9) can be conveniently followed by tracing the flow of electrons ($H = H^+ + e^-$) (reducing a compound is chemically equivalent to adding electrons, and oxidation is the removal of electrons). The oxidant involved with O_2 evolution (Eq. 5.8) is provided by Photosystem II. The primary oxidant is trap chl^+, which leads to the oxidation of water (see Eq. 5.6; Trap $chl^+ + D \rightarrow$ trap $chl + D^+$). The reductant required for carbon reduction (Eq. 5.9) is produced by the excited trap chl in Photosystem I (see Eq. 5.5; Trap $chl^* + A \rightarrow$ trap $chl^+ + A^-$). These two photosystems are linked by a chain of components along which a transfer of electrons occurs (Fig. 5-12).

For each photon absorbed by any of the accessory pigments or Chl a's whose excitations are funneled into a reaction center, one electron can be removed from its trap chl. Because four electrons are involved per O_2 derived from water, the evolution of this molecule of O_2

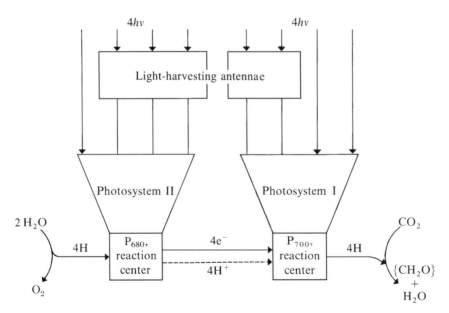

Figure 5-12. Schematic model for a series representation of the two photosystems of photosynthesis, indi-cating the stoichiometry of various factors involved in the reduction of CO_2 to a carbohydrate ($\{CH_2O\}$). Some of the photons ($h\nu$) are captured by the accessory pigments and Chl a in the light-harvesting antennae; these excitations are then fed into the two photosystems, but primarily to Photosystem II.

requires the absorption of four photons by Photosystem II or the light-harvesting antenna feeding into it (see Eq. 5.8 and Fig. 5-12). An additional four photons whose excitations arrive at the trap chl of Photosystem I are required for the reduction of the two molecules of $NADP^+$ necessary for the subsequent reduction of one CO_2 molecule (Eq. 5.9 and Fig. 5-12). Hence eight photons are needed for the evolution of one molecule of O_2 and the fixation of one molecule of CO_2. (In Chapter 6 we will consider how many photons are used to provide the ATPs required per CO_2 fixed.) Although there are alternatives to the series representation (Fig. 5-12) proposed by Hill and Fay Bendall in 1960, the scheme takes into consideration the results of many investigators and has generally become accepted as an overall description of electron flow in chloroplast lamellae. After introducing the concept of redox potential in Chapter 6, we will portray the energetics of the series representation (see Fig. 6-4, which includes many of the components that we will discuss next).

B. Components of the Electron Transfer Pathway

We shall now turn our attention to the specific molecules that act as electron acceptors or donors in chloroplasts. A summary of the characteristics of the better known components of the pathway is presented in Table 5-3. We will begin our discussion by considering the photochemistry at the reaction center of Photosystem II and then proceed with a consideration of the various substances in the sequence in which they are involved in electron transfer along the pathway from Photosystem II to Photosystem I. We will conclude by considering the fate of the excited electron in Photosystem I.

The electron removed from the excited trap chl in the reaction center of Photosystem II is replaced by one coming from water in the process leading to O_2 evolution (Eq. 5.8). Using very brief flashes of light, Pierre Joliot showed in the 1960s that essentially no O_2 evolution accompanied the first or even the second flash. If the four electrons involved come from four different Photosystem IIs, then a single intense flash should cause O_2 evolution because four Photosystem IIs would have been excited and the four electrons coming from each O_2 could be accepted. If two photosystems were involved, each one accepting two electrons sequentially, then two flashes should lead to O_2 evolution. In fact, every fourth flash leads to substantial O_2 evolution, indicating that a single Photosystem II is responsible for the four electrons involved in the evolution of each O_2 molecule, i.e., four consecutive photochemical acts in a particular Photosystem II are required before a molecule of O_2 can be evolved. The four steps, which result in the release of O_2 inside a thylakoid, require about 2 ms. (In an illuminated leaf, there are many O_2-evolving loci, and the four steps are in different stages at any one time, thus O_2 is continually evolved.) Manganese is apparently involved in the O_2-evolving process, presumably while bound to a protein. Four Mn atoms occur per reaction center in Photosystem II, and their progressive experimental removal leads to a stoichiometric reduction in the O_2-evolving ability. Chloride is also apparently necessary for O_2 evolution.

The electron from trap chl* of Photosystem II is transferred by a series of molecules making up the photosynthetic *electron transfer chain*, a term describing the pathway from Photosystem II to Photosystem I (see Fig. 5-12). The electron removed from P_{680} is very rapidly transferred to pheophytin (in about 2×10^{-12} s); *pheophytin* is chlorophyll (Fig. 5-2) without the central magnesium, indicating that it is a highly conjugated porphyrin absorbing in the blue region (Soret band) and the red region (another chlorophyll molecule may intervene between P_{680} and pheophytin). From pheophytin, the electron rapidly (in 1 or 2×10^{-10} s) moves to a quinone (generally referred to as Q_A), the latter step helping to

Table 5-3. Representative Properties of Some of the Components Involved with Electron Transfer in Chloroplasts[a]

Name	Molecular mass (kDa)	Approximate number per 600 chlorophylls	Numbers of electrons accepted or donated per molecule	Approximate midpoint redox potential (V)	Comment
P_{680}	2 × 0.893	1.6	1	1.10	Probably a Chl a dimer; acts as the trap of Photosystem II
Plastoquinone A	0.748	10	2	0.11	Located in membrane; acts as a mobile pool accepting electrons from Photosystem II and donating them to the Cyt b_6–f complex
Cyt b_6	23	2	1	−0.05 and −0.17	The two hemes are bound to the same polypeptide; part of the Cyt b_6–f complex
Cyt f	34	1	1	0.35	Part of the Cyt b_6–f complex; λ_{max} for α band at 554 nm
Plastocyanin	11	2	1	0.37	Blue protein (reduced form is colorless) that accepts electrons from the Cyt b_6–f complex and donates them to Photosystem I; contains one Cu; soluble in aqueous solutions, but occurs in the thylakoid lumen
P_{700}	2 × 0.893	1.0	1	0.48	A Chl a dimer; acts as the trap of Photosystem I
Ferredoxin	11	1–2	1	−0.42	Nonheme protein with two Fe and two S; accepts electrons from Photosystem I by way of intermediates; soluble in aqueous solutions
Ferredoxin–NADP$^+$ oxidoreductase	37	1	—	—	An enzyme containing one flavin adenine dinucleotide per molecule; bound to outside of lamellae
NADP$^+$–NADPH	0.743 or 0.744	30	2	−0.32	Soluble in aqueous solutions

[a]The frequency of components is per 600 chlorophylls for plants growing under moderate sunlight (see Table 5-1), and the order presented is in the sequence for electron flow. Redox potentials are discussed in Chapter 6.

stabilize the charge separation and thus preventing the electron from going back to P_{680}^+. The electron is later transferred to a second quinone (Q_B); when P_{680} is excited again, another electron is passed to this second quinone.

Chloroplast lamellae contain different types of *quinone*. A quinone becomes a semi-quinone when one hydrogen atom ($H^+ + e^-$) is added (a semiquinone is a free radical because of the presence of an unpaired electron) and a hydroquinone when two are added (a hydroquinone is also called a hydroquinol, or simply a quinol):

| quinone | semiquinone | hydroquinone |

There is great variation among quinones because of substituents attached to the ring, e.g., plastoquinone A is

Like chlorophyll, plastoquinone A has a nonpolar terpenoid or isoprenoid tail, which may stabilize the molecule at the proper location in the lamellar membranes of chloroplasts. When donating or accepting electrons, plastoquinones have characteristic absorption changes in the UV near 250–260, 290, and 320 nm that can be monitored to study their electron transfer reactions. (Plastoquinone refers to a quinone found in a plastid such as a chloroplast; these quinones have various numbers of isoprenoid residues, such as nine for plastoquinone A, the most common plastoquinone in higher plants.) The plastoquinones involved in photosynthetic electron transport are divided into two categories: (1) two plastoquinones, apparently bound to proteins, that rapidly receive single electrons from P_{680}; and (2) a mobile group or pool of about 10 plastoquinones that subsequently receives two electrons (plus two H^+) from the second of these plastoquinones (all of these quinones occur in the lamellar membranes). From the plastoquinone pool, electrons move to the cytochrome b_6-f complex.

Cytochromes are extremely important components of electron transfer pathways in chloroplasts and mitochondria. They have three absorption bands in the visible region: the α, β, and γ bands. (Absorption of light by cytochromes apparently plays no part in photosynthesis.) In 1925 David Keilin described three types of cytochrome based on the spectral position of their α (long) wavelength band. Cytochromes of the *a* type have a λ_{max} for the α band from 600 to 605 nm, *b* types near 560 nm, and *c* types near 550 nm *in vivo*. All cytochromes have β bands near 515–530 nm. The main short wavelength band, the γ or Soret band, has a λ_{max} between 415 and 430 nm. The first three cytochromes studied by

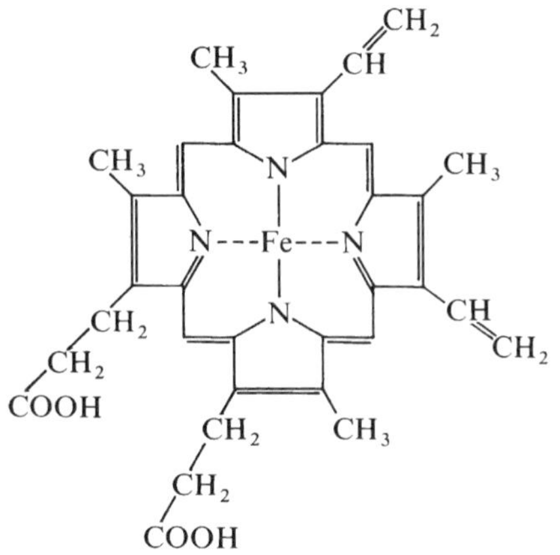

Figure 5-13. An iron-containing porphyrin known as heme, the chromophore for Cyt c.

Keilin were designated Cyt a, Cyt b, and Cyt c. Additional ones were indicated by subscripts. For instance, Cyt f (f from *frons*, the Latin for leaf) is also known as Cyt c_1 because it resembles the absorption properties of the first additional c-type cytochrome identified by Keilin. Cyt b_6, which is often referred to as Cyt b_{563}, has absorption properties similar to the sixth additional b-type cytochrome identified.

Cytochromes consist of an iron-containing tetrapyrrole or porphyrin known as *heme* (Fig. 5-13), which is bound to a protein. The various cytochromes differ in both the substituents around the periphery of the porphyrin ring and the protein to which the chromophore is attached. Cyt f occurs in chloroplasts and contains the chromophore indicated in Fig. 5-13. Many different hemoproteins of the Cyt b type are found in plants, two occurring in chloroplasts (Cyt b_{559}, which occurs associated with Photosystem II, and another one with a λ_{max} at about 563 nm, which is part of the Cyt b_6-f complex; Table 5-3). These various b cytochromes appear to have the same chromophore attached to different proteins, and thus their individual absorption properties are due to changes in the protein. Like the chlorophylls, cytochromes are tetrapyrroles (compare the structure of Chl a in Fig. 5-2), but they have an Fe atom in the center of the porphyrin ring, whereas chlorophylls have a Mg atom. Furthermore, the acceptance or donation of an electron by a cytochrome involves a transition between the two states of its iron, Fe^{2+} and Fe^{3+}, whereas the electron removed from chlorophyll in the photochemical reactions of photosynthesis is one of the π electrons in the conjugated system of the porphyrin ring.

Instead of the usual valence bonds, the metal atoms in chlorophylls and cytochromes should be presented in terms of the six coordinate bonds described by ligand–field (molecular–orbital) theory. Fe has one coordinate bond to each of the four N's in the porphyrin ring (the two solid and the two broken lines emanating from Fe in Fig. 5-13; also see Fig. 5-2), one to an N in the imidazole side chain of a histidine, and the sixth to another histidine for b-type cytochromes in chloroplasts and to a lysine or a methionine for Cyt f;[3] these amino acids occur in the protein to which the chromophore is bound (one bond is above and one is below the plane of the porphyrin in Fig. 5-13). The donation of

---

[3] For a-type cytochromes in mitochondria, the sixth position is not bound to an amino acid but rather interacts with O_2, Cu, or other ligands.

an electron by the ferrous form of Cyt c, ferrocytochrome c (Fig. 5-13), causes the iron to go to the ferric state, Fe^{3+}. The extra positive charge on the iron in ferricytochrome c can either attract anions such as OH^- or be delocalized to adjacent parts of the molecule. The six coordinate bonds remain in ferricytochrome c, and the conjugation in the porphyrin ring is only moderately changed from that in ferrocytochrome c. Hence, the extensive bleaching of chlorophyll following the loss of an electron from its porphyrin ring does not occur with electron donation by the iron atom in cytochrome because electrons in the ring conjugation are influenced only secondarily.

The removal of an electron from a cytochrome (oxidation) causes its three absorption bands to become less intense, broader, and to shift toward shorter wavelengths. The absorption coefficients at λ_{max} in the reduced form are about 3×10^3 m^2 mol^{-1} for the α band, somewhat less for the β band, and over 10^4 m^2 mol^{-1} for the γ (Soret) band. Upon oxidation, ε_λ for the α band decreases just over 50%, and smaller fractional changes generally occur in the absorption coefficients of the β and the γ bands (λ_{max} is also at shorter wavelengths in the oxidized form). Such absorption changes permit a study of the kinetics of electron transfer while the cytochrome molecules remain embedded in the internal membranes of chloroplasts or mitochondria. The spectral changes of chloroplast cytochromes indicate that the electron movement along the electron transfer chain between Photosystem II and Photosystem I (Figs. 5-12 and 6-4) occurs 2×10^{-3} s after light absorption by some pigment molecule in Photosystem II.

The Cyt b_6–f complex is a supramolecular protein unit embedded in the chloroplast lamellar membranes. It is composed of various polypeptides, including a polypeptide that binds two Cyt b_6's and another that binds Cyt f (Table 5-3), as well as an iron-plus-sulfur-containing protein (the Rieske Fe–S center) and a bound plastoquinone. The protein plastocyanin (Table 5-3) carries electrons from the Cyt b_6–f complex to Photosystem I. Plastocyanin occurs associated with the inner side of the thylakoid membranes, i.e., the side toward the lumen (see Fig. 1-9). The electrons from plastocyanin, which are apparently donated and accepted by copper atoms, can be accepted by the trap chl of Photosystem I, P_{700}, if the latter is in the oxidized form (see Eq. 5.6; Trap chl$^+$ + D $\rightarrow$ trap chl + D$^+$). The photochemical change in P_{700} can be followed spectrophotometrically because the loss of an electron causes a bleaching of both its Soret and red absorption bands; the subsequent acceptance of an electron restores the original spectral properties. Specifically, P_{700} is bleached (oxidized) by light absorbed by Photosystem I and then restored (reduced) following the absorption of photons by Photosystem II. Thus the electron removed from P_{700} is replaced by one coming from Photosystem II by means of the electron transfer chain (Figs. 5-12 and 6-4). Far-red light (above 690 nm) absorbed by Photosystem I leads to an oxidation of Cyt f, indicating that an electron from Cyt f can be donated to P_{700} in about 3×10^{-4} s.

Because Photosystem II tends to occur in the grana and Photosystem I in the stromal lamellae, the intervening components of the electron transport chain may have to diffuse over considerable distances in the lamellar membranes to link the two photosystems. We can examine such diffusion using the time–distance relationship derived in Chapter 1 (Eq. 1.6; $x_e^2 = 4D_j t_e$). In particular, the diffusion coefficient for plastocyanin in a membrane can be about 3×10^{-12} m^2 s^{-1} and about the same in the lumen of the thylakoids, unless diffusion of plastocyanin is physically restricted in the lumen by the appression of the membranes (Haehnel, 1984). For such a D_j, in 3×10^{-4} s (the time for electron transfer from the Cyt b_6–f complex to P_{700}), plastocyanin could diffuse only about $[(4)(3 \times 10^{-12}$ m^2 s$^{-1})$ $(3 \times 10^{-4}$ s$)]^{1/2}$ or 60 nm, indicating that this complex probably occurs in the lamellae in relatively close proximity to its electron acceptor, Photosystem I. Plastoquinone is smaller and hence would diffuse more readily than plastocyanin, and a longer time $(2 \times 10^{-3}$ s) is apparently necessary to move electrons from Photosystem II to the Cyt b_6–f complex; hence, these two components can be separated by much greater distances than are the Cyt b_6–f complex and Photosystem I.

The electron from P_{700}^* reduces the iron-containing protein, ferredoxin, via a series of at least five intermediates. Electron transfer to the first component, which may be a form of chlorophyll, is rapid (about 3×10^{-12} s); the next component may be a quinone (reached in about 35×10^{-12} s), followed by three centers containing iron and sulfur. Finally, the electron reaches ferredoxin, which contains two irons in the ferric state that interact with two sulfur atoms; the acceptance of an electron by ferredoxin reduces one of the ferric atoms to the ferrous form. By means of the enzyme ferredoxin–NADP$^+$ oxidoreductase, two molecules of ferredoxin reduce one molecule of NADP$^+$ to yield NADPH (we will discuss NADP$^+$ and NADPH in Chapter 6).

C. Types of Electron Flow

Three types of photosynthetic electron transfer, or flow, occur—noncyclic, pseudocyclic, and cyclic—each one depending on the compound to which electrons are transferred from ferredoxin (Fig. 5-14). In *noncyclic* electron flow, electrons coming from water reduce NADP$^+$: An electron from water goes to the trap chl$^+$ of Photosystem II; there it replaces

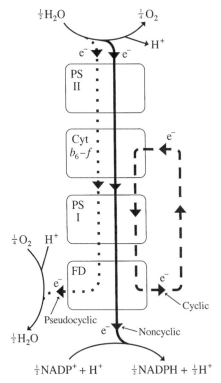

Figure 5-14. The three electron flow pathways in chloroplasts, showing the pivotal role played by ferredoxin (FD) in noncyclic (solid line), cyclic (dashed line), and pseudocyclic (dotted line) flow.

a donated electron, which moves along the electron transfer chain to the oxidized P_{700} in Photosystem I; the electron from P_{700} moves to ferredoxin and then to $NADP^+$. Such noncyclic electron flow follows essentially the same pathway as the reductant H moving from left to right in Fig. 5-12 (see also Fig. 6-4).

Electrons from ferredoxin may also reduce O_2, which yields H_2O_2 and eventually H_2O ($O_2 + 2e^- + 2H^+ \rightleftharpoons H_2O_2 \rightleftharpoons H_2O + \frac{1}{2}O_2$). (The light-dependent consumption of O_2, as occurs when electrons from ferredoxin or from one of the iron-plus-sulfur-containing intermediates of Photosystem I move to O_2, is termed the Mehler reaction.) Because equal amounts of O_2 are evolved at Photosystem II and then consumed using reduced ferredoxin in a separate reaction, such electron flow is termed *pseudocyclic* (see Figs. 5-14 and 6-4). No net O_2 change accompanies pseudocyclic electron flow, although it is not a cycle in the sense of having electrons cyclically traverse a certain pathway. To what extent pseudocyclic electron flow occurs *in vivo* is not clear, although such electron flow can be readily demonstrated with isolated chloroplasts.

For *cyclic* electron flow, an electron from the reduced form of ferredoxin moves back to the electron transfer chain between Photosystems I and II via the Cyt b_6–f complex (including the involvement of a plastoquinone) and eventually reduces an oxidized P_{700} (Fig. 5-14). Cyclic electron flow does not involve Photosystem II, so it can be caused by far-red light absorbed only by Photosystem I—a fact that is often exploited in experimental studies. When far-red light absorbed by Photosystem I is used, cyclic electron flow can occur but noncyclic does not, so no NADPH is formed and no O_2 is evolved (cyclic electron flow can lead to the formation of ATP, as is indicated in Chapter 6). When light absorbed by Photosystem II is added to cells exposed to far-red illumination, both CO_2 fixation and O_2 evolution can proceed, and photosynthetic enhancement is achieved. Treatment of chloroplasts or plant cells with the O_2-evolution inhibitor DCMU [3-(3,4-dichlorophenyl)-1,1-dimethyl urea] also leads to only cyclic electron flow; DCMU therefore has many applications in the laboratory and is also an effective herbicide because it markedly inhibits photosynthesis. Cyclic electron flow may be more common in stromal lamellae because they have predominantly Photosystem I activity.

As indicated in Table 5-3, P_{680}, P_{700}, the cytochromes, plastocyanin, and ferredoxin accept or donate only one electron per molecule. These electrons interact with $NADP^+$ and the plastoquinones, both of which transfer two electrons at a time. The two electrons that reduce plastoquinone apparently come sequentially from the same Photosystem II. It is less clear how the two electrons move from a plastoquinone to the one-electron carrier Cyt f. The enzyme ferredoxin–$NADP^+$ oxidoreductase matches the one-electron chemistry of ferredoxin to the two-electron chemistry of NADP. Both the pyridine nucleotides and the plastoquinones are considerably more numerous than are other molecules involved with photosynthetic electron flow (Table 5-3), which has important implications for the electron transfer reactions. Moreover, $NADP^+$ is soluble in aqueous solutions and so can diffuse to the ferredoxin–$NADP^+$ oxidoreductase, where two electrons are transferred to it to yield NADPH (ferredoxin and plastocyanin are also soluble in aqueous solutions).

D. Photophosphorylation

Three ATP molecules are generally required for the reductive fixation of one CO_2 molecule into a carbohydrate (see Fig. 5-1). Such ATP is produced by *photophosphorylation*; i.e., light absorbed by the photosynthetic pigments in the lamellar membranes leads to a flow of electrons, to which is coupled the phosphorylation of ADP. We will consider the energetics of this dehydration of ADP plus phosphate to yield ATP in Chapter 6.

Photophosphorylation was first demonstrated in cell-free systems in 1954. Albert Frenkel, working with bacterial chromatophores, and Daniel Arnon, Mary Belle Allen, and Robert Whatley, using broken spinach chloroplasts, observed ATP formation in the light. The enzymes are localized in or on the lamellar membranes, and the energy transfer steps are very sensitive to perturbation of the membranes. Moreover, none of the molecular species (ADP, ATP, and phosphate) can be readily determined nondestructively *in vivo*. For instance, the interconversions of these compounds in the chloroplasts cannot be monitored by measuring changes in spectral properties, a technique for studying the acceptance or donation of electrons by cytochromes and trap chls. Furthermore, all three molecules (ATP, ADP, and phosphate) take part in many different biochemical reactions. Nevertheless, considerable progress has been made in understanding the relationship between ATP formation and proton (H^+) transport across membranes. In Chapter 6 we will reconsider ATP formation coupled to electron flow in both chloroplasts and mitochondria, after some of the underlying energy concepts have been introduced.

E. Vectorial Aspects of Electron Flow

Because the electron flow components are associated with membranes, the possibility exists for chemical asymmetries to develop. In fact, the electrons and their associated protons are moved in specific directions in space by the processes that we have been considering, causing the flows to have a vectorial nature (see Fig. 5-15).

The chlorophyll–protein complexes are oriented in the lamellar membranes in such a way that the electron transfer steps at the reaction centers lead to an outward movement of

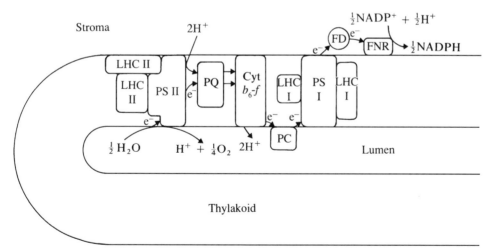

Figure 5-15. Schematic representation of reactions occurring at the photosystems and certain electron transfer components, emphasizing the vectorial or unidirectional flows developed in the thylakoids of a chloroplast. Outwardly directed electron movements occur in the two photosystems (PS I and PS II), where the electron donors are on the inner side and the electron acceptors are on the outer side of the membrane. Light-harvesting complexes (LHC) act as antennae for these photosystems. The plastoquinone pool (PQ) and the Cyt b_6–f complex occur in the membrane, whereas plastocyanin (PC) occurs on the lumen side and ferredoxin–NADP$^+$ oxidoreductase (FNR), which catalyzes electron flow from ferredoxin (FD) to NADP$^+$, occurs on the stromal side of the thylakoids. Protons (H^+) are produced in the lumen by the oxidation of water and also are transported into the lumen accompanying electron (e^-) movement along the electron transfer chain.

electrons. For instance, the electron donated by Photosystem II is moved from the lumen side to the stromal side of a thylakoid (see Figs. 1-9 and 5-15). The electron that is donated back to the trap chl (P_{680}) comes from H_2O, leading to the evolution of O_2 by Photosystem II (Eq. 5.8). The O_2 and the H^+ from this reaction are released inside the thylakoid (Fig. 5-15). Because O_2 is a small neutral molecule, it readily diffuses out across the lamellar membranes into the chloroplast stroma. However, the proton (H^+) carries a charge and hence has a low partition coefficient for the membrane, so it does not readily move out of the thylakoid lumen.

The electron excited away from P_{680} in Photosystem II eventually reaches a quinone in that photosystem that accepts two electrons and also picks up two protons (H^+) from the stroma (Fig. 5-15). This quinone transfers its two electrons and two protons to a mobile plastoquinone in the plastoquinone pool occurring in the lamellar membranes, and the mobile plastoquinone in turn interacts with the Cyt b_6-f complex. The Cyt b_6-f complex is responsible for the final step in the vectorial transport of protons from the stroma to the lumen of a thylakoid; it delivers electrons to plastocyanin, which occurs on the inner side of a lamellar membrane. Electrons from plastocyanin move to the reaction center of Photosystem I. Via a photochemical event, the trap chl of Photosystem I (P_{700}) donates an electron that eventually reaches ferredoxin, which occurs on the outer side of a thylakoid (Fig. 5-15). Ferredoxin, which is soluble in aqueous solutions, diffuses to ferredoxin–$NADP^+$ oxidoreductase (a flavoprotein; Table 5-3), where two electrons are accepted by $NADP^+$, yielding NADPH. The flavoprotein is bound on the outer side of a lamellar membrane, so NADPH is formed in a region where subsequent biochemical reactions can utilize this crucially important molecule.

We now recapitulate the accomplishments of the various processes described previously. O_2 is evolved inside a thylakoid and readily diffuses out. The protons from the O_2-evolving step plus those transported by the Cyt b_6-f complex are released in the thylakoid lumen, where the membranes prevent their ready escape. In cyclic electron flow (Fig. 5-14), electrons from P_{700} move to ferredoxin and thence to the Cyt b_6-f complex, which also causes protons to be delivered from the stroma to the lumen of a thylakoid. The accumulation of protons inside the thylakoid, together with the transfer of electrons out, raises the electrical potential inside with respect to outside and also increases the internal concentration of protons, thus setting up a chemical potential gradient capable of doing work. This proton chemical potential gradient is energetically coupled to the formation of ATP (photophosphorylation), as we will see in Chapter 6.

Problems

5.1 A spherical spongy mesophyll cell is 40 μm in diameter and contains 50 spherical chloroplasts that are 4 μm in diameter. Assume that such cells contain 1 g chlorophyll kg^{-1}, that the cell is 90% water by weight, and that the cellular density is 1000 kg m^{-3} ($= 1.00$ g cm^{-3}).

    a. What volume fraction of the cell is occupied by chloroplasts?

    b. If the CO_2 fixation rate is 100 mmol (g chlorophyll)$^{-1}$ h^{-1}, how long does it take to double the dry mass of the cell? Assume that CO_2 and H_2O are the only substances entering the cell.

    c. If the ratio Chl a/Chl b is 3, what is the mean molecular weight of chlorophyll?

d. Assuming that the chlorophyll is uniformly distributed throughout the cell, what is the maximum absorbance by one cell in the red and the blue regions? Use absorption coefficients given in Fig. 5-3.

5.2 Suppose that some pigment has eight double bonds in conjugation and has a single absorption band with a λ_{max} at 580 nm, which corresponds to a transition to the fourth vibrational sublevel of the excited state (see Fig. 4-8). A similar pigment has 10 double bonds in conjugation, which causes the lowest vibrational sublevel of the excited state to move down in energy by 20 kJ mol^{-1} and the lowest vibrational sublevel of the ground state to move up in energy by 20 kJ mol^{-1} compared with the corresponding levels in the other molecule. Assume that the splitting between vibrational sublevels remains at 10 kJ mol^{-1} and that the most likely transition predicted by the Franck–Condon principle for this second molecule is also to the fourth vibrational sublevel of the excited state.

a. What is the shortest λ_{max} for the main fluorescence by each of the two molecules?
b. Can either or both molecules readily pass their excitation on to Chl *a in vivo*?
c. Can the absorption of blue light by Chl *a* lead to excitation of either of the pigments? Give your reasoning.

5.3 Let us approximate chloroplasts by short cylinders 4 μm in diameter and 2 μm thick (i.e., 2 μm along the cylinder axis), which contain 20 mol chlorophyll m^{-3}. The chloroplasts are exposed to 40 W m^{-2} of 675 nm light parallel to the axis of the cylinder. Assume that ε_{675} is 0.60×10^4 m^2 mol^{-1} for the chlorophylls.

a. What is the absorbance at 675 nm for the chlorophyll in a single chloroplast? What is the fraction of the incident light absorbed?
b. How many μmol photons m^{-2} s^{-1} of 675-nm light will be absorbed in passing through a single chloroplast? How many chlorophyll molecules participate in this absorption?
c. Assume that a photosynthetic unit has 250 chlorophyll molecules and that 0.01 s is needed to process each excitation. How often are chlorophyll molecules excited on the average and what fraction of the absorbed photons can be processed?
d. How many moles of O_2 m^{-2} s^{-1} are evolved for each chloroplast? Assume that eight photons are needed to evolve one molecule of O_2.
e. What would be the answers to (c) for a chloroplast shaded by three overlying chloroplasts?

5.4 Chloroplasts corresponding to 10 mmol of chlorophyll m^{-3} of solution are suspended in a cuvette with a 10-mm light path. The rate of O_2 evolution is proportional to PPFD up to 10 μmol photons absorbed m^{-2} s^{-1}, which gives 10^{-4} mol m^{-3} (10^{-7} M) O_2 evolved s^{-1}. The maximum O_2 evolution rate under very high PPFD is 5×10^{-4} mol m^{-3} s^{-1}. For very brief and intense flashes of light, the O_2 evolution is 5×10^{-6} mol m^{-3} per flash.

a. Using the data given, how many photons are required per O_2 evolved?
b. How many chlorophyll molecules occur in a photosynthetic unit?
c. How much time is required for the processing of an excitation by a photosynthetic unit?
d. An "uncoupler" is a compound that decreases the ATP formation coupled to photosynthetic electron flow. When such a compound is added to chloroplasts

incubated at a high photon flux density, the O_2 evolution rate eventually becomes less than a control without the uncoupler. Explain.

5.5 Suppose that the absorbance of pea chloroplasts in a cuvette with a 10-mm light path is 0.1 at 710 nm and 1.0 at 550 nm. Assume that chlorophyll is the only species absorbing at 710 nm and that no chlorophyll absorbs at 550 nm. Suppose that no CO_2 is fixed when either 550- or 710-nm light is used alone, but that both together lead to CO_2 fixation.

  a. Is any ATP formation caused by the 550- or by the 710-nm light?
  b. What type of pigments are absorbing at 550 nm? Are they isoprenoids or tetrapyrroles?
  c. If equal but low incident photon flux densities are simultaneously used at both 550 and 710 nm, what is the maximum quantum yield for CO_2 fixation for each beam?
  d. The initial bleaching of P_{700} at 700 nm leads to a decrease in absorbance of 10^{-5} in 10^{-6} s. What is the minimum number of moles of photons per unit area absorbed by Photosystem I that could account for this? Assume that ε_{700} is 0.8×10^4 m^2 mol^{-1} for the trap chl.

References

Amesz, J., and Hoff, A. (Eds.) (1996). *Biophysical Techniques in Photosynthesis*. Kluwer, Dordrecht.

Anderson, J. M. (1986). Photoregulation of the composition, function, and structure of thylakoid membranes. *Annu. Rev. Plant Physiol.* **37**, 93–136.

Arnon, D. I., Allen, M. B., and Whatley, F. R. (1954). Photosynthesis by isolated chloroplasts. *Nature* **174**, 394–396.

Baker, N. R. (Ed.) (1996). *Photosynthesis and the Environment*. Kluwer, Dordrecht.

Barber, J. (Ed.) (1992). *The Photosystems: Structure, Function and Molecular Biology*. Elsevier, Amsterdam.

Cramer, W. A., Soriano, G. M., Ponomarev, M., Huang, D., Zhang, H., Martinez, S. E., and Smith, J. L. (1996). Some new structural aspects and old controversies concerning the cytochrome $b_6 f$ complex of oxygenic photosynthesis. *Annu. Rev. Plant Physiol. Plant Mol. Biol.* **47**, 477–508.

Deisenhofer, J., and Norris, J. R. (Eds.) (1993). *The Photosynthetic Reaction Center*, Vols. 1 and 2. Academic Press, San Diego.

Duysens, L. M. N., Amesz, J., and Kamp, B. M. (1961). Two photochemical systems in photosynthesis. *Nature* **190**, 510–511.

Emerson, R. (1957). Dependence of yield of photosynthesis in long-wave red on wavelength and intensity of supplementary light. *Science* **125**, 746.

Emerson, R., and Rabinowitch, E. (1960). Red drop and role of auxiliary pigments in photosynthesis. *Plant Physiol.* **35**, 477–485.

Falkowski, P. G., and Raven, J. A. (1997). *Aquatic Photosynthesis*. Blackwell Science, Malden, MA.

Frenkel, A. (1954). Light-induced phosphorylation by cell-free preparations of photosynthetic bacteria. *J. Am. Chem. Soc.* **76**, 5568–5569.

Glazer, A. N., and Melis, A. (1987). Photochemical reaction centers: Structure, organization, and function. *Annu. Rev. Plant Physiol.* **38**, 11–45.

Goodwin, T. W. (Ed.) (1976). *Chemistry and Biochemistry of Plant Pigments* (2nd ed.). Academic Press, London.

Green, B. R., and Durnford, D. G. (1996). The chlorophyll–carotenoid proteins of oxygenic photosynthesis. *Annu. Rev. Plant Physiol. Plant Mol. Biol.* **47**, 685–714.

Gregory, R. P. F. (1989). *Biochemistry of Photosynthesis* (3rd ed.). Wiley, Chichester, UK.

Gross, J. (1991). *Pigments in Vegetables: Chlorophylls and Carotenoids*. Van Nostrand–Reinhold, New York.

Haehnel, W. (1984). Photosynthetic electron transport in higher plants. *Annu. Rev. Plant Physiol.* **36**, 659–693.

Hall, D. O., and Rao, K. K. (1994). *Photosynthesis* (5th ed.). Cambridge Univ. Press, Cambridge, UK.

Hankamer, B., Barber, J., and Boekema, E. J. (1997). Structure and membrane organization of photosystem II in green plants. *Annu. Rev. Plant Physiol. Plant Mol. Biol.* **48**, 641–671.

Haxo, F. T., and Blinks, L. R. (1950). Photosynthetic action spectra of marine algae. *J. Gen. Physiol.* **33**, 389–422.

Hill, R., and Bendall, F. (1960). Function of the two cytochrome components in chloroplasts: A working hypothesis. *Nature* **186**, 136–137.

Holt, A. S., and Jacobs, E. E. (1954). Spectroscopy of plant pigments. I. Ethyl chlorophyllides *A* and *B* and their pheopher bides. *Am. J. Bot.* **41**, 710–717.

Jennings, R. C., Zucchelli, G., Ghetti, F., and Colombetti, G. (1996). *Light as an Energy Source and Information Carrier in Plant Physiology*. Plenum, New York.

Kirk, J. T. O. (1994). *Light and Photosynthesis in Aquatic Ecosystems* (2nd ed.). Cambridge Univ. Press, Cambridge, UK.

Lawlor, D. W. (1993). *Photosynthesis: Molecular, Physiological, and Environmental Processes* (2nd ed.). Longman, Harlow, UK.

Longhurst, A., Sathyendranath, S., Platt, T., and Caverhill. C. (1995). An estimate of global primary production in the ocean from satellite radiometer data. *J. Plankton Res.* **17**, 1245–1271.

Nugent, J. H. A. (1996). Oxygenic photosynthesis. Electron transfer in photosystem I and photosystem II. *Eur. J. Biochem.* **237**, 519–531.

Ó hEocha, C. (1965). Phycobilins. In *Chemistry and Biochemistry of Plant Pigments* (T. W. Goodwin, Ed.), pp. 175–196. Academic Press, London.

Ort, D. R., and Yocum, C. F. (Eds.) (1996). *Oxygenic Photosynthesis: The Light Reactions*. Kluwer, Dordrecht.

Pessarakli, M. (Ed.) (1997). *Handbook of Photosynthesis*. Dekker, New York.

Raghavendra, A. S. (Ed.) (1997). *Photosynthesis: A Comprehensive Treatise*. Cambridge Univ. Press, Cambridge, UK.

Scheer, H. (Ed.) (1991). *Chlorophylls*. CRC Press, Boca Raton, FL.

Wild, A., and Ball, R. (1997). *Photosynthetic Unit and Photosynthesis*. Backhuys, Leiden, The Netherlands.

Young, A., and Britton, G. (Eds.) (1993). *Carotenoids in Photosynthesis*. Chapman & Hall, London.

Zscheile, F. P., White, J. W., Jr., Beadle, B. W., and Roach, J. R. (1942). The preparation and absorption spectra of five pure carotenoid pigments. *Plant Physiol.* **17**, 331–346.

Zuber, H. (1987). The structure of light-harvesting pigment–protein complexes. In *The Light Reactions*, (J. Barber, Ed.), pp. 197–259. Elsevier, Amsterdam.

6

Bioenergetics

I.	**Gibbs Free Energy**	**223**
	A. Chemical Reactions and Equilibrium Constants	224
	B. Interconversion of Chemical and Electrical Energy	227
	C. Redox Potentials	228
II.	**Biological Energy Currencies**	**230**
	A. ATP—Structure and Reactions	230
	B. Gibbs Free Energy Change for ATP Formation	233
	C. $NADP^+$–NADPH Redox Couple	235
III.	**Chloroplast Bioenergetics**	**236**
	A. Redox Couples	237
	B. H^+ Chemical Potential Differences Caused by Electron Flow	240
	C. Evidence for Chemiosmotic Hypothesis	241
	D. Coupling of Flows	242
IV.	**Mitochondrial Bioenergetics**	**244**
	A. Electron Flow Components—Redox Potentials	245
	B. Oxidative Phosphorylation	247
V.	**Energy Flow in the Biosphere**	**249**
	A. Incident Light—Stefan–Boltzmann Law	250
	B. Absorbed Light and Photosynthetic Efficiency	251
	C. Food Chains and Material Cycles	253
	Problems	**253**
	References	**255**

THROUGHOUT THIS book we have considered various aspects of energy in biological systems. The concept of chemical potential was introduced in Chapter 2 and applied to the specific case of water. In Chapter 3 we used this thermodynamic approach to discuss the movement of ions. We also considered the use of energy for the active transport of substances toward higher chemical potentials. Chapter 4 dealt with the absorption of light, an event that is followed by various deexcitation reactions for the excited states of the molecules. The photochemistry of photosynthesis discussed in Chapter 5 involves the conversion of such electromagnetic energy into forms that are biologically useful. This last aspect—the production and use of various energy currencies in biological systems—is the topic of this chapter.

The two energy "currencies" produced in chloroplasts following the trapping of radiant energy are ATP and NADPH. These substances represent the two main classes of energy-storage compounds associated with the electron transfer pathways of photosynthesis and respiration. We can appreciate the importance of ATP by noting that about 100 mol of ATP ($\cong$ 50 kg because the molecular weight of ATP is 507.2) is hydrolyzed in the synthesis of 1 kg dry weight ($\cong$ 10 kg wet weight) of many microorganisms. Also, a typical adult human uses more than his or her total weight in ATP each day! Because only about 10^{-2} mol of ATP typically occurs per kilogram dry mass of cells, a great turnover of ATP is necessary to synthesize new tissue and to maintain mature cells in a state far from equilibrium.

In this chapter we will first examine energy storage in terms of the chemical potential changes accompanying the conversion of a set of reactants into their products. This consideration of the Gibbs free energy allows us to determine the amount of chemical energy that a particular reaction can store or release. We will then evaluate the energy-carrying capacity of ATP in terms of the energetics of its formation and hydrolysis. NADPH can be regarded as possessing electrical energy, with the particular amount depending on the oxidation-reduction potential of the system with which it interacts. After considering ATP and NADPH as individual molecules, we will place them in their biological context— namely, as part of the bioenergetic scheme of chloroplasts and mitochondria.

I. Gibbs Free Energy

Using the appropriate thermodynamic relations, we can calculate the energy changes that accompany biological reactions. Two conditions that are often met by physiological processes greatly simplify these calculations. First, most biological reactions take place at a constant temperature. Second, processes in cells or tissues generally take place at a constant pressure. These two special conditions make the Gibbs free energy, G, very convenient for describing energetics in biology because the decrease in G under these conditions equals the maximum amount of energy available for work. (See Chapter 2 for an introduction to G and Appendix IV for a mathematical presentation of the Gibbs free energy.) Biological use of free energy, or "work," takes a number of different forms—from muscular movement to chemical synthesis and active transport. The change in Gibbs free energy between two states predicts the direction for a spontaneous reaction and indicates how much energy the transition makes available for performing work. Biologists are generally more interested in such changes in free energy than in the absolute amount of energy, which must be defined relative to some arbitrary level.

A reaction at constant T and P spontaneously moves toward a minimum of the Gibbs free energy of the system; minimum Gibbs free energy is achieved at equilibrium (Fig. 6-1). In principle, a spontaneous process can be harnessed to do work; the reversal of a spontaneous reaction requires an input of free energy (see Fig. 2-5). As we will elaborate later, light can be harnessed to produce free energy that causes the phosphorylation of ADP and the reduction of $NADP^+$. These two processes are prime examples of energy-requiring reactions that are at the very heart of chloroplast bioenergetics.

The concept of free energy was introduced in Chapter 2 in presenting the chemical potential μ_j. The chemical potential is actually the partial molal Gibbs free energy with respect to that species, i.e., $\mu_j = (\partial G / \partial n_j)_{T,P,E,h,n_i}$ (Eq. IV.9, Appendix IV). We must

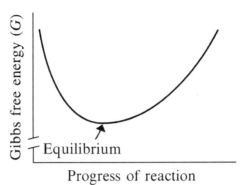

Figure 6-1. Relation between progress of a reaction and the Gibbs free energy of the system (G). When the product concentration is low (left-hand side of the abscissa), the reaction will spontaneously proceed in its forward direction toward lower values of G. The Gibbs free energy for a reaction attains a minimum at equilibrium; ΔG is then zero for the reaction proceeding a short distance in either direction. For high product concentrations (right-hand side of the abscissa), ΔG to drive the reaction further in the forward direction is positive, indicating that a free energy input is then needed. When ΔG in the forward direction is positive, ΔG for the reverse reaction is negative. The absolute value of G is arbitrary (e.g., $G = \sum_j n_j \mu_j$, where each μ_j contains an unknown constant, μ_j^*), so the ordinate is interrupted in the figure.

consider the Gibbs free energy of an entire system to define the chemical potential of species j. In turn, G depends on each of the species present, an appropriate expression being

$$G = \sum_j n_j \mu_j \tag{6.1}$$

where n_j is the number of moles of species j in some system, μ_j is the chemical potential of species j (Eq. 2.4), and the summation is over all of the species present (see Appendix IV, where Eq. 6.1 is derived). The Gibbs free energy is expressed relative to some arbitrary zero level—the arbitrariness in the baseline for G in Eq. (6.1) is a consequence of the μ_j^* included in each μ_j (see Fig. 6-1). The Gibbs free energy as represented by Eq. (6.1) is very useful for applying free energy relations to bioenergetics.

A. Chemical Reactions and Equilibrium Constants

From the thermodynamic point of view, we are interested in the overall change in free energy for an individual reaction—or perhaps a sequence of reactions. Let us consider a general chemical reaction for which A and B are the reactants and C and D are the products:

$$aA + bB \rightleftharpoons cC + dD \tag{6.2}$$

where a, b, c, and d are the numbers of moles of the various species taking part in the reaction.

How much energy is stored (or released) when the reaction proceeds a certain extent in either direction? More specifically, what is the change in Gibbs free energy for the reaction in Eq. (6.2) proceeding in the forward direction, with a moles of A and b moles of B reacting to give c moles of C and d moles of D? For most applications, we can consider that the chemical potentials of the species involved are constant; in other words, we are concerned with a hypothetical change in the Gibbs free energy when the reaction takes place at certain concentrations under fixed conditions. Using Eq. (6.1) $(G = \sum_j n_j \mu_j)$, we can express the change in the Gibbs free energy for such a reaction as

$$\Delta G = -a\mu_A - b\mu_B + c\mu_C + d\mu_D \tag{6.3}$$

where Δn_j is positive for a product and negative for a reactant $(\Delta G = \sum_j \Delta n_j \mu_j$ when all μ_j are constant). Equation (6.3) indicates that the free energy change for a chemical reaction is the Gibbs free energy of the products minus that of the reactants.

To transform Eq. (6.3) into a more useful form, we need to incorporate expressions for the chemical potentials of the species involved. The chemical potential of species j was presented in Chapter 2, where μ_j is a linear combination of various terms: $\mu_j = \mu_j^* + RT$ $\ln a_j + \bar{V}_j P + z_j FE + m_j gh$ (Eq. 2.4; μ_j^* is a constant, a_j is the activity of species j, $\bar{V}_j$ is its partial molal volume, P is the pressure in excess of atmospheric, z_j is its charge number, F is Faraday's constant, E is the electrical potential, m_j is its mass per mole, and h is the vertical position in the gravitational field). Substituting such chemical potentials of A, B, C, and D into Eq. (6.3), and collecting similar terms, we obtain

$$\begin{aligned} \Delta G = &-a\mu_A^* - b\mu_B^* + c\mu_C^* + d\mu_D^* \\ &+ RT(-a\ln a_A - b\ln a_B + c\ln a_C + d\ln a_D) \\ &+ P(-a\bar{V}_A - b\bar{V}_B + c\bar{V}_C + d\bar{V}_D) \\ &+ FE(-az_A - bz_B + cz_C + dz_D) \\ &+ gh(-am_A - bm_B + cm_C + dm_D) \end{aligned} \tag{6.4}$$

Let us next simplify this equation. The constant terms, $-a\mu_A^* - b\mu_B^* + c\mu_C^* + d\mu_D^*$, can be replaced by ΔG^*, a quantity that we will evaluate soon. If the volume of the products, $c\bar{V}_C + d\bar{V}_D$, is the same as that of the reactants, $a\bar{V}_A + b\bar{V}_B$, the factor multiplying P in Eq. (6.4) is zero. In any case, the value of $P\sum_j n_j \bar{V}_j$ for biochemical reactions usually is relatively small for pressures encountered in plant cells, so we will not retain these terms here.[1] The algebraic sum of the charge terms multiplying FE is zero because no charge is created or destroyed by the reaction given in Eq. (6.2); i.e., the total charge of the products, $cz_C + dz_D$, equals that of the reactants, $az_A + bz_B$. Likewise, no mass is created or destroyed by the reaction, so the mass factor multiplying gh is also zero.

We can now convert Eq. (6.4) into a relatively simple form. Using ΔG^*, the constancies of charge and mass, and the generally valid assumption that $P(-a\bar{V}_A - b\bar{V}_B + c\bar{V}_C + d\bar{V}_D)$ is negligible, the change in Gibbs free energy for the reaction in Eq. (6.2) proceeding in the forward direction becomes

$$\Delta G = \Delta G^* + RT\ln\frac{(a_C)^c(a_D)^d}{(a_A)^a(a_B)^b} \tag{6.5}$$

where various properties of logarithms (Appendix III) have been used to obtain the form indicated. Equation (6.5) expresses the Gibbs free energy stored or released by a chemical reaction.

---

[1] Ignoring $P\sum_j n_j \bar{V}_j$ may not be valid for reactions occurring under very high pressures (e.g., deep in the ocean), especially if volume changes are suspected, such as can occur when two oppositely charged species react to form a neutral one.

At equilibrium, the argument of the logarithm in Eq. (6.5), $[(a_C)^c(a_D)^d]/[(a_A)^a(a_B)^b]$, is the equilibrium constant K of the reaction given by Eq. (6.2). Furthermore, at constant T and P the Gibbs free energy achieves a minimum at equilibrium (Fig. 6-1), and it does not change for a conversion of reactants to products under such conditions, i.e., $\Delta G = 0$ for a chemical reaction proceeding a short distance in either direction at equilibrium under these conditions. Equation (6.5) thus indicates that $\Delta G^* + RT \ln K$ is zero at equilibrium, so

$$\Delta G^* = -RT \ln K \qquad (6.6)$$

where K equals the equilibrium value for $[(a_C)^c(a_D)^d]/[(a_A)^a(a_B)^b]$.

Let us next assume that the reactants, A and B, and the products, C and D, initially all have activities equal to 1. Thus $[(a_C)^c(a_D)^d]/[(a_A)^a(a_B)^b]$ is 1, so the logarithm of the activity term is zero. Equation (6.5) indicates that for a moles of A plus b moles of B reacting to form c moles of C plus d moles of D under these conditions, the Gibbs free energy change ΔG is ΔG^*. If the equilibrium constant K for the reaction is >1, ΔG^* is negative ($\Delta G^* = -RT \ln K$; Eq. 6.6), so the reaction in the forward direction is spontaneous for this case of unit activity of reactants and products. For $K > 1$ we would expect the reaction to proceed in the forward direction because at equilibrium the products are then favored over the reactants—in the sense that $[(a_C)^c(a_D)^d] > [(a_A)^a(a_B)^b]$ at equilibrium. On the other hand, if $K < 1$, ΔG^* is positive, and such a reaction would not proceed spontaneously in the forward direction for the given initial condition of unit activity of all reactants and products.

Let us now consider the units of G and ΔG. G is an extensive variable, i.e., it depends on the extent or size of the system and is obtained by summing its values throughout the whole system (see Appendix IV). Specifically, Eq. (6.1) ($G = \sum_j n_j \mu_j$) indicates that the Gibbs free energy is the sum, over all species present, of the number of moles of species j (n_j, an extensive variable) times the energy per mole of species j (μ_j, an intensive variable, i.e., a quantity that can be measured at some point in a system, such as is the case for T, P, $\bar{V}_j$, and μ_j). Hence G has the units of energy. On the other hand, Eq. (6.5) indicates that ΔG has the same units as RT—namely, energy per mole (the logarithm is dimensionless). To help us out of this apparent dilemma, let us reconsider the conventions used in Eq. (6.2), $aA + bB \rightleftharpoons cC + dD$. We usually write a chemical reaction using the smallest possible integers for a, b, c, and d, not the actual number of moles reacting. In fact, in the equations describing nearly all biochemical reactions, either or both a and c are unity, e.g., ADP + phosphate $\rightleftharpoons$ ATP $+ H_2O$. Another convention is to express the Gibbs free energy change per mole of a particular species, e.g., per mole of a certain reactant ($\Delta G/a$) or per mole of a certain product ($\Delta G/c$). Therefore, when a or c represents 1 mol, ΔG has the same magnitude, whether as energy or as energy per mole. When we use specific values for ΔG or ΔG^* to describe chemical reactions, we will indicate on what basis we are using the Gibbs free energy, e.g., "ΔG per mole of ATP formed."

What are the numerical values of ΔG^* per mole of reactant A or product C for Ks of 100 and 0.01? Because RT is 2.48 kJ mol^{-1} at 25°C and $\ln = 2.303 \log$, an equilibrium constant of 100 corresponds to a $\Delta G^*/a$ or $\Delta G^*/c$ given by Eq. (6.6) of $-(2.48$ kJ mol$^{-1})$ (2.303) log(100), or -11.4 kJ mol^{-1} for a moles of reactant A or c moles of product C, and a K of 0.01 leads to a $\Delta G^*/a$ or $\Delta G^*/c$ of 11.4 kJ mol^{-1}. In the former case Eq. (6.5) indicates that 11.4 kJ of energy per mole of the reactant or product (assuming a or c is 1 mol) is released, and in the latter case the same amount of energy per mole is required, when the reaction proceeds in the forward direction beginning with activities of 1 for all reactants and products.

B. Interconversion of Chemical and Electrical Energy

To help understand how chemical energy can be converted into electrical energy, and vice versa, we must reexamine the properties of both chemical reactions and the movement of charged species. Let us first consider a chemical reaction such as the dissociation of sodium chloride: $NaCl \rightleftharpoons Na^+ + Cl^-$. Although two charged species are produced upon dissociation of NaCl, no net change in the electrical components of the chemical potentials of Na^+ plus Cl^- occurs. In other words, the electrical term $z_j FE$ for Na^+ ($z_{Na} = 1$) is balanced by an opposite change in the electrical component of μ_{Cl} ($z_{Cl} = -1$). Next, let us consider the following type of reaction: $Ag_s \rightleftharpoons Ag^+ + e^-$, i.e., the dissociation of solid silver to an ion plus an electron. Again, no net change in the net electrical contribution to the two chemical potentials occurs for the dissociation as written. However, the production of an electron opens up other possibilities because electrons can be conducted to regions where the electrical potential may be different. Such reactions, in which electrons are produced in one region and then conducted to regions of different electrical potential, allow for the interconversion of chemical and electrical energy. The electron-producing and electron-consuming reactions are referred to as *electrode*, or *half-cell*, reactions and occur in batteries as well as in the electron transfer pathways located in chloroplast and mitochondrial membranes.

To elaborate on energy conversion, let us consider mixing ferrous (Fe^{2+}) and cupric (Cu^{2+}) ions in an aqueous solution; we will assume that the common anion is Cl^-. A chemical reaction occurs in which the products are ferric (Fe^{3+}) and cuprous (Cu^+) ions. Because this is a spontaneous process, the Gibbs free energy decreases; also, the reaction is exothermic, so heat is evolved. Next, let us consider the electrode reaction, $Fe^{2+} \rightleftharpoons Fe^{3+} + e^-$. If the electrons produced in such a half-cell initially containing only Fe^{2+} can be conducted by a wire to the Cu^{2+} ions (Fig. 6-2), another electrode reaction can occur

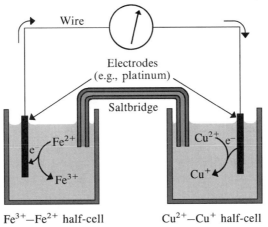

Figure 6-2. Two half-cells, or redox couples, connected by a wire and a saltbridge to complete the electrical circuit. Electrons donated by Fe^{2+} to one electrode are conducted by the wire to the other electrode where they reduce Cu^{2+}. Both electrodes (couples) are necessary before electrons can flow. A saltbridge, which provides a pathway along which ions can move and so helps maintain electroneutrality by avoiding the buildup of charge in either half-cell, often contains agar and KCl; the latter minimizes the diffusion potentials at the junctions between the saltbridge and the solutions in the beakers (see Chapter 3).

in a second beaker initially containing only Cu^{2+}—namely, $Cu^{2+} + e^- \rightleftharpoons Cu^+$. Except for heat evolution, the net result in the solutions is the same as occurs by mixing Fe^{2+} and Cu^{2+}. When the electrons move in the conductor, however, they can be used to do various types of electrical work, e.g., powering a direct-current electrical motor or a light bulb. Such an arrangement provides a way of converting the change in Gibbs free energy of the two spontaneous half-cell reactions into electrical energy that can be used for performing work. Indeed, the important aspect for obtaining electrical work from the two half-cell reactions is the conducting pathway between them.

When electrons are moved to a lower electrical potential ($\Delta E < 0$), e.g., by using the chemical energy in a battery, their electrical energy ($z_j F E$, where $z_j = -1$ for an electron) increases. We can use this increase in the electrical energy of the electrons to power a chemical reaction when the electrons subsequently move spontaneously to higher E. In photosynthesis, light energy is used to move electrons toward lower electrical potentials, thereby setting up a spontaneous flow of electrons in the opposite direction using the electron transfer components introduced in Chapter 5. This latter, energetically downhill, spontaneous electron movement is harnessed to drive the photophosphorylation reaction, $ADP + phosphate \rightleftharpoons ATP + H_2O$, in the forward direction, thereby storing chemical energy.

Now let us consider the interconversion of chemical and electrical energy in more formal terms. Suppose that n moles of electrons ($z_j = -1$) are transferred from one region to another where the electrical potential differs by ΔE, e.g., from one half-cell to another. As noted in Chapter 3, the charge carried by a mole of protons is Faraday's constant (F); hence the total charge moved in the current case is $-nF$. Electrical work equals the charge transported times the electrical potential difference through which it moves (ΔE). The change in the electrical energy of n moles of electrons therefore is $-nF\Delta E$. This can be converted to an equal change in Gibbs free energy (ΔG):

$$\Delta G = -nF\Delta E \qquad (6.7)$$

We note that n is usually expressed with respect to the compound of interest, for instance, 2 mol of electrons is used to reduce 1 mol of $NADP^+$; because n is dimensionless, ΔG has the units of energy mol^{-1}, just as for the ΔG of chemical reactions.

According to Eq. (6.7), the amount of Gibbs free energy stored or released is directly proportional to the difference in electrical potential across which the electrons move. Moreover, this equation indicates that the flow of electrons toward more positive electrical potentials ($\Delta E > 0$) corresponds to a decrease in the free energy ($\Delta G < 0$), and so the transfer proceeds spontaneously. We should emphasize that two half-cells are necessary to get a ΔE and thus a ΔG for electron transfer (see Fig. 6-2). We will apply these free energy considerations to the energetics of electrons moving from molecule to molecule in the electron transfer pathways of chloroplasts and mitochondria.

C. Redox Potentials

Many organic compounds involved in photosynthesis accept or donate electrons (see Table 5-3). The negatively charged electrons spontaneously flow toward more positive electrical potentials, which are measured by redox potentials for the components involved with electron flow in chloroplast lamellae or the inner membranes of mitochondria. Redox potentials are a measure of the relative chemical potential of electrons accepted or donated by a particular type of molecule. The oxidized form plus the reduced form of each electron transfer component can be regarded as an electrode, or half-cell. Such a half-cell can interact with other electron-accepting and electron-donating molecules in the membrane, in which case the electrons spontaneously move toward the component with the higher redox potential.

We can represent electron acceptance or donation by some species as a general electrode (half-cell) reaction:

$$\text{Oxidized form} + q\text{e}^- \rightleftharpoons \text{reduced form} \tag{6.8}$$

where q is a dimensionless parameter indicating the number of electrons transferred per molecule; oxidized and reduced are different forms of the same species—e.g., $NADP^+$ is an oxidized form and NADPH represents the corresponding reduced form (q is 2 in this case). Like any other chemical reaction, an oxidation–reduction reaction such as Eq. (6.8) has a change in Gibbs free energy associated with it when the reactants are converted to products. Thus oxidation–reduction, or "redox," reactions can be described by the relative tendency of the redox system, or *couple* (the oxidized plus the reduced forms of the compound), to proceed in the forward direction, which for Eq. (6.8) means accepting electrons.

Redox reactions are more conveniently described in terms of relative electrical potentials instead of the equivalent changes in Gibbs free energy. The electrons in Eq. (6.8) come from or go to some other redox couple, and whether or not the reaction in Eq. (6.8) proceeds in the forward direction depends on the relative electrical potentials of these two couples. Therefore, a specific electrical potential is assigned to a couple accepting or donating electrons, a value known as its *redox potential*. This oxidation–reduction potential is compared with that of another couple to predict the direction for spontaneous electron flow when the two couples interact—electrons spontaneously move toward higher redox potentials. The redox potential of species j, E_j, is defined as

$$E_j = E_j^* - \frac{RT}{qF} \ln \frac{(\text{reduced}_j)}{(\text{oxidized}_j)} \tag{6.9}$$

where E_j^* is an additive constant, q is the number of electrons transferred (the same q as in Eq. 6.8), and (reduced$_j$) and (oxidized$_j$) refer to the activities of the two different redox states of species j. Equation (6.9) indicates that the oxidation–reduction potential of a particular redox couple is determined by the ratio of the reduced form to the oxidized form plus an additive constant, a quantity that we consider next.

An electrical circuit is formed when two electrodes (half-cells) are connected and pathways for electron flow are provided (see Fig. 6-2). Because the sum of the electrical potential drops (voltage changes) around a circuit is zero, we can determine the half-cell potential on an absolute basis for a particular electrode if the potential of some standard reference electrode is known. By international agreement, the E_j^* of a hydrogen half-cell ($\frac{1}{2}H_{2gas} \rightleftharpoons H^+ + e^-$) is arbitrarily set equal to zero for an activity of hydrogen ions of 1 molal (m) equilibrated with hydrogen gas at a pressure of 1 atm, i.e., $E_H^* = 0$ V. Fixing the zero level of the electrical potential for the hydrogen half-cell removes the arbitrary nature of redox potentials for all half-cells because the redox potential for any species can be determined relative to that of the hydrogen electrode. We will replace E_j^* in Eq. (6.9) by E_j^{*H} to emphasize the convention of referring electrode potentials to the standard hydrogen electrode.[2]

---

[2] The hydrogen half-cell is not very convenient for routine laboratory usage—indeed, 1 m H^+ and 1 atm H_2 can be rather dangerous. Hence secondary standards are used, e.g., mercury/mercurous (calomel) or silver/silver chloride electrodes, which have midpoint redox potentials of 0.244 and 0.222 V, respectively.

According to Eq. (6.9), the larger the ratio (reduced$_j$)/(oxidized$_j$), the more negative the redox potential becomes. Because electrons are negatively charged, a lower E_j corresponds to higher energies for the electrons. Thus the further Eq. (6.8) is driven in the forward direction, the more the energy that is required to reduce species j. Likewise, the larger (reduced$_j$)/(oxidized$_j$), the higher the electrical energy of the electrons that the reduced form can donate. When (reduced$_j$) = (oxidized$_j$), $E_j = E_j^{*H}$ by Eq. (6.9). E_j^{*H} is hence commonly referred to as the *midpoint redox potential* (see the values in Table 5-3 for the E_j^{*H} of some components involved with electron transfer in chloroplasts). For certain purposes, such as estimating the energy available between redox couples, knowledge of the midpoint redox potentials may be sufficient, as we will show later in this chapter.

II. Biological Energy Currencies

In photosynthesis photons are captured, initiating an electron flow leading both to the production of NADPH and to a coupled process whereby ATP is formed. Light energy is thereby converted into chemical energy by the formation of a phosphoanhydride (ATP) in an aqueous environment. Light energy is also converted into electrical energy by providing a reduced compound (NADPH) under oxidizing conditions. ATP and NADPH are the two energy storage compounds, or "currencies," that we will consider in this section. Both occur as ions, both can readily diffuse around within a cell or organelle, and both can carry appreciable amounts of energy under biological conditions. In addition to the use of ATP in processes such as active transport and muscle contraction, the chemical energy stored in ATP is also used in certain biosynthetic reactions involving the formation of anhydrous links, or bonds, in the aqueous milieu of a cell. The relatively high atmospheric levels of O_2 ensure that appreciable amounts of this strong oxidizing agent will be present in most biological systems; a reduced compound such as NADPH is thus an important currency for energy storage. We will discuss these two compounds in turn, after briefly considering the difference between ATP and NADPH as energy currencies.

Redox couples are assigned a relative electrical energy, whereas chemical reactions have a specific chemical energy. In a chemical reaction certain reactants are transformed into products, and the accompanying change in Gibbs free energy can be calculated. This change in chemical energy does not need to depend on any other chemical species. For instance, if the concentrations—strictly speaking, the chemical activities—of ADP, phosphate, and ATP as well as certain other conditions (e.g., temperature, pH, Mg^{2+} concentration, and ionic strength) are the same in different parts of an organism, then the Gibbs free energy released upon the hydrolysis of a certain amount of ATP to ADP and phosphate will be the same in each of the different locations. However, an oxidation–reduction couple must donate electrons to, or accept electrons from, another redox system, and the change in electrical energy depends on the difference in the redox potential between the two couples. Thus the amount of electrical energy released when NADPH is oxidized to $NADP^+$ depends on the redox potential of the particular couple with which NADPH interacts.

A. ATP—Structure and Reactions

To help understand the bioenergetics of chloroplasts and mitochondria, we need to know how much energy is stored in ATP, that is, the difference between its chemical potential and that of the reactants (ADP and phosphate) used in its formation. We must then identify reactions that have a large enough free energy decrease to drive the ATP synthesis reaction in the energetically uphill direction; this will lead us to a consideration of the energetics of electron flow in organelles—topics that we will discuss in the next two sections. Our immediate concern is (1) the chemical reaction describing ATP formation, (2) the associated change in Gibbs free energy for that reaction, and (3) the implications of the substantial amount of energy storage in ATP.

ADP, ATP, and phosphate all occur in a number of different charge states in aqueous solutions. Moreover, all three compounds can interact with other species, notably Mg^{2+} and Ca^{2+}. Thus many different chemical reactions describe ATP formation. A predominant reaction occurring near neutral pH in the absence of divalent cations is

where adenosine is adenine (top portion) esterified to the 1′ position of the sugar ribose (bottom portion), i.e.,

The attachment of adenosine to the phosphates in ADP and ATP—and in $NADP^+$ as well as in FAD—is by means of an ester linkage with the hydroxymethyl group on the 5′ position of the ribose moiety (Fig. 6-3).

Equation (6.10) indicates a number of features of ATP production. For instance, the formation of ATP from ADP plus phosphate is a dehydration; the reversal of Eq. (6.10), in which the phosphoanhydride is split with the incorporation of water, is known as ATP hydrolysis. Because Eq. (6.10) contains H^+, the equilibrium constant depends on the pH, $-\log(a_{H^+})$. Moreover, the fractions of ADP, phosphate, and ATP in various states of ionization depend on the pH. Near pH 7 about half of the ADP molecules are doubly charged and half are triply charged, the latter form being indicated in Eq. (6.10) (for simplicity, we are ignoring

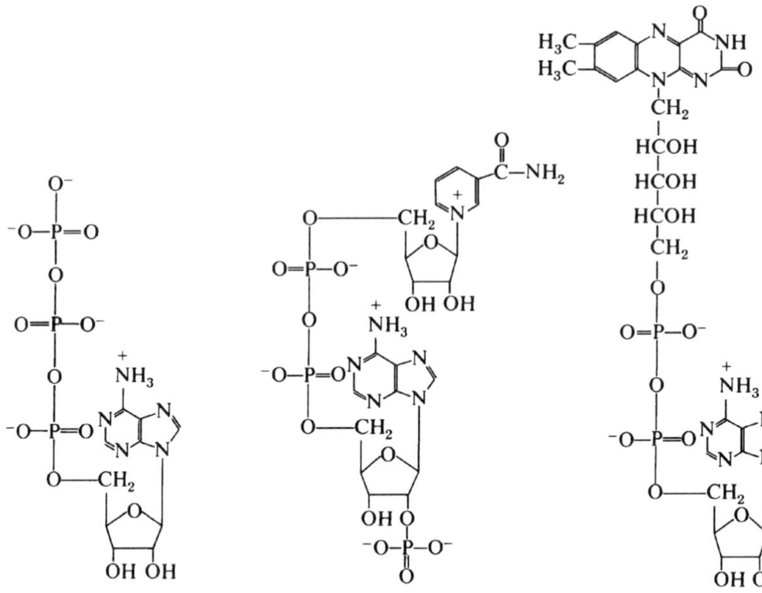

Figure 6-3. Structures of three molecules important in bioenergetics. The dissociations and bindings of protons indicated in the figure are appropriate near pH 7. Note the similarity among the molecules.

the charge due to the extra proton bound to an adenine nitrogen, which gives that part of the ADP and ATP molecules a single positive charge at pH 7). Likewise, ATP at pH 7 is about equally distributed between the forms with charges of -3 and -4. Because of their negative charges in aqueous solutions, both ADP and ATP can readily bind positive ions, especially divalent cations such as Mg^{2+} or Ca^{2+}. A chelate is formed such that Mg^{2+} or Ca^{2+} is held electrostatically between two negatively charged oxygen atoms on the same molecule (consider the many—O^-s occurring on the chemical structures indicated in Eq. 6.10). Also, inorganic phosphate can interact electrostatically with Mg^{2+} and other divalent cations, further increasing the number of complexed forms of ADP, ATP, and phosphate.

The activities (or concentrations) of a species in all of its ionization states and complexed forms are generally summed to obtain the total activity (or concentration) of that species. The number of relations and equilibrium constants needed to describe a reaction such as ATP formation is then reduced to one; i.e., a separate equilibrium constant is not needed for every possible combination of ionization states and complexed forms of all the reactants and products. Using this convention, we can replace Eq. (6.10) and many others like it, which also describe ATP formation, by the following general reaction for the phosphorylation of ADP:

$$ADP + phosphate \rightleftharpoons ATP + H_2O \tag{6.11}$$

We will return to ATP formation as represented by Eq. (6.11) after briefly commenting on two important conventions used in biochemistry. First, most equilibrium constants for biochemical reactions are defined at a specific pH, usually pH 7 ($a_{H^+} = 10^{-7}$ M). At a constant pH the activity of H^+ does not change. Thus H^+ does not need to be included as a reactant or a product in the expression for the change in Gibbs free energy (Eq. 6.5). In other words, the effect of H^+ in relations such as Eq. (6.10) is incorporated into the equilibrium constant, which itself generally depends on the pH. Second, biological reactions such as ATP formation usually take place in aqueous solutions for which the concentration of water does not change appreciably. (The concentrations of other possible reactants and products are much less than that of water.) Hence the a_{H_2O} term in relations such as Eq. (6.11) is also usually incorporated into the equilibrium constant. We can illustrate these points concerning a_{H^+} and a_{H_2O} by specifically considering ATP formation as described by Eq. (6.10). For the reaction as written, the equilibrium constant K is equal to $[(a_{ATP})(a_{H_2O})]/[(a_{ADP})(a_{phosphate})(a_{H^+})]$; hence $(a_{ATP})/[(a_{ADP})(a_{phosphate})] = (a_{H^+})K/(a_{H_2O})$. For a dilute aqueous solution, a_{H_2O} is essentially constant during the reaction; therefore, $(a_{H^+})K/(a_{H_2O})$—which is conventionally called the equilibrium constant in biochemistry—has a fixed value at a given pH, e.g., at pH 7, $(a_{H^+})K/(a_{H_2O}) = (10^{-7}M)K/(a_{H_2O}) = K_{pH7}$.

Next, we will specifically consider the equilibrium constant for ATP formation under biological conditions. Using the previous conventions for H^+ and H_2O, an equilibrium constant for Eq. (6.11) at pH 7 (K_{pH7}) is

$$K_{pH7} = \frac{[ATP]}{[ADP][phosphate]} \cong 5 \times 10^{-6}\,M^{-1} \quad \text{at } 25°C \qquad (6.12)$$

where the total concentration of each species involved is indicated in brackets; i.e., it is experimentally more convenient to measure K_{pH7} using concentrations (indicated by brackets) instead of activities (indicated by parentheses).[3]

When activities of ions $(a_j = \gamma_j c_j$; Eq. 2.5) are replaced by concentrations $[c_j]$, the resulting equations for equilibrium constants or free energy changes apply only to a particular ionic strength $(\frac{1}{2}\sum_j c_j z_j^2)$ because the activity coefficients of ions (γ_j) can markedly depend on ionic strength (see Eq. 3.3). The value for K_{pH7} given in Eq. (6.12) is suitable for ionic strengths close to 0.2 M (200 mol m^{-3}), which is an ionic strength that can occur *in vivo* (a 0.05 M increase or decrease in ionic strength changes K in the opposite direction by about 10%). The magnitude of the equilibrium constant for ATP formation also depends on the concentration of Mg^{2+}. The value of K_{pH7} in Eq. (6.12) is appropriate for 10 mM Mg^{2+}, a concentration similar to that in many plant cells (a 5 mM increase or decrease in Mg^{2+} changes K in the opposite direction by 10–20%). An equilibrium constant also depends on temperature, with K_{pH7} for ATP formation increasing by 2–4% per °C. A large effect on K is produced by pH, which may not be near 7 and often is unknown in a cell. The equilibrium constant for ATP formation increases about threefold as the pH is lowered by 1 unit from pH 7 and decreases sevenfold as it is raised by 1 unit.

B. Gibbs Free Energy Change for ATP Formation

The energetics of a reaction such as ATP formation is summarized by its Gibbs free energy change, ΔG. For a general chemical reaction, Eq. (6.5) indicates that $\Delta G = \Delta G^* + RT$ $\ln\{[(a_C)^c(a_D)^d]/[(a_A)^a(a_B)^b]\}$, where ΔG^* is $-RT \ln K$ (Eq. 6.6). For the current case, the reactant A is ADP, B is phosphate, the product C is ATP, and the equilibrium constant is given by Eq. (6.12). Therefore, our sought-after free energy relationship describing ATP

---

[3] Because the equilibrium constant for ATP formation is quite small (see Eq. 6.12) and is sensitive to temperature, pH, Mg^{2+}, and ionic strength, measured values of K vary considerably. Actually, instead of K, the standard Gibbs free energy, ΔG^*, is usually determined ($\Delta G^* = -RT \ln K$; Eq. 6.6). Values of ΔG^* for ATP formation range from 28 to 45 kJ mol^{-1}, and a value near 30 kJ mol^{-1} is appropriate at pH 7, 25°C, 10 mM Mg^{2+}, and an ionic strength of 0.2 M.

formation is

$$\Delta G = -RT \ln(K_{pH7}) + RT \ln \frac{[ATP]}{[ADP][phosphate]} \qquad (6.13a)$$

which at pH 7 and 25°C becomes

$$\Delta G \cong 30 + 5.71 \log \frac{[ATP]}{[ADP][phosphate]} \quad kJ \, (mol \, ATP)^{-1} \qquad (6.13b)$$

where $\ln = 2.303 \log$, $2.303 RT = 5.71 \, kJ \, mol^{-1}$ at 25°C (Appendix I), $K_{pH7} = 5 \times 10^{-6}$ M^{-1}, and $-(5.71 \, kJ \, mol^{-1}) \log(5 \times 10^{-6}) = 30 \, kJ \, mol^{-1}$.

ATP usually does not tend to form spontaneously because the Gibbs free energy change for the reaction is generally quite positive (Eq. 6.13). In fact, the energy required for the phosphorylation of ADP is large compared with the free energy changes for most biochemical reactions. Stated another way, much energy can be stored by converting ADP plus phosphate to ATP. Although ATP in an aqueous solution is thermodynamically unstable, meaning that its hydrolysis can release a considerable amount of Gibbs free energy, ATP can still last for a long enough time in cells to be an important energy currency, i.e., it is kinetically stable. In particular, ATP is generally not hydrolyzed very rapidly unless the appropriate enzymes necessary for its use in certain biosynthetic reactions or other energy-requiring processes are present (for long-term energy storage, plants use carbohydrates such as the polysaccharide starch).

We will next estimate the changes in Gibbs free energy that might be expected for photophosphorylation under physiological conditions. For purposes of calculation, we will assume that in unilluminated chloroplasts the concentration of ADP is 1.2 mM, that of phosphate is 5 mM, and that of ATP is 0.2 mM (1 mM = 1 mol m^{-3}). From Eq. (6.13b), the free energy change required to form ATP then is

$$\Delta G = 30 + 5.71 \log \frac{(0.2 \times 10^{-3} \, M)}{(1.2 \times 10^{-3} \, M)(5 \times 10^{-3} \, M)}$$
$$= 39 \, kJ \, (mol \, ATP)^{-1}$$

The change in Gibbs free energy required is positive, indicating that energy must be supplied to power photophosphorylation. Moreover, the energy necessary depends in a predictable way on the concentrations of the reactants and the product. After a period of time in the light, ATP may increase to 1.2 mM with a concomitant decrease in ADP to 0.2 mM and in phosphate to 4 mM. The free energy required for photophosphorylation under these conditions is 48 kJ (mol ATP)$^{-1}$. Consequently, the further that photophosphorylation goes to completion, the greater the energy required to form more ATP.

The high energy of ATP relative to ADP plus phosphate is not the property of a single bond but of the local configuration in the ATP molecule, which we can appreciate by considering the phosphorus atoms in ADP, ATP, and phosphate. Phosphorus is in group V of the third period of the periodic table and has five electrons in its outermost shell. It can enter into a total of five bonds with four oxygen atoms, the bonding to one O being a double bond (consider the structures in Eq. 6.10). In inorganic phosphate all four bonds are equivalent, so four different structures for phosphate exist in resonance with each other. The terminal P of ADP has only three resonating forms because one of the O's is connected to a second phosphorus atom and does not assume a double bond configuration (see Eq. 6.10). When inorganic phosphate is attached to this terminal P of ADP to form ATP, a resonating form is lost from both ADP and phosphate. Configurations having more resonating structures are in general more probable or stable (lower in energy), so energy must be supplied to form ATP from ADP plus phosphate with an accompanying loss of two resonating forms.

We next quantitatively examine some of the ways in which ATP can be used as a free energy currency. Each of our four examples will relate to a different variable term in the chemical potential ($\mu_j = \mu_j^* + RT \ln a_j + \bar{V}_j P + z_j FE + m_j gh$; Eq. 2.4). To transfer a mole of a neutral compound against a 10-fold increase in activity requires $2.303 RT \log 10$, or 5.7 kJ of Gibbs free energy at 25°C, and it takes 11.4 kJ for a 100-fold increase and 17.1 kJ for a 1000-fold increase in activity. To move a monovalent cation from one side of a membrane to the other, where it has the same concentration but the electrical potential is 0.1 V higher, requires 9.6 kJ mol^{-1}($F = 96.49$ kJ mol^{-1} V^{-1}; Appendix I). ATP usually supplies at least 40 kJ mol^{-1} when hydrolyzed and can act as the Gibbs free energy source for the active transport of solutes across membranes toward regions of higher chemical potential. ATP is also the free energy currency for the contraction of muscles. The ATP-driven contraction of the muscles surrounding the left ventricle of the human heart can increase the blood pressure within it by 20 kPa (0.2 bar or 150 mm Hg). This increases the chemical potential of the water in the blood (i.e., the $\bar{V}_w P$ term), which causes the blood to flow out to the aorta and then to the rest of the circulatory system toward lower hydrostatic pressures. Pressure-driven flow is an efficient way to move fluids; e.g., it takes only 0.02 kJ of Gibbs free energy to increase the pressure of 10^{-3} m^3 (1 liter) of water by 20 kPa. As an example of gravitational work that can be mediated by ATP, the increase in Gibbs free energy as a 50-kg person climbs 100 m is 49 kJ. ATP is an extremely useful cellular energy currency because of its large free energy release upon hydrolysis—about 40–50 kJ mol^{-1} (10–12 kcal mol^{-1})—and because of its convenient form as a relatively abundant ion.

C. NADP$^+$–NADPH Redox Couple

Another class of energy storage compounds consists of redox couples such as NADP$^+$– NADPH. The reduced form, NADPH, is produced by noncyclic electron flow in chloroplasts. Photosynthesis in bacteria makes use of a different redox couple, NAD$^+$–NADH. The reduced member of this latter couple causes an electron flow in mitochondria and an associated formation of ATP. NAD is nicotinamide adenine dinucleotide and differs from NADP by not having a phosphate esterified to the 2$'$ hydroxy group of the ribose in the adenosine part of the molecule (see Fig. 6-3). Our current discussion will focus on the NADP$^+$–NADPH couple, but the same arguments and also the same midpoint redox potential apply to the NAD$^+$–NADH couple.

The reduction of a molecule of NADP$^+$ involves its acceptance of two electrons. Only the nicotinamide portion (illustrated in Eq. 6.14) of NADP$^+$ is involved in accepting the electrons. The half-cell reaction describing this reduction is

NADP$^+$ + 2e$^-$ + 2H$^+$ $\rightleftharpoons$ NADPH + H$^+$ (6.14)

where R represents a ribose attached at its 1C position to nicotinamide and at its 5C position by a pyrophosphate bridge

$$-O-\overset{\underset{\parallel}{O}}{\underset{\underset{O^-}{|}}{P}}-O-\overset{\underset{\parallel}{O^-}}{\underset{\underset{O}{\parallel}}{P}}-O-$$

to an adenosine having the 2′ hydroxy group of its ribose moiety esterified to an additional phosphate (see Fig. 6-3). Adenosine minus the ribose is called adenine, hence the name of nicotinamide adenine dinucleotide phosphate, or NADP. The reduction of $NADP^+$ involves the transfer of two electrons to the nicotinamide ring, plus the attachment of one H^+ to the *para* position (top of the ring for the NADPH indicated in Eq. 6.14). That is, two electrons are accepted by the $NADP^+$ molecule during its reduction, although only one of the two accompanying protons is attached to the reduced form, as Eq. (6.14) indicates.

A particular half-cell reaction such as Eq. (6.14) can accept or donate electrons. We quantitatively describe this by the redox potential for that reaction, as expressed by Eq. (6.9) $\{E_j = E_j^{*H} - (RT/qF)\ln [(\text{reduced}_j)/(\text{oxidized}_j)]\}$. We will use (NADPH) to represent the activity of all of the various ionization states and complexed forms of the reduced nicotinamide adenine dinucleotide phosphate, and ($NADP^+$) has an analogous meaning for the oxidized species of the $NADP^+$–NADPH couple. For redox reactions of biological interest, the midpoint (standard) redox potential is generally determined at pH 7. By using Eq. (6.9), in which the number q of electrons transferred per molecule reduced is 2, we can express the oxidation–reduction potential of the $NADP^+$–NADPH couple (Eq. 6.14) as

$$E_{NADP^+-NADPH} = E_{pH\,7}^{*H} - \frac{RT}{2F}\ln\frac{(\text{NADPH})}{(\text{NADP}^+)} \tag{6.15a}$$

which at 25°C becomes

$$E_{NADP^+-NADPH} = -0.32 - 0.030\log\frac{(\text{NADPH})}{(\text{NADP}^+)} \quad \text{V} \tag{6.15b}$$

where in Eq. (6.15b) ln has been replaced by 2.303 log, and the numerical value for $2.303RT/F$ of 0.0592 V at 25°C (Appendix I) has been used. The midpoint redox potential, $E_{pH\,7}^{*H}$, for the $NADP^+$–NADPH couple is −0.32 V (Table 5-3), a value achieved when (NADPH) = ($NADP^+$).

To determine whether electrons spontaneously flow toward or away from the $NADP^+$–NADPH couple, we must compare its redox potential with that of some other redox couple. As indicated previously, electrons spontaneously flow toward more positive redox potentials, whereas energy must be supplied to move electrons in the energetically uphill direction of algebraically decreasing redox potentials. We will next examine the redox potentials of the various redox couples involved in electron flow in chloroplasts and then in mitochondria.

III. Chloroplast Bioenergetics

In Chapter 5 we introduced the various molecules involved with electron transfer in chloroplasts, together with a consideration of the sequence of electron flow between components. Now that the concept of redox potential has been presented, we will resume our discussion of electron transfer in chloroplasts. We will compare the midpoint redox potentials of the various redox couples not only to help understand the direction of spontaneous electron flow but also to see the important role of light absorption in changing the redox properties of the trap chl. Also, we will consider how ATP formation is coupled to electron flow.

A. Redox Couples

Although the ratio of reduced to oxidized forms of species j affects its redox potential $\{E_j = E_j^{*H} - (RT/qF)\ln[(\text{reduced}_j)/(\text{oxidized}_j)]$; Eq. 6.9}, the actual activities of the two forms are generally unknown *in vivo*. Moreover, the value of the local pH (which can affect E_j^{*H}) is also generally unknown. Consequently, midpoint (standard) redox potentials determined at pH 7 are usually compared to predict the direction for spontaneous electron flow in the lamellar membranes of chloroplasts. We will assume that free energy is required to transfer electrons to a compound with a more negative midpoint redox potential, whereas the reverse process can go on spontaneously.

The absorption of a photon markedly affects the redox properties of a pigment molecule. An excited molecule such as trap chl* has an electron in an antibonding orbital (see Chapter 4). Less energy is needed to remove such an electron from the excited molecule than when that molecule is in its ground state. Thus the electronically excited molecule is a better electron donor (reducing agent) than is one in the ground state because it has a more negative redox potential. Once the electron is removed, this oxidized molecule (trap chl$^+$) becomes a very good electron acceptor (oxidizing agent). Such electron acceptance, Trap chl$^+$ + D → trap chl + D$^+$ (Eq. 5.6), involves the ground state of the chlorophyll molecule, which has a positive redox potential (see Table 5-3 and Fig. 6-4). The electronic state of trap chl that donates an electron is therefore an excited state with a negative redox potential, and the ground state with its positive redox potential can readily accept an electron. In short, the absorption of light energy transforms trap chls from couples with positive redox potentials to couples with negative redox potentials. In this section we will estimate the redox potential spans at each of the two photosystems in chloroplasts and then diagram the overall pattern of photosynthetic electron flow.

Following light absorption by Chl a or an accessory pigment feeding into Photosystem II, the excitation migrates to P$_{680}$, where an electron transfer reaction takes place (Trap chl* + A → trap chl$^+$ + A$^-$; Eq. 5.5). The electron removed from P$_{680}^*$ is replaced by one coming from water, which results in O$_2$ evolution (for energetic considerations, we write Eq. 5.8 as $H_2O \rightleftharpoons \frac{1}{2}O_2 + 2H$). The water–oxygen couple has a midpoint redox potential of 0.82 V at 25°C, pH 7, and an O$_2$ pressure of 1 atm (Fig. 6-4). Water oxidation and the accompanying O$_2$ evolution follow spontaneously after the photochemistry at the reaction center of Photosystem II has led to P$_{680}^+$. Thus the required oxidant for water, i.e., P$_{680}^+$ or some intermediate oxidized by it, must have a redox potential more positive than 0.82 V for the electron to move energetically downhill from water to the trap chl$^+$ in the reaction center of Photosystem II. As indicated in Table 5-3, the redox potential of P$_{680}$ in the ground state is about 1.10 V.

The electron removed from P$_{680}^*$ goes to pheophytin, from which it moves to two intermediate quinones and then to the plastoquinone pool. The midpoint redox potentials are -0.61 V for pheophytin, -0.05 to -0.25 V for the first plastoquinone (Q$_A$), about 0.10 V for the second one (Q$_B$), and 0.11 V for plastoquinone A. The negatively charged electron spontaneously moves toward higher redox potentials, in this case from pheophytin to intermediate quinones to plastoquinone A. The electrical potential span in Photosystem II

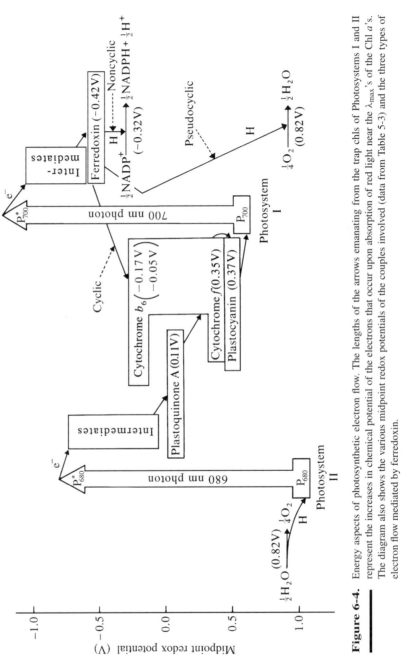

Figure 6-4. Energy aspects of photosynthetic electron flow. The lengths of the arrows emanating from the trap chls of Photosystems I and II represent the increases in chemical potential of the electrons that occur upon absorption of red light near the λ_{max}'s of the Chl a's. The diagram also shows the various midpoint redox potentials of the couples involved (data from Table 5-3) and the three types of electron flow mediated by ferredoxin.

is thus from a redox potential of 1.10 V for the ground state of P_{680} to about -0.70 V for the excited state of this trap chlorophyll, which must have a redox potential that is more negative than the -0.61 V of the pheophytin couple, or about 1.80 V overall.

The energy required to move an electron in the energetically uphill direction toward lower redox potentials in Photosystem II is supplied by a photon (Fig. 6-4). Photosystem II can be excited by 680-nm light (as well as by other wavelengths, this value being near the λ_{max} for the red band of its P_{680}). From Eq. (4.2a) ($E = hc/\lambda_{vacuum}$) and the numerical value of hc (1240 eV nm, Appendix I), the energy of 680-nm light is (1240 eV nm)/(680 nm), or 1.82 eV per photon. Such a photon could move an electron across 1.82 V, which is approximately the redox potential span of 1.80 V estimated for Photosystem II.[4]

We can similarly analyze the energetics for Photosystem I, where the trap chl is P_{700}. The redox potential span across which electrons are moved is from the midpoint redox potential of 0.48 V for the P_{700}–P_{700}^+ couple (Table 5-3) to about -1.20 V for the redox couple representing the excited state, which has a more negative redox potential than the -1.05 V estimated for the monomeric form of chlorophyll that acts as the first acceptor. Thus the electrical potential span in Photosystem I is about 1.68 V. A photon at 700 nm, which is the λ_{max} for the red band of P_{700} in Photosystem I, has an energy of 1.77 eV, which is ample energy to move an electron between the redox couples representing the ground state of P_{700} and its excited state, the latter having a more negative redox potential than the first acceptor in Photosystem I. From the first acceptor, the electron spontaneously moves to a quinone (midpoint redox potential possibly near -0.80 V), then to three Fe-S centers and then to ferredoxin (Table 5-3). From ferredoxin to the next component in the noncyclic electron flow sequence, $NADP^+$, electrons go from -0.42 to -0.32 V (midpoint redox potentials of the couples; Fig. 6-4). Again, moving toward higher redox potentials is energetically downhill for electrons, so this step leading to the reduction of the pyridine nucleotide follows spontaneously from the reduced ferredoxin—a step catalyzed by the enzyme ferredoxin–$NADP^+$ oxidoreductase (see Table 5-3).

In noncyclic electron flow, two electrons originating in the water–oxygen couple with a midpoint redox potential of 0.82 V are moved to the redox level of -0.32 V for the reduction of one molecule of the $NADP^+$–NADPH couple. Because a midpoint redox potential of -0.32 V is more negative than most encountered in biology, NADPH can spontaneously reduce most other redox systems—reduced pyridine nucleotides are therefore an important energy currency. Moving electrons from 0.82 V to -0.32 V requires considerable free energy, which helps explain why light, with its relatively large amount of energy (see Table 4-1), is needed. We can calculate the actual free energy change for the overall process using Eq. (6.7) ($\Delta G = -nF\Delta E$):

$$\Delta G = -(2)(96.5 \text{ kJ mol}^{-1} \text{ V}^{-1})(-0.32 \text{ V} - 0.82 \text{ V})$$

$$= 220 \text{ kJ mol}^{-1} \text{ NADPH}$$

for the overall movement of 2 mol of electrons along the pathway for noncyclic electron flow—a process leading to the reduction of 1 mol of $NADP^+$.

The incorporation of CO_2 into a carbohydrate during photosynthesis requires three ATPs and two NADPHs (see Fig. 5-1). Using these energy currencies, CO_2 fixation can

---

[4] Actually, the amount of absorbed light energy and the accompanying changes in free energy available for decreasing the redox potential are not the same, i.e., the increase in internal energy U upon light absorption is generally not equal to the change in Gibbs free energy G (see Eq. IV.4a, Appendix IV; $G = U + PV - TS$). The magnitude of the increase in G caused by the absorption of a photon by chlorophyll depends on the level of illumination as well as on the various pathways competing for the deexcitation of trap chl*.

energetically proceed in the absence of light, so the steps of the reductive pentose cycle are often referred to as the *dark reactions* of photosynthesis (actually, because several of the enzymes are light-activated, not much CO_2 fixation occurs in the dark). ATP formation in chloroplasts requires about 48 kJ mol^{-1}; we have just indicated that 220 kJ is required to reduce a mole of NADP$^+$ using electrons coming from water (this is actually the Gibbs free energy change between standard states because midpoint redox potentials were used in our calculations); and the increase in Gibbs free energy per mole of CO_2 incorporated into a carbohydrate during photosynthesis is 479 kJ. Using these numbers, we can estimate the efficiency for free energy storage by the dark reactions. Dividing the energy stored per mole of CO_2 fixed, 479 kJ, by the energy input (3 mol ATP)[48 kJ (mol ATP)$^{-1}$] + (2 mol NADPH) [220 kJ (mol NADPH)$^{-1}$], which is 584 kJ, we find that the efficiency is [(479 kJ)/(584 kJ)] (100), or 82%! The dark reactions of photosynthesis are indeed extremely efficient.

Figure 6-4 incorporates the midpoint redox potentials of various components involved with photosynthetic electron transfer discussed both here and in Chapter 5 (see Table 5-3). The direction for spontaneous electron flow is to higher midpoint redox potentials (downward in Fig. 6-4); the absorption of light quanta with their relatively large energies corresponds to moving electrons toward higher energy. The role played by ferredoxin at the crossroads of cyclic, noncyclic, and pseudocyclic electron flow (see Fig. 5-14) is also illustrated in Fig. 6-4.

B. H$^+$ Chemical Potential Differences Caused by Electron Flow

In the previous chapter we indicated that the components involved with electron flow are situated in the lamellar membranes of chloroplasts such that they lead to a vectorial or unidirectional movement of electrons and protons (see Fig. 5-15). We now return to this theme and focus on the gradients in H$^+$ (protons) thus created. The difference in the chemical potential of H$^+$ from the inside to the outside of a thylakoid in the light acts as the energy source to drive photophosphorylation. This was first clearly recognized in the 1960s by Peter Mitchell, who received the 1978 Nobel prize in chemistry for his enunciation of what has become known as the *chemiosmotic* hypothesis for interpreting the relationship between electron flow, proton movements, and ATP formation.

Figure 6.5 indicates that the O_2-evolution step and the electron flow mediated by the plastoquinones and the Cyt b_6–f complex lead to an accumulation of H$^+$ inside a thylakoid in the light. This causes the internal H$^+$ concentration c_H^i or activity a_H^i to increase. These steps depend on the light-driven electron flow, which leads to electron movement outward across the thylakoid in each of the photosystems (see Fig. 5-15). Such movements of electrons out and protons in can increase the electrical potential inside the thylakoid (E^i) relative to that outside (E^o), allowing an electrical potential difference to develop across a thylakoid membrane. By the definition of chemical potential ($\mu_j = \mu_j^* + RT \ln a_j + z_j FE$; Eq. 2.4 with the pressure and gravitational terms omitted; see Chapter 3, Section I), the difference in chemical potential of H$^+$ across a membrane is

$$\mu_H^i - \mu_H^o = RT \ln a_H^i + FE^i - RT \ln a_H^o - FE^o \qquad (6.16a)$$

Incorporating the definitions of pH (pH $= -\log a_H = -\ln a_H/2.303$) and $E_M (E_M = E^i - E^o)$ into Eq. (6.16a), we obtain

$$\mu_H^i - \mu_H^o = -2.303RT \, pH^i + 2.303RT \, pH^o + FE_M \qquad (6.16b)$$

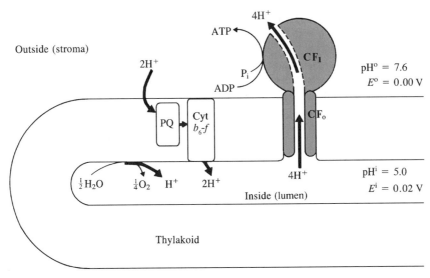

Figure 6-5. Energetics and directionality of the coupling between electron flow and ATP formation in chloroplasts, emphasizing the role played by H^+ (see also Fig. 5-15). The O_2 evolution from H_2O and
electron flow via plastoquinones (PQ) and the cytochrome b_6–f complex (Cyt b_6–f) lead to H^+
accumulation inside the thylakoid. This H^+ can move back out through a hydrophobic channel
(CF_o) and another protein factor (CF_1), leading to ATP formation.

which, using numerical values from Appendix I, at 25°C becomes

$$\mu_H^i - \mu_H^o = 5.71(pH^o - pH^i) + 96.5E_M \quad \text{kJ mol}^{-1} \tag{6.16c}$$

where E_M is in volts.

According to the chemiosmotic hypothesis (which might, more appropriately, be termed a
"transmembrane hypothesis"), the ATP reaction is driven in the energetically uphill direction
by protons moving out of the thylakoids in their energetically downhill direction. Although
a ratio of three H^+s per ATP had been widely accepted, the stoichiometry is apparently
four H^+s per ATP (Kobayashi *et al.*, 1995; Berry and Rumberg, 1996; van Walraven *et al.*,
1996). Because the formation of ATP in chloroplasts can require about 48 kJ (mol ATP)$^{-1}$,
the difference in chemical potential of H^+ must be at least (48 kJ mol^{-1})/(4) or 12 kJ
(mol H^+)$^{-1}$ to drive the reaction energetically using four H^+s per ATP. By Eq. (6.16c), such
an energy difference corresponds to a pH difference of (12 kJ mol^{-1})/(5.71 kJ mol^{-1}) or
2.1 pH units, or a difference in electrical potential of (12 kJ mol^{-1})/(96.5 kJ mol^{-1} V^{-1})
or 0.14 V. In turn, if the proton chemical potential gradient is established by coupling to
electron flow, then an energetically uphill movement of protons of at least 12 kJ mol^{-1}
requires an energetically even larger downhill flow of electrons. We will next examine the
evidence and the thermodynamic requirements for the various steps coupling electron flow
to ATP formation.

C. Evidence for Chemiosmotic Hypothesis

One of the most striking pieces of evidence in support of the chemiosmotic hypothesis was
obtained in the 1960s by Andre Jagendorf and Ernest Uribe. When chloroplast lamellae are

incubated in a solution at pH 4—in which case pH^i presumably attains a value near 4—and then rapidly transferred to a solution with a pH^o of 8 containing ADP and phosphate, the lamellae are capable of leading to ATP formation in the dark. Approximately 100 ATPs can be formed per Cyt f. When the difference in pH across the membranes is <2.5, essentially no ATP is formed by the chloroplast lamellae. This agrees with the energetic argument presented previously, where a minimum ΔpH of 2.1 pH units is required to lead to ATP formation. Also, if the pH of the external solution is gradually increased from 4 to 8 in the dark (over a period of tens of seconds), the protons "leak" out across the lamellar membranes, ΔpH is relatively small, and no ATP is formed.

The electrical term in the chemical potential of H$^+$ can also power ATP formation. For instance, when an E_M of 0.18 V is artificially created across lamellar membranes, ATP formation can be induced in the dark. This is consistent with our prediction that an electrical potential difference of at least 0.14 V is necessary. In chloroplast thylakoids, E_M in the light appears to be fairly low, e.g., near 0.02 V in the steady state (see Fig. 6-5). However, the electrical term seems to be the main contributor to $\Delta\mu_H$ for the first 1 or 2 s after chloroplasts are exposed to a high photosynthetic photon flux density (PPFD). The electrical component of the H$^+$ chemical potential difference can be large for the chromatophores of certain photosynthetic bacteria such as *Rhodopseudomonas spheroides*, for which E_M can be 0.20 V in the light in the steady state.

When chloroplasts are illuminated, electron flow commences, which causes μ_H^i within the thylakoids to increase relative to μ_H^o. We would expect a delay before $\Delta\mu_H$ given by Eq. (6.16) is large enough to lead to ATP formation. Indeed, a lag of a fraction of 1 s to a few seconds occurs before photophosphorylation commences at low PPFDs, and the lag is decreased by increasing the PPFD. We can also reason that a gradient in the chemical potential of H$^+$ will affect the movement of other ions. For instance, the light-induced uptake of H$^+$ into the thylakoids is accompanied by a release of Mg^{2+}, which can cause its stromal concentration to increase to 10 mM. This released Mg^{2+} can activate various enzymes involved with CO$_2$ fixation in the stroma, indicating that the ionic readjustments following light-dependent proton movements can act as a cellular control for biochemical reactions.

The chemical potential gradient of H$^+$ across the lamellar membranes can be dissipated in various ways, thus uncoupling electron flow from ATP formation. Compounds that accomplish this are called *uncouplers*. For instance, neutral weak bases [e.g., methyl amine (CH$_3$NH$_2$) or ammonia (NH$_3$)] can readily diffuse into the thylakoids and there combine with H$^+$. This lowers a_H^i and raises pH^i. Moreover, the protonated base (CH$_3$NH$_3^+$ or NH$_4^+$) cannot readily diffuse back out because it carries a charge. The uncoupler nigericin competitively binds H$^+$ and K$^+$; it can exchange K$^+$ outside for H$^+$ inside, which also tends to lower a_H^i (because nigericin can stoichiometrically lead to H$^+$ movement one way and K$^+$ the other, it acts as an H$^+$/K$^+$ antiporter). Detergents can remove certain membrane components, thus making the thylakoids leaky to H$^+$ and other ions, which also dissipates $\Delta\mu_H$. Such studies further show the importance of the H$^+$ chemical potential difference in leading to ATP formation.

D. Coupling of Flows

We will next reconsider the vectorial aspects of proton and electron flow (Figs. 5-15 and 6-5) and examine the associated energetics. We will discuss the structures involved in the coupling of ATP formation to proton flow. We will also consider the stoichiometry of the various flows with respect to the ATP and NADPH requirements of CO$_2$ fixation.

Let us start with the O_2-evolution step, $\frac{1}{2}H_2O \rightleftharpoons \frac{1}{4}O_2 + H^+ + e^-$ (essentially Eq. 5.8). To obtain H^+ inside thylakoids from the O_2-evolving step, we need (1) H_2O inside the thylakoids, which can readily diffuse in from the stroma; (2) an oxidant with a redox potential more positive than the midpoint redox potential of 0.82 V for the H_2O–O_2 couple, which is supplied by the P_{680}^+–P_{680} couple in Photosystem II with a midpoint redox potential of 1.10 V (Table 5-3); (3) a pathway for removing electrons, which is provided by Photosystem II, because it moves electrons from the inner to the outer side of the thylakoid membrane (see Fig. 5-15); and (4) removal of O_2 from inside the thylakoid, which readily occurs by outward diffusion of this small neutral molecule. Thus the asymmetrical nature of the reaction center of Photosystem II, together with the membrane property of allowing small neutral molecules to cross easily while retarding the penetration of charged species, leads to an accumulation of H^+ inside thylakoids from the light-dependent, O_2-evolution step.

H^+ is also transferred from the stroma to inside a thylakoid as electrons move along the electron transport pathway from Photosystem II to Cyt f (Figs. 5-15 and 6-5). Apparently two H^+s are moved as one electron spontaneously moves from Q_A (midpoint redox potential probably near −0.15 V) to Cyt f (0.35 V; Table 5-3 and Fig. 6-4). By Eq. (6.7) ($\Delta G = -nF\Delta E$), this 0.50 V increase in redox potential corresponds to a change in Gibbs free energy of $-(1)(96.49 \text{ kJ mol}^{-1} \text{ V}^{-1})(0.50 \text{ V})$ or -48 kJ mol^{-1} electrons. We argued previously that a decrease in the chemical potential of H^+ of at least $12 \text{ kJ mol}^{-1} H^+$ is needed if four H^+s are required per ATP, so 48 kJ mol^{-1} electrons is more than sufficient energy to move the two H^+s to higher energy inside a thylakoid per electron moving along the electron transfer pathway.

Coupling between the H^+ movements across the thylakoid membranes associated with electron flow and ATP formation occurs via an ATP synthetase, usually referred to as ATP synthase but also as an ATPase (because it can catalyze the reverse reaction leading to ATP hydrolysis) as well as a coupling factor. As illustrated in Fig. 6-5, the ATP synthase has two components: (1) a protein factor that occurs on the stromal side of a thylakoid, which can bind ADP, P_i, and ATP (labeled CF_1 in Fig. 6-5), and (2) another protein factor that is hydrophobic and hence occurs in the thylakoid membrane, through which H^+ can pass (labeled CF_0).[5] The chemical nature of these complexes is similar for ATP synthases in chloroplasts, mitochondria, and bacterial membranes. CF_1 is readily dislodged from the thylakoids and is soluble in water, where it can catalyze ATP hydrolysis. Approximately one CF_1, which comprises about 10% of the thylakoid protein, occurs per Photosystem I (just as for Photosystem I, CF_1 is absent where the thylakoids stack together to form grana). Upon removal of CF_1, CF_0 remains in the thylakoid membrane as a channel through which passive proton movement can readily occur. Although the H^+ moving through CF_0 can come from the lumen of a thylakoid (Fig. 6-5), protons (H^+) can also diffuse along domains on the inner side of the thylakoid membrane, such as via H^+-binding sites on proteins or phospholipids in the membrane.

A matter related to the coupling of various flows across the thylakoids is the relative amounts of ATP produced and $NADP^+$ reduced in chloroplasts. Three ATPs and two NADPHs are needed per CO_2 photosynthetically fixed in the majority of plants, which

---

[5] F_0 and F_1 were originally studied in mitochondria and when analogous complexes were found in chloroplasts, they were designated CF_0 and CF_1. F_1 was the first of a series of proteinaceous factors involved with oxidative phosphorylation that were isolated from mitochondria by Efraim Racker and others in the 1960s. Later the hydrophobic factor F_0, which makes the ATPase activity of F_1 sensitive to oligomycin and hence has "o" as a subscript, was isolated from mitochondrial membranes. For their work on the mechanism of action of ATP synthase, Paul Boyer and John Walker shared the Nobel prize in chemistry in 1997.

are referred to as C₃ plants because CO_2 is incorporated into ribulose-1,5-bisphosphate to yield two molecules of 3-phosphoglyceric acid, a three-carbon compound (see Fig. 8-10; about 92% of plant species use the C₃ pathway). Four or five ATPs (depending on which of three different pathways is involved) and two NADPHs are required per CO_2 fixed in C₄ plants, where the first photosynthetic products are four-carbon organic acids (e.g., oxaloacetic acid). The absorption of eight photons can lead to the processing of four excitations in each of the two photosystems, causing one O_2 to be evolved and four H^+s to be produced inside a thylakoid by Photosystem II, eight H^+s to be delivered from the stroma to the thylakoid lumen by the plastoquinones plus the Cyt b_6–f complex, and two NADPHs to be produced by the overall noncyclic electron flow (see Figs. 5-12, 5-15, 6-4, and 6-5). Assuming four H^+s are needed per ATP, the 12 H^+s lead to three ATPs that together with the two NADPHs are sufficient to fix one CO_2 in C₃ plants. C₄ plants require one or two more ATPs per CO_2 fixed; the additional ATP can be supplied by cyclic or possibly pseudocyclic electron flow (as we indicated in Chapter 5, no $NADP^+$ reduction accompanies either of these types of electron flow). For instance, cyclic electron flow takes electrons from ferredoxin to the Cyt b_6–f complex, from which the electrons move via plastocyanin back to Photosystem I; accompanying the electron movement, H^+ is transferred from the stroma to the thylakoid lumen, which can lead to the extra ATP needed as the H^+s move back to the stroma through the ATP synthase (Fig. 6-5).

IV. Mitochondrial Bioenergetics

The activities of chloroplasts and mitochondria are related in various ways (Fig. 6-6). For instance, the O_2 evolved by photosynthesis can be consumed during respiration, and the CO_2 produced by respiration can be fixed by photosynthesis. Moreover, ATP formation is coupled to electron flow in both organelles; in mitochondria the electron flow is from a

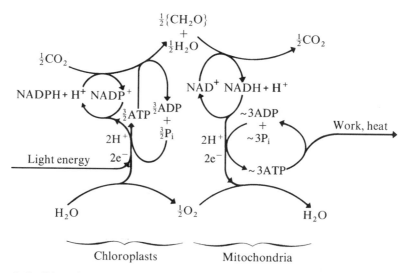

Figure 6-6. Schematic representation of the interrelationship between components involved in chloroplast and mitochondrial bioenergetics.

reduced pyridine nucleotide to the oxygen–water half-cell, and in chloroplasts it is in the opposite direction (Fig. 6-6). A few to many thousands of mitochondria occur in each plant cell, their frequency tending to be lower in cells in which chloroplasts are abundant. In photosynthetic tissue at night, and at all times in the nongreen tissues of a plant, oxidative phosphorylation in mitochondria is the supplier of ATP for the cells.

A. Electron Flow Components—Redox Potentials

As for chloroplast membranes, various compounds in mitochondrial membranes accept and donate electrons. These electrons originate from biochemical cycles in the cytosol as well as in the mitochondrial matrix (see Fig. 1-8)—most come from the tricarboxylic acid (Krebs) cycle, which leads to the oxidation of pyruvate and the reduction of NAD^+ within mitochondria. Certain principal components for mitochondrial electron transfer and their midpoint redox potentials are indicated in Fig. 6-7, in which the spontaneous electron flow

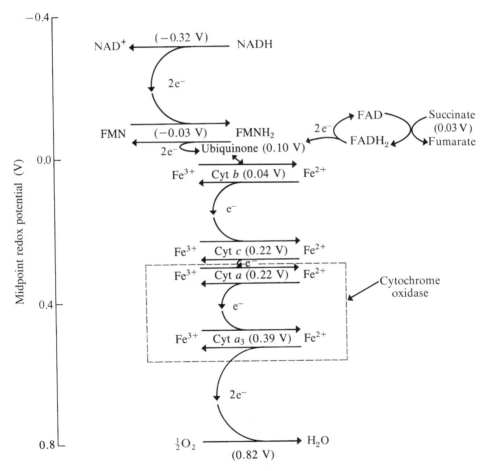

Figure 6-7. Components of the mitochondrial electron transport chain with midpoint redox potentials in parentheses.

to higher redox potentials is toward the bottom of the figure. As for photosynthetic electron flow, only a few types of compounds are involved in electron transfer in mitochondria—namely, pyridine nucleotides, flavoproteins, quinones, cytochromes, and the water–oxygen couple (some iron-plus-sulfur-containing centers or clusters also occur).

The reduced compounds that introduce electrons directly into the mitochondrial electron transfer chain are NADH and succinate ($COOHCH_2CH_2COOH$), the latter passing two hydrogens to flavin adenine dinucleotide (FAD) (Fig. 6-7). FAD consists of riboflavin (vitamin B_2) bound by a pyrophosphate bridge to adenosine (Fig. 6-3). Upon accepting two electrons and two protons—one H going to each N with a double bond in the riboflavin part of the molecule—FAD is reduced to $FADH_2$. The FAD–$FADH_2$ couple is bound to a protein, which is referred to as a *flavoprotein* [ferredoxin–NADP$^+$ oxidoreductase (Table 5-3) is a flavoprotein involved with photosynthetic electron transfer]. We note that a flavoprotein containing flavin mononucleotide (FMN) as the prosthetic group also occurs in mitochondria, where FMN is riboflavin phosphate:

Mitochondria contain *ubiquinone*, which differs from plastoquinone A (p. 212) by two methoxy groups in place of the methyl groups on the ring, and 10 instead of 9 isoprene units in the side chain. A c-type cytochrome, referred to as Cyt c_1 in animal mitochondria, intervenes just before Cyt c, and a b-type cytochrome occurs in plant mitochondria that is involved with an electron transfer that bypasses cytochrome oxidase on the way to O_2. The cytochrome oxidase complex contains two Cyt a plus two Cyt a_3 molecules and copper on an equimolar basis with the hemes (see Fig. 5-13). Both the Cu and the Fe of the heme of Cyt a_3 are involved with the reduction of O_2 to H_2O. Cytochromes a, b, and c are in approximately equal amounts in mitochondria (the ratios vary somewhat with plant species); flavoproteins are about 4 times, ubiquinones 7–10 times, and pyridine nucleotides 10–30 times more abundant than individual cytochromes. Likewise, in chloroplasts the quinones and pyridine nucleotides are much more abundant than are the cytochromes (see Table 5-3).

Most of the components involved in electron transport in mitochondria are contained in four supramolecular protein complexes that traverse the inner mitochondrial membrane. Complex I, which contains FMN and various iron–sulfur clusters as active sites, transfers electrons from NADH to ubiquinone (Fig. 6-7). Complex II, which contains FAD, various iron–sulfur clusters, and Cyt b_{560}, transfers electrons from succinate also to a ubiquinone. Ubiquinone functions as a pool of two-electron carriers, analogous to the function of plastoquinone A in the lamellar membranes of chloroplasts, which accepts electrons from complexes I and II and delivers them to the third protein complex.[6] This complex III, which

---

[6] Complex III is analogous to the Cyt b_6–f complex of chloroplasts, both with respect to contents (two Cyt b's, one Cyt c, an Fe–S protein, and a quinone) and function within the membranes (e.g., interaction with a quinol; the isolated Cyt b_6–f complex can also pass electrons to Cyt c as well as to its natural electron acceptor, plastocyanin). It is also structurally and functionally analogous to supramolecular protein complex in bacteria.

contains Cyt b_{562}, Cyt b_{566}, Cyt c_1, an iron–sulfur cluster, and a quinone, transfers electrons to Cyt c (Fig. 6-7). In turn, the pool of Cyt c molecules, which are soluble in aqueous solutions, passes electrons to complex IV (also called cytochrome oxidase), which contains Cyt a, Cyt a_3, and copper atoms as active sites. Complex IV delivers electrons to oxygen, which acts as the terminal electron acceptor in mitochondria.

B. Oxidative Phosphorylation

ATP formation coupled to electron flow in mitochondria is generally called *oxidative phosphorylation*. Because electron flow involves reduction as well as oxidation, more appropriate names are "respiratory phosphorylation" and "respiratory-chain phosphorylation," terms that are also more consistent with photophosphorylation for ATP formation in photosynthesis. As with photophosphorylation, the mechanism of oxidative phosphorylation is not yet fully understood in molecular terms. Processes like phosphorylation accompanying electron flow are intimately connected with membrane structure, so they are much more difficult to study than are the biochemical reactions taking place in solution. A chemiosmotic coupling mechanism between electron flow and ATP formation in mitochondria is generally accepted, and we will discuss some of its characteristics.

Accompanying electron flow in mitochondria, H^+ is transported from the matrix on the inner side of the inner membrane to the lumen between the limiting membranes (Fig. 6-8). Certain electron flow components are so situated in the membranes that they can carry out this vectorial movement. Protein complex I, which oxidizes NADH, apparently transfers

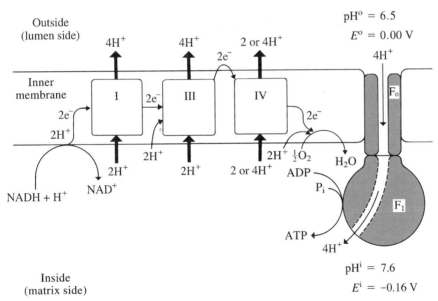

Figure 6-8. Schematic representation of certain electron flow and ATP synthesis components in the inner mitochondrial membrane, emphasizing the directional flows of H^+, various protein complexes, and the ATP synthase. The stoichiometry of H^+ per pair of electrons for the protein complexes is tentative (Bogachev *et al.*, 1996; Lorusso *et al.*, 1995; Papa *et al.*, 1994). The H^+, which is moved out toward higher μ_H accompanying electron flow along the respiratory chain, can move back through a hydrophobic channel (F_o) and another protein factor attached to the inside of the inner membrane (F_1), leading to ATP synthesis.

four H$^+$s across the inner membrane per pair of electrons from NADH. Complex II, which oxidizes FADH$_2$ and leads to the reduction of a ubiquinone whose two electrons also move to Complex III, apparently causes no H$^+$s to move from the matrix to the lumen. Transport of four H$^+$s outward from the matrix most likely occurs via protein complex III per pair of electrons traversing the electron transport chain. Complex IV (cytochrome oxidase) may also transport four H$^+$s out (Fig. 6-8 summarizes these possibilities). We also note that two H$^+$s are necessary for the reduction of $\frac{1}{2}O_2$ to H$_2$O, and these protons can also be taken up from the internal solution (Fig. 6-8). In any case, the pH is higher and the electrical potential is lower in the inner region (Fig. 6-8) than they are outside the inner mitochondrial membrane, opposite to the polarity found for chloroplasts (Fig. 6-5).

The transport of protons out of the matrix leads to a difference in the H$^+$ chemical potential across the inner mitochondrial membrane. Using Eq. (6.16c) and the values in Fig. 6-8, we calculate that $\mu_H^i - \mu_H^o = (5.71)(6.5 - 7.6) + (96.5)(-0.16 - 0.00) = -6.3 - 15.4$ or -22 kJ (mol H$^+$)$^{-1}$, indicating that the H$^+$ chemical potential is lower on the inside. For some cases in which ATP formation occurs, E_M is -0.14 V and pH^o $-$pH^i is -0.5, in which case 16 kJ (mol H$^+$)$^{-1}$ is available for ATP formation from the chemical potential difference of H$^+$ across the inner mitochondrial membrane. We indicated previously that at least 12 kJ per mole H$^+$ is required for ATP formation if four H$^+$s are used per ATP synthesized. We also note that for chloroplasts most of the $\Delta\mu_H$ is due to the pH term, whereas for mitochondria the electrical term is generally more important.

As with chloroplasts, we can uncouple ATP formation from electron flow by adding compounds that dissipate the H$^+$ chemical potential difference. Valinomycin acts like an organic ring with a hydrophilic center through which K$^+$ and NH$_4^+$ can readily pass but Na$^+$ and H$^+$ cannot, thus providing a selective channel when it is embedded in a membrane. If the antiporter nigericin, which facilitates K$^+$–H$^+$ exchange, is added together with the ionophore valinomycin, protons tend to move back into the matrix through the antiporter, thereby diminishing the ΔpH without affecting E_M, while the ionophore causes K$^+$ entry, thereby collapsing the electrical potential difference. Because $\Delta\mu_H$ is thus dissipated, ATP formation ceases.

ATP formation is coupled to the energetically downhill H$^+$ movement back into the mitochondrial matrix through a hydrophobic protein factor in the inner membrane (F$_o$; see footnote 5) and a protein factor (F$_1$) about 9 nm in diameter that protrudes from the inner membrane into the matrix (Fig. 6-8). Indeed, subunits of F$_1$ rotate during ATP formation, so this protein structure has been called a "rotary motor." When removed from F$_o$, F$_1$ can lead to the hydrolysis of ATP in an aqueous solution and the inner membrane becomes leaky to H$^+$, indicating that the hydrophobic F$_o$ is a channel or transporter for protons in mitochondria, just as CF$_o$ is in chloroplasts. Although a ratio of three H$^+$s crossing the inner mitochondrial membrane per ATP synthesized had been widely accepted, the formation of one ATP apparently requires the movement of four H$^+$s through the mitochondrial ATP synthase (F$_o$+F$_1$; Fig. 6-8), just as for the chloroplastic ATP synthase (CF$_o$+CF$_1$; Fig. 6-5; Boyer, 1997; Fillingame, 1997; Yasuda et al., 1997). The H$^+$ motion is inward for the inner mitochondrial membrane but outward for the chloroplast thylakoid.

ATP is produced in the mitochondrial matrix but is usually needed in the cytosol. As mentioned in Chapter 1, the outer mitochondrial membrane is readily permeable to solutes such as succinate, ADP, and ATP; a channel-forming protein is responsible for this high permeability. On the other hand, specific porters are necessary for moving charged solutes such as ATP across the inner mitochondrial membrane. In fact, an ADP/ATP antiporter exists in the inner membrane, which replenishes the internal ADP pool as well. Phosphate, which is also needed for ATP synthesis, enters by a P_i/OH^- antiporter (the concentration of OH^- is relatively high in the matrix) or, perhaps, an H^+/P_i symporter. For the usual state of ionization, an ADP/ATP antiporter taking ATP into the mitochondrial matrix would be electrogenic (one more negative charge brought in than taken out). Also, a P_i/OH^- antiporter (or an H^+/P_i symporter) transporting HPO_4^{2-} would be electrogenic but in the opposite direction. Because these two porters must operate at the same rate to replace ADP and P_i and thus to sustain ATP formation in the mitochondrial matrix, no overall effect on the membrane potential is expected. However, the extra proton required for these porters per ATP synthesized raises the proton requirement to five H^+s per ATP formed in the mitochondrial matrix. In contrast, the ATP synthesized during photophosphorylation is produced where it is mainly utilized, namely, in the chloroplast stroma where CO_2 fixation occurs.

As with chloroplasts, many questions concerning electron flow and the coupled ATP formation in mitochondria remain unanswered. The first part of the mitochondrial electron transfer chain has a number of two-electron carriers (NAD^+, FMN, and ubiquinone) that interact with the cytochromes (one-electron carriers). In this regard, the reduction of O_2 apparently involves four electrons coming sequentially from the same Cyt a_3. Of more interest is the stoichiometry between electron flow and proton movement. Twelve H^+s move out across the inner mitochondrial membrane when a pair of electrons moves from NADH to O_2 (see Fig. 6-8); this is consistent with four H^+s per ATP and the long-standing view of three ATPs formed per NADH oxidized. However, steady-state ATP formation apparently requires five H^+s per ATP and complex IV may transport only two H^+s from the mitochondrial matrix to the lumen per pair of electrons; these matters require further research.

V. Energy Flow in the Biosphere

The foregoing discussion of the way organisms interconvert energy on an organelle level sets the stage for a consideration of bioenergetics in a broader context. We will begin with certain biochemical aspects and then discuss the overall flow of energy from the sun through the biosphere. We will consider photosynthetic efficiency as well as transfer of energy from plants to animals. This material will serve as a transition between the molecular and cellular levels, considered up to now, and the organ and organism levels of the succeeding three chapters.

In Chapter 4 we indicated that the radiation input of the sun to the earth's atmosphere is $1367 \, W \, m^{-2}$ (the "solar constant"). Some of the radiant energy is used to form ATP and NADPH in chloroplasts. In turn, these energy currencies lead to the reductive fixation of CO_2 into a carbohydrate in photosynthesis (see Fig. 5-1). In the same photosynthetic cells, in other plant cells, and in animal cells, the carbohydrates formed during photosynthesis can serve as the energy source for respiration, which leads to the generation of ATP by oxidative phosphorylation.

When used as a fuel in respiration, the carbohydrate glucose is first broken down into two molecules of pyruvate in the cytosol. Pyruvate enters the mitochondria and is eventually oxidized to CO_2 and H_2O by the tricarboxylic acid cycle. One mole of glucose can lead to the formation of about 30 mol ATP. As we noted previously, the Gibbs free energy released by the complete oxidation of 1 mol of glucose is 2872 kJ, and about 48 kJ is required for the phosphorylation of 1 mol of ADP in mitochondria. Hence, the efficiency of the many-faceted conversion of Gibbs free energy from glucose to ATP can be [(30 mol ATP/mol glucose)(48 kJ/mol ATP)/(2872 kJ/mol glucose)](100), or 50%, indicating that the Gibbs free energy in glucose can be efficiently mobilized to produce ATP. Such ATP is used by the cells to transport ions, to synthesize proteins, and to provide for growth and maintenance in other ways. We can readily appreciate that, if free energy were not constantly supplied to their cells, plants and animals would drift toward equilibrium and die.

One of the consequences of the flux of energy through the biosphere is the formation of complex and energetically improbable molecules such as proteins and nucleic acids. Such compounds represent a considerably greater amount of Gibbs free energy than does an equilibrium mixture containing the same relative amounts of the various atoms. (Equilibrium corresponds to a minimum in Gibbs free energy; see Fig. 6-1.) For instance, the atoms in the nonaqueous components of cells have an average of about 26 kJ mol^{-1} more Gibbs free energy than do the same atoms at equilibrium (Morowitz, 1979). The Boltzmann energy distribution (Eq. 3.21b) predicts that at equilibrium the fraction of atoms with kinetic energy in excess of E is equal to $e^{-E/RT}$, which for 26 kJ mol^{-1} is only 0.000028 at 25°C. Thus, only a very small fraction of atoms would have a kinetic energy equal to the average enrichment in Gibbs free energy per atom of the nonaqueous components in cells. It is the flux of energy from the sun through plants and animals (see Fig. 6-6) that leads to such an energy enrichment in the molecules and that ensures that biological systems will be maintained in a state far from equilibrium, as is essential for life.

A. Incident Light—Stefan–Boltzmann Law

To help understand the energy available to the biosphere, we need to reconsider some properties of radiating bodies. In Chapter 4 we indicated that the distribution of radiant energy per unit wavelength interval is proportional to $\lambda^{-5}/(e^{hc/\lambda kT} - 1)$, where T is the surface temperature of the radiation source. This form of Planck's radiation distribution formula applies to an object that radiates maximally, a so-called blackbody. When blackbody radiation is integrated over all wavelengths, we can determine the maximum amount of energy radiated by an object. Appropriate integration[7] of Planck's radiation distribution formula leads to the following expression:

$$\text{Maximum radiant energy flux density} = \sigma T^4 \tag{6.17}$$

where σ is a constant and T is in kelvin units (temperature in °C + 273.15).

Although Eq. (6.17) can be derived from quantum-physical considerations developed by Max Planck in 1900, it was first proposed in the latter part of the 19th century. In 1879 Josef Stefan empirically determined that the maximum radiation was proportional to the fourth power of the absolute temperature; in 1884 Ludwig Boltzmann interpreted this in terms of classical physics. The coefficient of proportionality σ was deduced from measurements then available and has become known as the Stefan–Boltzmann constant,

---

[7] To integrate Planck's radiation distribution formula over all wavelengths, x can conveniently be substituted for $1/(\lambda T)$ and hence $dx = -(1/T)(1/\lambda^2)\,d\lambda$, so $d\lambda = -\lambda^2 T\,dx = -dx/(Tx^2)$. The total energy radiated is thus:

$$\text{Energy radiated} = \text{const} \int_0^\infty \lambda^{-5}(e^{hc/\lambda kT} - 1)^{-1}\,d\lambda$$

$$= \text{const} \int_\infty^0 (1/xT)^{-5}(e^{hcx/k} - 1)^{-1}(-dx/Tx^2)$$

$$= \text{const}\, T^4 \int_0^\infty x^3(e^{hcx/k} - 1)^{-1}dx$$

where the definite integral equals $(\pi^4/15)(k/hc)^4$.

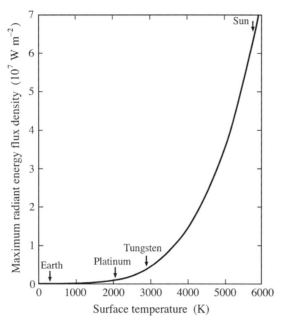

Figure 6-9. Dependence of energy radiated on surface temperature of object. Values are indicated for the earth (mean surface temperature of 290 K), the melting point of platinum (2042 K; formerly used to define a candle; see Chapter 4), a tungsten lamp (2900 K; also see Chapter 4), and the sun (5800 K).

which equals 5.670×10^{-8} W m^{-2} K^{-4} (Appendix I). For the case in which the object does not radiate as a blackbody, the radiant energy flux density at the surface of the radiator equals $e\sigma T^4$, where e is the *emissivity*. Emissivity depends on the surface material of the radiating body and achieves its maximum value of 1 for a blackbody. Equation (6.17), which is known as the Stefan–Boltzmann law, indicates that the amount of radiant energy emitted by an object increases extremely rapidly with its surface temperature (Fig. 6-9).

We will now estimate the amount of energy radiated from the sun's surface and how much of this is annually incident on the earth's atmosphere. Using Eq. (6.17) and the effective surface temperature of the sun, about 5800 K, the rate of energy radiation per unit area of the sun's surface is

$$J_{energy} = (5.670 \times 10^{-8} \text{ W m}^{-2} \text{ K}^{-4})(5800 \text{ K})^4 = 6.4 \times 10^7 \text{ W m}^{-2}$$

The entire output of the sun is about 3.84×10^{26} W, which leads to 1.21×10^{34} J year^{-1}. Because the amount incident on the earth's atmosphere is 1367 W m^{-2} and the projected area of the earth is 1.276×10^{14} m^2, the annual energy input into the earth's atmosphere from the sun is $(1367 \text{ J m}^{-2} \text{ s}^{-1})(1.276 \times 10^{14} \text{ m}^2)(3.156 \times 10^7 \text{ s year}^{-1})$, or 5.50×10^{24} J year^{-1} (Table 6-1).

B. Absorbed Light and Photosynthetic Efficiency

Only a small fraction of the sun's energy incident on the earth's atmosphere each year is actually absorbed by photosynthetic pigments, and only a small fraction of the absorbed

Table 6-1. Annual Energy Magnitudes

Quantity	J year^{-1}	Percentage of immediately above quantity
Sun's output	1.21×10^{34}	—
Energy into earth's atmosphere	5.50×10^{24}	4.6×10^{-8}
Energy absorbed by photosynthetic pigments	2.8×10^{23}	5.1
Energy stored in photosynthetic products	4.0×10^{21}	1.4
Human food energy consumption	2.6×10^{19}	0.7

energy is stored as chemical energy of the photosynthetic products. Specifically, approximately 5% of the 5.50×10^{24} J annually incident on the earth's atmosphere is absorbed by chlorophyll or other photosynthetic pigments, leading to a radiant energy input into this part of the biosphere of about 2.8×10^{23} J year^{-1} (Table 6-1). How much of this energy is annually stored in photosynthetic products? As we indicated in Chapter 5, a net of approximately 1.0×10^{14} kg of carbon is annually fixed by photosynthesis. For each mole of CO_2 (12 g carbon) incorporated into carbohydrate, approximately 479 kJ of Gibbs free energy is stored. The total amount of energy stored each year by photosynthesis is thus $(1.0 \times 10^{17}$ g year$^{-1})(1$ mol/12 g$)(4.79 \times 10^5$ J mol$^{-1})$, or 4.0×10^{21} J year^{-1}. Hence only about 1% of the radiant energy absorbed by photosynthetic pigments is ultimately stored by plant cells (Table 6-1).

The efficiency of photosynthesis can be represented in many different ways. If we express it on the basis of the total solar irradiation incident on the earth's atmosphere (5.50×10^{24} J year^{-1}), it is only 0.073%. This low figure takes into consideration many places of low productivity, e.g., cold noncoastal regions of oceans, polar icecaps, winter landscapes, and arid regions. Furthermore, not all solar radiation reaches the earth's surface, and much that does is in the infrared (see Fig. 4-3); the efficiency would be 0.127% if we considered only the solar irradiation reaching the ground. Nevertheless, even with the very low overall energy conversion, the trapping of solar energy by photosynthesis is the essential source of free energy used to sustain life.

What is the highest possible efficiency for photosynthesis? For low levels of red light, the conversion of radiant energy into the Gibbs free energy of photosynthetic products can be up to 34% in the laboratory (see Chapter 5). Solar irradiation includes many wavelengths; slightly less than half of this radiant energy is in the region that can be absorbed by photosynthetic pigments, 400–700 nm (see Figs. 4-3 and 7-2). If all the incident wavelengths from 400 to 700 nm (about half of the radiant energy) were absorbed by photosynthetic pigments, and eight photons were required per CO_2 fixed, the maximum photosynthetic efficiency for the use of low levels of incident solar irradiation would be just under half of 34%, e.g., 15%. However, some sunlight is reflected or transmitted by plants (see Fig. 7-3), and some is absorbed by nonphotosynthetic pigments in the plant cells. Thus the maximum photosynthetic efficiency for using incident solar energy under ideal conditions of temperature, water supply, and physiological status of plants in the field is closer to 10%.

Actual measurements of photosynthetic efficiency in the field have indicated that up to about 7% of the incident solar energy can be stored in photosynthetic products for a rapidly growing crop under ideal conditions. Usually, the photosynthetic photon flux density (PPFD) on the upper leaves of vegetation is too high for all excitations of the photosynthetic pigments to be used for the photochemistry of photosynthesis. The energy of many absorbed photons is therefore wasted as heat, especially when leaves of C_3 plants are exposed to a PPFD of more than 600 μmol m^{-2} s^{-1}. Hence, the maximum sustained efficiency for the conversion of solar energy into Gibbs free energy stored in photosynthetic products is often near 3% for crops when averaged over a day in the growing season. For all vegetation averaged over a year, the efficiency is about 0.5% of the incident solar irradiation. This is consistent with our previous statement that about 1% of the radiant energy absorbed by photosynthetic pigments is stored in the products of photosynthesis because only about half of the solar irradiation incident on plants or algae is absorbed by chlorophylls, carotenoids, and phycobilins.

C. Food Chains and Material Cycles

We will now consider the fate of the energy stored in photosynthetic products when animals enter the picture. We begin by noting that across each step in a food chain the free energy decreases, as required by the second law of thermodynamics. For instance, growing herbivores generally retain only 10–20% of the free energy of the ingested plant material, and a mature (nongrowing) animal uses essentially all of its Gibbs free energy consumption just to maintain a nonequilibrium state. Growing carnivores will store about 10–20% of the free energy content of herbivores or other animals that they eat. Thus a sizable loss in Gibbs free energy occurs for each link in a food chain, and as a consequence seldom do more than four links, or steps, occur in a chain.

Although modern agriculture tends to reduce the length of our food chain for meat, humans still make a large demand on the Gibbs free energy available in the biosphere. The global average intake of free energy is about $10 \, \text{MJ person}^{-1} \, \text{day}^{-1}$ (2400 kcal person^{-1} day^{-1}). For a world population of 7 billion, the annual consumption of Gibbs free energy for food is about $2.6 \times 10^{19} \, \text{J year}^{-1}$. (Humans also consume plants and animals for clothing, shelter, firewood, papermaking, and in many other ways.) Thus our food consumption alone amounts to almost 1% of the $4.0 \times 10^{21} \, \text{J year}^{-1}$ stored in photosynthetic products (Table 6-1). If we were to eat only carnivores that ate herbivores with a 10% retention in Gibbs free energy across each step in the food chain, we would indirectly be responsible for the consumption of the entire storage of energy by present-day photosynthesis. Fortunately, we obtain most of our free energy requirements directly from plants. The average daily consumption in the United States is about $12 \, \text{MJ person}^{-1}$, of which just over 70% comes from plants and just under 30% from animals. The animals consumed retain approximately 8% of the Gibbs free energy in the plant material that they eat.

The harnessing of solar radiation by photosynthesis starts the flow of Gibbs free energy through the biosphere. In addition to maintaining individual chemical reactions as well as entire plants and animals in a state far from equilibrium, the annual degradation of chemical energy to heat sets up various cycles. We have already indicated some of these, e.g., O_2 is evolved in photosynthesis and then consumed by respiration and CO_2 cycles in the reverse direction between these two processes (Fig. 6-6). There is also a cycling between ATP and ADP + phosphate at the cellular level, as well as the inevitable birth–death cycle of organisms. In addition, we can recognize a cycling of nitrogen, phosphorus, and sulfur in the biosphere. All these material cycles can be regarded as consequences of the unidirectional flow of Gibbs free energy, which becomes less after each step along the way.

Problems

6.1 A reaction $A + B \rightleftharpoons C$ has a ΔG^* of $-17.1 \, \text{kJ mol}^{-1}$ of reactant or product at 25°C (K in molality). Assume that activity coefficients are unity. In which direction will the reaction proceed under the following conditions?

a. The concentrations of A, B, and C are all $1\,m$.
b. The concentrations of A, B, and C are all $1\,mm$.
c. The concentrations of A, B, and C are all $1\,\mu m$.
d. What is the equilibrium constant for the reaction?

6.2 Consider the following two half-cell reactions at $25°C$:

$$A \rightleftharpoons A^+ + e^- \qquad \Delta G^* = 8.37 \text{ kJ mol}^{-1}$$
$$B \rightleftharpoons B^+ + e^- \qquad \Delta G^* = 2.93 \text{ kJ mol}^{-1}$$

Assume that the midpoint redox potential of the second reaction is 0.118 V and that all activity coefficients are 1.

a. If the redox potential of the B–B^+ couple is 0.000 V, what is the ratio of B^+ to B?
b. What is the midpoint redox potential of the A–A^+ couple?
c. Suppose that all reactants and products are initially $1\,m$ but that the couples are in separate solutions of equal volume. If the half-cells are electrically connected with a metal wire, what is the initial electrical potential difference between them, and in which direction do electrons flow?
d. If all reactants and products are initially $1\,m$, what is the concentration of each at equilibrium in a single solution?
e. Qualitatively, how would the answer to (d) change if the initial conditions were identical to (c), but as the electrons flow through the wire they do electrical work?

6.3 Suppose that isolated chloroplasts are suspended in an aqueous medium initially containing 2 mM ADP, 5 mM phosphate, and no ATP (ignore any of these solutes originally in the chloroplasts). Assume that the temperature is $25°C$ and that the pH is 7.

a. What is the ATP concentration at equilibrium?
b. When the chloroplasts are illuminated, the ADP concentration decreases to 1 mM. What is the new concentration of ATP, and what is the change in Gibbs free energy for continued photophosphorylation?
c. If ferredoxin has a redox potential of -0.580 V and the activity of NADPH is 3% of that of $NADP^+$, what is the difference in redox potential between the two couples?
d. How much Gibbs free energy is available between the two couples in (c) when a pair of electrons moves between them? Is this enough for the continued formation of ATP under the conditions of (b)?
e. What difference in pH across a thylakoid membrane whose E_M is zero represents enough energy for ATP synthesis under the conditions of (b) when two protons are used per ATP? What if three protons are required per ATP?

6.4 Consider the following two mitochondrial cytochromes:

$$\text{Cyt } b\,(\text{Fe}^{2+}) \rightleftharpoons \text{Cyt } b\,(\text{Fe}^{3+}) + e^- \qquad E_b^{*H} = 0.040 \text{ V}$$
$$\text{Cyt } c\,(\text{Fe}^{3+}) + e^- \rightleftharpoons \text{Cyt } c\,(\text{Fe}^{2+}) \qquad E_c^{*H} = 0.220 \text{ V}$$

Assume that the temperature is $25°C$, the chemical activity of Cyt b (Fe^{2+}) is 20% of that of the oxidized form, activity coefficients are equal to 1, and 40 kJ is required to form 1 mol of ATP in mitochondria.

a. What is the redox potential of Cyt b?

b. If the concentration of Cyt c (Fe^{2+}) is 1 mM, what is the concentration of ferricytochrome c such that the Cyt c couple can just transfer electrons back to Cyt b?

c. What is the redox potential of Cyt c such that one electron going from Cyt b to Cyt c represents the same energy as is required to form one ATP?

d. Assume that for each pair of electrons that ubiquinone delivers to Cyt b in a supramolecular protein complex, four protons are moved from the matrix side out across the inner mitochondrial membrane. If the proton concentration is the same on the two sides of the membrane, what difference in redox potential energetically corresponds to having the electrical potential 0.15 V higher on the inner side?

e. If the pH is the same on the two sides of the membrane and four protons move inward through the F_o–F_1 ATP synthase per ATP, what is the minimum electrical potential difference across the inner mitochondrial membrane required to synthesize ATP by proton movement?

f. What is E_M in (e) if three protons are required per ATP? What is E_M if three H^+ are required and the energy losses (inefficiencies) are 30%?

References

Abrol, Y. P., Mohanty, P., and Govindjee (Eds.) (1993). *Photosynthesis: Photoreactions to Plant Productivity.* Kluwer, Dordrecht.

Alberty, R. A., and Silbey, R. J. (1997). *Physical Chemistry*, 2nd ed. Wiley, New York.

Allen, J. F., and Holmes, N. G. (1986). Electron transport and redox titration. In *Photosynthesis Energy Transduction: A Practical Approach* (M. F. Hipkins and N. R. Baker, Eds.). IRL, Oxford.

Barrow, G. M. (1996). *Physical Chemistry*, 6th ed. McGraw-Hill, New York.

Berry, S., and Rumberg, B. (1996). H^+/ATP coupling ratio at the unmodulated CF_o–CF_1 ATP synthase determined by proton flux measurements. *Biochim. Biophys. Acta* **1276**, 51–56.

Bogachev, A. V., Murtazina, R. A., and Skulachev, V. P. (1996). H^+/e^- stoichiometry for NADH dehydrogenase I and dimethyl sulfoxide reductase in anaerobically grown *Escherichia coli* cells. *J. Bacteriol.* **178**, 6233–6237.

Boyer, P. D. (1997). The ATP synthase—A splendid molecular machine. *Annu. Rev. Biochem.* **66**, 717–749.

Brown, G. C., and Cooper, C. E. (Eds.) (1995). *Bioenergetics: A Practical Approach.* Oxford Univ. Press, New York.

Clayton, R. K. (1980). *Photosynthesis: Physical Mechanisms and Chemical Patterns.* Cambridge Univ. Press, Cambridge, UK.

Crow, D. R. (1994). *Principles and Applications of Electrochemistry*, 4th ed. Blackie, London.

Dilley, R. A., Theg, S. M., and Beard, W. A. (1987). Membrane–proton interactions in chloroplast bioenergetics: Localized proton domains. *Annu. Rev. Plant Physiol.* **38**, 347–389.

Douce, R., and Neuburger, M. (1989). The uniqueness of plant mitochondria. *Annu. Rev. Plant Physiol. Plant Mol. Biol.* **40**, 371–414.

Elston, T., Wang, H., and Oster, G. (1998). Energy transduction in ATP synthase. *Nature* **391**, 510–513.

Ernsfer, L. (Ed.) (1992). *Molecular Mechanisms in Bioenergetics.* Elsevier, Amsterdam.

Fillingame, R. H. (1997). Coupling H^+ transport and ATP synthesis in F_1F_o–ATP synthases: Glimpses of interacting parts in a dynamic molecular machine. *J. Exp. Biol.* **200**, 217–224.

Garby, L., and Larsen, P. S. (1995). *Bioenergetics: Its Thermodynamic Foundations.* Cambridge Univ. Press, New York.

Gates, D. M. (1980). *Biophysical Ecology.* Springer-Verlag, New York.

Graber, P., and Milazzo, G. (Eds.) (1997). *Bioenergetics.* Birkhauser, Basal.

Gregory, R. P. F. (1989). *Biochemistry of Photosynthesis*, 3rd ed. Wiley, Chichester, UK.

Jagendorf, A. T., and Uribe, E. (1966). ATP formation caused by acid-base transition of spinach chloroplasts. *Proc. Natl. Acad. Sci. USA.* **55**, 170–177.

Kobayashi, Y., Kaiser, W., and Heber, U. (1995). Bioenergetics of carbon assimilation in intact chloroplasts: Coupling of proton to electron transport at the ratio $H^+/e = 3$ is incompatible with $H^+/ATP = 3$ in ATP synthesis. *Plant Cell Physiol.* **38**, 1629–1637.

Koryta, J. (1993). *Principles of Electrochemistry*, 2nd ed. Wiley, Chichester, UK.

Ksenzhek, O. S., and Volkov, A. G. (1998). *Plant Energetics*. Academic Press, San Diego.

Lorusso, M., Cocco, T., Minuto, M., Capitanio, N., and Papa, S. (1995). Proton/electron stoichiometry of mitochondrial bc_1 complex. Influence of pH and transmembrane ΔpH. *J. Bioenerg. Biomembr.* **27**, 101–108.

Mitchell, P. (1979). Compartmentation and communication in living systems. Ligand conduction: A general catalytic principle in chemical, osmotic and chemiosmotic reaction systems. *Eur. J. Biochem.* **95**, 1–20.

Morowitz, H. J. (1979). *Energy Flow in Biology*. Ox Bow Press, Woodbridge, CT.

Nicholls, D. G., and Ferguson, S. J. (1992). *Bioenergetics 2*. Academic Press, San Diego.

Palmieri, F. (Ed.) (1995). *Thirty Years of Progress in Mitochondrial Bioenergetics and Molecular Biology*. Elsevier, Amsterdam.

Papa, S., Lorusso, M., and Capitanio, N. (1994). Mechanistic and phenomenological features of proton pumps in the respiratory chain of mitochondria. *J. Bioenerg. Biomembr.* **26**, 609–618.

Robinson, J. D. (1997). *Moving Questions: A History of Membrane Transport and Bioenergetics*. Oxford Univ. Press, New York.

Rosing, J., and Slater, E. C. (1972). The value of $\Delta G°$ for the hydrolysis of ATP. *Biochim. Biophys. Acta* **267**, 275–290.

Stryer, L. (1994). *Biochemistry*, 4th ed. Freeman, New York.

Tinoco, I., Jr., Sauer, K., and Wang, J. C. (1995). *Physical Chemistry: Principles and Applications in Biological Sciences*, 3rd ed. Prentice Hall, Upper Saddle River, NJ.

van Walraven, H. S., Strotmann, H., Schwarz, O., and Rumberg, B. (1996). The H^+/ATP coupling ratio of the ATP synthase from thiol-modulated chloroplasts and two cyanobacterial strains is four. *FEBS Lett.* **379**, 309–313.

Witt, H. T. (1979). Energy conversion in the functional membrane of photosynthesis. Analysis by light pulse and electric pulse methods. The central role of the electric field. *Biochim. Biophys. Acta* **505**, 355–427.

Yasuda, R., Noji, H., Kinosita, K., Motojima, F., and Yoshida, M. (1997). Rotation of the γ subunit in F_1-ATPase; Evidence that ATP synthase is a rotary motor enzyme. *J. Bioenerg. Biomembr.* **29**, 207–209.

7

Temperature—Energy Budgets

I. **Energy Budget—Radiation** **258**
   A. Solar Irradiation 260
   B. Absorbed Infrared Irradiation 263
   C. Emitted Infrared Radiation 263
   D. Values for a, a_{IR}, and e_{IR} 265
   E. Net Radiation 266
   F. Examples for Radiation Terms 266
II. **Wind–Heat Conduction and Convection** **269**
   A. Wind—General Comments 270
   B. Air Boundary Layers 271
   C. Boundary Layers for Bluff Bodies 273
   D. Heat Conduction/Convection Equations 274
   E. Dimensionless Numbers 276
   F. Examples of Heat Conduction /Convection 278
III. **Latent Heat—Transpiration** **279**
   A. Heat Flux Density Accompanying Transpiration 279
   B. Heat Flux Density for Dew or Frost Formation 279
   C. Examples of Frost and Dew Formation 281
IV. **Further Examples of Energy Budgets** **282**
   A. Leaf Shape and Orientation 282
   B. Shaded Leaves within Plant Communities 283
   C. Heat Storage 284
   D. Time Constants 285
V. **Soil** **286**
   A. Thermal Properties 286
   B. Soil Energy Balance 287
   C. Variations in Soil Temperature 288
   Problems **290**
   References **291**

WE HAVE already encountered many aspects of temperature, which affects essentially all processes in plants. When introducing the special properties of water in Chapter 2, we noted that physiological processes generally take place within a fairly narrow temperature range. The water vapor content of air at saturation, as is nearly the case inside a leaf, is very temperature dependent (see Appendix I). Biochemical reactions usually exhibit a temperature optimum, although specific reactions in plants can acclimate to different temperature regimes. In Chapter 3 we discussed the Boltzmann energy distribution, Arrhenius plots, and Q_{10}, all of which involve the thermal energy of molecular motion. Light absorption (see Chapter 4) causes molecules to attain states that are simply too improbable to be reached by collisions based on thermal energy. Transitions from an excited state to another one at a lower energy or to the ground state can occur by radiationless transitions, releasing energy as heat that is eventually shared by the surrounding molecules. The surface temperature of an object indicates both the wavelengths where radiation from it will be maximal (Wien's displacement law, Chapter 4) and also the total energy radiated (Stefan–Boltzmann law, Chapter 6).

The temperature of an object is the net result of all the ways that energy can enter or leave it. In this chapter we will examine these various ways, especially for leaves. We will then be able to predict the temperatures of leaves, as well as more massive plant parts, based on the ambient environmental conditions. We can also appreciate the consequences of certain adaptations of plants to their environment and identify the experimental data needed for future refinements of our calculations.

We should emphasize at the outset that individual plants as well as environmental conditions vary tremendously. Thus in this and the next two chapters we will indicate an approach to the study of plant physiology and physiological ecology rather than provide a compendium of facts suitable for all situations. Nevertheless, certain basic features should become clear. For instance, CO_2 uptake during photosynthesis is accompanied by a water efflux through the stomata. This water loss during transpiration cools a leaf. Also, energy influxes are balanced against effluxes by changes in leaf temperature, which affect the amount of radiation emitted by the leaf as well as the heat conducted to the surrounding air. Another generality is that the temperatures of small leaves tend to be closer to those of the air than do the temperatures of large leaves. To appreciate the relative contributions of the various factors, we will use representative values for leaf and environmental parameters to describe the gas fluxes and the energy balance of leaves.

I. Energy Budget—Radiation

The law of conservation of energy (the first law of thermodynamics) states that energy cannot be created or destroyed but only changed from one form to another. We will apply this principle to the energy balance of a leaf, which occurs in an environment with many energy fluxes. We can summarize the various contributors to the energy balance of a leaf as follows:

$$
\underbrace{
\begin{array}{c}
\text{absorbed solar} \\
\text{irradiation,} \\
\text{absorbed infrared} \\
\text{irradiation from} \\
\text{surroundings}
\end{array}
}_{\text{energy into leaf}}
\quad - \quad
\underbrace{
\begin{array}{c}
\text{emitted infrared} \\
\text{radiation,} \\
\text{heat convection,} \\
\text{heat conduction,} \\
\text{heat loss accompanying} \\
\text{water evaporation}
\end{array}
}_{\text{energy out of leaf}}
\quad = \quad
\underbrace{
\begin{array}{c}
\text{photosynthesis,} \\
\text{other metabolism,} \\
\text{leaf temperature} \\
\text{changes}
\end{array}
}_{\text{energy storage by leaf}}
\tag{7.1}
$$

Equation (7.1) describes the case in which the leaf temperature is greater than the temperature of the surroundings; when the leaf temperature is less than that of the surrounding turbulent air, heat moves into a leaf. Also, when water condenses onto a leaf, the leaf will gain heat. In such cases, the appropriate energy terms in Eq. (7.1) change sign.

The various terms in Eq. (7.1) differ greatly in magnitude. For instance, the energy storage terms generally are relatively small. As a basis for comparison, we will consider the average amount of solar irradiation incident on the earth's atmosphere (the term "irradiation" refers to incident radiation; see Chapter 4, footnote 3, for radiation terminology). This radiant flux density, the solar constant, is about 1367 W m^{-2}. The solar irradiation absorbed by an exposed leaf is often about half this during the daytime. Hence, any process under 7 W m^{-2} corresponds to <1% of the absorbed solar irradiation for full sunlight, which usually cannot be measured to an accuracy >1%. How much energy is stored by photosynthesis? A typical net rate of CO_2 fixation by a photosynthetically active leaf is 10 μmol m^{-2} s^{-1} (see Chapter 8). As indicated in Chapter 5, about 479 kJ of energy is stored per mole of CO_2 fixed into photosynthetic products. Hence, photosynthesis by leaves might store $(10 \times 10^{-6}$ mol m^{-2} s$^{-1}) (479 \times 10^3$ J mol^{-1}), or 5 W m^{-2}, which is only about 1% of the rate of absorption of solar irradiation under the same conditions. In some cases the rate of photosynthesis can be higher. Still, we can generally ignore the contribution of photosynthesis to the energy balance of a leaf. Other metabolic processes in a leaf, such as respiration and photorespiration, are usually even less important on an energy basis than is photosynthesis, so they too can generally be ignored.

We will now consider the amount of energy that can be stored because of changes in leaf temperature. For purposes of calculation, we will assume that a leaf has the high specific heat of water (4.19 kJ kg^{-1} °C^{-1} at 20°C), is 300 μm thick, and has an overall density of 700 kg m^{-3} (0.7 g cm^{-3})—a leaf is often 30% air by volume. Hence, the mass per unit leaf area in this case is $(300 \times 10^{-6}$ m)(700 kg m^{-3}), or 0.21 kg m^{-2}. If 7 W m^{-2} were stored by temperature changes in such a leaf, its temperature would rise at the rate of (7 J m^{-2} s^{-1})/[(4190 J kg^{-1} °C^{-1})(0.21 kg m^{-2})], or 0.008°C s^{-1} (0.5°C min^{-1}), i.e., energy absorbed per area per time = specific heat × mass per area × rate of temperature change, where the specific heat is the energy required to raise the temperature of unit mass by one degree. Because this is a faster temperature change than that sustained for very long periods by leaves, we can assume that very little energy is stored (or released) in the form of leaf temperature changes. Hence, all three energy storage terms in Eq. (7.1) usually are relatively small for a leaf.

When the energy storage terms in Eq. (7.1) are ignored, the remaining contributors to the energy balance of a leaf are either radiation or heat terms. We can then simplify our energy balance relation as follows:

$$
\underbrace{\begin{array}{c} \text{absorbed solar irradiation,} \\ \text{absorbed infrared irradiation} \\ \text{from surroundings} \end{array}}_{\text{energy into leaf}} \quad \cong \quad \underbrace{\begin{array}{c} \text{emitted infrared radiation,} \\ \text{heat convection,} \\ \text{heat conduction,} \\ \text{heat loss accompanying} \\ \text{water evaporation} \end{array}}_{\text{energy out of leaf}} \quad (7.2)
$$

The heat conducted and convected from leaves is sometimes referred to as *sensible* heat, and that associated with the evaporation or the condensation of water is known as *latent* heat. In this chapter we will consider each of the terms in Eq. (7.2), which have units of energy per unit area and per unit time (e.g., J m^{-2} s^{-1}, which is W m^{-2}).

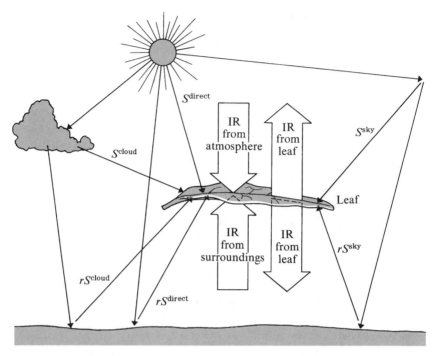

Figure 7-1. Schematic illustration of eight forms of radiant energy incident on an exposed leaf, including six that involve shortwave radiation from the sun and contain the letter S, and of the infrared radiation emitted from the two leaf surfaces.

A. Solar Irradiation

Solar irradiation can reach a leaf in many different ways, the most obvious being direct sunlight. Alternatively, sunlight can be scattered by molecules and particles in the atmosphere before striking a leaf. Also, both the direct and the scattered solar irradiation can be reflected by the surroundings toward a leaf (the term scattering usually denotes the irregular changes in the direction of light caused by small particles or molecules; reflection refers to the change in direction of irradiation at a surface). In Fig. 7-1 we summarize these various possibilities that lead to six different ways by which solar irradiation can impinge on a leaf. The individual energy fluxes can involve the upper surface of a leaf, its lower surface, or perhaps both surfaces—in Fig. 7-1 the direct solar irradiation (S^{direct}) is incident only on the upper surface of the leaf. To proceed with the analysis in a reasonable fashion, we will need to make many simplifying assumptions and approximations.

Some of the solar irradiation can be scattered or reflected from clouds before being incident on a leaf. On a cloudy day, the diffuse sunlight emanating from the clouds—S^{cloud}, or *cloudlight*—is substantial, whereas S^{direct} may be greatly reduced. For instance, a sky overcast by a fairly thin cloud layer 100 m thick might absorb or reflect away from the earth about 50% of the incident solar irradiation and diffusely scatter most of the rest toward the earth. A cloud 1 km thick will usually scatter somewhat less than 10% of the S^{direct} incident on it toward the earth (the actual percentages depend on the type and density of the clouds). Because the relative amounts of absorption, scattering, reflection, and transmission by clouds all depend on wavelength, cloudlight has a somewhat different wavelength

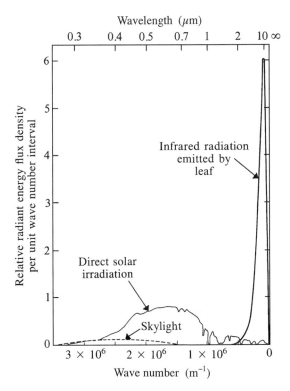

Figure 7-2. Wave number and wavelength distributions for direct solar irradiation, skylight, and the radiation emitted by a leaf at 25°C. Wave number, which we introduced in Problem 4.2, is the frequency of the radiation divided by the speed of light, ν/c, which equals the reciprocal of wavelength ($\lambda_{\text{vacuum}} \nu = c$ by Eq. 4.1); wave number is thus proportional to energy (see Eq. 4.2a; $E_\lambda = h\nu = hc/\lambda_{\text{vacuum}}$). The areas under the curves indicate the total energy radiated; S^{direct} is 840 W m^{-2}, S^{sky} is 80 W m^{-2}, and the IR emitted is 860 W m^{-2}.

distribution than does direct solar irradiation. Specifically, cloudlight is generally white or gray, whereas direct sunlight tends to be yellowish.

Solar irradiation scattered by air molecules and airborne particles leads to S^{sky}, or *skylight*. Such scattering is generally divided into two categories: Rayleigh scattering due to molecules, whose magnitude is approximately proportional to $1/\lambda^4$ (λ is the wavelength; see Chapter 4), and Mie scattering due to particles such as dust, which is approximately proportional to $1/\lambda$. Because of the greater scattering of the shorter wavelengths, skylight differs considerably from the wavelength distribution for S^{direct}. In particular, S^{sky} is enriched in the shorter wavelengths, because Rayleigh scattering causes most of the wavelengths below 500 nm to be scattered out of the direct solar beam (Fig. 7-2) and to become skylight. This explains the blue color of the sky (and its skylight) during the daytime—and the red color of the sun at sunset (S^{direct} must then travel a greater distance through the earth's atmosphere, leading to more scattering of the shorter wavelengths). In terms of energy, S^{sky} can be up to 10% of S^{direct} for a horizontal leaf (Fig. 7-1) on a cloudless day when the sun is overhead (Fig. 7-2) and can exceed S^{direct} when the sun is near the horizon.

As is customary, we refer to the direct sunlight plus the cloudlight and skylight as the *global irradiation*, S. Generally, $S^{\text{cloud}} + S^{\text{sky}}$ is referred to as diffuse shortwave irradiation, a readily measured quantity, whereas S^{cloud} and S^{sky} are difficult to measure separately. Thus the global irradiation equals the direct plus the diffuse shortwave irradiation. The value of the global irradiation varies widely with the time of day, the time of year, latitude, altitude, and atmospheric conditions, which must be taken into consideration when determining the energy balance of a leaf. As indicated previously, the maximum solar irradiation incident on the earth's atmosphere is about 1367 W m^{-2}. Because of scattering and absorption of solar irradiation by atmospheric gases (see Fig. 4-2), S on a cloudless day with the sun directly overhead in a dust-free sky is about 1000 W m^{-2} at sea level.

In the absence of clouds, S can be related to the solar constant (1367 W m^{-2}; S_c) and the atmospheric transmittance τ (the fraction of sunlight transmitted when the sun is directly overhead), which ranges from 0.5 under hazy conditions at sea level to 0.8 for clear skies at higher elevations. In particular, $S = S_c \tau^{1/\sin\gamma} \sin\gamma$, where γ is the sun's altitude, or angle above the horizon; γ depends on the time of day, the time of year, and the latitude.[1] When the sun is directly overhead, γ is 90° and $\sin\gamma = 1.00$. S then equals $S_c\tau$, where τ averages about 0.75 on clear days.

Sunlight may impinge on a leaf as direct solar irradiation, cloudlight, or skylight. These three components of global irradiation may first be reflected from the surroundings before striking a leaf (see Fig. 7-1). Although the reflected global irradiation can be incident on a leaf from all angles, for a horizontal exposed leaf it occurs primarily on the lower surface (Fig. 7-1). The reflected sunlight, cloudlight, and skylight generally are 10–30% of the global irradiation. A related quantity is the fraction of the incident shortwave irradiation reflected from the earth's surface, termed the *albedo*, which averages about 0.60 for snow, 0.35 for dry sandy soil, 0.25 for dry clay, 0.10 for peat soil, 0.25 for most crops, and 0.15 for forests. The albedo can vary with the angle of incidence of the direct solar beam, being greater at smaller angles of incidence.

Each of these six forms of solar irradiation (reflected as well as direct forms of sunlight, cloudlight, and skylight) can have a different wavelength distribution. Because absorption depends on wavelength, the fraction of each one absorbed can be different. Moreover, the fraction reflected also depends on wavelength. For simplicity, we will assume that the same absorptance applies to S^{direct}, S^{cloud}, and S^{sky}, as well as to the reflected forms of these irradiations. We will also assume that the same reflectance applies to each component of the global irradiation. We can then represent the absorption of all forms of solar irradiation by a leaf as follows:

Absorbed direct, scattered, and reflected solar irradiation

$$\cong a(S^{\text{direct}} + S^{\text{cloud}} + S^{\text{sky}}) + ar(S^{\text{direct}} + S^{\text{cloud}} + S^{\text{sky}})$$

$$= a(1+r)S \tag{7.3}$$

where the *absorptance* a is the fraction of the global radiant energy flux density S absorbed by the leaf, and the *reflectance* r is the fraction of S reflected from the surroundings onto the leaf. Absorptance is often called *absorptivity*, and reflectance is called *reflectivity*, especially when dealing with smooth surfaces of uniform composition.

---

[1] The dependence of γ on the time of day can be handled by the hour angle, $h = 15°(t - 12)$, where t is the solar time in hours and equals 12 at solar noon when the sun reaches its highest daily point in the sky. The time of year is handled by the solar declination δ, which equals $-23.5° \cos[(D + 10)360°/365.25]$, where D is the day of the year (January 1 = 1) and $-23.5°$ incorporates the tilt of the earth's axis relative to the plane of the earth's orbit. Using these parameters, $\sin\gamma = \sin\delta \sin\lambda + \cos h \cos\delta \cos\lambda$, where λ is the latitude in degrees (Gates, 1980; Nobel, 1988; Campbell and Norman, 1997).

B. Absorbed Infrared Irradiation

Besides the absorption of the various components of solar irradiation, additional infrared (IR), or thermal, radiation is also absorbed by a leaf (see Eq. 7.2). Any object with a temperature above 0 K ("absolute zero") emits such thermal radiation, including a leaf's surroundings as well as the sky (see Fig. 6-9). The peak in the spectral distribution of thermal radiation can be described by Wien's displacement law, which states that the wavelength for maximum emission of energy, λ_{max}, times the surface temperature of the emitting body, T, equals 2.90×10^6 nm K (Eq. 4.3b). Because the temperature of the surroundings is generally near 290 K, λ_{max} for radiation from them is close to $(2.90 \times 10^6$ nm K$)/(290$ K$)$, or 10,000 nm, which is 10 μm. Thus the emission of thermal radiation from the surroundings occurs predominantly at wavelengths far into the infrared. Because of its wavelength distribution, we will also refer to thermal radiation as infrared radiation and as *longwave* radiation (over 99% of the radiant energy from the surroundings occurs at wavelengths longer than 4 μm, and over 98% of the solar or *shortwave* irradiation occurs at wavelengths shorter than this).

Most of the thermal radiation from the sky comes from H_2O, CO_2, and other molecules in the atmosphere that emit considerable radiation from 5 to 8 μm and above 13 μm. Moreover, the concentration of these gases varies, so the effective temperature of the sky, T^{sky}, as judged from its radiation, also varies. For instance, clouds contain much water in the form of vapor, droplets, or crystals, which leads to a substantial emission of infrared radiation, so T^{sky} can be as high as 280 K on a cloudy day or night. On the other hand, a dry, cloudless, dust-free atmosphere might have a T^{sky} as low as 220 K.

We will now consider the amount of thermal irradiation absorbed by an unshaded leaf. We will suppose that the infrared irradiation from the surroundings, acting as a planar source at an effective temperature of T^{surr}, is incident on the lower surface of the leaf, and that the upper surface of the leaf is exposed to the sky, which acts as a planar source with an effective temperature of T^{sky} (Fig. 7-1). In Chapter 6 we introduced the Stefan–Boltzmann law, which indicates that the amount of radiation emitted by a body depends markedly on its surface temperature (Eq. 6.17; Maximum radiant energy flux density $= \sigma T^4$). The Stefan–Boltzmann law predicts the maximum rate of energy radiation, i.e., that emitted by a perfect radiator, a so-called blackbody (Figs. 6-9 and 7-3). Here we will use *effective temperature* in the sense that the actual radiant energy flux density equals $\sigma(T_{effective})^4$. For instance, T^{sky} is not the temperature we would measure at some particular location in the sky, although $\sigma(T^{sky})^4$ equals the actual amount of radiant energy from the sky. By the Stefan–Boltzmann law, with effective temperatures to give the radiation emitted by the surroundings and the sky, the IR absorbed by a leaf is

$$\text{IR irradiation absorbed} = a_{IR}\sigma[(T^{surr})^4 + (T^{sky})^4] \tag{7.4}$$

where the absorptance a_{IR} is the fraction of the energy of the incident IR irradiation absorbed by the leaf.

C. Emitted Infrared Radiation

Thermal radiation is also emitted by a leaf. Such radiation occurs at wavelengths far into the IR because leaf temperatures, like those of its surroundings, are near 300 K. This is illustrated in Fig. 7-2, in which the emission of radiant energy from a leaf at 25°C is plotted in terms of both wavelength and wave number. Using the wave number scale makes it easier

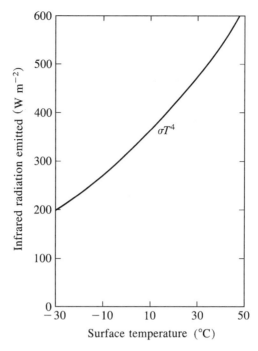

Figure 7-3. Rate of emission of infrared (longwave) radiation per unit area by a blackbody ($e_{IR} = 1.00$) as a function of its surface temperature as predicted by the Stefan–Boltzmann law (Eq. 6.17).

to illustrate the spectral distribution of radiation from the sun and a leaf in the same figure; moreover, the area under a curve is then proportional to the total radiant energy flux density. The λ_{max} for sunlight is in the visible region; for a leaf it is in the IR near 10 μm. Figure 7-2 also indicates that essentially all of the thermal radiation emitted by a leaf has wave numbers $< 0.5 \times 10^6$ m^{-1}, corresponding to IR wavelengths > 2 μm.

We will express the IR emitted by a leaf at a temperature T^{leaf} using the Stefan–Boltzmann law (Eq. 6.17), which describes the maximum rate of radiation per unit area. For the general emission case we incorporate a coefficient known as the *emissivity*, or *emittance*, e, which takes on its maximum value of 1 for a perfect, or blackbody, radiator. The actual radiant energy flux density equals $e\sigma(T_{actual})^4$, which is the same as $\sigma(T_{effective})^4$. We will use emissivities and actual temperatures to describe the energy radiated by leaves; effective temperatures are usually employed for thermal radiation from the surroundings and the sky because their temperatures are difficult to measure or, indeed, hypothetical (empirical equations incorporating the influence of air temperature, water vapor content, and clouds can also be used to predict the IR radiation from the sky). Because IR radiation is emitted by both sides of a leaf (see Fig. 7-1), the factor 2 is necessary to describe its energy loss by thermal radiation:

$$\text{IR radiation emitted} = 2e_{IR}\sigma(T^{leaf})^4 \tag{7.5}$$

As for our other flux density relations, Eq. (7.5) is expressed on the basis of unit area for one side of a leaf. The substantial temperature dependency of emitted IR is depicted in Fig. 7-3.

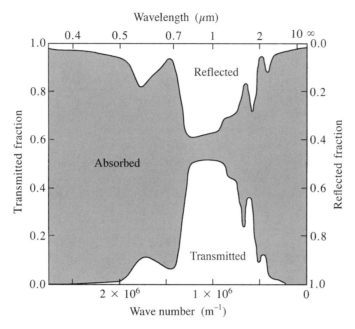

Figure 7-4. Representative fractions of irradiation absorbed (shaded region), reflected, and transmitted by a leaf as a function of wave number and wavelength. The sum $a_\lambda + r_\lambda + t_\lambda$ is unity.

D. Values for a, a_{IR}, and e_{IR}

The parameters a, a_{IR}, and e_{IR} help determine the energy balance of a leaf. We will first consider how the absorptance of a leaf depends on wavelength and then discuss the leaf emittance for infrared radiation.

Figure 7-4 indicates that the leaf absorptance at a particular wavelength, a_λ, varies considerably with the spectral region. For example, a_λ averages about 0.8 in the visible region (0.4–0.7 μm). Such relatively high fractional absorption by a leaf is due mainly to the photosynthetic pigments; the local minimum in a_λ near 0.55 μm is where chlorophyll absorption is relatively low (Fig. 5-3) and thus a larger fraction of the incident light is reflected or transmitted, leading to the green color (see Table 4-1) of leaves. Figure 7-4 indicates that a_λ is small from just beyond 0.7 μm up to nearly 1.2 μm. This is quite important for minimizing the energy input into a leaf because much global irradiation occurs in this interval of the IR. The fraction of irradiation absorbed becomes essentially 1 for IR irradiation beyond 2 μm. This does not lead to excessive heating of leaves from absorption of global irradiation because very little radiant emission from the sun occurs beyond 2 μm (see Fig. 7-2).

We have used two absorptances in our equations: a (in Eq. 7.3) and a_{IR} (in Eq. 7.4). In contrast to a_λ, these coefficients represent absorptances for a particular wavelength range. For example, a refers to the fraction of the incident solar energy absorbed (the wavelength distributions for direct sunlight and skylight are presented in Fig. 7-2). For most leaves, a is between 0.4 and 0.6. The shortwave absorptance can differ between the upper and the lower surfaces of a leaf, and a also tends to be lower for lower sun angles in the sky because the shorter wavelengths are then scattered more (Fig. 7-2) and the a_λs for the relatively enriched longer wavelengths are lower (Fig. 7-4). For wavelengths in the region 0.4–0.7 μm (designated "photosynthetic photon flux density," or PPFD in Chapter 4), the overall leaf absorptance is generally 0.75–0.90. Figure 7-4 shows that nearly all the IR irradiation beyond 2 μm is absorbed by a leaf. In fact, a_{IR} for leaves is usually from 0.95 to 0.98, and we will use 0.96 for purposes of calculation.

Because the emission of radiation is the reverse of its absorption, the same sort of electronic considerations that apply to the absorption of electromagnetic radiation (see Chapter 4) also apply to its emission. A good absorber of radiation is also a good emitter. In more precise language, the absorptance a_λ equals the emittance e_λ when they refer to the same wavelength (referred to as Kirchhoff's radiation law)—for a blackbody, $a_\lambda = e_\lambda = 1.00$ at all wavelengths. Because radiation from the surroundings and the sky occurs in essentially the same region of the IR as that emitted by a leaf, e_{IR} is about the same as a_{IR}, e.g., 0.96.[2]

E. Net Radiation

We have now considered each of the terms that involve radiation in the energy balance of a leaf (Eq. 7.2). These quantities comprise the *net radiation* balance for the leaf:

$$\text{Net radiation} = \begin{matrix} \text{absorbed solar irradiation} \\ + \\ \text{absorbed IR irradiation} \\ \text{from surroundings} \end{matrix} - \text{emitted IR radiation} \qquad (7.6a)$$

Using Eqs. (7.3)–(7.5), we can express the net radiation balance as

$$\text{Net radiation} = \begin{matrix} a(1+r)S \\ + \\ a_{IR}\sigma[(T^{\text{surr}})^4 + (T^{\text{sky}})^4] \end{matrix} - 2e_{IR}\sigma(T^{\text{leaf}})^4 \qquad (7.6b)$$

Before continuing with our analysis of the energy balance of a leaf, we will examine representative values for each of the terms in the net radiation.

F. Examples for Radiation Terms

We will consider a horizontal leaf exposed to full sunlight at sea level where the global irradiation S is 840 W m^{-2}. We will assume that the absorptance of the leaf for global irradiation a is 0.60, and that the reflectance of the surroundings r is 0.20. By Eq. (7.3), the

---

[2] We can relate e_λ and a_λ to radiation quantities introduced in Chapter 4. The amount of radiant energy emitted by a blackbody per unit wavelength interval is proportional to $\lambda^{-5}/(e^{hc/\lambda kT} - 1)$, as predicted by Planck's radiation distribution formula. When we multiply this maximum radiation by e_λ at each wavelength, we obtain the actual spectral distribution of the emitted thermal radiation. The absorptance a_λ is related to the absorption coefficient ε_λ. Equation (4.18) indicates that log $J_0/J_b = \varepsilon_\lambda cb$, where c is the concentration of the absorbing species, b is the optical path length, J_0 is the incident flux density, and J_b is the flux density of the emergent beam when only absorption takes place (i.e., in the absence of reflection and scattering). The fraction of irradiation absorbed at a particular wavelength, $(J_0 - J_b)/J_0$, is the absorptance, a_λ. Thus $a_\lambda = 1 - J_b/J_0$, which is $1 - 10^{-\varepsilon_\lambda cb}$, so the absorptance tends to be greater for wavelengths where the pigments absorb more (higher ε_λ), for leaves with higher pigment concentrations, and for thicker leaves.

direct plus the reflected sunlight, cloudlight, and skylight absorbed by the leaf is

$$a(1+r)S = (0.60)(1.00 + 0.20)(840 \text{ W m}^{-2})$$
$$= 605 \text{ W m}^{-2}$$

To calculate the IR irradiation absorbed by the leaf, we will let a_{IR} be 0.96, the temperature of the surroundings be 20°C, and the sky temperature be −20°C. Using Eq. (7.4) with a Stefan–Boltzmann constant of 5.67×10^{-8} W m^{-2} K^{-4} (see Appendix I), the absorbed IR is

$$a_{IR}\sigma[(T^{surr})^4 + (T^{sky})^4]$$
$$= (0.96)(5.67 \times 10^{-8} \text{ W m}^{-2} \text{ K}^{-4})[(293 \text{ K})^4 + (253 \text{ K})^4]$$
$$= 624 \text{ W m}^{-2}$$

Hence, the total irradiation load on the leaf is 605 W m^{-2} + 624 W m^{-2}, or 1229 W m^{-2}.

The rate of energy input per unit leaf area (1229 W m^{-2}) here is nearly the size of the solar constant (1367 W m^{-2}). About half is contributed by the various forms of irradiation from the sun (605 W m^{-2}) and half by IR irradiation from the surroundings plus the sky (624 W m^{-2}). Because the sky generally has a much lower effective temperature for radiation than does the surroundings, the upper surface of an exposed leaf usually receives less IR than does the lower one: $a_{IR}\sigma(T^{sky})^4$ here is 223 W m^{-2}, and the IR absorbed by the lower surface of the leaf, $a_{IR}\sigma(T^{surr})^4$, is 401 W m^{-2} (see Fig. 7-3). Changes in the angle of an exposed leaf generally have a major influence on the absorption of direct solar irradiation but less influence on the total irradiation load because the IR, the scattered, and the reflected irradiation received by a leaf come from all angles. In any case, leaf angle can have important implications for the interaction of certain plants with their environment.

To estimate the IR radiation emitted by a leaf, we will let e_{IR} be 0.96 and the leaf temperature be 25°C. By Eq. (7.5), the energy loss by thermal radiation then is

$$2e_{IR}\sigma(T^{leaf})^4 = (2)(0.96)(5.67 \times 10^{-8} \text{ W m}^{-2} \text{ K}^{-4})(298 \text{ K})^4$$
$$= 859 \text{ W m}^{-2}$$

In this case, the IR emitted by both sides of the exposed leaf is about 40% greater than the leaf's absorption of either solar or IR irradiation. Leaves shaded by other leaves are not fully exposed to T^{sky}, and for them the IR emitted can be approximately equal to the IR absorbed, a matter to which we will return later.

The net radiation balance (see Eq. 7.6) for our exposed leaf is 1229 W m^{-2} − 859 W m^{-2}, or 370 W m^{-2}. As we will discuss later, such excess energy is dissipated by conduction, convection, and the evaporation of water accompanying transpiration. However, most of the energy input into a leaf is balanced by the emission of IR radiation. The IR emitted in the current case, for example, amounts to (100) (859 W m^{-2})/(1229 W m^{-2}), or 70% of the energy input from all sources of incident irradiation. The radiation terms for this exposed leaf are summarized in the top line of Table 7-1.

Because many conditions affect the net radiation balance for leaves, let us now consider some other examples. At an elevation of 2000 m, S near noon on a cloudless day might be 920 W m^{-2}. Because of the higher incident global irradiation at 2000 m, the leaf there will absorb 57 W m^{-2} more direct, scattered, and reflected solar irradiation than the leaf at sea level. The effective temperature of the sky is generally slightly lower at the higher elevation

Table 7-1. Representative Values for the Various Terms in the Net Radiation Balance of an Exposed Leaf[a]

Condition	Global irradiation, S (W m^{-2})	Absorbed solar irradiation, $a(1+r)S$ (W m^{-2})	Temperature of surroundings (°C)	Sky temperature (°C)	Absorbed infrared, $a_{IR}\sigma[(T^{surr})^4 + (T^{sky})^4]$ (W m^{-2})	Leaf temperature (°C)	Emitted infrared, $2e_{IR}\sigma(T^{leaf})^4$ (W m^{-2})	Net radiation (W m^{-2})
Sea level on cloudless day	840	605	20	−20	624	25	859	370
2000 m on cloudless day	920	662	10	−25	555	24	847	370
Silvery leaf ($a = 0.50$) at 2000 m on cloudless day	920	552	10	−25	555	13	737	370
Sea level on cloudy night	0	0	1	1	614	1	614	0
Sea level on cloudless night	0	0	1	−20	530	−9	530	0
cloudless night	0	0	1	−20	530	−1	596	−66

[a]Equation (7.6) is used, taking a as 0.60 (except where indicated), r as 0.20, and both a_{IR} and e_{IR} as 0.96. (See text for interpretations.)

(-25 vs $-20°C$ in Table 7-1), and we will assume that the surroundings are $10°C$ lower in temperature than for our example at sea level. The total irradiation input is then 1217 W m^{-2}, which is 12 W m^{-2} less than that at sea level (see Table 7-1). If the net radiation were the same in the two cases (370 W m^{-2}), the leaf at the higher altitude must emit 12 W m^{-2} less thermal radiation than the leaf at sea level. As Table 7-1 indicates, this could be accomplished if T^{leaf} for the leaf at 2000 m were $1°C$ lower than the one at sea level.

Certain plants have silvery or shiny leaves, which increases the amount of solar irradiation reflected. The fraction of S reflected by the leaf may increase from typical values of 0.1 or 0.2 (see Fig. 7-4) to 0.3 for silvery leaves, with an accompanying reduction in the absorptance from 0.6 to 0.5 or lower. This reduction in absorptance can have a major influence on T^{leaf}. Other conditions remaining the same, a reduction of the absorptance a by only 0.1 can cause the leaf temperature to go from 24 to $13°C$ (Table 7-1). Such a reduction in leaf temperature can have substantial effects on transpiration and photosynthesis, which can be particularly important for desert plants in hot environments. For instance, seasonal variations in pubescense (epidermal hairs) can increase the shortwave reflectance of leaves of the desert shrub *Encelia farinosa* produced in dry, hot periods by about 0.2 over the reflectance for leaves from cool, wet periods. The higher reflectance decreases T^{leaf} by about $5°C$ for this species; this leads to photosynthetic rates that are $10–60\%$ higher during the dry, hot periods than they would otherwise be and at the same time conserves water because transpiration rates average 23% less (Ehleringer and Mooney, 1978).

We next consider the effect of clouds on leaf temperature at night. On a clear night at sea level the effective sky temperature might be $-20°C$, whereas for a heavy cloud cover it could be $1°C$ because of the IR emitted by the clouds (Table 7-1). If the temperature of the surroundings were $1°C$ in both cases, the IR absorbed by the leaf, $a_{IR}\sigma[(T^{surr})^4 + (T^{sky})^4]$, would be 84 W m^{-2} lower on the clear night because of less IR from the cloudless sky (Table 7-1). For there to be no net gain or loss of energy by a leaf from all forms of radiation, i.e., $a_{IR}\sigma[(T^{surr})^4 + (T^{sky})^4] = 2e_{IR}(T^{leaf})^4$, the leaf temperature would be $1°C$ on the cloudy night and $-9°C$ on the clear one (Table 7-1). Thus on the clear night with the surroundings at $1°C$, T^{leaf} could be considerably below freezing, so heat would be conducted from the surroundings to the leaf, raising its temperature to perhaps $-1°C$. Using this leaf temperature, we can calculate that there would be a net energy loss by radiation of 66 W m^{-2} on the clear night (see Table 7-1). A plant on a cloudless night may therefore freeze even though the temperatures of both the air and the surroundings are above $0°C$ because the excess of IR emitted over that absorbed can lower the leaf temperature below the freezing point. Such freezing of leaves on clear nights is a severe problem for certain agricultural crops. A method for avoiding freezing damage employed for certain citrus and other orchards is to spray water on the trees. The release of the heat of fusion as this water freezes maintains the plant tissues at $0°C$, which is above the freezing point of their cell sap.

II. Wind–Heat Conduction and Convection

Now that we have considered the net radiation balance of a leaf, we will examine other ways that energy may be exchanged with the environment. For instance, heat can be conducted from one body to an adjacent cooler one by molecular and/or electronic collisions. Of particular importance in this regard is heat conduction across layers of air surrounding plant parts. In particular, frictional interactions between a fluid and a solid phase moving with respect to each other lead to boundary layers of fluid adhering to the solid phase; heat and mass are exchanged across such boundary layers. In Chapter 1 we considered unstirred layers of water adjacent to membranes (see Fig. 1-10), and in this section we will consider boundary layers of air adjacent to leaves. We will assume that a temperature difference generally exists across such an air boundary layer adjacent to a leaf surface. Heat can be conducted across this air layer by the random thermal collisions of the gas molecules. Heat convection, on the other hand, involves turbulent movement of a fluid, brought about for example by differences in pressure.

There are two types of convection, *free* and *forced*. Free (natural) convection occurs when the heat transferred from a leaf causes the air outside the unstirred layer to warm, expand, and thus to decrease in density; this more buoyant warmer air then moves upward and thereby moves heat away from the leaf. Forced convection, caused by wind, can also remove the heated air outside the boundary layer. As the wind speed increases, more and more heat is dissipated by forced relative to free convection. However, even at a very low wind speed of 0.10 m s^{-1}, forced convection dominates free convection as a means of heat loss from most leaves (0.10 m s^{-1} $= 0.36$ km h^{-1} $= 0.22$ mile h^{-1}). We can therefore generally assume that heat is conducted across the boundary layer adjacent to the leaf and then removed by forced convection in the surrounding turbulent air. We will next examine some general characteristics of wind, paying particular attention to the air boundary layers adjacent to plant parts, and then introduce certain dimensionless numbers that can help indicate whether forced or free convection should prevail.

A. Wind—General Comments

Wind can influence plant growth, reproduction, and distribution and in some cases can be lethal. It can mechanically deform plants and can also disperse pollen, seeds, disease organisms, and gaseous substances such as CO_2 and pollutants. Many effects of wind depend on the air boundary layers next to the aerial surfaces of a plant across which mass and heat exchanges occur with the environment.

Wind is caused by differences in air pressure that result from differential absorption of shortwave irradiation by the earth's surface and by clouds. On a macro scale, wind speed is affected by land surface features such as mountains and canyons, whereas at about 10 m above a plant canopy wind can be predominantly influenced by the plants, becoming arrested at their surfaces. In coastal regions wind speed can annually average as high as 7 m s^{-1} at 10 m above the ground, and in topographically flat inland areas 1 m s^{-1} is a typical mean wind speed. Wind speeds are generally lower at night. Turbulent flow, where air movement is not parallel and orderly, characterizes the wind regime near vegetation. In fact, the standard deviation of wind speed divided by the mean wind speed, the *turbulence intensity*, is often about 0.4 near vegetation, indicating that wind has considerable temporal variation. In such turbulent regimes air can be described as moving in packets or "eddies" (to be further considered in Chapter 9).

Whenever air is decelerated by an object such as a leaf or a branch, *form drag* occurs. When the airflow is stopped, the force transmitted per unit area of surface perpendicular to the wind direction is $\frac{1}{2}\rho v^2$, where ρ is the air density (1.20 kg m^{-3} at 20°C, Appendix I) and v is the wind speed (the magnitude of the wind velocity, which is a vector having both a magnitude and a direction). However, much of the airflow is directed around most objects, so the actual form drag is reduced from its maximum value. For increasing wind speeds, leaves with flexible petioles (leaf stalks) generally change their aspect such that a smaller area is presented perpendicular to the wind direction. If we designate the original area projected in the wind direction by A_{orig}, we can define a dimensionless drag coefficient c_d as follows:

$$c_d = \frac{\text{actual drag force}}{\frac{1}{2}\rho v^2 A_{\text{orig}}} \qquad (7.7)$$

For relatively flat leaves that can align with the wind direction, c_d is generally 0.02–0.2, but it can be 0.5–0.9 for *bluff* bodies such as branches and tree trunks that block and hence substantially change the airflow. For flexible objects the drag coefficient defined by Eq. (7.7) can decrease as the wind speed increases. In particular, the areas of leaves and small branches perpendicular to the wind direction can decrease threefold as the wind speed increases, leading to an aerodynamically more streamlined shape, which occurs for many trees.

Sites experiencing greater wind speeds tend to have shorter vegetation, such as cushion plants in alpine tundra or the procumbent forms on coastal dunes. Also, stem elongation can be reduced two- or threefold by high winds. On the other hand, environmental chambers in which plants are grown for research purposes tend to have low wind speeds (generally below 0.5 m s^{-1}) and low light, often leading to spindly plants unlike their field-grown counterparts. The retardation of stem elongation and the increase in girth caused by wind are mainly due to the development of shorter cells with thicker cell walls. Agronomic implications of wind-induced sway have long been recognized, e.g., increasing the spacing between nursery stock leads to sturdier trees with larger trunk diameters. Mechanically shaking closely spaced young trees in a nursery for about 20 s per day can also increase stem diameter, avoiding the space requirements between plants necessary for movement to occur by frictional interactions with the wind.

Buttresses at the base of tree trunks and roots are more common on the windward side. Such a location more effectively resists upsetting forces caused by wind than if the buttresses or enhanced root growth were on the leeward side because the tensile strength of wood is greater than its compressional strength. A consequence of a prevailing wind direction are "flag trees," where branches occur mainly toward the leeward side. Most of these effects of wind on stem morphology are hormonally mediated. At the extreme of sporadic high winds such as occur in gales (wind speeds of 17–21 m s^{-1}, which corresponds to 61–76 km h^{-1} or 38–47 miles h^{-1}), form drag can cause stems to be permanently displaced from their upright position. This process is termed "lodging" for various cereal crops. To prevent lodging, breeding programs have developed rice, wheat, barley, and oat genotypes with shorter, sturdier stems. Wind is also one of the major factors in the ecology of forests, forming gaps in the canopy by uprooting trees, creating special microhabitats by distributing leaf litter, and influencing succession by dispersing pollen, spores, and certain seeds.

B. Air Boundary Layers

Wind speed affects the thickness of the air boundary layer next to a leaf or other aerial plant part. Because boundary layers influence heat exchange and hence the temperature of the shoot, any process in a shoot depending on temperature can be affected by wind speed. Also, any molecule entering or leaving a leaf in the gas phase must cross the boundary layer next to the surface.

The boundary layer is a region dominated by the shearing stresses originating at the surface; such layers arise for any solid immersed in a fluid, e.g., a leaf in air. Adjacent to the leaf is a laminar sublayer of air (Fig. 7-5), where movement is predominantly parallel

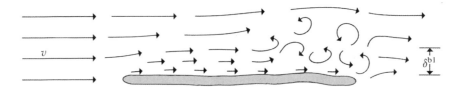

Figure 7-5. Schematic illustration of originally nonturbulent air flowing over the top of a flat leaf, indicating the laminar sublayer (shorter straight arrows), the turbulent region (curved arrows), and the effective boundary layer thickness δ^{bl}. The arrows indicate the relative speed and the direction of air movement.

to the leaf surface; air movement is arrested at the surface and has increasing speed at greater distances from the surface. Diffusion perpendicularly away from the leaf surface is by molecular motion in the laminar sublayer. Farther from the surface, especially on the downwind part of a leaf, the boundary layer becomes turbulent (Fig. 7-5). Here, heat and gas movements are eddy assisted, i.e., the air swirls around in vortices and behaves as if it were moving in little units or packets. Instead of describing the local transfer processes in the laminar and the turbulent portions, both of which change in thickness across a leaf's surface, we use an effective or equivalent boundary layer thickness, δ^{bl}, averaged over the whole leaf surface (this is often referred to as the displacement air boundary layer). The boundary layer is actually somewhat thinner at the upwind or leading edge of a leaf than at its center (Fig. 7-5); as a consequence, the temperature can differ slightly along a leaf. Because there is no sharp discontinuity of wind speed between the air adjacent to the leaf and the free airstream, the definition of boundary layer thickness is arbitrary, so it is generally defined operationally, e.g., in terms of the flux of heat across it, which we will consider later.

As we have indicated, δ^{bl} represents an average thickness of the unstirred air layer adjacent to a leaf (or to leaflets for a compound leaf). The main factors affecting δ^{bl} are the ambient wind speed and the leaf size, with leaf shape exerting a secondary influence. Partly for convenience, but mainly because it has proved experimentally justifiable, we will handle the effect of leaf size on boundary layer thickness by the characteristic dimension l, which is the mean length of a leaf in the direction of the wind. Based on hydrodynamic theory for laminar flow adjacent to a flat surface as modified by actual observations under field conditions, an approximate expression for the average thickness of the boundary layer next to a flat leaf is

$$\delta^{bl}_{(mm)} = 4.0 \sqrt{\frac{l_{(m)}}{v_{(m\ s^{-1})}}} \tag{7.8}$$

where $l_{(m)}$ is the mean length of the leaf in the downwind direction in m, $v_{(m\ s^{-1})}$ is the ambient wind speed in m s^{-1}, and $\delta^{bl}_{(mm)}$ is the average thickness of the boundary layer in mm; the factor 4.0 has units of mm s$^{-0.5}$, as we can deduce by using the indicated units for the three variables.

Instead of the numerical factor 4.0 in Eq. (7.8), hydrodynamic theory predicts a factor closer to 6.0 for the effective boundary layer thickness adjacent to a flat plate. However, wind tunnel measurements under an appropriate turbulence intensity, as well as field measurements, indicate that 4.0 is more suitable for leaves. This divergence from theory relates

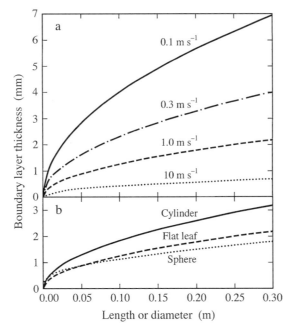

Figure 7-6. Mean thickness of the air boundary layer adjacent to a flat leaf at various wind speeds indicated next to the curves (a) and adjacent to objects of three different geometries at a wind speed of 1 m s^{-1} (b). The length for a flat leaf represents the mean distance across it in the direction of the wind; the diameter is used for the bluff bodies represented by cylinders and spheres. Note that 1.0 m s^{-1} = 3.6 km h^{-1} = 2.2 mile h^{-1}.

to the relatively small size of leaves, their irregular shape, leaf curl, leaf flutter, and, most important, the high turbulence intensity under field conditions. Moreover, the dependency of δ^{bl} on $l^{0.5}$, which applies to large flat surfaces, does not always best fit the data; wind tunnel measurements for various leaf shapes and sizes give values for the exponent of 0.3–0.5. Thus various shape considerations, including the leaf dimension perpendicular to the wind direction, can affect δ^{bl}. In addition, instead of the dependence on $v^{-0.5}$ indicated in Eq. (7.8), measurements with leaves are best described by exponents of -0.5 to -0.7. Consequently, Eq. (7.8) is only a useful approximation for indicating how the average boundary layer thickness varies with leaf size and wind speed.

Average daily wind speeds just above vegetation generally range from 0.1 to 10 m s^{-1}— exposed leaves commonly experience wind speeds near 1 m s^{-1}. Because the thickness of the air boundary layer enters into many calculations of heat and gas fluxes for leaves, the magnitudes of δ^{bl} are indicated in Fig. 7-6a versus leaf size for four wide-ranging wind speeds.

C. Boundary Layers for Bluff Bodies

Although relatively flat leaves can be described by the boundary layer considerations presented previously (Fig. 7-5 and Eq. 7.8), many plant parts such as stems, branches, inflorescences, fruits, and even certain leaves (e.g., the tubular leaves of onion, *Allium cepa*) represent three-dimensional objects. Airflow is intercepted by such bluff bodies and forced

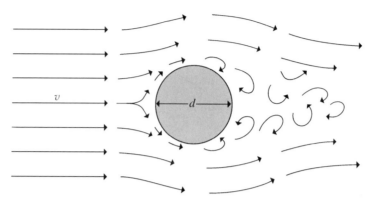

Figure 7-7. Schematic illustration of airflow around a cylinder. Flow can be laminar on the upwind side, but
turbulence develops on the downwind side.

to move around them. Here we will consider two shapes, cylinders and spheres. In the next
section we will present heat flux equations for objects of cylindrical and spherical symmetry
as well as for flat leaves.

A laminar boundary layer develops on the upwind side of a cylinder (Fig. 7-7). It is
analogous to the laminar sublayer for flat plates (Fig. 7-5), and air movements in it can be
described analytically. On the downwind side of the cylinder, the air flow becomes turbulent,
can be opposite in direction to the wind, and in general is quite difficult to analyze. Nevertheless, an effective boundary layer thickness can be estimated. For turbulence intensities
appropriate to field conditions, $\delta^{bl}_{(mm)}$ in mm can be represented as follows:

$$\delta^{bl}_{(mm)} = 5.8 \sqrt{\frac{d_{(m)}}{v_{(m\ s^{-1})}}} \qquad \text{cylinder} \qquad (7.9)$$

where $d_{(m)}$ is the cylinder diameter in m. As for flat leaves, the boundary layer is thinner
for smaller objects and at higher wind speeds (Fig. 7-6b).

A similar analysis for the effective average boundary layer around a sphere under turbulent
intensities appropriate to field conditions leads to

$$\delta^{bl}_{(mm)} = 2.8 \sqrt{\frac{d_{(m)}}{v_{(m\ s^{-1})}}} + \frac{0.25}{v_{(m\ s^{-1})}} \qquad \text{sphere} \qquad (7.10)$$

where $d_{(m)}$ is the diameter of the sphere in m (see Fig. 7-6b). Equation (7.10) has been
successfully used to predict boundary layer thicknesses for various spherical fruits.

D. Heat Conduction/Convection Equations

Now that we have considered the average boundary layer thickness, we return to a consideration of convective heat exchange, where heat is first conducted across the boundary layer
and then convected away in the moving airstream. For the one-dimensional case, heat flow
by conduction equals $-K\ \partial T/\partial x$, where K is the *thermal conductivity coefficient* (e.g.,
$W\ m^{-1}\ {}^{\circ}C^{-1}$) and $\partial T/\partial x$ is the temperature gradient (this relation is some times referred to
as Fourier's heat-transfer law). Because heat can be conducted across the boundary layers

on either side of a leaf, the factor 2 is needed to describe the total rate of heat flux by conduction per unit area of one side of a leaf. (For convenience we will assume that the boundary layers on the two sides are of equal thickness, δ^{bl}.) The heat conducted across the boundary layers and convected away from a leaf per unit time and area therefore is

$$J_H^C = -2K^{air}\frac{\partial T}{\partial x}$$

$$= 2K^{air}\frac{(T^{leaf} - T^{ta})}{\delta^{bl}} \qquad (7.11)$$

where J_H^C is the rate of heat conduction per unit area (e.g., W m^{-2}), K^{air} is the thermal conductivity coefficient of air, T^{leaf} is the leaf temperature, and T^{ta} is the temperature of the turbulent air outside a boundary layer of thickness δ^{bl}. Because heat is conducted from the solid surface of the leaf across the adjacent unstirred air, J_H^C does not depend on whether the stomata are open or closed—we have a planar surface as a source at a specific temperature, T^{leaf}, from which heat is conducted across a boundary layer to the outside turbulent air at T^{ta}. The heat flux density in Eq. (7.11) is considered to be positive when heat goes from the leaf to the surrounding air. Heat is conducted into the leaf when $T^{leaf} < T^{ta}$, in which case J_H^C is negative.

All of the flux equations used so far in this book have been for one-dimensional cases. Because we have introduced the average thickness of the boundary layer for cylinders (Eq. 7.9) and spheres (Eq. 7.10), let us also consider the appropriate fluxes for such cases, which can have many biological applications. For cylindrical symmetry, thermal properties can change in the radial direction perpendicularly away from the cylinder axis but not with angle around the cylinder or length along its axis. The heat flux density at the cylinder's surface for such cylindrical symmetry is

$$J_H^C = \frac{K^{air}(T^{surf} - T^{ta})}{r \ln\left(\frac{r+\delta^{bl}}{r}\right)} \qquad \text{cylinder} \qquad (7.12)$$

where r is the cylinder radius, T^{surf} is its surface temperature, δ^{bl} is calculated by Eq. (7.9), and the other quantities have the same meaning as for the one-dimensional case (Eq. 7.11). For spherical symmetry, properties vary only in the radial direction and not with any angle; the heat flux density at the sphere's surface for conduction across the boundary layer and convection in the surrounding turbulent air then is

$$J_H^C = \frac{(r+\delta^{bl})K^{air}(T^{surf} - T^{ta})}{r\delta^{bl}} \qquad \text{sphere} \qquad (7.13)$$

where r is the radius of the sphere and δ^{bl} is calculated by Eq. (7.10).

As can be seen, the flux for heat conduction across the air boundary layer is proportional to $K^{air}(T^{surf} - T^{ta})$ for all the shapes considered. Values of the thermal conductivity coefficient for dry air at various temperatures are given in W m^{-1} °C^{-1} in Appendix I. Because the conduction of heat in a gas phase is based on the random thermal motion of the molecules, the composition of air, such as the content of water vapor, can influence K^{air}. For instance, K^{air} at 20°C is 1% less for 100% relative humidity than it is for dry air. Air can hold considerably more water vapor at 40°C, where K^{air} is 2% lower for water-saturated air than that of 0% relative humidity.

E. Dimensionless Numbers

In many studies on heat and gas fluxes, relationships between parameters are expressed in terms of dimensionless numbers. This facilitates comparisons between objects of the same shape but different sizes and for different wind speeds, i.e., dimensionless numbers allow application of data to different but geometrically similar situations. For instance, dimensionless numbers were used to determine the effects of wind speed on δ^{bl} for the various shapes presented previously. Dimensionless numbers can also be used to study boundary layer phenomena and water flow characteristics—water speeds can exceed 10 m s^{-1} in intertidal regions due to wave action, although they are usually from 0.01 to 0.2 m s^{-1} in coastal regions of lakes and oceans. We will consider three dimensionless numbers that are particularly important for analyzing heat fluxes.

Before presenting these dimensionless numbers, we will indicate a common convention for describing heat conduction across boundary layers, a convention that is invariably used for objects of irregular shape and is often used for geometrically regular objects. Instead of expressions involving δ^{bl} (e.g., Eq. 7.11 for flat plates, Eq. 7.12 for cylinders, and Eq. 7.13 for spheres), the following relation is used to describe the heat flux density across the air boundary layer:

$$J_H^C = h_c(T^{surf} - T^{ta}) \tag{7.14}$$

where h_c is called the *heat convection coefficient* (or the convective heat-transfer coefficient; Eq. 7.14 is known as Newton's law of cooling). Upon comparing Eq. (7.14) with Eq. (7.11) and noting that h_c generally refers to a unit surface area of one side of a flat leaf [hence, the total heat flux density is $2h_c(T^{surf} - T^{ta})$], we find that $h_c = K^{air}/\delta^{bl}$ for flat leaves (slightly more complex relations hold for cylinders and spheres). Even when the boundary layer thickness cannot be determined analytically, e.g., for the irregular shapes of cacti with their surface ribbing and projecting spines, Eq. (7.14) can still be used to relate convective heat exchange to the temperature difference between the plant part and the air. Appropriate units of h_c are W m^{-2} °C^{-1}.

Now that we have introduced the heat convection coefficient, we will define our first dimensionless number, the *Nusselt number*. We represent the size of a particular plant part by a characteristic dimension d, which for a flat plate is the quantity l in Eq. (7.8) and for a cylinder or sphere is the diameter. This leads to

$$\text{Nusselt number} = \text{Nu} = \frac{h_c d}{K^{air}}$$

$$= \frac{d}{\delta^{bl}} \tag{7.15}$$

where in the second line we have used our comment on h_c for flat leaves ($h_c = K^{air}/\delta^{bl}$), i.e., Nu relates the characteristic dimension to the boundary layer thickness, and thus Nusselt numbers are useful for describing heat transfer across boundary layers.

Next, we introduce a dimensionless number that describes flow characteristics, e.g., it can help indicate whether the flow will be laminar or turbulent. This quantity indicates the ratio of inertial forces (due to momentum, which tends to keep things moving) to viscous forces (due to friction, which tends to slow things down) and is known as the *Reynolds number*:

$$\text{Reynolds number} = \text{Re} = \frac{vd}{\nu} \tag{7.16}$$

where v is the magnitude of the fluid velocity moving past an object of characteristic dimension d and ν is the kinematic viscosity (ordinary viscosity divided by density) for the fluid (1.51×10^{-5} m^2 s^{-1} for dry air at 20°C; Appendix I). At low Reynolds numbers viscous forces dominate inertial forces and the flow tends to be laminar, whereas at high Reynolds numbers (above about 10^4 for plant parts) the flow is mainly turbulent owing to the dominance of inertial forces. For a cylindrical branch 0.1 m in diameter, turbulence predominates for wind speeds above about 1.5 m s^{-1}. Although air movement is different below compared with above such critical Reynolds numbers, Eqs. (7.8)–(7.13) are satisfactory for most applications to plants in either flow regime.[3]

At very low wind speeds and large values of $T^{\text{surf}} - T^{\text{ta}}$, free convection can dominate forced convection for large objects. In such cases, the Reynolds number should be replaced in heat flow studies by the *Grashof number*, which takes into account buoyant forces. Specifically, the Grashof number (Gr) indicates the tendency of a parcel of air to rise or fall and thus describes the tendency for free convection. In fact, Gr represents the ratio of buoyant times inertial forces to the square of viscous forces:

$$\text{Grashof number} = \text{Gr} = \frac{g\,\beta\,\Delta T\,d^3}{\nu^2} \tag{7.17}$$

where g is the gravitational acceleration, β is the coefficient of volumetric thermal expansion (i.e., the fractional change in volume with temperature at constant pressure, which equals $1/T$ for an essentially perfect gas such as air, e.g., $\beta = 3.4 \times 10^{-3}$ °C^{-1} at 20°C), and ΔT is the temperature difference from the object's surface to the ambient air ($T^{\text{surf}} - T^{\text{ta}}$).

Using the Grashof and the Reynolds numbers, we can indicate whether forced or free convection dominates in a particular case. Because Re = inertial forces/viscous forces and Gr = buoyant × inertial forces/(viscous forces)^2, Re^2/Gr = inertial forces/buoyant forces. Thus Re^2/Gr reflects forced convection/free convection. Experiments reveal that forced convection accounts for nearly all heat transfer when Re^2/Gr > 10, free convection accounts for nearly all heat transfer when Re^2/Gr < 0.1, and the intervening region has mixed convection, i.e., both forced and free convection should then be considered, especially for Re^2/Gr near 1. Using Eqs. (7.16) and (7.17), we obtain

$$\frac{\text{Re}^2}{\text{Gr}} = \frac{\left(\frac{vd}{\nu}\right)^2}{\left(\frac{g\,\beta\,\Delta T\,d^3}{\nu^2}\right)} = \frac{v^2}{g\,\beta\,\Delta T\,d}$$

$$= \frac{v^2}{(9.8 \text{ m s}^{-2})(3.4 \times 10^{-3} \text{ °C}^{-1})\,\Delta T\,d}$$

$$= (30 \text{ s}^2\,°\text{C m}^{-1})\,\frac{v^2}{\Delta T\,d} \tag{7.18}$$

Considering $v^2/(\Delta T\,d)$ in Eq. (7.18), we note that inertial forces and hence forced convection become more important for higher wind speeds, smaller temperature differences, and smaller objects. For a ΔT of 5°C and a d of 0.1 m, Eq. (7.18) indicates that Re^2/Gr

---

[3] For flat leaves under field conditions, analysis by dimensionless numbers indicates that Nu $= 0.97$ Re$^{0.5}$. Hence, $d/\delta^{\text{bl}} = 0.97(vd/\nu)^{1/2}$ by Eqs. (7.15) and (7.16), so δ^{bl} is $(1/0.97)(\nu/vd)^{1/2}d = (1/0.97)(1.53 \times 10^{-5}$ m^2 s$^{-1})^{1/2}$ $(d/v)^{1/2} = 4.0 \times 10^{-3} \sqrt{d/v}$ m s$^{-1/2}$, which is essentially Eq. (7.8).

= 1 when the wind speed is

$$v = \left[\frac{(5°C)(0.1\text{ m})}{(30\text{ s}^2\text{ °C m}^{-1})} \right]^{1/2} = 0.1\text{ m s}^{-1}$$

Thus for wind speeds >0.1 m s^{-1}, forced convection dominates free convection when ΔT is 5°C and d is 0.1 m. This domination of forced convection over free convection occurs for most of our applications.

F. Examples of Heat Conduction/Convection

We now calculate the heat conduction across the air boundary layer for a leaf at 25°C when the surrounding turbulent air is at 20°C. We will consider a leaf with a mean length in the wind direction of 0.10 m (10 cm) and a wind speed of 0.8 m s^{-1}. From Eq. (7.8), the boundary layer thickness is

$$\delta^{bl}_{(mm)} = 4.0 \sqrt{\frac{(0.10\text{ m})}{(0.8\text{ m s}^{-1})}} = 1.4\text{ mm}$$

which is typical for a leaf. Using Eq. (7.11) and K^{air} of 0.0259 W m^{-1} °C^{-1} (appropriate for 20–25°C; see Appendix I), the heat flux density conducted across the boundary layer is

$$J^C_H = \frac{(2)(0.0259\text{ W m}^{-1}\text{ °C}^{-1})(25°C - 20°C)}{(1.4 \times 10^{-3}\text{ m})}$$

$$= 190\text{ W m}^{-2}$$

A leaf at 25°C at sea level on a sunny day can have a net radiation balance of 370 W m^{-2} (see top line of Table 7-1). In the current case, just over half of this energy input by net radiation is dissipated by conduction of heat across the boundary layers on each side of the leaf (190 W m^{-2} total), followed by forced convection in the surrounding turbulent air outside the boundary layers.

A leaf with a thick boundary layer can have a temperature quite different from that of the surrounding air because air is a relatively poor conductor of heat. Specifically, K^{air} is low compared with the thermal conductivity coefficients for liquids and most solids. A large leaf exposed to a low wind speed might have a boundary layer that is 4 mm thick (see Fig. 7-6a or Eq. 7.8 for the wind speeds and leaf sizes implied by this). If J^C_H and K^{air} are the same as in the previous paragraph, where the difference between leaf and turbulent air temperatures is 5°C when the boundary layer is 1.4 mm thick, then $T^{leaf} - T^{ta}$ will be (4 mm/1.4 mm) (5°C), or 14°C, for a δ^{bl} of 4 mm. Thus the combination of large leaves and low wind speeds favors a large drop in temperature across the boundary layers. On the other hand, a small leaf in a moderate wind can have a δ^{bl} of 0.2 mm. For the same J^C_H and K^{air} as given previously, $T^{leaf} - T^{ta}$ is (0.2 mm/1.4 mm)(5°C), or 0.7°C, for this thin boundary layer. Hence, small leaves tend to have temperatures quite close to those of the air, especially at moderate to high wind speeds. This close coupling between leaf and air temperatures for small leaves can keep the leaf temperature low enough for optimal photosynthesis (often 30–35°C) in hot, sunny climates. Also, the lower the leaf temperature, the lower the concentration of water vapor in the pores of the cell walls of mesophyll cells, and consequently less water then tends to be lost in transpiration (see Chapter 8), an important consideration in arid and semiarid regions.

III. Latent Heat—Transpiration

Evaporation of water is a cooling process. Water evaporates at the air–liquid interfaces along the pores in the cell walls of mesophyll, epidermal, and guard cells (see Fig. 1-2) and then diffuses out of a leaf. Thus transpiration represents a means of heat loss by a leaf (Eq. 7.2). A leaf can gain latent heat if dew or frost condenses onto it, as we will also discuss.

A. Heat Flux Density Accompanying Transpiration

We represent the flux density of water vapor diffusing out of a leaf during transpiration by J_{wv}. If we multiply the amount of water leaving per unit time and per unit leaf area, J_{wv}, by the energy necessary to evaporate a unit amount of water at the temperature of the leaf, H_{vap}, we obtain the heat flux density accompanying transpiration, J_H^T:

$$J_H^T = J_{wv} H_{vap} = \frac{H_{vap} D_{wv} \Delta c_{wv}^{total}}{\Delta x^{total}} = \frac{H_{vap} D_{wv} \left(c_{wv}^e - c_{wv}^{ta} \right)}{\Delta x^{total}} \qquad (7.19)$$

where Fick's first law (Eqs. 1.1 and 1.8) has been used to express J_{wv} in terms of the diffusion coefficient for water vapor, D_{wv}, and the total drop in water vapor concentration, Δc_{wv}^{total}, over some effective total distance, Δx^{total}. In turn, Δc_{wv}^{total} equals the water vapor concentration at the sites of evaporation within a leaf, c_{wv}^e, minus the value in the turbulent air just outside the boundary layer, c_{wv}^{ta}. J_H^T and J_{wv} in Eq. (7.19) are conventionally based on the area of one side of a leaf, and Δx^{total} is usually incorporated into a resistance or a conductance, which we will discuss in Chapter 8.

How much of the heat load on a leaf is dissipated by the evaporation of water during transpiration? For an exposed leaf of a typical mesophyte during the daytime, J_{wv} is about 4 mmol m^{-2} s^{-1}. In Chapter 2 we noted that water has a high heat of vaporization, e.g., 44.0 kJ mol^{-1} at 25°C (values at other temperatures are given in Appendix I). By Eq. (7.19), the heat flux density out of the leaf by transpiration then is

$$J_H^T = (4 \times 10^{-3} \text{ mol m}^{-2} \text{ s}^{-1})(44.0 \times 10^3 \text{ J mol}^{-1}) = 180 \text{ W m}^{-2}$$

For the leaf described in the top line of Table 7-1, a heat loss of 180 W m^{-2} by evaporation dissipates slightly under half of the net radiation balance (370 W m^{-2}), the rest of the energy input being removed by heat conduction across the boundary layer followed by forced convection (Fig. 7-8a). The heat loss accompanying transpiration reduces leaf temperatures during the daytime. Although such latent heat losses can benefit a plant by lowering leaf temperatures, evaporation and its associated cooling are simply inevitable consequences of gas exchange by leaves, for which opening of the stomata is necessary for substantial rates of CO_2 uptake.

B. Heat Flux Density for Dew or Frost Formation

So far we have regarded c_{wv}^e as greater than c_{wv}^{ta}, in which case a net loss of water occurs from the leaf, with a consequent dissipation of heat. When the turbulent air is warmer than the leaf, however, the water vapor concentration in the turbulent air may be greater than that in the leaf (in Chapter 2 we noted that the water vapor concentration and partial pressure at saturation increase rapidly with temperature, e.g., Fig. 2-13; also, see values for P_{wv}^* and c_{wv}^* in Appendix I). If c_{wv}^{ta} is greater than the water vapor concentration in the leaf, then a net diffusion of water vapor occurs toward the leaf. This can increase c_{wv} at the leaf surface,

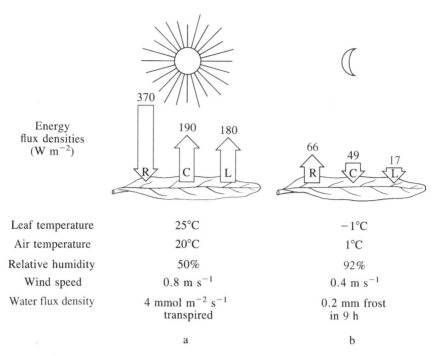

Figure 7-8. Energy budget for an exposed leaf (a) at midday and (b) at night with frost. The flux densities in W m^{-2} are indicated for net radiation (R); conduction/convection (C), also referred to as sensible heat; and latent heat (L).

and it may reach c_{wv}^*, the saturation value. If c_{wv}^{ta} is greater than this c_{wv}^*, dew—or frost, if the leaf temperature is below freezing—can form as water vapor diffuses toward the leaf and then condenses onto its surface, which is cooler than the turbulent air. Condensation resulting from water emanating from the soil is sometimes called "distillation," with the term "dew" being reserved for water coming from the air above.

Condensation of water vapor leads to a heat gain by a leaf. Because water condensation is the reverse of the energy-dissipating process of evaporation, the heat gain per unit amount of water condensed is the heat of vaporization of water at the temperature of the leaf, H_{vap}. Because the condensation is on the leaf surface, the diffusion is across the air boundary layers of thickness δ^{bl} that are present on either side of a leaf. To describe the rate of heat gain per unit area accompanying the water vapor condensation leading to dew formation, Eq. (7.19) becomes

$$J_H^{dew} = \frac{2 H_{vap} D_{wv} \left(c_{wv}^{ta} - c_{wv}^{leaf^*} \right)}{\delta^{bl}} \qquad (7.20)$$

where $c_{wv}^{leaf^*}$ is the saturation concentration of water vapor at the temperature of the leaf. The factor 2 is necessary because water vapor diffuses toward the leaf across the boundary layer on each side (as usual, we will assume that these layers are of equal average thickness δ^{bl}).

The temperature to which the turbulent air must be reduced at constant pressure for it to become saturated with water vapor is known as the *dew point*, or dew point temperature, $T_{dew\ point}^{ta}$. When $T^{leaf} < T_{dew\ point}^{ta}$, the turbulent air contains more water vapor (c_{wv}^{ta}) than the air at the leaf surface can hold ($c_{wv}^{leaf^*}$). Water vapor then diffuses toward the leaf, which can lead to dew formation (Eq. 7.20). If $T^{\text{leaf}} < 0°C$ and $< T^{\text{ta}}_{\text{dew point}}$, the water that condenses onto the leaf surface freezes. Under such conditions we must replace H_{vap} in Eq. (7.20) with the heat of sublimation, H_{sub}, at that particular leaf temperature to describe the heat gain by frost formation.

C. Examples of Frost and Dew Formation

As an example of nighttime frost formation, we will consider a leaf with a temperature of $-1°C$ when the surrounding air is at $1°C$ and 92% relative humidity. The leaf can be the same leaf considered under Section IIF, where we estimated a boundary layer thickness of 1.4 mm during the daytime; at night, the wind speed is generally lower—if v were halved, then the boundary layer would be 41% thicker, or 2.0 mm, by Eq. (7.8). D_{wv} is 2.13×10^{-5} m^2 s^{-1} at $0°C$, H_{sub} is 51.0 kJ mol^{-1} at $-1°C$, and c^*_{wv} is 0.249 mol m^{-3} at $-1°C$ and 0.288 mol m^{-3} at $1°C$ (Appendix I). By Eq. (7.20), the rate of heat gain per unit area by frost formation is

$$J_H^{\text{frost}} = (2)(51.0 \times 10^3 \text{ J mol}^{-1})(2.13 \times 10^{-5} \text{ m}^2 \text{ s}^{-1})$$
$$\times \frac{[(0.92)(0.288 \text{ mol m}^{-3}) - (0.249 \text{ mol m}^{-3})]}{(2.0 \times 10^{-3} \text{ m})}$$
$$= 17 \text{ W m}^{-2}$$

Because $T^{\text{leaf}} < T^{\text{ta}}$, heat is conducted into the leaf from the air. By Eq. (7.11) [$J_H^C = 2K^{\text{air}}(T^{\text{leaf}} - T^{\text{ta}})/\delta^{\text{bl}}$, where K^{air} is 0.0243 W m^{-1} °C^{-1} at $0°C$; Appendix I], the heat conduction across the boundary layer is $(2)(0.0243$ W m^{-1} °C$^{-1})(-1°C -1°C)/(2.0 \times 10^{-3}$ m$)$ or -49 W m^{-2}, where the minus sign indicates that heat is conducted into the leaf. The bottom line of Table 7-1 indicates that this leaf can lose 66 W m^{-2} by net radiation. Thus the heat inputs from frost formation (17 W m^{-2}) and heat conduction (49 W m^{-2}) balance the energy loss by net radiation (see Fig. 7-8b).

How much time is required to form a layer of frost 0.1 mm thick on each side of a leaf under the previous conditions? Because $J_H^{\text{frost}} = J_{wv}H_{\text{sub}}$ (see Eq. 7.19), the rate of water deposition per unit area (J_{wv}) in kg m^{-2} s^{-1} is J_H^{frost} divided by H_{sub} in J kg^{-1} (H_{sub} is 2.83×10^6 J kg^{-1} at $-1°C$; Appendix I). This J_{wv} (mass of water per unit area and per unit time) divided by the mass of ice per unit volume, which is the density of ice (ρ_{ice}, 917 kg m^{-3}), gives the thickness of frost accumulation per unit time in m s^{-1}. To accumulate 0.1 mm of ice on each side of a leaf therefore requires

$$t = \frac{\text{thickness}}{J_{wv}/\rho_{\text{ice}}} = \frac{(0.2 \text{ mm}) \rho_{\text{ice}}}{J_H^{\text{frost}}/H_{\text{sub}}}$$
$$= \frac{(0.2 \times 10^{-3} \text{ m})(917 \text{ kg m}^{-3})}{(17 \text{ J s}^{-1} \text{ m}^{-2})/(2.83 \times 10^6 \text{ J kg}^{-1})}$$
$$= 3.1 \times 10^4 \text{ s} \quad (8.6 \text{ h})$$

Dew formation can be as much as 0.5 mm per night, which can be an important source of water in certain regions for part of the year.[4] For the lichen *Ramalina maciformis* in the

---

[4] At $10°C$, a layer of liquid water 0.5 mm thick corresponds to the content of water vapor in a column of saturated air (see Appendix I) that is 53 m tall. Hence, considerable downward movement of water vapor is necessary to have dew formation over large areas, or much of the water vapor must emanate from the ground.

Negev Desert, dew sufficient to lead to photosynthesis the next day occurs on about half of the nights, and the annual dewfall can be 30 mm (Kappen et al., 1979). Dew or frost formation is favored on cloudless nights and when the relative humidity of the surrounding turbulent air is very high. As mentioned previously, T^{leaf} tends to be further from T^{ta} for large leaves than for small ones. On a cloudless night when $T^{\text{leaf}} < T^{\text{surr}}$ and T^{ta} (see Table 7-1), the larger exposed leaves generally dip below $T^{\text{ta}}_{\text{dew point}}$ sooner than do the small ones, and hence dew or frost tends to form on the larger leaves first. For convenience, we have assumed that the leaf has a uniform temperature. In fact, however, the boundary layer tends to be thinner at the edges of a leaf (Fig. 7-5), so T^{leaf} is closer to T^{ta} at the edges than at the center. Thus dew or frost generally forms first at the center of a leaf, where δ^{bl} is larger and the temperature slightly lower than at the leaf edges.

IV. Further Examples of Energy Budgets

Many observations and calculations indicate that exposed leaves in sunlight tend to be above air temperature for T^{ta} up to about 30°C and below T^{ta} for air temperatures above about 35°C. This primarily reflects the increasing importance of IR radiation emission as leaf temperature rises [see Eq. 7.5; $J_{\text{IR}} = 2e_{\text{IR}}\sigma(T^{\text{leaf}})^4$] and the increase with temperature of the water vapor concentration in the leaves, which affects transpiration (discussed in Chapter 8). Such influences often lead to temperatures for exposed leaves that are more favorable for photosynthesis than is the ambient air temperature. We can readily extend our analysis to include leaves shaded by overlying ones. For certain plant parts, heat storage and heat conduction within the tissues are important. We will conclude this section with some comments on the time constants for changes in leaf and stem temperature.

A. Leaf Shape and Orientation

Leaf sizes and shapes vary tremendously, which can have important consequences for leaf temperature. Leaves developing in full sunlight tend to have smaller areas when mature than do leaves on the same plant that develop in the shade—"sun" leaves generally have 20–80% less surface area than do "shade" leaves. When shade leaves are placed in exposed locations, their larger size leads to thicker boundary layers (Eq. 7.8), less convective heat loss (Eq. 7.11), and consequently greater differences from air temperature than for sun leaves. For leaves above air temperature, this can lead to high transpiration rates because the water vapor concentration at saturation depends more or less exponentially on temperature, as indicated in Chapter 2 (also see Fig. 8-5 and Appendix I). Moreover, the amount of CO_2 photosynthetically fixed per unit of water transpired can be higher for sun leaves than for shade leaves in exposed sunlit locations but higher for shade leaves than for sun leaves in shaded locations based on model calculations and observations on a desert shrub, Hyptis emoryi (Smith and Nobel, 1977).

Lobing and dissection tend to decrease the effective length across a leaf in the direction of the wind and hence to reduce δ^{bl} (Eq. 7.8), with a consequent enhancement of convective exchange. For instance, the heat convection coefficient h_c (Eq. 7.14) increases with the depth of leaf serrations. The greater lobing often observed for sun leaves compared with shade leaves reduces the heating of sun leaves above air temperature. Also, heat convection is greater for a pinnate leaf than for a simple leaf of the same area.

Certain plants, especially those exposed to intense shortwave irradiation, have vertical leaves, e.g., willow, many species of *Eucalyptus*, and certain chaparral and desert shrubs. Over the course of a day, vertical leaves can intercept nearly as much shortwave irradiation as horizontal leaves, but they intercept less at midday, when air temperatures tend to be high. Higher leaf temperatures at midday lead to greater transpiration for a given stomatal opening and possibly to temperatures above those photosynthetically optimal. Also, leaves generally become more vertical upon wilting, thereby reducing their interception of shortwave irradiation at higher sun angles. For the exposed horizontal leaf that we have considered (radiation terms in the top line of Table 7-1; convection and transpiration in Fig. 7-8a), rotation from horizontal to vertical decreases its temperature by 5°C. Consistent with this are observations that rotating the leaves of *Cercis canadensis* (redbud) and *Erythrina berteroana* from vertical to horizontal at midday increases their temperatures by 2–6°C (Gates, 1980).

Leaf temperature can be influenced by seasonal differences in leaf orientation and by diurnal solar "tracking" movements, as occurs for cotton, other species in the Malvaceae, many clovers, and certain desert annuals (Ehleringer and Forseth, 1989). Leaves of *Malvastrum rotundifolium*, a winter annual inhabiting warm deserts but growing during the cool part of the year, track the sun so well that the leaf surface is within 20° of perpendicular to the direct solar beam throughout the day. As well as leading to better light interception, this tracking raises the leaf temperature to values more optimal for photosynthesis. (Mooney and Ehleringer, 1978). The mechanism for solar tracking involves changes in hydrostatic pressure induced by the absorption of blue light by photoreceptors in the leaf veins or the pulvinus (large, thin-walled cells at base of petiole), perhaps via steps similar to those for stomatal opening (discussed in Chapter 8), which lead to changes in leaf orientation.

B. Shaded Leaves within Plant Communities

We next consider a shaded leaf at the same temperature as its surroundings. The IR radiation absorbed by such a leaf is the same as the IR emitted by it (when $a_{IR} = e_{IR}$ and $T^{leaf} = T^{surr}$). The net radiation is then due solely to the various forms of solar radiation that reach the leaf. Because the transmission by leaves is fairly high from 0.7 to 2 μm (Fig. 7-4), much of the solar radiation reaching a shaded leaf is in a region not useful for photosynthesis. The leaf absorptance in this range is lower than for the solar irradiation incident on an exposed leaf (Fig. 7-4). For purposes of calculation, we will assume that S on the shaded leaf is 70 W m^{-2}, its absorptance a is 0.30, and the reflectance r of the surroundings has the rather high value of 0.40 because of considerable reflection of radiation by the other leaves within the plant community. By Eqs. (7.3) and (7.6), the net radiation balance is then (0.30)(1.00 + 0.40)(70 W m^{-2}), or 30 W m^{-2}.

Such a shaded leaf has a low rate of photosynthesis because the amount of radiation in the visible region reaching it is fairly small. Moreover, its stomata are partially closed at the lower illumination levels, which decreases the stomatal conductance and hence photosynthesis (see Chapter 8). Three factors tend to reduce the flux of water vapor out of such a shaded leaf: the decrease in stomatal conductance at the lower light levels, which is the main factor reducing transpiration; a lower wind speed for a protected than for an exposed leaf, which leads to a thicker air boundary layer; and a concentration of water vapor in the turbulent air that is generally higher than it is at the top of the plant canopy. Instead of the water vapor flux density of 4 mmol m^{-2} s^{-1} for an exposed leaf of a typical mesophyte,

J_{wv} for the shaded leaf might be only 0.7 mmol m^{-2} s^{-1}. By Eq. (7.19) ($J_H^T = J_{wv}H_{vap}$), heat dissipation by the latent heat term is then 30 W m^{-2}. No heat is conducted across the boundary layers if the leaf is at the same temperature as the surrounding air. Thus the heat loss by water evaporation (30 W m^{-2}) balances the energy gain from the absorption of shortwave irradiation for this shaded leaf.

C. Heat Storage

We indicated at the beginning of this chapter that very little energy is stored in the temperature changes of leaves. However, massive plant parts, such as tree trunks, can store considerable amounts of energy. We will represent the heat storage rate (e.g., in J s^{-1}, or W) as follows:

$$\text{Heat storage rate} = C_P V \frac{\Delta T}{\Delta t} \qquad (7.21)$$

where C_P is the *volumetric heat capacity* (e.g., in J m^{-3} °C^{-1}), and V is the volume that undergoes a change in temperature ΔT in the time interval Δt. C_P thus indicates the amount of heat required to raise the temperature of unit volume by 1°C.

As indicated previously, heat storage can be important for any massive plant part. To model its energy balance, such a part can be divided into isothermal subvolumes, which are generally referred to as *nodes* in heat transfer studies. This approach has been applied to the energy balance of massive stems of cacti, such as the barrel-shaped *Ferocactus acanthodes* (Fig. 7-9) and the tall, columnar *Carnegiea gigantea* (saguaro). The stem is divided into surface nodes, which have no volume and hence no heat storage, plus interior nodes, which have a finite volume but no radiation, boundary-layer, or latent-heat terms (Fig. 7-9). The interior nodes are involved with changes in temperature leading to heat storage (Eq. 7.21) and heat conduction to or from surface nodes as well as to or from other interior nodes (describable by equations of the form of Eq. 7.11).

Once the stem has been divided into nodes, an energy balance for each node can be calculated—greater precision requires a greater number of nodes. The analysis is complicated, in part because the various contributors depend on temperature in different ways. Specifically, shortwave and longwave absorption is independent of temperature, longwave emission depends on T^4, conduction depends on a temperature gradient or difference (e.g., $T^{surf} - T^{ta}$), the saturation water vapor content—which can affect the latent heat term— varies approximately exponentially with temperature, and heat storage depends on $\Delta T/\Delta t$. When these various energy terms for all the nodes are incorporated into a simulation model, or when cactus temperatures are directly measured in the field, parts of the cactus stem facing the sun are found to range up to 15°C above air temperature for stems approximately 0.25 m in diameter. Parts facing away from the sun can be below air temperature, and time lags of a few hours are observed in the heating of the center of the stem.

Consumption or production of energy by metabolic processes can generally be ignored in the energy budget of a leaf. An interesting exception occurs for the inflorescences of many members of the Araceae (*Arum* family), for which high respiratory rates can substantially raise the inflorescence temperature and lead to considerable heat storage. An extreme example is presented by the 2- to 9-g inflorescence or spadix of *Symplocarpus foetidus* (eastern skunk cabbage). By consuming O_2 at the same rate as an active mammal of the same size (heat production of about 0.10 W g^{-1}), the tissue temperature can be 15–35°C above ambient temperatures of −15 to 10°C for over 2 weeks. This prevents freezing of

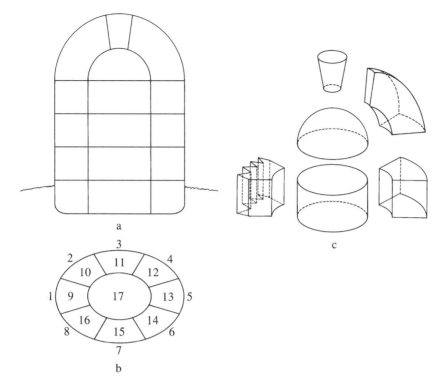

Figure 7-9. System of nodes or subvolumes used in energy balance studies on the barrel cactus, *Ferocac-tus acanthodes*: (a) vertical section indicating the division of the stem into various levels; (b) horizontal section indicating the surface nodes (1–8) and interior nodes (9–17); and (c) three-dimensional representation of certain nodes [adapted from Lewis and Nobel (1977); used by permission].

the inflorescence and also may lead to the volatilization of certain insect attractants, thus increasing the chances for pollination (Seymour, 1997).

D. Time Constants

A matter related to heat storage is the time constant for temperature changes. Analogous to our use of lifetimes in Chapter 4, we will define a time constant τ as the time required for the change of surface temperature from some initial value T_0^{surf} to within $1/e$ (37%) of the overall change to a final value approached asymptotically (T_∞^{surf}):

$$T^{\text{surf}} - T_\infty^{\text{surf}} = \left(T_0^{\text{surf}} - T_\infty^{\text{surf}}\right)e^{-t/\tau} \tag{7.22}$$

At $t = 0$, $e^{-t/\tau}$ is $e^{-0/\tau} = 1$, so $T^{\text{surf}} = T_0^{\text{surf}}$ by Eq. (7.22); at $t = \infty$, $e^{-t/\tau}$ is $e^{-\infty/\tau} = 0$, so $T^{\text{surf}} = T_\infty^{\text{surf}}$. If we ignore transpiration and assume uniform tissue temperatures, then the time constant is

$$\tau = \frac{(V/A)C_P}{4e_{\text{IR}}\sigma(T^{\text{surf}})^3 + K^{\text{air}}/\delta^{\text{bl}}} \tag{7.23}$$

where V is the volume of a plant part having total surface area A (V/A indicates the mean depth for heat storage, which is half its thickness for a leaf), C_P is the volumetric heat capacity, and T^{surf} is expressed in Kelvin units. For a rapidly transpiring leaf, the time constant is about 50% less than indicated by Eq. (7.23) (Gates, 1980; Monteith, 1981).

Let us estimate the time constants for a leaf and a cactus stem. If we consider a 300-μm-thick leaf at 25°C with a volumetric heat capacity 70% as large as that of water and a boundary layer thickness of 1.4 mm, by Eq. (7.25) the time constant is

$$\tau = \frac{(150 \times 10^{-6} \text{ m})(0.7)(4.2 \times 10^6 \text{ J m}^{-3} \text{ °C}^{-1})}{(4)(0.96)(5.67 \times 10^{-8} \text{ J s}^{-1} \text{ m}^{-2} \text{ K}^{-4})(298 \text{ K})^3 + \frac{(0.0259 \text{ J s}^{-1} \text{ m}^{-1} \text{ °C}^{-1})}{(1.4 \times 10^{-3} \text{ m})}} = 18 \text{ s}$$

Thus 63% of the overall change in leaf temperature occurs in only 18 s, indicating that such leaves respond rapidly to variations in environmental conditions, consistent with our statement that very little heat can be stored by means of temperature changes of such leaves. In particular, for a τ of 18 s, T^{surf} will go 96% of the way from T_0^{surf} to T_∞^{surf} in 1 min (see Eq. 7.22). On the other hand, stems of cacti can store appreciable amounts of heat. For the stem portrayed in Fig. 7-8, V/A is about 0.05 m, and C_P is about 90% of C_P^{water} (cactus stems generally have a much smaller volume fraction of intercellular air spaces than the 30% typical of leaves). Assuming the other factors are the same as in the previous calculation for a leaf, τ is then 7.8 × 10³ s (2.2 h) for the cactus stem. Indeed, massive stems do have long time constants for thermal changes.

Long thermal time lags help avoid overheating of tree trunks as well as of seeds and roots in the soil during rapidly moving fires. For instance, temperatures near the soil surface average about 300°C during fires in different ecosystems, but at 0.1 m below the soil surface they rarely exceed 50°C. Energy budget analyses can provide information on a multitude of physiological and ecological processes involving temperature—from frost to fire to photosynthesis.

V. Soil

Soil acts as an extremely important component in the energy balance of plants. For instance, shortwave irradiation can be reflected from its surface, it is the source of longwave radiation that can correspond to a temperature considerably different from that of the air, and heat can be conducted to or from stems in their region of contact with the soil. Also, considerable amounts of energy can be stored by the soil, contrary to the case for most leaves. Although the soil surface can have large daily oscillations in temperature, the soil temperature at moderate depths of 1 m can be extremely steady on a daily basis (variations less than 0.1°C) and fairly steady seasonally.

A. Thermal Properties

Soil has a relatively high heat capacity. To raise the temperature of 1 kg of dry sand by 1°C takes about 0.8 kJ, with similar values for dry clay or loam. Because the density of soil solids is about 2600 kg m^{-3} and soil is about half pores by volume, the overall density of dry soil is about 1300 kg m^{-3}—values range from 1200 kg m^{-3} for dry loam, slightly higher for clay, and up to 1500 kg m^{-3} for dry sand. Thus the volumetric heat capacity at constant pressure, C_P, of dry soil is about 1.1 MJ m^{-3} °C^{-1}. For a moist loam containing 20% water by volume (water has a C_P of 4.18 MJ m^{-3} °C^{-1} at 20°C), C_P^{soil} is about 1.9 MJ m^{-3} °C^{-1}.

The relatively high heat capacity of soil means that considerable energy can be involved with its temperature changes. For instance, if the temperature of the upper 0.4 m of a moist soil with a C_P^{soil} of 2.0 MJ m^{-3} °C^{-1} increases by an average of 2°C during the daytime, then (0.4 m)(2°C)(2.0 MJ m^{-3} °C^{-1}), or 1.6 MJ, can be stored per m^2 of ground. Nearly all the heat stored in the soil during the daytime is released that night. Generally, the soil temperature at depths below 0.4 m changes less than 0.5°C during a day or night, although annually it will vary considerably more, as we will indicate later.

Although soil has a substantial volumetric heat capacity, it does not have a very high thermal conductivity coefficient, K^{soil}. Heat is therefore not readily conducted in soil, where the heat flux density by conduction is

$$J_H^C = -K^{soil}\frac{\partial T}{\partial z} \tag{7.24}$$

Heat can be conducted in all directions in the soil, instead of only vertically, as we will consider here (see Eq. 7.24). J_H^C can be expressed in W m^{-2}, $\partial T/\partial z$ in °C m^{-1}, and therefore K^{soil} in W m^{-1} °C^{-1}, just as for K^{air}. K^{soil} depends on the soil water content; replacement of soil air (a relatively poor heat conductor) by water increases K^{soil}, as the water forms bridges between the soil particles (see Fig. 9-8) and thereby increases the thermal conductivity. For instance, K^{soil} can vary from 0.2 W m^{-1} °C^{-1} for a dry soil to 2 W m^{-1} °C^{-1} for a wet one. For comparison, K^{water} is 0.60 W m^{-1} °C^{-1} and K^{air} is 0.026 W m^{-1} °C^{-1} near 20°C (see Appendix I).

During the daytime the surface of the ground can be considerably warmer than the underlying layers of the soil, which leads to heat conduction into the soil. Because the soil exposed to the turbulent air tends to be drier than the underlying layers, the thermal conductivity coefficient can be lower near the soil surface. For the upper part of a fairly moist sandy loam, K^{soil} may be 0.6 W m^{-1} °C^{-1} and $\partial T/\partial z$ may be -100°C m^{-1} (at least for the upper 0.05 m or so; z is considered positive into the soil). Using Eq. (7.24), the heat flux density by conduction into the soil then is

$$J_H^C = -(0.6 \text{ W m}^{-1} \text{ °C}^{-1})(-100°C \text{ m}^{-1}) = 60 \text{ W m}^{-2}$$

This heat conducted into the soil could lead to the daytime heat storage calculated previously, 1.6 MJ m^{-2}, in (1.6 × 10^6 J m^{-2})/(60 J m^{-2} s^{-1}), or 2.67 × 10^4 s, which is about 7 h.

B. Soil Energy Balance

The components of the energy balance for the soil surface are similar to those for leaves (see Eqs. 7.1 and 7.2). However, we must also take into consideration the heat conducted into the soil (Eq. 7.24), which leads to a gradual temperature change of its upper layers.

The absorption and the emission of radiation usually takes place in the upper few millimeters of the soil. Using Eqs. (7.3)–(7.6), we can generally represent the net radiation balance for the soil surface by $aS + a_{IR}\sigma(T^{surr})^4 - e_{IR}\sigma(T^{soil})^4$, where the values of all parameters are those at the soil surface. (If the soil is exposed directly to the sky, T^{surr} should be replaced by T^{sky} for the incident IR.) For a soil exposed to direct sunlight, the net energy input by radiation can be quite large—in the desert the temperature of the soil surface can exceed 70°C.

Let us now consider heat fluxes for a soil. The heat conducted across the relatively still air next to the soil surface equals $-K^{air} \partial T / \partial z$, where K^{air} is the thermal conductivity coefficient of air at the local temperature (see Eq. 7.11, which indicates that the heat conducted from both sides of a leaf is $-2K^{air} \partial T / \partial x$). The heat conducted within the soil can be calculated using Eq. (7.24) ($J_H^C = -K^{soil} \partial T / \partial z$). For a layer of soil where the water vapor flux density changes by ΔJ_{wv}, the heat loss accompanying water evaporation (or heat gain accompanying condensation) equals $\Delta J_{wv} H_{vap}$, where H_{vap} is the heat necessary to evaporate a unit amount of water at the local soil temperature (see Eq. 7.19; $J_H^T = J_{wv} H_{vap}$). Except in the upper few millimeters, the main energy flux in the soil is for heat conduction, not for radiation or for phase changes of water. Because of the large heat capacity of water, its movement in the soil can also represent an important means of heat flow.

C. Variations in Soil Temperature

Because the energy flux in the soil is often mainly by heat conduction, we can estimate the soil temperature at various depths, although complications arise because of the heterogeneous nature of soil as well as the many types of plant cover. To obtain some idea of daily and annual temperature variations, we will assume that the volumetric heat capacity (C_P^{soil}) and the thermal conductivity coefficient (K^{soil}) are both constant with depth, and we will ignore water movement in the soil. Moreover, we will assume that the soil surface temperature varies sinusoidally around an average value $\bar{T}^{surf}$, with a daily or annual amplitude of ΔT^{surf}, a useful approximation that can readily be tested. We then obtain the following relation for the temperature T at a time t and depth z:

$$T = \bar{T}^{surf} + \Delta T^{surf} e^{-z/d} \cos \left(\frac{2\pi t}{p} - \frac{2\pi t_{max}}{p} - \frac{z}{d} \right) \qquad (7.25)$$

where the *damping depth d*, which is the depth in the soil where the variation in temperature has been damped to $1/e$ of the value at the soil surface, equals $(pK^{soil}/\pi C_P^{soil})^{1/2}$; and p is the period (24 h, or 8.64×10^4 s, for a daily variation and 365 times longer for an annual variation). We note that $z = 0$ at the surface and is considered positive downward, and that the soil surface has its maximum temperature, $\bar{T}^{surf} + \Delta T^{surf}$, when $t = t_{max}$. Equation (7.25) is of the form $y = A + B \cos \alpha$, where A is the average value of y and B is the amplitude of the variation about the mean.

We next estimate the depths where the variation in soil temperature is only $\pm 1°C$, in one case daily and in another annually. Again, we will use a C_P^{soil} of 2 MJ m^{-3} °C^{-1} and a K^{soil} of 0.6 W m^{-1} °C^{-1}. The damping depth for the daily case then is

$$d = \left[\frac{(8.64 \times 10^4 \text{ s})(0.6 \text{ J s}^{-1} \text{ m}^{-1} °C^{-1})}{(\pi)(2 \times 10^6 \text{ J m}^{-3} °C^{-1})} \right]^{1/2}$$
$$= 0.091 \text{ m}$$

and it is 1.7 m for the annual case. From Eq. (7.25) the daily variation in temperature at depth z is $\pm \Delta T^{surf} e^{-z/d}$. The amplitude of the daily variation in soil surface temperature about its mean for bare soil is often 15°C (i.e., $\bar{T}_{max}^{surf} - \bar{T}_{min}^{surf} = 30°C$), and the annual amplitude for variations in average daily surface temperatures is generally somewhat less (e.g., 10°C).

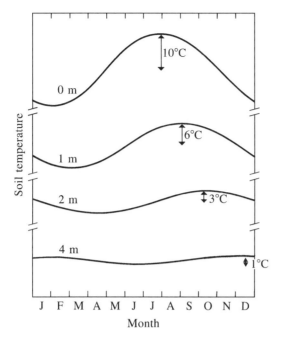

Figure 7-10. Simulated annual variation in soil temperatures at the indicated depths. The average daily temperature at the soil surface was assumed to vary sinusoidally, with a maximum on August 1 and an annual amplitude of 10°C; the damping depth d is 1.7 m.

The depth where the daily variation in temperature is $\pm1°C$ is then

$$z = -d \ln\left(\frac{1°C}{\Delta T^{surf}}\right) = -(0.091 \text{ m})\ln\left(\frac{1°C}{15°C}\right)$$

$$= 0.25 \text{ m}$$

which is consistent with our previous statement that soil temperatures generally change <0.5°C daily at about 0.4 m. On an annual basis, the $\pm1°C$ variation occurs at a depth of 3.9 m (Fig. 7-10).

The factor z/d in the cosine in Eq. (7.25) indicates that the peak of the temperature "wave" arrives later at greater depths in the soil. This peak occurs when cosine equals 1, which corresponds to $2\pi t/p - 2\pi t_{max}/p = z/d(\cos 0 = 1)$. In a time interval Δt, the peak moves a distance Δz, where $2\pi \Delta t/p = \Delta z/d$. The speed of movement of the wave into the soil, $\Delta z/\Delta t$, is thus $2\pi d/p$. Figure 7-10 shows this effect at various depths on an annual basis. For instance, at 4 m the peak temperature occurs $4\frac{1}{2}$ months later than at the surface, indicating a major seasonal displacement of maximum and minimum soil temperatures. To allow for the superposition of effects of seasonally changing temperatures with depth on daily temperatures, we can incorporate both daily and annual additive terms in Eq. (7.25), with each term containing its appropriate d and p. In any case, soil properties markedly affect the thermal environment of roots, which can represent about half of the plant's biomass, as well as animal burrows and even certain wine cellars.

Problems

7.1   An exposed leaf at 10°C has an a_{IR} and an e_{IR} of 0.96, whereas a is 0.60 and r is 0.10. Suppose that the effective temperature for radiation is 2°C for a cloudy sky and -40°C for a clear sky.

    a. If the absorbed IR equals the emitted IR for the leaf, what are the temperatures of the surroundings for a clear sky and for a cloudy one?

    b. What are the λ_{max}'s for emission of radiant energy by the leaf and the surroundings for (a)?

    c. If the global irradiation is 700 W m^{-2} and the temperature of the surroundings is 9°C on a clear day, how much radiation is absorbed by the leaf?

    d. Under the conditions of (c), what percentage of the energy input by absorbed irradiation is dissipated by the emission of thermal radiation?

    e. Assume that the clouds block out the sunlight and the skylight. Let S^{cloud} be 250 W m^{-2} on the upper surface of the leaf and 15% as much be reflected onto the lower surface. If the temperature of the surroundings is 9°C, what is the net radiation for the leaf?

7.2   Consider a circular leaf at 25°C that is 0.12 m in diameter. The ambient wind speed is 0.80 m s^{-1}, and the ambient air temperature is 20°C.

    a. What is the mean distance across the leaf in the direction of the wind?

    b. What is the boundary layer thickness? What would δ^{bl} be, assuming that the mean distance is the diameter?

    c. What is the heat flux density conducted across the boundary layer?

    d. If the net radiation balance for the leaf is 300 W m^{-2}, what is the transpiration rate such that the leaf temperature remains constant?

7.3   Let us consider a spherical cactus 0.2 m in diameter with essentially no stem mass below ground. The surface of the cactus averages 25.0°C and is 50% shaded by spines. Assume that the ambient wind speed is 1.0 m s^{-1}, the ambient air temperature is 20.0°C, the global irradiation with the sun overhead is 1000 W m^{-2}, the effective temperature of the surroundings (including the sky) is -20°C, $a_{IR} = e_{IR} = 0.97$, $r_{surr} = 0.00$, $a_{spine} = 0.70$, and the spines have no transpiration or heat storage.

    a. What is the mean boundary layer thickness for the stem and the heat conduction across it?

    b. What is the stem heat convection coefficient?

    c. Assuming that the spines can be represented by cylinders 1.2 mm in diameter, what is their boundary layer thickness?

    d. If the net radiation balance averaged over the spine surface is due entirely to shortwave irradiation (i.e., IR$_{absorbed}$ = IR$_{emitted}$), and if the maximum shortwave irradiation measured perpendicular to the cylinder lateral surface is 100 W m^{-2}, what is the mean T^{spine} to within 0.1°C? Ignore spine heat conduction to the stem and assume that spines do not transpire.

    e. What is the absorbed minus emitted longwave radiation at the stem surface in the presence and the absence of spines?

    f. Assume that 30% of the incident shortwave is absorbed by the stem surface for the plant with spines. What is the net energy balance averaged over the stem surface? What is the hourly change in mean tissue temperature? Assume that the volumetric heat capacity is 80% of that of water, and ignore transpiration.

7.4 Suppose that the global radiation absorbed by the ground below some vegetation averages 100 W m^{-2}. We will assume that the bulk of the vegetation is at 22°C, the top of the soil is at 20°C, and that both emit like ideal blackbodies.

    a. What is the net radiation balance for the soil?

    b. Suppose that there are four plants/m^2 of ground and that their average stem diameter is 3 cm. If the thermal conductivity coefficient of the stem is the same as that of water, and the temperature changes from that of the bulk of the vegetation to that of the ground in 0.8 m, what is the rate of heat conduction in W down each stem? What is the average value of such J_H^C per m^2 of the ground?

    c. Suppose that the 4 mm of air immediately above the ground acts like a boundary layer and that the air temperature at 4 mm is 21°C. What is the rate of heat conduction from the soil into the air?

    d. What is J_H^C into the soil in the steady state if 0.3 mmol m^{-2} s^{-1} of water evaporates from the upper part of the soil where the radiation is absorbed?

    e. If K^{soil} is the same as K^{water}, what is $\partial T/\partial z$ in the upper part of the soil?

References

Arya, S. P. S. (1988). *Introduction to Micrometeorology*. Academic Press, San Diego.

Campbell, G. S., and Norman, J. M. (1997). *An Introduction to Environmental Biophysics*, 2nd ed. Springer-Verlag, New York.

Dogniaux, R. (Ed.) (1994). *Prediction of Solar Radiation in Areas with a Specific Microclimate*. Kluwer, Dordrecht.

Ehleringer, J. R. (1989). Temperature and energy budgets. In *Plant Physiological Ecology: Field Methods and Instrumentation* (R. W. Pearcy, J. Ehleringer, H. A. Mooney, and P. W. Rundel, Eds.), pp. 117–135. Chapman & Hall, London.

Ehleringer, J. R., and Forseth, I. N. (1989). Diurnal leaf movements and productivity in canopies. In *Plant Canopies: Their Growth, Form and Function* (G. Russell, B. Marshall, and P. G. Jarvis, Eds.), pp. 129–142. Cambridge Univ. Press, Cambridge, UK.

Ehleringer, J. R., and Mooney, H. A. (1978). Leaf hairs: Effects on physiological activity and adaptive value to a desert shrub. *Oecologia* **37**, 183–200.

Gates, D. M. (1980). *Biophysical Ecology*. Springer-Verlag, New York.

Gates, D. M., and Papian, L. E. (1971). *Atlas of Energy Budgets of Plant Leaves*. Academic Press, New York.

Incropera, F. P., and DeWitt, D. P. (1996). *Fundamentals of Heat and Mass Transfer*, 4th ed. Wiley, New York.

Jaffe, M. J. (1985). Wind and other mechanical effects in the development and behavior of plants, with special emphasis on the role of hormones. In *Hormonal Regulation of Development III* (R. P. Pharis and D. M. Reid, Eds.) Encyclopedia of Plant Physiology, New Series, Vol. 11, pp. 444–484. Springer-Verlag, Berlin.

Kappen, L., Lange, O. L., Schulze, E.-D., Evenari, M., and Buschbom, U. (1979). Ecophysiological investigations on lichens of the Negev Desert. VI. Annual course of the photosynthetic production of *Ramalina maciformis* (Del.) Bory. *Flora* **168**, 85–108.

Kaviany, M. (1994). *Principles of Convective Heat Transfer*. Springer-Verlag, New York.

Koller, D. (1986). Yearly review: The control of leaf orientation by light. *Photochem. Photobiol.* **44**, 819–826.

Kreith, F., and Bohn, M. S. (1996). *Principles of Heat Transfer*, 5th rev. ed. Harper & Row, New York.

Leuning, R. (1987). Leaf temperatures during radiation frost part II. A steady state theory. *Agric. Forest Meteorol.* **42**, 135–155.

Lewis, D. A., and Nobel, P. S. (1977). Thermal energy exchange model and water loss of a barrel cactus, *Ferocactus acanthodes*. *Plant Physiol.* **60**, 609–616.

Monteith, J. L. (1981). Coupling of plants to the atmosphere. In *Plants and Their Atmospheric Environment* (J. Grace, E. D. Ford, and P. G. Jarvis, Eds.), pp. 1–29. Blackwell, Oxford.

Monteith, J. L., and Unsworth, M. H. (1990). *Principles of Environmental Physics*, 2nd ed. Edward Arnold, New York.

Mooney, H. A., and Ehleringer, J. R. (1978). The carbon gain benefits of solar tracking in a desert annual. *Plant Cell Environ.* **1**, 307–311.

Niklas, K. J. (1992). *Plant Biomechanics: An Engineering Approach to Plant Form and Function*. Univ. of Chicago Press, Chicago.

Nobel, P. S. (1974). Boundary layers of air adjacent to cylinders. Estimation of effective thickness and measurements on plant material. *Plant Physiol.* **54**, 177–181.

Nobel, P. S. (1975). Effective thickness and resistance of the air boundary layer adjacent to spherical plant parts. *J. Exp. Bot.* **26**, 120–130.

Nobel, P. S. (1981). Wind as an ecological factor. In *Physiological Plant Ecology* (O. L. Lange, P. S. Nobel, C. B. Osmond, and H. Ziegler, Eds.), Encyclopedia of Plant Physiology, New Series, Vol. 12A, pp. 475–500. Springer-Verlag, Berlin.

Nobel, P. S. (1988). *Environmental Biology of Agaves and Cacti*. Cambridge Univ. Press, New York.

Passerat de Silans, A., Monteny, B. A., and Lhomme, J. P. (1997). The correction of soil heat flux measurements to derive an accurate surface energy balance by the Bowen ratio method. *J. Hydrol.* **188**, 453–465.

Rahman, M. (Ed.) (1997). *Laminar and Turbulent Boundary Layers*. Computational Mechanics, Southampton, UK.

Rosenberg, N. J., Blad, B. L., and Verma, S. B. (1983). *Microclimate: The Biological Environment*, 2nd ed. Wiley, New York.

Schlichting, H. (1979). *Boundary-Layer Theory*, 7th ed. McGraw-Hill, New York.

Schuepp, P. H. (1993). Tansley review No. 59. Leaf boundary layers. *New Phytol.* **125**, 477–507.

Seymour, R. S. (1997). Plants that warm themselves. *Sci. Am.* **276**, 104–109.

Smith, W. K., and Nobel, P. S. (1977). Temperature and water relations for sun and shade leaves of a desert broadleaf, *Hyptis emoryi. J. Exp. Bot.* **28**, 169–183.

Suryanarayana, N. V. (1995). *Engineering Heat Transfer*. West, St. Paul, MN.

Vogel, S. (1989). Drag and reconfiguration of broad leaves in high winds. *J. Exp. Bot.* **40**, 941–948.

Vogel, S. (1994). *Life in Moving Fluids: The Physical Biology of Flow*, 2nd ed. Princeton Univ. Press, Princeton, NJ.

8

Leaves and Fluxes

I. Resistances and Conductances—Transpiration	**294**
A. Boundary Layer Adjacent to Leaf	296
B. Stomata	298
C. Stomatal Conductance and Resistance	301
D. Cuticle	303
E. Intercellular Air Spaces	303
F. Fick's First Law and Conductances	304
II. Water Vapor Fluxes Accompanying Transpiration	**306**
A. Conductance and Resistance Network	306
B. Values of Conductances	308
C. Effective Lengths and Resistance	309
D. Water Vapor Concentrations and Mole Fractions for Leaves	310
E. Examples of Water Vapor Levels in a Leaf	312
F. Water Vapor Fluxes	313
G. Control of Transpiration	314
III. CO_2 Conductances and Resistances	**315**
A. Resistance and Conductance Network	316
B. Mesophyll Area	317
C. Resistance Formulation for Cell Components	319
D. Partition Coefficient for CO_2	320
E. Cell Wall Resistance	321
F. Plasma Membrane Resistance	322
G. Cytosol Resistance	323
H. Mesophyll Resistance	323
I. Chloroplast Resistance	324
IV. CO_2 Fluxes Accompanying Photosynthesis	**324**
A. Photosynthesis	324
B. Respiration and Photorespiration	326
C. Comprehensive CO_2 Resistance Network	329
D. Compensation Points	330
E. Fluxes of CO_2	332
F. CO_2 Conductances	334
G. Range in Photosynthetic Rates	335
H. Environmental Productivity Index	336

V. **Water-Use Efficiency**   **337**
   A. Values of WUE   337
   B. Elevational Effects on WUE   339
   C. Stomatal Control of WUE   340
   D. C_3 versus C_4 Plants   343
   Problems   **345**
   References   **347**

IN THIS chapter we will reconsider the transpiration of water and the photosynthetic fixation of carbon dioxide by leaves. The driving forces for such fluxes are differences in CO_2 and H_2O concentrations or mole fractions. Resistances were first used to describe gas fluxes quantitatively for leaves by Henry Brown and Fergusson Escombe in 1900. Resistance networks were developed to specify which parts of the pathway are most limiting for photosynthesis or transpiration. Recently, the use of two different forms of conductance has become more popular, especially for describing transpiration, and we will consider both forms. We will see that greater stomatal opening may be an advantage for photosynthesis but can result in excessive transpiration. Consequently, a benefit/cost index, such as the amount of CO_2 fixed per unit of water lost, can be important for evaluating ecological aspects of gas exchange. Figure 8-1 indicates how the flux densities of water vapor and CO_2 can be measured for a leaf. Although we will usually be referring to leaves, the discussion of gas fluxes is also applicable to stems, flower petals, and other plant parts.

I. Resistances and Conductances—Transpiration

The resistances and conductances that we will discuss in this section are those encountered by water vapor as it diffuses from the pores in the cell walls of mesophyll cells or other sites of water evaporation out into the turbulent air surrounding a leaf. We will define these quantities for the intercellular air spaces, the stomata, the cuticle, and the boundary layer next to a leaf. As considered later in this chapter, CO_2 diffuses across the same gaseous-phase resistances or conductances as does water vapor and also across a number of other components in the liquid phases of mesophyll cells.

Throughout this book we have used equations of the following general form: flux density = proportionality coefficient × force (as indicated in Chapter 1, a flux density, indicating amount area^{-1} time^{-1}, is often simply referred to as a flux). The proportionality coefficients in such flux density expressions are measures of *conductivity*. To represent force, instead of using the gradient in chemical potential—which, as we noted in Chapter 3, is a very general force—we often use quantities that are more convenient experimentally, such as differences in concentration (consider Fick's first law; Eq. 1.1). Such a "force," however, does not have the proper units for force: Therefore, to be correct, the proportionality coefficient is not the conductivity, but rather the *conductance*:

$$\text{Flux density} = \text{conductivity} \times \text{force} \qquad (8.1a)$$

$$\text{Flux density} = \text{conductance} \times \text{``force''} \qquad (8.1b)$$

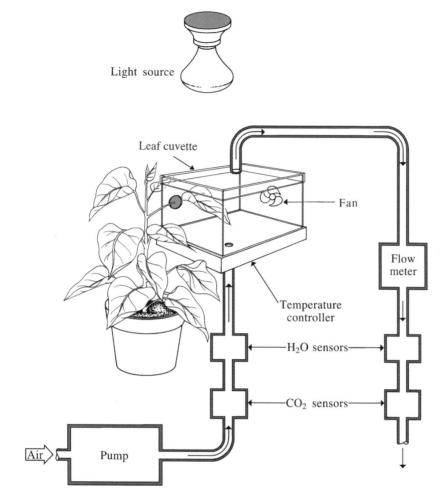

Light source

Leaf cuvette

Fan

Flow
meter

Temperature
controller

—H₂O sensors—

—CO₂ sensors—

Air

Pump

Figure 8-1. Experimental arrangement for measuring leaf transpiration and photosynthesis. The water vapor and CO_2 content of the gas entering a transparent chamber enclosing a leaf is compared with that leaving. A fan is necessary to mix the air in the leaf cuvette. If the water vapor concentration increases from 0.6 mol m^{-3} for the air entering to 1.0 mol m^{-3} for that leaving for a gas flow rate of 1.0×10^{-5} m^3 s^{-1} (10 cm^3 s^{-1}), then J_{wv} for a leaf of area 1.0×10^{-3} m^2 (10 cm^2) would be (1.0 mol m^{-3} − 0.6 mol m^{-3})(1.0×10^{-5} m^3 s^{-1})/(1.0×10^{-3} m^2), or 0.004 mol m^{-2} s^{-1}.

We might equally well choose an alternative form for the relation between forces and fluxes: Flux density = force/resistivity, where *resistivity* is the reciprocal of conductivity. In turn, we can define a *resistance,* which is the reciprocal of conductance:

$$\text{Flux density} = \frac{\text{force}}{\text{resistivity}} \tag{8.1c}$$

$$\text{Flux density} = \frac{\text{``force''}}{\text{resistance}} \tag{8.1d}$$

To help understand the difference between resistivity and resistance, we will consider the electrical usage of these terms. Electrical resistivity is a fundamental physical property of a material—such resistivities are tabulated in handbooks. Electrical resistance, on the other hand, describes a particular component in an electrical circuit, i.e., a particular piece of material. We can readily measure the magnitude of resistance in the laboratory or purchase a resistor of known resistance in an electronics store. Electrical resistivity, ρ, generally has the units of ohm m, whereas electrical resistance, $R = \rho \Delta x / A$, is expressed in ohms. Besides differing conceptually (compare Eqs. 8.1c and 8.1d), resistance and resistivity thus differ in their units. In fact, much of our attention in this chapter will be devoted to the units for variables such as resistance because many different systems are in use in plant physiology, ecology, agronomy, meteorology, soil science, and other related disciplines.

We will consider both the resistance and the conductance of a particular component, and we will present the expressions for both resistance and conductance for series versus parallel arrangements of components. Because a flux density is directly proportional to conductance (Eq. 8.1b) but inversely proportional to resistance (Eq. 8.1d), conductance terminology is often more convenient for discussing plant responses to environmental factors. Also, resistance has no upper limit—i.e., it varies from some minimal value to infinity—whereas conductance varies between zero and some maximum value. However, resistance terminology can be easier to use when a substance must cross a series of components in sequence, such as CO_2 diffusing across the cell wall, plasma membrane, cytosol, and chloroplast membranes. To help become familiar with the various conventions in the scientific literature, we will use both systems, emphasizing conductances for transpiration and resistances for photosynthesis.

A. Boundary Layer Adjacent to Leaf

As a starting point for our discussion of gas fluxes across air boundary layers, let us consider the one-dimensional form of Fick's first law of diffusion, $J_j = -D_j \partial c_j / \partial x$ (Eq. 1.1). As in Chapter 1, we will replace the concentration gradient by the difference in concentration across some distance. In effect, we are considering cases not too far from equilibrium, so the flux density depends linearly on the force, and the force can be represented by the difference in concentration. The distance is across the air boundary layer adjacent to the surface of a leaf, δ^{bl} (Chapter 7 discusses boundary layers and equations for their thickness). Consequently, Fick's first law assumes the following form for the diffusion of species j across the boundary layer:

$$J_j = -D_j \frac{\partial c_j}{\partial x} = D_j \frac{\Delta c_j}{\Delta x} = D_j \frac{\Delta c_j^{bl}}{\delta^{bl}}$$

$$= g_j^{bl} \Delta c_j^{bl} = \frac{\Delta c_j^{bl}}{r_j^{bl}} \tag{8.2}$$

Equation (8.2) shows how the net flux density of substance j depends on its diffusion coefficient D_j and on the difference in its concentration Δc_j^{bl} across a distance δ^{bl} of the air. The net flux density J_j is toward regions of lower c_j, which requires the negative sign associated with the concentration gradient and otherwise is incorporated into the definition of Δc_j in Eq. (8.2); we will specifically consider the diffusion of water vapor and CO_2 toward lower concentrations in this chapter. Also, we will assume that the same boundary layer thickness δ^{bl} derived for heat transfer (Eqs. 7.8–7.13) applies for mass transfer, an example of the similarity principle. Outside δ^{bl} is a region of air turbulence, where we will assume that the concentrations of gases are the same as in the bulk atmosphere (an assumption that we will remove in Chapter 9). Equation (8.2) indicates that J_j equals Δc_j^{bl} multiplied by a conductance, g_j^{bl}, or divided by a resistance, r_j^{bl}.

The air boundary layers on both sides of a leaf influence the entry of CO_2 and the exit of H_2O, as was clearly shown by Klaus Raschke in the 1950s. Movement of gas molecules across these layers is by diffusion in response to differences in concentration. Using Eq. (8.2), we can represent the conductance and the resistance of a boundary layer of air as follows:

$$g_j^{bl} = \frac{J_j}{\Delta c_j^{bl}} = \frac{D_j}{\delta^{bl}} = \frac{1}{r_j^{bl}} \tag{8.3}$$

The SI units for the diffusion coefficient D_j are $m^2\ s^{-1}$ (see Chapter 1) and the thickness of the boundary layer δ^{bl} is in m, so g_j^{bl} is in $(m^2\ s^{-1})/(m)$, or $m\ s^{-1}$ (values are often expressed in $mm\ s^{-1}$), and r_j^{bl} is in $s\ m^{-1}$. J_j is expressed per unit leaf area, so g_j^{bl} and r_j^{bl} also relate to unit area of a leaf.

D_j is a fundamental measure of conductivity (values available in suitable handbooks) describing the diffusion of species j in a given medium. On the other hand, δ^{bl} characterizes a particular situation because the thickness of the boundary layer depends on wind speed and leaf size (see Chapter 7, e.g., Eq. 7.8). Thus r_j^{bl} as defined by Eq. (8.3) describes a particular component of the path, analogous to the resistance (R) used in Ohm's law. Recalling Eq. (1.9), $P_j = D_j K_j/\Delta x$, we recognize that D_j/δ^{bl} represents the permeability coefficient for substance j, P_j, as it diffuses across an air boundary layer of thickness δ^{bl}. When something readily diffuses across a boundary layer, P_j and g_j^{bl} are large and r_j^{bl} is small (see Eq. 8.3). Using resistances and conductances, we can describe gas fluxes into and out of leaves employing a number of relations originally developed for the analysis of electrical circuits.

We will next estimate values for g_{wv}^{bl} and r_{wv}^{bl}—the conductance and the resistance, respectively—for water vapor diffusing across the boundary layer of air next to a leaf. In Chapter 7 we indicated that the boundary layer thickness in mm ($\delta_{(mm)}^{bl}$) for a flat leaf under field conditions is $4.0\sqrt{l_{(m)}/v_{(m\ s^{-1})}}$ (Eq. 7.8), where $l_{(m)}$ is the mean length of the leaf in the direction of the wind in m and $v_{(m\ s^{-1})}$ is the ambient wind speed in $m\ s^{-1}$. Let us consider a relatively thin boundary layer of 0.3 mm and a thick one of 3 mm (see Fig. 7-6a for the values of $l_{(m)}$ and $v_{(m\ s^{-1})}$ implied). For water vapor diffusing in air at 20° C, D_{wv} is $2.4 \times 10^{-5}\ m^2\ s^{-1}$ (see Appendix I). Using Eq. (8.3), for the thin boundary layer we obtain

$$g_{wv}^{bl} = \frac{(2.4 \times 10^{-5}\ m^2\ s^{-1})}{(0.3 \times 10^{-3}\ m)} = 8 \times 10^{-2}\ m\ s^{-1} = 80\ mm\ s^{-1}$$

$$r_{wv}^{bl} = \frac{(0.3 \times 10^{-3}\ m)}{(2.4 \times 10^{-5}\ m^2\ s^{-1})} = 13\ s\ m^{-1}$$

For the thick boundary layer, $g_{wv}^{bl} = 8\ mm\ s^{-1}$ and $r_{wv}^{bl} = 130\ s\ m^{-1}$ (Table 8-1).[1] Boundary layer conductances usually are larger and resistances are smaller than their respective values for diffusion along the stomatal pores, which we will examine next.

---

[1] Instead of estimating the boundary layer conductance or resistance based on δ^{bl} calculated by Eq. (7.8), which cannot account for all the intricacies of different leaf shapes, it is often more expedient to construct a filter-paper replica of the leaf; if this "leaf" is then moistened, the observed J_{wv} from it for a certain Δc_{wv}^{bl} will indicate g_{wv}^{bl} or r_{wv}^{bl} (Eq. 8.2) because water vapor will then cross only a boundary layer.

Table 8-1. Summary of Representative Values of Conductances and Resistances for Water Vapor Diffusing out of Leave[a]

Component condition	Conductance		Resistance	
	$(mm\ s^{-1})$	$(mmol\ m^{-2}\ s^{-1})$	$(s\ m^{-1})$	$(m^2\ s\ mol^{-1})$
Boundary layer				
Thin	80	3,200	13	0.3
Thick	8	320	130	3
Stomata				
Large area—open	19	760	53	1.3
Small area—open	1.7	70	600	14
Closed	0	0	∞	∞
Mesophytes—open	4–20	160–800	50–250	1.3–6
Xerophytes and trees—open	1–4	40–160	250–1,000	6–25
Cuticle				
Crops	0.1–0.4	4–16	2,500–10,000	60–250
Many trees	0.05–0.2	2–8	5,000–20,000	125–500
Many xerophytes	0.01–0.1	0.4–4	10,000–100,000	250–2,500
Intercellular air spaces				
Calculation	24–240	1,000–10,000	4.2–42	0.1–1
Waxy layer				
Typical	50–200	2,000–8,000	5–20	0.1–0.5
Certain xerophytes	10	400	100	2.5
Typical	40–100	1,600–4,000	10–25	0.2–0.6
Leaf (lower surface)				
Crops—open stomata	2–10	80–400	100–500	2.5–13
Trees—open stomata	0.5–3	20–120	300–2,000	8–50

[a] See text for specific calculations (conductances in mmol $m^{-2}\ s^{-1}$ are from Eq. 8.8) as well as Farquhar and von Caemmerer (1982) and Körner *et al.* (1979).

B. Stomata

As we indicated in Chapter 1, stomata control the exit of water vapor from leaves and the entry of CO_2 into them. Although the epidermal cells occupy a much greater fraction of the leaf surface area than do the stomatal pores, the waxy cuticle covering the outer surface of the epidermal cells greatly reduces the water loss from their cell walls to the turbulent air surrounding a leaf, the *cuticular* transpiration. The usual pathway for water vapor leaving a leaf during transpiration is therefore through the stomata.

The stomatal aperture is controlled by the conformation of the two guard cells surrounding a pore (see Figs. 1-2 and 8-2). These cells are generally kidney-shaped (dumbbell shaped in grasses), may be 40 μm long, and, unlike ordinary epidermal cells, usually contain chloroplasts. When the guard cells are relatively flaccid, the stomatal pore is nearly closed, as is the case for most plants at night. Upon illumination, guard cells take up K^+, which may increase in concentration by 0.2–0.6 M. The K^+ uptake raises the internal osmotic pressure, thus lowering the internal water potential; water then spontaneously flows from the epidermal cells into the guard cells by osmosis (Fig. 8-2). This water entry leads to an increase in the internal hydrostatic pressure of a pair of guard cells, causing them to expand and their cell walls on either side of the pore to become concave. As the kidney-shaped guard cells bow outward, an elliptical pore develops between the two cells. The formation of this pore is a consequence of the anisotropic properties of the cell wall surrounding each

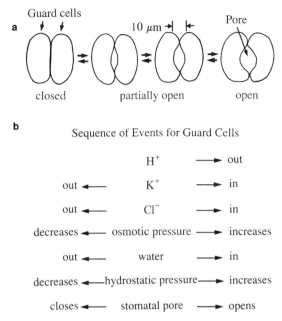

Figure 8-2. Schematic representation of the opening and closing movements for stomatal pores: (a) Pair of guard cells as viewed toward leaf surface and (b) cellular events involved. Arrows to the right are for stomatal opening and those to the left are for closing. [Adapted by permission of the publisher from THE CACTUS PRIMER by A. C. Gibson and P. S. Nobel, Cambridge, Mass.: Harvard University Press, Copyright ©1986 by A. C. Gibson and P. S. Nobel.]

guard cell. The distance between the guard cells across the open pore (the pore "width") is generally 5–15 μm, and the major axis of the elliptical pore can be about 20 μm.

The water relations of the special epidermal cells immediately surrounding the guard cells, which are referred to as subsidiary cells, are also crucial for stomatal opening. For instance, solutes such as K^+ can move from the subsidiary cells to the guard cells, causing water to leave the subsidiary cells and their internal hydrostatic pressure P^i to decrease while P^i of the guard cells increases. Once the stomatal pore has opened, water evaporates from the inner side of the guard cells and subsidiary cells and then diffuses out of the leaf. This lowers the water potential Ψ in the cell walls, which can cause water to leave the protoplasts of the guard cells and subsidiary cells, thereby lowering their P^i and causing some stomatal closure. The ensuing partial stomatal closure reduces the water loss from the leaf and hence reduces evaporation, thereby allowing P^i to build back up in the guard cells and subsidiary cells, leading to reopening of the stomata. The resulting oscillation of stomatal pore aperture can have a period of 30–60 min, although it is often masked by changes in environmental factors such as wind and irradiation.

What controls the opening of stomatal pores? This question is difficult to answer, in part because a number of factors are involved. An initial event apparently is the active H^+ extrusion from the guard cells, which lowers the electrical potential inside relative to outside (i.e., hyperpolarizes the membrane potential, E_M; see Chapter 3) as well as lowers the internal concentration of H^+ (i.e., raises the internal pH). The lowered E_M favors passive K^+ uptake, the latter most likely by K^+ channels that are opened by the hyperpolarization of the plasma membrane. After stomatal opening is initiated, the added K^+ within the guard cells is electrically balanced—partly by a Cl^- influx (perhaps via an OH^- antiporter or more likely by an H^+ symporter) and partly by the production of organic anions such as malate in the guard cells. For most plants, malate is the main counterion for K^+, but Cl^- appears to be the main anion in those few species whose guard cells lack chloroplasts. For some plants stomatal opening increases with the light level only up to a photosynthetic photon flux density (PPFD) of about 200 μmol m^{-2} s^{-1}, but for others it may increase all the way up to full sunlight (2000 μmol m^{-2} s^{-1}).

The degree of stomatal opening often depends on the CO_2 concentration in the guard cells, which reflects their own carbohydrate metabolism as well as the CO_2 level in the air within the leaf. For instance, upon illumination the CO_2 concentration in the leaf intercellular air spaces is decreased by photosynthesis, resulting in decreased CO_2 levels in the guard cells, which somehow triggers stomatal opening. CO_2 can then enter the leaf and photosynthesis can continue. In the dark, respiration generally leads to relatively high CO_2 levels in the leaves, which triggers stomata to close. Lowering the CO_2 concentration in the ambient air can induce stomatal opening in the dark (again as a response to the low CO_2 level in the guard cells), which indicates that the energy for opening can be supplied by respiration. Stomata can also respond directly to light, which stimulates proton extrusion from guard cells, independent of the response to CO_2; the light responses involve the absorption of PPFD by chloroplasts in the guard cells as well as another system that absorbs in the blue region and that is sensitive to low photon flux densities. In addition, stomata often tend to close partially as the relative humidity of the ambient air decreases, another way of regulating water loss. If a constant water vapor concentration difference from inside the leaf to the surrounding air is maintained, stomata in the light generally tend to open with increasing temperature up to that optimal for photophosphorylation, about 35°C; in the dark, the temperature-induced stomatal opening continues up to higher temperatures, possibly reflecting the higher optimal temperature for oxidative phosphorylation than for photophosphorylation. Stomatal movements can also be controlled by hormones, including those produced in the roots as well as in the leaves. For instance, abscisic acid increases in illuminated leaves during water stress. This leads to stomatal closure, which conserves water, although at the expense of a decrease in photosynthesis. The halftimes for stomatal movements are generally 5–20 min, with closing usually more rapid than opening.

Less is currently known about what triggers stomatal closure compared with opening. An initial event may be the opening of Ca^{2+} channels in the plasma membrane of the guard cells, causing Ca^{2+} entry into the cytosol and a depolarization of the plasma membrane. This depolarization and the increased cytosolic Ca^{2+} concentration apparently cause anion channels to open, allowing Cl^- and possibly malate to move out across the plasma membrane, leading to further depolarization. Such depolarization opens K^+ channels, so K^+ also passively moves out of the guard cells. The decreased osmotic pressure causes water to move out, which in turn causes the hydrostatic pressure to decrease, leading to stomatal closure (Fig. 8-2).

For leaves of dicots, stomata are usually more numerous on the lower surface than on the upper one. In many dicots, stomata may even be nearly absent from the upper epidermis. On the other hand, many monocots and certain other plants with vertically oriented leaves have approximately equal numbers of stomata per unit area on each side. A frequency of 40–300 stomata per mm^2 is representative for the lower surface of most leaves. The pores of the open stomata generally occupy 0.2–2% of the leaf surface area. Thus the area for diffusion of gases through the stomatal pores in the upper or lower epidermis of a leaf, A^{st}, is much less than the leaf surface area, A.

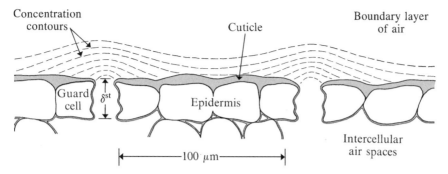

Figure 8-3. Anatomy near leaf surface showing the concentration contours of water vapor outside open stomata.

C. Stomatal Conductance and Resistance

We can apply Fick's first law in the form $J_j = D_j \Delta c_j / \Delta x$ (Eq. 8.2) to describe the diffusion of gases through stomatal pores. We will let the depth of the stomatal pore be δ^{st} (Fig. 8-3); the concentration of substance j changes by Δc_j^{st} along the distance δ^{st}. For the steady state and ignoring the cuticular pathway, the amount of substance j moving per unit time toward or away from the leaf (J_j times the leaf area A) must equal the amount of substance j moving per unit time through the stomata (the flux density within the stomata, $D_j \Delta c_j^{st} / \delta^{st}$, times the stomatal area A^{st} that occurs for leaf area A). Recognition of this constricting or bottleneck effect that stomata have on the area available for the diffusion of gas molecules leads to the following relations:

$$J_j A = D_j \frac{\Delta c_j^{st}}{\delta^{st}} A^{st} \tag{8.4a}$$

or

$$J_j = D_j \frac{\Delta c_j^{st}}{\delta^{st}} \frac{A^{st}}{A} = \frac{D_j n a^{st}}{\delta^{st}} \Delta c_j^{st} \tag{8.4b}$$

where n is the number of stomata per unit area of the leaf and a^{st} is the average area per stomatal pore; thus na^{st} equals the fraction of the leaf surface area occupied by stomatal pores, A^{st}/A. The flux density J_j in Eq. (8.4) refers to the rate of movement of substance j per unit area of the leaf, a quantity that is considerably easier to measure than the flux density within a stomatal pore.

The area available for water vapor diffusion abruptly changes from A^{st} to A at the leaf surface. On the other hand, the three-dimensional surfaces of equal concentration fan out from each stomatal pore (Fig. 8-3). This geometrical aspect could introduce considerable complications, but fortunately a one-dimensional form for Fick's first law can still describe gases moving across the boundary layer next to a leaf, although we need to make an "end correction" to allow for the diffusion pattern at the end of the stomatal pores. The distance between stomata is often about 100 μm, which is considerably less than the thicknesses of nearly all boundary layers (see Fig. 7-6a). The three-dimensional concentration contours from adjacent stomata therefore tend to overlap in the boundary layer (Fig. 8-3). Because of this, the concentration of water vapor varies only slightly in planes parallel to the leaf surface but changes substantially in the direction perpendicular to the leaf surface, so we can generally use a one-dimensional form of Fick's first law. However, the concentration patterns on both ends of a stomatal pore cause the pore to have an effective depth $> \delta^{st}$ by about the mean "radius" of the pore, r^{st}. We will define r^{st} as $\pi(r^{st})^2 = a^{st}$, where a^{st} is the area of the pore. Formulas for the end correction due to the three-dimensional nature of the concentration gradients at each end of a stomatal pore vary and are more complicated than that used here, r^{st}, although differences among the various correction formulas are generally relatively small compared with δ^{st}.

For our applications to transpiration and photosynthesis, we will define a stomatal conductance, g_j^{st}, and resistance, r_j^{st}, for the diffusion of species j using Eq. (8.4):

$$g_j^{st} = \frac{J_j}{\Delta c_j^{st}} = \frac{D_j na^{st}}{\delta^{st} + r^{st}} = \frac{1}{r_j^{st}} \tag{8.5}$$

Equation (8.5) incorporates the effective depth of a stomatal pore, $\delta^{st} + r^{st}$, where r^{st} is the mean pore radius just introduced. When the width of a stomatal pore is 0.1–0.3 μm, as occurs when the pores are nearly closed, the mean free path for molecules diffusing in air is of the same magnitude as the dimensions of the opening. In particular, the mean free path of a gas molecule (the average distance that such a molecule travels before colliding with another gas molecule) is about 0.07 μm at pressures and temperatures normally experienced by plants. The molecular interactions with the sides of the stomatal pore are then important, which affects the value of D_j. Even though we will ignore this interaction and will make certain geometrical approximations for δ^{st}, r^{st}, and a^{st}, Eq. (8.5) is still quite useful for estimating stomatal conductances and resistances [for a general treatment, including the effects of water vapor movements on the stomatal CO_2 conductance, see Field *et al.* (1989) and Leuning (1983)].

We now calculate some values of the stomatal conductance for the diffusion of water vapor. We will consider air at 20°C for which D_{wv} is 2.4×10^{-5} m^2 s^{-1}. The stomatal conductance tends to be high when a large portion of the leaf surface area (e.g., $na^{st} = 0.02$) is occupied by open stomata of relatively short pore depth (e.g., $\delta^{st} = 20 \, \mu$m). Using a mean radius r^{st} of 5 μm, Eq. (8.5) predicts that the conductance is

$$g_{wv}^{st} = \frac{(2.4 \times 10^{-5} \text{ m}^2 \text{ s}^{-1})(0.02)}{(20 \times 10^{-6} \text{ m} + 5 \times 10^{-6} \text{ m})}$$

$$= 1.9 \times 10^{-2} \text{ m s}^{-1} = 19 \text{ mm s}^{-1}$$

At the other extreme, the open stomata may occupy only 0.4% of the lower surface of a leaf ($na^{st} = 0.004$), and the pore depth may be relatively great, e.g., 50 μm. Again assuming that r^{st} is 5 μm, the stomatal conductance calculated using Eq. (8.5) is 1.7 mm s^{-1}, a small value for open stomata (Table 8-1).

The stomatal conductance is usually 2–20 mm s^{-1} for water vapor diffusing out through the open stomata of most mesophytes (r_{wv}^{st} of 50–500 s m^{-1}). In general, crops tend to have high values of g_{wv}^{st}, whereas it may be only 1 mm s^{-1} for certain xerophytes and many trees with open stomata (Table 8-1). Some xerophytes have sunken stomata leading to another conductance in series with g_{wv}^{st}, which slightly decreases the overall conductance for water vapor loss by transpiration. As the stomatal pores close, the conductance decreases accordingly because g_{wv}^{st} is proportional to the stomatal area by Eq. (8.5). Because g_{wv}^{st} is the only conductance in the whole diffusion pathway that is variable over a wide range, changes in the openings of the stomatal pores can regulate the movement of gases into and out of the leaves.

D. Cuticle

Some water molecules diffuse across the waxy cuticle of the epidermal cells, a process called *cuticular transpiration*, which probably involves both liquid water and water vapor. We will identify a cuticular conductance, g_j^c, and a resistance, r_j^c, for the diffusion of species j across the cuticle. These quantities are in parallel with the analogous quantities for the stomata, i.e., substance j can leave the leaf either by crossing the cuticle or by going through the stomatal pores (see Fig. 1-2). The conductance for cuticular transpiration, g_{wv}^c, generally ranges from 0.05 to 0.3 mm s^{-1} for different species, although it can be even lower for xerophytes with thick cuticles that greatly restrict water loss (Table 8-1). Thus g_{wv}^c is usually much smaller than g_{wv}^{st} for open stomata. When the stomata are nearly closed (low g_{wv}^{st}), cuticular transpiration can exceed the loss of water through the stomata. If the cuticle is mechanically damaged or develops cracks, as can occur for older leaves, g_{wv}^c can be raised, and more water will then move out of the leaf via this pathway.

E. Intercellular Air Spaces

Another conductance encountered by the diffusion of substance j in plant leaves is that of the intercellular air spaces, g_j^{ias}. The irregular shape of these air spaces, which generally account for about 30% of the leaf volume, makes g_j^{ias} difficult to estimate accurately. However, the intercellular air spaces do act as an unstirred air layer across which substances must diffuse; hence we will describe g_j^{ias} by a relation similar to Eq. (8.3) ($g_j^{bl} = D_j/\delta^{bl}$):

$$g_j^{ias} = \frac{D_j}{\delta^{ias}} = \frac{1}{r_j^{ias}} \tag{8.6}$$

For convenience, we will combine a number of factors to get the effective distance, δ^{ias}. For instance, we will let δ^{ias} include the effective length of the air-filled part of the cell wall pores from the cell wall surface to the sites where the evaporation of water or the dissolving of CO_2 takes place. This length is greater than the actual distance along the pores because we must correct for the decrease in the cross-sectional area available for diffusion, a decrease caused by the nongaseous parts of the cell wall. We will also include in δ^{ias} the effective length of the thin waxy layer that generally occurs on the surfaces of mesophyll cells. This thin waxy layer has a conductance of 50–200 mm s^{-1} for mesophytes but may be much lower for certain xerophytes for which cutinization of the mesophyll cells is appreciable (Table 8-1). In addition, δ^{ias} incorporates the fact that the entire cross section of the mesophyll region is not available for diffusion of gases, as the flow is constricted to the intercellular air spaces, which have a smaller cross-sectional area than does the corresponding leaf area, i.e., $A^{ias}/A < 1$. Thus that part of δ^{ias} referring to the intercellular air spaces per se equals the actual distance involved times A/A^{ias}. If the mesophyll region were one-third air by volume, A/A^{ias} would be 3. CO_2 and water vapor can enter or leave cells along the length of the intercellular air spaces, which further complicates the analysis.

The effective length δ^{ias}, including all the factors just enumerated, ranges from 100 μm to 1 mm for most leaves. Equation (8.6) indicates that the water vapor conductance across the intercellular air spaces then ranges from an upper limit of

$$g_{wv}^{\text{ias}} = \frac{(2.4 \times 10^{-5}\,\text{m}^2\,\text{s}^{-1})}{(100 \times 10^{-6}\,\text{m})} = 0.24\,\text{m s}^{-1} = 240\,\text{mm s}^{-1}$$

for a δ^{ias} of 100 μm down to about 24 mm s^{-1} for a δ^{ias} of 1 mm ($D_{wv} = 2.4 \times 10^{-5}\,\text{m}^2\,\text{s}^{-1}$ at 20°C; see Appendix I). Thus the conductance of the intercellular air spaces is relatively large (the resistance is relatively small) compared with the other conductances encountered by gases diffusing into or out of leaves (Table 8-1).

F. Fick's First Law and Conductances

The form of Fick's first law used to describe the flux density of water vapor and CO_2 for a leaf is appropriate for a one-dimensional situation (e.g., $J_j = D_j \Delta c_j / \Delta x = g_j \Delta c_j = \Delta c_j / r_j$). The H_2O lost from a leaf during transpiration evaporates from the cell walls of mesophyll cells, the inner sides of guard cells, and the adjacent subsidiary cells. If the cell walls were uniform and wet, then most of the water would evaporate from the immediate vicinity of the stomatal pores. However, the waxy material that occurs on the cell walls within a leaf, especially on guard cells and other nearby cells, causes much of the water to evaporate from the mesophyll cells in the leaf interior. We can imagine that the water vapor moves in the intercellular air spaces toward the leaf surface by diffusing down planar fronts of successively lower concentration. Our imaginary planar fronts are parallel to the leaf surface, so the direction for the fluxes is perpendicular to the leaf surface. When we reach the inner side of a stomatal pore, the area for diffusion is reduced from A^{ias} to A^{st} (see Fig. 8-3). In other words, we are still discussing the advance of planar fronts in one dimension, but the flux is now constricted to the stomatal pores. (A small amount of water constituting the cuticular transpiration diffuses across the cuticle in parallel with the stomatal fluxes.) Finally, the movement of water vapor across the boundary layer at the leaf surface is again a one-dimensional diffusion process—in this case across a distance δ^{bl}. Thus all fluxes of gases that we will consider here are moving perpendicular to the leaf surface, and thus a one-dimensional form of Fick's first law is usually appropriate. We generally express gas fluxes on the basis of unit leaf area. All conductances and resistances are also based on unit leaf area, as mentioned previously.

We next ask whether Δc_j is an accurate representation of the force for diffusion. Also, do the coefficients g_j and r_j relating flux densities to Δc_j change greatly as other quantities vary over physiological ranges? For instance, the coefficients could depend on the concentration of substance j or on temperature. Finally, are other quantities—such as changes in chemical potential $\Delta \mu_j$, water potential $\Delta \Psi_{wv}$, or partial pressure ΔP_j—more appropriate than Δc_j for flux calculations?

We will begin by writing the equation for the chemical potential of substance j in a gas phase, μ_j^{vapor}. By Eqs. (2.20) and (IV.10) (in Appendix IV), $\mu_j^{\text{vapor}} = \mu_j^* + RT \ln(P_j / P_j^*)$, where P_j is the partial pressure of substance j in the gas phase and P_j^* is its saturation partial pressure at that temperature and 1 atm of total pressure (we are ignoring the gravitational term of μ_j^{vapor} because over the short distances involved it does not influence the fluxes of gases into or out of leaves). At constant T the most general way to represent the force promoting the movement of substance j in the x-direction is $-\partial \mu_j^{\text{vapor}} / \partial x$, which becomes $-RT\,\partial[\ln(P_j / P_j^*)]/\partial x$ for our representation of the chemical potential, and in turn equals $-(RT/P_j)\partial P_j / \partial x$. For an ideal or perfect gas, $P_j V = n_j RT$, so $P_j = n_j RT / V$. The total number of moles of substance j divided by the volume (n_j/V) is the concentration of substance j, c_j; hence P_j can be replaced by $c_j RT$. Making this substitution for P_j, and using the very general flux relation given by Eq. (3.6) $[J_j = u_j c_j(-\partial \mu_j/\partial x)]$, we can express the flux density of gaseous substance j as follows:

$$J_j = u_j c_j \left(\frac{-\partial \mu_j^{\text{vapor}}}{\partial x} \right) = -u_j c_j \frac{RT}{P_j} \frac{\partial P_j}{\partial x}$$

$$= -u_j \frac{\partial P_j}{\partial x} = -u_j RT \frac{\partial c_j}{\partial x} \tag{8.7}$$

$$= \frac{u_j RT}{\Delta x} \Delta c_j = \frac{D_j}{\Delta x} \Delta c_j = g_j \Delta c_j$$

where in the last line we have replaced the negative concentration gradient, $-\partial c_j/\partial x$, by an average concentration gradient, $\Delta c_j/\Delta x$. Also, we have replaced $u_j RT$ by D_j (see Chapter 3, Section IIA) and then used Eq. (8.2), which indicates that $g_j = D_j/\Delta x$. In essence, Eq. (8.7) represents a thermodynamic derivation of Fick's first law for a gas phase.

Let us next examine the coefficients multiplying the various driving forces in Eq. (8.7). First, we note that $g_j = u_j RT/\Delta x$, so g_j is essentially independent of concentration—the very slight dependence of u_j or D_j on concentration can be ignored for gases. On the other hand, if we replace $\partial \mu_j^{\text{vapor}}/\partial x$ by $\Delta \mu_j^{\text{vapor}}/\Delta x$, Eq. (8.7) becomes $J_j = u_j c_j \Delta \mu_j^{\text{vapor}}/\Delta x$. The factor multiplying $\Delta \mu_j^{\text{vapor}}$ depends directly on concentration, so in addition to knowing $\Delta \mu_j^{\text{vapor}}$, we also need to specify c_j before calculating J_j. Moreover, c_j can vary near a leaf. Thus $\Delta \mu_j^{\text{vapor}}$ is not used to represent the force in the relations for transpiration or photosynthesis. Because $\Delta \Psi_{wv} = \Delta \mu_{wv}/\bar{V}_w$ (Eq. 2.21), Eq. (8.7) becomes $J_{wv} = (u_{wv} c_{wv} \bar{V}_w/\Delta x)\Delta \Psi_{wv}$ if differences in water potential are used to represent the force. Again, the coefficient depends on concentration, so using $\Delta \Psi_{wv}$ is not appropriate for describing transpiration. However, Eq. (8.7) also indicates that J_j could equal $(u_j/\Delta x)$ ΔP_j. This can lead to an alternative formulation for flux relations in which conductance has less dependence on temperature and ambient pressure.

Instead of using partial pressures, this alternative formulation uses mole fractions to represent the driving force. We have already encountered a difference in mole fraction representing a driving force when we discussed water flow ($J_{V_w} = P_w \Delta N_w$; see Chapter 2, Section IVF). For the case of ideal gases, which approximates situations of biological interest, Dalton's law of partial pressures indicates that the mole fraction of species j, N_j, equals P_j/P, where P is the total pressure. We can then modify Eq. (8.7) as follows:

$$J_j = -u_j \frac{\partial P_j}{\partial x} = u_j \frac{\Delta P_j}{\Delta x} = \frac{u_j RT}{RT} \frac{P}{\Delta x} \frac{\Delta P_j}{P}$$

$$= \frac{D_j P}{RT \Delta x} \Delta N_j = g_j' \Delta N_j \tag{8.8}$$

where we have again incorporated the relationship between diffusion coefficients and mobility ($D_j = u_j RT$) to lead to the conductance g_j'.

Let us next see how the conductance in Eqs. (8.2) and (8.7) ($g_j = D_j/\Delta x$) and that introduced in Eq. (8.8) [$g_j' = D_j P/(RT \Delta x) = g_j P/(RT)$] depend on pressure and temperature. We have noted that, for gases, D_j depends inversely on pressure (see Chapter 1,

Section IIC) and on temperature raised to the power 1.8 (see Chapter 3, Section IIA):

$$D_j \cong D_{j_0} \left(\frac{T}{273}\right)^{1.8} \frac{P_0}{P} \tag{8.9}$$

where D_{j_0} is the diffusion coefficient of substance j at 273 K (0°C) and an ambient pressure of P_0 (often taken as 1 atm). As the pressure decreases, D_j increases proportionally (Eq. 8.9), as does g_j ($g_j = D_j/\Delta x$; Eq. 8.2). On the other hand, as pressure changes, g_j' is unchanged [$g_j' = D_j P/(RT\Delta x)$; Eq. 8.8]. As temperature increases, D_j and hence g_j increase approximately as $T^{1.8}$ (Eq. 8.9), whereas g_j' increases approximately as $T^{0.8}$. Thus g_j' has no dependence on pressure and much less dependence on temperature than does g_j, and thus the former conductance is more appropriate for describing gas fluxes.

We can illustrate the difference between the two types of conductance by comparing a leaf with the same anatomical properties but under various environmental conditions. If a leaf at sea level and 10°C were heated to 40°C, g_j would increase 20%. If the leaf were then transferred at constant temperature to 2000 m, where the ambient air pressure is 22% lower than that at sea level, g_j would increase 28% more. On the other hand, heating from 10 to 40°C would increase g_j' only 8%, and transferring to 2000 m would not affect g_j'.

Because ΔN_j is dimensionless, g_j' has the same units as J_j, e.g., mol m^{-2} s^{-1}. At 1 atm and 20°C, $P/RT = 41.6$ mol m^{-3}. Thus a conductance g_j of 1 mm s^{-1} then corresponds to a conductance g_j' of (1 mm s^{-1})(41.6 mol m^{-3}), or 41.6 mmol m^{-2} s^{-1}. In our discussion of transpiration and photosynthesis, we will use both forms of conductance and resistance (see Table 8-1) but will emphasize g_j'.

II. Water Vapor Fluxes Accompanying Transpiration

The flux of water vapor out of a leaf during transpiration can be quantified using the conductances and the resistances involved. We will represent the conductances and the resistances using symbols (namely, ⎓⎓⎓) borrowed from electrical circuit diagrams. Typical values for the components will be presented along with the resulting differences in water vapor concentration and mole fraction across them. Our analysis of water vapor fluxes will indicate the important control of transpiration exercised by stomata.

A. Conductance and Resistance Network

Water vapor that evaporates from cell walls of mesophyll cells or the inner side of leaf epidermal cells diffuses through the intercellular air spaces to the stomata and then into the outside air. We have already introduced the four components involved—two are strictly anatomical (intercellular air spaces and cuticle), one depends on anatomy and yet responds to metabolic as well as environmental factors (stomata), and one depends on leaf morphology and wind speed (boundary layer). Figure 8-4 summarizes the symbols and arranges them into an electrical circuit analogy. We will analyze resistances and conductances for these components, some of which occur in series and some in parallel.

The conductance g_{wv}^{ias} and the resistance r_{wv}^{ias} include all parts of the pathway from the site of water evaporation to the leaf epidermis. Water can evaporate at the air–water interfaces of mesophyll cells, the inner side of epidermal cells (including guard cells), and even cells of the vascular tissue in a leaf before diffusing in the tortuous pathways of the

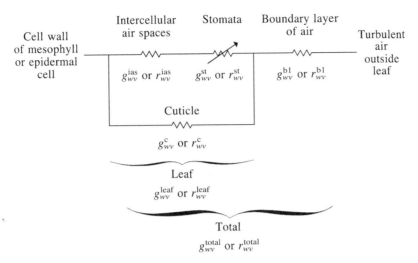

Figure 8-4. Conductances and resistances involved in water vapor flow accompanying transpiration, as arranged into an electrical circuit. The symbol for the stomatal component indicates that it is variable.

intercellular air spaces. The water generally has to cross a thin waxy layer on the cell walls of certain cells within a leaf. After crossing the waxy layer, which can be up to 0.1 μm thick, the water vapor diffuses through the intercellular air spaces and then through the stomata (conductance $= g_{wv}^{st}$, resistance $= r_{wv}^{st}$) to reach the boundary layer adjacent to the leaf surface. Alternatively, water in the cell walls of interior leaf cells might move as a liquid to the cell walls on the cuticle side of epidermal cells, where it could evaporate and the vapor then diffuse across the cuticle (or water may move across the cuticle as a liquid) before reaching the boundary layer at the leaf surface. The pathway for such cuticular transpiration (conductance $= g_{wv}^{c}$, resistance $= r_{wv}^{c}$) is in parallel with the pathway for transpiration through the stomatal pores. For simplicity, we have not included in Fig. 8-4 the intermediate case of water evaporating from the cell walls of mesophyll cells and then moving through the intercellular air spaces before crossing the waxy cuticle (the resistance of this pathway is indistinguishable from r_{wv}^{c} because $r_{wv}^{c} \gg r_{wv}^{ias}$). The final component encountered by the diffusing water vapor is the boundary layer just outside the leaf (conductance $= g_{wv}^{bl}$, resistance $= r_{wv}^{bl}$).

We will analyze the electrical circuit presented in Fig. 8-4 first in terms of resistances and then in terms of conductances. We begin by noting that the resistances r_{wv}^{ias} and r_{wv}^{st} occur in series. The total resistance of a group of resistors in series, r_{series}, is the sum of the individual resistances ($r_{series} = \sum_{i} r_{i}$, where r_{i} is the resistance of series resistor i). Thus the resistance of the pathway from the site of water evaporation across the intercellular air spaces and through the stomata is $r_{wv}^{ias} + r_{wv}^{st}$. This resistance is in parallel with r_{wv}^{c}, the cuticular resistance. The reciprocal of the total resistance of a group of resistors in parallel is the sum of the reciprocals of the individual resistances ($1/r_{parallel} = \sum_{i} 1/r_{i}$, where r_{i} is the resistance of parallel resistor i). For two resistors in parallel, $r_{parallel} = r_{1}r_{2}/(r_{1} + r_{2})$. The combined resistance for ($r_{wv}^{ias} + r_{wv}^{st}$) and r_{wv}^{c} in parallel is the leaf resistance for water vapor, r_{wv}^{leaf}:

$$r_{wv}^{leaf} = \frac{\left(r_{wv}^{ias} + r_{wv}^{st}\right)\left(r_{wv}^{c}\right)}{r_{wv}^{ias} + r_{wv}^{st} + r_{wv}^{c}} \tag{8.10}$$

When the stomata are open, the cuticular resistance (r_{wv}^c) is generally much larger than is the resistance in parallel with it, $r_{wv}^{ias} + r_{wv}^{st}$. In that case, the leaf resistance as given by Eq. (8.10) is approximately equal to $r_{wv}^{ias} + r_{wv}^{st}$.

The leaf resistance is in series with that of the boundary layer, r_{wv}^{bl}. Thus the total resistance for the flow of water vapor from the site of evaporation to the turbulent air surrounding a leaf (r_{wv}^{total}) is $r_{wv}^{leaf} + r_{wv}^{bl}$ (see Fig. 8-4). We now must face the complication created by the two leaf surfaces representing parallel pathways for the diffusion of water vapor from the interior of a leaf to the surrounding turbulent air. We will represent the leaf resistance for the pathway through the upper (adaxial) surface by $r_{wv}^{leaf_u}$ and that for the lower (abaxial) surface by $r_{wv}^{leaf_l}$. Each of these resistances is in series with that of an air boundary layer—$r_{wv}^{bl_u}$ and $r_{wv}^{bl_l}$ for the upper and the lower surfaces, respectively. The parallel arrangement of the pathways through the upper and the lower surfaces of a leaf leads to the following total resistance to the diffusion of water vapor from a leaf:

$$r_{wv}^{total} = \frac{\left(r_{wv}^{leaf_u} + r_{wv}^{bl_u}\right)\left(r_{wv}^{leaf_l} + r_{wv}^{bl_l}\right)}{r_{wv}^{leaf_u} + r_{wv}^{bl_u} + r_{wv}^{leaf_l} + r_{wv}^{bl_l}} \tag{8.11}$$

Now let us reconsider the analysis of the electrical circuit using conductances. For a group of conductances in series, the reciprocal of the total conductance ($1/g_{series}$) is the sum of the reciprocals of the individual conductances ($1/g_{series} = \sum_i 1/g_i$), whereas the conductance for a group of conductances in parallel is the sum of the individual conductances ($g_{parallel} = \sum_i g_i$). Thus the water vapor conductance of the intercellular air spaces and the stomata in series is $g_{wv}^{ias} g_{wv}^{st}/(g_{wv}^{ias} + g_{wv}^{st})$, and the water vapor conductance of the leaf is

$$g_{wv}^{leaf} = \frac{g_{wv}^{ias} g_{wv}^{st}}{g_{wv}^{ias} + g_{wv}^{st}} + g_{wv}^c \tag{8.12}$$

Next we note that g_{wv}^{leaf} is in series with a boundary layer conductance g_{wv}^{bl} and that the two sides of a leaf act as parallel conductances for water vapor diffusing from the interior of a leaf. We therefore obtain the following expression for the total conductance of a leaf with air boundary layers on each side:

$$g_{wv}^{total} = \frac{g_{wv}^{leaf_u} g_{wv}^{bl_u}}{g_{wv}^{leaf_u} + g_{wv}^{bl_u}} + \frac{g_{wv}^{leaf_l} g_{wv}^{bl_l}}{g_{wv}^{leaf_l} + g_{wv}^{bl_l}} \tag{8.13}$$

B. Values of Conductances

We generally assume that the average thickness of the boundary layer is the same on the two sides of a leaf, in which case $g_{wv}^{bl_u} = g_{wv}^{bl_l}$. However, this does not simplify our analysis of gas fluxes very much. Usually the boundary layer conductance is much greater than that of the leaf. If, in addition, $g_{wv}^{leaf_l} \gg g_{wv}^{leaf_u}$, as can occur for leaves with stomata primarily on the lower surface, then the diffusion of water vapor is mainly out through the lower surface of the leaf. This is often the case for the leaves of deciduous trees. The total conductance for water vapor diffusion from the sites of evaporation inside a leaf to the turbulent air surrounding the leaf (g_{wv}^{total}) is then approximately $g_{wv}^{leaf_l} g_{wv}^{bl_l}/(g_{wv}^{leaf_l} + g_{wv}^{bl_l})$ (see Eq. 8.13). For the symmetrical case in which $g_{wv}^{leaf_l} = g_{wv}^{leaf_u}$ and the boundary layers are of equal thickness (hence, $g_{wv}^{bl_l} = g_{wv}^{bl_u}$), the conductance is the same through either surface; g_{wv}^{total} is then twice as large as that through either surface, e.g., $2g_{wv}^{leaf_l} g_{wv}^{bl_l}/(g_{wv}^{leaf_l} + g_{wv}^{bl_l})$ by Eq. (8.13). This can occur for certain monocots—e.g., certain grasses, oats, barley, wheat, and corn—which can have fairly equal stomatal frequencies on the two leaf surfaces. For many plants, two to four times more stomata occur on the lower surface of the leaves than on the upper one, so A^{st}/A is considerably larger for the lower surface.

We next examine some representative values for the conductances encountered by water vapor as it diffuses out of leaves. For crop plants such as beet, spinach, tomato, and pea, g_{wv}^{ias} is generally 1600–4000 mmol m^{-2} s^{-1}, and for open stomata g_{wv}^{st} usually ranges from 200 to 800 mmol m^{-2} s^{-1}. Both conductances can be somewhat lower for other crops, but the maximum g_{wv}^{leaf} is still 80–400 mmol m^{-2} s^{-1} for most cultivated plants. Compared with crop plants, the maximum g_{wv}^{leaf} is usually somewhat lower for leaves of deciduous trees and conifer needles (see Table 8-1). As the stomata close, g_{wv}^{st} and hence g_{wv}^{leaf} decrease accordingly. The minimum value for leaf water vapor conductance occurs for fully closed stomata and essentially equals the cuticular conductance (see Eq. 8.12). Measured values of g_{wv}^{c} vary considerably, in part because of the difficulty in telling when stomata are fully closed, and may be as low as 4 mmol m^{-2} s^{-1} for cultivated plants, 2-fold lower for trees, and 10-fold lower for certain xerophytes (Table 8-1).

C. Effective Lengths and Resistance

We next examine a simplified expression that often adequately describes diffusion of water vapor from the sites of evaporation in cell walls to the turbulent air surrounding a leaf. We will consider the case in which nearly all the water vapor moves out across the lower epidermis and when cuticular transpiration is negligible. By Eqs. (8.10) and (8.11), the total resistance then is

$$r_{wv}^{total} \cong r_{wv}^{leaf_l} + r_{wv}^{bl_l}$$

$$= r_{wv}^{ias} + r_{wv}^{st_l} + r_{wv}^{bl_l}$$

$$= \frac{1}{D_{wv}}\left(\delta^{ias} + \frac{\delta^{st} + r^{st}}{na^{st}} + \delta^{bl}\right) \tag{8.14}$$

where Eqs. (8.3), (8.5), and (8.6) have been used to obtain the bottom line.

When diffusion is described by a one-dimensional form of Fick's first law, Eq. (8.14) indicates that each part of the pathway contributes its own effective length influencing the movement of water vapor out of the leaf. For instance, the effective length of the boundary layer is its thickness, δ^{bl}. The distance δ^{ias} includes the effective depth of the cell wall pores, the effective thickness of the waxy layer on the mesophyll cells, and the constriction on the region available for diffusion of gases within a leaf caused by the presence of mesophyll cells. A similar constricting effect greatly increases the effective length of the stomatal pores over the value of $\delta^{st} + r^{st}$. Specifically, Eq. (8.14) indicates that the effective length of the stomatal pores is $(\delta^{st} + r^{st})/(na^{st})$, which is considerably greater than $\delta^{st} + r^{st}$ because na^{st} is much less than 1 (recall that $na^{st} = A^{st}/A$, so na^{st} is a dimensionless number indicating the fraction of the surface area occupied by stomatal pores). The large effective length of the stomatal pores caused by the factor $1/na^{st}$ is an alternative way of viewing the constricting effect of the stomata.

If we need to incorporate another series resistance for gaseous diffusion, we can include its effective length within the parentheses in Eq. (8.14). For instance, the stomata in some xerophytes and conifers occur sunken beneath the leaf surface in cavities or crypts. We can estimate the effective length of this additional part of the pathway by using appropriate geometrical approximations. The effective length divided by the diffusion coefficient gives the resistance. Many leaves are covered by epidermal "hairs," which can be unicellular projections from the epidermal cells or multicellular appendages with the cells occurring in a row. Such *trichomes* cause an additional air layer to be held next to the leaf surface. The thickness of this layer may equal the average distance that the hairs project perpendicularly away from the leaf surface (more than 1 mm for some leaves). This length can be added within the parentheses in Eq. (8.14) to calculate the overall resistance. The extra resistance for water vapor diffusion caused by the hairs equals the thickness of the additional air layer divided by D_{wv}. Some water may evaporate from the surfaces of the epidermal hairs; such evaporation can lead to a substantial complication in the analysis of water vapor fluxes for leaves with numerous trichomes.

D. Water Vapor Concentrations and Mole Fractions for Leaves

As can be seen from relations such as Eq. (8.2) ($J_j = g_j \Delta c_j = \Delta c_j/r_j$), the conductances or resistances of the various parts of the pathway determine the drop in concentration across each component when the flux density is constant. Here we will apply this condition to a consideration of water vapor concentration and mole fraction in a leaf. We will also discuss the important effect of temperature on the water vapor content of air.

Let us represent the difference in water vapor concentration across the intercellular air spaces by Δc_{wv}^{ias}, that across the stomatal pores by Δc_{wv}^{st}, and that across the boundary layer adjacent to the leaf surface by Δc_{wv}^{bl}. Then the overall drop in water vapor concentration from the cell walls where the water evaporates to the turbulent air surrounding a leaf, Δc_{wv}^{total}, is

$$\Delta c_{wv}^{total} = c_{wv}^{e} - c_{wv}^{ta}$$
$$= \Delta c_{wv}^{ias} + \Delta c_{wv}^{st} + \Delta c_{wv}^{bl} \qquad (8.15)$$

where c_{wv}^{e} is the concentration of water vapor at the evaporation sites, presumably in the cell wall pores, and c_{wv}^{ta} is its value in the turbulent air surrounding the leaf. We also note that $\Delta c_{wv}^{ias} + \Delta c_{wv}^{st} = \Delta c_{wv}^{c}$, the difference in water vapor concentration from the cell walls of epidermal cells across the cuticle to the leaf surface, because the end points of the pathway are the same in each case (Fig. 8-4).

The rate of water vapor diffusion per unit leaf area, J_{wv}, equals the difference in water vapor concentration multiplied by the conductance across which Δc_{wv} occurs ($J_j = g_j \Delta c_j$; Eq. 8.2). In the steady state, when the flux density of water vapor and the conductance of each component are constant, this relation holds for both the overall pathway and any individual segment of it. Because some water evaporates from the cell walls of mesophyll cells along the pathway within the leaf, J_{wv} is actually not spatially constant in the intercellular air spaces. For simplicity, however, we generally assume that J_{wv} is unchanging from the mesophyll cell walls out to the turbulent air outside the leaf. When water vapor moves out only across the lower epidermis of the leaf, and when cuticular transpiration is negligible, we obtain the following relations in the steady state:

$$J_{wv} = g_{wv}^{total} \Delta c_{wv}^{total} = g_{wv}^{ias} \Delta c_{wv}^{ias}$$
$$= g_{wv}^{st} \Delta c_{wv}^{st} = g_{wv}^{bl} \Delta c_{wv}^{bl} \qquad (8.16)$$

Equation (8.16) indicates that the drop in water vapor concentration across a particular component is inversely proportional to its conductance when J_{wv} is constant (Δc_{wv} is then directly proportional to the resistance because $J_{wv} = \Delta c_{wv}/r_{wv}$). Because g_{wv}^{ias} and g_{wv}^{bl} are

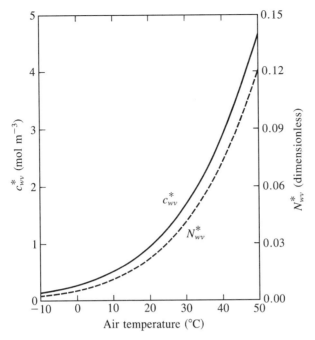

Figure 8-5. Variation in saturation values of water vapor concentration (c_{wv}^*) and mole fraction (N_{wv}^*) with temperature. Ambient pressure (needed to determine N_{wv}^*) was assumed to be 1 atm (0.1013 MPa). Digital values are presented in Appendix I.

generally larger than g_{wv}^{st} when the stomata are open (Table 8-1), by Eq. (8.16) the largest Δc_{wv} then occurs across the stomata.

The quantity c_{wv} is sometimes referred to as the *absolute humidity*. Its value in the turbulent air generally does not change very much during a day unless there is precipitation or other marked changes in the weather, whereas the *relative humidity* can vary greatly as the air temperature changes. For instance, air that is saturated with water vapor at 20°C (100% relative humidity) drops to 57% relative humidity when heated at constant pressure to 30°C. Most data for c_{wv} are expressed as mass of water/unit volume of air, and therefore we will use g m^{-3} as one of our units for water vapor concentration. A unit for c_{wv} that is often more appropriate is mol m^{-3}, and we will use it as well (Fig. 8-5).

Equations (8.15) and (8.16) can be adapted to water vapor expressed as a mole fraction, N_{wv}, simply by replacing c with N throughout. For instance, $\Delta N_{wv}^{total} = N_{wv}^{e} - N_{wv}^{ta}$, and $J_{wv} = g_{wv}^{total} \Delta N_{wv}^{total}$ (we will use the same symbol for conductance in either system, the units being dictated by whether changes in concentration or mole fraction represent the driving force). Also, $\Delta N_{wv}^{total} = \Delta N_{wv}^{ias} + \Delta N_{wv}^{st} + \Delta N_{wv}^{bl}$ (see Eq. 8.15). The magnitude of each of these differences in water vapor mole fraction is inversely proportional to the conductance across which the drop occurs—e.g., $g_{wv}^{x} \Delta N_{wv}^{x} = g_{wv}^{total} \Delta N_{wv}^{total}$—and so ΔN_{wv}^{x} is larger when g_{wv}^{x} is smaller, where x refers to any series component in the pathway (see Eq. 8.16).

Water vapor is generally the third most prevalent gas in air, although its mole fraction is relatively low compared with mole fractions for O_2 or N_2. Its saturation value is quite temperature dependent, e.g., $N_{wv}^* = 0.0231$ at 20°C and 0.0419 at 30°C (Fig. 8-5). In fact, the water vapor content of air at saturation (represented by c_{wv}^*, N_{wv}^*, or P_{wv}^*) increases nearly exponentially with temperature [Fig. 8-5; it may be helpful for understanding this temperature dependence to note that the fraction of molecules in the upper part of the Boltzmann energy distribution (see Eq. 3.21) that have a high enough energy to escape from the liquid increases exponentially with temperature]. Appendix I lists c_{wv}^*, N_{wv}^*, and P_{wv}^* at various temperatures from -30 to $60°C$.

E. Examples of Water Vapor Levels in a Leaf

We will now consider the actual relative humidities, concentrations, and mole fractions of water vapor in different parts of the transpiration pathway associated with a leaf (Fig. 8-6). We will suppose that the leaf temperature is $25°C$ and that the turbulent air surrounding the leaf is at $20°C$ with a relative humidity of 50%, which correspond to a c_{wv}^{ta} of 0.48 mol m^{-3} and an N_{wv}^{ta} of 0.0115 (see Fig. 8-6 and Appendix I).

We will assume that the water vapor in the pores of the cell walls of mesophyll cells is in equilibrium with the water in the cell wall and that $\Psi^{cell\,wall}$ is -1.0 MPa. Using Eq. (2.21) and omitting the gravity term, the corresponding % relative humidity (RH) at $25°C$ is

$$\Psi^{cell\,wall} = \Psi_{wv}^e = \frac{RT}{\bar{V}_w} \ln\left(\frac{RH}{100}\right)$$

or

$$RH = 100\, e^{\bar{V}_w \Psi^{cell\,wall}/RT}$$

$$= 100\, e^{(-1.0\,MPa/137.3\,MPa)} = 99.3\%$$

Thus the air in the pores of the mesophyll cell walls is nearly saturated with water vapor. From Appendix I, $c_{wv}^* = 1.28$ mol m^{-3} and $N_{wv}^* = 0.0313$ at $25°C$; therefore 99.3% relative humidity corresponds to a c_{wv}^e of (0.993)(1.28 mol m^{-3}), or 1.27 mol m^{-3}, and an N_{wv}^e of 0.0311. Hence, the difference in water vapor concentration from the mesophyll cell walls to the turbulent air is 1.27 mol m^{-3} $-$ 0.48 mol m^{-3}, or 0.79 mol m^{-3} ($\Delta c_{wv}^{total} = c_{wv}^e - c_{wv}^{ta}$; Eq. 8.15), and $N_{wv}^e - N_{wv}^{ta}$ is 0.0311 $-$ 0.0115, or 0.0196 (Fig. 8-6).[2]

Figure 8-6 also indicates specific values of the conductances as well as the overall series conductance for the diffusion of water vapor from the sites of evaporation to the turbulent air. The largest drop in water vapor level occurs across the stomatal pores because they have the smallest conductance in the current case. For instance, $\Delta N_{wv}^{st} = g_{wv}^{total} \Delta N_{wv}^{total}/g_{wv}^{st}$, which is (154 mmol m^{-2} s^{-1})(0.0196)/(200 mmol m^{-2} s^{-1}), or 0.0151 (Fig. 8-6).

We note that a small drop in relative humidity, here from 99% in the cell walls of mesophyll cells to 95% at the inner side of the stomata, is necessary for the diffusion of water vapor across the intercellular air spaces. The greater the fraction of water evaporating from near the guard cells, the smaller the drop across the intercellular air spaces. A large humidity drop occurs across the stomatal pores, such that the relative humidity is 47% at the leaf surface (Fig. 8-6). After crossing the stomatal pores, water vapor moves energetically downhill as

---

[2] The condition $N_{wv}^e > N_{wv}^{ta}$ necessary for the transpirational efflux of water vapor from a leaf leads to $N_{air}^{ta} > N_{air}^e$, where N_{air} represents the mole fraction of everything but water vapor—i.e., $N_{wv} + N_{air} = 1$ (there is a relatively small ΔN_{CO_2} resulting from CO_2 uptake during photosynthesis, which is mostly compensated for by a small ΔN_{O_2} acting in the opposite direction). Thus we expect a diffusion of air (mainly N_2) into a leaf, which can lead to slightly higher pressures in leaves (generally 0.1–1 kPa higher) compared to the ambient pressure outside (the continual supply of water vapor by evaporation inside the leaf and its removal in the outside turbulent air is necessary for the maintenance of this pressure difference). The internally elevated pressure can lead to mass flow of air within a plant, such as along the stems of the yellow waterlily, *Nuphar luteum*, where a ΔN_{wv} of 0.01 is accompanied by an air pressure 0.2 kPa higher inside a young leaf (Dacey, 1981).

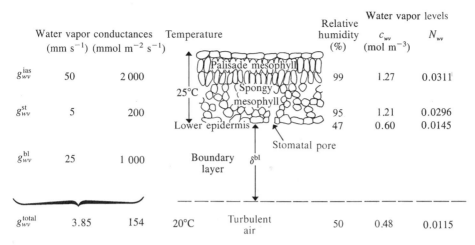

Figure 8-6. Representative values of quantities influencing the diffusion of water vapor out of a leaf. Conductances are given for the indicated parts of the pathway assuming that water moves out only through the lower surface and ignoring cuticular transpiration.

it diffuses across the boundary layer from 47% relative humidity at 25°C to 50% relative humidity at 20°C in the turbulent air surrounding the leaf (Fig. 8-6). Because a temperature change occurs in this part of the pathway, the driving force for the diffusion of water vapor must be expressed in terms of the difference in concentration or mole fraction and not the change in relative humidity.

In this text we assume that a leaf has a uniform temperature. However, temperature differences of a few degrees Celsius can develop across the width of a moderate-sized leaf at a wind speed of 1 m s^{-1}, reflecting differences in boundary layer thickness and spatial stomatal variation. The turbulent air outside a leaf generally has a different temperature. Fick's first law (e.g., $J_j = D_j \Delta c_j / \Delta x$; Eq. 8.2) strictly applies only to isothermal situations. For instance, D_j depends on the absolute temperature (D_j is proportional to $T^{1.8}$; Eq. 8.9), so $\Delta(D_j c_j)$ may be nonzero and lead to a flux even when Δc_j is zero. Fortunately, even when there are temperature differences between leaves and the turbulent air, Fick's first law in the form of Eq. (8.2) generally proves adequate for describing the fluxes of H_2O and CO_2, our primary concern in this chapter. However, flux relations based on differences in mole fraction (e.g., Eq. 8.8), which have a much lower dependence on temperature for their conductance, are preferred when there is a temperature difference from leaf to air, as is usually the case.

F. Water Vapor Fluxes

Based on quantities in Fig. 8-6, we can readily calculate the flux density of water vapor moving out of the lower side of a leaf. Specifically, $J_{wv}^l = g_{wv}^{total} \Delta c_{wv}^{total}$ or $g_{wv}^{total} \Delta N_{wv}^{total}$, e.g.,

$$J_{wv}^l = (154 \text{ mmol m}^{-2} \text{ s}^{-1})(0.0311 - 0.0115)$$
$$= 3.0 \text{ mmol m}^{-2} \text{ s}^{-1}$$

For simplicity, we have been considering the movement of water vapor across only the lower surface of a leaf, and we have ignored cuticular transpiration. Cuticular transpiration is generally small compared with transpiration through open stomata that is in parallel with it; i.e., the drop in water vapor level across the cuticle is essentially the

same as the drop across the stomata, whereas the stomatal conductance for open stomata is much greater than the cuticular conductance (see Table 8-1). We can add the cuticular transpiration to that through the stomata to get the total transpiration through one side of
a leaf. To obtain the overall rate of water vapor diffusing out of both sides of a leaf, we can
scale up J^l_{wv} calculated for the lower surface by an appropriate factor—the reciprocal of the
fraction of transpiration through the lower surface. For instance, about 70% of the water
loss in transpiration might be through the lower surface for a representative mesophyte. A
J^l_{wv} of 3.0 mmol m^{-2} s^{-1} then becomes (1/0.70)(3.0 mmol m^{-2} s^{-1}), or 4.3 mmol m^{-2} s^{-1},
when both leaf surfaces are considered, i.e., $J^l_{wv} = 0.70 J_{wv}$ implies that $J_{wv} = J^l_{wv}/0.70$.
Alternatively, we could use the actual g^{total}_{wv} as given by Eq. (8.13), which considers the two
leaf surfaces in parallel, or we could measure g^{total}_{wv} experimentally. This conductance times
Δc_{wv} or ΔN_{wv}, as appropriate, gives J_{wv} through both surfaces, but expressed per unit area
of one side of the leaf, which is the usual convention.

Let us next relate the rate of water loss by transpiration to the water content of a leaf.
For a 300-μm-thick leaf containing 30% intercellular air spaces by volume, the nongaseous
material corresponds to a thickness of (0.70)(300 μm), or 210 μm. Water typically comprises about 90% of a leaf's mass, so the water thickness would be about (0.90)(210 μm) or
190 μm. The density of water is 1000 kg m^{-3}, which corresponds to 56 kmol m^{-3}, so the
leaf has (190 $\times 10^{-6}$ m) (56 $\times 10^3$ mol m^{-3}) or 11 mol water m^{-2}. For the calculated J_{wv} of
4.3 mmol m^{-2} s^{-1}, this amount of water could be transpired in (11 mol m^{-2})/(4.3 $\times 10^{-3}$
mol m^{-2} s^{-1}) or 2600 s, which is 43 min. Hence, such a transpiring leaf must be continually supplied with water. For many cultivated plants and other mesophytes under these
conditions ($T^{leaf} = 25°C$, $T^{ta} = 20°C$, relative humidity$^{ta} = 50\%$), J_{wv} for open stomata is 2–5 mmol m^{-2} s^{-1}, so our example represents a slightly above-average transpiration rate. Many systems of units are used for transpiration rates; conversion factors for the
more common ones are summarized in Table 8-2.

G. Control of Transpiration

We now reconsider the values of the various conductances affecting the diffusion of water
vapor through the intercellular air spaces, out the stomata, and across the boundary layer at
the leaf surface. Usually g^{ias}_{wv} is relatively large, g^{bl}_{wv} is rarely less than 500 mmol m^{-2} s^{-1}, but
g^{st}_{wv} is generally less than this and decreases as the stomata close. Consequently, control for
limiting transpiration rests with the stomata, not with the boundary layer or the intercellular
air spaces. When g^{st}_{wv} is at least a few times smaller than is g^{bl}_{wv}, as generally occurs under
field conditions, moderate changes in the ambient wind speed have relatively little effect
on g^{total}_{wv}. However, the boundary layer conductance can be the main determinant of J_{wv}
for the fruiting bodies of Basidiomycetes (fungi), which have no stomata. For instance,
near the ground, where these fruiting bodies occur and the wind speed is relatively low
(generally <0.2 m s^{-1}), g^{bl}_{wv} exerts the main control on water loss for *Lycoperdon perlatum*
and *Scleroderma australe* (Nobel, 1975).

Stomata tend to close as a leaf wilts, a common response of plants to water strees. Assuming that g^{st}_{wv} decreases 20-fold, and using values presented in Fig. 8-6, g^{total}_{wv} decreases
about 15-fold from 154 mmol m^{-2} s^{-1} for a g^{st}_{wv} of 200 mmol m^{-2} s^{-1} to 9.9 mmol m^{-2} s^{-1}
for a g^{st}_{wv} of 10 mmol m^{-2} s^{-1}. The decrease in g^{total}_{wv} causes transpiration to decrease to only
about 6% of its former value, if we ignore the parallel pathway across the cuticle. When the
stomata close tightly, only cuticular transpiration remains, which can have a conductance
of 1–10 mmol m^{-2} s^{-1} (Table 8-1). In this case, cuticular transpiration accounts for all the
loss of water vapor from the leaves.

Table 8-2. Conversion Factors for Some of the More Common Units Used for Transpiration, CO_2 Levels, and Photosynthesis[a]

Transpiration unit	$(mmol\ m^{-2}\ s^{-1})$	$(mg\ m^{-2}\ s^{-1})$
$\mu mol\ H_2O\ cm^{-2}\ s^{-1}$	10	180.2
$mol\ H_2O\ m^{-2}\ h^{-1}$	0.278	5.01
$\mu g\ H_2O\ cm^{-2}\ s^{-1}$	0.555	10
$\mu g\ H_2O\ cm^{-2}\ min^{-1}$	0.00925	0.1667
$mg\ H_2O\ dm^{-2}\ min^{-1}$	0.0925	1.667
$g\ H_2O\ dm^{-2}\ h^{-1}$	1.542	27.8
$kg\ H_2O\ m^{-2}\ h^{-1}$	15.42	278

CO_2 level	$(mmol\ m^{-3})$	$(mg\ m^{-3})$	(Pa)	Mole fraction $\times 10^6$ (ppm by volume) $(\mu liter\ liter^{-1})$ $(\mu bar\ bar^{-1})$ $(\mu mol\ mol^{-1})$
$mmol\ CO_2\ m^{-3}$, $nmol\ CO_2\ cm^{-3}$, μM	1	44.0	2.44	24.4
$mg\ CO_2\ m^{-3}$, $ng\ CO_2\ cm^{-3}$	0.0227	1	0.0554	0.554
$\mu liter\ CO_2\ liter^{-1}$,[b] ppm CO_2 by volume	0.0410	1.806	0.1	1
μbar[b]	0.0410	1.806	0.1	1
Pa[b]	0.410	18.06	1	10

Photosynthesis unit	$(\mu mol\ m^{-2}\ s^{-1})$
$ng\ CO_2\ cm^{-2}\ s^{-1}$	0.227
$nmol\ CO_2\ cm^{-2}\ s^{-1}$	10
$mg\ CO_2\ m^{-2}\ s^{-1}$	22.7
$mg\ CO_2\ dm^{-2}\ h^{-1}$, $kg\ CO_2\ hectare^{-1}\ h^{-1}$	0.631
$kg\ carbohydrate\ hectare^{-1}\ h^{-1}$	0.92
$mm^3\ CO_2\ cm^{-2}\ h^{-1}$	0.114

[a] Values were determined using quantities in Appendices I and II. To convert from one set to another, a quantity expressed in the units in the left column should be multiplied by the factor in the column of the desired units.
[b] To convert a volume/volume number like ppm or mole fraction to a mole/volume or a pressure unit, or vice versa, we need to know the temperature and the pressure; an air temperature of 20°C and a pressure of 0.1 MPa (1 bar) were used here. To adjust for other temperatures and pressures, the ideal gas law ($PV = nRT$) must be employed, e.g., 1 $\mu mol\ mol^{-1}$ of CO_2 at temperature T_x in K and pressure P_x in MPa is equal to 0.410 $(293.15/T_x)$ (P_x) mmol m^{-3}.

III. CO_2 Conductances and Resistances

We will next consider the main function of a leaf, photosynthesis, in terms of the conductances and the resistances encountered by CO_2 as it diffuses from the turbulent air, across the boundary layers next to a leaf, through the stomata, across the intercellular air spaces, into the mesophyll cells, and eventually into the chloroplasts. The situation is obviously more complicated than the movement of water vapor during transpiration because CO_2 not only must diffuse across the same components encountered by water vapor moving in the opposite direction[3] but also must cross the cell wall of a mesophyll cell, the plasma membrane, part of the cytosol, the membranes surrounding a chloroplast, and some of the chloroplast stroma. We will specifically indicate the resistance of each component, resistances being

---

[3] Additional complications relate to the mass flows that occur, particularly through the stomata where the diffusing species can also interact with the walls of the pores. The flux of water vapor thus affects the CO_2 concentration gradient because of molecular collisions between H_2O and CO_2 moving in opposite directions and also because of pressure driven (bulk) flow [see footnote 2; also see Field *et al.* (1989) and Leuning (1983)].

somewhat easier to deal with than conductances for the series of components involved in the pathway for CO_2 movement.

A. Resistance and Conductance Network

Figure 8-7 illustrates the various conductances and resistances affecting CO_2 as it diffuses from the turbulent air surrounding a leaf up to the sites in the chloroplasts where it is incorporated into photosynthetic products. For simplicity, we will initially restrict our attention to the diffusion of CO_2 into a leaf across its lower epidermis only. When the parallel pathways through the upper and the lower surfaces of a leaf are both important, we can readily modify our equations to handle the reduction in resistance, or increase in conductance, encountered between the turbulent air surrounding the leaf and the cell walls of its mesophyll cells (see Eqs. 8.11 and 8.13). We will also ignore the cuticular path for CO_2 entry into the leaf for the same type of reason that we have neglected this part of the pathway when discussing transpiration–namely, $r^c_{CO_2}$ is generally considerably greater than $r^{ias}_{CO_2} + r^{st}_{CO_2}$, the resistance in parallel with it.

The first three resistances encountered by CO_2 entering a leaf through the lower epidermis ($r^{bl_1}_{CO_2}, r^{st_1}_{CO_2}$, and $r^{ias}_{CO_2}$) have analogs in the case of transpiration. We can thus transfer Eq. (8.14) to our current discussion, changing only the subscripts:

$$r^{leaf_1}_{CO_2} + r^{bl_1}_{CO_2} = \frac{1}{D_{CO_2}} \left(\delta^{ias} + \frac{\delta^{st} + r^{st}}{na^{st}} + \delta^{bl} \right) \tag{8.17}$$

where $r^{leaf_1}_{CO_2}$ is the resistance of the intercellular air spaces plus the stomatal pores of the lower leaf surface to the diffusion of CO_2. We also note that $r^{leaf_1}_{CO_2} + r^{bl_1}_{CO_2}$ is analogous to r^{total}_{wv}, the total resistance encountered by water vapor (see Figs. 8.4 and 8.7).

The diffusion coefficient for CO_2, D_{CO_2}, is 1.51×10^{-5} m^2 s^{-1} in air at 20°C (see Appendix I). This is smaller than D_{wv} (2.42×10^{-5} m^2 s^{-1}) because CO_2 molecules are heavier and thus diffuse more slowly than do H_2O molecules. By Eqs. (8.14) and (8.17), $(r^{leaf_1}_{CO_2} + r^{bl_1}_{CO_2})/r^{total}_{wv} = D_{wv}/D_{CO_2}$, which is $(2.42 \times 10^{-5}$ m^2 s$^{-1})/(1.51 \times 10^{-5}$ m^2 s$^{-1})$, or 1.60. Consequently, CO_2 diffusing from the turbulent air up to the cell walls of mesophyll cells encounters a resistance that is 60% higher than does water vapor diffusing in the opposite direction over the same pathway. Likewise, the gas phase conductance is $(100\%)/(1.60)$ or only 62% as great for CO_2 as for water vapor.[4]

As indicated in Fig. 8-7, five additional resistances are involved in CO_2 movement compared with water vapor movement. The new components of the pathway are the nongaseous parts of the cell wall of a mesophyll cell (resistance $= r^{cw}_{CO_2}$), a plasma membrane ($r^{pm}_{CO_2}$), the cytosol ($r^{cyt}_{CO_2}$), the chloroplast limiting membranes ($r^{clm}_{CO_2}$), and the interior of the chloroplasts ($r^{stroma}_{CO_2}$). For convenience we will divide these five resistances into two parts, the

---

[4] Actually, movement across the boundary layer is partly by diffusion, where the ratio D_{wv}/D_{CO_2} applies, and partly by turbulent mixing (see Fig. 7-5), where molecular differences are obliterated. Thus $r^{bl}_{CO_2}/r^{bl}_{wv}$ is intermediate between the extremes of 1.60 and 1.00, and indeed it is found to be $(D_{wv}/D_{CO_2})^{2/3}$, which is 1.37. Although Eq. (8.17) is generally satisfactory, certain situations may warrant replacing δ^{bl}/D_{CO_2} by 1.37 δ^{bl}/D_{wv}, which equals 0.86 δ^{bl}/D_{CO_2}. We also note that the effective boundary layer thicknesses for water vapor and heat transfer are quite similar, being within 7% for flat plates (Monteith and Unsworth, 1990) and within 5% for cylinders and spheres (Nobel, 1974, 1975).

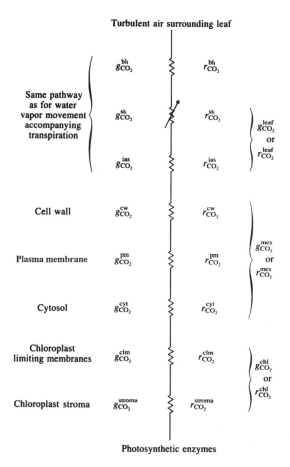

Figure 8-7. Principal conductances and resistances involved in the movement of CO$_2$ from the turbulent air surrounding a leaf, across the lower epidermis, and then to the enzymes involved in the fixation of CO$_2$ into photosynthetic products in the chloroplasts of mesophyll cells.

mesophyll resistance, $r_{CO_2}^{mes}$, and the chloroplast resistance, $r_{CO_2}^{chl}$:

$$r_{CO_2}^{mes} = r_{CO_2}^{cw} + r_{CO_2}^{pm} + r_{CO_2}^{cyt} \tag{8.18a}$$

$$r_{CO_2}^{chl} = r_{CO_2}^{clm} + r_{CO_2}^{stroma} \tag{8.18b}$$

B. Mesophyll Area

The area of the mesophyll cell walls across which CO$_2$ can diffuse is considerably larger than the surface area of the leaf. For the constricting effect caused by the stomata, we used A^{st}/A, the fraction of the leaf surface area occupied by stomatal pores. Here we will use the ratio A^{mes}/A to indicate the increase in area available for CO$_2$ diffusion, where A^{mes} is the total area of the cell walls of mesophyll cells that is exposed to the intercellular air spaces, and A is the area of one side of the same leaf.

Although A^{mes}/A varies with plant species as well as with leaf development, it is generally between 10 and 40 for mesophytes (Björkman, 1981; Nobel and Walker, 1985). We can

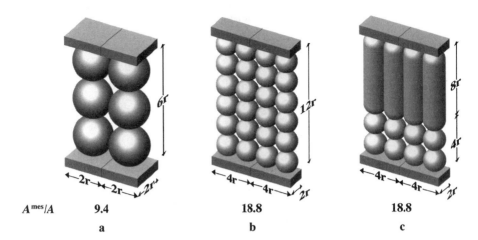

Figure 8-8. Representations of mesophyll cells showing how geometry affects A^{mes}/A. Spheres or cylinders with hemispherical ends in an orthogonal (right-angled) array lead to the indicated A^{mes}/A. The length of the lateral walls of the "palisade" cells in c is six times the radius.

appreciate the large values of A^{mes}/A by examining Figs. 1-2 and 8-6, which indicate that a tremendous amount of cell wall area is exposed to air within a leaf; e.g., the palisade mesophyll is generally 15–40% air by volume, and the spongy mesophyll is 50–70% air. Although the spongy mesophyll region generally has a greater volume fraction of air, the palisade region usually has a greater total cell wall area exposed to the intercellular air spaces. Xerophytes tend to have a more highly developed palisade region than do mesophytes (in some cases, the spongy mesophyll cells are even absent in xerophytes), which leads to values of 20–50 for A^{mes}/A of many xerophytes.

To help appreciate the magnitude of A^{mes}/A, we will consider some geometrical idealizations. For a single layer of uniform spheres in an orthogonal array, A^{mes}/A is $4\pi r^2/(2r \times 2r)$, or π. Hence, three layers of spherical cells have an A^{mes}/A of 3π, or 9.4 (Fig. 8-8a). If the radius of the spheres were halved but the thickness of the array remained unchanged, then A^{mes}/A doubles (Fig. 8-8a versus Fig. 8-8b). In a more realistic representation of a leaf with a single palisade layer having cells four times longer than wide and two layers of spherical spongy cells, A^{mes}/A is 18.8, two-thirds of which is contributed by the palisade cells (Fig. 8-8c). Moreover, three-quarters of the exposed cell wall area of the palisade cells occurs on their lateral walls. To show that the relatively large area contributed by the lateral walls is typical, we will represent the palisade cells as cylinders of radius r and length l with hemispheres on each end. The total area of the lateral surface of the cylinder $(2\pi rl)$ is generally greater than that of the two hemispherical ends $(4\pi r^2)$ because l is usually considerably greater than r for such palisade cells—e.g., r may be 10–20 μm, whereas l is 30–100 μm for representative palisade cells. When such cylinders are packed together to form a layer of palisade cells, nearly the entire surface area of the lateral walls is exposed to the intercellular air spaces. This surface area and most of that of both ends of a palisade cell are available for the inward diffusion of CO_2.

The illumination under which a leaf develops can greatly influence the anatomy of its mesophyll. Development in a dark or shaded environment can lead to a shade leaf, and differentiation under moderate to high illumination can lead to a sun leaf. Besides being smaller in area, sun leaves usually are thicker and have a higher ratio of palisade to spongy mesophyll cells than do shade leaves on the same plant. Moreover, their palisade cells are generally longer (larger l for the cylinders). Consequently, A^{mes}/A can be two to four times higher for sun leaves than for shade leaves on the same plant. For example, growing *Plectranthus parviflorus* under low light levels (a photosynthetic photon flux density of 17 μmol m^{-2} s^{-1} for 12-h days) leads to thin leaves with an A^{mes}/A of 11, whereas high average light levels (a PPFD of 810 μmol m^{-2} s^{-1}) lead to thick leaves with an A^{mes}/A of 50 (Nobel *et al.*, 1975). The columnar nature of palisade cells and their abundance in sun leaves developing under high light cause internal reflections that allow the light to penetrate further into a leaf, especially collimated direct sunlight (Vogelmann, 1993).

Although light generally has the greatest influence, A^{mes}/A can also be influenced by changes in other environmental factors during leaf development (Nobel and Walker, 1985). For instance, higher temperatures generally induce smaller cells and increase A^{mes}/A by up to 40%. Reduced cell size generally accompanies water stress, but the influences on A^{mes}/A vary with species, ranging from no change to a 50% increase in A^{mes}/A. Higher salinities during leaf development generally lead to thicker leaves, which can be accompanied by a corresponding increase in cell dimensions with no change in A^{mes}/A or sometimes by an increase in A^{mes}/A.

C. Resistance Formulation for Cell Components

The resistance to diffusion of a molecular species across a barrier equals the reciprocal of its permeability coefficient. We will let $P^j_{CO_2}$ be the permeability coefficient for CO_2 diffusion across barrier j. To express the resistance of a particular mesophyll or chloroplast component on a leaf area basis, we must also incorporate A^{mes}/A to allow for the actual area available for diffusion—the greater internal leaf area acts like more pathways in parallel and thus reduces the effective resistance. Because the area of the plasma membrane is about the same as that of the cell wall, and the chloroplasts generally occupy a single layer around the periphery of the cytosol (Fig. 8-9), the factor A^{mes}/A applies to all the diffusion steps of CO_2 in mesophyll cells (all five individual resistances in Eq. 8.18). In other words, we are imagining for simplicity that the cell wall, the plasma membrane, the cytosol, and the chloroplasts are all in layers having essentially equal areas (Fig. 8-9). Thus the resistance of any of the mesophyll or chloroplast components for CO_2 diffusion, $r^j_{CO_2}$, is reduced from $1/P^j_{CO_2}$ by the reciprocal of the same factor, A^{mes}/A:

$$r^j_{CO_2} = \frac{1}{A^{mes}/A} \frac{1}{P^j_{CO_2}} = \frac{A}{A^{mes} P^j_{CO_2}}$$

$$= \frac{A \, \Delta x^j}{A^{mes} D^j_{CO_2} K^j_{CO_2}} = \frac{1}{g^j_{CO_2}} \tag{8.19}$$

where the second line follows from the definition of a permeability coefficient, $P_j = D_j K_j / \Delta x$ (Eq. 1.9). In particular, Δx^j is the thickness of the jth barrier, $D^j_{CO_2}$ is the diffusion coefficient of CO_2 in it, and $K^j_{CO_2}$ is a suitably defined partition coefficient. If the chloroplasts do not occur all around a mesophyll cell (Fig. 8-9), then their area can be $<A^{mes}$. In such a case, A^{mes} in Eq. (8.19) can be replaced by $A^{chloroplast}$, which is the chloroplast area corresponding to leaf area A. This recognizes the importance of the diffusion of CO_2 into chloroplasts for photosynthesis.

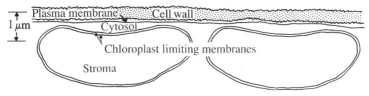

Intercellular air spaces

Figure 8-9. Schematic cross section near the periphery of a mesophyll cell, indicating the sequential occurrence of anatomical components across which CO_2 diffuses from the intercellular air spaces to
the carboxylation enzymes in the chloroplast stroma.

D. Partition Coefficient for CO$_2$

In Chapter 1 we introduced a partition coefficient to describe the ratio of the concentrations of
some species in two adjacent phases. For instance, $K_{CO_2}c_{CO_2}$ can be the actual concentration
of CO_2 in some region where concentrations are difficult to measure, c_{CO_2} is the equilibrium
concentration of CO_2 in an adjacent region where it is readily measured, and K_{CO_2} is the
partition coefficient. Similarly, we will express every $K^j_{CO_2}$ in the mesophyll cells as the
actual concentration of all forms of CO_2 in component j divided by the concentration of
CO_2 that occurs in an adjacent air phase at equilibrium, $c^{air}_{CO_2}$:

$$K^j_{CO_2} = \frac{\text{concentration of all forms of "CO}_2\text{" in barrier } j}{\text{equilibrium CO}_2 \text{ concentration in adjacent air phase}}$$

$$= \frac{c^j_{CO_2} + c^j_{H_2CO_3} + c^j_{HCO_3^-} + c^j_{CO_3^{2-}}}{c^{air}_{CO_2}} \qquad (8.20)$$

A concentration referred to as $c^{air}_{CO_2}$ thus equals the actual concentration of all forms of
CO_2 in component j divided by $K^j_{CO_2}$. This convention allows us to discuss fluxes in a
straightforward manner because CO_2 then diffuses toward regions of lower $c^{air}_{CO_2}$ regardless
of the actual concentrations and partition coefficients involved. For instance, to discuss the
diffusion of "CO_2" across a cell wall, we need to consider the partitioning of CO_2 between
the air in the cell wall pores and the various types of CO_2 in the adjacent water within the
cell wall interstices. $K^{cw}_{CO_2}$ is thus the actual concentration of CO_2 plus H_2CO_3, HCO_3^-, and
CO_3^{2-} in the cell wall water divided by the concentration of CO_2 in air in equilibrium with
the cell wall water.

The concentrations of the various forms of "CO_2" present in an aqueous phase are
temperature dependent and extremely sensitive to pH (the concentrations also depend on
the presence of other solutes, which presumably is a small effect for the cell wall water). For
instance, the equilibrium concentration of CO_2 dissolved in water divided by that of CO_2 in
an adjacent gas phase, $c^{water}_{CO_2}/c^{air}_{CO_2}$, decreases over twofold from 10 to 40°C (Table 8-3; the
decreased solubility of CO_2 at higher temperatures is a characteristic of dissolved gases,
which fit into the interstices of water, such space becoming less available as molecular
motion increases with temperature). This partition coefficient is not very pH dependent,
but the equilibrium concentration of HCO_3^- in water relative to that of dissolved CO_2 is

Table 8-3. Influence of Temperature and pH on Partitioning of "CO_2" between an Aqueous Solution and an Adjacent Air Phase[a]

Temperature (°C)	$\dfrac{c_{CO_2}^{water}}{c_{CO_2}^{air}}$	pH	$\dfrac{c_{CO_2}^{water} + c_{H_2CO_3}^{water} + c_{HCO_3}^{water}}{c_{CO_2}^{air}}$
0	1.65	4	0.91
10	1.19	5	0.96
20	0.91	6	1.48
30	0.71	7	6.6
40	0.58	8	58
50	0.50		

[a] The partition coefficient for all three forms of "CO_2" at various pHs is for 20°C. Data are an ionic strength of about 200 mol m^{-3} and an air pressure of 0.1 MPa (source: Stumm and Morgan, 1996).

markedly affected by pH. In particular, CO_2 dissolved in an aqueous solution may interact with OH^- to form bicarbonate, which then associates with H^+ to form H_2CO_3: $CO_2 + OH^- + H^+ \rightleftharpoons HCO_3^- + H^+ \rightleftharpoons H_2CO_3$. Alternatively, CO_2 in water might form H_2CO_3, which then dissociates to HCO_3^- and H^+. The interconversions of CO_2 and H_2CO_3 are actually relatively slow unless a suitable catalyst, such as the enzyme carbonic anhydrase, is present. Because H^+ is involved in these reactions, the pH will affect the amount of HCO_3^- in solution, which in turn depends on the CO_2 concentration. The equilibrium concentration of H_2CO_3 is only about 1/400 of that of the dissolved CO_2, so our main concern will be with CO_2 and HCO_3^- (CO_3^{2-} is also not a major type until the pH exceeds 8, and at pH 8 it is only 1% of HCO_3^- at an ionic strength of 200 mol m^{-3}, so CO_3^{2-} will be ignored here).

E. Cell Wall Resistance

We begin our discussion of the newly introduced resistances by evaluating the components of $r_{CO_2}^{cw}$, the resistance encountered by CO_2 as it diffuses through the water-filled interstices of the cell wall from the interface with the intercellular air spaces on one side to the plasma membrane on the other side (Fig. 8-9). We will use Eq. (8.19) to describe this resistance:
$r_{CO_2}^{cw} = A \Delta x^{cw}/(A^{mes} D_{CO_2}^{cw} K_{CO_2}^{cw})$.
The distance across the barrier Δx^{cw} is the average thickness of the cell walls of the mesophyll cells. The diffusion coefficient for the gas CO_2 dissolved in water is 1.7×10^{-9} m^2 s^{-1} at 20°C (see Table 1-1). However, the effective $D_{CO_2}^{cw}$ is lower by a factor of about three or four because the water-filled interstices represent slightly less than half the cell wall and their course through the cell wall is rather tortuous. Thus $D_{CO_2}^{cw}$ may be about 5×10^{10} m^2 s^{-1}. Besides moving as the dissolved gas, "CO_2" may also diffuse across the cell wall as H_2CO_3 or HCO_3^-. The diffusion coefficients of these two species in the cell wall are also most likely about 5×10^{-10} m^2 s^{-1}. However, the possible presence of H_2CO_3 and HCO_3^- makes the effective concentration of "CO_2" in the cell wall uncertain, which is why it is convenient to introduce $K_{CO_2}^{cw}$, the partition coefficient for CO_2 in the cell wall.

At 20°C, $(c_{CO_2}^{water} + c_{H_2CO_3}^{water} + c_{HCO_3}^{water})/c_{CO_2}^{air}$ (our definition for $K_{CO_2}^{cw}$; see Eq. 8.20) is about 1 from pH 4 to 6 but increases markedly above pH 7 (Table 8-3). The equilibrium value is

Table 8-4. Summary of Representative Values of Conductances and Resistances for CO_2 Diffusing into Leaves[a]

	Conductance		Resistance	
Component	$(mm\ s^{-1})$	$(mmol\ m^{-2}\ s^{-1})$	$(s\ m^{-1})$	$(m^2\ s\ mol^{-1})$
Leaf (lower surface)—gas phase				
Crops—open stomata	1.2–6	50–250	160–800	4–20
Trees—open stomata	0.3–2	12–75	500–2500	13–80
Cell wall	30	1200	30	0.8
Plasma membrane	10	400	100	2.5
Cytosol	100	4000	10	0.25
Mesophyll				
Estimation	7	300	140	3.5
Measurements—mesophytes	2.5–25	100–1000	40–400	1–10
Chloroplast				
Estimation	10	400	100	2.5
Measurements	>5	>200	<200	<5

[a] Certain values are calculated in the text (also see Evans *et al.*, 1986; Longstreth *et al.*, 1980; and references for Table 8-1).

affected by temperature in approximately the same way as the partition coefficient $c_{CO_2}^{water}/c_{CO_2}^{air}$ cited previously, e.g., from 20 to 30°C it decreases by about 22%. We note that this appreciable temperature dependence of K_{CO_2} results in a similar temperature dependence of $r_{CO_2}^j$ (see Eq. 8.19). Although the pH in the cell walls of mesophyll cells within a leaf is not known with certainty, it is probably <6. Thus at usual leaf temperatures, $K_{CO_2}^{cw}$ will be close to 1, the value we will use for calculation.

We now estimate $r_{CO_2}^{cw}$. We will assume that the mesophyll cells have a typical cell wall thickness, Δx^{cw}, of 0.3 μm, that the diffusion coefficient in the cell walls for CO_2 or HCO_3^-, $D_{CO_2}^{cw}$, is $5 \times 10^{-10}\ m^2\ s^{-1}$, and that $K_{CO_2}^{cw}$ is 1. The magnitude of $r_{CO_2}^{cw}$ also depends on the relative surface area of the mesophyll cells compared with the leaf area. We will let A^{mes}/A be 20, a representative value for mesophytes. Using Eq. (8.19), the resistance of the cell walls to the diffusion of CO_2 is

$$r_{CO_2}^{cw} = \frac{(0.3 \times 10^{-6}\ m)}{(20)(5 \times 10^{-10}\ m^2\ s^{-1})(1)} = 30\ s\ m^{-1}$$

This is a small value for a CO_2 resistance (Table 8-4) and indicates that the cell walls of the mesophyll cells generally do not represent a major barrier to the diffusion of the various species of CO_2.

F. Plasma Membrane Resistance

We now examine $r_{CO_2}^{pm}$, the resistance of the plasma membrane of mesophyll cells to diffusion of the various types of CO_2. Although we do not know the actual permeability coefficient of the plasma membrane of mesophyll cells for CO_2 or HCO_3^-, we expect it to be much lower for a charged species such as HCO_3^-. For instance, $P_{HCO_3^-}^{pm}$ might be about $10^{-8}\ m\ s^{-1}$. On the other hand, CO_2 is a small neutral linear molecule that enters cells extremely easily. The permeability coefficient of CO_2 entering plant cells is probably at least $2 \times 10^{-4}\ m\ s^{-1}$, which is somewhat higher than $P_{H_2O}^{pm}$, and may be $1 \times 10^{-3}\ m\ s^{-1}$—we will use $5 \times 10^{-4}\ m\ s^{-1}$ for purposes of calculation. Using Eq. (8.19) and a value of 20 for A^{mes}/A, the CO_2 resistance of the plasma membrane is

$$r_{CO_2}^{pm} = \frac{1}{(20)(5 \times 10^{-4} \text{ m s}^{-1})} = 100 \text{ s m}^{-1}$$

Based on the relative values for the two permeability coefficients, the resistance to the diffusion of HCO_3^- across the plasma membrane is over 10^4 times higher than that for CO_2, namely 5×10^6 s m^{-1}. Because of the extremely high resistance for HCO_3^-, we conclude that bicarbonate does not diffuse across the plasma membrane at the rates necessary to sustain photosynthesis. The resistance calculated for CO_2 (100 s m^{-1}) is relatively small, which suggests that diffusion of CO_2 can be adequate for moving this substrate of photosynthesis across the plasma membrane.

HCO_3^- or CO_2 could be actively transported across the plasma membrane or perhaps could cross by facilitated diffusion. Facilitated diffusion would act as a low-resistance pathway in parallel with the ordinary diffusion pathway and thus would reduce the effective resistance of the plasma membrane. Unfortunately, the actual mechanism for CO_2 or HCO_3^- movement across the plasma membrane of mesophyll cells is not known with certainty, although $r_{CO_2}^{pl}$ for diffusion of CO_2 is apparently low enough to account for the observed CO_2 fluxes.

G. Cytosol Resistance

The resistance of the cytosol of mesophyll cells to the diffusion of CO_2 is small because the distance is short—the chloroplasts are located around the periphery of most mesophyll cells (Figs. 1-1 and 8-9). In particular, the average distance from the plasma membrane to the chloroplasts, Δx^{cyt}, is only 0.1–0.3 μm. We will use a value of 0.2 μm for Δx^{cyt} when estimating $r_{CO_2}^{cyt}$ using Eq. (8.19). Because of the presence of fibrous proteins, the diffusion coefficient of CO_2 in this region of the cytosol is somewhat less than that in water at 20°C, 1.7×10^{-9} m^2 s^{-1}. For purposes of calculation, we will let $D_{CO_2}^{cyt}$ be 1.0×10^{-9} m^2 s^{-1} for the various types of CO_2 diffusing from the plasma membrane to the chloroplasts. The value of $K_{CO_2}^{cyt}$ for mesophyll cells is not known, primarily because the cytosolic pH is not known with certainty. In any case, $K_{CO_2}^{cyt}$ cannot be much less than 1 and values >1—the magnitude we will assume here—will not change our conclusions about the relative importance of $r_{CO_2}^{cyt}$. As before, we will let A^{mes}/A be 20. Using Eq. (8.19), the cytosol resistance for CO_2 is $(0.2 \times 10^{-6}$ m$)/[(20)(1.0 \times 10^{-9}$ m^2 s$^{-1})(1)]$ or only 10 s m^{-1}, a very small resistance for CO_2 diffusion (Table 8-4). Thus the location of the chloroplasts around the periphery of a mesophyll cell is a "good" design that causes the resistance of this part of the pathway for CO_2 diffusion to be low.

H. Mesophyll Resistance

Most of the measurements of $r_{CO_2}^{mes}$ are indirect, but they indicate that this resistance is generally 40–400 s m^{-1} for mesophytes (Table 8-4). Based on our estimates of 30 s m^{-1} for $r_{CO_2}^{cw}$, 100 s m^{-1} for $r_{CO_2}^{pm}$, and 10 s m^{-1} for $r_{CO_2}^{cyt}$, we predict 140 s m^{-1} for $r_{CO_2}^{mes}$ (by Eq. 8.18a; $r_{CO_2}^{mes} = r_{CO_2}^{cw} + r_{CO_2}^{pm} + r_{CO_2}^{cyt}$). Thus our estimate based on the diffusion of CO_2 across each barrier is consistent with the measured mesophyll resistance.

There are many assumptions and parameter choices involved in the current calculation of $r_{CO_2}^{mes}$. For instance, we let A^{mes}/A be 20, whereas many leaves have values from 30 to 40; the latter ratios would reduce $r_{CO_2}^{mes}$ to 70–90 s m^{-1}. The cell walls of some mesophyll cells are only 0.07 μm thick, which would decrease $r_{CO_2}^{cw}$ to less than 10 s m^{-1}. Permeability coefficients of the plasma membrane of mesophyll cells for CO_2 have not been adequately measured. In this regard, $P_j = D_j K_j / \Delta x$ (Eq. 1.9), where the diffusion coefficients of H_2O and CO_2 in the plasma membrane are probably about the same (within a factor of 2 of each other), the partition coefficient for CO_2 is most likely at least 10 times higher than $K_{H_2O}^{pm}$, and Δx^{pm} is the same for H_2O and CO_2. Because $P_{H_2O}^{pm}$ can be 10^{-4} m s^{-1} (see Chapter 1, Section IVB), our assumed value of 5×10^{-4} m s^{-1} for $P_{CO_2}^{pm}$ may be too low—we noted in Chapter 1 (Section IVB) that P_j for another small molecule, O_2, crossing erythrocyte membranes has an extremely high value of 0.3 m s^{-1}. A higher value for $P_{CO_2}^{pm}$ would decrease our estimate for $r_{CO_2}^{pm}$ and thus for $r_{CO_2}^{mes}$.

I. Chloroplast Resistance

The last two structural resistances encountered by the CO_2 in photosynthesis are due to the chloroplasts ($r_{CO_2}^{chl} = r_{CO_2}^{clm} + r_{CO_2}^{stroma}$; Eq. 8.18b; see Fig. 8-9). As for the plasma membrane, the resistance of the chloroplast limiting membranes, $r_{CO_2}^{clm}$, is extremely large for the diffusion of HCO_3^- and relatively low for the diffusion of CO_2. The chloroplast-limiting membranes are actually permeable to small solutes, so $r_{CO_2}^{clm}$ may be lower than $r_{CO_2}^{pm}$, which we estimated to be 100 s m^{-1}. Because Δx^{stroma} averages nearly 1 μm, whereas Δx^{cyt} is about 0.2 μm and the other pertinent parameters (A^{mes}/A, D_{CO_2}, K_{CO_2}) are approximately the same in the two cases, $r_{CO_2}^{stroma}$ is a few times larger than $r_{CO_2}^{cyt}$, which we calculated to be only 10 s m^{-1} (if the pH in the chloroplast stroma were near or above 7, this would increase $K_{CO_2}^{chl}$ above 1 and reduce $r_{CO_2}^{stroma}$ accordingly). In any case, $r_{CO_2}^{stroma}$ is a relatively small resistance.

Currently, we can only estimate a value of about 100 s m^{-1} for the resistance to the diffusion of CO_2 into chloroplasts and across their stroma. Measurement of $r_{CO_2}^{chl}$ *in vivo* is also difficult—analysis of available data indicates that it is most likely < 200 s m^{-1}. Although active transport or facilitated diffusion of CO_2 or HCO_3^- into chloroplasts may lower the effective resistance, the experimental values for $r_{CO_2}^{chl}$ are compatible with diffusion of CO_2 across the chloroplast limiting membranes. All the resistances that we have just discussed and their corresponding conductances are summarized in Table 8-4.

IV. CO$_2$ Fluxes Accompanying Photosynthesis

Now that we have discussed the CO_2 resistances and conductances, we are ready to examine CO_2 fluxes. We will do this for photosynthesis, a process that consumes CO_2, as well as for respiration and photorespiration, processes that evolve CO_2. Our analysis will use an electrical circuit analogy so that we can represent the interrelationships between the various factors influencing net CO_2 uptake by a leaf.

A. Photosynthesis

The flux density of CO_2 into chloroplasts represents the gross rate of photosynthesis per unit leaf area, $J_{CO_2}^{ps}$. In the steady state, $J_{CO_2}^{ps}$ is the same as the flux density entering the leaf, J_{CO_2}, corrected for any other reactions evolving or consuming CO_2, as we will consider later (most evidence indicates that CO_2, not HCO_3^-, is the substrate for photosynthesis in chloroplasts, and we will therefore focus our attention on CO_2). We will represent the average rate of photosynthesis per unit volume of the chloroplasts by v_{CO_2}, which can have units of mol

CO$_2$ fixed m^{-3} s^{-1}. If chloroplasts have an average or effective thickness Δx^{chl} in a direction perpendicular to the plasma membrane (see Fig. 8-9), the rate of photosynthesis per unit leaf area $J^{\text{ps}}_{\text{CO}_2}$ can be represented as

$$J^{\text{ps}}_{\text{CO}_2} = v_{\text{CO}_2} \Delta x^{\text{chl}} A^{\text{mes}} / A \qquad (8.21)$$

where A^{mes}/A relates the chloroplast area to the leaf area. We can also derive Eq. (8.21) by regarding the chloroplasts as a flat layer of average thickness Δx^{chl} occupying some internal leaf area, e.g., A^{mes}. The rate of photosynthesis in this volume ($v_{\text{CO}_2} \Delta x^{\text{chl}} A^{\text{mes}}$) equals the gross CO$_2$ flux per unit leaf area times the leaf area corresponding to A^{mes}, namely $J^{\text{ps}}_{\text{CO}_2} A$.

The gross rates of CO$_2$ fixation by leaves and isolated chloroplasts are proportional to the CO$_2$ concentration over the lower part of its range and eventually reach an upper limit when the CO$_2$ concentration is sufficiently high. One way to describe such behavior is with a Michaelis–Menten type of expression:

$$v_{\text{CO}_2} = \frac{V_{\text{max}} \, c^{\text{chl}}_{\text{CO}_2}}{K_{\text{CO}_2} + c^{\text{chl}}_{\text{CO}_2}} \qquad (8.22)$$

where V_{max} is the maximum rate of CO$_2$ fixation per unit volume and K_{CO_2} is essentially a Michaelis constant for CO$_2$ fixation, i.e., the value of $c^{\text{chl}}_{\text{CO}_2}$ at which $v_{\text{CO}_2} = \frac{1}{2} V_{\text{max}}$ [see Eq. 3.27a, $J^{\text{in}}_j = J^{\text{in}}_{j\,\text{max}} c^o_j / (K_j + c^o_j)$, and the discussion that follows it]. Although convenient, using a Michaelis–Menten type of expression for the photosynthetic rate per unit volume may not always be justified for a complicated series of reactions such as photosynthesis.

In Eq. (8.22), V_{max} and, to some extent, K_{CO_2} depend on PPFD, temperature, and nutrient status. For instance, V_{max} is zero in the dark because photosynthesis then ceases, and it is directly proportional to PPFD up to about 50 μmol m^{-2} s^{-1}. If we continually increase the PPFD, V_{max} can reach an upper limit, its value for light saturation. This generally occurs at about 600 μmol m^{-2} s^{-1} for most C$_3$ plants, whereas photosynthesis for C$_4$ plants is usually not light saturated even at full sunlight, 2000 μmol m^{-2} s^{-1} (see Chapter 6, Section IIID for comments on C$_3$ and C$_4$ plants; see also Fig. 8-14 for responses of leaves of C$_3$ plants and a C$_4$ plant to PPFD). Photosynthesis is maximal at certain temperatures, often from 30 to 40°C. V_{max} increases as the leaf temperature is raised to the optimum and then decreases with a further temperature increase.

The values of V_{max} at light saturation and at the optimal temperatures for photosynthesis vary with plant species, but they are generally from 2 to 10 mol m^{-3} s^{-1}. We can also estimate V_{max} from measurements of the maximum rates of CO$_2$ fixation by isolated chloroplasts. These maximum rates—which are sustained for short periods and are for optimal conditions—can be 100 mmol of CO$_2$ fixed (kg chlorophyll)$^{-1}$ s^{-1} [360 μmol (mg chlorophyll)$^{-1}$ h^{-1} in another common unit], which is approximately 3 mol m^{-3} s^{-1} (1 kg chlorophyll is contained in about 0.035 m^3 of chloroplasts *in vivo*). *In vitro*, the key enzyme for CO$_2$ fixation, ribulose-1,5-bisphosphate carboxylase/oxygenase, can have rates equivalent to 200 mmol (kg chlorophyll)$^{-1}$ s^{-1}. The estimates of V_{max} using isolated chloroplasts or enzymes usually are somewhat lower than its values determined for a leaf. Measurements using leaves generally indicate that K_{CO_2} is 5–15 mmol m^{-3}. For instance, K_{CO_2} can be 9 mmol m^{-3} at 25°C with a Q_{10} (see Chapter 3, Section IIIB) of 1.8 (Woodrow and Berry, 1988).

In Chapter 5 (Section IVB) we noted that the processing time per CO$_2$ fixed is about 5 ms. Eight photons are required, which are absorbed by approximately 2000 chlorophyll molecules in chloroplasts where the chlorophyll concentration is about 30 mol m^{-3} (see Chapter 4, Section IVD). Hence, the photosynthetic rate per unit volume of chloroplasts is

$$v_{CO_2} = \frac{\text{(concentration of } CO_2 \text{ fixation sites)}}{\text{(time for } CO_2 \text{ fixation at a site)}}$$

$$= \frac{(30 \text{ mol chlorophyll m}^{-3})/(2.0 \times 10^3 \text{ chlorophylls per site})}{(5 \times 10^{-3} \text{s}/CO_2 \text{ at a site})}$$

$$= 3 \text{ mol } CO_2 \text{ m}^{-3} \text{ s}^{-1}$$

Under high PPFD, $c_{CO_2}^{chl}$ is generally limiting for photosynthesis, so this v_{CO_2} may be only 40–70% of the V^{max} at a particular temperature. Conversely, the 5 ms processing time may in large measure reflect the suboptimal CO_2 levels in the chloroplasts (these comments apply to C_3 plants; for C_4 plants $c_{CO_2}^{chl}$ is generally near the saturation value for photosynthesis, a topic that we will return to at the end of this chapter; also see Fig. 8-13). A related factor is the increase in atmospheric CO_2, primarily due to the burning of fossil fuels, from about 300 μmol mol^{-1} in 1900 to 365 μmol mol^{-1} in 1998, when the annual rate of CO_2 increase was nearly 2 μmol mol^{-1}.[5] Other things being equal, this increase in $c_{CO_2}^{ta}$ will raise $c_{CO_2}^{chl}$ and hence v_{CO_2} (Eq. 8.22).

B. Respiration and Photorespiration

So far, the only process involving CO_2 that we have considered in this chapter is photosynthesis. However, we cannot neglect the CO_2 produced within mesophyll (and other) cells by respiration and photorespiration. If mitochondrial respiration in leaf cells were the same in the light as in the dark, when it can be readily measured, respiration would produce about 5% as much CO_2 as is consumed by photosynthesis at a moderate PPFD. In C_3 plants, the rate of light-stimulated production of CO_2 by photorespiration at moderate temperatures is often about 30% (range, 20–50%) of the rate of CO_2 fixation into photosynthetic products.

We have already considered respiration in Chapter 6, so we will briefly comment on photorespiration. Photorespiration is the uptake of O_2 and the evolution of CO_2 in the light resulting from glycolate synthesis in chloroplasts and subsequent glycolate and glycine metabolism in peroxisomes and mitochondria (Fig. 8-10). A crucial role is played by the enzyme ribulose-1,5-bisphosphate carboxylase/oxygenase (Rubisco), which has a molecular mass of about 540 kDa. Rubisco constitutes up to half of the soluble protein in the leaves of C_3 plants and perhaps one-sixth of the soluble protein in C_4 plants, making it the most abundant protein in the world (approximately 10 kg of Rubisco occurs per person). This carboxylase/oxygenase can interact with CO_2, leading to photosynthesis, or with O_2,

---

[5] CO_2 levels are expressed in many units, parts per million by volume (ppm or ppmv) being commonly used, concise, clearly understood, but no longer generally accepted in the biological literature. Some data are published on a volume basis using units of μliter liter^{-1} or cm^3 m^{-3} (numerically equivalent to ppm). However, most data are currently expressed on a mole fraction basis, either as a dimensionless number or in μmol mol^{-1} (e.g., 350×10^{-6} and 350 μmol mol^{-1}, respectively) or on the basis of partial pressure, e.g., Pa MPa^{-1} and Pa. If pressure is used for CO_2 levels, then the ambient atmospheric pressure should also be stated. By the ideal or perfect gas law ($PV = nRT$, where $P = \sum_i P_i$ and $V = \sum_i V_i$, the summations being over all gaseous species present), $V_{CO_2}/V_{total} = P_{CO_2}/P_{total} = n_{CO_2}/n_{total} = N_{CO_2}$, where N_{CO_2} is the mole fraction of CO_2, the unit that we will adopt. Another unit is mmol m^{-3}, the concentration of CO_2 (which we will also use), although both temperature and pressure should then be specified.

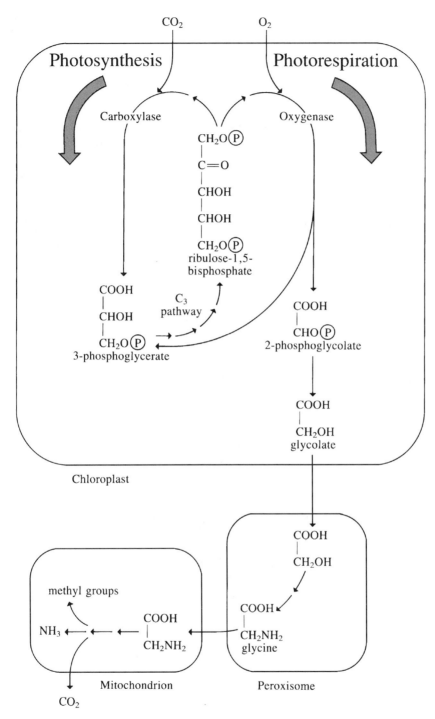

Figure 8-10. Schematic illustration of Rubisco (ribulose-1,5-bisphosphate carboxylase/oxygenase) acting as the branch point for photosynthesis and photorespiration. All three of the organelles involved, but only a few of the biochemical steps, are indicated (Ⓟ represents phosphate).

leading to photorespiration (Fig. 8-10). The competition for the same active site on Rubisco by O_2 and CO_2, which thus act as alternative substrates, can be modeled by a modification of Eq. (8.22) in which K_{CO_2} is replaced by $K_{CO_2}(1 + c_{O_2}^{chl}/K_{O_2})$ (Farquhar and von Caemmerer, 1982; Woodrow and Berry, 1988).

In photorespiration, ribulose-1,5-bisphosphate is split into 3-phosphogylcerate and 2-phosphoglycolate, the latter undergoing dephosphorylation in the chloroplasts and then entering the peroxisomes. Although CO_2 can be released by a decarboxylation of glycolate in the peroxisomes, the main product of glycolate metabolism is glycine, which then moves to the mitochondria where the CO_2 is released (Fig. 8-10). Because the generation of ribulose-1,5-bisphosphate depends on the C_3 photosynthetic pathway, photorespiration is influenced by PPFD and by temperature, although not quite in the same manner as is photosynthesis. For instance, photosynthesis usually doubles in going from 20 to 30°C, whereas photorespiration often triples over this interval of leaf temperature; the oxygenase activity is apparently favored over the carboxylase activity with increasing temperature, and thus photorespiration increases at the expense of photosynthesis at higher temperatures. Regarding CO_2 fixation, photorespiration apparently undoes what photosynthesis has done. We might then ask whether photorespiration benefits a plant—a question that so far has no convincing answer.

Not all plants photorespire significantly. Many of these "nonphotorespiring" plants are tropical monocots, and all tend to have a leaf anatomy (referred to as Kranz anatomy) differing from that of plants with high rates of photorespiration (Fig. 1-2 versus Fig. 8-11). Nonphotorespires have a conspicuous group of chloroplast-containing cells surrounding the vascular bundles known as the *bundle sheaths* (the bundle sheaths in photorespirers tend to have smaller cells with few or no chloroplasts). Outside a bundle sheath in nonphotorespirers

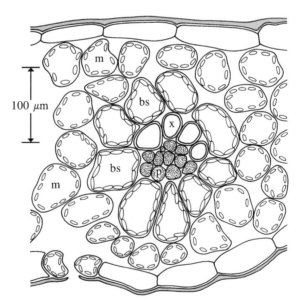

Figure 8-11. Schematic transverse section through a leaf of a C_4 plant, indicating a vascular bundle containing xylem (x) and phloem (p) cells, a concentric layer of bundle sheath cells (bs), and the surrounding mesophyll cells (m). Bundle sheath cells of C_4 plants appear more conspicuously green than do mesophyll cells because the former generally contain more and/or larger chloroplasts (which are granaless in some types of C_4 plants).

are mesophyll cells where CO_2 is fixed into four-carbon dicarboxylic acids via a C_4 pathway elaborated by Hal Hatch and Charles Slack in the 1960s. In this C_4 pathway, CO_2 in the form of HCO_3^- reacts with phosphoenolpyruvate via the enzyme phosphoenolpyruvate carboxylase located in the cytosol of the mesophyll cells[6]. The initial product is oxaloacetate, which is rapidly converted to malate and aspartate. For all chloroplasts in photorespiring (C_3) plants, and for the chloroplasts in the bundle sheath cells of C_4 plants, photosynthesis uses the ordinary C_3 pathway of the Calvin cycle—known since the 1940s—where CO_2 is incorporated into ribulose-1,5-bisphosphate, yielding two molecules of the three-carbon compound, 3-phosphoglycerate (Fig. 8-10). Biochemical shuttles in C_4 plants move the four-carbon compounds initially produced in the light, such as malate, from the mesophyll cells into the bundle sheath cells (Fig. 8-11). Decarboxylation of these compounds raises the CO_2 level in the bundle sheath cells to much higher levels than expected based on diffusion of CO_2 in from the atmosphere (to over 1500 μmol CO_2 mol^{-1} in one estimate; Ehleringer and Björkman, 1977). Because of the high CO_2 level, the carboxylase activity of Rubisco is dominant over (i.e., outcompetes) the oxygenase activity, so photosynthesis takes place in the bundle sheath cells with very little photorespiration. Similarly, raising the external CO_2 level to 1500 μmol mol^{-1} virtually eliminates photorespiration in C_3 plants (see Fig. 8-13); such CO_2 enrichment is common in greenhouses containing commercially valuable C_3 crops.

C. Comprehensive CO$_2$ Resistance Network

We can now develop an analytical framework to represent CO_2 fixation in photosynthesis and its evolution in respiration and photorespiration. The net flux of CO_2 into a leaf, J_{CO_2}, indicates the apparent (net) CO_2 assimilation rate by photosynthesis (see Fig. 8-1). The gross or "true" rate of photosynthesis, $J_{CO_2}^{ps}$, minus the rate of CO_2 evolution by respiration and photorespiration per unit leaf area, $J_{CO_2}^{r+pr}$, equals J_{CO_2}:

$$J_{CO_2} = J_{CO_2}^{ps} - J_{CO_2}^{r+pr} \qquad (8.23)$$

Equation (8.23) summarizes the overall steady-state balance of CO_2 fluxes for leaves.

We will consider CO_2 fluxes and resistances for photosynthesis, respiration, and photorespiration using an electrical circuit analogy (Fig. 8-12). The sources of CO_2 for photosynthesis are the turbulent air surrounding a leaf (represented by the E battery in Fig. 8-12) and respiration plus photorespiration (the e battery). The E battery corresponds to the drop in CO_2 concentration (or mole fraction) from the turbulent air surrounding a leaf to the enzymes of photosynthesis inside chloroplasts, $c_{CO_2}^{ta} - c_{CO_2}^{chl}$, which represents the driving force for CO_2 diffusion. The batteries lead to currents that correspond to fluxes of CO_2; e.g., I, the current from the E battery, corresponds to J_{CO_2}, and i represents the flux density of CO_2 emanating from respiration and photorespiration, $J_{CO_2}^{r+pr}$. The current I crosses the resistances $r_{CO_2}^{bl_1}$, $r_{CO_2}^{leaf_1}$, and $r_{CO_2}^{mes}$ before being joined by i (Fig. 8-12). The current i encounters $r_{CO_2}^i$, which is the resistance to the movement of CO_2 out of mitochondria and then across a short distance in the cytosol. Both I and i cross the resistance $r_{CO_2}^{chl}$ because CO_2 coming from the surrounding air, as well as that evolved in mitochondria by respiration

---

[6] CO_2 can diffuse across the plasma membrane and become hydrated to HCO_3^- by carbonic anhydrase, which occurs in the cytosol of mesophyll cells of C_4 plants. The Michaelis constant of phosphoenolpyruvate carboxylase for HCO_3^- is about 200 mmol m^{-3} (0.2 mM), which suggests that the cytosolic pH must be above 7 to get sufficient HCO_3^- formation (see Table 8-3) to match the CO_2 uptake rates of C_4 plants.

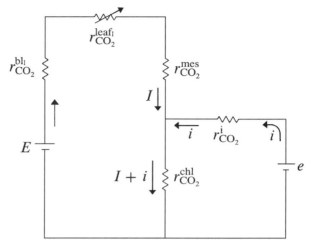

Figure 8-12. Electrical circuit indicating the resistances affecting photosynthesis, respiration, and photorespiration. The sources of CO_2 are the turbulent air surrounding the leaf (represented by the
battery of electromotive force E) and respiration plus photorespiration (the e battery). The
current I corresponds to the net CO_2 influx into the leaf (J_{CO_2}), i represents CO_2 evolution by
respiration and photorespiration ($J_{CO_2}^{r+pr}$), and $I+i$ corresponds to gross photosynthesis ($J_{CO_2}^{ps}$).
The voltages at various locations correspond to specific CO_2 concentrations, e.g., the voltage
in the upper right corner (between $r_{CO_2}^{leaf_l}$ and $r_{CO_2}^{mes}$) corresponds to $c_{CO_2}^{ias}$, that in the middle right
corner (between $r_{CO_2}^{i}$ and e) to $c_{CO_2}^{mito}$, and that along the lower line to $c_{CO_2}^{chl}$.

and photorespiration, can be used for photosynthesis in the chloroplasts (this is a way of
paraphrasing Eq. 8.23).

To analyze the electrical circuit in Fig. 8-12, we will use Ohm's law ($\Delta E = IR$) and
Kirchhoff's laws. Kirchhoff's first law for electrical circuits states that the algebraic sum
of the currents at any junction equals zero. For instance, at the junction in Fig. 8-12 where
current I meets current i, the current leaving that point equals $I + i$. Kirchhoff's second
law, also known as the loop theorem and a consequence of the conservation of energy, states
that the overall change in electrical potential in going completely around a closed loop is
zero. By considering a complete pathway around the left-hand part of the electrical circuit
in Fig. 8-12, we obtain the following relationship:

$$E - I\left(r_{CO_2}^{bl_l} + r_{CO_2}^{leaf_l} + r_{CO_2}^{mes}\right) - (I+i)\left(r_{CO_2}^{chl}\right) = 0 \tag{8.24}$$

For CO_2 exchange, E in Eq. (8.24) can be replaced by $c_{CO_2}^{ta} - c_{CO_2}^{chl}$, I by J_{CO_2}, and i by
$J_{CO_2}^{r+pr}$. Upon moving the resistance terms to the right-hand side of the equation, we obtain

$$c_{CO_2}^{ta} - c_{CO_2}^{chl} = J_{CO_2}\left(r_{CO_2}^{bl_l} + r_{CO_2}^{leaf_l} + r_{CO_2}^{mes} + r_{CO_2}^{chl}\right) + J_{CO_2}^{r+pr} r_{CO_2}^{chl} \tag{8.25}$$

If respiration or photorespiration increase, Eq. (8.25) indicates that the net photosynthetic
rate will decrease when other factors are unchanged.

D. Compensation Points

The atmospheric CO_2 concentration at which the CO_2 evolved by respiration and photorespiration is exactly compensated by CO_2 consumption in photosynthesis is known as the

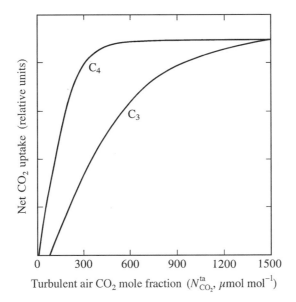

Figure 8-13. Dependence of net CO$_2$ uptake on external CO$_2$ level for leaves of representative C$_3$ and C$_4$ plants. C$_3$ plants require a higher $N_{CO_2}^{ta}$ at the CO$_2$ compensation point ($J_{CO_2} = 0$) and for CO$_2$ saturation than do C$_4$ plants. We note that because photosynthesis for C$_4$ plants is already saturated at current atmospheric CO$_2$ levels, higher CO$_2$ levels generally will not enhance their photosynthetic rates, whereas the increasing atmospheric CO$_2$ levels will progressively increase photosynthesis for C$_3$ plants.

CO$_2$ compensation point. We can use Fig. 8-12 and Eq. (8.25) to demonstrate the CO$_2$ compensation point for photosynthesis in terms of forces and fluxes. If we steadily decrease $c_{CO_2}^{ta}$ for a leaf initially having a net uptake of CO$_2$, J_{CO_2} will decrease and eventually will become zero when $c_{CO_2}^{ta} - c_{CO_2}^{chl} = J_{CO_2}^{r+pr} r_{CO_2}^{chl}$ (see Eq. 8.25). Thus reducing the concentration of CO$_2$ in the turbulent air surrounding an illuminated leaf will eliminate net CO$_2$ fixation at the CO$_2$ compensation point.

The CO$_2$ compensation point is much higher for a C$_3$ plant than for a C$_4$ plant. At the compensation point, $c_{CO_2}^{ta} - c_{CO_2}^{chl} = J_{CO_2}^{r+pr} r_{CO_2}^{chl}$, and $J_{CO_2}^{r+pr}$ is larger when photorespiration is appreciable. Most C$_4$ plants (e.g., sugarcane, sorghum, maize, bermuda grass, Sudan grass, and *Amaranthus*) have CO$_2$ compensation points of 3–10 μmol CO$_2$ mol^{-1} in the turbulent air (Fig. 8-13; CO$_2$ levels in the gas phase are usually expressed as mole fractions, as we will do here). Most dicotyledons and temperate monocots are C$_3$ plants (e.g., cotton, tobacco, tomato, lettuce, oaks, maples, roses, wheat, and orchard grass) and have CO$_2$ compensation points of 40–100 μmol CO$_2$ mol^{-1} (Fig. 8-13). A few species (e.g., some species of *Mollugo, Moricandia,* and *Panicum*) have intermediate CO$_2$ compensation points near 25 μmol CO$_2$ mol^{-1}, and shifts between C$_3$ and C$_4$ patterns can even occur during leaf development. The CO$_2$ compensation points generally increase with increasing temperature and decreasing PPFD, the values given being appropriate at 25°C when PPFD is not limiting for photosynthesis.

If we reduce the amount of light incident on a leaf from the value for direct sunlight, we eventually reach a PPFD for which there is no net photosynthesis. This PPFD for which J_{CO_2} is zero is known as the *light compensation point* for photosynthesis. Because photorespiration depends on photosynthetic products, both photorespiration and gross photosynthesis decrease as the PPFD is lowered. Hence, the light compensation point for leaves is approximately the same for C_3 and C_4 plants—at $20°C$ and $350\ \mu mol\ CO_2\ mol^{-1}$, light compensation usually occurs at a PPFD of about $8–16\ \mu mol\ m^{-2}\ s^{-1}$ for C_3 plants and $6–14\ \mu mol\ m^{-2}\ s^{-1}$ for C_4 plants (the lower values are for shade leaves or shade plants). The light compensation point, which occurs at $<1\%$ of a full sunlight PPFD of about $2000\ \mu mol\ m^{-2}\ s^{-1}$, is important in our consideration of plant canopies in Chapter 9. For example, leaves shaded by many overlying leaves can actually be at (or below) the light compensation point when the exposed leaves have appreciable net rates of photosynthesis. Also, even the uppermost leaves can reach the light compensation point on cloudy days or near sunset.

At either compensation point, J_{CO_2} is zero when $c_{CO_2}^{ta} - c_{CO_2}^{chl} = J_{CO_2}^{r+pr} r_{CO_2}^{chl}$ (Eq. 8.25). For the light compensation point, $c_{CO_2}^{ta}$ is unchanged but $c_{CO_2}^{chl}$ increases owing to a decrease in CO_2 fixation by photosynthesis at the low PPFD. If we lower the PPFD below the light compensation point, J_{CO_2} reverses direction, meaning there is a net flux of CO_2 out of the leaf. For instance, when $c_{CO_2}^{ta} = c_{CO_2}^{chl}$, $J_{CO_2} = -J_{CO_2}^{r+pr} r_{CO_2}^{chl}/(r_{CO_2}^{bl_i} + r_{CO_2}^{leaf_i} + r_{CO_2}^{mes} + r_{CO_2}^{chl})$ by Eq. (8.25), a conclusion that can also be reached by applying the loop theorem to the left-hand side of Fig. 8-12 (note that E is zero when $c_{CO_2}^{ta} = c_{CO_2}^{chl}$). Thus part of the respiratory plus photorespiratory flux density is then refixed in the chloroplasts and part comes out of the leaf. At night, $J_{CO_2} = -J_{CO_2}^{r+pr}$ because $J_{CO_2}^{ps}$ is zero as gross photosynthesis stops upon cessation of illumination (Eq. 8.23), as does photorespiration. The respiratory flow of CO_2 out of the leaf is then driven by the higher CO_2 concentration in the mitochondria than in the turbulent air, encountering the resistances $r_{CO_2}^i, r_{CO_2}^{mes}, r_{CO_2}^{leaf_i}$ and $r_{CO_2}^{bl_i}$, in that order (see Fig. 8-12). In particular, the E battery still corresponds to $c_{CO_2}^{ta} - c_{CO_2}^{chl}$, but $c_{CO_2}^{chl} > c_{CO_2}^{ta}$ at night, so the battery reverses its polarity. The condition $J_{CO_2}^{ps}$ equaling zero means that there is no current through $r_{CO_2}^{chl}$ ($I = -i$), so applying the loop theorem to the left-hand side of Fig. 8-12 yields $c_{CO_2}^{chl} - c_{CO_2}^{ta} = J_{CO_2}^{r+pr}(r_{CO_2}^{bl_i} + r_{CO_2}^{leaf_i} + r_{CO_2}^{mes})$ and to the right-hand side yields $c_{CO_2}^{mito} - c_{CO_2}^{ta} = J_{CO_2}^{r+pr} r_{CO_2}^i$. Adding these two relations yields $c_{CO_2}^{mito} - c_{CO_2}^{ta} = J_{CO_2}^{r+pr}(r_{CO_2}^{bl_i} + r_{CO_2}^{leaf_i} + r_{CO_2}^{mes} + r_{CO_2}^i)$, which can be used to describe the efflux of respiratory CO_2 from leaves at night. Our electrical circuit in Fig. 8-12, and Eq. (8.25) derived from it, is able to portray the CO_2 compensation point, the light compensation point, as well as the general interrelations of the fluxes of CO_2 for photosynthesis, photorespiration, and respiration in the light and the dark. Our discussion and Fig. 8-12 are for C_3 plants—to apply an electrical circuit analog to C_4 and Crassulacean acid metabolism (CAM) plants, we need to consider the cytosolic location of the initial CO_2 fixing enzymes as well as the fate of the CO_2 released upon decarboxylation of the four-carbon acids.

E. Fluxes of CO_2

Using Eq. (8.25), we can relate the apparent or net rate of photosynthesis, J_{CO_2}, to the various resistances, the rate of respiration plus photorespiration, and the overall drop in CO_2 concentration, $\Delta c_{CO_2}^{total} = c_{CO_2}^{ta} - c_{CO_2}^{chl}$. Let us first rearrange Eq. (8.25) into the following form:

$$J_{CO_2} = \frac{c_{CO_2}^{ta} - c_{CO_2}^{chl}}{r_{CO_2}^{bl_i} + r_{CO_2}^{leaf_i} + r_{CO_2}^{mes} + \left(1 + \frac{J_{CO_2}^{r+pr}}{J_{CO_2}}\right) r_{CO_2}^{chl}}$$

$$= \frac{\Delta c_{CO_2}^{total}}{r_{CO_2}^{total}}$$

$$(8.26)$$

where we have used our customary definition of resistance to obtain $r_{CO_2}^{total}$, the total effective resistance for CO$_2$ fixation—namely, resistance equals the concentration drop divided by the flux density (see Eq. 8.1d). We note that $r_{CO_2}^{total}$ depends on $J_{CO_2}^{r+pr}$, which is a consequence of the complicated electrical circuit (Fig. 8-12) needed to represent the various CO$_2$ components. Sometimes it may be convenient to rearrange Eq. (8.25) in other ways, e.g., $J_{CO_2} = (\Delta c_{CO_2}^{total} - J_{CO_2}^{r+pr} r_{CO_2}^{chl})/(r_{CO_2}^{bl_1} + r_{CO_2}^{leaf_1} + r_{CO_2}^{mes} + r_{CO_2}^{chl})$. This form clearly shows that J_{CO_2} is zero at the compensation points ($\Delta c_{CO_2}^{total} = J_{CO_2}^{r+pr} r_{CO_2}^{chl}$). Also, using Eq. (8.23) we can manipulate the factor in Eq. (8.26) containing $J_{CO_2}^{r+pr}$ as follows:

$$1 + \frac{J_{CO_2}^{r+pr}}{J_{CO_2}} = \frac{J_{CO_2} + J_{CO_2}^{r+pr}}{J_{CO_2}} = \frac{J_{CO_2}^{ps}}{J_{CO_2}} = \frac{J_{CO_2}^{ps}}{J_{CO_2}^{ps} - J_{CO_2}^{r+pr}} = \frac{1}{1 - \frac{J_{CO_2}^{r+pr}}{J_{CO_2}^{ps}}}$$

The appropriate form of this factor to use in Eq. (8.26) depends on which ratio of fluxes is known.

We next consider J_{CO_2} for specific values of the various parameters affecting net CO$_2$ uptake. We will use a CO$_2$ mole fraction of 360 μmol mol^{-1} in the turbulent air around a leaf within a plant canopy, which corresponds to a concentration of (360)(0.0410), or 14.8 mmol CO$_2$ m^{-3} at 20°C and a pressure of 0.1 MPa (see conversion factor in Table 8-2). Although we do not have reliable measurements of $c_{CO_2}^{chl}$, it may be about 8.0 mmol m^{-3} for a photorespiring plant at saturating PPFD. At 20°C, respiration plus photorespiration might be 30% as large as net photosynthesis. For purposes of calculation, we will let the gas phase resistance $r_{CO_2}^{bl_1} + r_{CO_2}^{leaf_1}$ be 400 s m^{-1}, $r_{CO_2}^{mes}$ be 140 s m^{-1}, and $r_{CO_2}^{chl}$ be 100 s m^{-1} (see Table 8-4). Using Eq. (8.26), we then calculate the net photosynthesis to be

$$J_{CO_2} = \frac{(14.8 \times 10^{-3} \text{ mol m}^{-3} - 8.0 \times 10^{-3} \text{ mol m}^{-3})}{400 \text{ s m}^{-1} + 140 \text{ s m}^{-1} + (1.0 + 0.3)(100 \text{ s m}^{-1})}$$

$$= \frac{(6.8 \times 10^{-3} \text{ mol m}^{-3})}{(670 \text{ s m}^{-1})} = 10.1 \ \mu\text{mol m}^{-2} \text{ s}^{-1}$$

We have so far considered CO$_2$ diffusing into a leaf only across its lower surface. In the general case, CO$_2$ can move in across its upper surface as well, which we can incorporate into our considerations by appropriately reducing the resistance $r_{CO_2}^{bl_1} + r_{CO_2}^{leaf_1}$ ($r_{CO_2}^{mes}$ and $r_{CO_2}^{chl}$ are unaffected when CO$_2$ diffuses in through both sides of a leaf). If 30% of the CO$_2$ diffused in through the upper side of a leaf, the effective resistance between the turbulent air and the surfaces of the mesophyll cells would be 70% as great as $r_{CO_2}^{bl_1} + r_{CO_2}^{leaf_1}$.[7] For instance, $r_{CO_2}^{bl_1} + r_{CO_2}^{leaf_1}$ is 400 s m^{-1}, so the resistance of this part of the pathway is reduced to (0.70)(400 s m^{-1}), or 280 s m^{-1}, if 30% of the CO$_2$ enters through the upper surface. This reduces $r_{CO_2}^{total}$ from 670 to 550 s m^{-1}, which would raise J_{CO_2} to 12.4 μmol m^{-2} s^{-1} if $c_{CO_2}^{chl}$ were unchanged. Actually, $c_{CO_2}^{chl}$ must increase to lead to the higher photosynthetic rate (Eq. 8.22), offsetting most of the possible increase in J_{CO_2} caused by the lower gas-phase resistance (see Eq. 8.26). For example, if K_{CO_2} is 10 mmol m^{-3}, then simultaneously solving Eqs. (8.22) and (8.26) for $c_{CO_2}^{chl}$ (see Appendix III for the solution of a quadratic

---

[7] See Section IIF for discussion of the analogous situation in transpiration. Instead of knowing the relative flux densities through the two sides, we might know the resistances. We could then use Eq. (8.11) (with CO$_2$ replacing wv as subscripts) to determine the resistance for CO$_2$ movement of the two leaf surfaces in parallel. Quite often the gas phase resistance for water vapor is measured for the two leaf surfaces together, in which case the gas-phase CO$_2$ resistance is obtained by multiplying r_{wv}^{total} by D_{wv}/D_{CO_2}.

equation) indicates that $c_{CO_2}^{chl}$ will increase to 8.9 mmol m^{-3} (equivalent to about 220 μmol CO_2 mol^{-1}) and J_{CO_2} will be 10.7 μmol m^{-2} s^{-1} for the lower resistance condition.

We next consider what would happen to CO_2 uptake if the stomata provided no resistance to CO_2 entry. Instead of a gas-phase resistance for CO_2 of 280 s m^{-1}, it might then be only 80 s m^{-1} for the two leaf surfaces acting in parallel. The $r_{CO_2}^{total}$ would thus be lowered to 350 s m^{-1}. Again simultaneously solving Eqs. (8.22) and (8.26) for $c_{CO_2}^{chl}$ with $K_{CO_2} = 10$ mmol m^{-3}, we find that J_{CO_2} is 11.8 μmol m^{-2} s^{-1}. Thus removing both epidermises completely will enhance CO_2 uptake by only 10%, indicating that the stomata do not greatly restrict the photosynthetic rate in this case, although their pores occupy only a very small fraction of the leaf surface area. In summary, stomata exert major control on transpiratory water loss while restricting net CO_2 uptake relatively little.

F. CO_2 Conductances

Our analysis for CO_2 fluxes could be carried out using conductances and mole fractions. Also, we could divide the CO_2 pathway into a gas-phase component from the turbulent air up to the mesophyll cells and a liquid-phase component representing the mesophyll cells. The drop in CO_2 mole fraction across the gas phase $\Delta N_{CO_2}^{gas}$ can be related to the CO_2 conductance for the gas phase $g_{CO_2}^{gas}$ and CO_2 mole fractions as follows:

$$J_{CO_2} = g_{CO_2}^{gas} \Delta N_{CO_2}^{gas} = g_{CO_2}^{gas} \left(N_{CO_2}^{ta} - N_{CO_2}^{ias} \right) \qquad (8.27)$$

where $N_{CO_2}^{ta}$ is the CO_2 mole fraction in the turbulent air and $N_{CO_2}^{ias}$ is that in the intercellular air spaces. We note that $g_{CO_2}^{gas} = g_{wv}^{total}/1.60$, where g_{wv}^{total} can be defined by Eq. (8.13) and 1.60 is the value of D_{wv}/D_{CO_2}. Under optimal photosynthetic conditions, $N_{CO_2}^{ias}$ is nearly twice as high for leaves of C_3 compared with C_4 plants (discussed later). Indeed, $N_{CO_2}^{ias}$ is an important parameter for evaluating photosynthesis at the mesophyll or liquid-phase level as well as for studying the regulation of stomatal opening.

We can identify a liquid-phase CO_2 conductance for the part of the pathway from the mesophyll cell walls up to the CO_2-fixation enzymes:

$$J_{CO_2} = g_{CO_2}^{liquid} \left(N_{CO_2}^{ias} - N_{CO_2}^{chl} \right) \qquad (8.28)$$

If respiration and photorespiration can be ignored, then $1/g_{CO_2}^{liquid} = 1/g_{CO_2}^{mes} + 1/g_{CO_2}^{chl}$ (equivalently, $r_{CO_2}^{liquid} = r_{CO_2}^{mes} + r_{CO_2}^{chl}$). Otherwise, we could return to Fig. 8-12 and note that $N_{CO_2}^{ias} - N_{CO_2}^{chl} = J_{CO_2}/g_{CO_2}^{mes} + J_{CO_2}^{ps}/g_{CO_2}^{chl} = J_{CO_2}[1/g_{CO_2}^{mes} + (J_{CO_2}^{ps}/J_{CO_2})/g_{CO_2}^{chl}]$, which in turn equals $J_{CO_2}/g_{CO_2}^{liquid}$ by Eq. (8.28). This intermingling of conductances and fluxes again reflects the complication of having more than one source of CO_2 that can be fixed photosynthetically (see Eq. 8.25). If we are interested in the photosynthetic properties of the mesophyll cells, we might wish to express the CO_2 conductance on the basis of mesophyll cell surface area:

$$g_{CO_2}^{cell} = \frac{g_{CO_2}^{liquid}}{A^{mes}/A} \qquad (8.29)$$

where $g_{CO_2}^{cell}$ is the cellular conductance for CO_2.

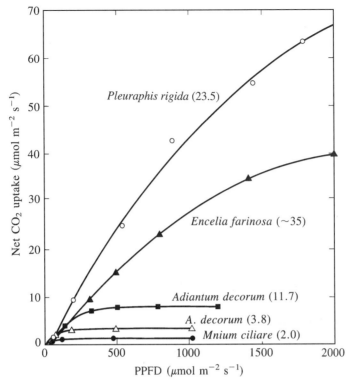

Figure 8-14. Photosynthetic responses to photosynthetic photon flux density for species differing in mesophyll surface area per unit leaf area. Curves were obtained at ambient CO$_2$ and O$_2$ concentrations, optimal temperatures, and the A^{mes}/A indicated in parentheses. [Sources: for the C$_4$ desert grass *Pleuraphis rigida*, Nobel (1980); for the C$_3$ desert composite *Encelia farinosa*, Ehleringer *et al.* (1976); and for the C$_3$ maidenhair fern *Adiantum decorum* and the C$_3$ moss *Mnium ciliare*, Nobel (1977).]

G. Range in Photosynthetic Rates

The net rates of photosynthesis vary considerably with plant species, temperature, PPFD, and other conditions (Fig. 8-14). For instance, the maximum J_{CO_2} is often 5–10 μmol m^{-2} s^{-1} for the leaves of trees. Certain C$_3$ crop plants, such as sugar beet, soybean, and tobacco, can have a J_{CO_2} of 20–25 μmol m^{-2} s^{-1} at saturating PPFD and temperatures near 30°C. For C$_4$ plants, J_{CO_2} tends to be higher because $J_{CO_2}^{r+pr}$ is small and the liquid-phase resistances ($r_{CO_2}^{mes}$ and $r_{CO_2}^{chl}$) are often relatively small. Under optimal conditions of high PPFD and a leaf temperature of 35°C, J_{CO_2} can exceed 40 μmol m^{-2} s^{-1} for bermuda grass, maize, sorghum, sugarcane, and certain other C$_4$ plants (as well as a few C$_3$ species). An extremely high value of 67 μmol m^{-2} s^{-1} can occur for the C$_4$ *Pleuraphis rigida* at full sunlight (Fig. 8-14) and for a few other species.

The influence of A^{mes}/A on J_{CO_2} deserves special comment. A^{mes}/A can be two or more times larger for sun leaves than for shade leaves on the same plant; this reduces $r_{CO_2}^{total}$ and thus enhances the maximal photosynthetic rates of sun leaves compared with shade leaves. Maximal rates of CO$_2$ uptake per unit area for C$_3$ plants can thus vary with A^{mes}/A for leaves on a single plant, on different plants of the same species, and sometimes even between species (Fig. 8-14). If the mesophyll cells were tightly packed into a layer with no intervening air spaces, A^{mes}/A could equal 2.0. This occurs for the moss *Mnium ciliare*, whose leaves are one cell thick with the lateral walls completely touching; thus the only area available for CO_2 to diffuse from the gas phase into the cells is their end walls, which have a total area twice that of one side of the leaf. Instead of a mesophyll resistance of 140 s m^{-1} that we calculated for an A^{mes}/A of 20 (Table 8-4), $r_{CO_2}^{mes}$ is 1400 s m^{-1} for an A^{mes}/A of 2. The evolution of a leaf anatomy with abundant mesophyll cell surface area leading to a large value for A^{mes}/A allows $r_{CO_2}^{mes}$ and $r_{CO_2}^{chl}$ to be fairly low, with a correspondingly high value for J_{CO_2} (see Fig. 8-14).

Many units are used to express photosynthesis and CO_2 fluxes for leaves. Conversion factors for some of the more common units are summarized in Table 8-2. For example, 16 mg CO_2 dm^{-2} h^{-1} corresponds to $(16)(0.631)$, or 10 μmol m^{-2} s^{-1}. Chlorophyll per unit leaf area generally ranges from 0.2 to 0.8 g m^{-2}, with 0.4–0.5 g m^{-2} being typical. Thus 10 μmol m^{-2} s^{-1} might correspond to $(10\ \mu\text{mol m}^{-2}\text{ s}^{-1})/(0.4\text{ g chlorophyll m}^{-2})$, or 25 μmol (g chlorophyll)$^{-1}$ s^{-1}, which equals 90 μmol CO_2 fixed (mg chlorophyll)$^{-1}$ h^{-1}.

H. Environmental Productivity Index

Water status, temperature, and PPFD all affect stomatal opening and the photosynthetic rates of leaves. Sometimes such environmental effects are incorporated into photosynthetic models by their influences on $g_{CO_2}^{st}$ (Eq. 8.5) or V_{max} (Eq. 8.22). Also, indices have been proposed relating J_{CO_2} and the associated plant productivity to rainfall and the water status of the plants (Le Houérou, 1984); to the ambient temperature, including the accumulated time that the temperature is above some minimum value (Long and Woodward, 1988); and to the intercepted radiation (Monteith, 1977; Jones, 1992). In the latter case, net CO_2 uptake depends on the PPFD absorbed times a conversion efficiency that in turn depends on water status and temperature.

Recognizing that various environmental factors can simultaneously limit net CO_2 uptake, an environmental productivity index (EPI) has been proposed to help predict daily net CO_2 uptake in the field based on J_{CO_2} for individual leaves (Nobel, 1984, 1988):

$$\text{EPI} = \text{water index} \times \text{temperature index} \times \text{PPFD index} \qquad (8.30)$$

where each of the component indices ranges from zero, when limitations by that factor abolish net CO_2 uptake, to 1, when that factor is optimal for net CO_2 uptake. To determine values for each component index, effects of individual factors on J_{CO_2} are determined over 24-h periods in the laboratory when the other environmental conditions are held constant. Then the values of each component index are calculated for the environmental conditions prevailing in the field to predict EPI, which hence represents the fraction of maximum net CO_2 uptake expected over the course of a day for a leaf under specific environmental conditions. This dimensionless EPI times the maximum net CO_2 uptake measured in the laboratory for a 24-h period when the soil is wet (water index $= 1.00$), the air temperature is optimal for net CO_2 uptake (temperature index $= 1.00$), and the PPFD is saturating for photosynthesis (PPFD index $= 1.00$) gives the daily CO_2 uptake under field conditions. Such measurements of maximum net CO_2 uptake can be made at any ambient CO_2 level and under the nutrient status pertinent to the field situation. Alternatively, a nutrient index can be incorporated into EPI to take into consideration the effect of soil elements on leaf net CO_2 uptake. Indeed, J_{CO_2} increases nearly linearly with leaf nitrogen content for many plant species.

Using EPI, net CO_2 uptake can be predicted at various times of the year, at various locations, and under various environmental conditions, such as under elevated atmospheric CO_2 levels and its associated suite of environmental changes or as a function of elevation. For instance, net CO_2 uptake over 24-h periods is known for leaves of a common succulent of the Sonoran Desert, *Agave deserti*, for wet conditions as well as various durations of drought, for day/night temperatures encompassing the entire temperature range occurring in its native habitat, and from darkness up to full sunlight, so all three component indices of EPI (Eq. 8.30) can be calculated (Nobel, 1984). In the summer, EPI for *A. deserti* increases fourfold from an elevation of 300 m to one of 1300 m, primarily reflecting a threefold increase in rainfall, which raises the water index, and a nearly 10°C decrease in temperature, which raises the temperature index because the warm temperatures at low elevations at this time of the year are considerably higher than those optimal for net CO_2 uptake (Nobel and Hartsock, 1986). In the winter, EPI increases to mid-elevations, reflecting an increase in the water index, and then decreases at higher elevations, reflecting the overriding importance of a decreasing temperature index at this cold time of the year. The seasonal changes in EPI with elevation correlate well with changes in productivity measured independently, indicating that a productivity index based on net CO_2 uptake determined in the laboratory, in which one factor is varied at a time, can help interpret field CO_2 uptake and productivity when various environmental factors are changing simultaneously. Although secondary interactions do occur, such as a lower PPFD required for saturation of J_{CO_2} at suboptimal temperatures, EPI can be used to predict the major influences of climate on net CO_2 uptake in the field.

V. Water-Use Efficiency

Stomatal opening leading to CO_2 uptake necessary for photosynthesis results in an inevitable loss of water. A useful parameter relating the two fluxes and showing the total CO_2 fixed (benefit) per unit water lost (cost) is the *water-use efficiency* (WUE):

$$\text{WUE} = \frac{\text{mass } CO_2 \text{ fixed}}{\text{mass } H_2O \text{ transpired}} \qquad \text{mass basis} \qquad (8.31a)$$

$$\text{WUE} = \frac{\text{mol } CO_2 \text{ fixed}}{\text{mol } H_2O \text{ transpired}} \qquad \text{mole basis} \qquad (8.31b)$$

A related quantity is the *transpiration ratio*, which is the reciprocal of the water-use efficiency and hence represents the water lost per CO_2 fixed.

A. Values of WUE

From the J_{CO_2} and J_{wv} calculated in this chapter, we can determine a WUE for the leaf of a representative C_3 mesophyte. Specifically, we obtained a J_{CO_2} of 10.7 μmol CO_2 fixed $m^{-2} s^{-1}$ (see Section IVE) and a J_{wv} of 4.3 mmol H_2O transpired $m^{-2} s^{-1}$ (see Section IIF). By Eq. (8.31b), the WUE then is

$$\text{WUE} = \frac{(10.7 \times 10^{-6} \text{ mol } CO_2 \text{ m}^{-2} \text{ s}^{-1})}{(4.3 \times 10^{-3} \text{ mol } H_2O \text{ m}^{-2} \text{ s}^{-1})} = 0.0025 \, CO_2/H_2O$$

On a mass basis, this corresponds to a WUE of 6.1 g CO_2 (kg $H_2O)^{-1}$ (the molar masses of CO_2 and H_2O are 44.0 and 18.0 g mol^{-1}, respectively). We also note that the transpiration ratio in this case is 400 H_2O/CO_2. This substantial water loss per CO_2 fixed is generally not a problem when plenty of water is available for transpiration. Plants in such environments often have a high g_{wv}^{total}, which leads to a somewhat higher $g_{CO_2}^{total}$ and somewhat higher rates of photosynthesis than for plants with a moderate g_{wv}^{total}.

Any loss of water can potentially be harmful for plants growing in arid regions, many of which have evolved a novel way of fixing CO_2 in a manner leading to a high WUE. For example, many species in the family Crassulaceae, as well as other desert succulents, have their stomata closed during the daytime. This greatly reduces transpiration but also essentially eliminates the net influx of CO_2 at that time. When the stomata open at night, CO_2 diffuses in and is fixed into malate (e.g., by carboxylation of phosphoenolpyruvate) and other organic acids. During the next daytime these organic acids are decarboxylated, and the released CO_2 is retained within the plant because of the closed stomata. This CO_2 is then fixed into photosynthetic products by means of the C_3 pathway. Plants with this CO_2 fixation mechanism are referred to as Crassulacean acid metabolism (CAM) plants because such reactions were initially studied extensively in the Crassulaceae, although apparently first observed in the Cactaceae in 1804. About 6–7% of the nearly 300,000 species of vascular plants use the CAM pathway, most of which are tropical or subtropical epiphytes.

As just indicated, stomata for CAM plants tend to open at night, when leaf and air temperatures are lower than daytime values. The concentration of water vapor in the pores of the cell walls of chlorenchyma cells (c_{wv}^e) is then much lower, markedly reducing the rate of transpiration. For example, leaf temperatures of *Agave deserti* can be 25°C in the afternoon and 5°C at night (Fig. 8-15), leading to saturation water vapor concentrations of 23.1 and 6.8 g m^{-3}, respectively (see Appendix I). For air with a water vapor content of 4.0 g m^{-3}, which is fairly typical during the wintertime in the native habitat of *A. deserti*, Δc_{wv}^{total} is $(23.1 - 4.0)/(6.8 - 4.0)$ or 7 times higher at 25°C than at 5°C, and therefore so is J_{wv} for the same degree of stomatal opening. Clearly, nocturnal stomatal opening can result in water conservation and hence a higher WUE. For the CAM plant *A. deserti* on the day depicted in Fig. 8-15, the WUE is 56 g CO_2 (kg H_2O)$^{-1}$, and it can be 40 g CO_2 (kg H_2O)$^{-1}$ when averaged over a whole year (Nobel, 1976)—both very high values.

Changes in the thickness of the air boundary layers adjacent to a leaf have a greater influence on the flux of water vapor than on the flux of CO_2. For instance, the total resistance for water vapor diffusion can equal $r_{wv}^{bl_1} + r_{wv}^{st_1} + r_{wv}^{ias}$ (Eq. 8.14), whereas $r_{CO_2}^{bl_1} + r_{CO_2}^{st_1} + r_{CO_2}^{ias}$ (Eq. 8.17) is generally just over half of the total resistance for CO_2 diffusion. Thus changes in wind speed have a smaller fractional effect on $r_{CO_2}^{total}$ than on r_{wv}^{total}. Similarly, partial stomatal closure can appreciably reduce g_{wv}^{total} but lead to smaller fractional reductions in $g_{CO_2}^{total}$. Certain xerophytes have fairly low maximal values for g_{wv}^{st}—the maximal stomatal conductance is generally <80 mmol m^{-2} s^{-1} for *A. deserti* (see Fig. 8-15) compared with over 400 mmol m^{-2} s^{-1} for many mesophytes (Table 8-1). Such a low maximal stomatal conductance reduces transpiration to a greater degree than CO_2 uptake, with a consequent enhancement in WUE (Eq. 8.27).

Even though photosynthesis and transpiration depend on environmental conditions, we can still make some generalizations about WUE for different types of plants. Specifically, WUEs averaged over a day for mature leaves are usually 1–3 g CO_2 (kg H_2O)$^{-1}$ for C_3 plants, 2–5 g CO_2 (kg H_2O)$^{-1}$ for C_4 plants, and 10–40 g CO_2 (kg H_2O)$^{-1}$ for CAM plants (Osmond *et al.*, 1980; Szarek and Ting, 1975). C_4 plants have approximately double the WUE of C_3 plants because C_4 plants tend to have lower gas-phase conductances (which conserves water with a relatively small negative effect on photosynthesis) and higher liquid-phase conductances (which affects photosynthesis positively) than do C_3 plants. However,

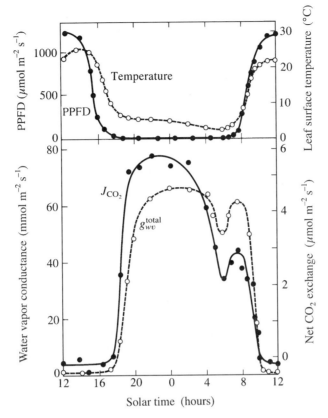

Figure 8-15. Photosynthetic photon flux density (on a horizontal surface), leaf surface temperature, water vapor conductance, and net CO_2 exchange for *Agave deserti* on clear winter days in the northwestern Sonoran Desert [modified from Nobel (1976); used by permission].

maximizing WUE may not always be adaptive, e.g., water may not be limiting for an aquatic plant.

B. Elevational Effects on WUE

Both transpiration and photosynthesis are affected by elevation because diffusion coefficients depend inversely on ambient (barometric) pressure [$D_j = D_{j_0}(T/273)^{1.8}(P_0/P)$; Eq. 8.9] and the partial pressures of water vapor and CO_2 generally decrease with elevation. Barometric pressure averages 0.101 MPa at sea level, 0.079 MPa at 2000 m, and about 0.054 MPa at 5000 m. Thus diffusion coefficients are nearly twice as large at 5000 m as at sea level owing to the pressure change, which correspondingly increases the gas-phase conductances based on Δc (e.g., Eq. 8.2), whereas those based on ΔN (Eq. 8.8) are unchanged. For a typical lapse rate of $-5°C$ per kilometer of elevation,[8] temperatures can decrease from $30°C$ at sea level to $5°C$ at 5000 m, which by itself decreases diffusion coefficients by 14%

---

[8] At the dry adiabatic lapse rate (9.8°C decrease in temperature per kilometer increase in altitude), a rising parcel of dry air will cool by expansion due to the decrease in air pressure and will achieve the same temperature as the surrounding air—a case of neutral stability. That is, air movement is then neither favored nor retarded by buoyancy. Observed lapse rates are usually -5 to $-6°C$ km^{-1}.

according to Eq. (8.9). The partial pressure of CO_2 is reduced more or less in concert with the reduced barometric pressure, i.e., the mole fraction of CO_2 is approximately constant with elevation. The partial pressure of water vapor in the air also tends to decrease with elevation, so under isothermal conditions the driving force for water loss (both Δc_{wv} and ΔN_{wv}) increases, as does transpiration. Because P_{wv}^* is essentially independent of P and $N_{wv}^* = P_{wv}^*/P$, the mole fraction of water vapor in the leaf increases as ambient pressure decreases, i.e., at higher elevations; however, the temperature decreases with increasing elevation generally more than offset the effects of P changes on N_{wv}^*.

Because of the interaction of many factors, especially the numerous temperature effects on both transpiration and photosynthesis, the effect of elevation on WUE is difficult to predict. When the turbulent mixing aspect in the leaf boundary layer is ignored (see footnote 4), the higher D_{CO_2} in the gas phase at higher elevations tends to offset the lower P_{CO_2} as far as the CO_2 level in the intercellular air spaces is concerned. If we ignore changes in stomatal aperture and temperature, the reduction in P_{CO_2} with elevation thus translates into smaller fractional decreases in $P_{CO_2}^{ias}$ and $c_{CO_2}^{ias}$, and $N_{CO_2}^{ias}$ is unchanged if J_{CO_2} is unchanged. Diffusion coefficients in the liquid phase are unaffected by barometric pressure. Hence, liquid-phase conductances expressed in mm s^{-1} are unaffected by the pressure changes with elevation, but those expressed in mmol m^{-2} s^{-1} are proportional to P (see Section IF) and hence decrease with elevation. Based on these changes in gas-phase and liquid-phase conductances, J_{CO_2} tends to decrease with elevation—the greater the liquid-phase conductance relative to that of the gas phase, the less the elevational effect. The lowering of temperature with elevation could have an even larger effect on photosynthesis, although the optimal temperature for photosynthesis can acclimate (usually by 2–15°C) to match the average ambient temperature of the environment.

Many factors affect leaf temperature, and an energy budget analysis can be used to calculate T^{leaf} and thus to indicate the effects of elevation on transpiration. Leaves at higher elevations generally experience a higher net radiation balance (more incident shortwave irradiation although somewhat less incident longwave irradiation; Eq. 7.6) and lower air temperatures. For small leaves, which tend to be close to air temperature, transpiration generally decreases with elevation. For a 5 × 5-cm leaf in full sunlight, transpiration is unaffected at a lapse rate of -1°C km^{-1} but decreases 25% at 1000 m compared to sea level at a lapse rate of -5°C km^{-1} (Gale, 1972). For large leaves in full sunlight, transpiration at typical lapse rates may decrease only slightly with elevation because large leaves are further above air temperature at higher elevations. For temperature inversion conditions (increasing temperature with elevation), transpiration can actually increase with elevation. Because photosynthesis and transpiration can vary in so many ways with elevation, effects of elevation on WUE must be judged case by case.

C. Stomatal Control of WUE

To maximize WUE, stomatal opening must be synchronized with the capability for CO_2 fixation. Stomatal opening can be regulated by the CO_2 level in the intercellular air spaces, a depletion of $N_{CO_2}^{ias}$ by photosynthesis leading to an increase in $g_{CO_2}^{st}$, which then lets more CO_2 into the leaf under a PPFD and other conditions favorable for photosynthesis. This is an example of a *feedback* system, as $N_{CO_2}^{ias}$ feeds a signal back to the stomata, which in turn leads to a change in $N_{CO_2}^{ias}$. Also, the PPFD may directly affect the metabolism of guard cells, which indeed contain chloroplasts and hence can utilize such radiation. This is an example of a *feedforward* system, as changes in stomatal aperture due to photosynthetic responses of guard cell chloroplasts feed forward (or anticipate) and adjust CO_2 entry into the leaf, thereby matching photosynthesis of the mesophyll region to environmental conditions.

Stomatal opening is also affected by the leaf water status. For instance, stomata tend to close as a leaf begins to wilt, especially after the leaf water potential drops below some threshold level and abscisic acid (ABA) is produced. In fact, ABA can induce stomatal closure even when $N_{CO_2}^{ias}$ favors opening. The water status thus affects stomatal opening and hence transpiration, which in turn feeds back onto the leaf water status. Stomatal opening is usually increased by higher N_{wv}^{ta}. This is another example of a feedforward system, as it anticipates the effect of the ambient water vapor concentration on transpiration; e.g., higher N_{wv}^{ta} means less "force" leading to water loss from a leaf, so the stomata can open wider without leading to excessive transpiration. These various processes regulating stomatal movements interact with each other—we will examine the consequences of this for gas exchange by leaves.

What is the optimal behavior of stomata over the course of a day? WUE is maximized by minimal stomatal opening because transpiration is decreased more than photosynthesis by partial stomatal closure (see Eq. 8.31), i.e., J_{wv} changes proportionally more than does J_{CO_2} as g_{wv}^{st} changes (Fig. 8-16). However, this can lead to very little CO_2 uptake. Thus a more pertinent consideration might be the maximal amount of CO_2 that can be taken up for a certain amount of water transpired. The amount of water lost depends on plant condition and environmental factors and should be considered over the course of a whole day. To help analyze the relationship between gas fluxes, curves showing J_{CO_2} versus J_{wv} can be drawn for any PPFD, temperature, wind speed, or relative humidity occurring for a particular leaf during the day, the location of the curves varying but still conforming to the general shape illustrated in Fig. 8-16. In addition to environmental factors, the location of the curves is influenced by leaf properties such as size, age, A^{mes}/A, and shortwave absorptance. Nevertheless, Ian Cowan and Graham Farquhar hypothesized in the 1970s that stomata will open or close depending on the various feedback and feedforward processes in such a way that the following relation is obeyed:

$$\frac{\partial J_{wv}/\partial g_{wv}^{st}}{\partial J_{CO_2}/\partial g_{wv}^{st}} = \frac{\partial J_{wv}}{\partial J_{CO_2}} = \lambda = \text{constant} \qquad (8.32)$$

where $\partial J_{wv}/\partial g_{wv}^{st}$ and $\partial J_{CO_2}/\partial g_{wv}^{st}$ represent the sensitivity of transpiration and photosynthesis, respectively, to changes in stomatal conductance.

The solid circles in Fig. 8-16 indicate the CO_2 and H_2O fluxes that can occur at different times of the day for a leaf with a λ of 1000 H_2O/CO_2. At low PPFD, little stomatal opening occurs, so little water is used under conditions in which the rate of photosynthesis inherently cannot be very high. Stomatal opening is much greater at high PPFD and thus both transpiration and photosynthesis are then greater, but the local slope of the J_{CO_2} versus J_{wv} curve is still the same for points of the same λ (slope $= \partial y/\partial x$). We next consider what happens to the fluxes if we move along a curve away from one of the solid circles, e.g., the one for medium PPFD (Fig. 8-16). If the stomatal opening changes so that J_{wv} increases by 1.0 mmol m^{-2} s^{-1}, then J_{CO_2} increases by 0.8 μmol m^{-2} s^{-1}. To lead to the same total transpiration for the day, we must decrease J_{wv} by the same amount at another time, which for simplicity we can also consider for the medium PPFD curve (note that all the solid circles in Fig. 8-16 occur for the same slope, 1000 H_2O/CO_2). A decrease in J_{wv} of 1.0 mmol m^{-2} s^{-1} is accompanied by a decrease in J_{CO_2} of 1.4 μmol m^{-2} s^{-1}. When the effects of both changes are considered together, we get the same total transpiration but a

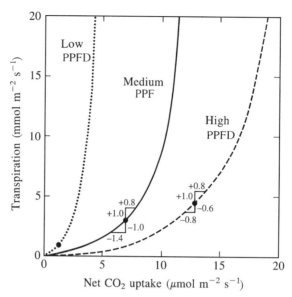

Figure 8-16. Relation between net photosynthesis (J_{CO_2}) and transpiration (J_{wv}) as stomatal conductance is varied. The three curves depict various PPFD levels, indicated as "low," "medium," and "high." The circles indicate where the slope $\partial J_{wv}/\partial J_{CO_2}$ is 1000 H_2O/CO_2. Cuticular transpiration is ignored. The numbers indicate changes in the fluxes, on the medium PPFD curve for no net change in transpiration and at the same slope on the high PPFD curve for no net change in CO_2 uptake. Curve shapes indicate that J_{wv} increases faster than J_{CO_2} as stomates open.

lower net CO_2 uptake. Similarly, if we move along a curve (e.g., the one for high PPFD; Fig. 8-16) so that increased stomatal opening increases J_{wv} by 1.0 mmol m^{-2} s^{-1} and increases J_{CO_2} by 0.8 μmol m^{-2} s^{-1} as before but now move along a curve to decrease J_{CO_2} by 0.8 μmol m^{-2} s^{-1}, we find that J_{wv} decreases by 0.6 mmol m^{-2} s^{-1}. When both of these changes are considered, we get the same net CO_2 uptake but more transpiration. In fact, the criterion expressed in Eq. (8.32) leads to the maximal amount of CO_2 fixed for a particular amount of water transpired as well as to the minimal amount of water transpired for a particular amount of CO_2 fixed in a day. Thus if stomata respond to keep $\partial J_{wv}/\partial J_{CO_2}$ constant, then the WUE of the leaf for the entire day is maximized.

The value of λ can change during the growth of a plant. For instance, λ can be small (e.g., 300 H_2O/CO_2) when water is in short supply. In such cases, constancy of λ requires that stomata close partially near midday, when temperatures are the highest and transpiration is potentially the greatest. Also, water stress generally leads to higher ABA levels in the leaves and a lower λ. On the other hand, a large λ (e.g., 1300 H_2O/CO_2) occurs when the plant is not under water stress, and no midday stomatal closure then takes place.

The proposed constancy of $\partial J_{wv}/\partial J_{CO_2}$ helps us interpret the partial stomatal closure at midday when water is limiting, as well as the nocturnal stomatal closure when PPFD is limiting. We can also use the constancy of $\partial J_{wv}/\partial J_{CO_2}$ to help interpret experiments in which a single environmental factor is varied, such as the driving force for water vapor loss, ΔN_{wv}. If the stomata maintained a constant J_{wv}, then changes in g_{wv}^{st} would be the inverse of changes in ΔN_{wv} (when cuticular transpiration is ignored). On the other hand, maintenance of constant J_{CO_2} as ΔN_{wv} is varied requires constancy of $g_{CO_2}^{st}$, which equals $g_{wv}^{st}/1.60$. In fact, varying ΔN_{wv} over a fourfold range for *Nicotiana glauca* (tobacco),

Corylus avellana (hazel; Farquhar *et al.*, 1980), and *Vigna unguiculata* (cowpea; Hall and Schulze, 1980) leads to stomatal behavior resulting in variation of both J_{wv} and J_{CO_2}, but $\partial J_{wv}/\partial J_{CO_2}$ is approximately constant.

D. C$_3$ versus C$_4$ Plants

We will next recapitulate some of the characteristics of C$_3$ and C$_4$ plants previously introduced. After examining the influence of stomata on maximal photosynthetic rates under optimal conditions, we will predict effects on WUE for elevated levels of atmospheric CO_2.

The ecological advantages of the C$_4$ pathway are most apparent for plants in environments having high PPFD, high temperature, and limited water supply. C$_4$ plants are effective at high PPFD because photosynthesis for leaves of many C$_3$ plants saturates near 600 μmol m^{-2} s^{-1}, but most C$_4$ plants have an increasing J_{CO_2} as the PPFD is raised up to 2000 μmol m^{-2} s^{-1} (Fig. 8-14). Optimal temperatures for net CO_2 uptake are usually 20–35°C for C$_3$ plants but 30–45°C for C$_4$ plants, which can be interpreted by considering Rubisco; the CO_2-evolving photorespiration, which has very low rates in C$_4$ plants, becomes proportionally more important at higher temperatures and thus reduces the net CO_2 uptake at the higher temperatures for C$_3$ plants. The optimal temperature for photosynthesis is actually variable, and it can change by 10°C in a matter of days (even for mature leaves), allowing for seasonal acclimation of photosynthetic performance. C$_4$ plants can more readily cope with limited water supply because generally their gas-phase conductance for CO_2 is lower and their liquid-phase CO_2 conductance is higher than those for C$_3$ plants. Thus C$_4$ plants tend to become dominant in deserts, grasslands, and certain subtropical regions, i.e., areas of high PPFD, high temperature, and limited water supply, as indicated previously. Low rates of photorespiration and the associated high WUE allow C$_4$ plants to become very successful weeds—in fact, 8 of the 10 agriculturally most noxious weeds use the C$_4$ pathway (Holm *et al.*, 1977), although only about 1% of the nearly 300,000 species of vascular plants are C$_4$. CAM plants have an even higher WUE than do C$_4$ plants and achieve their greatest relative importance in regions of high PPFD and very limited water supply.

At 30°C and for an absorbed PPFD up to about 100 μmol m^{-2} s^{-1}, leaves of C$_3$ and C$_4$ plants can have a very similar quantum yield (approximately 0.053 mol CO_2/mol photons at an $N_{CO_2}^{ta}$ of 325 μmol mol^{-1}; Ehleringer and Björkman, 1977). As the temperature is raised, however, photorespiration increases relative to photosynthesis, so the quantum yield declines for C$_3$ plants but is essentially unchanged for C$_4$ plants. On the other hand, lowering the ambient O$_2$ level raises the quantum yield for C$_3$ (photorespiring) plants because the oxygenase activity of Rubisco (see Fig. 8-10) is then suppressed; such changes have little effect on C$_4$ plants until the O$_2$ level falls below about 2%, where mitochondrial respiration is affected.

CO_2 uptake by C$_4$ plants is CO_2 saturated at a relatively low CO_2 level in the intercellular air spaces. For instance, an $N_{CO_2}^{ias}$ equivalent to 150 μmol mol^{-1} usually leads to over 90% of the maximum J_{CO_2} for C$_4$ plants, so increasing the ambient CO_2 level usually has little effect on their quantum yield (Fig. 8-13). However, the quantum yield for CO_2 fixation by C$_3$ plants progressively increases as the ambient CO_2 level is raised, and at an $N_{CO_2}^{ias}$ corresponding to 700 μmol mol^{-1} it approaches to within 10% of the value occurring when the ambient O$_2$ level is reduced 10-fold (about 0.081 mol CO_2/mol photons). Such raising of the CO_2 level is another way of favoring the carboxylase activity of Rubisco. Also, the requirement for a high $N_{CO_2}^{ias}$ for C$_3$ plants is consistent with the K_{CO_2} of 5–15 mmol m^{-3} for CO_2 fixation by Rubisco.

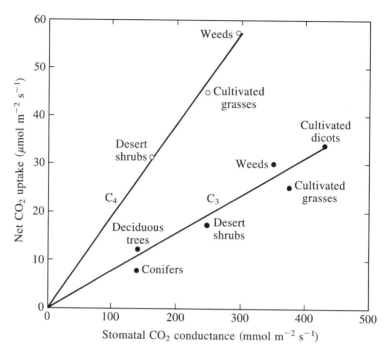

Figure 8-17. Relation between stomatal CO_2 conductance ($g_{CO_2}^{st}$) and net CO_2 uptake (J_{CO_2}) for various categories of C_3 and C_4 plants under optimal conditions and 340 μmol CO_2 mol^{-1} (data are from references cited in Tables 8-1 and 8-4).

For a series of C_3 and C_4 plants, stomata open to a degree that gives an approximately constant CO_2 level in the intercellular air spaces, with the level differing between plants from the two photosynthetic pathways (Fig. 8-17). A similar adjustment in stomatal conductance occurs as the PPFD increases for a particular plant (see Fig. 8-16). We must therefore conclude that stomata regulate the entry of CO_2 to match the photosynthetic capability of the mesophyll region. In particular, the slope of $g_{CO_2}^{st}$ versus J_{CO_2} gives the drop in CO_2 across the stomata, i.e., $J_{CO_2} = g_{CO_2}^{st} \Delta N_{CO_2}^{st}$ (the equation for a straight line is $y = mx + b$, where the slope m is $\partial y/\partial x$). For the C_3 plants in Fig. 8-17, the slope is 74×10^{-6} (i.e., $74 \,\mu$mol mol^{-1}), and for the C_4 plants it is 189×10^{-6}. For an ambient CO_2 level of 340 μmol mol^{-1}, and ignoring the CO_2 drop across the boundary layer (about 10–30 μmol mol^{-1}), $N_{CO_2}^{ias}$ corresponds to 270 μmol mol^{-1} for C_3 plants and 150 μmol mol^{-1} for C_4 plants. [Changes in $N_{CO_2}^{ias}$ can occur, such as higher values at lower PPFD in the lower parts of a canopy (Farquhar *et al.*, 1989).] Even though $N_{CO_2}^{ias}$ is lower for C_4 plants, it is still high enough to saturate their CO_2 fixation pathway. A higher $N_{CO_2}^{ias}$ brought about by a higher $g_{CO_2}^{st}$ does not benefit photosynthesis for C_4 plants, but the accompanying greater stomatal conductance would lead to more water loss. For a C_3 plant, photosynthesis does not approach saturation until $N_{CO_2}^{ias}$ exceeds 700 μmol mol^{-1} (Fig. 8-13). However, opening stomata further than required to maintain an $N_{CO_2}^{ias}$ of about 270 μmol mol^{-1} would not enhance photosynthesis very much, but it would considerably increase transpiration (see Fig. 8-16). Thus the adjustment of stomatal opening to meet the conflicting demands of photosynthesis and transpiration, using feedback and feedforward control by microclimatic and leaf parameters, leads to a remarkable regulation that minimizes water loss while maximizing CO_2 uptake.

What will happen to the WUE of C_3 and C_4 plants as the atmospheric CO_2 level increases? We will assume that other conditions, such as temperature, water status, and nutrients, are not severely limiting net CO_2 uptake. As the CO_2 level in the turbulent air increases, $N_{CO_2}^{ias}$ will increase, which will decrease stomatal opening and hence transpiration. For a doubling of atmospheric CO_2, transpiration should decrease 30–40% for leaves of both C_3 and C_4 plants. On the other hand, elevated CO_2 will increase the rate of photosynthesis for C_3 plants, approximately 30–60% for a doubling in atmospheric CO_2 levels, but should have no major effect on the photosynthetic rate of C_4 plants (Fig. 8-13). Thus as atmospheric CO_2 levels become twice as high as the 1990 level in the latter half of the 21st century, the WUE should increase about 35% for C_4 plants and 75% for C_3 plants. CAM plants show more variation in their responses to increased atmospheric CO_2 levels and often exhibit even greater enhancements than do C_3 plants.

Problems

8.1 Consider a leaf that is 0.5 mm thick with 64 stomata per mm². Approximate the stomatal opening by a rectangle that is 6×20 μm with a depth of 25 μm. Assume that the leaf and air temperatures are both 20°C and that the ambient air pressure is 1 atm.

a. What are g_{wv}^{bl} (in mm s^{-1}) and r_{wv}^{bl} if the boundary layer is 0.8 mm thick?

b. What are na^{st} and the effective r^{st}?

c. What is the average flux of water vapor within the stomatal pores compared with that across the boundary layer?

d. What is g_{wv}^{st} in mm s^{-1} and mmol m^{-2} s^{-1}? What are the values if the ambient air pressure is reduced to 0.9 atm?

e. What is g_{wv}^{ias} in the two units if the effective path length in the intercellular air spaces equals the leaf thickness?

f. Suppose that each stoma is sunken in a cylindrical cavity 50 μm across and 100 μm deep. What additional resistance to water vapor diffusion does this provide?

8.2 Suppose that $g_{wv}^{bl_i}$ is 20 mm s^{-1}, $g_{wv}^{st_i}$ is 6 mm s^{-1}, g_{wv}^{c} is 0.1 mm s^{-1}, and g_{wv}^{ias} is 40 mm s^{-1}.

a. What is g_{wv}^{total} if water vapor diffuses out only across the lower epidermis of the leaf?

b. What are the three g_{wv}^{total}'s in (a) if the cuticular pathway is ignored, if the intercellular air spaces are ignored, and if both g_{wv}^{c} and g_{wv}^{ias} are ignored?

c. What is g_{wv}^{total} if the stomata in the upper epidermis have the same conductance as those in the lower one?

d. What is g_{wv}^{total} if 28% of J_{wv} is through the upper epidermis?

e. Suppose that the leaf temperature is 30°C, the air in the cell wall pores where the water evaporates is at 99% relative humidity, and c_{wv}^{ta} is 7.5 g m^{-3}. What is J_{wv} through the lower epidermis?

f. What are g_{wv}^{total} in mmol m^{-2} s^{-1} and J_{wv} (in mmol m^{-2} s^{-1}) under the conditions of (e)? Assume that the air pressure is 1 atm and that N_{wv}^{ta} is 0.0103.

g. Under the conditions of (e) and (f), what is the drop in water vapor concentration and mole fraction along the stomatal pores (ignore cuticular transpiration)?

8.3 Suppose that a shade leaf has a layer of tightly packed palisade mesophyll cells with rectangular sides that are externally $30 \times 100 \, \mu m$ and with square ends $30 \times 30 \, \mu m$ (the long dimension is perpendicular to the leaf surface). Suppose that there are two spherical spongy mesophyll cells ($30 \, \mu m$ in diameter) under each palisade cell. Let the cell wall thickness of mesophyll cells be $0.2 \, \mu m$, the mean distance from the plasma membrane to the chloroplasts be $0.1 \, \mu m$, and the average distance that CO_2 diffuses in the chloroplasts before reaching the photosynthetic enzymes be $0.5 \, \mu m$.

a. What is A^{mes}/A if essentially the entire surface area of the mesophyll cells is exposed to the intercellular air spaces?

b. Assume that a sun leaf on the same plant has two layers of palisade cells and half as many spongy mesophyll cells. If the dimensions of the cells are the same as for the shade leaf, what is A^{mes}/A for the sun leaf?

c. If $D^{cw}_{CO_2}$ is 5.0×10^{-10} m^2 s^{-1}, what is the maximum value for $r^{cw}_{CO_2}$ at 20°C for the shade leaf?

d. If P_{CO_2} is 1.0×10^{-3} m s^{-1} for the plasma membrane and the chloroplast limiting membranes, what are $r^{pl}_{CO_2}$ and $r^{clm}_{CO_2}$ (shade leaf)?

e. If $D^{cyt}_{CO_2}$ and $D^{stroma}_{CO_2}$ are 1.0×10^{-9} m^2 s^{-1}, what are $r^{cyt}_{CO_2}$ and $r^{stroma}_{CO_2}$ (shade leaf)? Assume that the relevant partition coefficients for the various forms of CO_2 are 1.

f. What is the resistance to CO_2 diffusion from the intercellular air spaces to the photosynthetic enzymes for the sun leaf? Assume that $J^{r+pr}_{CO_2}$ is negligible and that $r^{cw}_{CO_2}$ has its maximal 20°C value.

8.4 Let us suppose that $c^{ta}_{CO_2}$ is 13 mmol m^{-3}, K_{CO_2} is 5 μM, $r^{bl}_{CO_2}$ is 60 s m^{-1}, $r^{leaf_1}_{CO_2}$ is 250 s m^{-1}, $r^{mes}_{CO_2}$ is 150 s m^{-1}, and $r^{chl}_{CO_2}$ is 100 s m^{-1} for a leaf that CO_2 enters only across the lower epidermis.

a. If the rate of gross photosynthesis is 4 mol CO_2 fixed m^{-3} s^{-1} when $c^{chl}_{CO_2}$ is 9 μM, what is V_{max}?

b. What is $c^{chl}_{CO_2}$ when v_{CO_2} is 90% of V_{max}?

c. If the rate of respiration plus photorespiration is 45% of that of gross photosynthesis, what are $r^{total}_{CO_2}$ and J_{CO_2}? Assume that $c^{chl}_{CO_2}$ is 9 μM.

d. Repeat (c) for a nonphotorespiring plant where the rate of respiration is 5% of $J^{ps}_{CO_2}$. Assume that $c^{chl}_{CO_2}$ is 7 μM.

e. Let us place a small transparent bag completely around a leaf of the nonphotorespiring plant. What would $c^{chl}_{CO_2}$ be if the CO_2 concentration in the bag in the steady state were 10 μmol mol^{-1}? Assume that all resistances and the rate of respiration are unchanged.

f. What is the concentration of CO_2 in the mitochondria at night for the nonphotorespiring plant? Let $r^i_{CO_2}$ be 500 s m^{-1}, and assume that the rate of respiration as well as the resistances remain the same as the daytime values. What is the mitochondrial c_{CO_2} at night if stomatal closure causes $r^{leaf_1}_{CO_2}$ to become 5000 s m^{-1}?

8.5 Consider a sunlit leaf at 35°C with a g^{bl}_{wv} of 15 mm s^{-1}, stomata only in the lower epidermis, a g^{ias}_{wv} of 30 mm s^{-1}, and a $g^{total}_{CO_2}$ of 0.70 mm s^{-1}. Assume that the ambient air is at 30°C and 32% relative humidity, that cuticular transpiration is negligible and total transpiration is 5.0 mmol m^{-2} s^{-1}, and that the air in the intercellular air spaces reaches 100% relative humidity.

a. What are g_{wv}^{total} and g_{wv}^{st}?

b. What is the essentially immediate effect on g_{wv}^{total} and J_{wv} of decreasing the stomatal opening fourfold, as can occur during wilting?

c. What is the qualitative effect of the action in (b) on T^{leaf}?

d. Neglecting effects caused by leaf temperature, what are the percentage changes of photosynthesis and WUE caused by the action in (b)? Assume that $c_{CO_2}^{chl}$ is unchanged.

e. What is the essentially immediate effect on $g_{CO_2}^{total}$ and J_{wv} of increasing the wind speed by fourfold?

f. What is the qualitative effect of the action in (e) on heat conduction across the boundary layer $[J_H^C = 2K^{air}(T^{leaf} - T^{air})/\delta^{bl}$; Eq. 7.11] and on T^{leaf}?

References

Björkman, O. (1981). Responses to different quantum flux densities. In *Physiological Plant Ecology* (O. L. Lange, P. S. Nobel, C. B. Osmond, and H. Ziegler, Eds.), *Encyclopedia of Plant Physiology, New Series*, Vol. 12A, pp. 57–107. Springer-Verlag, Berlin.

Botre, F., Gros, G., and Storey, B. T. (Eds.) (1991). *Carbonic Anhydrase: From Biochemistry and Genetics to Physiology and Clinical Medicine*. VCH, New York.

Brown, H. T., and Escombe, F. (1900). Static diffusion of gases and liquids in relation to the assimilation of carbon and translocation in plants. *Philos. Trans. R. Soc. London Ser. B* **193**, 223–291.

Colman, B., and Espie, G. S. (1985). CO_2 uptake and transport in leaf mesophyll cells. *Plant Cell Environ.* **8**, 449–457.

Cowan, I. R. (1977). Stomatal behavior and environment. *Adv. Bot. Res.* **4**, 117–227.

Cowan, I. R., and Farquhar, G. D. (1977). Stomatal function in relation to leaf metabolism and environment. In *Integration of Activity in the Higher Plant* (D. H. Jennings, Ed.), pp. 471–505. Cambridge Univ. Press, Cambridge, UK.

Dacey, J. W. H. (1981). Pressurized ventilation in the yellow waterlily. *Ecology* **62**, 1137–1147.

Ehleringer, J., and Björkman, O. (1977). Quantum yields for CO_2 uptake in C_3 and C_4 plants: Dependence on temperature, CO_2, and O_2 concentration. *Plant Physiol.* **59**, 86–90.

Ehleringer, J., Björkman, O., and Mooney, H. A. (1976). Leaf pubescence: Effects on absorptance and photosynthesis in a desert shrub. *Science* **192**, 376–377.

Evans, J. R., Sharkey, T. D., Berry, J. A., and Farquhar, G. D. (1986). Carbon isotope discrimination measured concurrently with gas exchange to investigate CO_2 diffusion in leaves of higher plants. *Aust. J. Plant Physiol.* **13**, 281–292.

Farquhar, G. D., and Sharkey, T. D. (1982). Stomatal conductance and photosynthesis. *Annu. Rev. Plant Physiol.* **33**, 317–345.

Farquhar, G. D., and von Caemmerer, S. (1982). Modelling of photosynthetic response to environmental conditions. In *Physiological Plant Ecology* (O. L. Lange, P. S. Nobel, C. B. Osmond, and H. Ziegler, Eds.), *Encyclopedia of Plant Physiology, New Series*, Vol. 12B, pp. 549–587. Springer-Verlag, Berlin.

Farquhar, G. D., Schulze, E.-D., and Küppers, M. (1980). Responses to humidity by stomata of *Nicotiana glauca* L. and *Corylus avellana* L. are consistent with the optimization of carbon dioxide uptake with respect to water loss. *Aust. J. Plant Physiol.* **7**, 315–327.

Farquhar, G. D., Ehleringer, J. R., and Hubick, K. T. (1989). Carbon isotope discrimination and photosynthesis. *Annu. Rev. Plant Physiol. Plant Mol. Biol.* **40**, 503–507.

Field, C., and Mooney, H. A. (1986). The photosynthesis–nitrogen relationship in wild plants. In *On the Economy of Plant Form and Function* (T. J. Givnish, Ed.), pp. 25–55. Cambridge Univ. Press, Cambridge, UK.

Field, C. B., Ball, J. T., and Berry, J. A. (1989). Photosynthesis: Principles and field techniques. In *Plant Physiological Ecology: Field Methods and Instrumentation* (R. W. Pearcy, J. Ehleringer, H. A. Mooney, and P. W. Rundel, Eds.), pp. 209–253. Chapman & Hall, London.

Gale, J. (1972). Elevation and transpiration: Some theoretical considerations with special reference to Mediterranean-type climate. *J. Appl. Ecol.* **9**, 691–702.

Gibson, A. C., and Nobel, P. S. (1986). *The Cactus Primer*. Harvard Univ. Press, Cambridge, MA.

Haefner, J. W., Buckley, T. N., and Mott, K. A. (1997). A spatially explicit model of patchy stomatal responses to humidity. *Plant Cell Environ.* **20**, 1087–1097.

Hall, A. E., and Schulze, E.-D. (1980). Stomatal response to environment and a possible interrelation between stomatal effects on transpiration and CO_2 assimilation. *Plant Cell Environ.* **3**, 467–474.

Holm, L. G., Plucknett, D. L., Pancho, J. V., and Herberger, J. P. (1977). *The World's Worst Weeds: Distribution and Biology.* Univ. of Hawaii Press, Honolulu.

Jones, H. G. (1992). *Plants and Microclimate: A Quantitative Approach to Environmental Plant Physiology*, 2nd ed. Cambridge Univ. Press, Cambridge, UK.

Kerstiens, G. (1996a). Cuticular water permeability and its physiological significance. *J. Exp. Bot.* **47**, 1813–1832.

Kerstiens, G. (Ed.) (1996b). *Plant Cuticles: An Integrated Functional Approach.* BIOS Scientific, Oxford.

Körner, C. H., Scheel, J. A., and Bauer, H. (1979). Maximum leaf diffusive conductance in vascular plants. *Photosynthetica* **13**, 45–82.

Krömer, S. (1995). Respiration during photosynthesis. *Annu. Rev. Plant Physiol. Plant Mol. Biol.* **46**, 45–70.

Lake, J. V. (1967). Respiration of leaves during photosynthesis. *Aust. J. Biol. Sci.* **20**, 487–499.

Le Houérou, H. N. (1984). Rain use efficiency: A unifying concept in arid-land ecology. *J. Arid Environ.* **7**, 213–247.

Leuning, R. (1983). Transport of gases into leaves. *Plant Cell Environ.* **6**, 181–194.

Long, S. P., and Woodward, F. I. (Eds.) (1988). *Plants and Temperature.* Company of Biologists, Cambridge, UK.

Longstreth, D. J., Hartsock, T. L., and Nobel, P. S. (1980). Mesophyll cell properties for some C_3 and C_4 species with high photosynthetic rates. *Physiol. Plant.* **48**, 494–498.

MacRobbie, E. A. C. (1988). Control of ion fluxes in stomatal guard cells. *Bot. Acta* **101**, 140–148.

Monteith, J. L. (1977). Climate and the efficiency of crop production in Britain. *Philos. Trans. R. Soc. London Ser. B* **281**, 277–294.

Monteith, J. L., and Unsworth, M. H. (1990). *Principles of Environmental Physics*, 2nd ed. Edward Arnold, New York.

Nikolov, N. T., Massman, W. J., and Schoettle, A. W. (1995). Coupling biochemical and biophysical processes at the leaf level: An equilibrium photosynthesis model for leaves of C_3 plants. *Ecol. Modeling* **80**, 205–235.

Nobel, P. S. (1974). Boundary layers of air adjacent to cylinders. Estimation of effective thickness and measurements on plant material. *Plant Physiol.* **54**, 177–181.

Nobel, P. S. (1975). Effective thickness and resistance of the air boundary layer adjacent to spherical plant parts. *J. Exp. Bot.* **26**, 120–130.

Nobel, P. S. (1976). Water relations and photosynthesis of a desert CAM plant, *Agave deserti. Plant Physiol.* **58**, 576–582.

Nobel, P. S. (1977). Internal leaf area and CO_2 resistance: Photosynthetic implications of variations with growth conditions and plant species. *Physiol. Plant.* **40**, 137–144.

Nobel, P. S. (1980). Water vapor conductance and CO_2 uptake for leaves of a C_4 desert grass, *Hilaria rigida. Ecology* **61**, 252–258.

Nobel, P. S. (1984). Productivity of *Agave deserti*: Measurement by dry weight and monthly prediction using physiological responses to environmental parameters. *Oecologia* **64**, 1–7.

Nobel, P. S. (1988). *Environmental Biology of Agaves and Cacti.* Cambridge Univ. Press, New York.

Nobel, P. S. (1989). A nutrient index quantifying productivity of agaves and cacti. *J. Appl. Ecol.* **26**, 635–645.

Nobel, P. S., and Hartsock, T. L. (1986). Temperature, water, and PAR influences on predicted and measured productivity of *Agave deserti* at various elevations. *Oecologia* **68**, 181–185.

Nobel, P. S., and Walker, D. B. (1985). Structure of leaf photosynthetic tissue. In *Photosynthetic Mechanisms and the Environment* (J. Barber and N. R. Baker, Eds.), pp. 501–536. Elsevier, Amsterdam.

Nobel, P. S., Zaragoza, L. J., and Smith, W. K. (1975). Relation between mesophyll surface area, photosynthetic rate, and illumination level during development for leaves of *Plectranthus parviflorus* Henckel. *Plant Physiol.* **55**, 1067–1070.

Oke, T. R. (1987). *Boundary Layer Climates*, 2nd ed. Methuen, London.

Osmond, C. B., Björkman, O., and Anderson, D. J. (1980). *Physiological Processes in Plant Ecology. Toward a Synthesis with Atriplex.* Springer-Verlag, Berlin.

Parkhurst, D. F. (1994). Tansley review No. 65. Diffusion of CO_2 and other gases inside leaves. *New Phytol.* **126**, 449–479.

Raschke, K. (1956). Über die physikalischen Beziehungen zwischen Wärmeübergangszahl, Strahlungsaustausch, Temperatur und Transpiration eines Blattes. *Planta* **48**, 200–238.

Raskin, I., and Kende, H. (1985). Mechanism of aeration in rice. *Science* **228**, 327–329.

Reid, R. C., Prausnitz, J. M., and Poling, B. E. (1987). *The Properties of Gases and Liquids*, 4th ed. McGraw-Hill, New York.

Schlichting, H. (1979). *Boundary-Layer Theory*, 7th ed. McGraw-Hill, New York.

Schroeder, J. I., and Hedrich, R. (1989). Involvement of ion channels and active transport in osmoregulation and signaling of higher plant cells. *Trends Biochem. Sci.* **14**, 187–192.

Schulze, E.-D (1994). *Flux Control in Biological Systems: From Enzymes to Populations and Ecosystems*. Academic Press, San Diego.

Schulze, E.-D., and Caldwell, M. M. (Eds.) (1994). *Ecophysiology of Photosynthesis*. Springer-Verlag, Berlin.

Sharkey, T. D. (1985). Photosynthesis in intact leaves of C_3 plants: Physics, physiology and rate limitations. *Bot. Rev.* **51**, 53–105.

Smirnoff, N. (Ed.) (1995). *Environment and Plant Metabolism: Flexibility and Acclimation*. BIOS Scientific, Oxford.

Smith, J.A.C., and Griffiths, H. (1993). *Water Deficits: Plant Responses from Cell to Community*. BIOS Scientific, Oxford.

Stumm, W., and Morgan, J. J. (1996). *Aquatic Chemistry: Chemical Equilibria and Rates in Natural Waters*, 3rd ed. Wiley, New York.

Syvertsen, J. P., Lloyd, J., McConchie, C., Kriedmann, P. E., and Farquhar, G. D. (1995). On the relationship between leaf anatomy and CO_2 diffusion through the mesophyll of hypostomatous leaves. *Plant Cell Environ.* **18**, 149–157.

Szarek, S. R., and Ting, I. P. (1975). Photosynthetic efficiency of CAM plants in relation to C_3 and C_4 plants. In *Environmental and Biological Control of Photosynthesis* (R. Marcelle, Ed.) pp. 289–297. Junk, The Hague.

Vogelmann, T. C. (1993). Plant tissue optics. *Annu. Rev. Plant Physiol. Plant Mol. Biol.* **44**, 231–251.

Willmer, C., and Fricker, M. (1996). *Stomata*, 2nd ed. Chapman & Hall, London.

Winter, K., and Smith, J. A. C. (Eds.) (1996). *Crassulacean Acid Metabolism: Biochemistry, Ecophysiology and Evolution*. Springer-Verlag, Berlin.

Woodrow, I. E., and Berry, J. A. (1988). Enzymatic regulation of photosynthetic CO_2 fixation in C_3 plants. *Annu. Rev. Plant Physiol. Plant Mol. Biol.* **39**, 533–594.

Wullschleger, S. D. (1993). Biochemical limitations to carbon assimilation in C_3 plants—A retrospective analysis of A/C_i curves from 109 species. *J. Exp. Bot.* **44**, 907–920.

Yong, J. W. H., Wong, S. C., and Farquhar, G. D. (1997). Stomatal responses to changes in vapor pressure differences between leaf and air. *Plant Cell Environ.* **20**, 1213–1216.

Zeiger, E., Farquhar, G. D., and Cowan, I. R. (1987). *Stomatal Function*. Stanford Univ. Press, Stanford, CA.

9

Plants and Fluxes

I.	**Gas Fluxes above the Leaf Canopy**	**351**
	A. Wind Speed Profiles	352
	B. Flux Densities	353
	C. Eddy Diffusion Coefficients	353
	D. Resistance of Air above the Canopy	355
	E. Transpiration and Photosynthesis	356
	F. Values for Fluxes and Concentrations	356
	G. Condensation	358
II.	**Gas Fluxes within Plant Communities**	**359**
	A. Eddy Diffusion Coefficient and Resistance	359
	B. Water Vapor	360
	C. Attenuation of Photosynthetic Photon Flux Density	361
	D. Values of Foliar Absorption Coefficients	362
	E. Light Compensation Point	363
	F. CO_2 Concentrations and Fluxes	364
	G. CO_2 at Night	365
III.	**Soil**	**366**
	A. Soil Water Potential	367
	B. Darcy's Law	368
	C. Soil Hydraulic Conductivity Coefficient	369
	D. Flux for Cylindrical Symmetry	371
	E. Flux for Spherical Symmetry	372
IV.	**Water Movement in the Xylem and the Phloem**	**373**
	A. Root Tissues	374
	B. The Xylem	374
	C. Poiseuille's Law	375
	D. Applications of Poiseuille's Law	376
	E. The Phloem	378
	F. Phloem Contents and Speed of Movement	379
	G. Mechanism of Phloem Flow	380
	H. Values for Components of the Phloem Water Potential	381
V.	**The Soil–Plant–Atmosphere Continuum**	**384**
	A. Values of Water Potential Components	384
	B. Resistances and Areas	386
	C. Specific Resistances and Conductances	389

D. Capacitance and Time Constants   391
E. Daily Changes   393
F. Global Climate Change   394
Problems   **397**
References   **399**

IN CHAPTER 8 we analyzed gas fluxes for single leaves. We repeatedly used Fick's first law in the following form: Flux density equals concentration (or mole fraction) difference divided by resistance, or, equivalently, conductance times concentration difference. This approach can be extended to an entire plant community. We first describe fluxes in the air above the plants. Although the fluxes of water vapor and CO_2 in the air above vegetation resemble diffusion, because net migration of these gases is toward regions of lower concentration, we are not dealing with the random thermal motion of molecules but rather with the random motion of relatively large packets of air in the turbulent region above the plants.

Our next task is to discuss concentrations and fluxes within a plant community. When we analyze water vapor and CO_2 fluxes from the soil up to the top of plants, we are confronted by the great structural diversity among different types of vegetation. Each plant community has its own unique spatial patterns for water vapor and CO_2 concentration. The possibility of many layers of leaves and the constantly changing illumination also greatly complicate the analysis. Even approximate descriptions of the gas fluxes within carefully selected plant communities involve complex calculations based on models incorporating numerous simplifying assumptions. We will consider a cornfield as a specific example.

Three-fourths of the water vaporized on land is transpired by plants. This water comes from the soil, which also affects the CO_2 fluxes for vegetation. Therefore, after we consider gas fluxes within a plant community, we will examine some of the hydraulic properties of soil. For instance, water in the soil is removed from larger pores before smaller ones. This removal decreases the soil conductivity for subsequent water movement, and a greater drop in water potential from the bulk soil up to a root is therefore necessary for a particular water flux density.

Our final topic will be the flow of water as a continuous stream from the soil, to the root, into the root xylem, up to the leaves, and eventually out through the stomata into the atmosphere. As a useful first approximation, the negative gradient of the water potential represents the driving force for the flux across any segment where water moves as a liquid. We usually replace $-\partial\Psi/\partial x$ by $\Delta\Psi/\Delta x$. The greater the resistance—or, alternatively, the lower the conductance—the larger the $\Delta\Psi$ required to maintain a given water flux across a particular component. However, $\Delta\Psi$ does not always represent the driving force on water. Furthermore, water movement in the xylem interacts with that in the other major transport system in plants, the phloem.

I. Gas Fluxes above the Leaf Canopy

When we considered the fluxes of H_2O and CO_2 for individual leaves in Chapter 8, we assumed that outside the boundary layers on each side of a leaf occurs a turbulent region

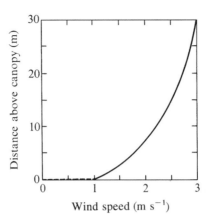

Figure 9-1. Change in wind speed with distance above a leaf at the top of a canopy. At the leaf surface air
motion is arrested and at a distance δ^{bl} (on the order of mm) v is 1.0 m s^{-1}. At 0.5 m into the
turbulent air, v increases to 1.1 m s^{-1}, and it can triple at 30 m above the canopy.

where both water vapor and CO_2 have specific concentrations. Actually, gradients in both
CO_2 and H_2O exist within this turbulent region around plants. We will also find that the
ambient wind speed is not constant but instead varies with distance above the vegetation.

A. Wind Speed Profiles

Because of frictional interactions between moving air and a leaf, the air immediately adjacent
to a leaf is stationary. As we move short distances away from the leaf surface, a transition
occurs from laminar flow parallel to the leaf in the lower part of the boundary layer to
turbulent flow with eddying motion (see Fig. 7-5). The wind speed increases as we move
further away from the leaf (Fig. 9-1), increasing approximately logarithmically for a few
hundred meters above the leaf canopy (the turbulent region generally extends 0.5–1.0 km
above the earth's surface, above which more or less laminar flow occurs in the direction of
the prevailing wind). Because wind speed varies in the turbulent air above vegetation, where
to measure the ambient wind speed is ambiguous. However, wind speed generally does not
increase substantially until we are at least 1 m above the foliage (Fig. 9-1). Thus wind speed
measured about 0.2 m above vegetation may be used as the ambient value, which is needed
to calculate the boundary layer thickness for an exposed leaf at the top of the canopy (e.g.,
using Eq. 7.8).

Plants exert a frictional drag on moving air masses and thereby modify the local wind
patterns (Fig. 9-1). The frictional interaction between trees and wind is different from that
of a flexible crop such as wheat, which leads to different form drag (Eq. 7.7) and different
wind patterns in the overlying turbulent air. Topographical features such as canyons or cliffs
also affect the local wind speed profile.

As just indicated, wind speed can increase approximately logarithmically with distance
above a canopy (Fig. 9-1) and it is also influenced by properties of the plants. The variation
in wind speed v with distance above a large, horizontal, uniform canopy under stable
atmospheric conditions can be

$$v = \frac{v^*}{k} \ln \frac{z - d}{z_0} \tag{9.1}$$

where v^* is termed the shearing or friction velocity, k is the von Karman constant (about 0.41), z is the height above the ground, d is the zero plane displacement, and z_0 is the roughness length. Generally, d is about 70% of the canopy height because most of the plant parts that produce form drag occur near that level. Although z_0 depends on the length of protrusions above the general canopy surface, it is often about 10% of the canopy height for dense vegetation and less for sparse vegetation. Equation (9.1) indicates that the extrapolated wind speed is zero at a height of $z_0 + d$ [the argument of the logarithm is then $(z_0 + d - d)/z_0$ or 1], although the actual wind speed at this height in the vegetation is nonzero, as we will discuss later (Eq. 9.1 is applicable only above the vegetation).

B. Flux Densities

During the daytime, a transpiring and photosynthesizing plant community as a whole can have a net vertical flux density of CO_2 (J_{CO_2}) toward it and a net vertical flux density of water vapor (J_{wv}) away from it into the turbulent air above the canopy. These flux densities are expressed per unit area of the ground or, equivalently, per unit area of the (horizontal) plant canopy. Each of the flux densities depends on the appropriate gradient. The vertical flux density of water vapor, for example, depends on the rate of change of water vapor concentration in the turbulent air, c_{wv}^{ta}, with respect to distance, z, above the vegetation:

$$J_{wv} = -K_{wv}\frac{\partial c_{wv}^{ta}}{\partial z} \tag{9.2}$$

In Eq. (9.2) we again employ the relation, flux density equals a proportionality coefficient times a force, where the force is the negative gradient of water vapor concentration. Because J_{wv} from the plant community can be expressed in mol m^{-2} s^{-1} and $\partial c_{wv}^{ta}/\partial z$ in mol m^{-4}, the coefficient K_{wv} in Eq. (9.2) can have units of m^2 s^{-1}, the same as for diffusion coefficients. In fact, K_{wv} in Eq. (9.2) is analogous to D_j in Fick's first law ($J_j = -D_j\partial c_j/\partial x$; Eqs. 1.1 and 8.2), except that it does not reflect the random thermal motion of water vapor molecules but rather the irregular swirling motion of packets, or eddies, of air in the turbulent region (Fig. 9-2). This makes the coefficient much larger than for molecular motion.

By analogy with Eq. (9.2), we can represent the flux density of CO_2 above a canopy as

$$J_{CO_2} = -K_{CO_2}\frac{\partial c_{CO_2}^{ta}}{\partial z} \tag{9.3}$$

where K_{CO_2} is the "air packet" or *eddy* diffusion coefficient for CO_2 (K_j is also referred to as a transfer coefficient, an exchange coefficient, or a diffusivity coefficient). Similarly, the vertical flux density of O_2 in the turbulent air, J_{O_2}, can be equated to $-K_{O_2}\partial c_{O_2}^{ta}/\partial z$. The positive direction for z is increasing altitude; consequently, the positive direction for a net flux density is from the plant canopy upward into the turbulent air.

C. Eddy Diffusion Coefficients

The eddy diffusion coefficients, K_{wv} and K_{CO_2}, unlike the ordinary diffusion coefficients, D_{wv} and D_{CO_2}, have the same value in a given situation. A small packet of air moves more or less as a unit, carrying with it all the H_2O, CO_2, and other molecules that it contains (see Fig. 9-2). Although we cannot assign actual volumes to these eddies—which are constantly changing in size and shape because of shearing effects or coalescence with neighboring packets—they are very large compared with intermolecular distances and contain enormous

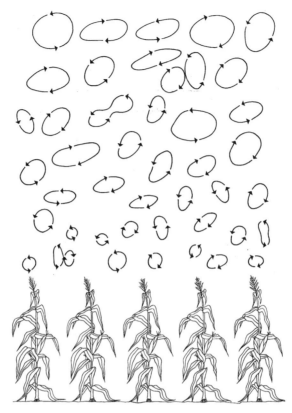

Figure 9-2. Schematic illustration of small packets or eddies of air swirling about in the turbulent region above vegetation. The eddies, which tend to increase in size with height, carry all the molecules that they contain more or less as a unit. They are continually changing—breaking up or coalescing with other eddies—making their actual size somewhat hypothetical.

numbers of molecules. The mean eddy size above vegetation has been approximated by $k(z - d)$ (parameters in Eq. 9.1). The random motion of an air packet is caused by random fluctuations in pressure in local regions of the turbulent air. The eddying motions of the air packets promote a mixing, formally like the mixing due to diffusion, and thus lead to relations such as Eqs. (9.2) and (9.3). Besides their eddying motion, air packets have an average drift velocity represented by the local wind velocity. Pressure gradients over large distances cause the winds and the resulting horizontal drift of the air packets.

Values for K_{wv} and K_{CO_2} describing the "diffusion" of air packets vary with the wind speed above the canopy. Also, eddy diffusion coefficients are affected by the rates of change of both wind speed and air temperature with altitude. For instance, hot air tends to rise and become replaced by cooler air. Such buoyancy effects, which are encouraged when $\partial T/\partial z$ is steeply negative above the canopy, lead to more rapid mixing and higher values for K_j. The eddy diffusion coefficients usually are approximately proportional to the local wind speed. As the wind speed increases, turbulent mixing of the air is more likely, and thus K_j becomes larger. Because wind speed varies with height (see Fig. 9-1), and because K_j also depends on the gradient in wind speed, we often employ an eddy diffusion coefficient averaged over an appropriate distance to describe the vertical fluxes in some region of the turbulent air above the canopy. Moreover, the wind speed, its gradient, and the vertical temperature gradient all vary during the day. Consequently, K_j should also be averaged over a suitable time interval, e.g., an hour.

For a moderate wind speed of 2 m s^{-1}, the eddy diffusion coefficient is generally 0.05–0.2 m^2 s^{-1} just above the plant canopy. Under these conditions, K_j might be about 2 m^2 s^{-1} at 30 m above the canopy and can exceed 5 m^2 s^{-1} at or above 300 m, where turbulent mixing is even greater. By comparison, D_{wv} is 2.4×10^{-5} m^2 s^{-1} and D_{CO_2} is 1.5×10^{-5} m^2 s^{-1} in air at 20°C. Thus K_j is 10^4 to 10^5 times larger in the turbulent air above the canopy than are these D_j's. The random motion of air packets is indeed much more effective than the random thermal motion of molecules in moving H_2O and CO_2.

Because K_j increases with altitude as we move into turbulent regions with higher wind speeds, the steady-state concentration gradients become less steep. Specifically, $J_{wv} = -K_{wv}\partial c_{wv}^{ta}/\partial z$ by Eq. (9.2), and because K_{wv} increases with altitude, the absolute value of $\partial c_{wv}^{ta}/\partial z$ becomes smaller at greater heights above the canopy. K_j may increase by a factor of 10 in the first 20 m above the vegetation, in which case the gradient in water vapor concentration becomes one-tenth as large over this interval.

The equality of eddy diffusion coefficients for different gaseous species results from an air packet and the molecules within it moving as a unit. In fact, K_j is often assumed to be the same for the transfer of gases, heat, and momentum (expressed in the same units), a relation that is referred to as the *similarity principle*. K_j is therefore generally measured for the most convenient quantity in some situation and is then assumed to be the same (or at least similar) for all others.

D. Resistance of Air above the Canopy

As for gaseous diffusion resistances in Chapter 8, we will identify a resistance to the flow of water vapor in the turbulent air by r_{wv}^{ta} and that for CO_2 by $r_{CO_2}^{ta}$. To derive such quantities, we will replace the negative gradient by the difference in concentration of species j, Δc_j^{ta}, across a distance, Δz, in the turbulent air, i.e., Eq. (9.2) becomes $J_{wv} = -K_{wv}\partial c_{wv}^{ta}/\partial z = K_{wv}\Delta c_{wv}^{ta}/\Delta z$. By analogy with our previous definition of resistance ($J_j/\Delta c_j^{bl} = D_j/\delta^{bl} = 1/r_j^{bl}$; Eq. 8.3), we can identify resistances from the flux density expressions in Eqs. (9.2) and (9.3). Because $K_{wv} = K_{CO_2}$, we obtain the following equalities:

$$r_{wv}^{ta} = \frac{\Delta c_{wv}^{ta}}{J_{wv}} = \frac{\Delta z}{K_{wv}} = \frac{\Delta z}{K_{CO_2}} = \frac{\Delta c_{CO_2}^{ta}}{J_{CO_2}} = r_{CO_2}^{ta} \qquad (9.4)$$

As with the analogous relations in Chapter 8 (e.g., Eqs. 8.3 and 8.5), Eq. (9.4) describes the steady-state condition. Equation (9.4) indicates that r_{wv}^{ta} has the same value as $r_{CO_2}^{ta}$, as we would indeed expect based on the random motions of whole packets of air (also, $g_{wv}^{ta} = 1/r_{wv}^{ta} = 1/r_{CO_2}^{ta} = g_{CO_2}^{ta}$).

We now estimate the resistance of the turbulent air immediately above a plant canopy. We will let K_j average 1.0 m^2 s^{-1} for the first 30 m above the plants, a typical value in a moderate wind during the daytime. By Eq. (9.4), the resistance over this 30-m interval then is

$$r_{wv}^{ta} = r_{CO_2}^{ta} = \frac{(30\,\text{m})}{(1\,\text{m}^2\,\text{s}^{-1})} = 30\,\text{s m}^{-1}$$

Measured values for this resistance generally range from 20 to 40 s m^{-1} for moderate wind speeds, as do predicted values from computer analyses of r_j^{ta} using models incorporating the variation of K_j with altitude. Wind speeds, and therefore K_j, tend to be lower at night, so r_j^{ta} tends to be somewhat higher then than during the daytime.

E. Transpiration and Photosynthesis

We express flux densities above plants per unit area of the ground or, equivalently, per unit area of the canopy. For many agricultural and ecological considerations, such a measure of the average transpiration or average photosynthesis of the whole plant community is more useful than is the water vapor or CO_2 flux densities of an individual leaf. Environmental measurements in the turbulent air above vegetation can thus indicate the overall rates of transpiration and photosynthesis, especially if the extent of similar plants is fairly large, as might occur for a cornfield or a grassland. Moreover, such measurements can generally be made without disturbing the plants or their leaves. On this large scale, however, we lose sight of certain factors, such as the effect of stomatal opening or leaf size on the gas fluxes. Also, the turbulent air above the canopy is greatly influenced by the terrain and by the vegetation, so we must reckon with other factors not involved in our study of leaves. For example, both K_j and the gradients in water vapor or CO_2 depend on whether we are at the edge or the center of a field, whether and what types of trees are present, and whether the region is flat or hilly.

For simplicity, we are considering a one-dimensional situation in which the net fluxes of water vapor and CO_2 occur only in the vertical direction above the canopy, as occurs near the center of a large uniform plant community. Just as our assumption of a boundary layer of uniform thickness breaks down at the leading and the trailing edges of a leaf, we must also consider air packets transferring H_2O and CO_2 horizontally in and out at the sides of vegetation. Such net horizontal transfer of various gases is referred to as *advection*. Instead of using Eqs. (9.2)–(9.4) to analyze net gas flux densities, we may then have to use much more cumbersome three-dimensional equations to handle advection for small fields or individual plants.

As we considered in Chapter 8 (Section VA), J_{wv}/J_{CO_2} can be about 400 H_2O/CO_2 for a sunlit mesophytic leaf and 200 for a photosynthetically efficient C_4 species such as corn (maize). For an entire plant community, however, the water lost per CO_2 fixed is generally considerably higher than that for a single, well-illuminated leaf. In particular, J_{wv} measured above the canopy also includes water vapor coming from the soil and from leaves that are not well illuminated and that therefore contribute little to the net photosynthesis. Some of the CO_2 taken up by the plant community is evolved by soil microorganisms, roots, and leaves that do not receive much sunlight and so are below light compensation—all of which decrease the amount of CO_2 that needs to be supplied from above the plant canopy. These effects tend to raise J_{wv}/J_{CO_2} above the values for an exposed leaf. Although the absolute value of J_{wv}/J_{CO_2} above a leaf canopy depends on the ambient relative humidity and the physiological status of the plants, it is generally between 400 and 2000 H_2O/CO_2 when averaged over a day in the growing season. Moreover, mainly because J_{CO_2} for C_4 species is often about twice as large as that for C_3 species, the absolute value of J_{wv}/J_{CO_2} is lower and daily growth tends to be greater for C_4 species.

F. Values for Fluxes and Concentrations

We will use representative values of J_{CO_2} to calculate the decreases in concentration of CO_2 that can occur over a certain vertical distance in the turbulent atmosphere. When a net CO_2

uptake is occurring, the flux density of CO_2 is directed from the turbulent air down into the canopy. J_{CO_2} above the vegetation is then negative by our sign convention, which means that $c_{CO_2}^{ta}$ increases as we go vertically upward ($J_{CO_2} = -K_{CO_2} \partial c_{CO_2}^{ta}/\partial z$; Eq. 9.3). This is as we would expect if CO_2 is to be transferred downward toward the plants by the random motion of eddies in the turbulent air.

J_{CO_2} above a plant canopy can be -20 μmol m^{-2} s^{-1} at midday. For comparison, the flux density of CO_2 into an exposed leaf of a mesophyte at a moderate light level can be 11 μmol m^{-2} s^{-1} (Chapter 8, Section IVE). Using Eq. (9.4), an analogy with Ohm's law ($\Delta E = IR$; namely, $\Delta c_j^{ta} = J_j r_j^{ta}$), and a resistance of 30 s m^{-1} for the lower 30 m of the turbulent air ($r_{CO_2}^{ta}$), we calculate that the decrease in CO_2 concentration across this region is

$$\Delta c_{CO_2}^{ta} = J_{CO_2} r_{CO_2}^{ta} = (-20 \times 10^{-6} \text{ mol m}^{-2} \text{ s}^{-1})(30 \text{ s m}^{-1})$$

$$= -0.60 \text{ mmol m}^{-3}$$

Employing a conversion factor from Table 8-2, this $\Delta c_{CO_2}^{ta}$ corresponds to a CO_2 mole fraction difference of $(0.60)(24.4)$, or 15 μmol mol^{-1} at 20°C and an air pressure of 0.1 MPa. Thus CO_2, which might have a mole fraction near 380 μmol mol^{-1} well into the turbulent air, e.g., 30 m above vegetation, could be at $380 - 15$, or 365 μmol mol^{-1} just above the canopy.

For a rapidly photosynthesizing corn crop at noon, J_{CO_2} can be -60 μmol m^{-2} s^{-1}. For the previous $r_{CO_2}^{ta}$, $\Delta c_{CO_2}^{ta}$ corresponds to -44 μmol mol^{-1}, so the CO_2 mole fraction at the top of the canopy can be 44 μmol mol^{-1} lower than the 380 μmol mol^{-1} in the turbulent air tens of meters above the corn plants. In fact, measurement of $c_{CO_2}^{ta}$ at the canopy level indicates the net rate of photosynthesis by the plants. At night respiration occurs, but not photosynthesis, so vegetation then acts as a source of CO_2. Thus the mole fraction of CO_2 just above the canopy at night is usually a few μmol mol^{-1} greater than it is higher up in the turbulent air, as we will discuss later.

The flux density of water vapor just above the canopy, which includes transpiration from the leaves plus evaporation from the soil, is often termed *evapotranspiration*. For fairly dense vegetation and a moist soil, evapotranspiration is appreciable, generally amounting to 60–90% of the flux density of water vapor from an exposed water surface, such as a lake, at the ambient air temperature. The daily evapotranspiration from a forest is often equivalent to a layer of water 3–5 mm thick, which averages 2–3 mmol m^{-2} s^{-1}. At noon on a sunny day with a moderate wind, J_{wv} above a leaf canopy can be 7 mmol m^{-2} s^{-1}. Using Eq. (9.4) ($\Delta c_{wv}^{ta} = J_{wv} r_{wv}^{ta}$) and employing our value for r_{wv}^{ta} of 30 s m^{-1}, we note that the water vapor concentration decreases by $(7 \times 10^{-3}$ mol m^{-2} s$^{-1})(30$ s m$^{-1})$ or 0.21 mol m^{-3} over the first 30 m of the turbulent air. We indicated in Chapter 8 that the turbulent air immediately outside the boundary layer adjacent to a leaf contains 0.48 mol water m^{-3} when it is at 20°C and 50% relative humidity (see Fig. 8-6). Our calculation indicates that c_{wv}^{ta} could drop by 0.21 mol m^{-3}, which means from 0.48 to 0.27 mol m^{-3} (28% relative humidity at 20°C), as we move 30 m upward into the turbulent air above the canopy. Such a decrease in absolute and relative humidity is generally observed in the turbulent air above vegetation.

Because evapotranspiration is important both ecologically and agronomically, various methods have been developed for estimating it, including approaches based on measurements made above the plant canopy. Just as for a leaf (see Chapter 7), the main energy input into the plant canopy generally is from net radiation (see Eq. 7.6), and the main energy losses are from sensible heat (heat conduction across a boundary layer followed by convection to the turbulent air; see Eqs. 7.11–7.14) and latent heat loss accompanying water evaporation

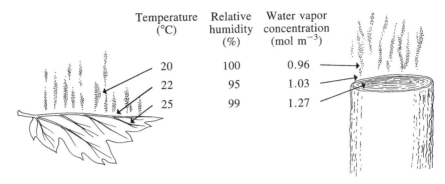

Temperature (°C)	Relative humidity (%)	Water vapor concentration (mol m^{-3})
20	100	0.96
22	95	1.03
25	99	1.27

Figure 9-3. "Steam" rising from a leaf and a wooden fencepost that are rapidly heated by the sun after a rainstorm. Moisture-laden air next to the objects is swept in an eddying motion into a cooler region, where the water vapor condenses.

(Eq. 7.19). The ratio of the flux density of sensible heat to the air to the flux density of latent heat loss from the soil and the plant canopy is referred to as the *Bowen ratio*, based on work by Ira Bowen in the 1920s. When the evaporation rate is low, such as when water availability is limited, the Bowen ratio, β, tends to be high. For instance, β is about 0.1 for tropical oceans, 0.2 for tropical rain forests, 0.4–0.8 for temperate forests and grasslands, 2–6 for semiarid regions, and 10 for deserts. Incorporating β into the energy budget relation for an entire plant canopy allows the determination of the rate of evapotranspiration from measurements of the net radiation, the heat flux density into the soil, and the gradients in temperature and water vapor concentration above the canopy. Evapotranspiration can thus be estimated to within a few percent of measured values for grasses and other short crops as well as for forests using an energy budget analysis incorporating the Bowen ratio.

G. Condensation

What appears to be steam is often seen rising from leaves or other surfaces (Fig. 9-3) when the sun breaks through the clouds following a rainstorm or at sunrise following a night with a heavy dew. To analyze this phenomenon, we will assume that the sun warms the leaves at the top of the canopy to 25°C and that the concentration of water vapor in the cell wall pores of their mesophyll cells is then 1.27 mol m^{-3} (see Chapter 8, Section IIE). Suppose that the air just outside the boundary layer adjacent to a leaf is at 22°C and has a high relative humidity of 95% just after the rainstorm; c_{wv}^{ta} is then 1.03 mol m^{-3} ($c_{wv}^* = 1.08$ mol m^{-3} at 22°C). Hence, water vapor will diffuse from the leaf, across the boundary layer, and into the turbulent air (Fig. 9-3). Now suppose that the air at a greater distance from the leaf is somewhat cooler, e.g., 20°C at 10 mm from the leaf. At 20°C c_{wv}^* is 0.96 mol m^{-3}. Thus as the air with 1.03 mol water m^{-3} moves away from the unstirred layer adjacent to the leaf in an eddy, or air parcel, it will be cooled and some of its water vapor will condense because c_{wv}^{ta} cannot exceed c_{wv}^* for the local air temperature. This condensation leads to the fog, or "steam," seen moving away from plants or a fence post into the surrounding cooler turbulent air (Fig. 9-3). As we move even further away to regions of lower water vapor concentration, the condensed water evaporates and so the steam disappears.

II. Gas Fluxes within Plant Communities

The pattern of concentrations and fluxes within vegetation varies with the plant community. Airflow within communities also depends on the three-dimensional architecture of the plants. For instance, wind speed does not necessarily decrease as we move toward the ground—air in an open forest can "tunnel" under the branches and hence the wind speed can be greater there than further up in the canopy. We will not attempt to examine all types of vegetation but instead will focus on a cornfield representing a monospecific stand of high productivity. The same general principles apply to other fairly uniform plant communities, but isolated plants provide special difficulty for analysis because the gas concentrations and fluxes then vary in three dimensions.

A. Eddy Diffusion Coefficient and Resistance

We begin by considering how the eddy diffusion coefficient might vary within a plant community. Near the ground a thick boundary layer can occur because the air there is generally quite still. In fact, K_j often averages 5×10^{-5} m^2 s^{-1} over the first 10 mm above the ground, a value only two or three times larger than the diffusion coefficients of water vapor and CO_2 in air. When K_j is of the same order of magnitude as diffusion coefficients, differences in movement between molecular species can become apparent, which are ignored here. As we move to the top of the canopy, the eddy diffusion coefficient increases, often more or less logarithmically with height in the upper part of many plant communities. It may reach 0.2 m^2 s^{-1} at the top of a canopy in a moderate wind (Fig. 9-4). Because K_j is approximately proportional to wind speed, v^{wind} within the plant community varies in a manner similar to the variation described for the eddy diffusion coefficient (Fig. 9-4). For instance, the wind speed about 0.2 m above the ground might be 0.1 m s^{-1}, increasing to 2 m s^{-1} at the top of the canopy.

Two aspects concerning K_j within a plant community deserve special emphasis. First, transfer of gaseous substances within the vegetation takes place by the random motion of relatively large parcels or eddies of air, just as in the turbulent region above the canopy.

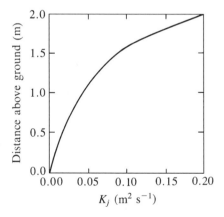

Figure 9-4. Idealized representation of the variation in the eddy diffusion coefficient within a uniform corn crop 2 m in height. The wind speed is 2 m s^{-1} at the top of the canopy.

Table 9-1. Summary of Gas Exchange Parameters and Flux Densities within a 2-m-Tall Corn Crop at Noon on a Sunny Day[a]

Height above ground (m)	$\bar{K}_j$ (m² s⁻¹)	r_j^{ta} (s m⁻¹)	$\bar{J}_{wv}$ (mmol m⁻² s⁻¹)	Δc_{wv} (mol m⁻³)	$\bar{J}_{CO_2}$ (μmol m⁻² s⁻¹)	Δc_{CO_2} (mmol m⁻³)	ΔN_{CO_2} (μmol mol⁻¹)
Above							
canopy			7		−60		
1–2	1×10^{-1}	10	4	0.04	−30	−0.30	−7
0.1–1	2×10^{-2}	50	1	0.05	3	0.15	4
0.01–0.1	1×10^{-3}	90	0.5	0.05	2	0.18	4
0.00–0.01	5×10^{-5}	200	0.5	0.10	2	0.40	10
Ground							
level			0.5		2		

[a] Bars over K_j, J_{wv}, and J_{CO_2} indicate values averaged over the height increment involved (source: Fig. 9-4; Lemon et al., 1971).

Second, because of frictional drag with the many leaves, branches, and other plant parts, the eddy diffusion coefficient within the vegetation is considerably less than in the air above the canopy.

To illustrate the relative contributions of various air layers from the ground to the top of a corn corp 2 m in height (Fig. 9-4), we will let K_j have specific average values for various height intervals above the ground (Table 9-1), consistent with the plot of K_j versus height in Fig. 9-4. We can use Eq. (9.4), $r_{wv}^{ta} = \Delta z / K_{wv} = r_{CO_2}^{ta}$, to estimate the resistance of each of the four air layers in series, e.g., r_j^{ta} for the lowermost layer is $(0.01 \text{ m})/(5 \times 10^{-5}$ m² s⁻¹), or 200 s m⁻¹. The total resistance from the ground up to the top of the canopy equals $200 + 90 + 50 + 10$, or 350 s m⁻¹ (summarized in Table 9-1). (In a sense, we are performing a numerical integration to determine the resistance.) Computer analyses using models describing the turbulent air within such a crop also indicate that the resistance is generally 300–400 s m⁻¹. Most of the resistance within a plant community is generally due to the relatively still air next to the ground. For instance, just over half (200 of 350 s m⁻¹) of the resistance for the 2-m pathway is provided by the lowest 0.01 m, whereas the entire upper half of the corn crop accounts for a resistance of only 10 s m⁻¹ (see Table 9-1).

B. Water Vapor

A considerable amount of water can evaporate from the soil and move by air packets up through the vegetation. For instance, J_{wv} from a moist, intermittently illuminated soil, such as commonly occurs in a temperate forest, can be 0.2–1.0 mmol m⁻² s⁻¹. (For comparison, 0.5 mmol m⁻² s⁻¹ corresponds to a depth of water of 0.8 mm/day or 280 mm/year.) If a flow of 0.5 mmol m⁻² s⁻¹ occurs across a resistance of 290 s m⁻¹ to reach a distance 0.1 m above the ground, using Eq. (9.4) (by which $\Delta c_{wv}^{ta} = J_{wv} r_{wv}^{ta}$) we calculate that the drop in water vapor from the ground to this level is $(0.5 \times 10^{-3}$ mol m⁻² s⁻¹) (290 s m⁻¹) or 0.15 mol m⁻³. At 20°C the saturation water vapor concentration is 0.96 mol m⁻³ (see Appendix I), so a water vapor drop of 0.15 mol m⁻³ then corresponds to a 16% decrease in relative humidity. Thus an appreciable drop in water vapor concentration can occur across the relatively still air near a moist soil under a canopy (Fig. 9-5).

Because of water vapor transpired by the leaves, J_{wv} increases as we move from the ground up through a corn crop. On a sunny day the water vapor flux density might be 1 mmol m⁻² s⁻¹ at 0.5 m, 2 mmol m⁻² s⁻¹ at 1.0 m, 4 mmol m⁻² s⁻¹ at 1.5 m, and

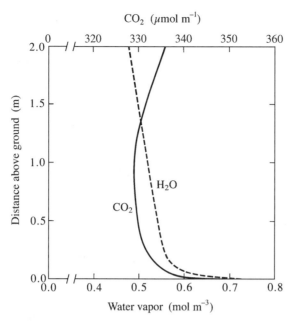

Figure 9-5. Possible variation of water vapor and CO_2 concentrations within a 2-m-high corn crop at noon on a sunny day. At the top of the canopy, the wind speed is 2 m s^{-1}. In the turbulent air 30 m above the vegetation, c_{wv}^{ta} is 0.27 mol m^{-3} and $N_{CO_2}^{ta}$ is 380 μmol mol^{-1} (for values under field conditions, see Lemon *et al.*, 1971). Such variations of atmospheric CO_2 and H_2O with height need to be taken into account in canopy gas-exchange models that deal with leaves in various layers.

7 mmol m^{-2} s^{-1} at 2.0 m, the top of the canopy. (On a cloudy humid day, J_{wv} for a corn crop might be only 1 mmol m^{-2} s^{-1} at the top of the canopy.) If J_{wv} averages 1 mmol m^{-2} s^{-1} from 0.1 to 1.0 m above the ground, where the resistance is 50 s m^{-1} (Table 9-1), then by Eq. (9.4) Δc_{wv}^{ta} for this part of the pathway is (1 mmol m^{-2} s^{-1}) (50 s m^{-1}) or 0.05 mol m^{-3}. For the upper 1 m the resistance is 10 s m^{-1}, and so, for an average J_{wv} of 4 mmol m^{-2} s^{-1}, the decrease in water vapor concentration is 0.04 mol m^{-3} (Table 9-1). Thus Δc_{wv}^{ta} is 0.15 mol m^{-3} over the 0.1 m just above the ground, 0.05 mol m^{-3} from 0.1 to 1.0 m, and 0.04 mol m^{-3} from 1.0 to 2.0 m, or 0.24 mol m^{-3} overall (see Table 9-1 and Fig. 9-5). If the turbulent air at the top of the canopy is at 20°C and 50% relative humidity, it contains 0.48 mol water m^{-3}. The air near the soil then contains approximately 0.48 + 0.24, or 0.72, mol H_2O m^{-3}, which corresponds to 75% relative humidity at 20°C (c_{wv}^* = 0.96 mol m^{-3} at 20°C; see Appendix I). In summary, we note that (1) air close to the soil under a (fairly dense) plant canopy can have a high relative humidity, (2) c_{wv}^{ta} continuously decreases as we move upward from the ground, (3) most of the overall drop in water vapor concentration occurs near the ground, and (4) most of the water vapor comes from the upper half of the corn crop in the current example.

C. Attenuation of Photosynthetic Photon Flux Density

Before discussing J_{CO_2} within a plant community, we need to consider how the amount of light varies down through the various layers of vegetation because the photosynthetic photon flux density (PPFD; wavelengths of 400–700 nm) at each level helps determine the rate of photosynthesis there. The net rate of CO_2 fixation approaches light saturation near a PPFD of 600 μmol m^{-2} s^{-1} for leaves of many C_3 plants, and it decreases to zero at light compensation. A comprehensive formulation—including effects of leaf angle, sun elevation in the sky, the finite width of the sun's disc, changes in spectral distribution of PPFD at various levels within the plant community, multiple reflections from leaves and other surfaces, and clumping versus uniform arrangement of leaves—leads to a nearly hopeless complication of the algebra. Instead, we will assume that the PPFD decreases due to absorption by the foliage in a manner analogous to Beer's law, $\ln(J_b/J_0) = -k_\lambda cb$ (Eq. 4.17). This approximation is particularly useful when the leaves are randomly distributed horizontally, as can occur in certain moderately dense plant communities.

As we move downward into the vegetation, the PPFD decreases approximately exponentially with the amount of absorbing material encountered. For some canopies the greatest leaf area per interval of height occurs near the center (e.g., many grasses), and for others it occurs about three-fourths of the way up from the ground (e.g., many crops and trees). We will let F be the average cumulative total leaf area per unit ground area as we move down through the plant community. The dimensionless parameter F uses the area of only one side of a leaf and thus is expressed on the same basis as are our flux densities of H_2O and CO_2. F is zero at the top of a canopy and takes on its maximum value at ground level, a value generally referred to as the *leaf area index*. If the leaves in a particular plant community were all horizontal, the leaf area index would equal the average number of leaves above any point on the ground. In any case, the leaf area index equals the leaf area per plant divided by the ground area per plant.

We will represent the PPFD, J, incident on a leaf canopy by J_0. Primarily because of absorption by photosynthetic pigments, PPFD is attenuated down through the plant community. At any level in the vegetation, J is related to J_0 and F as follows:

$$\ln \frac{J_0}{J} = kF \qquad (9.5)$$

where k is a dimensionless parameter describing the absorption properties of a particular type of foliage and is referred to as the *foliar absorption coefficient*. Because we are ignoring changes in spectral distribution at different levels in the vegetation, J_0 and J in Eq. (9.5) can represent the flux density of photons from 400 to 700 nm or an energy flux density for these photons. Equation (9.5) was introduced into plant studies by Masami Monsi and Toshiro Saeki in 1953.

D. Values of Foliar Absorption Coefficients

The foliar absorption coefficient k ranges from 0.3 to 1.3 for most plant canopies. Light penetrates the vertically oriented blades of grasses rather easily; in such cases, k can be near 0.4. What cumulative leaf area per ground area reduces the incident PPFD by 95% for grasses with such a foliar absorption coefficient? By Eq. (9.5), the accumulated leaf area per unit ground area in this case is

$$F = \frac{\ln\left(\frac{J_0}{0.05 J_0}\right)}{0.4} = 7.5$$

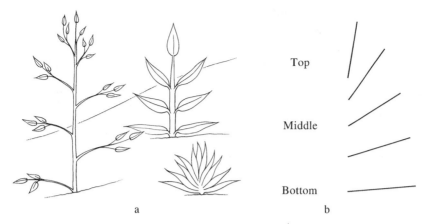

Figure 9-6. Variation in leaf angle and hence foliar absorption coefficient with distance above the ground for (a) various idealized plants and (b) sugar beet measured at various canopy positions (Hodáňová, 1979). The greater erectness of the uppermost leaves leads to a lower k for them, and hence better penetration of PPFD down to the lower leaves.

When the average leaf area index is 7.5 for such grasses, 5% of the PPFD reaches the soil surface. For 95% of the PPFD to be absorbed for a leaf area index of 3, k equals $\ln(1/0.05)/3$ or 1.0 by Eq. (9.5). Such a high foliar absorption coefficient applies to horizontal leaves with at least 0.5 g chlorophyll m^{-2}, which can occur for crops such as potato, soybean, sunflower, and white clover.

When the sun is overhead, vertical leaves absorb less sunlight and reflect more of it down into the vegetation per unit leaf area than do horizontal leaves. This accounts for the low values of k for grasses because their leaves are generally erect. For certain plants, leaves tend to be vertical near the top, e.g., sugar beet, becoming on average more horizontal toward the ground (Fig. 9-6; in other cases, leaves at the top of the canopy are more pendant, again becoming more horizontal toward the ground). This orientation reduces the foliar absorption coefficient of the upper leaves, so more of the light incident on the plants is available for the lower leaves. In fact, optimal light utilization for photosynthesis generally occurs when the incident PPFD is distributed as uniformly as possible over the leaves because the fraction of leaves exposed to PPFD levels approaching light saturation or below light compensation is then usually minimized. Our arguments about the effect of leaf orientation on k presuppose that essentially all the light is incident on the top of the canopy. When much PPFD comes in from the sides, as for an isolated tree, or in the early morning or late afternoon, foliar absorption coefficients determined for vertically incident light should not be used in Eq. (9.5)—indeed, k can be determined for other sun angles. Also, a foliar absorption coefficient can be determined for shortwave irradiation, instead of just for PPFD.

E. Light Compensation Point

We next consider the light compensation point for CO_2 fixation by leaves. As we mentioned in Chapter 8 (Section IVD), light compensation generally occurs at a PPFD of about 10 μmol m^{-2} s^{-1} for a leaf temperature near 20°C and a CO_2 concentration of 350 μmol mol^{-1} ($c_{CO_2}^{ta}$ generally is somewhat below this level within a plant community, as we will see later). Suppose that a moderate PPFD of 400 μmol m^{-2} s^{-1} occurs on trees whose leaves have a foliar absorption coefficient of 0.8. At what cumulative area of leaves per unit ground area is a light compensation point of 10 μmol m^{-2} s^{-1} reached? By Eq. (9.5), F is then $\ln(400/10)/0.8$, or 4.6. Thus only the upper five "layers" of leaves in a dense forest might be above light compensation for that part of the day when the PPFD on the canopy is 400 μmol m^{-2} s^{-1}. Of course, for a lower PPFD on the plant canopy, more leaves are below light compensation. Occasional sunflecks of high PPFD reach the lower parts of the vegetation, complicating our analysis of where light compensation occurs.

Leaves that are below light compensation for most of the day do not contribute to the net photosynthesis of the plant. Such leaves generally lose 20–50% of their dry weight before dying and abscising. Following this loss of leaves on the lower branches of trees, the branches themselves die and eventually fall off or are blown off by the wind. Thus tall trees in a dense forest often have few or no branches on the lower part of their trunks.

F. CO$_2$ Concentrations and Fluxes

In contrast to the concentration of water vapor, which continuously decreases with increasing distance above the ground, on a sunny day the CO$_2$ concentration generally achieves a minimum somewhere within the plant community (Fig. 9-5). This occurs because both the turbulent air above the canopy and the soil can serve as sources of CO$_2$. During the day CO$_2$ diffuses toward lower concentrations from the soil up into the vegetation and from the overlying turbulent air down into the leaf canopy.

Respiration in root cells and in soil microorganisms can lead to a net upward CO$_2$ flux density from the ground of 1–3 μmol m^{-2} s^{-1} during the growing season. (An O$_2$ flux density of similar magnitude occurs in the opposite direction.) J_{CO_2} from the soil varies in phase with the soil temperature, which is higher during the daytime (see Chapter 7). We have already estimated that $r_{CO_2}^{ta}$, which is the same as r_{wv}^{ta}, might be 200 s m^{-1} for the first 0.01 m and 90 s m^{-1} for the next 0.09 m above the ground for a corn crop that is 2 m tall (Table 9-1). Using Eq. (9.4) ($\Delta c_{CO_2}^{ta} = J_{CO_2} r_{CO_2}^{ta}$), we calculate, for a moderate CO$_2$ flux density of 2 μmol m^{-2} s^{-1} emanating from the soil, that the decrease in CO$_2$ concentration across the first 0.01 m above the ground is $(2 \times 10^{-3}$ mmol m^{-2} s$^{-1})$ $(200$ s m$^{-1})$ or 0.40 mmol m^{-3}, which by the conversion factor in Table 8-2 represents a mole fraction drop of 10 μmol CO$_2$ mol^{-1} at 20° C and 0.1 MPa air pressure. In the next 0.09 m, the CO$_2$ concentration might decrease by 0.18 mmol m^{-3}, which corresponds to about 4 μmol mol^{-1} (Table 9-1). Thus the CO$_2$ level might decrease by 14 μmol mol^{-1} from 347 μmol mol^{-1} at the soil surface to 333 μmol mol^{-1} at 0.1 m above the ground (Fig. 9-5).

As we move further upward from the ground, the flux density of CO$_2$ initially increases as we encounter leaves that are below light compensation and thus have a net evolution of CO$_2$. For instance, J_{CO_2} directed upward may increase from 2 μmol m^{-2} s^{-1} at 0.1 m to 5 μmol m^{-2} s^{-1} at 0.5 m. The maximum upward J_{CO_2} occurs at the canopy level where light compensation occurs. As we move even higher and encounter leaves with net photosynthesis, the net flux density of CO$_2$ in the turbulent air decreases, and it may become zero at 1.0 m above the ground in a 2-m-tall corn crop with a high photosynthetic rate. Because $J_{CO_2} = -K_{CO_2} \partial c_{CO_2}^{ta}/\partial z$ (Eq. 9.3), J_{CO_2} is zero (no net CO$_2$ flux upward or downward) when $\partial c_{CO_2}^{ta}/\partial z$ is zero, which corresponds to the local minimum in CO$_2$ concentration (Fig. 9-5). J_{CO_2} may average 3 μmol m^{-2} s^{-1} from 0.1 to 1.0 m above the ground, an interval that has a resistance of 50 s m^{-1} (Table 9-1). This would lead to a $\Delta c_{CO_2}^{ta}$ of 0.15 mmol m^{-3}, which corresponds to a CO$_2$ mole fraction decrease of 4 μmol mol^{-1}.

Hence, the CO_2 concentration may reach its lowest value of 329 μmol mol^{-1} midway through the crop (see Fig. 9-5).

At noon on a sunny day, J_{CO_2} down into a cornfield might be 60 μmol m^{-2} s^{-1}. Essentially all the net CO_2 flux from the turbulent air above the canopy is directed into the leaves in the upper half of the corn crop; e.g., J_{CO_2} may become -30 μmol m^{-2} s^{-1} at 1.5 m above the ground and zero at 1.0 m. Thus the average CO_2 flux density in the upper half of the vegetation is about -30 μmol m^{-2} s^{-1}, and the resistance is 10 s m^{-1} (Table 9-1). Conseqently, $\Delta c_{CO_2}^{ta}$ for this upper portion of an actively photosynthesizing cornfield might be -0.30 mmol m^{-3}, which corresponds to a 7 μmol mol^{-1} decrease in CO_2 from the top of the canopy to 1.0 m below (Table 9-1; Fig. 9-5). We indicated in Section IF that $N_{CO_2}^{ta}$ at the top of this canopy might be 336 μmol mol^{-1}. The CO_2 mole fraction at 1.0 m above the ground then is 329 μmol mol^{-1}, the same value that we estimated by working our way up from the ground (see Fig. 9-5). Two-thirds or more of the net photosynthesis generally occurs in the upper one-third of most canopies, as it does here for corn.

CO_2 concentrations in the air can vary over a wide range for different plant communities. For a cornfield exposed to a low wind speed (<0.3 m s^{-1} at the top of the canopy), for a rapidly growing plant community, or for other dense vegetation where the eddy diffusion coefficient may be relatively small, the CO_2 mole fraction in the turbulent air within the plant stand can drop below 200 μmol mol^{-1} during a sunny day. On the other hand, for sparse desert vegetation, especially on windy or overcast days, $N_{CO_2}^{ta}$ generally does not decrease even 1 μmol mol^{-1} from the value at the top of the canopy.

G. CO_2 at Night

The CO_2 concentration at night is highest near the ground and continuously decreases as we go upward through the plants into the turbulent air above. J_{CO_2} emanating from the soil might be 1 μmol m^{-2} s^{-1}, and the respiratory flux density of CO_2 from the above-ground parts of plants can be 3 μmol m^{-2} s^{-1} at night. Respiration averaged over a 24-h period can be 20% of gross photosynthesis for a rapidly growing plant community and can increase to over 50% as the community matures. For certain climax communities, respiration can become nearly 100% of gross photosynthesis. When considered over a growing season, respiration for an entire plant is generally 30–50% of gross photosynthesis for crops.

We will next estimate the decrease in CO_2 concentration that might occur in a 2-m-high corn crop at night. For purposes of calculation, we will assume that J_{CO_2} vertically upward increases from 1 μmol m^{-2} s^{-1} near the ground to 4 μmol m^{-2} s^{-1} at the top of the cornfield and that the lower wind speeds at night lead to somewhat higher resistances than the daytime values (see Fig. 9-7 and Table 9-1). Using Eq. (9.4) ($\Delta c_{CO_2}^{ta} = J_{CO_2} r_{CO_2}^{ta}$), the drop in CO_2 concentration from the ground to the top of the canopy is $0.41 + 0.16 + 0.05$ or 0.62 mmol m^{-3}, which corresponds to 15 μmol CO_2 mol^{-1} at 20°C (Fig. 9-7; conversion factor in Table 8-2). The resistance of the first 30 m of turbulent air above the plants might be 50 s m^{-1} at night (compare the value of 30 s m^{-1} during the daytime that we used earlier). If we let the total CO_2 flux density from the canopy be 4 μmol m^{-2} s^{-1}, then $\Delta c_{CO_2}^{ta}$ for the first 30 m above the canopy would be 0.20 mmol m^{-3}, which corresponds to 5 μmol CO_2 mol^{-1}. Therefore, assuming that $N_{CO_2}^{ta}$ is 380 μmol mol^{-1} 30 m up in the turbulent air, the CO_2 mole fraction at night would be 385 μmol mol^{-1} at the top of the canopy and 400 μmol mol^{-1} at the soil surface (Fig. 9-7).

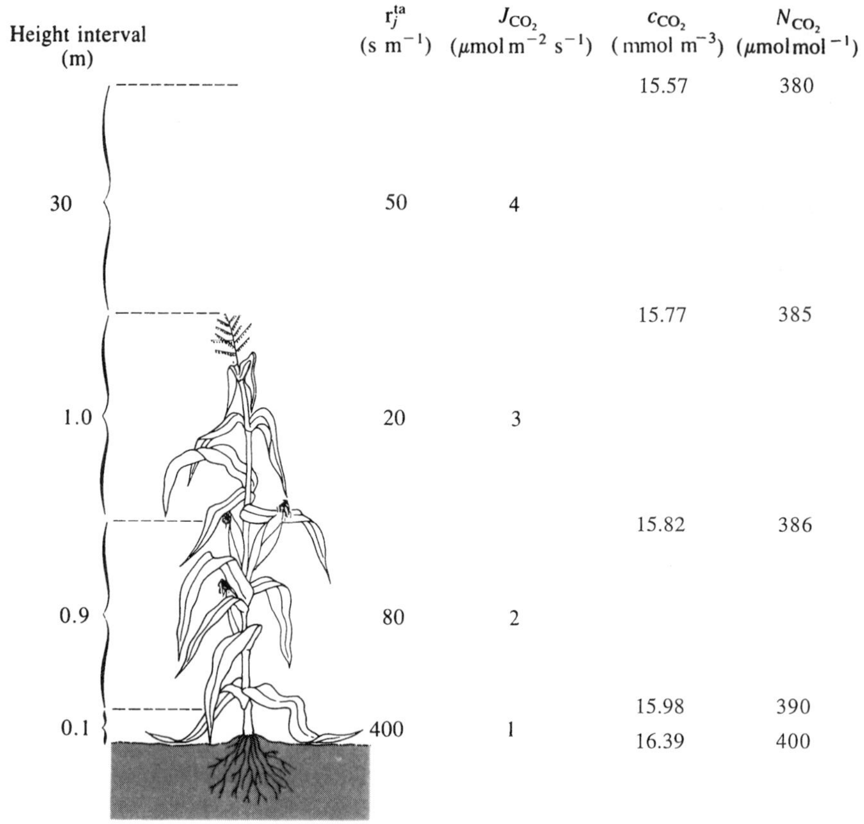
Height interval (m)
r_j^{ta} (s m^{-1})
J_{CO_2} (μmol m^{-2} s^{-1})
c_{CO_2} (mmol m^{-3})
N_{CO_2} (μmol mol^{-1})
15.57 380
30 50 4
15.77 385
1.0 20 3
15.82 386
0.9 80 2
15.98 390
0.1 400 1
16.39 400

Figure 9-7. CO_2 resistances, flux densities, and concentrations within and above a cornfield at night.

III. Soil

Soils vary tremendously in their physical properties, such as the size of individual particles. In a sandy soil many particles are over 1 mm in diameter, but in a clay soil most particles are $< 2\ \mu$m in diameter. In particular, clay refers to particles $< 2\ \mu$m across, silt to particle sizes from 2 to 50 μm, and sand to larger particles up to 2 mm across; after removing even larger particles, loosely termed gravel, the particle size distribution determines the soil textural class, e.g., "loam" is about 20% clay, 40% silt, and 40% sand by mass. Small particles have a much greater surface area per unit mass than do large particles. Sand, for example, can have under 1 m^2 of surface area/g, whereas most clays have 100–1000 m^2 of surface area/g. Most soil minerals are aluminosilicates, with negatively charged surfaces that act as Donnan phases (see Chapter 3, Section IIF) with the mobile cations in the adjacent soil water. Because of their large surface areas per unit mass, clays dominate the ion exchage properties of many soils. For example, the clay soil montmorillonite has 800 m^2 of surface area/g and can hold nearly 1 mmol of monovalent cations/g. Nutrient concentrations vary tremendously with soil type, which affects the ion exchange capacity and water content; for moist agricultural soil, phosphate can be about 2 μM and K^+ and NO_3^- about 2 mM (Ca^{2+}, Mg^{2+}, Na^+, and Cl^- can be even higher in concentration).

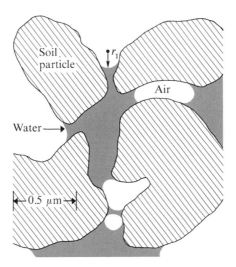

Figure 9-8. Schematic indication of the gas, liquid, and solid phases in a soil. The radii of curvature at the air–water interfaces can be related to the negative hydrostatic pressures or tensions in the liquid phase.

The irregularly shaped pores between soil particles contain both air and water (Fig. 9-8). The soil pores, or voids, vary from just under 40% to about 60% of the soil by volume. Thus a soil whose pores are completely filled with water contains 40–60% water by volume. In the vicinity of most roots, moist soil contains 8–30% water by volume, and the rest of the pore space is filled with air. The pores therefore provide many air–liquid interfaces where surface tension effects can lead to a negative hydrostatic pressure in the soil water. Such a negative P is generally the main contributor to the water potential in the soil, especially as the soil dries. The thermal properties of soil were discussed in Chapter 7, so we will focus here on soil water relations.

A. Soil Water Potential

As just indicated, the predominant influence on the soil water potential is usually the many air–liquid interfaces in the soil (Fig. 9-8). The soil water also contains dissolved solutes, which generally lead to an osmotic pressure (Π^{soil}) of 0.01–0.2 MPa for moist soil; the magnitude of Π^{soil} depends on the water content of the soil, which varies greatly. Because of the relatively low osmotic pressures in wet soil, we will often refer to the soil solution as water. In contrast to the generally small values for Π^{soil}, the many interfaces present in the small soil pores can lead to extremely negative hydrostatic pressures as the soil dries. A few days after being saturated by rainfall, a wet clay soil might retain 40% water by volume ("field capacity") and have a soil water potential (Ψ^{soil}) of -0.01 MPa, whereas permanent wilting from which many crops will not recover occurs when Ψ^{soil} is about -1.5 MPa and the volumetric water content of the clay soil is 15%. Field capacity and permanent wilting occur at about 30 and 10% water by volume, respectively, for loam, and 10 and 3% for sand.

If the arc formed by the intersection of a surface and a plane perpendicular to it is continued around to form a circle, the radius of the circle is the radius of curvature of the surface,

r. A slightly curved surface has a large r, and r becomes infinite if the surface is flat in a particular direction. By convention, r is positive for a concave surface as viewed from the air side (see r_1 in Fig. 9-8) and negative for a convex surface. In particular, the surfaces of the air–water interfaces in the pores between the soil particles are usually concave when viewed from the air side, just as for capillaries (see Fig. 2-3). However, the surfaces generally are not spherical or otherwise regularly shaped (the same restriction applies to the air–liquid interfaces in cell wall pores). We can nevertheless define two principal radii of curvature for a surface. Let us designate these radii, which occur in planes perpendicular to each other and to the liquid surface, by r_1 and r_2. The hydrostatic pressure in the soil water then is

$$P = -\sigma \left(\frac{1}{r_1} + \frac{1}{r_2} \right) \tag{9.6}$$

where σ is the surface tension at an air–liquid interface [to connect Eq. 9.6 with Eq. 2.23 ($P = -2\sigma \cos\alpha/r$, where r is the radius of the capillary), note that for a cylindrical capillary the two principal radii of the surface are the same ($r_1 = r_2$), and hence the factor $(1/r_1 + 1/r_2)$ is $2/r_1$, which equals $2 \cos\alpha/r$]. The negative hydrostatic pressure or positive tension described by Eq. (9.6) that results from the presence of air–water interfaces is often called the *soil matric potential*. Equation (9.6) was derived by Thomas Young and by Pierre Laplace in 1805 and is sometimes called the Young–Laplace equation.

Instead of being concave, the water surface extending between adjacent soil particles may assume a semicylindrical shape, i.e., like a trough or channel. One of the radii of curvature then becomes infinite, e.g., $r_2 = \infty$; in such a case, the pressure is $-\sigma/r_1$ by Eq. (9.6). If the air–liquid surface is convex when viewed from the air side, the radii as defined are negative; we would then have a positive hydrostatic pressure in the water. In the intermediate case—one radius positive and one negative (a so-called "saddle-shaped" surface)—whether the pressure is positive or negative depends on the relative sizes of the two radii of curvature.

We now estimate the hydrostatic pressure in the soil water within a wedge-shaped crevice between two adjacent soil particles, as is illustrated at the top of Fig. 9-8. We will assume that the air–liquid surface is cylindrical, so $r_2 = \infty$, and that $r_1 = 0.1$ μm. Letting σ be 0.0728 Pa m, the value for water at 20°C (see Appendix I), by Eq. (9.6) the hydrostatic pressure is

$$P = \frac{-(0.0728 \text{ Pa m})}{(1 \times 10^{-7} \text{ m})} = -7 \times 10^5 \text{ Pa}$$

$$= -0.7 \text{ MPa}$$

As the amount of soil water decreases, the air–water surface retreats into the crevice between the particles, the radius of curvature decreases, and the pressure accordingly becomes more negative. Because $\Psi^{\text{soil}} = P^{\text{soil}} - \Pi^{\text{soil}} + \rho_w g h$ (Eq. 2.13a), the soil water potential also becomes more negative as water is lost from such crevices.

B. Darcy's Law

Henri Darcy in 1856 recognized that the flow of water through soil is driven by a gradient in hydrostatic pressure. We can represent this relation, known as Darcy's law, by the following expression:

$$J_V = -L^{\text{soil}} \frac{\partial P^{\text{soil}}}{\partial x} \qquad (9.7)$$

where J_V is the volume of solution crossing unit area in unit time and L^{soil} is the soil hydraulic conductivity coefficient.

Although Eq. (9.7) is in a familiar form (flux density equals a proportionality coefficient times a force), we have used $-\partial P^{\text{soil}}/\partial x$ instead of the possibly more general force, $-\partial \Psi^{\text{soil}}/\partial x (\Psi^{\text{soil}} = P^{\text{soil}} - \Pi^{\text{soil}} + \rho_w g h$; Eq. 2.13a). In Chapter 3 we derived an expression for J_V that incorporated a reflection coefficient: $J_V = L_P(\Delta P - \sigma \Delta \Pi)$ (Eq. 3.38). When σ is zero, as occurs for a porous barrier such as soil, $\Delta \Pi$ does not lead to a volume flux density. Hence, we do not expect $\partial \Pi^{\text{soil}}/\partial x$ to influence the movement of water in the soil, and it is not included in Darcy's law. Because $\partial(\rho_w g h)/\partial x$ is not incorporated into Eq. (9.7), the indicated form of Darcy's law applies only to horizontal flow in the soil ($\partial h/\partial x = 0$ when x is in a horizontal direction). Actually, $\rho_w g$ is only 0.01 MPa m^{-1} (see Appendix I), so changes of $\rho_w g h$ in the vicinity of a root are relatively small. On the other hand, for percolation of water down appreciable distances into the soil, P^{soil} in Eq. (9.7) should be replaced by $P^{\text{soil}} + \rho_w g h$. In fact, for drainage of very wet soil, the gravitational term can be the dominant factor in such a formulation of Darcy's law.

C. Soil Hydraulic Conductivity Coefficient

L^{soil} in Eq. (9.7) has different units than $L_P[J_V = L_P(\Delta P - \sigma \Delta \pi)$; Eq. 3.38] or $L_w[J_{V_w} = L_w(\Psi^{\circ} - \Psi^i)$; Eq. 2.23]. L_P and L_w are volume flux densities/pressure drop, e.g., m s^{-1} Pa^{-1}; L^{soil} has units of volume flux density/pressure gradient, e.g., (m s^{-1})/(Pa m^{-1}), or m^2 s^{-1} Pa^{-1}. We can use other self-consistent sets of units for Darcy's law; indeed, the soil hydraulic conductivity coefficient is expressed in many different ways in the literature.[1]

The soil hydraulic conductivity coefficient depends on the geometry of the pores in the soil. For geometrically similar pore shapes, and ignoring certain surface effects, the hydraulic conductivity coefficient is approximately proportional to the square of the pore width. However, the pores are so complex in shape that, in general, we cannot directly calculate L^{soil}. As soil dries, its water potential decreases and P^{soil} becomes less. When P^{soil} decreases below the minimum hydrostatic pressure that can occur in fairly large pores, water flows out of them, but water will remain in pores that have smaller dimensions and can therefore have even more negative pressures [see Eq. 9.6; $P = -\sigma(1/r_1 + 1/r_2)$]. Not only is the higher conductivity of the larger pores lost but also the remaining pathway for water flow becomes more tortuous, so L^{soil} decreases as the soil dries. The soil hydraulic conductivity coefficient can be 1×10^{-17} m^2 s^{-1} Pa^{-1} or lower for a dry nonporous soil and 1×10^{-13} m^2 s^{-1} Pa^{-1} or higher for a wet porous one. In particular, L^{soil} is usually 10^{-13}–10^{-10} m^2 s^{-1} Pa^{-1} for a water-saturated porous clay and 10^{-8}–10^{-7} m^2 s^{-1} Pa^{-1} for a water-saturated sandy soil.

---

[1] Instead of using $-\partial P/\partial x$ [or $-\partial(P + \rho_w g h)/\partial x$], most soil literature expresses the force in Darcy's law as the negative gradient in the hydraulic head of water, the latter being the length of a vertical column of water yielding the same pressure, in which case the units for L^{soil} are the same as those for J_V. Because a 1-m height of water exerts a pressure of $\rho_w g \times 1$ m or 0.00979 MPa ($= 9.79 \times 10^3$ Pa) at sea level, 45° latitude, and 20°C (see Appendix I), an L^{soil} of 10^{-10} m^2 s^{-1} Pa^{-1} corresponds to one of 9.79×10^{-7} m s^{-1}.

The ground is often covered by a dry crust in which the soil hydraulic conductivity coefficient is low. Specifically, L^{soil} may average 1×10^{-16} m^2 s^{-1} Pa^{-1} in the upper 5 mm of the soil. If P^{soil} is -2.0 MPa near the surface and -1.5 MPa at 5 mm beneath it (Fig. 9-9), then by Darcy's law (Eq. 9.7) the volume flux density of water is

$$J_V = -(1 \times 10^{-16}\,\text{m}^2\,\text{s}^{-1}\,\text{Pa}^{-1})\left[\frac{(-2.0 \times 10^6\,\text{Pa}) - (-1.5 \times 10^6\,\text{Pa})}{(5 \times 10^{-3}\,\text{m})}\right]$$

$$= 1 \times 10^{-8}\,\text{m s}^{-1} \qquad (0.9\,\text{mm day}^{-1})$$

This water flux density directed upward at the soil surface equals $(1 \times 10^{-8}$ m^3 m^{-2} s^{-1}) $(1 \text{ mol}/18 \times 10^{-6}$ m^3), or 0.6×10^{-3} mol m^{-2} s^{-1}. When discussing water vapor movement in the previous section, we indicated that J_{wv} emanating from a moist shaded soil is generally 0.2–1.0 mmol m^{-2} s^{-1}, so our calculated flux density is consistent with the range of measured values. The calculation also indicates that a fairly large gradient in hydrostatic pressure can exist near the soil surface.

During a rainstorm the upper part of the soil can become nearly saturated with water. As a result, L^{soil} there might increase 10^6-fold, becoming 1×10^{-10} m^2 s^{-1} Pa^{-1} (Fig. 9-9). This facilitates the entry or infiltration of water into the soil, which initially can have a volume flux density of about 5×10^{-6} m s^{-1}. Such an infiltration rate, which equals an 18 mm depth of water per hour, is maintained for only short times because within minutes for a clay and after an hour or so for a sandy soil, the upper part of the soil becomes saturated with water and the infiltration rate decreases. In any case, we note that the uppermost layer or crust acts somewhat like a valve, retarding the outward movement of water when the soil is fairly dry (low L^{soil}) but promoting the infiltration of water upon moistening (high L^{soil}), as is indicated in Fig. 9-9.

Besides moving as a liquid, water can also move as a vapor in the soil. Because water is continually evaporating from and condensing onto the many air–liquid interfaces in the soil,

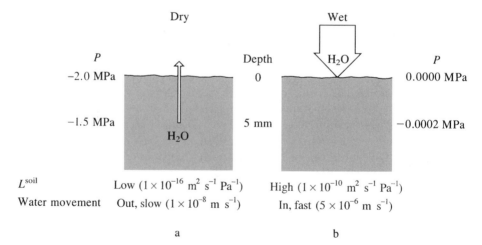

Figure 9-9. Summary of valve-like properties of upper layer of loamy soil: (a) a fairly dry state and (b) while being wet by rain. Water readily enters wet soil but is lost only gradually across a dry crust. Water movement is also influenced by gravity, the relevant component of the water potential ($\rho_w gh$) decreasing by 0.01 MPa per meter depth in the soil. For a depth of 5 mm, the gravitational contribution decreases Ψ by 0.00005 MPa compared with the soil surface, which increases the influx rate by 25% for the wet condition compared to considering only ΔP.

such movement can be relatively important, especially for dry soils in which the liquid phases are discontinuous. The saturation value for water vapor partial pressure or concentration increases nearly exponetially with temperature (see Fig. 8-5). Because the air between the soil particles generally is nearly saturated with water vapor (except near the soil surface), the amount of water vapor in the soil air increases rapidly with soil temperature. Consequently, the movement of water in the form of vapor tends to be greater at higher soil temperatures.

D. Flux for Cylindrical Symmetry

Although we have been mainly considering one-dimensional cases in Cartesian coordinates, the flow of soil water toward a root may be more appropriately described using cylindrical coordinates because the length of a root is much greater than is its diameter. We will restrict our attention to the steady-state condition in which the fluxes do not change with time. Also, we will consider the cylindrically symmetric case for which the fluxes and forces depend only on the radial distance r from the axis of the cylinder and not on any angle around it or location along its axis.

For a cylindrical surface of length l along the axis, the total volume of solution crossing per unit time is $J_V 2\pi r l$, where J_V is the volume flux density directed radially at a distance r from the axis of the cylinder. In the steady state, $J_V 2\pi r l$ is constant, so the magnitude of the flux density depends inversely on the radial distance. When L^{soil} is constant, we can represent the flux density for the cylindrically symmetric case by

$$J_V = \frac{1}{r} \frac{L^{soil}(P_a - P_b)}{\ln(r_a/r_b)} \tag{9.8}$$

where P_a is the hydrostatic pressure at a distance r_a from the axis of the cylinder, and P_b is the value at r_b. J_V is positive when the net flux density is directed into the root, as occurs when the hydrostatic pressure is higher ($P_a > P_b$) the further we are from the root ($r_a > r_b$). Equation (9.8) represents a general form for steady-state cases with cylindrical symmetry; e.g., Fick's first law then is $J_j = (1/r)D_j(c_j^a - c_j^b)/\ln(r_a/r_b)$, and Eq. (7.12) gives the heat flux density for cylindrical symmetry.

The uptake of water by a young root 1 mm in diameter is usually 1×10^{-5} to 5×10^{-5} m^3 day^{-1} per meter of root length (0.1–0.5 cm^3 day^{-1} per centimeter). This uptake occurs over a root surface area of $2\pi r l$, so the volume flux density of water at the root surface for a moderate water uptake rate of 3×10^{-5} m^3 day^{-1} per meter of root length is

$$J_V = \frac{(3 \times 10^{-5} \, m^3 \, day^{-1})}{(8.64 \times 10^4 \, s \, day^{-1})[(2\pi)(0.5 \times 10^{-3} \, m)(1 \, m)]}$$

$$= 1.1 \times 10^{-7} \, m \, s^{-1}$$

The influx of water is enhanced by root hairs, which protrude from the epidermal cells (see Fig. 1-4). They are often about 12 μm in diameter, up to 1 mm long, and generally vary in frequency from 0.5 to 50 per mm^2. J_V can be $\frac{1}{10}$ to $\frac{1}{100}$ as much for older roots because their outer surfaces lack root hairs and generally become extensively cutinized and suberized.

For representative root spacing in a soil, water may move toward a root over a radial distance of about 10 mm. We next estimate the decrease in hydrostatic pressure that might occur over this interval. We will assume that Ψ^{soil} at an r_a of 10.5 mm is -0.3 MPa, made up of an osmotic pressure of 0.1 MPa and a P_a^{soil} of -0.2 MPa ($\Psi = P - \Pi + \rho_w gh$; Eq. 2.13a). We will let L^{soil} be 1×10^{-15} m^2 s^{-1} Pa^{-1}, as might apply to a loam of moderately low water content, and we will suppose that J_V at the surface of a root 1 mm in diameter (i.e., $r_b = 0.5$

mm) is 1.1×10^{-7} m s^{-1}. Using Eq. (9.8), we calculate that the hydrostatic pressure near the root surface is

$$P_b^{\text{soil}} = -\frac{r J_V \ln(r_a/r_b)}{L^{\text{soil}}} + P_a^{\text{soil}}$$

$$= -\frac{(0.5 \times 10^{-3} \text{ m})(1.1 \times 10^{-7} \text{ m s}^{-1}) \ln(10.5 \text{ mm}/0.5 \text{ mm})}{(1 \times 10^{-15} \text{ m}^2 \text{ s}^{-1} \text{ Pa}^{-1})} - 0.2 \text{ MPa}$$

$$= -0.4 \text{ MPa}$$

Thus the hydrostatic pressure decreases 0.2 MPa across a distance of 10 mm in the soil next to the root. Assuming that the solute content of the soil water does not change appreciably over this interval, Π_b^{soil} is 0.1 MPa, and thus Ψ_b^{soil} adjacent to the root is -0.4 MPa -0.1 MPa, or -0.5 MPa (see Table 9-2). As the soil dries, L^{soil} decreases; therefore, the decreases in hydrostatic pressure and water potential adjacent to a root must then be larger to maintain a given volume flux density of solution toward a root.

E. Flux for Spherical Symmetry

Sometimes the fluxes of water in the soil toward plant parts can approximate spherical symmetry. For instance, water movement can occur radially over a 10-mm interval toward a seed or a recently initiated root. For some roots most water and nutrient uptake apparently takes place over the region containing the root hairs, which is proximal to the elongation region (see Fig. 1-4) and distal to where the periderm begins. The external cell layers of the periderm have suberin in their cell walls, which greatly limits water uptake. *Lenticels* occur in the suberized regions, consisting of a loose aggregation of cells facilitating gas exchange (lenticels also occur on stems). Lenticels and other interruptions of the suberized periderm can act as local sites toward which the flux of water converges. In such a case, the water movement from the surrounding soil toward the root can also be approximately spherically symmetric.

For spherical symmetry we note that $J_V 4\pi r^2$ is constant in the steady state. Thus we obtain the following steady-state relation describing the volume flux density J_V at distance r from the center of a sphere when J_V varies only in the radial direction and L^{soil} is constant:

$$J_V = \frac{1}{r^2}\left(\frac{r_a r_b}{r_a - r_b}\right) L^{\text{soil}}(P_a - P_b) \tag{9.9}$$

Equation (9.9) is similar to Eq. (7.13) $[J_H^C = (r + \delta^{\text{bl}}) K^{\text{air}}(T^{\text{surf}} - T^{\text{ta}})/(r\delta^{\text{bl}})]$ describing the heat flux density across an air boundary layer for spherical symmetry ($r_a = r + \delta^{\text{bl}}$ and $r_b = r$). A steady state is often not achieved in soils, so a "steady rate" is sometimes used, where the rate of volumetric water depletion is constant, leading to relations considerably more complicated than Eqs. (9.8) and (9.9).[2] We note that the equations describing water flow in a soil for the one-dimensional case (Eq. 9.7), for cylindrical symmetry (Eq. 9.8),

---

[2] In a steady state the water content of the soil does not change with time, whereas for a steady-rate situation the rate of change is constant. In the general time-dependent case, a relation similar to $\frac{\partial c_j}{\partial t} = -\frac{\partial}{\partial x}(-D_j \frac{\partial c_j}{\partial x})$ (Eq. 1.4), but in the proper coordinate system, must be satisfied to desribe the water flow in the soil for cylindrical or spherical symmetry; D_j is replaced by a quantity analogous to L^{soil} that varies with water content and hence location in the soil.

and for spherical symmetry (Eq. 9.9) all indicate that the volume flux density of water is proportional to L^{soil} times a difference in hydrostatic pressure.

During germination the volume flux density of water into a seed is often limited by a seed coat (testa) of thickness δ^{sc}. The seed coat can be two to four cell layers thick and thus is relatively thin compared with the radius of the seed. For the volume flux density at the seed surface ($r = r_s$, $r_a = r_s$, and $r_b = r_s - \delta^{\text{sc}}$), Eq. (9.9) becomes

$$
\begin{aligned}
J_V &= \frac{1}{r_s^2} \frac{(r_s)(r_s - \delta^{\text{sc}})}{[r_s - (r_s - \delta^{\text{sc}})]} L^{\text{sc}} \Delta P^{\text{sc}} \\
&= \frac{r_s - \delta^{\text{sc}}}{r_s} \frac{L^{\text{sc}}}{\delta^{\text{sc}}} \Delta P^{\text{sc}} \\
&\cong \frac{L^{\text{sc}}}{\delta^{\text{sc}}} \Delta P^{\text{sc}}
\end{aligned}
\tag{9.10}
$$

where L^{sc} is the hydraulic conductivity coefficient of a seed coat of thickness δ^{sc}, and the bottom line incorporates the supposition that $\delta^{\text{sc}} \ll r_s$. Equation (9.10) indicates that, when the region of interest is thin compared with the radial distance, the flow can be approximated by an equation appropriate for a one-dimensional movement such as across a seed coat of conductance $L^{\text{sc}}/\delta^{\text{sc}}$ (see Eq. 8.1b). Usually ΔP^{sc} in Eq. (9.10) is replaced by $\Delta \Psi^{\text{sc}}$ because osmotic pressures can also influence water uptake by seeds.

As we have indicated for L^{soil}, L^{sc} depends on water content, increasing more than 10^6-fold upon water uptake by the seed and subsequent rupture of the seed coat. L^{sc} then approaches values found for the L^{soil} of a moist soil (Shaykewich and Williams, 1971). When dry, seeds can have a very negative water potential, e.g., -100 to -200 MPa. As seeds imbibe water, their internal water potential rises; germination is initiated only when Ψ^{soil} and hence Ψ^{seed} are above about -1.0 MPa. Water uptake is not uniform over the whole seed surface, initially being higher near the micropyle (a small opening in the integument of an ovule through which the pollen tube enters and which remains as a minute pore in the testa).

IV. Water Movement in the Xylem and the Phloem

Under usual conditions, essentially all of the water entering a land plant comes from the soil by way of the roots. The water is conducted to other parts of the plant mainly in the xylem. To reach the xylem in the mature part of a root, water must cross the root epidermis, the cortex, and then the endodermis. We will briefly describe the various cells encountered by water as it moves from the soil to the root xylem (see Fig. 1-4) and then discuss the dependence of water flow in the xylem on the gradient in hydrostatic pressure. Flow in the phloem likewise depends on $\partial P/\partial x$, although osmotic pressures are also important for interpreting water movement in this other "circulatory" system. The divisions and subdivisions of the vascular tissue in a leaf generally result in individual mesophyll cells being no more than three or four cells away from the xylem or the phloem. This proximity facilitates the movement of photosynthetic products into the phloem, such "loading" being necessary for the distribution of sugars such as sucrose to "sinks" located in different parts of the plant.

A. Root Tissues

Figure 1-4 presents longitudinal and cross-sectional views near the tip of a root, indicating the types of cells that can act as barriers to flow. Water may fairly easily traverse the single-cell layer of the root epidermis to reach the cortex. The root cortex often consists of 5–10 cell layers, with the cytoplasm of adjacent cells being continuous because of plasmodesmata. The collective protoplasm of interconnected cells is referred to as the *symplasm* (see Chapter 1). In the symplasm, permeability barriers in the form of plasma membranes and cell walls do not have to be surmounted for diffusion to occur from cell to cell, which facilitates the movement of water and solutes across the cortex. Much of the water flow across the root cortex occurs in an alternative pathway, the cell walls, which form the *apoplast*.

Roots of most woody and some herbaceous plants have mycorrhizal associations. The fungal hyphae invade the cortical region of the young roots and can extend over 10 mm away from the root surface. The relation is mutualistic because the fungus obtains organic matter from the plant and in return increases the effective area of contact between roots and soil particles. The hyphae can remain active for older roots whose outer surface has become suberized and hence less conductive. The fungus thus increases the availability of certain nutrients such as N and especially P to the host plant.

Along that part of a root where cells have differentiated but the roots have not thickened much due to secondary conducting tissue, an *endodermis* occurs inside the cortex (see Fig. 1-4). At the endodermis, the cell wall pathway for water movement is blocked by the casparian strip, as realized by Robert Caspary in 1865. This is a continuous hydrophobic band around the radial walls of the endodermal cells that blocks the apoplastic pathway along which water and solutes could have crossed the endodermal layer. The casparian strip is impregnated with suberin (a polymer of fatty acids) and lignin. Besides the primary cell wall, suberin and lignin are also deposited in the secondary wall (which often becomes quite thick) and across the middle lamella. Water and ions therefore cannot flow across the endodermal cells in their cell walls but rather must move through their cytoplasm. Thus the endodermis, with its casparian strip, regulates the passage of solutes and water from the cortex to the xylem in the affected parts of the root. Inside the endodermis is a layer of parenchyma cells known as the *pericycle*, which surrounds the vascular tissue containing the xylem and the phloem (see Fig. 1-4). During water movement through a plant accompanying transpiration, the hydrostatic pressure in the root xylem is reduced and can become quite negative. This decreases the water potential ($\Psi = P - \Pi + \rho_w gh$; Eq. 2.13a) in the xylem and promotes water movement from the soil down a water potential gradient to the root xylem.

B. The Xylem

Before discussing the characteristics of flow in the xylem, we will briefly review some of its anatomical features. In general, the conducting xylem elements have thick, lignified secondary cell walls and contain no protoplasts, i.e., the xylem cells serve their special function of providing the plant with a low-resistance conduit for water flow only when they are dead.

Two types of conducting cells are distinguished in the xylem: the *vessel members* (also called vessel elements; found in angiosperms) and the phylogenetically more primitive *tracheids* (in angiosperms, gymnosperms, and the lower vascular plants). Tracheids typically are tapered at their ends, whereas the generally shorter and broader vessel members often abut each other with blunt ends. The end cell walls of the vessel members are perforated, so the vessel members arranged end to end form a continuous tube called a *vessel*. The end cell wall of a vessel member bearing the holes is referred to as a *perforation plate*; a simple perforation plate (see Fig. 1-3) essentially eliminates the end walls between the individual members in a vessel. Although xylem elements vary considerably in their widths (from about 8 μm up to 500 μm), we will represent them by cylinders with radii of 20 μm for purposes of calculation. Conducting cells of the xylem generally range in length from 1 to 10 mm for tracheids and from 0.2 to 3 mm for vessel members; vessels vary greatly in length, even within the same plant, ranging from about 10 mm to 10 m.

Besides vessel members and tracheids, parenchyma cells and fibers also occur in the xylem (see Fig. 1-3). Xylem fibers, which contribute to the structural support of a plant, are long thin cells with lignified cell walls; they are generally devoid of protoplasts at maturity but are nonconducting. The living parenchyma cells in the xylem are important for the storage of carbohydrates and for the lateral movement of water and solutes into and out of the conducting cells.

C. Poiseuille's Law

To describe fluid movement in the xylem quantitatively, we need to relate the flow to the driving force causing the motion. For cylindrical tubes, an appropriate relationship was determined experimentally by Gotthilf Hagen in 1839 and independently by Jan Poiseuille in 1840. They found that the volume of fluid moving in unit time along a cylinder is proportional to the fourth power of its radius and that the movement depends linearly on the drop in hydrostatic pressure. Hans Wiedemann in 1856 showed that this rate of volume movement per tube could be represented as follows:

$$\text{Volume flow rate per tube} = -\frac{\pi r^4}{8\eta}\frac{\partial P}{\partial x} \tag{9.11a}$$

where r is the cylinder radius, η is the solution viscosity, and $-\partial P/\partial x$ is the negative gradient of the hydrostatic pressure. Throughout this text we have been concerned with the volume flowing per unit time and area, J_V. For flow in a cylinder of radius r, and hence area πr^2, J_V is

$$J_V = -\frac{r^2}{8\eta}\frac{\partial P}{\partial x} \tag{9.11b}$$

The negative sign is necessary in Eq. (9.11) because positive flow ($J_V > 0$) occurs in the direction of decreasing hydrostatic pressure ($\partial P/\partial x < 0$).

Equation (9.11) is generally referred to as Poiseuille's law and sometimes as the Hagen–Poiseuille law. It assumes that the fluid in the cylinder moves in layers, or laminae, with each layer gliding over the adjacent one. Such laminar movement occurs only if the flow is slow enough to meet a criterion deduced by Osborne Reynolds in 1883. Specifically, the Reynolds number $\rho J_V d/\eta$ must be less than 2000 [ρ is the solution density and d is the cylinder diameter; Eq. 7.16 indicates that Re $= vd/\nu$, where the kinematic viscosity $\nu = \eta/\rho$; because J_V equals the mean velocity of fluid movement v, Re can be represented by $J_V d/(\eta/\rho)$, or $\rho J_V d/\eta$]. Otherwise, a transition to turbulent flow occurs, and Eq. (9.11) is no longer valid. The fluid in Poiseuille (laminar) flow is stationary at the wall of the cylinder; the speed of solution flow increases in a parabolic fashion to a maximum value in the center of the tube, where it is twice the average speed, J_V. Thus the flows in Eq. (9.11) are actually the mean flows averaged over the entire cross section of cylinders of radius r. The dimensions for the viscosity η are pressure $\times$ time, e.g., Pa s (the cgs unit, dyne s cm^{-2}, is termed a *poise*, which equals 0.1 Pa s); the viscosity of water at 20°C is 1.002×10^{-3} Pa s (values of η_w at other temperatures are given in Appendix I).

D. Applications of Poiseuille's Law

As we have already indicated, the volume flux densities described by both Poiseuille's law [$J_V = -(r^2/8\eta)\partial P/\partial x$; Eq. 9.11b] and Darcy's law ($J_V = -L^{soil}\partial P^{soil}/\partial x$; Eq. 9.7) depend on the negative gradient of the hydrostatic pressure. Can we establish any correspondence between these two equations? If the soil pores were cylinders of radius r, all aligned in the same direction, we could have Poiseuille flow in the soil, in which case L^{soil} would equal $r^2/8\eta$—we are assuming for the moment that the pores occupy the entire soil volume. When $r = 1$ μm and $\eta = 1.002 \times 10^{-3}$ Pa s, we obtain

$$\frac{r^2}{8\eta} = \frac{(1 \times 10^{-6}\,\text{m})^2}{(8)(1.002 \times 10^{-3}\,\text{Pa s})} = 1.2 \times 10^{-10}\,\text{m}^2\,\text{s}^{-1}\,\text{Pa}^{-1}$$

which corresponds to the L^{soil} for a very wet soil (Fig. 9-9). Actually, L^{soil} for water-filled porous clay with average pore radii of 1 μm is about 10^{-11} m^2 s^{-1} Pa^{-1}, about 10-fold smaller than $r^2/8\eta$ calculated using the average pore radius. As a result, J_V in the soil is considerably less than that for Poiseuille flow through pores of the same average radius because the soil pores are not in the shape of cylinders aligned in the direction of flow and the pores between soil particles occupy only about half of the soil volume.

Next, we will use Poiseuille's law to estimate the pressure gradient necessary to cause a specified volume flux density in the conducting cells of the xylem.[3] The speed of sap ascent in the xylem of a transpiring tree can be about 3.6 m h^{-1}, which equals 1 mm s^{-1}. We note that v_w equals the volume flux density of water, J_{V_w} (see Chapter 2, Section IVF); the average speed of the solution equals J_V, the volume flux density of the solution, which for a dilute aqueous solution such as occurs in the xylem $\cong J_{V_w}$ (see Chapter 3, Section VC). Thus J_V in the xylem of a transpiring tree might be 1 mm s^{-1}. For a viscosity of 1.0 mPa s and a xylem element with a lumen radius of 20 μm, the pressure gradient required to satisfy Eq. (9.11b) is

$$\frac{\partial P}{\partial x} = \frac{-8\eta J_V}{r^2} = \frac{-(8)(1.0 \times 10^{-3}\,\text{Pa s})(1.0 \times 10^{-3}\,\text{m s}^{-1})}{(20 \times 10^{-6}\,\text{m})^2}$$

$$= -2 \times 10^4\,\text{Pa m}^{-1} = -0.02\,\text{MPa m}^{-1}$$

As expected, the pressure decreases along the direction of flow (Fig. 9-10).

The estimate of -0.02 MPa m^{-1} is consistent with most experimental observations of the $\partial P/\partial x$ accompanying solution flow in horizontal xylem vessels. For the vertical vessels in a tree, however, a static hydrostatic pressure gradient of -0.01 MPa due to gravity exists even

---

[3] For application of Poiseuille's law to a complex tissue such as the xylem, care must be taken to ensure that particular vessel elements or tracheids are conducting (e.g., not blocked by embolisms), the actual radii must be determined (note the r^4 dependence in Eq. 9.11a), and corrections may be necessary for lumen shape, tracheid taper, and cell wall characteristics including pits (Calkin *et al.*, 1986; Schulte *et al.*, 1989a). For instance, if the lumen is elliptical with major and minor axes of a and b, respectively, then r^4 in Eq. (9.11a) should be replaced by $a^3 b^3/(8a^2 + 8b^2)$.

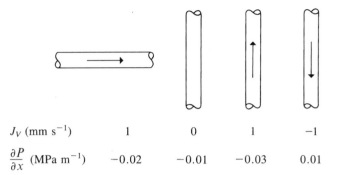

J_V (mm s^{-1})	1	0	1	-1
$\dfrac{\partial P}{\partial x}$ (MPa m^{-1})	-0.02	-0.01	-0.03	0.01

Figure 9-10. Flows and pressure gradients in cylindrical tubes. The tube radius is 20 μm and $\partial P/\partial x$ is calculated from Poiseuille's law (Eq. 9.11b) as modified by gravitational effects. For vertical tubes, x is considered positive upward. Arrows indicate the direction of flow.

in the absence of flow. We can appreciate this by considering a vertical column of pure water ($\Pi = 0$) at equilibrium, where the water potential Ψ is $P + \rho_w gh$; $\rho_w gh$ increases vertically upward, so P must decrease by the same amount for Ψ to remain unchanged, as it does at equilibrium. Because $\rho_w g = 0.0098$ MPa m^{-1} (see Appendix I), the additional pressure gradient caused by gravity amounts to -0.01 MPa m^{-1} (Fig. 9-10). For transpiring plants, the total $\partial P/\partial x$ in the vertical sections of the xylem is often about -0.03 MPa m^{-1}. Based on the previous calculations, this pressure gradient is sufficient to overcome gravity (-0.01 MPa m^{-1}) and to cause Poiseuille flow in the xylem vessels (-0.02 MPa m^{-1} is needed in the current example in which $r = 20\ \mu$m). The gradients refer to vessel members joined by simple perforation plates that offer little obstruction to flow (see Fig. 1-3). Gradients tend to be greater for tracheids because they are connected by numerous small pits containing locally thin regions of the cell wall. These pits can greatly restrict flow because some cell wall material must be traversed to move from one trachied to another. For instance, for fern tracheids with small radii ($<10\ \mu$m) the pressure drop is mainly along the lumen, reflecting the major influence of the lumen radius on Poiseuille flow, whereas for larger diameter tracheids most of the pressure drop is across the pits. In conifers half of the pressure drop along the xylem may occur across the pits, and hence the overall $\partial P/\partial x$ for the same lumen diameter and flow rate is about twice as large as that calculated here for the lumen.

We have been approximating xylem vessels by cylinders that are 20 μm in radius, as might be appropriate for diffuse-porous (small-porous) trees. For ring-porous (large-porous) trees, the mean radii of xylem vessels are often about 100 μm. For a given pressure gradient, Poiseuille's law (Eq. 9.11b) indicates that xylem sap moves faster in ring-porous trees, as is indeed observed. Using our previous values and a water density of 1000 kg m^{-3}, the Reynolds number for the diffuse-porous case is

$$\text{Re} = \frac{\rho J_V d}{\eta} = \frac{(1000\ \text{kg m}^{-3})(1 \times 10^{-3}\ \text{m s}^{-1})(40 \times 10^{-6}\ \text{m})}{(1.0 \times 10^{-3}\ \text{Pa s})}$$

$$= 0.04\ \text{kg m}^{-1}\,\text{s}^{-2}\,\text{Pa}^{-1} = 0.04$$

Such a low value indicates that no turbulence is expected. Even if d were 5-fold larger and J_V were 10-fold greater, as can occur in a ring-porous tree, the Reynolds number is still far less than the value of 2000 at which turbulence generally sets in.

What pressure gradient is necessary to promote a given flow through a cell wall? Because the interfibrillar spaces, or interstices, in a cell wall have diameters near 10 nm (100 Å), we will let r be 5 nm for purposes of calculation. (Complications due to the tortuosity of the aqueous channels through the interstices will be omitted here. We will also assume that the pores occupy the entire cell wall.) For $J_V = 1$ mm s^{-1}, Eq. (9.11b) indicates that the pressure gradient required for solution flow through the cell walls is $-(1 \times 10^{-3}$ m s$^{-1})$ $(8)(1.0 \times 10^{-3}$ Pa s$)/(5 \times 10^{-9}$ m$)^2$ or -3×10^5 MPa m^{-1}. In contrast, a $\partial P/\partial x$ of only -0.02 MPa m^{-1} is needed for the same J_V in the xylem element with a 20-μm radius. Thus the $\partial P/\partial x$ for Poiseuille flow through the small interstices of a cell wall is over 10^7 times greater than that for the same flux density through the lumen of the xylem element. Because of the tremendous pressure gradients required to force water through the small interstices available for solution conduction in the cell wall, fluid could not flow rapidly enough up a tree in the cell walls—as has been suggested—to account for the observed rates of water movement.

To compare the relative effects on flow of a cell wall, a plasma membrane, and the lumen of a xylem vessel, we will calculate the hydrostatic pressure drops across each of them in a hypothetical case. We consider a xylem vessel member that is 1-mm long and 20 μm in radius. We have just calculated that the pressure gradient necessary for a Poiseuille flow of 1 mm s^{-1} in the lumen of this vessel member is -0.02 MPa m^{-1}, which amounts to $(-0.02$ MPa m$^{-1})$ $(10^{-3}$ m), or -2×10^{-5} MPa for the 1-mm length of the cell. Let us suppose that a cell wall 1 μm thick is placed across both ends of the cell. In the previous paragraph we indicated that a pressure gradient of -3×10^5 MPa m^{-1} is necessary for a J_V of 1 mm s^{-1} through the interstices of a cell wall. To cross this cell wall (a total distance of 2 μm for both ends) would require a pressure change of $(-3 \times 10^5$ MPa m$^{-1})(2 \times 10^{-6}$ m), or -0.6 MPa. Finally, let us imagine that a plasma membrane with a typical hydraulic conductivity coefficient of 7×10^{-13} m s^{-1} Pa^{-1} is placed around the cell. Using Eq. (3.38) $[J_V = L_P(\Delta P - \sigma \Delta \Pi)]$ with $J_V = 1$ mm s^{-1}, the difference in hydrostatic pressure required for such flow out the ends is $(1 \times 10^{-3}$ m s$^{-1})/(7 \times 10^{-13}$ m s^{-1} Pa$^{-1})$, or 1.4×10^9 Pa, which means a total pressure drop of 3×10^3 MPa is necessary to traverse the plasma membrane at each end of the cell. In summary, the pressure drops needed in this hypothetical case are 2×10^{-5} MPa to flow through the lumen, 0.6 MPa to cross the cell walls, and 3×10^3 MPa to cross the membranes.

In fact, membranes generally serve as the main barrier to water flow into or out of plant cells. The interstices of the cell walls provide a much easier pathway for such flow, and hollow xylem vessels present the least impediment to flow (such as up a stem). Consequently, xylem provides a plant with tubes, or conduits, well suited for moving water over long distances. The region of a plant made up of cell walls and the hollow xylem vessels is often called the apoplast, as noted in Chapter 1. Water and the solutes that it contains can move fairly readily in the apoplast, but they must cross a membrane to enter the symplast (symplasm), the interconnected cytoplasm of the cells.

E. The Phloem

Water and solute movement in the phloem involves cooperative interactions among several types of cells. The conducting cells of the phloem, which generally have a high internal hydrostatic pressure, are the sieve cells in lower vascular plants and gymnosperms and the generally shorter, wider, and less tapered sieve-tube members in angiosperms (see

Fig. 1-3). Both types of cells are collectively called *sieve elements*. Mature sieve elements almost invariably have lost their nuclei, and the tonoplast has broken down, so no large central vacuole is present, although the plasma membrane remains intact. However, unlike the conducting elements of the xylem, the sieve elements contain cytoplasm and are living. The sieve elements in most plants range from 0.1 to 3 mm in length and tend to be longer in gymnosperms than in angiosperms. Typical radii of the lumens are 6–25 μm. Specialized cells are generally found adjacent to and in close association with the sieve elements, called *companion cells* in angiosperms (see Fig. 1-3) and *albuminous cells* in gymnosperms. These cells have nuclei and generally contain many mitochondria. Such cells are metabolically related to the conducting cells and may supply them with carbohydrates, ATP, and other materials. Moreover, companion and albuminous cells accumulate sugars and other solutes, which may either passively diffuse (e.g., through the plasmodesmata that are present) or be actively transported into the sieve elements.

Sieve-tube members usually abut end to end to form the *sieve tubes*. The pair of generally inclined end walls between two sequential sieve-tube members forms the *sieve plate* (see Fig. 1-3; sieve cells are joined at less specialized sieve areas). Sieve plates have many pores, ranging in diameter from <1 μm up to about 5 μm, and ordinarily have strands of cytoplasmic material passing through them. The pores are lined, not crossed, by the plasma membranes. Therefore, solution moving in the phloem does not have to cross any mebranes as it flows from cell to cell along a sieve tube. A distinguishing feature of sieve-tube members in dicots and some monocots is the presence of *phloem (P) protein*, which often occurs in tubular as well as fibrillar form. P protein can plug the sieve plate pores upon injury of the sieve tube. Specifically, when the phloem is opened by cutting a stem, the positive hydrostatic pressure in the phloem forces the contents of the conducting cells toward the incision, carrying the P protein into the sieve plate pores. As a further wound response, callose (a glucose polysaccharide) can also be deposited in the sieve plate pores, thereby closing them within minutes after injury.

The products of photosynthesis, *photosynthates*, are distributed throughout a plant by the phloem. Changes in phloem flow can affect the distribution of photosynthates, which in turn can affect the metabolism of mesophyll cells. For moderate water stress leading to reduced leaf water potentials, translocation of photosynthates can decrease more than photosynthesis in some plants, with the consequence that leaves and nearby parts of the stem accumulate more starch. In C_4 plants the profuse vascularization and close proximity of bundle sheath cells to the phloem (see Fig. 8-11) can lead to rapid removal of photosynthates from the leaves and little deposition of starch in the bundle sheath cells over the course of a day, whereas certain C_3 plants can have a large increase of starch in the mesophyll cells during the daytime. The number of sieve elements in a petiole can reflect the productivity of the leaf, e.g., petioles of sun leaves tend to have more sieve elements per unit leaf area than do petioles of shade leaves. The practice of "ringing" or "girdling" a branch (cutting away a band of phloem-containing bark all the way around) can stop the export of photosynthate, with the consequence that fruit on such a branch can become considerably larger than it otherwise would. The practice of girdling the trunk of a tree abolishes the export of photosynthate to the roots, ultimately leading to death of the tree.

F. Phloem Contents and Speed of Movement

An elegant way of studying the contents of certain sieve elements is by means of aphid stylets. An aphid feeds on the phloem by inserting its stylet into an individual sieve element.

After the aphid has been anesthetized and its body removed, the remaining mouthpart forms a tube that leads the phloem solution from the sieve element to the outside. Solutes in the phloem solution extracted using this technique are often over 90% carbohydrates, mainly sucrose and some other oligosaccharides. The concentration of sucrose is generally 0.2–0.7 M; values near 0.3 M are typical for small plants, whereas some tall trees have sucrose concentrations near the upper limit. The types and concentrations of the solutes exhibit daily and seasonal variations and also depend on the tissues that the phloem solution is flowing toward or away from. For example, the solution in the phloem moving out of senescing leaves is low in sucrose but contains an appreciable concentration of amino acids and amides, sometimes as high as 0.5 M—an amino acid concentration near 0.05 M is representative of phloem solution in general. The initial movement of ions from the root to the rest of the plant occurs mainly in the xylem; subsequent recirculation can take place in the phloem, such as the movement of ions out of leaves just before their abscission.

Solutes can move over long distances in the phloem, with flow being toward regions of lower osmotic pressure. Thus photosynthetic products move from the leaves to storage tissues in the stem and the root, where they are generally converted to starch. At other times, sugars produced from the hydrolysis of such starch may move in the opposite direction, from the storage tissue to meristematic areas at the top of the plant. The speed of solute movement is ordinarily 0.2–2 m h^{-1}, the rate varying, among other things, with the plant species and vigor of growth. Although a particular sugar generally moves in the phloem in the direction of a decrease in its concentration, diffusion is not the mechanism. First, the rate of solute movement far exceeds that of diffusion (see Chapter 1). Second, when a radioactive solute, a dye, or a heat pulse is introduced into conducting sieve elements, the "front" moves with a fairly constant speed (distance moved is proportional to time), whereas in a one-dimensional diffusional process, distance moved is proportional to the square root of time in such a case ($x_e^2 = 4D_j t_e$; Eq. 1.6).

Rather than the solute speed in the phloem, we are often more interested in how much matter is translocated. For example, if the sieve elements contain 0.5 M (500 mol m^{-3}) sucrose moving at an average speed of 0.6 m h^{-1}, what is the transfer rate of sucrose in kg m^{-2} h^{-1}? By Eq. (3.6) ($J_j = \bar{v}_j c_j$), the flux density of sucrose is

$$J_{\text{sucrose}} = (0.6\,\text{m h}^{-1})(500\,\text{mol m}^{-3})$$

$$= 300\,\text{mol m}^{-2}\,\text{h}^{-1}$$

Because sucrose has a mass of 0.342 kg mol^{-1}, this flux density corresponds to 103 kg m^{-2} h^{-1}. In the current example, the flow is per m^2 of sieve-tube lumens; the rate of flow per unit area of phloem tissue is less by the ratio of the lumen cross-sectional area to the total phloem cross-sectional area, which is generally 0.2–0.5.

G. Mechanism of Phloem Flow

What causes the movement of solutes in the phloem? This question has proved difficult to answer, primarily because of observational problems. Another complication is that water may readily enter and leave the various types of cells in the phloem and the surrounding tissue. The phloem therefore cannot be viewed as an isolated independent system. For example, when the water potential in the xylem decreases, as occurs during rapid transpiration, fluid in the phloem generally moves more slowly. Some water may move upward in the xylem and, later, downward in the phloem; however, this is not the whole story because movement in the phloem can be in either direction. Moreover, the phloem can sometimes be the main supplier of water to certain regions of a plant, such as for fruits and various young organs.

In 1930 Ernst Münch proposed that fluid movement in the phloem is caused by a gradient in hydrostatic pressure. This leads to flow analogous to that in the xylem as described by Poiseuille's law (Eq. 9.11b), $J_V = -(r^2/8\eta)(\partial P/\partial x)$. To examine this hypothesis, we will assume that the average speed of flow in the lumen of the conducting cells of the phloem, J_V, is 0.6 m h^{-1} (0.17 mm s^{-1}). Our sieve-tube members will be 12 μm in radius and 1 mm long, and the sieve plates will be 5 μm thick, with pores 1.2 μm in radius covering one-third of their surface area. Based on the relative areas available for conduction, J_V is three times as high in the sieve plates as in the lumen of the sieve tube—namely, 0.51 mm s^{-1} across the sieve plates. The viscosity of the solution in a sieve tube is greater than that of water; e.g., η^{phloem} may be about 1.7 mPa s at 20°C, the value for 0.5 M sucrose. The pressure change per cell expected from Poiseuille's law is then

$$\Delta P = \frac{-8\eta J_V}{r^2}\Delta x = \frac{-(8)(1.7 \times 10^{-3}\text{ Pa s})(0.17 \times 10^{-3}\text{ m s}^{-1})}{(12 \times 10^{-6}\text{ m})^2}(1 \times 10^{-3}\text{ m})$$

$$= -16\text{ Pa}$$

in the lumen and -24 Pa across the sieve plate. For cells 1 mm long, the pressure gradient needed to cause Poiseuille flow is then -40×10^3 Pa m^{-1}, or -0.040 MPa m^{-1}.

In the current case, slightly more than half of the hydrostatic pressure drop along the phloem is necessary to overcome the resistance of the sieve plate pores. When the end walls are steeply inclined to the axis of the sieve element, the pores of the sieve plate can occupy an area greater than that of the cross section of the sieve tube. This causes J_V in the pores to be less than that in the lumen, which tends to reduce the resistance to flow in the phloem.

H. Values for Components of the Phloem Water Potential

We will now examine some of the consequences of a pressure-driven flow in the phloem. We will let the water potential in the xylem be -0.6 MPa at ground level and -0.8 MPa at 10 m above the ground (see Fig. 9-11 and Table 9-2). Because there is no evidence for active transport of water across membranes into the phloem, the water potential in the phloem 10 m above the ground, Ψ^{phloem}_{10m}, must be lower than -0.8 MPa if water is to enter passively—water entering the phloem in the leaves can come from the xylem. The water potential there might be -1.0 MPa, for example. By analogous reasoning, Ψ^{phloem} in the root must be higher than Ψ^{xylem} if water is to leave the phloem spontaneously in the region where the solutes, such as sucrose, are removed; an appropriate value for the water potential in the phloem at ground level, Ψ^{phloem}_{0m}, might be -0.4 MPa. Water may then flow down the phloem from a water potential of -1.0 MPa to one of -0.4 MPa. This appears to be an energetically uphill movement. However, as in the xylem, the term representing the force on solution moving within the phloem is ΔP, not $\Delta\Psi$.

The fact that the decrease in hydrostatic pressure, not the change in water potential, represents the driving force along the phloem deserves special emphasis. By Eq. (3.38), $J_V = L_P(\Delta P - \sigma\Delta\Pi)$, where the reflection coefficient σ along the phloem is zero because no membranes intervene between sequential members of a sieve tube. Differences in the osmotic pressure at various locations in the phloem, which help determine the local water potential, therefore do not directly affect the movement of solution along the phloem. On the other hand, substances entering or leaving the phloem pass across a plasma membrane, which most likely has a mean reflection coefficient close to 1 for solutes such as sucrose. J_V given by Eq. (3.38) is then $L_P(\Delta P - \Delta\Pi)$, which equals $L_P\Delta\Psi$ by Eq. (2.13a) ($\Psi = P - \Pi + \rho_w gh$). Consequently, the change in water potential does represent the driving force on water moving into or out of the phloem (Fig. 9-11).

Next we will estimate the various components of the water potential at the upper end of the phloem tissue under consideration (Fig. 9-11). The osmotic pressure in a sieve tube in the phloem of a leaf 10 m above the ground, $\Pi^{phloem_{10m}}$, might be due to the following solutes: 0.5 M sucrose, 0.1 M other sugars, 0.05 M amino acids, and 0.05 M inorganic ions;

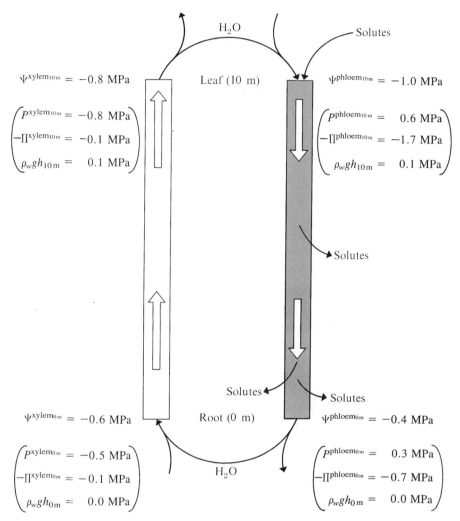

Figure 9-11. Idealized representation of xylem and phloem flow driven by gradients in hydrostatic pressure. Water enters and leaves the phloem by passively moving toward regions of lower water potential. The conducting cells of the xylem generally have a low and relatively constant osmotic pressure (here 0.1 MPa). Solutes either diffuse or are actively transported into and out of the sieve elements, leading to a decrease in the phloem osmotic pressure from 1.7 MPa in the leaf to 0.7 MPa in the root.

i.e., $\sum_j c_j = 0.5 + 0.1 + 0.05 + 0.05 = 0.7$ M, which is 700 mol m^{-3}. Using the Van't Hoff relation—$\Pi_s = RT \sum_j c_j$ (Eq. 2.10), where RT is 2.437×10^{-3} m^3 MPa mol^{-1} at 20°C (see Appendix I)—we estimate that the osmotic pressure in the phloem 10 m above the ground is $(2.437 \times 10^{-3}$ m^3 MPa mol$^{-1})(700$ mol m$^{-3})$, or 1.7 MPa. Such a large osmotic pressure, caused by the high concentrations of sucrose and other solutes, suggests that active transport is necessary at some stage to move certain photosynthetic products from leaf mesophyll cells to the sieve elements of the phloem. From the definition of water potential, $\Psi = P - \Pi + \rho_w gh$ (Eq. 2.13a), we conclude that the hydrostatic pressure in the phloem of a leaf 10 m above the ground is

$$P^{\text{phloem}_{10\,m}} = \Psi^{\text{phloem}_{10\,m}} + \Pi^{\text{phloem}_{10\,m}} - \rho_w gh_{10\,m}$$

$$= -1.0\,\text{MPa} + 1.7\,\text{MPa} - 0.1\,\text{MPa}$$

$$= 0.6\,\text{MPa}$$

(see Fig. 9-11; because $\rho_w g$ is 0.01 MPa m^{-1}, $\rho_w gh_{10\,m}$ is 0.1 MPa and $\rho_w gh_{0\,m}$ is 0.0 MPa). This appreciable hydrostatic pressure causes sieve-tube members to exude when cut.

As we have indicated, the hydrostatic pressure gradient can be -0.04 MPa m^{-1} for Poiseuille flow in representative (horizontal) conducting cells of the phloem. P would therefore decrease 0.4 MPa for a distance of 10 m. Gravity also leads to a pressure gradient in a column of water (0.1 MPa increase in P when descending 10 m), so in the current case the ΔP needed to cause Poiseuille flow vertically downward is -0.3 MPa. Thus the hydrostatic pressure in the phloem at ground level is $0.6 - 0.3$ or 0.3 MPa. The osmotic pressure there equals $P^{\text{phloem}_{0\,m}} + \rho_w gh_{0\,m} - \Psi^{\text{phloem}_{0\,m}}$ (Eq. 2.13a), which is 0.7 MPa. The values of Ψ^{phloem} and its components at $h = 0$ m and $h = 10$ m are summarized in Fig. 9-11.

In our current example, the osmotic pressure of the phloem solution decreases from 1.7 MPa in the leaf to 0.7 MPa in the root (Fig. 9-11). Such a decrease in Π is consistent with the phloem's function of delivering photosynthetic products to different parts of a plant. Moreover, our calculations indicate that flow is in the direction of decreasing concentration but that diffusion is not the mechanism. (Although the total concentration decreases in the direction of flow, the c_j of every solute does not necessarily do so.) Finally, we note the importance of removing solutes, either by active transport or by diffusion into the cells near the conducting cells of the phloem.

The known involvement of metabolism in translocation in the phloem could be due to active transport of solutes into the phloem of a leaf or other *source*, which is often referred to as *loading*, and/or to their removal, or *unloading*, in a root or other *sink*, such as a fruit. Indeed, loading often involves proton–sucrose cotransport via a carrier located in the plasma membrane of sieve elements, which moves this sugar from the apoplast into the conducting cells (movement of photosynthates among mesophyll cells and possibly to companion cells is mainly in the symplast). In sink regions, low concentrations of translocated solutes in the cells outside the phloem, which favor the diffusion of these solutes out of the sieve elements, could be maintained by metabolic conversions of phloem solutes, such as by starch formation in the cells adjacent to the conducting cells of the phloem. In any case, active loading or unloading of the phloem sets up a gradient in osmotic pressure, which in turn generates the hydrostatic pressure gradient that can lead to Poiseuille flow along the phloem.

Table 9-2. Representative Values for the Various Components of the Water Potential in the Soil–Plant–Atmosphere Continuum[a]

Location	Ψ (MPa)	P (MPa)	$-\Pi$ (MPa)	$\rho_w gh$ (MPa)	$\dfrac{RT}{\bar{V}_w}\ln\left(\dfrac{RH}{100}\right)$ (MPa)
Soil 0.1 m below ground and 10 mm from root	−0.3	−0.2	−0.1	0.0	
Soil adjacent to root	−0.5	−0.4	−0.1	0.0	
Xylem of root near ground surface	−0.6	−0.5	−0.1	0.0	
Xylem in leaf at 10 m above ground	−0.8	−0.8	−0.1	0.1	
Vacuole of mesophyll cell in leaf at 10 m	−0.8	0.2	−1.1	0.1	
Cell wall of mesophyll cell at 10 m	−0.8	−0.4	−0.5	0.1	
Air in cell wall pores at 10 m (water vapor assumed to be in equilibrium with water in cell wall)	−0.8			0.1	−0.9
Air just inside stomata at 95% relative humidity	−6.9			0.1	−7.0
Air just outside stomata at 60% relative humidity	−70.0			0.1	−70.1
Air just across boundary layer at 50% relative humidity	−95.1			0.1	−95.2

[a] $\Psi = P - \Pi + \rho_w gh$ in the liquid phases (Eq. 2.13a) and $= (RT/\bar{V}_w)\ln$ (% relative humidity/100) $+ \rho_w gh$ in the gas phases (Eq. 2.21), all at 25°C.

V. The Soil–Plant–Atmosphere Continuum

Water moves from the soil, through a plant, out into the surrounding atmosphere. During a growing season, about 100 times more water is transpired by a plant than remains in it for growth. Thus the amount of transpiration is a fairly accurate estimate of water uptake by the roots. Although the rate at which water crosses each section of the pathway is essentially the same in the steady state, the resistances and the areas across which water flows differ markedly for the various components. The generally but certainly not universally accepted mechanism for water movement through a plant is the "cohesion theory" proposed by Henry Dixon and Charles Joly at the end of the nineteenth century. In particular, water evaporating from the leaves creates a tension in the xylem where hydrogen bonds provide an intermolecular attraction and continuity between water molecules. Thus the column of water in the lumen of the xylem is drawn upward toward regions of lower hydrostatic pressure.

A. Values of Water Potential Components

Possible values for the water potential and its components in various parts of the soil–plant–atmosphere continuum are given in Table 9-2. The values do not apply to all plants, nor even to the same plant at all times. Rather, they serve to indicate representative contributions of P, Π, $\rho_w gh$, and relative humidity to Ψ in various parts of the soil–plant–atmosphere continuum.

First, we will consider the soil water potential, Ψ^{soil}. As indicated earlier in this chapter, Ψ^{soil} is usually dominated by P^{soil}, which is negative because of surface tension effects at the numerous air–liquid interfaces in the soil. The magnitude of the water potential varies with environmental conditions and the type of soil. After a rainfall or in freshly irrigated soil, Ψ^{soil} may be about -0.01 MPa, whereas permanent wilting of plants often occurs when it decreases below about -1.5 MPa.

The value of the soil water potential at which wilting of a plant occurs depends on the osmotic pressure in the vacuoles of its leaf cells ($\Pi^{vacuole}$ is generally the same as $\Pi^{cytosol}$; see Chapter 2, Section IIIA). Let us consider the case in Table 9-2, where the water potential in the vacuole of a leaf cell 10 m above the ground is initially -0.8 MPa. As the soil dries, Ψ^{soil} decreases and eventually becomes -1.0 MPa. When Ψ^{soil} becomes -1.0 MPa, Ψ^{leaf} must be less than this for water movement to continue from the soil to the leaf. Ψ^{leaf} could be -1.0 MPa when $P^{vacuole_{10m}}$ is 0.0 MPa, $\rho_w g h^{vacuole_{10m}}$ remains 0.1 MPa, and $\Pi^{vacuole_{10m}}$ is 1.1 MPa (the latter two are values in Table 9-2; actually, as the hydrostatic pressure in a leaf cell decreases, the cell will shrink somewhat because of the elastic properties of the cell wall, so $\Pi^{vacuole}$ will increase slightly; see Chapter 2, Section IVA). Zero hydrostatic pressure in the vacuole means that the cell has lost turgor, and the leaf thus wilts in response to this low Ψ^{soil}. For certain xerophytes in arid areas, the osmotic pressure in the leaves can be 2.5–5.0 MPa under normal physiological conditions. The value of the soil water potential at which wilting occurs for such plants is considerably lower (i.e., more negative) than that for the plant indicated in Table 9-2. A high osmotic pressure in the vacuoles of the leaf cells can therefore be viewed as an adaptation to low soil water potentials in arid regions.

Let us suppose that the soil dries even further from the level causing wilting. Because at least some transpiration still occurs due to the very low Ψ^{air}, Ψ^{leaf} will be less than Ψ^{soil}, and some cellular water will be lost from the leaf. The vacuolar contents then become more concentrated, and $\Pi^{vacuole}$ increases. For instance, if Ψ^{leaf} became -2.0 MPa for a Ψ^{soil} of -1.8 MPa, the osmotic pressure in the vacuole of a leaf cell 10 m above the ground would be 2.1 MPa, which represents a loss of nearly half of the cellular water ($\Pi^{vacuole_{10m}}$ originally was 1.1 MPa for this leaf cell; Table 9-2).

As we have indicated, the driving force for water movement in the xylem is the negative gradient in hydrostatic pressure, which leads to a flow describable by Poiseuille's law (Eq. 9.11). In Table 9-2, P^{xylem} decreases by 0.3 MPa from the root to a leaf 10 m above the ground. The xylary sap, which contains chiefly water plus some minerals absorbed from the soil, usually does not have an osmotic pressure in excess of 0.2 MPa. The hydrostatic pressure, on the other hand, can have much larger absolute values and generally changes markedly during the day. When there is extremely rapid transpiration, large tensions (negative hydrostatic pressures) can develop in the xylem. These tensions are maintained by the cohesion of water molecules resulting from the intermolecular hydrogen bonding (see Chapter 2). When transpiration essentially ceases, as it can at night or under conditions of very high relative humidity in the air surrounding the plant, the tension in the xylem becomes very small—in fact, the hydrostatic pressure can even become positive (reflecting water movement from the surrounding cells into the root xylem in response to Π^{xylem}). Such a positive hydrostatic pressure (termed *root pressure*) can cause guttation as xylem fluid is then exuded through specialized structures called *hydathodes* located near the ends of veins on the margins of leaves.

Water is conducted to and across the leaves in the xylem. It then moves to the individual leaf cells by flowing partly apoplastically in the cell walls and partly symplastically (only short distances are involved because the xylem ramifies extensively in a leaf). The water potential is usually about the same in the vacuole, the cytosol, and the cell wall of a particular mesophyll cell (see values in Table 9-2). If this were not the case, water would redistribute by flowing energetically downhill toward lower water potentials. The water in the cell wall pores is in contact with air, where evaporation can take place, leading to a flow along the cell wall interstices to replace the lost water. This flow can be approximately described by Poiseuille's law (Eq. 9.11), which indicates that a (very small) hydrostatic pressure gradient exists across such cell walls.

As water continually evaporates from the cell walls of mesophyll cells, the accompanying solutes originally in the xylary sap are left behind and can accumulate in the cell wall water. Some solutes are of course needed for cell growth. For halophytes and xerophytes growing in soils of high salinity, excess inorganic ions can be actively excreted from a leaf by salt glands on its surface. The periodic abscission of leaves is another way of "preventing" an excessive buildup of solutes in the cell wall water, as well as for returning mineral nutrients to the soil.

B. Resistances and Areas

We will now consider the resistances to water flow in those parts of the soil–plant–atmosphere continuum where water moves as a liquid. (We have already considered the gaseous parts of the pathway in Chapter 8.) If we let the flux density of water equal the drop in water potential across some component divided by its resistance, we would have only part of the story because we should also consider the relative areas of each component as well as whether $\Delta\Psi$ represents the driving force. Moreover, $\Delta\Psi$ represents the relative energies of water only at constant T, and thus ideally we should compare water potentials only between locations at the same temperature. The root and adjacent soil are usually at the same temperature, as are the mesophyll cells and air in the intercellular air spaces of a leaf. However, roots and leaves generally are not at the same temperature. Nevertheless, using the pressure gradient (Eq. 9.11) generally is sufficiently accurate for describing the flow in the xylem or the phloem, even when a temperature difference exists along the pathway.

Let us designate the average volume flux density of water across area A^j of component j by $J_{V_w}^j$. A^j could be the root surface area, the effective cross-sectional area of the xylem, or the area of one side of the leaves. In the steady state, the product $J_{V_w}^j A^j$ is essentially constant because nearly all the water taken up by the root is lost by transpiration, and thus the same volume of water moves across each component along the pathway per unit time. We will represent the drop in water potential across component j by $\Delta\Psi^j$, defining the resistance of component j (R^j) as follows:

$$J_{V_w}^j A^j = \frac{\Delta\Psi^j}{R^j} \cong \text{constant} \tag{9.12}$$

where $J_{V_w}^j A^j$ is the volume of water crossing component j in unit time (e.g., m^3 s^{-1}). A relation similar to Eq. (9.12) was proposed by Tako van den Honert in 1948.

To illustrate the constancy of water flow through a plant, let us reconsider its volumetric flux densities for various components of the pathway (Table 9-3). In Chapter 8 (Section IIF) we indicated that transpiration by an exposed leaf of a C$_3$ mesophyte might be 4.3 mmol m^{-2} s^{-1}, which corresponds to a volume flux density of water of 0.77×10^{-7} m s^{-1} (1 mol H$_2$O $= 0.018$ kg $= 18 \times 10^{-6}$ m^3). In this chapter (Section IIID) we calculated that J_{V_w} of a young root can be 1.1×10^{-7} m s^{-1}. For a plant having leaves and roots with such volume flux densities, equality of water flow across each component (Eq. 9.12) indicates that the area of one side of the leaves must be $(1.1 \times 10^{-7}$ m s$^{-1})/(0.77 \times 10^{-7}$ m s$^{-1})$, or 1.4

Table 9-3. Values for the Volume Flux Density of Water and Relative Areas for Its Flow along a Plant, Illustrating the Constancy of $J^j_{V_w} A^j$ (Eq. 9.12)

Component	$J^j_{V_w}$ (m s^{-1})	Relative A^j
Young roots	1.1×10^{-7}	1.0
Stem xylem	1.0×10^{-3}	1.1×10^{-4}
Leaves	0.77×10^{-7}	1.4

times larger than the surface area of the young roots (Table 9-3). $J^{root}_{V_w}$ depends markedly on root age and hence varies along the length of a root, e.g, it is usually considerably lower for older roots. Because the root systems of many perennials are quite extensive, A^{root} including the relatively nonconducting regions can be 20 times larger than the surface area of mature leaves. In the stem xylem J_V can be 1 mm s^{-1} (see Section IVD). Again using Eq. (9.12), we conclude that the cross-sectional area of the conducting parts of the stem xylem is $(1.1 \times 10^{-7}$ m s$^{-1})/(1.0 \times 10^{-3}$ m s$^{-1})$ or 1.1×10^{-4} times the surface area of the young roots (Table 9-3), i.e., a much smaller area is involved in plant water conduction than in water absorption or in water release.

Although Eq. (9.12) can be used to describe certain overall characteristics of water flow in the soil–plant–atmosphere continuum, $\Delta\Psi^j$ does not always represent the driving force on water. For instance, a change in the osmotic pressure component of Ψ has no direct effect on the flow along the xylem or the phloem. Also, such a relation is not useful for a gas phase because the resistance R^j depends on the concentration of water vapor (see Chapter 8, Section IF). When $J^j_{V_w}$ is in m s^{-1}, A^j in m^2, and $\Delta\Psi^j$ in MPa, Eq. (9.12) indicates that the units of R^j are MPa s m^{-3}. For young sunflower and tomato plants approximately 0.3 m tall, the resistance from the root surface to the leaf mesophyll cells, R^{plant}, is about 1.0×10^8 MPa s m^{-3} and $\Delta\Psi^{plant}$ is about 0.2 MPa (Kramer and Boyer, 1995). Using Eq. (9.12), we obtain

$$J^j_{V_w} A^j = \frac{(0.2\,\text{MPa})}{(1.0 \times 10^8\,\text{MPa s m}^{-3})} = 2 \times 10^{-9}\,\text{m}^3\,\text{s}^{-1}$$

which gives the volume of water flowing across each component per unit time. For sunflower, bean, and tomato, $R^{root}:R^{stem}:R^{leaves}$ is about 2:1:1.5, whereas R^{root} is relatively higher for soybean and R^{leaves} is relatively higher for safflower. For the previous R^{plant}, R^{stem} is about $(1/4.5)(1.0 \times 10^8$ MPa s m$^{-3})$ or 2.2×10^7 MPa s m^{-3} for the young sunflower or tomato plants. Resistances can be considerably higher for other plants, e.g., R^{plant} can be 10^{10} MPa s m^{-3} for wheat and even higher for barley.

The resistance for water movement along the stem can be separated into (1) a quantity expressing some inherent flow properties of the xylem and (2) the geometrical aspects of the conduits. By analogy with Ohm's law, where $R = \rho \Delta x / A$, we thus obtain

$$R^j = \frac{\rho^j \Delta x^j}{A^j} \tag{9.13}$$

where ρ^j is the hydraulic resistivity of the xylem tissue of length Δx^j and cross-sectional area A^j.

For many plants containing xylem vessels, $\rho^{stem} = 100$–500 MPa s m^{-2}; the hydraulic resistivity can be somewhat higher for plants with tracheids, such as conifers, e.g., it can be 1600 MPa s m^{-2} for certain ferns (Woodhouse and Nobel, 1982; Zimmerman, 1983). Just as for the gas fluxes discussed in Chapter 8, Eqs. (9.12) and (9.13) can be recast in terms of conductivities or conductances, which are the reciprocals of resistivities and resistances, respectively.[4] Also, the hydraulic conductance per unit length, $K_h (= \Delta x^j / R^j; \text{m}^4 \text{MPa}^{-1} \text{s}^{-1})$ is often determined for the xylem. By Eq. (9.12), $K_h = J_{V_w}^j A^j /(\Delta \Psi^j / \Delta x^j)$, which is the volume flow rate per unit water potential gradient. K_h is useful for comparing xylem characteristics among species.

Assuming a stem xylem cross-sectional area of 5 mm^2 for the young sunflower and tomato plants with a stem length of 0.3 m and an R^{stem} of 2.2×10^7 MPa s m^{-3}, Eq. (9.13) indicates that the hydraulic resistivity for the stem xylem is

$$\rho^{\text{stem xylem}} = \frac{R^{\text{stem}} A^{\text{stem xylem}}}{\Delta x^{\text{stem}}}$$

$$= \frac{(2.2 \times 10^7 \text{ MPa s m}^{-3})(5 \times 10^{-6} \text{ m}^2)}{(0.3 \text{ m})}$$

$$= 370 \text{ MPa s m}^{-2}$$

A tree 10 m tall might have 500 times more xylem area and many more xylem vessels with an overall xylem length much greater than those for the tomato or sunflower plant, leading to an R^{stem} of 1.5×10^6 MPa s m^{-3}. The tree has a much higher hydraulic conductance per unit length, $(10 \text{ m})/(1.5 \times 10^6 \text{ MPa s m}^{-3})$ or 6.7×10^{-6} m^4 MPa^{-1} s^{-1}, than do the tomato or sunflower, whose $K_h(1.4 \times 10^{-8}$ m^4 MPa^{-1} s$^{-1})$ is at the lower end of the range measured for a series of young stems and leaf petioles (Schulte et al., 1989a).

We can relate the hydraulic resistivity of the xylem to flow characteristics predicted by Poiseuille's law [Eq. 9.11b; $J_V = -(r^2/8\eta)\partial P/\partial x$]. Specifically, we note that $J_{V_w}^j = (1/\rho_j)(\Delta \Psi^j / \Delta x^j)$ by the definition of resistivity in Eq. (8.1c). By comparing Poiseuille's law with this form and identifying $-\partial P/\partial x$ with $\Delta \Psi^j / \Delta x^j$, we can equate ρ^j to $8\eta/r^2$. Using a representative value for $\rho^{\text{stem xylem}}$ and the viscosity of water at 20°C (see Appendix I), we then have

$$r = \sqrt{\frac{8\eta}{\rho^{\text{stem xylem}}}} = \sqrt{\frac{(8)(1.0 \times 10^{-3} \text{ Pa s})}{(370 \times 10^6 \text{ Pa s m}^{-2})}}$$

$$= 5 \times 10^{-6} \text{ m}$$

which is a possible effective radius of xylem elements. Owing to the presence of cell walls and nonconducting cells, the lumen of the xylem vessels might correspond to only about one-fourth of the cross-sectional area of the xylem tissue in the stem, $A^{\text{stem xylem}}$. A lumen radius of 10 μm for one-fourth the area has the same J_V and $\partial P/\partial x$ as pores 5 μm in radius occupying the entire area (consider the r^2 factor in Eq. 9.11b). Other complications, such as the resistance of the perforation plates, cause the effective radius to be even smaller than the actual radius. We should also note that, because of the inverse relationship between $\rho^{\text{stem xylem}}$ and r^2, the larger xylem vessels tend to conduct proportionally more than do the smaller ones in a given stem. Finally, recall that we used Poiseuille's law to equate the soil

---

[4] Hydraulic conductivities $(1/\rho^j)$ also can be used for the phloem; e.g., the phloem hydraulic conductivity can be about 1×10^{-4} m^2 MPa^{-1} s^{-1} for herbaceous species and 4×10^{-3} m^2 MPa^{-1} s^{-1} for trees; pressure gradients along the phloem are generally much higher in small herbaceous species than in tall trees.

hydraulic conductivity coefficient L^{soil} to $r^2/8\eta$ (see Section IVD), which is analogous to our current consideration of a reciprocally related quantity, the xylem hydraulic resistivity.

C. Specific Resistances and Conductances

For the example in Table 9-2, the drop in water potential is 0.2 MPa across the soil part of the pathway, 0.1 MPa from the root surface to the root xylem, and 0.2 MPa along the xylem—values that suggest the relative magnitudes of the three resistances involved ($\Delta \Psi^j \cong$ constant $\times R^j$; Eq. 9.12). As the soil dries, its hydraulic conductivity decreases, and the relative size of the water potential drop in the soil usually becomes larger. For soybean in pots R^{soil} becomes greater than R^{plant} below a Ψ^{soil} of -1.1 MPa (Blizzard and Boyer, 1980). R^{plant} can also increase as Ψ^{leaf} decreases, perhaps because the entry of air breaks the water continuity (cavitation) in some of the xylem vessels, which thus become nonconducting. The rapid uptake of soil water by the root during those parts of the day when transpiration is particularly high can also lead to a large hydrostatic pressure gradient [consider Darcy's law (Eq. 9.7)], and $\Delta \Psi^{soil}$ from the bulk soil to the root then increases. In fact, R^{soil} is often the largest resistance for that part of the soil–plant–atmosphere continuum in which water moves predominantly as a liquid (water can also move as a vapor in the soil). The resistance of the root epidermis, cortex, and endodermis is generally somewhat less than that of R^{soil}. The resistance of the xylem is proportional to its length. In conifers and other plants, the conducting area of the stem xylem is often proportional to the leaf area. For the same transpiration rate per unit leaf area and xylem element dimensions, the pressure gradient for Poiseuille flow is then the same; i.e., a higher $J_V^{leaves} A^{leaves}$ is compensated by a higher $A^{stem\,xylem}$ (see Eqs. 9.12 and 9.13), so $\Delta \Psi / \Delta x$ along the stem need not change.

As soil dries, roots often shrink in the radial direction, leading to the development of a root–soil air gap. Hence less contact occurs between a root and the water adjacent to soil particles, leading to a hydraulic resistance at the root–soil interface. Such a resistance can decrease water movement from a root to a drying soil and thereby help prevent excessive water loss from plants during the initial phases of drought (Fig. 9-12). In particular, let us designate the water potential of the bulk soil (at a distance $r_{distant}$ from the center of a root) by $\Psi_{distant}$, that in the soil at the root–soil gap (a distance of r_{gap} from the center of the root) by Ψ_{gap}, that at the root surface (r_{root}) by $\Psi_{surface}$, and that in the root xylem by Ψ_{xylem}. Using Darcy's law for cylindrical symmetry [Eq. 9.8, where $L_{eff}^{soil} = L^{soil}/(r_{root} \ln r_{distant}/r_{gap})$] to represent the soil part of the pathway, L_{gap} for the conductance of water vapor across the root–soil air gap, and L_P^{root} for the hydraulic conductivity of the root (see Eq. 3.39), we obtain

$$
\begin{aligned}
J_V &= L_{eff}^{soil}(\Psi_{distant} - \Psi_{gap}) \\
&= L_{gap}(\Psi_{gap} - \Psi_{surface}) \\
&= L_P^{root}(\Psi_{surface} - \Psi_{xylem}) \\
&= L_{overall}(\Psi_{distant} - \Psi_{xylem})
\end{aligned}
\qquad (9.14)
$$

where the volume flux density J_V is expressed at the root surface. For the three conductances in series, the reciprocal of the overall conductance from the bulk soil to the root xylem, $1/L_{overall}$, equals the sum of the reciprocals of the individual conductances, $1/L_{eff}^{soil} + 1/L_{gap} + 1/L_P^{root}$. Assuming that water vapor diffuses across a root–soil air gap

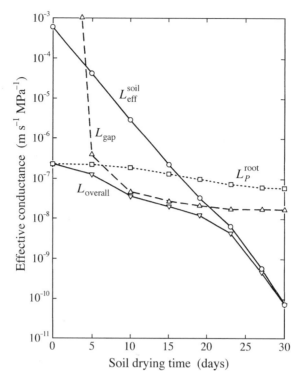

Figure 9-12. Changes in the hydraulic conductances of the root (L_P^{root}), the root–soil air gap (L_{gap}), the soil ($L_{\text{eff}}^{\text{soil}}$), and the overall pathway (L_{overall}; see Eq. 9.14) as the soil dries over a 30-day period. [Data for young roots of desert succulents are adapted from Nobel and Cui (1992a). *J. Exp. Bot.* **43**, 319–326; by permission of Oxford University Press.]

for a distance of Δx_{gap}, L_{gap} can be estimated from

$$L_{\text{gap}} = \frac{\bar{V}_w^2 D_{wv} P_{wv}^*}{(RT)^2 \Delta x_{\text{gap}}} \tag{9.15}$$

where $\bar{V}_w$ is the partial molal volume of water, D_{wv} is the diffusion coefficient of water vapor, P_{wv}^* is the water vapor pressure in air at saturation, R is the gas constant, and T is the absolute temperature. Equation (9.15) assumes isothermal conditions and roots located concentrically in the root–soil air gap but can approximate more realistic conditions (Nobel and Cui, 1992a,b).

Under wet conditions and for young roots of desert succulents, L_{overall} is determined essentially only by L_P^{root}, the root hydraulic conductivity (Fig. 9-12). In particular, L_{gap} is infinite before any root shrinkage occurs and $L_{\text{eff}}^{\text{soil}}$ is over 1000-fold larger than L_P^{root}. As the soil dries, the roots shrink radially and a root–soil air gap develops, which causes L_{gap} to decrease and to become the main limiter for water movement (Fig. 9-12). This occurs when Ψ_{distant} has become less than Ψ_{xylem}, so water flow is out of the root. In this phase L_{gap} therefore helps prevent water loss from the plant. As the soil continues to dry, root shrinkage eventually ceases so L_{gap} becomes constant, as does L_P^{root} after a few fold decrease, but L^{soil} and hence $L_{\text{eff}}^{\text{soil}}$ continue to decrease (Fig. 9-12). This causes L_{overall} to continue to decrease, as consequently does water loss during prolonged drought, which can last 6 months or longer for desert succulents.

D. Capacitance and Time Constants

The daily changes in hydrostatic pressure in the xylem can cause fluctuations in stem diameter. When the transpiration rate is high, the large tension within the xylem vessels is transmitted to the water in the cell walls of the xylem vessels, then to communicating water in adjacent cells, and eventually all the way across the stem. The decrease in hydrostatic pressure can thereby cause a tree trunk to contract during the daytime. At night, the hydrostatic pressure in the xylem increases and may even become positive, and the trunk diameter increases, generally by 0.3–1%. Such changes in diameter, and therefore volume, represent net release of water during the day and net storage at night. The daily change in water content of a plant can equal the amount of water transpired in a few minutes to a few hours during the daytime. These changes in water content correspond to a capacitance effect that is superimposed on the resistance network for water flow.

We begin by comparing the average daily evapotranspiration per unit ground area to the daily fluctuations in water content of a tree trunk. We will suppose that the trunks average 0.2 m in diameter, 10 m in height, and 4 m apart in a rectangular grid, and transpire a depth of water of 4 mm day^{-1}, nearly all of which occurs during the daytime. If a 1% diurnal change in diameter reflects volumetric changes in water content, each trunk would vary daily by $(10\,\text{m})(\pi)[(0.100\,\text{m})^2 - (0.099\,\text{m})^2]$ or $0.006\,\text{m}^3$ in water volume; each tree transpires $0.06\,\text{m}^3$ water daily ($4 \times 4 \times 0.004\,\text{m day}^{-1}$). Thus the change in water content of the trunk could supply about 1 h's worth of transpired water during the daytime. Absorption of water by the roots can hence lag behind the loss of water by leaf transpiration, a consequence of water coming from the storage capacity or *capacitance* of the trunk.

As transpiration increases following stomatal opening at dawn, the leaf water content is lowered, the water potential decreases, and the hydrostatic pressure decreases; concomitantly the average osmotic pressure increases slightly, thereby further decreasing the leaf water potential ($\Psi = P - \Pi + \rho_w gh$; Eq. 2.13a). During this period, water uptake from the soil does not balance transpiration by the leaves, so the steady-state relation embodied in Eq. (9.12) is not obeyed. The water content of the leaves may decrease 10% early in the daytime. For a leaf area index of 5 and leaves $300\,\mu\text{m}$ thick consisting of 70% water by volume, the change in water content corresponds to a water thickness of $(5)(300 \times 10^{-6}\,\text{m})(0.70)(0.10)$, or $1.1 \times 10^{-4}\,\text{m}$ (0.11 mm). This represents 3% of the daily transpiration of 4 mm or about 20 min of daytime water loss. Roots can also exhibit daily changes in water content. For some species, roots may have more biomass than do the leaves and trunk together; in such cases, roots may store the equivalent of a few hours' worth of transpiration.

Capacitance effects can also be seen on a longer time scale. For instance, the sapwood between the vascular cambium on the outside and the heartwood on the inside can represent 20–40% of the radial dimension of a mature tree and even more for a young tree or sapling (heartwood, which represents the central xylem and often is darkly pigmented, has no living cells or conduction capacity). The sapwood can store about 1 week's worth of water at moderate transpiration rates (even longer in drought periods with lowered rates of transpiration). A cactus stem (see Fig. 7-9) can store many months' worth of transpired water during drought periods when stomatal opening is severely limited.

We will define the water capacitance C^j of plant part j as follows:

$$C^j = \frac{\text{change in water content of component } j}{\text{change in average water potential along component } j}$$

$$= \frac{\Delta V_w^j}{\Delta \bar{\Psi}^j} \tag{9.16}$$

where V_w^j is the volume of water in component j. We estimated that a tree trunk might change its water content by $0.006\,\text{m}^3$ each day. This can be accompanied by a change in average xylem water potential from $-0.3\,\text{MPa}$ at dawn (the value of Ψ^{soil} 10 mm from the roots; Table 9-2) to $-0.7\,\text{MPa}$ (average of -0.6 and $-0.8\,\text{MPa}$, the values of Ψ at the two ends of the xylem indicated in Table 9-2), or $-0.4\,\text{MPa}$ overall. By Eq. (9.16), C^{trunk} is then

$$C^{\text{trunk}} = \frac{(-0.006\,\text{m}^3)}{(-0.4\,\text{MPa})} = 1.5 \times 10^{-2}\,\text{m}^3\,(\text{MPa})^{-1}$$

The much smaller stem of a young tomato or sunflower has a much lower water capacitance, e.g., its C^{stem} might be 1000-fold less than that for a tree trunk.

Upon comparing Eq. (9.16) with Eq. (3.1) ($Q = C\Delta E$, where Q is the net charge accumulated that leads to an electrical potential change ΔE across a region of capacitance C), we note that $\Delta \bar{\Psi}^j$ in Eq. (9.16) takes the place of ΔE in electrical circuits. In fact, we can again borrow from electrical circuit analysis to indicate how the initial average water potential along some component $\bar{\Psi}_0^j$ will approach a final average water potential $\bar{\Psi}_\infty^j$:

$$\bar{\Psi}^j - \bar{\Psi}_\infty^j = \left(\bar{\Psi}_0^j - \bar{\Psi}_\infty^j\right)e^{-t/\tau^j} \tag{9.17}$$

which is identical to Eq. (7.22) except that $\bar{\Psi}^j$ replaces T^{surf}. Similar to the time constant for thermal changes (Eq. 7.23), we identify a time constant τ^j for changes in the average water potential of component j:

$$\tau^j = R_s^j C^j \tag{9.18}$$

where C^j is defined by Eq. (9.16) and R_s^j represents the effective resistance from the water storage region to the main transpiration pathway (Fig. 9-13). This τ^j indicates the time required for the water potential to change to within $1/e$ or 37% of its final value. As C^j decreases, the changes in Ψ^j occur faster (Eqs. 9.17 and 9.18). In the limit of no water

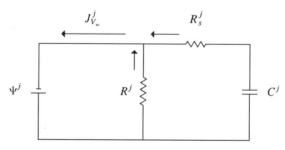

Figure 9-13. Electrical circuit portraying a plant component that can store water. The battery represents the drop in water potential along component j. The currents (arrows) represent the flux densities of water, which can come across R^j as well as from storage in capacitor C^j and then across R_s^j.

storage in a plant (all C^j's equal to zero; see Eq. 9.16), the water potentials at various locations in the plant instantaneously adjust to the new steady-state situation.

Because water can be stored all along the stem, C^{stem} equals the sum of the capacitances for individual parts of the pathway arranged in parallel ($C^j = \Sigma_i C_i^j$ for capacitances in parallel). When $\Delta\Psi^j$ is the same all along the pathway, then we simply add the component capacitances; otherwise, we can let $C^j = \Sigma_i \Delta\bar{\Psi}_i^j C_i^j / \Delta\bar{\Psi}^j$, where $\Delta\bar{\Psi}_i^j$ is the average water potential change along component i that has capacitance C_i^j. The accompanying resistances of individual parts of the pathway are greater than the overall resistance ($1/R_s^j = \Sigma_i 1/R_{s_i}^j$ for resistances in parallel). The overall resistance for water movement from the water storage region to the xylem in the stem can be about 20% of the stem xylem resistance defined by Eq. (9.12). Such movement can be via apoplastic or symplastic pathways, with the symplastic pathway dominating in many cases.

The time constant $R_s^j C^j$ indicates how rapidly the average water potential along component j changes following changes in water storage in C^j. Using the values introduced previously for the young tomato or sunflower and Eq. (9.16), we obtain a time constant for the stem part of the pathway:

$$\tau^{stem} = (0.2)(2.2 \times 10^7 \text{ MPa s m}^{-3})[1.5 \times 10^{-5} \text{ m}^3 \text{ (MPa)}^{-1}] = 66 \text{ s}$$

indicating that such stem water potentials respond rapidly to environmentally induced changes in xylary water flow. For the 10-m-tall tree and again assuming that $R_s^{stem} = 0.2 R^{stem}$, the time constant is 4.5×10^3 s (75 min). Thus considerable time is required to move water from the sapwood into the trunk xylem of a tree. In other words, because the trunk has a relatively high water capacitance, the peak xylary sap flow at the base of the trunk can lag several hours behind the peak transpiration of a tree. For the barrel cactus *Ferocactus acanthodes*, lags of 4 h can occur between the time of maximum transpiration and the time of maximum water uptake from the soil (Schulte *et al.*, 1989b). The osmotic pressure increases (represented in electrical circuit analogs by voltage sources) that accompany nocturnal CO_2 uptake by the stem of this CAM plant also affect its internal redistribution of water, leading to a tendency for the hydrostatic pressure in the outer, chlorophyll-containing tissue to increase during the night and to decrease during the daytime.

E. Daily Changes

We have already indicated daily changes that take place in the soil–plant–atmosphere continuum; e.g., the soil temperature changes, which affects the CO_2 evolution by respiration in soil microorganisms and in root cells. The hydrostatic pressure in the xylem is more negative when the rate of transpiration is high, causing plants to decrease slightly in diameter during the daytime. All these changes are direct or indirect consequences of the daily variation in sunlight. Of course, the rate of photosynthesis also changes during a day. Photosynthesis is affected not only by the PPFD but also by the leaf temperature, which depends on the varying air temperature and net radiation balance for a leaf (see Chapter 7).

Daily changes also occur in the water potentials in the soil–plant–atmosphere continuum. Let us consider a plant in a well-watered soil (Fig. 9-14). At night transpiration essentially ceases because the stomata close; the water potentials in the soil, root, and leaf may then all become nearly equal (and close to zero). At dawn, the stomata open. Transpiration then removes water from the leaf and Ψ^{leaf} decreases (Fig. 9-14). After a short lag, the length of which depends on the capacitances involved and the transpiration rate, Ψ^{root} begins to

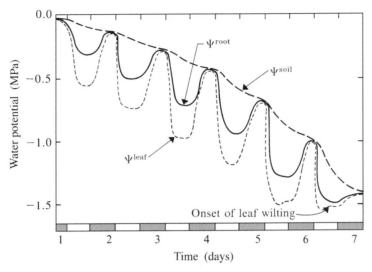

Figure 9-14. Schematic representation of daily changes in the water potentials in the soil, root, and leaf of a plant in an initially wet soil that dries over a 1-week period. Ψ^{soil} is the water potential in the bulk soil, Ψ^{root} is that in the root xylem, and Ψ^{leaf} is the value in a mesophyll cell. Shaded regions indicate night [adapted from Slatyer (1967); used by permission].

decrease, but only to about -0.3 MPa (Fig. 9-14) because plenty of water is initially available in the wet soil. At dusk, these changes in Ψ^{leaf} and Ψ^{root} are reversed. As soil moisture is lost, Ψ^{soil} becomes more negative day by day, so Ψ^{root} and Ψ^{leaf} also become more negative. (We are assuming that Ψ^{air} here remains essentially unchanged and is much lower than the other water potentials). As the soil becomes drier, a steeper gradient in water potential is necessary to sustain the water flow up to the root, and therefore the difference between Ψ^{soil} and Ψ^{root} becomes larger day by day (Fig. 9-14). On the other hand, $\Psi^{leaf} - \Psi^{root}$, to which the transpiration rate is proportional by Eq. (9.12), is similar for the first 4 days. As Ψ^{leaf} becomes more negative on ensuing days, the leaf turgor pressure decreases, and eventually the hydrostatic pressure in the mesophyll cells of the leaf becomes zero some time during the day, e.g., when Ψ^{leaf} becomes -1.5 MPa in response to the decreasing Ψ^{soil}. The leaf thus wilts but recovers at night. Permanent wilting and damage to the leaf may result when Ψ^{soil} becomes lower on subsequent days. This portrayal of successive daily changes of Ψ^{soil}, Ψ^{root} and Ψ^{leaf} illustrates that even such nonequilibrium processes in the soil–plant–atmosphere continuum can be analyzed in terms of the water potential.

F. Global Climate Change

In Chapter 8 we indicated that the mole fraction of atmospheric CO_2 is currently increasing by nearly $2 \, \mu$mol mol^{-1} annually, reflecting primarily the burning of fossil fuels and secondarily other anthropogenic causes (such as cement manufacture, land clearing, and other land-use changes), and at the end of Chapter 8 we indicated how the increasing atmospheric CO_2 level will affect leaf gas exchange. For instance, a doubling of the 1990 atmospheric CO_2 mole fraction may occur by the end of the twenty-first century, which will increase net CO_2 uptake for leaves of C_3 (see Fig. 1-2) and CAM (see Fig. 8-15) plants by enhancing the carboxylase activity of Rubisco and reducing its oxygenase activity (see Fig. 8-10), with little effect on C_4 plants (see Fig. 8-11). Partial stomatal closure generally occurs under elevated CO_2 levels, reducing transpiration and raising leaf temperatures somewhat (see Chapters 7 and 8). The water-use efficiency (Eq. 8.31) is thus predicted to increase substantially for leaves of plants representing all three photosynthetic pathways, with no penalty in water consumption (see Fig. 8-16). What is expected at the plant level and at the ecosystem level? Are other atmospheric gases that affect plant performance also expected to increase as part of global climate change?

CO_2 is called a "greenhouse" gas because its presence in the atmosphere leads to the absorption of longwave (infrared) radiation, whose energy is thereby trapped, analogous to the trapping of solar energy in a greenhouse. In particular, atmospheric greenhouse gases allow shortwave radiation from the sun (see Fig. 4-3) to pass through but absorb the longwave radiation emanating from the earth's surface, radiation whose temperature dependency is quantified by the Stefan–Boltzmann law (Eq. 6.17; also see Figs. 7-2 and 7-3). Based on calculations from computer models, an average worldwide increase of about 3°C is predicted to accompany a doubling of the atmospheric CO_2 mole fraction. A 3°C increase in temperature would extend frost-free regions an average of 330 km poleward at a given elevation, which has major agricultural and ecological implications. The changing atmospheric temperatures will also affect air circulation patterns, leading to changes in the seasonality of precipitation and changes in its overall amount. Any accompanying increases in cloud cover can reduce the PPFD, which by itself would decrease leaf net CO_2 uptake (see Fig. 8-14). In any case, the resultant effects of changes in temperature, soil water, and PPFD on net CO_2 uptake can be quantified using an environmental productivity index (Eq. 8.30).

Most proposed ecosystem effects of global climate change are based on computer models using experimental CO_2 uptake and water loss data under current ambient conditions along with leaf responses to elevated atmospheric CO_2 mole fractions measured in environmental chambers, field enclosures, and field sites with releases of gaseous CO_2. However, errors induced by scaling can occur, e.g., leaf-level measurements must be scaled up to the whole plant and plant responses must be scaled up to an ecosystem level (the scaling hierarchy is approximately subcellular → cell → organ → plant → canopy → community → ecosystem → biosphere). Complex processes such as photosynthesis may not scale linearly between levels. Also, the species composition of ecosystems may change as the climate changes. Although forests occupy wide regions throughout the world, the large sizes of mature trees limit the experiments on such ecosystems; however, studies on selected tree seedlings or saplings may not correctly predict responses of mature trees in native forests, leading to further uncertainties.

Although CO_2 is the most important greenhouse gas, leading to about 60% of the projected atmospheric temperature increase, atmospheric levels of methane (CH_4), choloroflurocarbons, and oxides of nitrogen are also rising. Methane, which is produced from many sources, including the processing and burning of fossil fuels, municipal landfills, ruminants, rice paddies, and natural wetlands, accounts for nearly 20% of the projected atmospheric temperature increase worldwide. Other changes that are occurring in the atmosphere include the destruction of stratospheric ozone, which leads to an increase in ultraviolet radiation (UV; see Table 4-1) reaching the earth's surface. The increase in UV (especially 290–330 nm) can decrease photosynthesis, as has been clearly demonstrated for short-term exposures of plants to high UV levels, and also can have deleterious effects on animals.

A rising atmospheric CO_2 level can change plants in many ways, such as changes in carbon partitioning and hence in canopy development (topics relevant to this chapter). For instance, the root:shoot ratio (usually defined as the mass of the root system after drying in an oven divided by the dry mass of the shoot) will increase for most species, especially when rooting volume is sufficient. The greater allocation of photosynthates to roots under elevated atmospheric CO_2 levels will cause greater exploration of the soil, which can lead to greater nitrogen uptake. However, increased atmospheric CO_2 mole fractions generally lead to lower leaf N concentrations, reflecting lower Rubisco concentrations. Leaves developing under elevated CO_2 levels tend to be thicker with a larger A^{mes}/A (see Fig. 8-8), which by itself leads to a higher photosynthetic capacity. On the other hand, the amount of chlorophyll (see Fig. 5-2) per unit leaf dry mass tends to decrease as the atmospheric CO_2 level increases. The stimulation of leaf net CO_2 uptake when plants are shifted from the current ambient CO_2 levels to elevated ones (see Fig. 8-13) often decreases over a period of months, especially when other factors such as sink strength or nutrients become limiting. Species having indeterminate growth generally demonstrate greater flexibility to altered resource availability than do those having determinate growth. For instance, crops such as cotton (*Gossypium hirsutum*) and faba bean (*Vicia faba*) respond more to elevated CO_2 levels than does wheat (*Triticum aestivum*), which can have important agronomic implications. Where they are currently cultivated, rice (*Oryza sativa*) and wheat, currently the two most important grain crops, will have less net CO_2 uptake and less growth as air temperatures increase. The interactions among temperature, light, and soil water status with respect to photosynthesis, transpiration, plant growth, and productivity are usually quantified using computer simulations, again underscoring the importance of the experimental input data for predictions of plant changes for the twenty-first century.

Although the gas exchange responses of ecosystems to elevated CO_2 levels are less dramatic than those of leaves of C_3 species, the rapid rise in the atmospheric CO_2 mole fraction should also cause major changes in ecosystems. For instance, the rate of change induced by humans is more rapid than is plant adaptation, so biodiversity as represented by the number of species in a particular ecosystem is expected to decrease (increases in atmospheric CO_2, other greenhouse gases, and nitrogen will favor some species over others), representing an irreversible change. The rates of litter decomposition, an important aspect of carbon and nutrient cycling at the ecosystem level, depend on soil temperature and moisture content, both of which are expected to increase accompanying global climate change. However, the lower N concentrations in leaves developing under elevated atmospheric CO_2 levels will tend to reduce the decomposition rates. Also, phytophageous insects are the major consumers in forests, especially four species in the order Lepidoptera, and their herbivory decreases as leaf N content decreases. These lowerings are in part compensated for by an increased release of N due to human activities ranging from automotive exhausts to the manufacture of N fertilizers. If plant growth is already severely nutrient limited, then plants may have little response to increases in the atmospheric CO_2 mole fraction but may respond to increases in the airborne nitrogen.

Remote sensing of optical reflections from leaves (see Fig. 7-4), measured from satellites and airplanes, can quantify the regions occupied by plants and can monitor their seasonal changes in biomass per unit ground area. In this regard, forests cover approximately one-third of the earth's land area but are responsible for just over 60% of terrestrial net CO_2 uptake and contain nearly 90% of terrestrial biomass, so they are crucial at the biosphere level. When the atmospheric CO_2 mole fraction is increased, evapotranspiration (Eq. 9.2) tends to decrease, the growth season is extended, and senescence is delayed. Many interacting, sometimes compensating, factors must all be considered for accurate prediction of the consequences of rising atmospheric CO_2 levels and possible mitigation policies—a difficult but challenging task considering our current less than quantitative understanding of plants. However, an inescapable conclusion is that overall plant productivity will increase and ecosystem composition will change because of the rapid, human-induced increases in atmospheric mole fractions of CO_2, other greenhouse gases, and nitrogen, all of which affect plant performance.

What new applications of physics and chemistry might we expect in plant physiology in the future? Our quantitative approach should be expanded to include the interdependence of forces and fluxes, such as for gas exchange. The permeability coefficient for CO_2 crossing cellular and organelle membranes needs to be determined. Nonisothermal conditions must be adequately handled. Of even greater potential impact is the application of a mathematical approach to the very complex field of plant growth and development, including hormone action, differentiation, photomorphogenesis, reproduction, and senescence. Many aspects of plant ecology are also ripe for physicochemical explanations, especially regarding root function. Consideration of the multiple effects of global climate change on photosynthesis and plant growth from the cell level to ecosystems underscores the complexity of nature that we seek to understand. Mastery of the basic principles emphasized in this book as well as the adoption of a problem-solving approach will help deal quantitatively with unresolved questions in these areas and also allow us to predict plant responses to new environmental situations. Progress in any field requires some good fortune. However, in the words of Louis Pasteur, "Chance favors only the prepared mind."

Problems

9.1 Suppose that J_{wv} above some canopy reaches a peak value equivalent to 1.0 mm of water h^{-1} during the daytime when the air temperature is 30°C and is 0.10 mm h^{-1} at night when T^{ta} is 20°C. Assume that during the daytime the relative humidity decreases by 20% across the first 30 m of turbulent air and that the eddy diffusion coefficient halves at night because of an ambient wind speed lower than during the daytime.

    a. What are J_{wv} in mmol $m^{-2} s^{-1}$ and Δc_{wv}^{ta} in mol m^{-3} during peak transpiration?

    b. What is r_{wv}^{ta} (over the first 30 m) during peak transpiration and at night?

    c. What is Δc_{wv}^{ta} at night in mol m^{-3}? To what drop in relative humidity does this correspond?

    d. If J_{wv}/J_{CO_2} is -700 H_2O/CO_2 during peak transpiration, what are J_{CO_2} and $\Delta c_{CO_2}^{ta}$ then?

    e. What is K_{CO_2} for the first 30 m of turbulent air at night?

    f. How long would it take for water vapor to diffuse 1 m at night by eddy diffusion and by ordinary diffusion? Assume that Eq. (1.6) ($x_e^2 = 4D_j t_e$) applies, where D_{wv} is 2.4×10^{-5} $m^2 s^{-1}$ at 20°C, and K_{wv} has the value averaged over the first 30 m of turbulent air.

9.2 Suppose that the foliar absorption coefficient is 0.7 for trees with an average leaf area index of 8.0.

    a. If the light compensation point for CO_2 fixation is at a PPFD of 8 μmol $m^{-2} s^{-1}$, what are the cumulative leaf areas for light compensation when 2000, 200, 20, and 0 μmol $m^{-2} s^{-1}$ are incident on the canopy?

b. Suppose that J_{wv}/J_{CO_2} for the soil is 200 H_2O/CO_2, that 90% of the water vapor passing out of the canopy comes from the leaves, and that the net photosynthetic rate for the forest (using CO_2 from above the canopy as well as from the soil) corresponds to 20 kg of carbohydrate ha^{-1} h^{-1}. If J_{wv} from the soil is 0.6 mmol m^{-2} s^{-1}, what are the J_{CO_2}'s up from the soil and down into the canopy?

c. What is the absolute value of J_{wv}/J_{CO_2} above the canopy under the conditions of (b)?

d. When the leaves are randomly distributed with respect to distance above the ground and the trees are 16 m tall, at what level does the maximum upward flux of CO_2 occur when 200 μmol m^{-2} s^{-1} is incident on the canopy?

e. If the net rate of photosynthesis is proportional to PPFD, about where does $c_{CO_2}^{ta}$ achieve a minimum? Assume that the conditions are as in (b) and (d) and that the maximal upward J_{CO_2} is 5.0 μmol m^{-2} s^{-1}. Note that in the current case the PPFD halves every 2 m downward in the vegetation.

9.3 A horizontal xylem element has a cross-sectional area of 0.004 mm^2 and conducts water at 20°C at a rate of 20 mm^3 h^{-1}.

a. What is the mean speed of the fluid in the xylem element?

b. What pressure gradient is necessary to cause such a flow?

c. If the pressure gradient remained the same, but cell walls with interstices 20 nm across filled the xylem element, what would be the mean speed of fluid movement? Assume that the entire area of the cell walls is available for conduction.

9.4 A horizontal sieve tube of the phloem has an effective radius of 10 μm. Assume that $D_{sucrose}$ in the pholem solution, which has a viscosity of 1.5 mPa s, is 0.3×10^{-9} m^2 s^{-1} at 20°C.

a. If there is no flow in the sieve tube and a thin layer of [^{14}C]-sucrose is inserted, how long would it take by diffusion for the radioactive label at 10 mm and at 1 m to be 37% of the value at the plane of insertion?

b. If there is a pressure gradient of -0.02 MPa m^{-1} that leads to Poiseuille flow in the phloem, what is J_V there?

c. For the conditions of (b), how long does it take on average for sucrose to move 10 mm and 1 m in the sieve tube (ignore concomitant diffusion)? Compare your values with answers to (a).

d. For the pressure gradient in (b), what would J_V be in a vertical sieve tube?

9.5 Suppose that L^{soil} is 1×10^{-11} and 2×10^{-16} m^2 s^{-1} Pa^{-1} when Ψ^{soil} is -0.01 and -1.4 MPa, respectively. For a plant in the wet soil, J_V^{xylem} is 2 mm s^{-1} and Ψ^{leaf} is -0.2 MPa for a leaf 3 m above the ground. Assume that the roots are 3 mm in diameter and that their surface area is five times that of one side of the leaves and 10^5 times larger than the conducting area of the xylem. Assume that all temperatures are 20°C and ignore all osmotic pressures.

a. What are the radii of curvature at hemispherical air–liquid interfaces in the wet soil and in the dry soil?

b. If water moves a distance of 8 mm to reach the root, what is the drop in hydrostatic pressure over that interval in the wet soil?

c. Suppose that the root xylem is arranged concentrically in a ring 500 μm below the root surface. If the average conductivity of the epidermis, cortex, endodermis, and xylem cell walls in the root is like that of the dry soil, what is the drop in hydrostatic pressure across them?

d. If the water in the leaf xylem is in equilibrium with that in the vacuoles of mesophyll cells, what is the average hydrostatic pressure gradient in the xylem for the wet soil case (assume that A^{xylem} is constant throughout the plant)?

e. If A^{mes}/A is 20 and the cell wall pores are 10 nm in diameter and 1 μm long, what ΔP along them will account for the rate of transpiration for the plant in the wet soil?

f. If the relative humidity above the dry soil is increased to 99%, what is J_V^{xylem} then?

9.6 Consider a tree with a leaf area index of 6 and a crown diameter of 6 m. The trunk is 3 m tall, has a mean cross-sectional area of $0.10\,\text{m}^2$ of which 5% is xylem tissue, and varies from an average water potential along its length of -0.1 MPa at dawn to -0.5 MPa in the steady state during the daytime.

a. If the average water vapor flux density of the leaves is 1 mmol $\text{m}^{-2}\,\text{s}^{-1}$, what is the transpiration rate of the tree in $\text{m}^3\,\text{s}^{-1}$?

b. How long could such a transpiration rate be supported by water from the leaves if the volume of water in the leaves per unit ground area changes from an equivalent depth of 1.0 to 0.8 mm?

c. What is R^{trunk} if the water potential at the base is -0.3 MPa during the daytime?

d. What is the hydraulic resistivity and hydraulic conductivity of the trunk?

e. What is the capacitance and the time constant for water release from the trunk, if water equivalent to 0.8% of the trunk volume enters the transpiration stream? Assume that the resistance involved is the same as R^{trunk}.

References

Adamson, A. W. (1990). *Physical Chemistry of Surfaces*, 5th ed. Wiley, New York.

Baker, N. R., and Thomas, H. (Eds.) (1992). *Crop Photosynthesis: Spatial and Temporal Determinants*. Elsevier, Amsterdam.

Bewley, J. D., and Black, M. (1978). *Physiology and Biochemistry of Seeds in Relation to Germination*, Vol. 1. Springer-Verlag, Berlin.

Blizzard, W. E., and Boyer, J. S. (1980). Comparative resistance of the soil and the plant to water transport. *Plant Physiol.* **66**, 809–814.

Boeckx, P., and Van Cleemput, O. (1996). Flux estimates from soil methanogenesis and methanotrophy: Landfills, rice paddies, natural wetlands and aerobic soils. *Environ. Monitoring Assessment* **42**, 189–207.

Borghetti, M., Grace, J., and Raschi, A. (1993). *Water Transport in Plants under Climatic Stress*. Cambridge Univ. Press, Cambridge, UK.

Bowes, G. (1993). Facing the inevitable: Plants and increasing atmospheric CO_2. *Annu. Rev. Plant Physiol. Plant Mol. Biol.* **44**, 309–332.

Caldwell, M. M. (1976). Root extension and water absorption. In *Water and Plant Life: Problems and Modern Approaches* (O. L. Lange, L. Kappen, and E.-D. Schulze, Eds.), Ecological Studies 19, p. 63–85. Springer-Verlag, Berlin.

Calkin, H. W., Gibson, A. C., and Nobel, P. S. (1986). Biophysical model of xylem conductance in tracheids of the fern *Pteris vittata*. *J. Exp. Bot.* **37**, 1054–1064.

Canny, M. J. (1995). A new theory for the ascent of sap—Cohesion supported by tissue pressure. *Ann. Bot.* **75**, 343–357.

Carlquist, S. (1988). *Comparative Wood Anatomy: Systematic, Ecological, and Evolutionary Aspects of Dicotyledon Wood*. Springer-Verlag, Berlin.

Carlson, T. N., and Lynn, B. (1991). The effects of plant water storage on transpiration and radiometric surface temperature. *Agric. Forest Meterol.* **57**, 171–186.

Cronshaw, J., Lucas, W. J., and Giaquinta, R. T. (1985). *Phloem Transport*. Liss, New York.

Darcy, H. (1856). *Les Fontaines Publiques de la Ville de Dijon*. Dalmont, Paris.

De Pury, D. G. G., and Farquhar, G. D. (1997). Simple scaling of photosynthesis from leaves to canopies without the errors of big-leaf models. *Plant Cell Environ*. **20**, 537–557.

Drake, B. G., González-Meler, M. A., and Long, S. P. (1997). More efficient plants: Consequence of rising atmospheric CO_2? *Annu. Rev. Plant Physiol. Plant Mol. Biol*. **48**, 609–639.

Ehleringer, J. R., and Field, C. B. (Eds.) (1993). *Scaling Physiological Processes: Leaf to Globe*. Academic Press, San Diego.

Ehleringer, J. R., Hall, A. E., and Farquhar, G. D. (Eds.) (1993). *Stable Isotopes and Plant Carbon–Water Relations*. Academic Press, San Diego.

Gartner, B. L. (Ed.) (1995). *Plant Stems: Physiology and Functional Morphology*. Academic Press, San Diego.

Hillel, D. (1982). *Introduction to Soil Physics*. Academic Press, New York.

Hodáňová, D. (1979). Sugar beet canopy photosynthesis as limited by leaf age and irradiance. Estimation by models. *Photosynthetica* **13**, 376–385.

Jarvis, P. G., Edwards, W. R. N., and Talbot, H. (1981). Models of plant and crop water use. In *Mathematics and Plant Physiology* (D. A. Rose and D. A. Charles-Edwards, Eds.), pp. 151–194. Academic Press, London.

Johnson, D. E., and Ward, G. M. (1996). Estimates of animal methane emissions. *Environ. Monitoring Assessment* **42**, 133–141.

Jones, H. G. (1992). *Plants and Microclimate: A Quantitative Approach to Environmental Plant Physiology*, 2nd ed. Cambridge Univ. Press, Cambridge, UK.

Koch, G. W., and Mooney, H. A. (Eds.) (1996). *Carbon Dioxide and Terrestrial Ecosystems*. Academic Press, San Diego.

Kozlowski, T. T., and Pallardy, S. G. (1996). *Physiology of Woody Plants*, 2nd ed. Academic Press, San Diego.

Kramer, P. J., and Boyer, J. S. (1995). *Water Relations of Plants and Soils*. Academic Press, San Diego.

Larcher, W. (1995). *Physiological Plant Ecology*. Springer-Verlag, Berlin.

Lemon, E., Stewart, D. W., and Shawcroft, R. W. (1971). The sun's work in a cornfield. *Science* **174**, 374–378.

Marschner, H. (1995). *Mineral Nutrition of Higher Plants*, 2nd ed. Academic Press, San Diego.

Marshall, T. J., Holmes, J. W. and Rose, C. W. (1996). *Soil Physics*, 3rd ed. Cambridge Univ. Press, Cambridge, UK.

Monsi, M., and Saeki, T. (1953). Über den Lichtfaktor in den Pflanzengesellschaften und seine Bedeutung für die Stoffproduktion. *Jpn. J. Bot*. **14**, 22–52.

Münch, E. (1930). *Die Stoffbewegungen in der Pflanze*. Fischer, Jena, Germany.

Myneni, R. B., and Ross, J. (Eds.) (1991). *Photon–Vegetation Interaction: Applications in Optical Remote Sensing and Plant Ecology*. Springer-Verlag, Berlin.

Niklas, K. J. (1994). *Plant Allometry: The Scaling of Form and Process*. Univ. of Chicago Press, Chicago.

Nobel, P. S., and Cui, M. (1992a) Hydraulic conductances of the soil, the root–soil air gap, and the root: Changes for desert succulents in drying soil. *J. Exp. Bot*. **43**, 319–326.

Nobel, P. S., and Cui, M. (1992b). Prediction and measurement of gap water vapor conductance for roots located concentrically and eccentrically in air gaps. *Plant Soil* **145**, 157–166.

Nobel, P. S., and Long, S. P. (1985). Canopy structure and light interception. In *Techniques in Bioproductivity and Photosynthesis* (J. Coombs, D. O. Hall, S. P. Long, and J. M. O. Scurlock, Eds.), 2nd ed., pp. 41–49. Pergamon, Oxford.

Nobel, P. S., Andrade, J. L., Wang, N., and North, G. (1994). Water potentials for developing cladodes and fruits of a succulent plant, including xylem-versus-phloem implications for water movement. *J. Exp. Bot*. **281**, 1801–1807.

Oke, T. R. (1987). *Boundary Layer Climates*, 2nd ed. Methuen, London.

Passioura, J. B. (1988). Water transport in and to roots. *Annu. Rev. Plant Physiol. Plant Mol. Biol*. **39**, 245–265.

Patrick, J. W. (1997). Phloem unloading: Sieve element unloading and post-sieve element transport. *Annu. Rev. Plant Physiol. Plant Mol. Biol*. **48**, 191–222.

Rendig, V. V., and Taylor, H. M. (1989). *Principles of Soil–Plant Interrelationships*. McGraw-Hill, New York.

Rochette, P., Desjardin, R. L., Pattey, E., and Lessard, R. (1996). Instantaneous measurement of radiation and water-use efficiencies of a maize crop. *Agron. J*. **88**, 627–635.

Rosenberg, N. J., Blad, B. L. and Verma, S. B. (1983). *Microclimate: The Biological Environment*, 2nd ed. Wiley, New York.

Russell, G., Marshall, B., and Jarvis, P. G. (Eds.) (1989). *Plant Canopies: Their Growth, Form and Function*. Cambridge Univ. Press, Cambridge, UK.

Schulte, P. J., Gibson, A. C., and Nobel, P. S. (1989a). Water flow in vessels with simple or compound perforation plates. *Ann. Bot*. **64**, 171–178.

Schulte, P. J., Smith, J. A. C., and Nobel, P. S. (1989b). Water storage and osmotic pressure influences on the water relations of a dicotyledonous desert succulent. *Plant Cell Environ.* **12**, 831–842.

Shaykewich, C. F., and Williams, J. (1971). Resistance to water absorption in germinating rapeseed (*Brassica napus* L.). *J. Exp. Bot.* **22**, 19–24.

Slatyer, R. O. (1967). *Plant–Water Relationships*. Academic Press, New York.

Smith, J. A. C., and Nobel, P. S. (1986). Water movement and storage in a desert succulent: Anatomy and rehydration kinetics for leaves of *Agave deserti. J. Exp. Bot.* **37**, 1044–1053.

Sperry, J. S., Saliendra, N. Z., Pockman, W. T., Cochard, H., Cruiziat, P., Davis, S. D., Ewers, F. W., and Tyree, M. T. (1996). New evidence for large negative xylem pressures and their measurement by the pressure chamber method. *Plant Cell Environ.* **19**, 427–436.

Streeter, V. L., Wylie, E. B., and Bedford, K. W. (1988). *Fluid Mechanics*, 9th ed. McGraw-Hill, New York.

Tyree, M. T. (1997). The Cohesion–Tension theory of sap ascent: Current controversies. *J. Exp. Bot.* **48**, 1753–1765.

Van Bel. A. J. E. (1993). Strategies of phloem loading. *Annu. Rev. Plant Physiol. Plant Mol. Biol.* **44**, 253–281.

van den Honert, T. H. (1948). Water transport as a catenary process. *Discussions Faraday Soc.* **3**, 146–153.

Veroustraete, F., Ceulemans, R. J. M., Impens, I. I. P., and van Rensbergen, J. B. H. F. (1994). *Vegetation, Modeling and Climate Change Effects*. SPB Academic, The Hague.

Vitousek, P. M. (1994). Beyond global warming: Ecology and global change. *Ecology* **75**, 1861–1876.

Vogel, S. (1994). *Moving Fluids: The Physical Biology of Flow*, 2nd ed. Princeton Univ. Press, Princeton, NJ.

Waisel, Y., Eschel, A., and Kafkafi, U. (1996). *Plant Roots: The Hidden Half*, 2nd ed. Dekker, New York.

Weatherley, P. E. (1982). Water uptake and flow in roots. In *Physiological Plant Ecology* (O. L. Lange, P. S. Nobel, C. B. Osmond, and H. Ziegler, Eds.), *Encyclopedia of Plant Physiology, New Series*, Vol. 12B, pp. 79–109. Springer-Verlag, Berlin.

Woodhouse, R. M., and Nobel, P. S. (1982). Stipe anatomy, water potentials, and xylem conductances in seven species of ferns (Filicopsida). *Am. J. Bot.* **69**, 135–140.

Zimmermann, M. H. (1983). *Xylem Structure and the Ascent of Sap*. Springer-Verlag, Berlin.

SOLUTIONS TO PROBLEMS

Chapter 1

1.1 **(a)** We note that c_j at $x = 3$ mm is 37% (i.e., $\sim e^{-1}$) of c_j at the origin ($x = 0$), so we can use Eq. (1.6) to calculate D_j:

$$D_j = \frac{x_e^2}{4t_e} = \frac{(0.003 \text{ m})^2}{(4)(3600 \text{ s})} = 0.63 \times 10^{-9} \text{ m}^2 \text{ s}^{-1}$$

(b) Similar to (a), but now we solve Eq. (1.6) for t_e:

$$t_e = \frac{x_e^2}{4D_j} = \frac{(0.090 \text{ m})^2}{(4)(0.63 \times 10^{-9} \text{ m}^2 \text{ s}^{-1})} = 3.2 \times 10^6 \text{ s} \qquad (37 \text{ days})$$

(c) We will use Eq. (1.5) and solve at the origin ($x = 0$, so $e^{-x^2/4D_j t} = 1$), leading to

$$M_j = 2c_j(\pi D_j t)^{1/2} = 2(100 \text{ mol m}^{-3})[(\pi)(0.63 \times 10^{-9} \text{ m}^2 \text{ s}^{-1})(3600 \text{ s})]^{1/2}$$
$$= 0.53 \text{ mol m}^{-2}$$

(d) If t_e is the same and D_j is 1/100 of the value in (b), Eq. (1.6) indicates that x_e is then 1/10 as large, or 9 mm.

1.2 **(a)** The mass of a mitochondrion is its volume times the density (mass/volume):

$$\text{Mass of mito} = [0.30 \times (10^{-6} \text{ m})^3](1110 \text{ kg m}^{-3}) = 3.33 \times 10^{-16} \text{ kg}$$

Hence, the mass of Avogadro's number of mitochondria is

$$\text{Mass/mol mito} = (3.33 \times 10^{-16} \text{ kg})(6.022 \times 10^{23} \text{ mol}^{-1})$$
$$= 2.0 \times 10^8 \text{ kg mol}^{-1}$$

Therefore, the "molecular weight" of mitochondria is 2.0×10^{11}.

(b) If D_j is proportional to $1/(MW)^{1/3}$, then

$$D_{\text{mito}} = D_x \frac{(MW_x)^{1/3}}{(MW_{\text{mito}})^{1/3}} = (0.5 \times 10^{-9} \text{ m}^2 \text{ s}^{-1})\frac{(200)^{1/3}}{(2 \times 10^{11})^{1/3}}$$
$$= 5 \times 10^{-13} \text{ m}^2 \text{ s}^{-1}$$

(c) Using Eq. (1.6), $t_e = \frac{x_e^2}{4D_j} = \frac{(0.2 \times 10^{-6}\,\text{m})^2}{(4)(5 \times 10^{-13}\,\text{m}^2\,\text{s}^{-1})} = 0.020\,\text{s}$

For x_e of 50 μm, $t_e = \dfrac{(50 \times 10^{-6}\,\text{m})^2}{(4)(5 \times 10^{-13}\,\text{m}^2\,\text{s}^{-1})} = 1250\,\text{s}$ (21 min)

(d) Again using Eq. (1.6), $t_e = \dfrac{(50 \times 10^{-6}\,\text{m})^2}{(4)(0.3 \times 10^{-9}\,\text{m}^2\,\text{s}^{-1})} = 2.1\,\text{s}$

Because ATP diffuses much faster than do mitochondria across such cellular distances, it is much more expedient for ATP to diffuse to where it is to be used than for mitochondria to so diffuse.

1.3 (a) The times for CO_2 to diffuse across the two barriers can be compared using Eq. (1.6):

$$\frac{t_e^{\text{air}}}{t_e^{\text{cell wall}}} = \frac{(1 \times 10^{-3}\,\text{m})^2(4)\left(D_{CO_2}^{cw}\right)}{(2 \times 10^{-6}\,\text{m})^2(4)\left(10^6 \times D_{CO_2}^{cw}\right)} = 0.25$$

(b) Again using Eq. (1.6),

$$\frac{D_{CO_2}^{\text{plasma membrane}}}{D_{CO_2}^{\text{cell wall}}} = \frac{(8 \times 10^{-9}\,\text{m})^2(4)(t_e)}{(2 \times 10^{-6}\,\text{m})^2(4)(t_e)} = 1.6 \times 10^{-5}$$

(c) Using Eq. (1.9) to define P_j, we obtain

$$\frac{P_{CO_2}^{\text{cell wall}}}{P_{CO_2}^{\text{plasma membrane}}} = \frac{(8 \times 10^{-9}\,\text{m})\left(D_{CO_2}^{\text{cell wall}}\right)\left(100 \times K_{CO_2}^{\text{plasma membrane}}\right)}{(2 \times 10^{-6}\,\text{m})\left(1.6 \times 10^{-5}D_{CO_2}^{\text{cell wall}}\right)\left(K_{CO_2}^{\text{plasma membrane}}\right)}$$

$$= 2.5 \times 10^4$$

1.4 (a) For the unstirred boundary layer, K_j is 1. Using Eq. (1.9), P_j then is

$$P_{D_2O}^{\text{bl}} = \frac{(2.6 \times 10^{-9}\,\text{m}^2\,\text{s}^{-1})(1)}{(20 \times 10^{-6}\,\text{m})} = 13 \times 10^{-5}\,\text{m}\,\text{s}^{-1}$$

$$P_{\text{methanol}}^{\text{bl}} = \frac{(0.8 \times 10^{-9}\,\text{m}^2\,\text{s}^{-1})(1)}{(20 \times 10^{-6}\,\text{m})} = 4.0 \times 10^{-5}\,\text{m}\,\text{s}^{-1}$$

$$P_{\text{L-leucine}}^{\text{bl}} = \frac{(0.2 \times 10^{-9}\,\text{m}^2\,\text{s}^{-1})(1)}{(20 \times 10^{-6}\,\text{m})} = 1.0 \times 10^{-5}\,\text{m}\,\text{s}^{-1}$$

(b) For the two barriers in series,

$$\frac{1}{P_j^{\text{total}}} = \frac{1}{P_j^{\text{bl}}} + \frac{1}{P_j^{\text{memb}}}$$

and so

$$P_j^{\text{memb}} = \frac{P_j^{\text{bl}}\, P_j^{\text{total}}}{P_j^{\text{bl}} - P_j^{\text{total}}}$$

Using measured values of P_j^{total} and P_j^{bl} from (a), we have

$$P_{D_2O}^{\text{memb}} = \frac{(1.3 \times 10^{-4}\,\text{m}\,\text{s}^{-1})(1.0 \times 10^{-4}\,\text{m}\,\text{s}^{-1})}{(1.3 \times 10^{-4}\,\text{m}\,\text{s}^{-1}) - (1.0 \times 10^{-4}\,\text{m}\,\text{s}^{-1})} = 4.3 \times 10^{-4}\,\text{m}\,\text{s}^{-1}$$

$$P_{\text{methanol}}^{\text{memb}} = \frac{(4.0 \times 10^{-5}\,\text{m s}^{-1})(2.0 \times 10^{-5}\,\text{m s}^{-1})}{(4.0 \times 10^{-5}\,\text{m s}^{-1}) - (2.0 \times 10^{-5}\,\text{m s}^{-1})} = 4.0 \times 10^{-5}\,\text{m s}^{-1}$$

$$P_{\text{L-leucine}}^{\text{memb}} = \frac{(1.0 \times 10^{-5}\,\text{m s}^{-1})(3.0 \times 10^{-8}\,\text{m s}^{-1})}{(1.0 \times 10^{-5}\,\text{m s}^{-1}) - (3.0 \times 10^{-8}\,\text{m s}^{-1})} = 3.0 \times 10^{-8}\,\text{m s}^{-1}$$

(c) The unstirred layer is the bigger barrier for D_2O because $P_{D_2O}^{\text{memb}} > P_{D_2O}^{\text{bl}}$; both barriers are of similar resistance for methanol because $P_{\text{methanol}}^{\text{memb}} = P_{\text{methanol}}^{\text{bl}}$; the membrane is the bigger barrier for L-leucine because $P_{\text{L-leucine}}^{\text{memb}} \ll P_{\text{L-leucine}}^{\text{bl}}$.

(d) If P_j^{memb} becomes large, then $1/P_j^{\text{memb}}$ approaches 0, and $1/P_j^{\text{total}} \cong 1/P_j^{\text{bl}}$. Thus the maximum value of P_j^{total} is P_j^{bl} [given in (a)].

1.5 (a) The volume to surface area ratio of a cylinder is $rl/[2(r+l)]$. As $c_j^o = 0$ and $c_j^i(t) = 0.1c_j^i(0)$, Eq. (1.12) becomes (after solving for t):

$$t = \frac{rl \ln 10}{2(r+l)P_j} = \frac{(0.5 \times 10^{-3}\,\text{m})(100 \times 10^{-3}\,\text{m})(2.303)}{2[(0.5+100) \times 10^{-3}\,\text{m}](10^{-6}\,\text{m s}^{-1})} = 573\,\text{s} \quad (9.6\,\text{min})$$

(b) Here, $\frac{V}{A}$ becomes $\pi r^2 l/(2\pi r^2)$ or $l/2$, and so

$$t = \frac{l \ln 10}{2P_j} = \frac{(100 \times 10^{-3}\,\text{m})(2.30)}{(2)(10^{-6}\,\text{m s}^{-1})} = 1.2 \times 10^5\,\text{s} \quad (32\,\text{h})$$

(c) Here, $c_j^i(t)$ is $0.01c_j^i(0)$, so the rearranged Eq. (1.12) yields

$$\frac{t_{99\%}}{t_{90\%}} = \frac{\frac{V}{P_jA}\ln 100}{\frac{V}{P_jA}\ln 10} = \frac{\ln 10^2}{\ln 10} = 2$$

(d) Again using the rearranged Eq. (1.12),

$$\frac{t_{\text{new}}}{t_{\text{original}}} = \frac{P_j^{\text{original}}}{P_j^{\text{new}}} = \frac{(10^{-6}\,\text{m s}^{-1})}{(10^{-8}\,\text{m s}^{-1})} = 100$$

1.6 (a) The surface area to volume ratio (A/V) for a sphere is $3/r$ and for a cylinder is $2(r+l)/rl$. Thus, A/V for *Nitella* is $\frac{2(0.5 \times 10^{-3}\,\text{m} + 100 \times 10^{-3}\,\text{m})}{(0.5 \times 10^{-3}\,\text{m})(100 \times 10^{-3}\,\text{m})}$ or $4020\,\text{m}^{-1}$; for *Valonia* is $\frac{3}{(5 \times 10^{-3}\,\text{m})}$ or $600\,\text{m}^{-1}$; and for *Chlorella* is $\frac{3}{(2 \times 10^{-6}\,\text{m})}$ or $1.5 \times 10^6\,\text{m}^{-1}$.

(b) *Chlorella* [see (a)].

(c) Because the concentration criterion and P_j are the same in all cases, Eq. (1.12) leads to

$$\frac{t_a}{t_b} = \frac{A_b/V_b}{A_a/V_a}$$

and so

$$t_{\text{Nitella}} = (1\,\text{s})\frac{(1.5 \times 10^6\,\text{m}^{-1})}{(4020\,\text{m}^{-1})} = 370\,\text{s} \quad (6.3\,\text{min})$$

$$t_{\text{Valonia}} = (1\,\text{s})\frac{(1.5 \times 10^6\,\text{m}^{-1})}{(600\,\text{m}^{-1})} = 2500\,\text{s} \quad (42\,\text{min})$$

(d) For a cylindrical cell like *Nitella*, Eq. (1.16) gives a tangential stress of rP/t, which is twice as large as the longitudinal stress. For a spherical cell, drawing a section through the center and analyzing as for Fig. 1.13 leads to

$$P\pi r^2 = 2\pi r t\sigma, \quad \text{so} \quad \sigma = \frac{rP}{2t}$$

Here, P/t is the same in all cases, so the maximum stress is $(0.5 \times 10^{-3} \text{ m})P/t$ for *Nitella*, $(2.5 \times 10^{-3} \text{ m})P/t$ for *Valonia*, and $(1.0 \times 10^{-6} \text{ m})P/t$ for *Chlorella*. Therefore, *Valonia* will experience the highest cell wall stress.

Chapter 2

2.1 (a) Still 15 mm, because the tilt will not affect the vertical rise, although the length of the water column in the capillary will be greater when tilted.

(b) By Eq. (2.2a) ($h = 2\sigma \cos \alpha / r\rho g$), the product $h\rho$ is the same in each case, so

$$h_{\text{sucrose}} = h_{\text{water}} \frac{\rho_{\text{water}}}{\rho_{\text{sucrose}}} = (15 \text{ mm}) \frac{(998 \text{ kg m}^{-3})}{(1200 \text{ kg m}^{-3})} = 12.5 \text{ mm}$$

(c) We can rearrange Eq. (2.2a) and solve:

$$\cos \alpha = \frac{hr\rho g}{2\sigma} = \frac{(7.5 \times 10^{-3} \text{ m})(1 \times 10^{-3} \text{ m})(998 \text{ kg m}^{-3})(9.8 \text{ m s}^{-2})}{(2)(0.0728 \text{ kg s}^{-2})}$$

$$= 0.50$$

$$\text{so } \alpha = 60°$$

(d) Because all the parameters are the same as in (c), $h = 7.5$ mm.

(e) Because the height of the rise is inversely proportional to the radius (Eqs. 2.2a or 2.2b), we can appropriately scale the rise given:

$$h = (15 \times 10^{-3} \text{ m}) \frac{(1 \times 10^{-3} \text{ m})}{(1 \times 10^{-6} \text{ m})} = 15 \text{ m}$$

(f) The capillary wall is not fully wettable in (c) and (d), and surface tension acts over a much smaller circumference in (e). The weight of the fluid supported ($mg = \pi r^2 h\rho$) is the same and greatest in (a) and (b).

2.2 (a) Because the system is at equilibrium, the water potential is the same on the two sides of the barrier. From Eq. (2.13a) ($\Psi = P - \Pi + \rho_w gh$) and recognizing that there is no change in vertical position, we obtain

$$P^A - \Pi^A = P^B - \Pi^B$$

and so

$$\Delta P = P^B - P^A = \Pi^B - \Pi^A$$

Using Eq. (2.10) ($\Pi_s \cong RT \sum_j c_j$), we obtain

$$\Pi^A = (2.44 \times 10^3 \text{ m}^3 \text{ Pa mol}^{-1})(0.1 \times 10^3 \text{ mol m}^{-3})$$

$$= 0.244 \times 10^6 \text{ Pa} = 0.244 \text{ MPa}$$

and

$$\Pi^B = 10 \, \Pi^A = 2.44 \text{ MPa}$$

$$\Delta P = 2.44 \text{ MPa} - 0.244 \text{ MPa} = 2.20 \text{ MPa, higher on 1-}m \text{ side}$$

(b) We cannot use Eq. (2.13a) because water is not what moves across the barrier—species j does. We will use Eq. (2.4) ($\mu_j = \mu_j^* + RT \ln a_j + \bar{V}_j P$, ignoring electrical and gravitational terms). At equilibrium, $\mu^A = \mu^B$ and so we obtain

$$\Delta P = P^B - P^A = \frac{RT}{\bar{V}_j}\left(\ln a_j^A - \ln a_j^B\right) = \frac{RT}{\bar{V}_j} \ln \frac{a_j^A}{a_j^B} = \frac{RT}{\bar{V}_j} \ln \frac{c_j^A}{c_j^B}$$

because activity coefficients are 1.

Now insert values:

$$\Delta P = \frac{(2.44 \times 10^3 \text{ m}^3 \text{ Pa mol}^{-1})}{(40 \times 10^{-6} \text{ m}^3 \text{ mol})} \ln\left(\frac{0.1\,m}{1\,m}\right)$$

$$= -140 \text{ MPa, higher on 0.1-}m \text{ side}$$

(c) We must change γ_{solute}^B:

$$P^A - P^B = \frac{RT}{\bar{V}_j} \ln \frac{c_j^A}{(0.5)c_j^B} = \frac{(2.44 \times 10^3 \text{ m}^3 \text{ Pa mol}^{-1})}{(40 \times 10^{-6} \text{ m}^3 \text{ mol}^{-1})} \ln \frac{(0.1\,m)}{(0.5)(1\,m)}$$

$$= 98 \text{ MPa, higher on 0.1-}m \text{ side}$$

(d) Side A would have the same composition as side B ($\Delta c_{\text{solute}} = 0\,m$), so $\Delta P = 0$ MPa.

(e) The value of μ_j is never known in absolute terms because μ_j^* is arbitrary.

2.3 **(a)** For an ideal solution, $\gamma_w = 1$. Therefore

$$a_w = N_w = \frac{\frac{(1.00 \text{ kg water})}{(0.018 \text{ kg mol}^{-1}\text{water})}}{\frac{(1.00 \text{ kg water})}{(0.018 \text{ kg mol}^{-1}\text{water})} + \frac{(0.080 \text{ kg sorbitol})}{(0.182 \text{ kg mol}^{-1}\text{sorbitol})}} = 0.992$$

(b) Rearrange Eq. (2.7) and solve:

$$\Pi = -(135.0 \text{ MPa})(\ln 0.992) = 1.08 \text{ MPa}$$

(c) Again using Eq. (2.7), we find

$$a_w = e^{-(1.0\,\text{MPa}/135.0\,\text{MPa})} = 0.9926$$

which is a 0.74% reduction from the value for pure water ($1.0000 \cdots$).

(d) We can rearrange Eq. (2.10) and solve:

$$\sum_j c_j = \frac{(1.0 \text{ MPa})}{(2.437 \times 10^3 \text{ m}^3 \text{ Pa mol}^{-1})} = 410 \text{ mol m}^{-3}$$

(e) Equation (2.10) predicts that

$$\Pi_s = (2.437 \times 10^3 \text{ m}^3 \text{ Pa mol}^{-1})(0.25 \text{ mol m}^{-3}) = 6 \times 10^{-4} \text{ MPa}$$

Hence, the polymer must be influencing Π mainly by decreasing γ_w (see Eq. 2.11).

(f) Using Eqs. (2.7) and (2.13a) (with the gravitational term omitted), the water potential of the cell is

$$\Psi = (0.8 \text{ MPa}) - (-135.0 \text{ MPa})(\ln 0.98) = -1.9 \text{ MPa}$$

2.4 **(a)** Because there are no barriers to the movement of water within the tank, the water potential will be the same at 0.1 m and at the bottom of the tank (-0.600 MPa).

(b) There are no barriers to diffusion, so Π is the same throughout the tank. Rearrange Eq. (2.13a) and solve for conditions at surface of tank:

$$\Pi = P + \rho_w g h - \Psi = 0 + (\rho_w g)(0) - (-0.600\,\text{MPa}) = 0.600\,\text{MPa}$$

(c) Because $\Psi = -\Pi$ [see (b)], $P = -\rho_w g h$ and $\rho_w g$ is $0.0098\,\text{MPa m}^{-1}$: at $h = 0\,\text{m}$, $\rho_w g h = 0\,\text{MPa}$ and $P = 0\,\text{MPa}$; at $h = -0.1\,\text{m}$, $\rho_w g h = -0.001\,\text{MPa}$ and $P = 0.001\,\text{MPa}$; and at $h = -10\,\text{m}$, $\rho_w g h = -0.098\,\text{MPa}$, and $P = 0.098\,\text{MPa}$.

(d) At equilibrium $\Psi_w = \Psi_{wv}$. Rearrange Eq. (2.21) and solve:

$$\% \text{ relative humidity} = (100)(e^{(-0.600\,\text{MPa})/(135.0\,\text{MPa})})\% = 99.6\%$$

2.5 **(a)** From Eq. (2.15), volume (V) is inversely related to external osmotic pressure (Π°). Plot volume versus $1/\Pi^\circ$ and determine that $V = 32\,\mu\text{m}^3$ for $\Pi^\circ = 0.4\,\text{MPa}$.

(b) Intercept in (a) on y-axis is the nonaqueous volume per chloroplast ($12\,\mu\text{m}^3$).

(c) Using values from (a) and (b), $(32\,\mu\text{m}^3 - 12\,\mu\text{m}^3)/(32\,\mu\text{m}^3) = 0.63$.

(d) Rearrange Eq. (2.15) and solve for n in a chloroplast:

$$n = \frac{(32\,\mu\text{m}^3 - 12\,\mu\text{m}^3)(0.4\,\text{MPa})}{(2.437 \times 10^3\,\text{m}^3\,\text{Pa}\,\text{mol}^{-1})} = 3.3 \times 10^{-15}\,\text{mol}$$

2.6 **(a)** Using Eqs. (2.23) and (2.13a),

$$J_{V_w} = L_w[\Psi^\circ - (P^i - \Pi^i)] = (10^{-12}\,\text{m s}^{-1}\,\text{Pa}^{-1})[0 - (0.6\,\text{MPa} - 1.0\,\text{MPa})]$$
$$= 4 \times 10^{-7}\,\text{m s}^{-1} \quad \text{(flow is inward)}$$

(b) Using Eq. (2.28) with $V/A = r/3$,

$$t_e = \frac{(0.5 \times 10^{-3}\,\text{m})}{(3)(10^{-12}\,\text{m s}^{-1}\,\text{Pa}^{-1})(5\,\text{MPa} + 1\,\text{MPa})} = 28\,\text{s}$$

(c) At point of incipient plasmolysis $\Psi^\circ = \Psi^i$, so $J_{V_w} = 0$ (Eq. 2.23).

(d) At equilibrium $\Psi_{wv} = \Psi^\circ$. Using Eq. (2.21), $\Psi_{wv} = 135.0\,\text{MPa}\,\ln(97/100) = -4.1\,\text{MPa}$. Hence, by Eq. (2.23):

$$J_{V_w} = (10^{-12}\,\text{m s}^{-1}\,\text{Pa}^{-1})[-4.1\,\text{MPa} - (-0.4\,\text{MPa})]$$
$$= -3.7 \times 10^{-6}\,\text{m s}^{-1} \quad \text{(flow is outward)}$$

(e) Because $\Pi^{\text{cell wall}}$ is 0, $P^{\text{cell wall}} = \Psi^i$ or $-0.4\,\text{MPa}$ (Eq. 2.13a). Rearrange Eq. (2.22) and solve:

$$\cos \alpha = \frac{-rP}{2\sigma} = -\frac{(10 \times 10^{-9}\,\text{m})(-0.4\,\text{MPa})}{(2)(7.28 \times 10^{-8}\,\text{MPa}\,\text{m})}$$

Hence, $\alpha = 88^\circ$.

Chapter 3

3.1 **(a)** The concentration of charge is Q/V, where the volume of such a thin layer (thickness t) is $4\pi r^2 \times t$, and so

$$\frac{\frac{4}{3}\pi(30 \times 10^{-6}\,\text{m})^3\,(10^{-3}\,\text{mol m}^{-3})}{4\pi(30 \times 10^{-6}\,\text{m})^2(3 \times 10^{-9}\,\text{m})} = 3.3\,\text{mol m}^{-3}$$

(b) Because a sulfate ion has a charge number of -2, the additional negative charge corresponds to

$$\frac{(2 \times 10^7)}{(6.022 \times 10^{23} \, \text{mol}^{-1})} = 3.32 \times 10^{-17} \, \text{mol}$$

The concentration of additional charge is

$$\frac{(3.32 \times 10^{-17} \, \text{mol})}{\frac{4}{3}\pi(30 \times 10^{-6} \, \text{m})^3} = 0.29 \times 10^{-3} \, \text{mol m}^{-3}$$

Hence, C increases by 29%, so ΔE becomes -129 mV by Eq. (3.2).

(c) The 10^7 sulfate ions carry charge (Q) equal to

$$Q = (-3.32 \times 10^{-17} \, \text{mol})(9.649 \times 10^4 \, \text{C mol}^{-1}) = -3.2 \times 10^{-12} \, \text{C}$$

The membrane potential changes from -100 to -129 mV during the time that the work is done, so the average difference in electrical potential across the membrane is -115 mV. Hence, the electrical work is about

$$Q \, \Delta E = (-3.2 \times 10^{-12} \, \text{C})(-115 \times 10^{-3} \, \text{V}) = 3.7 \times 10^{-13} \, \text{J}$$

3.2 **(a)** At equilibrium and with $\gamma = 1$, Eq. (3.5) indicates that

$$E_M = E_{N_K} = \frac{RT}{z_K F} \ln \frac{a_K^o}{a_K^i} = \frac{(25.3 \, \text{mV})}{(1)} \ln \frac{(1 \, \text{mol m}^{-3})}{(160 \, \text{mol m}^{-3})} = -128 \, \text{mV}$$

(b) The ionic strength of the solution inside the cell is

$$\frac{1}{2}\sum_j c_j z_j^2 = \frac{1}{2}(160 \, \text{mol m}^{-3})(+1)^2 + (160 \, \text{mol m}^{-3})(-1)^2] = 160 \, \text{mol m}^{-3}$$

and that of the external solution is $1 \, \text{mol m}^{-3}$. From Eq. (3.3),

$$\ln \gamma_{\pm}^i = \frac{(1.17)(+1)(-1)\sqrt{160 \, \text{mol m}^{-3}}}{32 + \sqrt{160 \, \text{mol m}^{-3}}} = -0.331$$

and so $\gamma_{\pm}^i = 0.718$, and $\gamma_{\pm}^o$ is 0.965.
Using Eq. (3.5a),

$$E_{N_K} = \frac{(25.3 \, \text{mV})}{(1)} \ln \frac{(0.965)}{(0.718)} \frac{(1 \, \text{mol m}^{-3})}{(160 \, \text{mol m}^{-3})} = -121 \, \text{mV}$$

(c) The concentration of ions inside the cell is as follows: $c_K = 160 \, \text{mol m}^{-3} + 3 \, \text{mol m}^{-3} = 163 \, \text{mol m}^{-3}$; $c_{Cl} = 160 \, \text{mol m}^{-3}$; and $c_{ATP} = 1 \, \text{mol m}^{-3}$. The ionic strength therefore is

$$\frac{1}{2}(163 \, \text{mol m}^{-3})(+1)^2 + (160 \, \text{mol m}^{-3})(-1)^2 + (1 \, \text{mol m}^{-3})(-3)^2]$$
$$= 166 \, \text{mol m}^{-3}$$

From Eq. (3.3),

$$\ln \gamma_{K-ATP} = \frac{(1.17)(+1)(-3)\sqrt{166}}{32 + \sqrt{166}} = -1.008$$

and so $\gamma_{K-ATP} = 0.37$, and a_{ATP} is $(0.37)(1 \, \text{mol m}^{-3}) = 0.37 \, \text{mol m}^{-3}$.

3.3 (a) $E_{N_j} = E_M$ when species j is in equilibrium. Using Eq. (3.5b),

$$E_{N_K} = \frac{(25.7\,\text{mV})}{(+1)}\ln\frac{(1\,\text{mM})}{(100\,\text{mM})} = -118\,\text{mV} \quad K^+ \text{ is in equilibrium}$$

$$E_{N_{Mg}} = \frac{(25.7\,\text{mV})}{(+2)}\ln\frac{(0.1\,\text{mM})}{(10\,\text{mM})} = -59\,\text{mV} \quad Mg^{2+} \text{ is not in equilibrium}$$

(b) Note that $E_{N_{Na}} = E_{N_{Ca}} = E_M$ and rearrange Eq. (3.5b):

$$a_{Na}^i = \frac{a_{Na}^o}{e^{-z_j E_M/25.7}} = \frac{(0.1\,\text{mM})}{\left(e^{(-118\,\text{mV})/(25.7\,\text{mV})}\right)} = 10\,\text{mM} = c_{Na}^i \quad (\gamma_{Na} = 1)$$

$$a_{Ca}^o = (1.0\,\text{mM})e^{(-118)(+2)/(25.7)} = 1.0 \times 10^{-4}\,\text{mM} = 0.1\,\mu\text{M} = c_{Ca}^o \quad (\gamma_{Ca} = 1)$$

(c) Here $E_M = E_{N_{Cl}} - 177\,\text{mV}$, so $E_{N_{Cl}} = 177\,\text{mV} - 118\,\text{mV}$ or $59\,\text{mV}$. Using Eq. (3.5b) ($\gamma_{Cl} = 1$, $c_{Cl}^o = 1.0\,\text{mM} + 0.1\,\text{mM} + 0.2\,\text{mM} = 1.3\,\text{mM}$), we obtain

$$c_{Cl}^i = \frac{(1.3\,\text{mM})}{\left(e^{(59\,\text{mV})(-1)/(25.7\,\text{mV})}\right)} = 13\,\text{mM}$$

(d) Using Eq. (3.24),

$$\frac{J_{Cl}^{in}}{J_{Cl}^{out}} = \frac{c_{Cl}^o}{c_{Cl}^i e^{(-1)(FE_M/RT)}} = \frac{(1.3\,\text{mM})}{(13\,\text{mM})e^{-(-118\,\text{mV})/(25.7\,\text{mV})}} = 0.0010$$

(e) Using Eq. (3.25),

$$\mu_{Cl}^o - \mu_{Cl}^i = (2.479 \times 10^3\,\text{J mol}^{-1})[\ln(0.0010)] = -17\,\text{kJ mol}^{-1}$$

From (a), $E_{N_{Mg}} = -59\,\text{mV}$ and so Eq. (3.26a) yields

$$\mu_{Mg}^o - \mu_{Mg}^i = 2F(E_{N_{Mg}} - E_M)$$
$$= (2)(9.65 \times 10^{-2}\,\text{kJ mol}^{-1}\,\text{mV}^{-1})(-59\,\text{mV} + 118\,\text{mV})$$
$$= 11\,\text{kJ mol}^{-1}$$

3.4 (a) Using Eq. (3.10b) ($u_- = 1.04\,u_+$),

$$E^{old} - E^{new} = (59.2\,\text{mV})\frac{(1.04\,u_K - u_K)}{(1.04\,u_K + u_K)}\log\left(\frac{1\,\text{mM}}{10\,\text{mM}}\right) = -1\,\text{mV}$$

(b) A Donnan potential that can be calculated using Eq. (3.10b) ($u_- = 0$):

$$E^{membrane} - E^{solution} = (59.2\,\text{mV})\frac{-u_+}{+u_+}\log\left(\frac{200\,\text{mM}}{10\,\text{mM}}\right) = -77\,\text{mV}$$

(c) Using the Goldman equation (Eq. 3.19) with $c_{Cl}^i = 20\,\text{mM}$, $c_K^o = c_{Na}^o = 10\,\text{mM}$, $c_{Cl}^i = c_K^i = 100\,\text{mM}$, $c_{Na}^i = 10\,\text{mM}$, $P_{Na} = 0.2\,P_K$, and $P_{Cl} = 0.01\,P_K$:

$$E_M = (25.7\,\text{mV})\ln\left[\frac{P_K(10\,\text{mM}) + (0.2\,P_K)(10\,\text{mM}) + (0.01\,P_K)(100\,\text{mM})}{P_K(100\,\text{mM}) + (0.2\,P_K)(10\,\text{mM}) + (0.01\,P_K)(20\,\text{mM})}\right]$$
$$= -53\,\text{mV}$$

(d) If $P_{Cl}/P_K = 0$, then using the Goldman equation we obtain

$$E_M = (25.7\,\text{mV})\ln\left[\frac{(10\,\text{mM}) + (0.2)(10\,\text{mM})}{(100\,\text{mM}) + (0.2)(10\,\text{mM})}\right] = -55\,\text{mV}$$

If $P_{Na}/P_K = P_{Cl}/P_K = 0$, then

$$E_M = (25.7\,\text{mV})\ln\frac{(10\,\text{mM})}{(100\,\text{mM})} = -59\,\text{mV}$$

3.5 **(a)** For the steady state for the cell (with no change in anion j in the cellular compartments),

$$J_j^{\text{out}} = (10\,\text{nmol}\,\text{m}^{-2}\,\text{s}^{-1})\frac{(50)(4\pi)(2\times 10^{-6}\,\text{m})^2}{(4\pi)(20\times 10^{-6}\,\text{m})^2} = 5\,\text{nmol}\,\text{m}^{-2}\,\text{s}^{-1}$$

(b) Using Eq. (3.25),

$$\mu_j^o - \mu_j^i = (2.479\times 10^3\,\text{J}\,\text{mol}^{-1})\ln\frac{(1\,\text{nmol}\,\text{m}^{-2}\,\text{s}^{-1})}{(5\,\text{nmol}\,\text{m}^{-2}\,\text{s}^{-1})} = -4.0\,\text{kJ}\,\text{mol}^{-1}$$

(c) If the concentration is the same on both sides of the membrane, then $E_{N_j} = 0$. Using Eq. (3.26b),

$$\mu_j^i - \mu_j^o = (-1)(9.65\times 10^{-2}\,\text{kJ}\,\text{mol}^{-1}\,\text{mV}^{-1})(-118\,\text{mV} - 0\,\text{mV})$$
$$= 11.4\,\text{kJ}\,\text{mol}^{-1}$$

Because the chemical potential of the anion is higher inside than outside the cell, the efflux is passive. The influx, however, requires energy of at least $11.4\,\text{kJ}\,\text{mol}^{-1}$.

(d) If one ATP is required per ion transported

$$\text{ATP consumed} = \frac{(0.1\times 10^{-9}\,\text{mol}\,\text{m}^{-2}\,\text{s}^{-1})(1\,\text{ATP}\,\text{ion}^{-1})(4\pi)(20\times 10^{-6}\,\text{m})^2}{\frac{4\pi}{3}(20\times 10^{-6}\,\text{m})^3}$$

$$= 15\,\mu\text{mol}\,\text{m}^{-3}\,\text{s}^{-1}$$

3.6 **(a)** Using Eq. (3.22) with an energy requirement of 20 kJ mol^{-1} to break 1 mol of hydrogen bonds (U_{min}), we obtain

$$Q_{10} = \sqrt{\frac{(283\,\text{K} + 10\,\text{K})}{(283\,\text{K})}}\,e^{(10\,\text{K})(20\,\text{kJ}\,\text{mol}^{-1})/[(8.314\,\text{J}\,\text{mol}^{-1}\,\text{K}^{-1})(283\,\text{K})(293\,\text{K})]} = 1.36$$

(b) Here we rearrange Eq. (3.22) to obtain

$$U_{\text{min}} = \frac{(8.3143\,\text{J}\,\text{mol}^{-1}\,\text{K}^{-1})(283\,\text{K})(293\,\text{K})}{(10\,\text{K})}\ln\left(\frac{3.2}{\sqrt{\frac{293}{283}}}\right) = 79\,\text{kJ}\,\text{mol}^{-1}$$

Hence, the number of hydrogen bonds per molecule that must be broken to account for a U_{min} of 79 kJ mol^{-1} is approximately 4.

3.7 **(a)** Because $J_{K\,\text{max}}^{\text{in}} = J_{Na\,\text{max}}^{\text{in}}$ and $c_K^o = c_{Na}^o$, we can use Eq. (3.27a) to obtain

$$\frac{J_K^{\text{in}}}{J_{Na}^{\text{in}}} = \frac{K_{Na} + c_{Na}^o}{K_K + c_K^o} = \frac{(1.0\,\text{mol}\,\text{m}^{-3}) + (0.01\,\text{mol}\,\text{m}^{-3})}{(0.01\,\text{mol}\,\text{m}^{-3}) + (0.01\,\text{mol}\,\text{m}^{-3})} = 50.5$$

(b) Again using Eq. (3.27a), we obtain

$$\frac{J_K^{\text{in}}}{J_{Na}^{\text{in}}} = \frac{(1.0\,\text{mol}\,\text{m}^{-3}) + (100\,\text{mol}\,\text{m}^{-3})}{(0.01\,\text{mol}\,\text{m}^{-3}) + (100\,\text{mol}\,\text{m}^{-3})} = 1.01$$

(c) Facilitated diffusion requires no energy so ATP is not involved. Using Eq. (3.27a), we obtain

$$J_K^{in} = \frac{(10\,\text{nmol m}^{-2}\,\text{s}^{-1})(0.1\,\text{mol m}^{-3})}{(0.01\,\text{mol m}^{-3}) + (0.1\,\text{mol m}^{-3})} = 9.1\,\text{nmol m}^{-2}\,\text{s}^{-1}$$

3.8 (a) By Eq. (3.40) we obtain

$$\sigma^\circ = \frac{\sum_j \sigma_j \Pi_j^\circ}{\Pi^\circ}$$

$$= \frac{(1.00)(0.2\,\text{MPa}) + (0.30)(0.1\,\text{MPa}) + (0.80)(0.1\,\text{MPa})}{(0.2\,\text{MPa}) + (0.1\,\text{MPa}) + (0.1\,\text{MPa})} = 0.78$$

(b) Because J_V is zero, the stationary state applies, so Eq. (3.40) with $\sum_j \sigma_j \Pi_j^i = \sigma_{glycerol}\Pi_{glycerol}^i + \sigma_{rest}\Pi_{rest}^i$ and $\tau^i = 0$ leads to

$$\sigma_{rest}^i = \frac{(0.78)(0.4\,\text{MPa}) - (0.80)(0.2\,\text{MPa}) + (0.5\,\text{MPa})}{(1.0\,\text{MPa})} = 0.65$$

(c) If the membrane is nonselective, then all σ_j's are zero. From Eq. (3.39) with $\Delta P = P^\circ - P^i$, we obtain

$$J_V = L_P\left(\Delta P - \sum_j \sigma_j \Delta \Pi_j\right)$$

$$= (10^{-12}\,\text{m s}^{-1}\,\text{Pa}^{-1})(0.0\,\text{MPa} - 0.5\,\text{MPa}) = -5 \times 10^{-7}\,\text{m s}^{-1}$$

(d) If the membrane is impermeable, then all σ_j's equal 1. Using Eq. (3.39), we obtain

$$J_V = (10^{-12}\,\text{m s}^{-1}\,\text{Pa}^{-1})[(0.0\,\text{MPa} - 0.5\,\text{MPa}) - (0.4\,\text{MPa} - 1.2\,\text{MPa})]$$

$$= 3 \times 10^{-7}\,\text{m s}^{-1}$$

3.9 (a) Because osmotic pressures are proportional to concentrations and sucrose is impermeant, we can use Eq. (3.42):

$$\sigma_{glycine} = \frac{(0.3\,m)}{(0.4\,m)} = 0.75$$

(b) At the point of incipient plasmolysis, $\sigma^\circ \Pi^\circ = \sigma^i \Pi^i$. Using Eq. (2.10) ($\Pi_s^\circ = RTc$) for sucrose ($\sigma = 1.00$), the effective osmotic pressure is

$$\sigma^\circ \Pi^\circ = (2.437 \times 10^3\,\text{m}^3\,\text{Pa mol}^{-1})(0.3 \times 10^3\,\text{mol m}^{-3}) = 0.73 \times 10^6\,\text{Pa}$$

Upon rearranging Eq. (3.44), we note that V is proportional to $1/\Pi^\circ$. From Problem 2.5, V is $20\,\mu\text{m}^3$ at $1/\Pi^\circ$ of $1.0\,\text{MPa}^{-1}$, $28\,\mu\text{m}^3$ at $2.0\,\text{MPa}^{-1}$, and so at $1/(0.73\,\text{MPa}) = 1.37\,\text{MPa}^{-1}$, V is $23\,\mu\text{m}^3$.

(c) The chloroplast volume did not change with the addition of glycine to the external solution; hence $\sigma_{glycine}$ is zero (see Eq. 3.42).

(d) Using Eq. (3.42) and noting that osmotic pressures are proportional to concentrations, we obtain

$$\text{Actual concentration} = \frac{\text{effective concentration}}{\sigma_{glycine}}$$

$$= \frac{(0.3\,m)}{(0.6)} = 0.5\,m$$

Glycerol is permeant and so enters the chloroplasts, which leads to water entry. If there were no other solutes inside the chloroplast, the concentration of glycerol inside would eventually equal that outside. However, because there are at least some impermeant solutes inside chloroplasts, Ψ^i will be lower than Ψ^o even for equal concentrations of glycerol inside and outside. Consequently, water (and glycerol) will continue to move in until the chloroplasts burst.

Chapter 4

4.1 (a) Using Eq. (4.2a) ($E_\lambda = hc/\lambda_{vacuum}$ for a single photon), we obtain

$$E_{400} = \frac{(10^{20})(1.986 \times 10^{-25}\,\text{J m})}{(400 \times 10^{-9}\,\text{m})} = 50\,\text{J}$$

(b) Using Eq. (4.2b), the energy absorbed is

$$E_{1800} = \frac{(1\,\text{mol})(119{,}600\,\text{kJ mol}^{-1}\,\text{nm})}{(1800\,\text{nm})} = 66.4\,\text{kJ}$$

Using the appropriate volumetric heat capacity of water, the final temperature will be

$$T + \Delta T = 0°\text{C} + \frac{(66.4 \times 10^3\,\text{J})}{(10^{-3}\,\text{m}^3)(4.19 \times 10^6\,\text{J m}^{-3}\,°\text{C}^{-1})} = 16°\text{C}$$

(c) The maximum photon flux density will occur when all photons have the lowest energy and hence the longest wavelength possible (600 nm). By Eq. (4.2b), E_{600} is

$$E_{600} = \frac{(119{,}600\,\text{kJ mol}^{-1}\,\text{nm})}{(600\,\text{nm})} = 199\,\text{kJ mol}^{-1}$$

Hence, an energy flux density of 1 W m^{-2} corresponds to

$$\frac{(1\,\text{J m}^{-2}\,\text{s}^{-1})}{(199 \times 10^3\,\text{J mol}^{-1})} = 5.0 \times 10^{-6}\,\text{mol m}^{-2}\,\text{s}^{-1}$$

(d) Cannot be interconverted, unless the sensitivity of photometric device at 600 nm is known.

4.2 (a) From Eq. (4.1) ($\lambda v = v$) or Eq. (4.2a), we note that $\lambda_{vacuum} = c/v$, and so

$$\lambda_{vacuum} = \frac{(3.0 \times 10^8\,\text{m s}^{-1})}{(0.9 \times 10^{15}\,\text{s}^{-1})} = 3.33 \times 10^{-7}\,\text{m} = 333\,\text{nm}$$

From Section IA and Eq. (4.1), v_{air} is about $0.9997c$, so

$$\lambda_{air} = (0.9997)(333\,\text{nm}) = 333\,\text{nm} \quad (\text{decrease is } 0.1\,\text{nm})$$

In the flint glass, we obtain

$$\lambda_{glass} = \frac{(2.0 \times 10^8\,\text{m s}^{-1})}{(0.9 \times 10^{15}\,\text{s}^{-1})} = 2.22 \times 10^{-7}\,\text{m} = 222\,\text{nm}$$

(b) To go from $S_{(\pi,\pi)}$ to $T_{(\pi,\pi^*)}$ for chlorophyll requires energy equivalent to the absorption of a photon with a wavelength a little longer than 680 nm, so a

333-nm photon would generally have enough energy. However, because a change in spin direction is also required, this transition is extremely improbable.

(c) A transition from π to π^* for a straight-chain hydrocarbon with six double bonds in conjugation can be caused by the absorption of a photon with a wavelength of about 350 nm (Section IVE), so the 333-nm photon has enough energy.

(d) Wave number $= \frac{(0.9 \times 10^{15}\,\mathrm{s}^{-1})}{(3.0 \times 10^{8}\,\mathrm{m\,s}^{-1})} = 3.0 \times 10^{6}\,\mathrm{m}^{-1}$

4.3 (a) The quantum yield for ATP production is equal to the rate of ATP production divided by the rate at which photons are absorbed. If all incident 680-nm photons are absorbed,

$$\text{Minimum photon flux density} = \frac{(0.2\,\mathrm{mol\,m^{-2}\,h^{-1}})}{(0.4)(3600\,\mathrm{s\,h^{-1}})} = 139\,\mu\mathrm{mol\,m^{-2}\,s^{-1}}$$

(b) Using energy/mole from Eq. (4.2b) times the photon flux density, we obtain

$$\text{Energy flux density} = \frac{(0.1196\,\mathrm{J\,mol^{-1}\,m})}{(680 \times 10^{-9}\,\mathrm{m})}(139 \times 10^{-6}\,\mathrm{mol\,m^{-2}\,s^{-1}})$$
$$= 24\,\mathrm{J\,m^{-2}\,s^{-1}} = 24\,\mathrm{W\,m^{-2}}$$

(c) ATP is synthesized using energy derived from $S^{a}_{(\pi,\pi^*)}$, which can be reached by all 680-nm photons but only by 95% of the 430-nm photons, which also represent more energy. Hence, the ratio of energy conversion is

$$\text{Ratio} = \frac{0.95/(hc/430\,\mathrm{nm})}{1.00/(hc/680\,\mathrm{nm})} = 0.60$$

(d) Using Eq. (4.2b), we obtain

$$\lambda_{\text{vacuum}} = \frac{(119{,}600\,\mathrm{kJ\,mol^{-1}\,nm})}{(45\,\mathrm{kJ\,mol^{-1}})} = 2600\,\mathrm{nm} = 2.6\,\mu\mathrm{m}$$

4.4 (a) Using Eq. (4.13) $\left(1/\tau = \sum_{j} 1/\tau_{j}\right)$, we obtain

$$\tau = \frac{1}{\left(\frac{1}{10^{-8}\,\mathrm{s}} + \frac{1}{5 \times 10^{-9}\,\mathrm{s}} + \frac{1}{10^{-8}\,\mathrm{s}}\right)} = 2.5 \times 10^{-9}\,\mathrm{s}$$

(b) Electromagnetic radiation comes directly as fluorescence and also can come as phosphorescence from the triplet state. Using Eq. (4.15), we obtain

$$\Phi_{\text{em}} = \frac{k_{\text{fluor}} + k_{\text{phos}}}{\sum_{j} k_{j}} = \left(\frac{1}{\tau_{\text{fluor}}} + \frac{1}{\tau_{\text{phos}}}\right)\tau$$
$$= \left(\frac{1}{10^{-8}\,\mathrm{s}} + \frac{1}{5 \times 10^{-9}\,\mathrm{s}}\right)(2.5 \times 10^{-9}\,\mathrm{s}) = 0.75$$

(c) By Eq. (4.13) we obtain

$$\frac{1}{\tau} = k = k_{\text{transfer}} + k_{\text{previous}}$$
$$= (10^{12}\,\mathrm{s^{-1}}) + \frac{1}{(2.5 \times 10^{-9}\,\mathrm{s})} \cong 10^{12}\,\mathrm{s^{-1}}$$

hence $\tau \cong 10^{-12}\,\mathrm{s}$.

4.5 **(a)** Rearranging Beer's law (Eq. 4.18), we obtain

$$c_{cis} = \frac{\log J_0/J_b}{\varepsilon_\lambda b} = \frac{\log(1.00/0.35)}{(2.0 \times 10^3 \text{ m}^2 \text{ mol}^{-1})(10 \times 10^{-3} \text{ m})} = 23 \times 10^{-3} \text{ mol m}^{-3}$$

(b)

$$A_{450}^{cis} = \log\left(\frac{1.00}{0.35}\right) = 0.46$$

The absorbance is unaffected by changes in the incident photon flux density.
(c) Assuming that each photon absorbed excites a molecule and using Eq. (4.15), we obtain

$$\text{Initial rate} = (0.5)(0.65)(10^{17} \text{ molecules m}^{-2} \text{ s}^{-1})$$

$$= 3.3 \times 10^{16} \text{ molecules m}^{-2} \text{ s}^{-1}$$

For the cuvette path length, this corresponds to

$$\frac{(3.3 \times 10^{16} \text{ molecules m}^{-2} \text{ s}^{-1})}{(10 \times 10^{-3} \text{ m})} = 3.3 \times 10^{18} \text{ molecules m}^{-3} \text{ s}^{-1}$$

or

$$\frac{(3.3 \times 10^{18} \text{ molecules m}^{-3} \text{ s}^{-1})}{(6.02 \times 10^{23} \text{ molecules mol}^{-1})} = 5.5 \times 10^{-6} \text{ mol m}^{-3} \text{ s}^{-1}$$

Using Eq. (4.18), we obtain

$$\Delta A_{450} = \varepsilon_{450}^{cis} \Delta c^{cis} b = (2 \times 10^3 \text{ m}^2 \text{ mol}^{-1})(5.5 \times 10^{-6} \text{ mol m}^{-3} \text{ s}^{-1})(10^{-2} \text{ m})$$

$$= 1.1 \times 10^{-4} \text{ s}^{-1}$$

(d) The quantum yields for the *cis* to *trans* and the *trans* to *cis* isomerizations are here the same (0.50), and hence after a long time (a "photostationary state") the relative amounts of the two forms depend inversely on the relative absorption coefficients:

$$cis/trans \text{ ratio} = \frac{(10^3 \text{ m}^2 \text{ mol}^{-1})}{(2.0 \times 10^3 \text{ m}^2 \text{ mol}^{-1})} = 0.5$$

The same conclusion can be deduced by noting that the same number of photons are absorbed by each isomer and using Eq. (4.18).

4.6 **(a)** By Problem 4.2, wave number equals v/c or $1/\lambda_{vacuum}$. Here $1/\lambda = \frac{1}{(500 \text{ nm})} \pm 1.2 \times 10^5 \text{ m}^{-1}$, so

$$\lambda_{below} = \frac{1}{\left[\frac{1}{(500 \times 10^{-9} \text{ m})} - (1.2 \times 10^5 \text{ m}^{-1})\right]} = 5.32 \times 10^{-7} \text{ m} = 532 \text{ nm}$$

$$\lambda_{above} = \frac{1}{\left[\frac{1}{(500 \times 10^{-9} \text{ m})} + (1.2 \times 10^5 \text{ m}^{-1})\right]} = 4.72 \times 10^{-7} \text{ m} = 472 \text{ nm}$$

(b) The percentage absorbed depends on ε_λ, c, and b (Eq. 4.18). Here, $\varepsilon_{satellite} cb = \log(J_0/J_b) = \log(1.00/0.80)$, or 0.097. Because $\varepsilon_{main} = 5\varepsilon_{satellite}$, $\varepsilon_{main} cb = 0.485$, indicating that $J_0/J_b = 3.05$, which means that $1 - (J_b/J_0)$ is 0.67, i.e., 67% is absorbed by the main band.

(c) Based on Eq. (4.18), we obtain (maximum absorptance occurs for the maximum absorption coefficient)

$$c = \frac{A_\lambda}{\varepsilon_\lambda b} = \frac{(0.3)}{(5 \times 10^3 \, \text{m}^2 \, \text{mol}^{-1})(5 \times 10^{-3} \, \text{m})} = 0.012 \, \text{mol m}^{-3}$$

4.7 (a) The energy for each band is calculated using Eq. (4.2b) ($E_\lambda = Nhc/\lambda_{\text{vacuum}}$). The splitting between vibrational sublevels equals the energy difference between adjacent bands:

$$E_{431} - E_{450} = \frac{(119{,}600 \, \text{kJ mol}^{-1} \, \text{nm})}{(431 \, \text{nm})} - \frac{(119{,}600 \, \text{kJ mol}^{-1} \, \text{nm})}{(450 \, \text{nm})} = 12 \, \text{kJ mol}^{-1}$$

(b) The temperature dependence of the minor band indicates a transition from an excited sublevel of the ground state. The Boltzmann factor, $e^{-E/RT}$, is 0.007 for an E of 12 kJ mol^{-1} at 20°C, so the transition is only 0.7% as probable as the 450-nm transition; compare the indicated ε_λ's. (Also, the population of the first excited vibrational sublevel of the ground state decreases as the temperature decreases.)

(c) By Eq. (4.2b) the energy difference between the 450-nm band and the 494-nm band is

$$E_{450} - E_{494} = \frac{(119{,}600 \, \text{kJ mol}^{-1} \, \text{nm})}{(450 \, \text{nm})} - \frac{(119{,}600 \, \text{kJ mol}^{-1} \, \text{nm})}{(494 \, \text{nm})}$$

$$= 24 \, \text{kJ mol}^{-1}$$

which is the energy difference between two vibrational sublevels. Because emission of fluorescence generally occurs from the lowest vibrational sublevel of the excited state, the 450-nm band apparently represents a transition from the first vibrational sublevel of the ground state (the most populated one according to the Boltzmann factor) to the third vibrational sublevel of the excited state.

(d) Originally the molecule had 11 double bonds in a single conjugated system. Therefore, using Eq. (4.2b) we obtain

$$E_{\text{max}}^{\text{original}} = \frac{(119{,}600 \, \text{kJ mol}^{-1} \, \text{nm})}{(450 \, \text{nm})} = 266 \, \text{kJ mol}^{-1}$$

The new molecule has five double bonds in each of two conjugated systems. Therefore, Eq. (4.2b) indicates that

$$\lambda_{\text{max}}^{\text{new}} = \frac{Nhc}{E_{\text{max}}^{\text{original}} + (6)(25 \, \text{kJ mol}^{-1})}$$

$$= \frac{(119{,}600 \, \text{kJ mol}^{-1} \, \text{nm})}{(266 \, \text{kJ mol}^{-1}) + (150 \, \text{kJ mol}^{-1})} = 288 \, \text{nm}$$

If ε_λ is proportional to the number of double bonds in a conjugated system and the new molecule has two conjugated systems per molecule, then

$$\varepsilon_{288} = \frac{(2)(5)(1.0 \times 10^4 \, \text{m}^2 \, \text{mol}^{-1})}{(11)} = 0.91 \times 10^4 \, \text{m}^2 \, \text{mol}^{-1}$$

Chapter 5

5.1 (a) The volume fraction of the cell occupied by chloroplasts is

$$\frac{(50)\left(\frac{4}{3}\pi\right)(2 \times 10^{-6}\,\text{m})^3}{\left(\frac{4}{3}\pi\right)(20 \times 10^{-6}\,\text{m})^3} = 0.05$$

(b) Cell volume times density equals the cell mass:

$$\tfrac{4}{3}\pi(20 \times 10^{-6}\,\text{m})^3 \times 1000\,\text{kg m}^{-3} = 3.35 \times 10^{-11}\,\text{kg cell}^{-1}$$

of which 10% or 3.35×10^{-12} kg cell^{-1} is dry mass. CO_2 fixation leads to the formation of carbohydrate (see Fig. 5.1), which has a mass of 30 g/mol C. Hence, the rate of dry mass production is

$$[0.1\,\text{mol } CO_2(\text{g chlorophyll})^{-1}\,\text{h}^{-1}](1\,\text{g chlorophyll kg}^{-1})$$
$$\times (3.35 \times 10^{-11}\,\text{kg cell}^{-1})[30\,\text{g(mol } CO_2)^{-1}] = 10.05 \times 10^{-11}\,\text{g h}^{-1}\text{cell}^{-1}$$

Hence, the time is $\dfrac{(3.35 \times 10^{-12}\,\text{kg cell}^{-1})}{(10.05 \times 10^{-14}\,\text{kg h}^{-1}\,\text{cell}^{-1})} = 33\,\text{h}$

(c) The molecular weight of Chl a is 893.5 (see Chapter 5, Section IA). For Chl b, a formyl group $(-CHO)$ replaces a methyl group $(-CH_3)$ of Chl a, so its molecular weight is 907.5. Using a weighted mean, the mean molecular weight of chlorophyll in the cell thus is

$$\frac{(3)(893.5) + (1)(907.5)}{(4)} = 897.0$$

(d) The concentration of chlorophyll is

$$(1\,\text{g chlorophyll kg}^{-1})[1\,\text{mol } (897\,\text{g})^{-1}](1000\,\text{kg m}^{-3})$$
$$= 1.11\,\text{mol chlorophyll m}^{-3}$$

Maximum absorptance occurs along a cell diameter (40 μm), and so using Beer's law ($A_\lambda = \varepsilon_\lambda cb$; Eq. 4.18) we obtain

$$A_{\text{blue}} = (1.21 \times 10^4\,\text{m}^2\,\text{mol}^{-1})(1.11\,\text{mol m}^{-3})(40 \times 10^{-6}\,\text{m}) = 0.54$$
$$A_{\text{red}} = (0.90 \times 10^4\,\text{m}^2\,\text{mol}^{-1})(1.11\,\text{mol m}^{-3})(40 \times 10^{-6}\,\text{m}) = 0.40$$

5.2 (a) The λ_{max} of fluorescence corresponds to a transition from the lowest sublevel of the excited state to some sublevel of the ground state, which is the lowest sublevel for the current case of the shortest λ_{max}. By Eq. (4.2b) ($E_\lambda = Nhc/\lambda_{\text{vacuum}}$), λ_{580} corresponds to an energy of (119,600 kJ mol^{-1} nm)/ (580 nm), or 206 kJ mol^{-1}. For the first pigment, fluorescence is maximum at wavelengths corresponding to $(3)(10\,\text{kJ mol}^{-1})$ less energy, so

$$\lambda_{\text{max}} = \frac{(119{,}600\,\text{kJ mol}^{-1}\,\text{nm})}{(206\,\text{kJ mol}^{-1}) - (30\,\text{kJ mol}^{-1})} = 680\,\text{nm}$$

The transition for the pigment with 10 double bonds in conjugation occurs $(2)(20\,\text{kJ mol}^{-1})$ lower in energy, so

$$\lambda_{\text{max}} = \frac{(119{,}600\,\text{kJ mol}^{-1}\,\text{nm})}{(166\,\text{kJ mol}^{-1}) - (30\,\text{kJ mol}^{-1})} = 879\,\text{nm}$$

(b) To pass energy on by resonance transfer, the available energy in the donating molecule (indicated by its fluorescence emission spectrum) must match the energy that can be accepted by the receiving molecule (indicated by its absorption spectrum). Because λ_{max} for fluorescence by the pigment with eight bonds in conjugation closely matches λ_{max} for absorption by the red band of Chl a, it can readily pass its excitation to Chl a, whereas the other molecule cannot.

(c) Blue light excites the Soret band of Chl a, which rapidly becomes deexcited to the lower excited singlet (see Fig. 4.5). Therefore, the energy available is given by the red fluorescence of Chl a ($\lambda_{max} = 666$ nm; see Fig. 5.3). Consequently, the first compound ($\lambda_{max} = 580$ nm) cannot be excited by Chl a. For the pigment with 10 double bonds in conjunction, Eq. (4.2) indicates that the absorption maximum occurs at

$$\lambda_{max} = \frac{(119{,}600 \text{ kJ mol}^{-1} \text{ nm})}{(166 \text{ kJ mol}^{-1})} = 720 \text{ nm}$$

Thus this pigment could become excited by resonance transfer from Chl a.

5.3 (a) Using Eq. (4.18) $[A_\lambda = \log(J_0/J_b) = \varepsilon_\lambda cb]$, we obtain

$$A_{675} = (0.6 \times 10^4 \text{ m}^2 \text{ mol}^{-1})(2 \times 10^{-6} \text{ m})(20 \text{ mol m}^{-3}) = 0.24$$

The fraction transmitted by the chloroplast (J_b/J_0) is $10^{-0.24}$ or 0.58, so the fraction absorbed is 0.42.

(b) The incident photon flux density is calculated by dividing the radiant energy flux density (40 W m^{-2} or 40 J s^{-1} m^{-2}) by the energy per mole of 675-nm photons, which by Eq. (4.2b) ($E_\lambda = Nhc/\lambda$) is 177 kJ mol^{-1}. If 0.42 of these photons are absorbed, then the rate at which photons are absorbed by a single chloroplast is

$$(0.42)\frac{(40 \text{ J s}^{-1} \text{ m}^{-2})}{(177 \times 10^3 \text{ J mol}^{-1})} = 95 \times 10^{-6} \text{ mol m}^{-2} \text{ s}^{-1} = 95 \text{ } \mu\text{mol m}^{-2} \text{ s}^{-1}$$

The number of chlorophyll molecules participating in this absorption for the single chloroplast is

$$(2 \times 10^{-6} \text{ m})(20 \text{ mol m}^{-3}) = 40 \times 10^{-6} \text{ mol m}^{-2} = 40 \text{ } \mu\text{mol m}^{-2}$$

(c) Each absorbed photon excites a chlorophyll, so the rate at which chlorophyll molecules are excited is

$$\frac{(95 \text{ } \mu\text{mol photon m}^{-2} \text{ s}^{-1})}{(40 \text{ } \mu\text{mol chlorophyll m}^{-2})} = 2.4 \text{ excitations chlorophyll}^{-1} \text{ s}^{-1}$$

The excitation frequency for a photosynthetic unit (PSU) is (2.4 excitations chlorophyll^{-1} s^{-1}) (250 chlorophylls PSU^{-1}), or 600 excitations s^{-1} PSU^{-1}. Because only 100 excitations s^{-1} PSU^{-1} can be processed, the processable fraction is (100/600), or 0.17.

(d) Because only 0.17 of excitations from the absorbed radiation are processable and eight photons are required /O_2, the O_2 evolution rate is

$$(0.17)\frac{(95 \text{ } \mu\text{mol photons m}^{-2} \text{ s}^{-1})}{(8 \text{ photons}/O_2)} = 2.0 \text{ } \mu\text{mol } O_2 \text{ m}^{-2} \text{ s}^{-1}$$

(e) Because the absorptance of three chloroplasts is three times that of the one chloroplast considered in (a), Eq. (4.18) indicates that

$$J_{transmitted} = J^{incident} 10^{-0.72} = \frac{(40\,J\,s^{-1}\,m^{-2})}{(177 \times 10^3\,J\,mol^{-1})}(0.191)$$

$$= 43 \times 10^{-6} \text{ mol photons m}^{-2}\,s^{-1}$$

The fraction absorbed by the single chloroplast is again 0.42, so (0.42) (43 μmol m^{-2} s^{-1}), or 18 μmol photons are absorbed m^{-2} s^{-1}. Thus the excitation frequency is

$$\frac{(18\,\mu\text{mol photons m}^{-2}\,s^{-1})}{(40\,\mu\text{mol chlorophyll m}^{-2})} = 0.45 \text{ excitations chlorophyll}^{-1}\,s^{-1}$$

which corresponds to 113 excitations (250 chlorophylls)$^{-1}$ s^{-1}, of which 100 excitations s^{-1} can be processed. Thus, (100/113) or 0.88 (nearly all) of the excitations are processed in the shaded chloroplast, which is a much higher fraction than for the unshaded chloroplast in (c).

5.4 (a) The number of photons required per oxygen evolved is derived from the initial linear part of the curve describing the relationship between incident photon flux density and oxygen evolution:

$$\frac{(10 \times 10^{-6} \text{ mol photon absorbed m}^{-2}\,s^{-1})}{(10^{-4} \text{ mol O}_2\,m^{-3})(10^{-2}\,m)}$$

$$= 10 \text{ photons absorbed (O}_2 \text{ evolved)}^{-1}$$

(b) During an intense, brief flash of light almost all chlorophyll molecules become excited, although only 1 excitation can be processed per PSU. If 10 excitations need to be processed before a molecule of oxygen can be evolved [see (a)], then the concentration of PSUs will be equal to 10 times the concentration of oxygen evolved during a flash. Thus the ratio of chlorophyll to PSUs is

$$\frac{(10 \times 10^{-3}\text{mol chlorophyll m}^{-3})}{(10\,PSU/O_2)(5 \times 10^{-6}\text{ mol O}_2\,m^{-3})} = 200 \text{ chlorophylls PSU}^{-1}$$

(c) Here, 10^{-2} mol chlorophyll m^{-3} corresponds to 10^{-4} mol chlorophyll m^{-2}, which by (b) is 0.50×10^{-6} mol PSU m^{-2}. This processed 10 μmol photons m^{-2} s^{-1}, so it could process 20 photons PSU^{-1} s^{-1} (0.05 s per photon).

(d) The uncoupling of ATP formation from electron flow is analogous to disengaging a clutch, which allows a motor to run faster; the O_2 evolution therefore initially speeds up. After a while, the lack of ATP formation causes no CO_2 to be fixed; the electron acceptors therefore stay reduced and electron flow to them is curtailed. Moreover, electron flow may eventually switch over to the pseudocyclic type, which involves no net O_2 evolution.

5.5 (a) The 710-nm light absorbed by Photosystem I leads to cyclic electron flow and accompanying ATP formation. In the idealized example given, 550-nm light by itself leads to excitation of Photosystem II only, and no photophosphorylation occurs.

(b) Accessory (or auxiliary) pigments, presumably carotenoids, which are isoprenoids.

(c) More photons are absorbed at 550 nm [$A_{550} = 1.0$ indicates that $J_b = J_0 \times 10^{-1.0} = 0.10 J_0$ by Eq. 4.18 ($A_\lambda = \log J_0/J_b$), so 90% are absorbed] than at 710 nm ($A_{710} = 0.1$ indicates that $J_b = J_0 \times 10^{-0.1} = 0.79 J_0$, so 21% are absorbed). Thus, for every 90 photons absorbed at 550 nm, 21 will be absorbed at 710 nm. Because CO_2 fixation requires an equal number of photons to be absorbed by each of the two photosystems, all 21 of the 710-nm photons can potentially be used (maximum quantum efficiency of 1.00), whereas only 21 of the 90 or 0.23 of the photons at 550 nm can be used.

(d) The minimum photon flux density occurs if each absorbed photon bleaches one P_{700}. Again using Eq. (4.18), we obtain

$$A_{700} = 10^{-5} = \varepsilon_{700} cb = (0.8 \times 10^4 \, m^2 \, mol^{-1})(cb)$$

Hence,

$$cb = \frac{A_{700}}{\varepsilon_{700}} = \frac{(10^{-5})}{(0.8 \times 10^4 \, m^2 \, mol^{-1})} = 1.3 \times 10^{-9} \, mol \, m^{-2}$$

Chapter 6

6.1 (a) By Eq. (6.5), we have

$$\Delta G = -17.1 \, kJ \, mol^{-1} + (2.479 \, kJ \, mol^{-1}) \ln \frac{(1 \, m)}{(1 \, m)(1 \, m)} = -17.1 \, kJ \, mol^{-1}$$

Because ΔG for the reaction is negative, the reaction proceeds in the forward direction.

(b) Using Eq. (6.5) as in (a), we calculate

$$\Delta G = -17.1 \, kJ \, mol^{-1} + 2.479 \, kJ \, mol^{-1} \ln \frac{(10^{-3} \, m)}{(10^{-3} \, m)(10^{-3} \, m)}$$

$$= -17.1 \, kJ \, mol^{-1} + 17.1 \, kJ \, mol^{-1} = 0 \, kJ \, mol^{-1}$$

Hence, the reactants and the product are in equilibrium, and thus the reaction does not proceed in either direction.

(c) As in (a) and (b) we use Eq. (6.5) to calculate $\Delta G = 17.1 \, kJ \, mol^{-1}$, indicating that the reaction proceeds in the backward direction.

(d) Equilibrium is attained when $\Delta G = 0$ [see (b)]. The equilibrium constant therefore is

$$\frac{(10^{-3} \, m)}{(10^{-3} \, m)(10^{-3} \, m)} = 10^3 \, m^{-1}$$

6.2 (a) Rearranging Eq. (6.9), we obtain

$$\ln \frac{(B)}{(B^+)} = \frac{E^*_{B^+ - B} - E_{B^+ - B}}{RT/F} = \frac{(0.118 \, V) - (0.000 \, V)}{(0.0257 \, V)} = 4.59$$

Hence, $(B^+)/(B) = e^{-4.59} = 0.010$.

(b) Based on the ΔG^*s, the first reaction has more of a tendency to go to the reduced form (A), meaning that it has a higher midpoint redox potential (e.g., consider beginning with unit activity of all species). Using Eq. (6.7) ($\Delta G = -nF\Delta E$), we calculate that E_A^{*H} is higher than E_B^{*H} by

$$\frac{(8.37 \text{ kJ mol}^{-1}) - (2.93 \text{ kJ mol}^{-1})}{(96.49 \text{ kJ mol}^{-1} \text{ V}^{-1})} = 0.056 \text{ V}$$

and so $E_A^{*H} = 0.056 \text{ V} + 0.118 \text{ V} = 0.174 \text{ V}$.

(c) When $(\text{oxidized})_j = (\text{reduced})_j$, $E_j = E_j^{*H}$. Thus the difference in electrical potential initially is the difference between the two midpoint redox potentials:

$$\Delta E = E_A^{*H} - E_B^{*H} = 0.174 \text{ V} - 0.118 \text{ V} = 0.056 \text{ V}$$

Electrons spontaneously flow toward higher redox potentials, which is toward the A–A^+ half cell (couple).

(d) Consider the reaction $A + B^+ \rightleftharpoons A^+ + B$, which is obtained by subtracting the second reaction from the first. For this composite reaction, $\Delta G^* = (8.37 \text{ kJ mol}^{-1}) - (2.93 \text{ kJ mol}^{-1}) = 5.44 \text{ kJ mol}^{-1}$. By Eq. (6.6) ($\Delta G^* = -RT \ln K$),

$$K = e^{-(5.44 \text{ kJ mol}^{-1})/(2.479 \text{ kJ mol}^{-1})} = 0.111$$

At equilibrium, A^+ and B have the same concentration, as do A and B^+. Thus, $K = \frac{(A^+)(B)}{(A)(B^+)} = \frac{(A^+)^2}{(A)^2}$, and so $(A^+) = \sqrt{K}(A) = \sqrt{0.111}(A) = 0.33(A)$. Because all reactants and products were initially $1\,m$, we have $(A) + (A^+) = 2\,m = 1.33(A)$, and so

$$(A) = (B^+) = 1.50\,m$$
$$(A^+) = (B) = 0.50\,m$$

(e) If instead of the chemical reaction in (d) we had two half-cell reactions, we could turn the difference in electron free energy into work instead of heat. However, the final concentrations at equilibrium would be the same in the two cases.

6.3 (a) Rearranging the expression for the equilibrium constant for ATP formation (Eq. 6.12), we obtain

$$[\text{ATP}] = (K_{\text{pH 7}})[\text{ADP}][\text{phosphate}]$$
$$= (5.0 \times 10^{-6} \text{ M}^{-1})(2.0 \times 10^{-3} \text{ M})(5.0 \times 10^{-3} \text{ M}) = 5.0 \times 10^{-11} \text{ M}$$

(b) Because $\text{ADP} + \text{P}_i \rightleftharpoons \text{ATP} + \text{H}_2\text{O}$ (Eq. 6.11) and there was negligible ATP to begin with [see (a)], a decrease in ADP from 2 to 1 mM causes the ATP concentration to become 1 mM. Using Eq. (6.13b) for the ATP formation energy, we obtain

$$\Delta G = 30 + 5.71 \log \frac{(1.0 \times 10^{-3} \text{ M})}{(1.0 \times 10^{-3} \text{ M})(4.0 \times 10^{-3} \text{ M})} \text{kJ (mol ATP)}^{-1}$$
$$= 44 \text{ kJ (mol ATP)}^{-1}$$

(c) Using Eq. (6.15b), we can calculate the redox potential of the NADP^+–NADPH couple:

$$E_{\text{NADP}^+\text{–NADPH}} = -0.32 - 0.030 \log \left(\frac{(0.03)}{(1.00)}\right) \quad \text{V} = -0.274 \text{ V}$$

Therefore, the difference in redox potential between the NADP$^+$–NADPH couple and ferredoxin is

$$(-0.274 \text{ V}) - (-0.580 \text{ V}) = 0.306 \text{ V}$$

(d) When 2 mol of electrons is transferred (NADPH can donate two electrons), Eq. (6.7) indicates that

$$\Delta G = - (2)(96.5 \text{ kJ mol}^{-1} \text{ V}^{-1})(0.306 \text{ V}) = - 59 \text{ kJ per 2 moles of electrons}$$

which is a large enough free energy decrease to form 1 mol of ATP [$\Delta G = 44$ kJ (mol ATP)$^{-1}$ in (b)].

(e) Assuming $E_M = 0$ V, Eq. (6.16c) indicates that

$$\text{pH}^\circ - \text{pH}^\text{i} = \frac{\mu_H^\text{i} - \mu_H^\text{o}}{5.71} = \frac{(44/2)}{(5.71)} = 3.9$$

when two H$^+$s are involved and $(44/3)/(5.71) = 2.6$ when three H$^+$s are involved.

6.4 (a) Using Eq. (6.9), we obtain

$$E_b = (0.040 \text{ V}) - \frac{(0.0257 \text{ V})}{(1)} \ln \left(\frac{0.20}{1.00}\right) = 0.081 \text{ V}$$

(b) For Cyt c to be just able to pass electrons to Cyt b [see (a)], E_c has to be equal to E_b (or be slightly lower). Thus by Eq. (6.9) we obtain

$$E_c = 0.220 \text{ V} - \frac{(0.0257 \text{ V})}{(1)} \ln \frac{(1 \text{ mM})}{(\text{Fe}^{3+})} = 0.081 \text{ V}$$

Hence

$$\ln \frac{(1 \text{ mM})}{(\text{Fe}^{3+})} = \frac{(0.220 \text{ V} - 0.081 \text{ V})}{(0.0257 \text{ V})} = 5.41$$

$$(\text{Fe}^{3+}) = \frac{(1 \text{ mM})}{(e^{5.41})} = \frac{(10^{-3} \text{ M})}{(224)} = 4.5 \times 10^{-6} \text{ M}$$

(c) The ΔE equivalent to a drop in free energy of 40 kJ mol^{-1} is calculated using Eq. (6.7):

$$\Delta E = \frac{(-40 \text{ kJ mol}^{-1})}{-(1)(96.49 \text{ kJ mol}^{-1} \text{ V}^{-1})} = 0.415 \text{ V}$$

Thus E_c must be 0.415 V more positive than E_b, or $E_c = 0.081$ V $+ 0.415$ V $= 0.496$ V.

(d) Because two H$^+$s move per electron and the activity (or pH) term for H$^+$ is the same on the two sides of the membrane, the redox potential difference is twice the electrical potential difference, or 0.30 V.

(e) Assuming pH$^\circ$ = pH$^\text{i}$, Eq. (6.16c) indicates that

$$E_M = \frac{\mu_H^\text{i} - \mu_H^\text{o} \text{ kJ mol}^{-1}}{96.5 \text{ kJ mol}^{-1} \text{ V}^{-1}} = \frac{(-40/2)}{(96.5 \text{ V}^{-1})} = -0.21 \text{ V}$$

(f) When three H$^+$s are required, the minimum E_M is $(-40/3)/(96.5) = -0.14$ V. When 30% of the energy is lost, E_M is $(-0.14 \text{ V})/(0.70) = -0.20$ V.

Chapter 7

7.1 (a) We know that absorbed infrared (IR) is here equal to emitted IR for the leaf, and therefore we can set Eq. (7.4) equal to Eq. (7.5) and solve for T^{surr}:

$$a_{IR}\sigma[(T^{surr})^4 + (T^{sky})^4] = 2\,e_{IR}\sigma(T^{leaf})^4$$

$$T^{surr} = \left[\frac{2\,e_{IR}\sigma(T^{leaf})^4}{a_{IR}\sigma} - (T^{sky})^4\right]^{1/4}$$

$$= [2(T^{leaf})^4 - (T^{sky})^4]^{1/4}$$

because a_{IR} here equals e_{IR}. For a clear sky we therefore have

$$T^{surr} = [2(283\,\text{K})^4 - (233\,\text{K})^4]^{1/4} = 315\,\text{K} = 42°\text{C}$$

and for a cloudy sky we obtain

$$T^{surr} = [2(283\,\text{K})^4 - (275\,\text{K})^4]^{1/4} = 290\,\text{K} = 17°\text{C}$$

(b) For emitted radiant energy, Wien's displacement law (Eq. 4.3) becomes $\lambda_{max}T = 2.90 \times 10^6$ nm K. Therefore, for the leaf

$$\lambda_{max} = \frac{(2.90 \times 10^6\,\text{nm K})}{(283\,\text{K})} = 10,200\,\text{nm} = 10.2\,\mu\text{m}$$

Emitted radiant energy for the surroundings with the clear sky has

$$\lambda_{max} = \frac{(2.90 \times 10^6\,\text{nm K})}{(315\,\text{K})} = 9200\,\text{nm} = 9.2\,\mu\text{m}$$

Emitted radiant energy for the surroundings with the cloudy sky has

$$\lambda_{max} = \frac{(2.90 \times 10^6\,\text{nm K})}{(290\,\text{K})} = 10,000\,\text{nm} = 10.0\,\mu\text{m}$$

(c) The amount of absorbed radiation is equal to the amount of absorbed solar radiation plus absorbed IR radiation. Using Eqs. (7.3) [for $a(1+r)S$] and (7.4), total absorbed radiation equals

$$\text{Absorbed shortwave and longwave} = (0.6)(1+0.10)(700\,\text{W m}^{-2})$$
$$+ (0.96)(5.67 \times 10^{-8}\,\text{W m}^{-2}\,\text{K}^4)[(282\,\text{K})^4 + (233\,\text{K})^4]$$
$$= 462\,\text{W m}^{-2} + 505\,\text{W m}^{-2} = 967\,\text{W m}^{-2}$$

(d) The thermal (i.e., longwave or IR) radiation emitted is calculated using Eq. (7.5):

$$\text{Emitted IR} = (2)(0.96)(5.67 \times 10^{-8}\,\text{W m}^{-2}\,\text{K}^{-4})(283\,\text{K})^4 = 698\,\text{W m}^{-2}$$

Thus, 100 (698 W m^{-2})/(967 W m^{-2}) or 72% of the absorbed radiation calculated in (c) is emitted as thermal radiation.

(e) Using Eq. (7.3), the solar (i.e., shortwave) radiation absorbed is

$$(0.60)(1+0.15)(250\,\text{W m}^{-2}) = 173\,\text{W m}^{-2}$$

and by Eq. (7.4) the absorbed IR radiation is

$$(0.96)(5.67 \times 10^{-8}\,\text{W m}^{-2}\,\text{K}^{-4})[(282\,\text{K})^4 + (275\,\text{K})^4] = 656\,\text{W m}^{-2}$$

Therefore, the total absorbed radiation is $173\ \mathrm{W\,m^{-2}} + 656\ \mathrm{W\,m^{-2}}$, or $829\ \mathrm{W\,m^{-2}}$, and the outgoing radiation is $698\ \mathrm{W\,m^{-2}}$ from (d), so the net radiation is

$$829\ \mathrm{W\,m^{-2}} - 698\ \mathrm{W\,m^{-2}} = 131\ \mathrm{W\,m^{-2}}$$

7.2 **(a)** The average value of a function is $\frac{1}{b-a}\int_a^b f(x)\,dx$, i.e., the area under the function divided by the distance between the two endpoints. For a circle, we can apply similar reasoning, thus

$$\bar{l} = \frac{A}{d} = \frac{\pi r^2}{2r} = \frac{\pi r}{2} = \frac{\pi(0.06\,\mathrm{m})}{2} = 0.094\,\mathrm{m}$$

(b) Using Eq. (7.8), we have

$$\delta^{bl}_{(mm)} = 4.0\sqrt{\frac{l_{(m)}}{v_{(m\,s^{-1})}}} = 4.0\sqrt{\frac{(0.094\,\mathrm{m})}{(0.80\,\mathrm{m\,s^{-1}})}} = 1.37\,\mathrm{mm}$$

Assuming that the mean distance across a leaf is the diameter, we have

$$\delta^{bl}_{mm} = 4.0\sqrt{\frac{(0.12\,\mathrm{m})}{(0.80\,\mathrm{m\,s^{-1}})}} = 1.55\,\mathrm{mm}$$

which is similar to the value determined using mean length.

(c) Using Eq. (7.11), the heat flux density is

$$J^C_H = \frac{2\,K^{air}(T^{leaf} - T^{ta})}{\delta^{bl}} = \frac{(2)(0.0259\ \mathrm{W\,m^{-1}\,^\circ C^{-1}})(25^\circ C - 20^\circ C)}{(1.37 \times 10^{-3}\,\mathrm{m})}$$

$$= 189\ \mathrm{W\,m^{-2}}$$

(d) From (c) we know that conductive/convective heat loss is $189\ \mathrm{W\,m^{-2}}$, and we also know that the total heat load from net radiation is $300\ \mathrm{W\,m^{-2}}$. Therefore, the amount of energy that must be dissipated as latent heat through transpiration is

$$300\ \mathrm{W\,m^{-2}} - 189\ \mathrm{W\,m^{-2}} = 111\ \mathrm{W\,m^{-2}}$$

Rearranging Eq. (7.19) we thus find

$$J_{wv} = \frac{J^T_H}{H_{vap}} = \frac{(111\ \mathrm{J\,s^{-1}\,m^{-2}})}{(44.0 \times 10^3\ \mathrm{J\,mol^{-1}})} = 2.5 \times 10^{-3}\ \mathrm{mol\,m^{-2}\,s^{-1}}$$

7.3 **(a)** Using Eq. (7.10) for a sphere, we obtain

$$\delta^{bl}_{(mm)} = 2.8\sqrt{\frac{d_{(m)}}{v_{(m\,s^{-1})}}} + \frac{0.25}{v_{(m\,s^{-1})}} = 2.8\sqrt{\frac{(0.2\,\mathrm{m})}{(1.0\,\mathrm{m\,s^{-1}})}} + \frac{(0.25)}{(1.0\,\mathrm{m\,s^{-1}})} = 1.50\,\mathrm{mm}$$

The amount of heat conducted across the boundary layer of a sphere can be calculated using Eq. (7.13):

$$J^C_H = \frac{(r + \delta^{bl})K^{air}(T^{surf} - T^{ta})}{r\delta^{bl}}$$

$$= \frac{(0.1\,\mathrm{m} + 0.0015\,\mathrm{m})(0.0259\ \mathrm{W\,m^{-1}\,^\circ C^{-1}})(25^\circ C - 20^\circ C)}{(0.1\,\mathrm{m})(0.0015\,\mathrm{m})} = 88\ \mathrm{W\,m^{-2}}$$

(b) Rearranging Eq. (7.14), we find

$$h_c = \frac{J_H^C}{T^{surf} - T^{ta}} = \frac{(88 \, \text{W m}^{-2})}{(25°\text{C} - 20°\text{C})} = 18 \, \text{W m}^{-2} \, °\text{C}^{-1}$$

(c) Using Eq. (7.9) for a cylinder, we obtain

$$\delta_{(mm)}^{bl} = 5.8\sqrt{\frac{d_{(m)}}{v_{(m\,s^{-1})}}} = 5.8\sqrt{\frac{(1.2 \times 10^{-3} \, \text{m})}{(1.0 \, \text{m s}^{-1})}} = 0.20 \, \text{mm}$$

(d) Given that net radiation is due entirely to absorbed shortwave radiation and that heat loss does not occur through evaporation or conduction to the stem, Eq. (7.2) indicates that absorbed solar irradiation must be balanced by convective heat loss (Eq. 7.12):

$$a\bar{S} = J_H^C = \frac{K^{air}(T^{spine} - T^{ta})}{r \ln\left(\frac{r+\delta^{bl}}{r}\right)}$$

The mean solar flux density on the spine, $\bar{S}$, is less than the maximum S:

$$\bar{S} = \frac{\text{projected area}}{\text{actual area}} S = \frac{dl}{\pi \, dl} S = \frac{S}{\pi} = \frac{100 \, \text{W m}^{-2}}{\pi} = 32 \, \text{W m}^{-2}$$

Therefore, we calculate

$$T^{spine} = \frac{(0.7)(32 \, \text{W m}^{-2})(0.6 \times 10^{-3} \, \text{m}) \ln\left(\frac{0.6\,\text{mm} + 0.2\,\text{mm}}{0.6\,\text{mm}}\right)}{(0.0257 \, \text{W m}^{-1} \, °\text{C}^{-1})} + 20.0°\text{C} = 20.2°\text{C}$$

(e) In the absence of spines, IR is absorbed from one source (surroundings) and emitted from one source (cactus surface). Combining Eqs. (7.4) and (7.5), we obtain

$$\text{Net IR} = a_{IR}\sigma(T^{surr})^4 - e_{IR}\sigma(T^{surf})^4$$
$$= (0.97)(5.67 \times 10^{-8} \, \text{W m}^{-2} \, \text{K}^{-4})[(253 \, \text{K})^4 - (298 \, \text{K})^4]$$
$$= -208 \, \text{W m}^{-2}$$

In the presence of spines, IR is absorbed from two sources (surroundings and spines) and emitted from one source (cactus surface). Therefore

$$\text{Net IR} = a_{IR}\sigma[(T^{surr})^4 + (T^{spine})^4](\tfrac{1}{2}) - e_{IR}\sigma(T^{surf})^4$$
$$= (0.97)(5.67 \times 10^{-8} \, \text{W m}^{-2} \, \text{K}^{-4})[\tfrac{1}{2}(253 \, \text{K})^4 + \tfrac{1}{2}(293 \, \text{K})^4 - (298 \, \text{K})^4]$$
$$= -118 \, \text{W m}^{-2}$$

(f) Because the plant is spherical, the total solar radiation absorbed equals its projected area (πr^2) times the solar flux density perpendicular to the solar beam $(1000 \, \text{W m}^{-2})$ times the absorbance (0.30). To obtain the average solar radiation absorbed over the plant surface, we divide by the plant's surface area $(4\pi r^2)$. Thus

$$a\bar{S} = \frac{\pi r^2 (1000 \, \text{W m}^{-2})(0.30)}{4\pi r^2} = 75 \, \text{W m}^{-2}$$

Using answers from (e) and (a), the average net energy balance is

Absorbed solar + net IR − net heat convection
$$= 75 \, \text{W m}^{-2} - 118 \, \text{W m}^{-2} - 88 \, \text{W m}^{-2} = -131 \, \text{W m}^{-2}$$

Considering the whole plant surface area, the net energy balance equals $(-131 \, \text{W m}^{-2}) \, (4\pi)(0.1 \, \text{m})^2 = -16.5 \, \text{W}$. Given that the volumetric heat capacity is $(0.80)(4.175 \, \text{MJ m}^{-3} \, ^\circ\text{C}^{-1}) = 3.34 \, \text{MJ m}^{-3} \, ^\circ\text{C}^{-1}$, the change in tissue temperature can be calculated (Eq. 7.21):

$$\frac{\Delta T}{\Delta t} = \frac{\text{heat storage rate}}{C_P V} = \frac{(16.5 \, \text{J s}^{-1})}{(3.34 \times 10^6 \, \text{J m}^{-3} \, ^\circ\text{C}^{-1})(4/3)(\pi)(0.1 \, \text{m})^3}$$
$$= -0.00118^\circ\text{C s}^{-1} = -4.2^\circ\text{C h}^{-1}$$

7.4 **(a)** Using the equation for net radiation balance in the soil (Section VB), we find that net radiation equals

$$aS + a_{IR}\sigma(T^{\text{surr}})^4 - \varepsilon_{IR}\sigma(T^{\text{soil}})^4$$
$$= 100 \, \text{W m}^{-2} + (1)(5.67 \times 10^{-8} \, \text{W m}^{-2} \, \text{K}^{-4})(295 \, \text{K})^4$$
$$- (1)(5.67 \times 10^{-8} \, \text{W m}^{-2} \, \text{K}^4)(293 \, \text{K})^4$$
$$= 100 \, \text{W m}^{-2} + 429 \, \text{W m}^{-2} - 418 \, \text{W m}^{-2}$$
$$= 111 \, \text{W m}^{-2}$$

(b) Modifying Eq. (7.24), we find the heat flux density down each stem to be

$$J_H^C = \frac{K^{\text{water}}(T^{\text{veg}} - T^{\text{soil}})}{\Delta z}$$

Because flux density is a rate per unit area, multiplying by the stem area gives the rate of heat conduction down a stem:

$$A^{\text{stem}} J_H^C = \frac{A^{\text{stem}} K^{\text{water}}(T^{\text{veg}} - T^{\text{soil}})}{\Delta z}$$
$$= \frac{\pi(0.015 \, \text{m})^2(0.60 \, \text{W m}^{-1} \, ^\circ\text{C}^{-1})(22^\circ\text{C} - 20^\circ\text{C})}{(0.8 \, \text{m})}$$
$$= 0.00106 \, \text{W} = 1.06 \, \text{mW}$$

The average value of J_H^C per m^2 of ground is $(1.06 \, \text{mW plant}^{-1})$ (4 plants m^{-2} of ground) $= 4.2 \, \text{mW m}^{-2}$ of ground.

(c) Modifying Eq. (7.11) by removing the factor 2 because the soil conducts to the air only from one surface, we find

$$J_H^C = \frac{K^{\text{air}}(T^{\text{soil}} - T^{\text{air}})}{\delta^{bl}} = \frac{(0.0257 \, \text{W m}^{-1} \, ^\circ\text{C}^{-1})(20^\circ\text{C} - 21^\circ\text{C})}{(4 \times 10^{-3} \, \text{m})}$$
$$= -6 \, \text{W m}^{-2}$$

(d) Using Eq. (7.19), latent heat loss is

$$J_H^T = J_{wv} H_{vap} = (0.3 \times 10^{-3} \, \text{mol m}^{-2} \, \text{s}^{-1})(44.2 \times 10^3 \, \text{J mol}^{-1}) = 13 \, \text{W m}^{-2}$$

The flux density of heat conducted into the soil can be calculated from the overall energy balance (see Eq. 7.2):

$$J_H^C = \text{(net radiation)} - \text{(conduction/convection)} - \text{(latent heat)}$$
$$= 111 \text{ W m}^{-2} - (-6 \text{ W m}^{-2}) - 13 \text{ W m}^{-2} = 104 \text{ W m}^{-2}$$

(e) Rearranging Eq. (7.24), we find

$$\frac{\partial T}{\partial z} = \frac{J_H^C}{-K^{\text{soil}}} = -\frac{(104 \text{ W m}^{-2})}{(0.60 \text{ W m}^{-1}\,^{\circ}\text{C}^{-1})} = -173^{\circ}\text{C m}^{-1}$$

Chapter 8

8.1 (a) Using Eq. (8.3) we find

$$g_{wv}^{bl} = \frac{D_{wv}}{\delta^{bl}} = \frac{(2.42 \times 10^{-5} \text{ m}^2 \text{ s}^{-1})}{(0.8 \times 10^{-3} \text{ m})} = 0.030 \text{ m s}^{-1} = 30 \text{ mm s}^{-1}$$

If we rearrange Eq. (8.3),

$$r_{wv}^{bl} = \frac{1}{g_{wv}^{bl}} = \frac{1}{(0.030 \text{ m s}^{-1})} = 33 \text{ s m}^{-1}$$

(b) The fraction of the leaf surface area occupied by stomatal pores is

$$na^{\text{st}} = (64 \text{ stomata mm}^{-2})(6 \times 10^{-3} \text{ mm})(20 \times 10^{-3} \text{ mm}) = 0.0077$$

The effective radius of a stomate (see Chapter 8, Section IC) is

$$r^{\text{st}} = \left(\frac{a^{\text{st}}}{\pi}\right)^{1/2} = \left(\frac{(6\,\mu\text{m})(20\,\mu\text{m})}{\pi}\right)^{1/2} = 6\,\mu\text{m}$$

(c) The average flux density within an individual stomatal pore will be greater than that in the boundary layer by the ratio of the areas available for gas diffusion:

$$J_{wv}^{\text{st}} = \frac{A}{A^{\text{st}}} J_{wv}^{bl} = \frac{1}{(0.0077)} J_{wv}^{bl} = 130 J_{wv}^{bl}$$

(d) Using Eq. (8.5) we find

$$g_{wv}^{\text{st}} = \frac{D_{wv}na^{\text{st}}}{\delta^{\text{st}} + r^{\text{st}}} = \frac{(2.42 \times 10^{-5} \text{ m}^2 \text{ s}^{-1})(0.0077)}{(25 \times 10^{-6} \text{ m}) + (6 \times 10^{-6} \text{ m})}$$
$$= 0.0060 \text{ m s}^{-1} = 6.0 \text{ mm s}^{-1}$$

Considering the discussion following Eq. (8.8), we find

$$g_{wv}^{\prime\text{st}} = g_{wv}^{\text{st}} \frac{P}{RT} = (6.0 \times 10^{-3} \text{ m s}^{-1})(41.6 \text{ mol m}^{-3})$$
$$= 0.25 \text{ mol m}^{-2} \text{ s}^{-1} = 250 \text{ mmol m}^{-2} \text{ s}^{-1}$$

If the air pressure is reduced to 0.9 atm, then D_{wv} increases by (1.0 atm / 0.9 atm) (see Eq. 8.9); therefore g_{wv}^{st} increases by that amount (see Eq. 8.5):

$$g_{gv}^{\text{st, 0.9 atm}} = \left(\frac{1 \text{ atm}}{0.9 \text{ atm}}\right)(6.0 \text{ mm s}^{-1}) = 6.7 \text{ mm s}^{-1}$$

Because $g_{wv}^{\prime\,\mathrm{st}}$ is independent of pressure (see Chapter 8, Section IF), it remains at 250 mmol m^{-2} s^{-1}.

(e) Using Eq. (8.6) we find

$$g_{wv}^{\mathrm{ias}} = \frac{D_{wv}}{\delta^{\mathrm{ias}}} = \frac{(2.42 \times 10^{-5}\,\mathrm{m^2\,s^{-1}})}{(0.5 \times 10^{-3}\,\mathrm{m})} = 0.048\,\mathrm{m\,s^{-1}} = 48\,\mathrm{mm\,s^{-1}}$$

For the other conductance, we have

$$g_{wv}^{\prime\,\mathrm{ias}} = g_{wv}^{\mathrm{ias}}\frac{P}{RT} = (48 \times 10^{-3}\,\mathrm{m\,s^{-1}})(41.6\,\mathrm{mol\,m^{-3}}) = 2.0\,\mathrm{mol\,m^{-2}\,s^{-1}}$$

(f) Because the cavity is cylindrical, the effective radius equals the actual radius. Using Eq. (8.5)(see also Eq. 8.14), we thus obtain

$$r_{wv}^{\mathrm{cavity}} = \frac{\delta^{\mathrm{cav}} + r^{\mathrm{cav}}}{D_j na^{\mathrm{cav}}}$$

$$= \frac{(100 \times 10^{-6}\,\mathrm{m}) + (25 \times 10^{-6}\,\mathrm{m})}{(2.42 \times 10^{-5}\,\mathrm{m^2\,s^{-1}})(64 \times 10^{6}\,\mathrm{m^{-2}})(\pi)(25 \times 10^{-6}\,\mathrm{m})^2} = 41\,\mathrm{s\,m^{-1}}$$

8.2 (a) Using Eq. (8.12), we can calculate the water vapor conductance of the leaf:

$$g_{wv}^{\mathrm{leaf}} = \frac{g_{wv}^{\mathrm{ias}}\,g_{wv}^{\mathrm{st}}}{g_{wv}^{\mathrm{ias}} + g_{wv}^{\mathrm{st}}} + g_{wv}^{\mathrm{c}} = \frac{(40\,\mathrm{mm\,s^{-1}})(6\,\mathrm{mm\,s^{-1}})}{(40\,\mathrm{mm\,s^{-1}}) + (6\,\mathrm{mm\,s^{-1}})} + (0.1\,\mathrm{mm\,s^{-1}})$$

$$= 5.3\,\mathrm{mm\,s^{-1}}$$

Then using Eq. (8.13), we can incorporate the conductance through the boundary layer with that of the leaf to find the total water vapor conductance:

$$g_{wv}^{\mathrm{total}} = \frac{(5.3\,\mathrm{mm\,s^{-1}})(20\,\mathrm{mm\,s^{-1}})}{(5.3\,\mathrm{mm\,s^{-1}}) + (20\,\mathrm{mm\,s^{-1}})} = 4.2\,\mathrm{mm\,s^{-1}}$$

(b) If the cuticular pathway is ignored, we have three conductances in series:

$$g_{wv}^{\mathrm{total}} = \frac{1}{\frac{1}{(40\,\mathrm{mm\,s^{-1}})} + \frac{1}{(6\,\mathrm{mm\,s^{-1}})} + \frac{1}{(20\,\mathrm{mm\,s^{-1}})}} = 4.1\,\mathrm{mm\,s^{-1}}$$

If the intercellular air spaces are ignored, stomatal and cuticular conductances are in parallel ($g_{wv}^{\mathrm{leaf}} = g_{wv}^{\mathrm{st}} + g_{wv}^{\mathrm{c}}$) and in series with the boundary layer conductance:

$$g_{wv}^{\mathrm{total}} = \frac{1}{\frac{1}{(6\,\mathrm{mm\,s^{-1}}) + (0.1\,\mathrm{mm\,s^{-1}})} + \frac{1}{(20\,\mathrm{mm\,s^{-1}})}} = 4.7\,\mathrm{mm\,s^{-1}}$$

If both the cuticle and the intercellular air spaces are ignored, then the stomatal and boundary layer conductances are in series:

$$g_{wv}^{\mathrm{total}} = \frac{1}{\frac{1}{(6\,\mathrm{mm\,s^{-1}})} + \frac{1}{(20\,\mathrm{mm\,s^{-1}})}} = 4.6\,\mathrm{mm\,s^{-1}}$$

(c) The conductance of the upper surface here equals the conductance of the lower surface [determined in (a)] and is in parallel with it:

$$g_{wv}^{\mathrm{total}} = g_{wv}^{\mathrm{upper}} + g_{wv}^{\mathrm{lower}} = 2g_{wv}^{\mathrm{lower}} = (2)(4.2\,\mathrm{mm\,s^{-1}}) = 8.4\,\mathrm{mm\,s^{-1}}$$

(d) Here 0.72 of g_{wv}^{total} comes from the lower surface, whose conductance is determined in (a), so

$$g_{wv}^{total} = \frac{g_{wv}^l}{(0.72)} = \frac{(4.2 \text{ mm s}^{-1})}{(0.72)} = 5.8 \text{ mm s}^{-1}$$

(e) Using Eq. (8.16) and $c_{wv}^{leaf} = (\text{RH}^{leaf}/100)c_{wv}^{*,\,leaf}$, we find

$$J_{wv}^{lower} = g_{wv}^{total} \Delta c_{wv}^{total} = g_{wv}^{total}\left(c_{wv}^{leaf} - c_{wv}^{ta}\right)$$
$$= (4.2 \text{ mm s}^{-1})[(0.99)(30.40 \text{ g m}^{-3}) - (7.5 \text{ g m}^{-3})] = 95 \text{ mg m}^{-2} \text{ s}^{-1}$$

(f) Considering the discussion following Eq. (8.8), we obtain

$$g_{wv}^{\prime total} = g_{wv}^{total} \frac{P}{RT}$$

$$= (4.2 \times 10^{-3} \text{ m s}^{-1})\frac{(1 \text{ atm})}{(0.08205 \text{ liter atm mol}^{-1} \text{ K}^{-1})(303 \text{ K})}$$

$$\times \left(\frac{1000 \text{ liter}}{\text{m}^3}\right) = 0.169 \text{ mol m}^{-2} \text{ s}^{-1} = 169 \text{ mmol m}^{-2} \text{ s}^{-1}$$

Equation (8.16) was modified as described in the text (see Section IID), so that

$$J_{wv} = g_{wv}^{\prime total} \Delta N_{wv}^{total} = g_{wv}^{\prime total}\left(N_{wv}^{leaf} - N_{wv}^{ta}\right)$$
$$= (169 \text{ mmol m}^{-2} \text{ s}^{-1})[(0.99)(0.04190) - (0.0103)] = 5.3 \text{ mmol m}^{-2} \text{ s}^{-1}$$

(g) Rearranging Eq. (8.16), the drop in water vapor concentration is

$$\Delta c_{wv}^{st} = \frac{J_{wv}}{g_{wv}^{st}} = \frac{(0.095 \text{ g m}^{-2} \text{ s}^{-1})}{(6.0 \times 10^{-3} \text{ m s}^{-1})} = 15.8 \text{ g m}^{-3}$$

Similarly, rearranging Eq. (8.16) for the mole fraction case we obtain

$$\Delta N_{wv}^{st} = \frac{J_{wv}}{g_{wv}^{\prime st}} = \frac{J_{wv}}{g_{wv}^{st} \frac{P}{RT}} = \frac{(5.3 \times 10^{-3} \text{ mol m}^{-2} \text{ s}^{-1})}{(6.0 \times 10^{-3} \text{ m s}^{-1})(40.2 \text{ mol m}^{-3})} = 0.022$$

8.3 (a) We will calculate the total surface area of the palisade mesophyll cells (rectangular parallelepipeds) and spongy mesophyll cells (spheres) and express it per unit area of the leaf surface:

$$\frac{A^{mes}}{A} = \frac{[(4)(100 \,\mu\text{m})(30 \,\mu\text{m}) + (2)(30 \,\mu\text{m})(30 \,\mu\text{m})] + (2)(4\pi)(15 \,\mu\text{m})^2}{(30 \,\mu\text{m})(30 \,\mu\text{m})} = 22$$

(b) The value for the sun leaf is

$$\frac{A^{mes}}{A} = \frac{(2)[(4)(100 \,\mu\text{m})(30 \,\mu\text{m}) + (2)(30 \,\mu\text{m})(30 \,\mu\text{m})] + (4\pi)(15 \,\mu\text{m})^2}{(30 \,\mu\text{m})(30 \,\mu\text{m})} = 34$$

(c) Using Eq. (8.19), the maximum $r_{CO_2}^{cw}$ will be achieved when $K_{CO_2}^{cw}$ is minimal, namely 0.91 (i.e., no HCO_3^- or H_2CO_3; see Table 8-3):

$$r_{CO_2}^{cw} = \frac{A \, \Delta x^{cw}}{A^{mes} D_{CO_2}^{cw} K_{CO_2}^{cw}} = \frac{(1)(0.2 \times 10^{-6} \text{ m})}{(22)(5.0 \times 10^{-10} \text{ m}^2 \text{ s}^{-1})(0.91)} = 20 \text{ s m}^{-1}$$

(d) Using Eq. (8.19), we find

$$r_{CO_2}^{pm} = r_{CO_2}^{clm} = \frac{A}{A^{mes} P_{CO_2}} = \frac{1}{(22)(1.0 \times 10^{-3} \text{ m s}^{-1})} = 45 \text{ s m}^{-1}$$

(e) Using Eq. (8.19), we find

$$r_{CO_2}^{cyt} = \frac{A \Delta x^{cyt}}{A^{mes} D_{CO_2}^{cyt} K_{CO_2}^{cyt}} = \frac{(1)(0.1 \times 10^{-6} \text{ m})}{(22)(1.0 \times 10^{-9} \text{ m}^2 \text{ s}^{-1})(1)} = 5 \text{ s m}^{-1}$$

$$r_{CO_2}^{stroma} = \frac{A \Delta x^{stroma}}{A^{mes} D_{CO_2}^{stroma} K_{CO_2}^{stroma}} = \frac{(1)(0.2 \times 10^{-6} \text{ m})}{(22)(1.0 \times 10^{-9} \text{ m}^2 \text{ s}^{-1})(1)} = 23 \text{ s m}^{-1}$$

(f) Using Eqs. (8.18a) and (8.18b), the total liquid-phase resistance to CO_2 diffusion for the shade leaf is

$$r_{CO_2}^{liquid} = r_{CO_2}^{cw} + r_{CO_2}^{pm} + r_{CO_2}^{cyt} + r_{CO_2}^{clm} + r_{CO_2}^{stroma}$$

$$= 20 \text{ s m}^{-1} + 45 \text{ s m}^{-1} + 5 \text{ s m}^{-1} + 45 \text{ s m}^{-1} + 23 \text{ s m}^{-1}$$

$$= 138 \text{ s m}^{-1}$$

The total liquid-phase resistance to CO_2 diffusion for the sun leaf is smaller than that of the shade leaf because the sun leaf has a higher A_{mes}/A:

$$r_{CO_2}^{liquid} = \left(\frac{22}{34}\right)(138 \text{ s m}^{-1}) = 89 \text{ s m}^{-1}$$

8.4 (a) Rearranging Eq. (8.22), we find

$$V_{max} = \frac{v_{CO_2}\left(K_{CO_2} + c_{CO_2}^{chl}\right)}{c_{CO_2}^{chl}} = \frac{(4 \text{ mol m}^{-3} \text{ s}^{-1})(5 \, \mu M + 9 \, \mu M)}{(9 \, \mu M)}$$

$$= 6.2 \text{ mol m}^{-3} \text{ s}^{-1}$$

(b) Again using Eq. (8.22), we have

$$0.9 V_{max} = \frac{V_{max} c_{CO_2}^{chl}}{5 \, \mu M + c_{CO_2}^{chl}} \quad \text{or} \quad 4.5 \, \mu M + 0.9 \, c_{CO_2}^{chl} = c_{CO_2}^{chl}$$

hence $c_{CO_2}^{chl} = \dfrac{4.5 \, \mu M}{0.1} = 45 \, \mu M$

(c) Utilizing Eq. (8.26) with the flux densities expressed as in Section IVE, we find that the total resistance to CO_2 diffusion is

$$r_{CO_2}^{total} = r_{CO_2}^{bl_1} + r_{CO_2}^{leaf_1} + r_{CO_2}^{mes} + \left(\frac{1}{1 - \frac{J_{CO_2}^{r+pr}}{J_{CO_2}^{ps}}}\right) r_{CO_2}^{chl}$$

$$= 60 \text{ s m}^{-1} + 250 \text{ s m}^{-1} + 150 \text{ s m}^{-1} + \left(\frac{1}{1 - 0.45}\right)(100 \text{ s m}^{-1})$$

$$= 642 \text{ s m}^{-1}$$

Using Eq. (8.26), we find

$$J_{CO_2} = \frac{c_{CO_2}^{ta} - c_{CO_2}^{chl}}{r_{CO_2}^{total}} = \frac{(13 \times 10^{-3}\,\text{mol m}^{-3}) - (9 \times 10^{-3}\,\text{mol m}^{-3})}{(642\,\text{s m}^{-1})}$$

$$= 6.2 \times 10^{-6}\,\text{mol m}^{-2}\,\text{s}^{-1}$$

(d) We use the same equations as in (c) and find

$$r_{CO_2}^{total} = 60\,\text{s m}^{-1} + 250\,\text{s m}^{-1} + 150\,\text{s m}^{-1} + \left(\frac{1}{1 - 0.05}\right)(100\,\text{s m}^{-1})$$

$$= 565\,\text{s m}^{-1}$$

$$J_{CO_2} = \frac{(13 \times 10^{-3}\,\text{mol m}^{-3}) - (7 \times 10^{-3}\,\text{mol m}^{-3})}{(565\,\text{s m}^{-1})}$$

$$= 10.6 \times 10^{-6}\,\text{mol m}^{-2}\,\text{s}^{-1}$$

(e) We need to determine $J_{CO_2}^{r+pr}$, e.g., using Eq. (8.23) and J_{CO_2} from (d) and knowing that $J_{CO_2}^{r+pr} = 0.05 J_{CO_2}^{ps}$:

$$J_{CO_2} = J_{CO_2}^{ps} - J_{CO_2}^{r+pr} = 20\,J_{CO_2}^{r+pr} - J_{CO_2}^{r+pr} = 19\,J_{CO_2}^{r+pr}$$

$$J_{CO_2}^{r+pr} = J_{CO_2}/19 = (10.6 \times 10^{-6}\,\text{mol m}^{-2}\,\text{s}^{-1})/(19)$$

$$= 0.56 \times 10^{-6}\,\text{mol m}^{-2}\,\text{s}^{-1}$$

We also note that 10 μmol mol^{-1} CO_2 is about 0.41 mmol m^{-3} (see Table 8-2). Thus, rearranging Eq. (8.25) and noting that in the steady state for a confined bag J_{CO_2} is zero, we find

$$c_{CO_2}^{chl} = c_{CO_2}^{ta} - \left(J_{CO_2}^{r+pr}\right)\left(r_{CO_2}^{chl}\right)$$

$$= (0.41\,\text{mmol m}^{-3}) - (0.56 \times 10^{-3}\,\text{mmol m}^{-2}\,\text{s}^{-1})(100\,\text{s m}^{-1})$$

$$= 0.35\,\text{mmol m}^{-3} = 0.35\,\mu\text{M}$$

(f) By considering Fig. 8.12 and the discussion in Section IVE, we can obtain a relation for the CO_2 efflux at night:

$$c_{CO_2}^{mito} - c_{CO_2}^{ta} = J_{CO_2}^{r}\left(r_{CO_2}^{i} + r_{CO_2}^{mes} + r_{CO_2}^{leaf_1} + r_{CO_2}^{bl_1}\right)$$

$$\text{so } c_{CO_2}^{mito} = (0.56 \times 10^{-3}\,\text{mmol m}^{-2}\,\text{s}^{-1})(500\,\text{s m}^{-1} + 150\,\text{s m}^{-1}$$

$$+ 250\,\text{s m}^{-1} + 60\,\text{s m}^{-1}) + 13\,\text{mmol m}^{-3}$$

$$= 14\,\text{mmol m}^{-3} = 14\,\mu\text{M}$$

If stomatal closure causes $r_{CO_2}^{leaf}$ to become 5000 s m^{-1}, then

$$c_{CO_2}^{mito} = (0.56 \times 10^{-3}\,\text{mmol m}^{-2}\,\text{s}^{-1})(500\,\text{s m}^{-1} + 150\,\text{s m}^{-1} + 5000\,\text{s m}^{-1}$$

$$+ 60\,\text{s m}^{-1}) + 13\,\text{mmol m}^{-3} = 16\,\text{mmol m}^{-3} = 16\,\mu\text{M}$$

8.5 (a) We can use Eq. (8.16) to find g_{wv}^{total}:

$$g_{wv}^{total} = \frac{J_{wv}}{c_{wv}^{leaf} - c_{wv}^{air}} = \frac{(5.0 \times 10^{-3}\,\text{mol m}^{-2}\,\text{s}^{-1})}{(2.20\,\text{mol m}^{-3}) - (0.32)(1.69\,\text{mol m}^{-3})}$$

$$= 3.0 \times 10^{-3}\,\text{m s}^{-1} = 3.0\,\text{mm s}^{-1}$$

For conductances in series, the reciprocal of the total conductance $(1/g^{\text{total}})$ is the sum of the reciprocals of the individual conductances (see Section IIA), so

$$\frac{1}{g^{\text{st}}_{wv}} = \frac{1}{g^{\text{total}}_{wv}} - \frac{1}{g^{\text{bl}}_{wv}} - \frac{1}{g^{\text{ias}}_{wv}}$$

$$= \frac{1}{(3.0\,\text{mm s}^{-1})} - \frac{1}{(15\,\text{mm s}^{-1})} - \frac{1}{(30\,\text{mm s}^{-1})} = 0.23\,\text{s mm}^{-1}$$

$$g^{\text{st}}_{wv} = \frac{1}{(0.23\,\text{s mm}^{-1})} = 4.3\,\text{mm s}^{-1}$$

(b) Stomatal conductance will decrease fourfold, so total conductance is

$$g^{\text{total}}_{wv} = \frac{1}{\frac{1}{(15\,\text{mm s}^{-1})} + \frac{1}{(0.25)(4.3\,\text{mm s}^{-1})} + \frac{1}{(30\,\text{mm s}^{-1})}} = 1.0\,\text{mm s}^{-1}$$

Because g^{total}_{wv} is one-third of that in (a), transpiration will be reduced threefold (to 1.7 mmol m^{-2} s^{-1}).

(c) Leaf temperature will increase because of the decrease in latent heat loss accompanying transpiration.

(d) First we find the original gas-phase CO_2 conductance (see Section IVF):

$$g^{\text{gas}}_{CO_2} = \frac{g^{\text{total}}_{wv}}{1.60} = \frac{(3.0\,\text{mm s}^{-1})}{(1.60)} = 1.88\,\text{mm s}^{-1}$$

The new gas-phase conductance is

$$g^{\text{gas}}_{CO_2} = \frac{g^{\text{total}}_{wv}}{1.60} = \frac{(1.0\,\text{mm s}^{-1})}{(1.60)} = 0.63\,\text{mm s}^{-1}$$

If we use the original gas-phase conductance, we can find the liquid-phase conductance to CO_2, which is unchanged:

$$g^{\text{liquid}}_{CO_2} = \frac{1}{\frac{1}{g^{\text{total}}_{CO_2}} - \frac{1}{g^{\text{gas}}_{CO_2}}} = \frac{1}{\frac{1}{(0.70\,\text{mm s}^{-1})} - \frac{1}{(1.88\,\text{mm s}^{-1})}} = 1.12\,\text{mm s}^{-1}$$

$$\frac{g^{\text{total}}_{CO_2}\,_{\text{new}}}{g^{\text{total}}_{CO_2}\,_{\text{original}}} = \frac{\frac{g^{\text{gasnew}}_{CO_2}\,g^{\text{liq}}_{CO_2}}{g^{\text{gasnew}}_{CO_2} + g^{\text{liq}}_{CO_2}}}{0.70\,\text{mm s}^{-1}} = \frac{\frac{(0.63\,\text{mm s}^{-1})(1.12\,\text{mm s}^{-1})}{(0.63\,\text{mm s}^{-1})+(1.12\,\text{mm s}^{-1})}}{(0.70\,\text{mm s}^{-1})}$$

$$= \frac{(0.40\,\text{mm s}^{-1})}{(0.70\,\text{mm s}^{-1})} = 0.57$$

and hence photosynthesis is decreased 43%.

By Eq. (8.29), water-use efficiency (WUE) is proportional to photosynthesis/transpiration, so

$$\frac{\text{WUE}^{\text{new}}}{\text{WUE}^{\text{original}}} = \frac{(0.57)/(0.33)}{(1.00)/(1.00)} = 1.73$$

and WUE is increased by 73%.

(e) Given that $\delta^{bl} = 4.0\sqrt{\frac{l}{v}}$ (Eq. 7.8), a fourfold increase in windspeed will change δ^{bl} by a factor of $\sqrt{\frac{1}{4}} = \frac{1}{2}$. Because $g_{wv}^{bl} = \frac{D_{wv}}{\delta^{bl}}$ (Eq. 8.3), g_{wv}^{bl} will be doubled and so becomes $(15 \text{ mm s}^{-1})(2)$ or 30 mm s^{-1}. Therefore, total water vapor conductance becomes

$$g_{wv}^{\text{total}} = \cfrac{1}{\cfrac{1}{g_{wv}^{bl}} + \cfrac{1}{g_{wv}^{st}} + \cfrac{1}{g_{wv}^{ias}}} = \cfrac{1}{\cfrac{1}{(30 \text{ mm s}^{-1})} + \cfrac{1}{(4.3 \text{ mm s}^{-1})} + \cfrac{1}{(30 \text{ mm s}^{-1})}}$$

$$= 3.3 \text{ mm s}^{-1}$$

Because g_{wv}^{total} was originally 3.0 mm s^{-1}, J_{wv} will increase by 10%.

(f) Because δ^{bl} is halved, J_H^C initially doubles; T^{leaf} becomes closer to T^{ta}. If T^{leaf} is above T^{ta}, as generally occurs during the daytime, the decrease in T^{leaf} in response to a higher wind will reduce c_{wv}^e (which is nearly at the saturation value). If the accompanying fractional reduction in $c_{wv}^e - c_{wv}^{ta}$ is more than the fractional increase in g_{wv}^{total}, the increase in wind speed will lead to a decrease in transpiration because $J_{wv} = g_{wv}^{\text{total}}(c_{wv}^e - c_{wv}^{ta})$ by Eqs. (8.15) and (8.16).

Chapter 9

9.1 **(a)** We must convert the transpiration value from mm H_2O h^{-1} to the more common units of J_{wv}, mmol m^{-2} s^{-1}, noting that 1 m^3 H_2O corresponds to 996 kg at 30°C and 18.0 g corresponds to 1 mol:

$$1 \text{ mm h}^{-1} = \frac{10^{-3} \text{ m}^3 \text{ m}^{-2}}{3600 \text{ s}} = \frac{996 \text{ g m}^{-2}}{3600 \text{ s}} = \frac{(996/18.0) \text{ mol m}^{-2}}{3600 \text{ s}}$$

$$= 0.0154 \text{ mol m}^{-2} \text{ s}^{-1}$$

The change in water vapor concentration during peak transpiration is

$$\Delta c_{wv}^{ta} = (c_{wv}^*)\left(\frac{\Delta RH}{100}\right) = (1.69 \text{ mol m}^{-3})\left(\frac{20}{100}\right) = 0.34 \text{ mol m}^{-3}$$

(b) Using Eq. (9.4), r_{wv}^{ta} during peak transpiration is

$$r_{wv}^{ta} = \frac{\Delta c_{wv}^{ta}}{J_{wv}} = \frac{(0.34 \text{ mol m}^{-3})}{(15.4 \times 10^{-3} \text{ mol m}^{-2} \text{ s}^{-1})} = 22 \text{ s m}^{-1}$$

Because the value of K_{wv} at night is half the daytime value, by Eq. (9.4) r_{wv}^{ta} at night must be twice the daytime r_{wv}^{ta}:

$$r_{wv}^{ta} = \frac{\Delta z}{K_{wv}} = (22 \text{ s m}^{-1})(2) = 44 \text{ s m}^{-1}$$

(c) At night, J_{wv} is only 10% of the peak value and thus is $(15.4 \text{ mmol m}^{-2} \text{ s}^{-1})(0.10)$, or $1.54 \text{ mmol m}^{-2} \text{ s}^{-1}$. Therefore, using Eq. (9.4) we find

$$\Delta c_{wv}^{ta} = J_{wv}r_{wv}^{ta} = (1.54 \text{ mmol m}^{-2} \text{ s}^{-1})(44 \text{ s m}^{-1}) = 68 \text{ mmol m}^{-3}$$

At an air temperature of 20°C, the drop in relative humidity is

$$100 \frac{\Delta c_{wv}^{ta}}{c_{wv}^*} = (100)\frac{(0.068 \, \text{mol m}^{-3})}{(0.96 \, \text{mol m}^{-3})} = 7\%$$

(d) Because we know J_{wv} from (a), we can calculate J_{CO_2}:

$$J_{CO_2} = \frac{J_{wv}}{-700} = \frac{(15.4 \, \text{mmol m}^{-2} \, \text{s}^{-1})}{(-700)} = -0.022 \, \text{mmol m}^{-2} \, \text{s}^{-1}$$

$$= -22 \, \mu\text{mol m}^{-2} \, \text{s}^{-1}$$

Using Eq. (9.4), we find

$$\Delta c_{CO_2}^{ta} = J_{CO_2} r_{CO_2}^{ta} = (-0.022 \, \text{mmol m}^{-2} \, \text{s}^{-1})(22 \, \text{s m}^{-1}) = -0.48 \, \text{mmol m}^{-3}$$

(e) Rearranging Eq. (9.4), we find

$$K_{CO_2} = \frac{\Delta z}{r_{CO_2}^{ta}} = \frac{(30 \, \text{m})}{(44 \, \text{s m}^{-1})} = 0.68 \, \text{m}^2 \, \text{s}^{-1}$$

(f) Using Eq. (1.6), we can calculate the time for eddy diffusion at night:

$$t_e = \frac{x_e^2}{4K_{wv}} = \frac{(1 \, \text{m})^2}{(4)(0.68 \, \text{m}^2 \, \text{s}^{-1})} = 0.37 \, \text{s}$$

Also using Eq. (1.6), we can calculate the time for ordinary diffusion:

$$t_e = \frac{x_e^2}{4D_{wv}} = \frac{(1 \, \text{m})^2}{(4)(2.4 \times 10^{-5} \, \text{m}^2 \, \text{s}^{-1})} = 10,400 \, \text{s} = 2.9 \, \text{h}$$

9.2 (a) We use Eq. (9.5) to calculate the cumulative leaf areas at different incident light levels, where $F = (\ln J_0/J)/K$:

For $J_0 = 2000 \, \mu\text{mol m}^{-2} \, \text{s}^{-1}$, $F = \frac{\ln(2000/8)}{(0.7)} = 7.9$ (i.e., essentially none of the leaves is then below light compensation)

For $J_0 = 200 \, \mu\text{mol m}^{-2} \, \text{s}^{-1}$, $F = \frac{\ln(200/8)}{(0.7)} = 4.6$

For $J_0 = 20 \, \mu\text{mol m}^{-2} \, \text{s}^{-1}$, $F = \frac{\ln(20/8)}{(0.7)} = 1.3$

For $J_0 = 0 \, \mu\text{mol m}^{-2} \, \text{s}^{-1}$, all the leaves are below light compensation (i.e., $F = 0.0$)

(b) We can calculate $J_{CO_2}^{soil}$ coming up from the soil as follows:

$$J_{CO_2}^{soil} = \frac{J_{wv}^{soil}}{200} = \frac{(0.6 \, \text{mmol m}^{-2} \, \text{s}^{-1})}{(200)} = 0.0030 \, \text{mmol m}^{-2} \, \text{s}^{-1}$$

$$= 3.0 \, \mu\text{mol m}^{-2} \, \text{s}^{-1}$$

We can calculate $J_{CO_2}^{total}$ (which includes CO_2 from above the canopy and from the soil) by conversion of the net photosynthetic rate (see Table 8-2):

$$J_{CO_2}^{total} = (20 \, \text{kg ha}^{-1} \, \text{h}^{-1}) \left(0.92 \frac{\mu\text{mol m}^{-2} \, \text{s}^{-1}}{\text{kg ha}^{-1} \, \text{h}^{-1}} \right) = 18.4 \, \mu\text{mol m}^{-2} \, \text{s}^{-1}$$

The flux from above the canopy is then $18.4 \, \mu\text{mol m}^{-2} \, \text{s}^{-1} - 3.0 \, \mu\text{mol m}^{-2} \, \text{s}^{-1}$, or $15.4 \, \mu\text{mol m}^{-2} \, \text{s}^{-1}$.

(c) We are given that 10% of the water comes from the soil ($J_{wv}^{soil} = 0.6 \, \text{mmol m}^{-2}$ s^{-1}), so J_{wv} above the canopy is $6.0 \, \text{mmol m}^{-2} \, \text{s}^{-1}$. From (b) we calculated J_{CO_2} above the canopy to be $-15.4 \, \mu\text{mol m}^{-2} \, \text{s}^{-1}$, so

$$\frac{J_{wv}}{J_{CO_2}} = \left| \frac{(6000 \, \mu\text{mol m}^{-2} \, \text{s}^{-1})}{(-15.4 \, \mu\text{mol m}^{-2} \, \text{s}^{-1})} \right| = 390 \, \text{H}_2\text{O}/\text{CO}_2$$

(d) The maximum upward flux of CO_2 will occur for the level at the light compensation point (leaves below this level have a net evolution of CO_2). At $200 \, \mu\text{mol m}^{-2} \, \text{s}^{-1}$ incident on the canopy, light compensation occurs at $F = 4.6$ [calculated in (a)]. Because the leaf area index is 8.0, F increases by 1.0 every 2.0 m (16 m/8.0). Therefore, an F of 4.6 occurs at 9.2 m from the top of the canopy, or 6.8 m above the ground.

(e) The minimum in $c_{CO_2}^{ta}$ occurs at the level corresponding to the cumulative uptake of all the CO_2 coming down into the canopy ($15.4 \, \mu\text{mol m}^{-2} \, \text{s}^{-1}$); net uptake of CO_2 occurs only between 6.8 and 16 m in the canopy, and the magnitude of this uptake is $20.4 \, \mu\text{mol m}^{-2} \, \text{s}^{-1}$ (i.e., $15.4 + 5.0$, the latter from below 6.8 m). We will divide the canopy into 2-m-thick layers across which F increases by 1.0, and we will let x equal that part of the total flux taken up by the first layer of leaves. Because the PPFD is halved every 2 m [by Eq. 9.5; $\ln(J_0/J) = kF = (0.7)(1.0) = 0.7$, or $J_0/J = 2.0$] and because J_{CO_2} is proportional to PPFD by supposition, CO_2 uptake is also halved every 2 m. Therefore we obtain

$$\text{Total uptake} = 20.4 \, \mu\text{mol m}^{-2} \, \text{s}^{-1} = x + \tfrac{1}{2}x + \left(\tfrac{1}{2}\right)^2 x + \left(\tfrac{1}{2}\right)^3 x$$
$$+ \left(\frac{8.0 - 6.8 \, \text{m}}{2.0 \, \text{m}} \right) \left(\tfrac{1}{2}\right)^4 x = 1.91x$$
$$x = 10.7 \, \mu\text{mol m}^{-2} \, \text{s}^{-1}$$

The first 2-m layer thus takes up $10.7 \, \mu\text{mol m}^{-2} \, \text{s}^{-1}$ and the second layer one-half of this, or $5.3 \, \mu\text{mol m}^{-2} \, \text{s}^{-1}$. Thus, at 4 m from the top of the canopy, the total uptake is $16.0 \, \mu\text{mol m}^{-2} \, \text{s}^{-1}$, which is slightly more than the flux of CO_2 from the air above the canopy. Therefore, $c_{CO_2}^{ta}$ will be minimal at slightly above 12 m from the ground.

9.3 (a) The mean speed of the fluid in the xylem element is

$$J_V = \frac{\text{volume flow rate per tube}}{\text{tube area}} = \frac{(20 \, \text{mm}^3 \, \text{h}^{-1})}{(0.004 \, \text{m}^2)(3600 \, \text{s h}^{-1})} = 1.4 \, \text{mm s}^{-1}$$

(b) Rearranging Poiseuille's law (Eq. 9.11b), we obtain

$$\frac{\Delta P}{\Delta x} = \frac{-J_V 8\eta}{r^2} = \frac{-J_V 8\eta}{A/\pi} = \frac{-(1.4 \text{ mm s}^{-1})(8)(1.002 \times 10^{-3} \text{ Pa s})}{(0.004 \text{ mm}^2/\pi)}$$

$$= -8.8 \text{ Pa mm}^{-1} = -0.0088 \text{ MPa m}^{-1}$$

(c) Again using Poiseuille's law, we find

$$J_V = -\frac{r^2}{8\eta}\frac{\Delta P}{\Delta x} = \frac{-(10 \times 10^{-9} \text{ m})^2}{(8)(1.002 \times 10^{-3} \text{ Pa s})}(-0.0088 \times 10^6 \text{ Pa m}^{-1})$$

$$= 1.1 \times 10^{-10} \text{ m s}^{-1}$$

9.4 **(a)** We can rearrange Eq. (1.6) to find the time required for diffusion:

$$\text{For } 10 \text{ mm:} t = \frac{x_e^2}{4D_{\text{sucrose}}} = \frac{(1 \times 10^{-2} \text{ m})^2}{(4)(0.3 \times 10^{-9} \text{ m}^2 \text{ s}^{-1})} = 8.3 \times 10^4 \text{ s} = 23 \text{ h}$$

$$\text{For } 1 \text{ m:} t = \frac{x_e^2}{4D_{\text{sucrose}}} = \frac{(1 \text{ m})^2}{(4)(0.3 \times 10^{-9} \text{ m}^2 \text{ s}^{-1})} = 8.3 \times 10^8 \text{ s} = 26 \text{ years}$$

(b) We use Poiseuille's law (Eq. 9.11b) to find

$$J_V = -\frac{r^2}{8\eta}\frac{\Delta P}{\Delta x} = -\frac{(10 \times 10^{-6} \text{ m})^2}{(8)(1.5 \times 10^{-3} \text{ Pa s})}(-0.02 \times 10^6 \text{ Pa m}^{-1})$$

$$= 1.7 \times 10^{-4} \text{ m s}^{-1} = 0.17 \text{ mm s}^{-1} \quad (0.6 \text{ m h}^{-1})$$

(c) Because J_V is the mean velocity, to travel 10 mm takes $[10 \text{ mm }/(0.17 \text{ mm s}^{-1})]$ or 59 s and to travel 1 m takes $[1 \text{ m }/(0.17 \times 10^{-3} \text{ m s}^{-1})]$ or 5900 s, which is 1.6 h, considerably less than that calculated for diffusion in (a). Also note that the distance traveled in the phloem vessel is proportional to time rather than to the square root of time, as it is for diffusion (see Eq. 1.6).

(d) A vertical sieve tube has a static gradient in the absence of flow and due to gravity of -0.01 MPa m^{-1}. The gradient leading to flow would be either -0.01 MPa m^{-1} (upward) or -0.03 MPa m^{-1} (downward). Thus, J_V is either half that in (b), namely, 0.3 m h^{-1} upward, or 0.9 m h^{-1} downward.

9.5 **(a)** If we ignore Π^{soil}, then $\Psi^{\text{soil}} = P^{\text{soil}}$, which equals $-\sigma(\frac{1}{r_1} + \frac{1}{r_2})$ by Eq. (9.5). For hemispherical interfaces $r_1 = r_2$, so $\Psi^{\text{soil}} = \frac{-2\sigma}{r}$. Therefore, for the wet soil we obtain

$$r = \frac{-2\sigma}{\Psi^{\text{soil}}} = \frac{-(2)(0.0728 \text{ Pa m})}{(-0.01 \times 10^6 \text{ Pa})} = 1.5 \times 10^{-5} \text{ m} = 15 \text{ }\mu\text{m}$$

For the dry soil we obtain

$$r = \frac{-2\sigma}{\Psi^{\text{soil}}} = \frac{-(2)(0.0728 \text{ Pa m})}{(-1.4 \times 10^6 \text{ Pa})} = 1.0 \times 10^{-7} \text{ m} = 0.10 \text{ }\mu\text{m}$$

(b) First, we must find J_V of the root from the following relationship based on Eq. (9.12):

$$J_V^{\text{root}} = J_V^{\text{xylem}}\frac{A^{\text{xylem}}}{A^{\text{root}}} = (2 \times 10^{-3} \text{ m s}^{-1})\left(\frac{1}{10^5}\right) = 2 \times 10^{-8} \text{ m s}^{-1}$$

Then assuming cylindrical symmetry we rearrange Eq. (9.8) to find the drop in hydrostatic pressure in the wet soil:

$$\Delta P = P_a - P_b = \frac{J_V r \ln(r_a/r_b)}{L^{\text{soil}}}$$

$$= \frac{(2 \times 10^{-8}\,\text{m s}^{-1})(1.5 \times 10^{-3}\,\text{m}) \ln\left(\frac{8\,\text{mm}+1.5\,\text{mm}}{1.5\,\text{mm}}\right)}{(1 \times 10^{-11}\,\text{m}^2\,\text{s}^{-1}\,\text{Pa}^{-1})} = 5.5\,\text{Pa}$$

(c) Again, by rearranging Eq. (9.8) we find the drop in hydrostatic pressure to be

$$\Delta P = \frac{J_V r \ln(r_a/r_b)}{L^{\text{soil}}}$$

$$= \frac{(2 \times 10^{-8}\,\text{m s}^{-1})(1.5 \times 10^{-3}\,\text{m}) \ln\left(\frac{1.5\,\text{mm}}{1.5\,\text{mm}-0.5\,\text{mm}}\right)}{(2 \times 10^{-16}\,\text{m}^2\,\text{s}^{-1}\,\text{Pa}^{-1})}$$

$$= 6.1 \times 10^4\,\text{Pa} = 0.061\,\text{MPa}$$

(d) First, we should determine the water potential in the root xylem:

$$\Psi^{\text{root xylem}} = \Psi^{\text{soil}} - \Delta\Psi^{\text{soil}} - \Delta\Psi^{\text{root}}$$

$$= (-0.01\,\text{MPa}) - (5.5 \times 10^{-6}\,\text{MPa}) - (0.061\,\text{MPa}) = -0.071\,\text{MPa}$$

$$= P^{\text{root xylem}} - \Pi^{\text{root xylem}} + \rho_w g h = P^{\text{root xylem}}$$

In the leaf 3 m above the ground, $\Psi^{\text{leaf}}(-0.20\,\text{MPa})$ is $P^{\text{leaf}} - \Pi^{\text{leaf}} + \rho_w g h$, which is $P^{\text{leaf}} - 0\,\text{MPa} + (0.01\,\text{MPa m}^{-1})(3\,\text{m})$ or $P^{\text{leaf}} + 0.03\,\text{MPa}$, so P^{leaf} is $-0.20\,\text{MPa} - 0.03\,\text{MPa}$, or $-0.23\,\text{MPa}$. Thus the gradient is

$$\frac{\Delta P}{\Delta x} = \frac{P^{\text{leaf}} - P^{\text{root xylem}}}{\Delta x} = \frac{(-0.23\,\text{MPa}) - (0.07\,\text{MPa})}{(3\,\text{m})}$$

$$= -0.053\,\text{MPa m}^{-1}$$

(e) We must first find J_V of the leaf based on Eq. (9.12):

$$J_V^{\text{leaf}} = J_V^{\text{xylem}} \frac{A^{\text{xylem}}}{A^{\text{root}}} \frac{A^{\text{root}}}{A^{\text{leaf}}} = (2 \times 10^{-3}\,\text{m s}^{-1})(10^{-5})(5) = 1.0 \times 10^{-7}\,\text{m s}^{-1}$$

We must then determine J_V of the mesophyll:

$$J_V^{\text{mes}} = J_V^{\text{leaf}} \frac{A^{\text{leaf}}}{A^{\text{mes}}} = (1.0 \times 10^{-7}\,\text{m s}^{-1})\left(\tfrac{1}{20}\right) = 5 \times 10^{-9}\,\text{m s}^{-1}$$

Now we apply Poiseuille's law (Eq. 9.11b) to find the drop in hydrostatic pressure that could account for the rate of transpiration for the plant in the wet soil:

$$\Delta P = \frac{-J_V 8\eta \Delta x}{r^2} = \frac{-(5 \times 10^{-9}\,\text{m s}^{-1})(8)(1.002 \times 10^{-3}\,\text{Pa s})(1 \times 10^{-6}\,\text{m})}{(5 \times 10^{-9}\,\text{m})^2}$$

$$= -1.6\,\text{Pa, which is very small}$$

(f) If we use Eq. (2.21), we find that at 99% relative humidity

$$\Psi^{\text{ta}} = \frac{RT}{V_w} \ln\frac{RH}{100} = (135\,\text{MPa}) \ln\left(\tfrac{99}{100}\right) = -1.4\,\text{MPa}$$

Because Ψ^{ta} then equals Ψ^{soil}, water will not move from the dry soil to the air through the plant, so J_V^{xylem} is zero.

9.6 **(a)** The total leaf area is equal to the leaf area index times the projected area of one leaf layer, which is $(6)(\pi)(3\,m)^2$ or $170\,m^2$. Noting that 1 mol of H_2O is 18 g, which occupies $18 \times 10^{-6}\,m^3$, the transpiration rate is

$$\text{Transpiration rate} = J_{wv}(18 \times 10^{-6}\,m^3\,mol^{-1})A^{leaf}$$
$$= (1 \times 10^{-3}\,mol\,m^{-2}\,s^{-1})(18 \times 10^{-6}\,m^3\,mol^{-1})(170\,m^2)$$
$$= 3.1 \times 10^{-6}\,m^3\,s^{-1}$$

(b) We must first determine the amount of water available for transpiration:

$$(1.0 \times 10^{-3}\,m - 0.8 \times 10^{-3}\,m)(\pi)(3\,m)^2 = 5.7 \times 10^{-3}\,m^3$$

We can now calculate how long this will last:

$$\frac{(5.7 \times 10^{-3}\,m^3)}{(3.1 \times 10^{-6}\,m^3\,s^{-1})} = 1840\,s = 31\,min$$

(c) We use Eq. (9.12) to find R^{trunk} using Ψ^{trunk} of -0.3 MPa at the base and -0.5 MPa at midheight:

$$R^{trunk} = \frac{\Delta\Psi}{J_{wv}A^{leaf}} = \frac{2(\Psi^{base} - \Psi^{mid})}{(J_{wv}A^{leaf})} = \frac{2[(-0.3\,MPa) - (-0.5\,MPa)]}{(3.1 \times 10^{-6}\,m^3\,s^{-1})}$$
$$= 1.3 \times 10^5\,MPa\,s\,m^{-3}$$

(d) Using Eq. (9.13), we obtain

$$\rho = \frac{R^{trunk}A^{trunk}}{\Delta x} = \frac{(1.3 \times 10^5\,MPa\,s\,m^{-3})(0.10\,m^2)(0.05)}{(3\,m)}$$
$$= 220\,MPa\,s\,m^{-2}$$

Hydraulic conductivity is the inverse of hydraulic resistivity:

$$\frac{1}{\rho} = \frac{1}{(220\,MPa\,s\,m^{-2})} = 4.6 \times 10^{-3}\,m^2\,s^{-1}\,MPa^{-1}$$

(e) We use Eq. (9.14) to find the capacitance:

$$C^{trunk} = \frac{\Delta V^{trunk}}{\Delta\Psi^{day}} = \frac{(0.10\,m^2)(3\,m)(0.008)}{(-0.1\,MPa) - (-0.5\,MPa)}$$
$$= 6.0 \times 10^{-3}\,m^3\,MPa^{-1}$$

We use Eq. (9.15) to calculate the time constant for water release from the trunk:

$$t = R^{trunk}C^{trunk} = (1.3 \times 10^5\,MPa\,s\,m^{-3})(6 \times 10^{-3}\,m^3\,MPa^{-1}) = 780\,s = 13\,min$$

APPENDIX I

Numerical Values of Constants and Coefficients

Symbol	Description	Magnitude
c	Speed of light in vacuum	2.998×10^8 m s^{-1}
c_{wv}^*	Saturation concentration of water vapor (i.e., at 100% relative humidity)	See pp. 443–444 for values from -30 to $60°$C
C_P^{water}	Volumetric heat capacity of water at constant pressure (1 atm, 0.1013 MPa)	4.217 MJ m^{-3} °C^{-1} at $0°$C
		4.175 MJ m^{-3} °C^{-1} at $20°$C
		4.146 MJ m^{-3} °C^{-1} at $40°$C
C_P^{air}	Volumetric heat capacity of dry air at constant pressure (1 atm)	1.300 kJ m^{-3} °C^{-1} at $0°$C
		1.212 kJ m^{-3} °C^{-1} at $20°$C
		1.136 kJ m^{-3} °C^{-1} at $40°$C
D_{CO_2}	Diffusion coefficient of CO_2 in air (1 atm, 0.1013 MPa)	1.33×10^{-5} m^2 s^{-1} at $0°$C
		1.42×10^{-5} m^2 s^{-1} at $10°$C
		1.51×10^{-5} m^2 s^{-1} at $20°$C
		1.60×10^{-5} m^2 s^{-1} at $30°$C
		1.70×10^{-5} m^2 s^{-1} at $40°$C
D_{O_2}	Diffusion coefficient of O_2 in air (1 atm, 0.1013 MPa)	1.95×10^{-5} m^2 s^{-1} at $20°$C
D_{wv}	Diffusion coefficient of water vapor in air (1 atm, 0.1013 MPa)	2.13×10^{-5} m^2 s^{-1} at $0°$C
		2.27×10^{-5} m^2 s^{-1} at $10°$C
		2.42×10^{-5} m^2 s^{-1} at $20°$C
		2.57×10^{-5} m^2 s^{-1} at $30°$C
		2.72×10^{-5} m^2 s^{-1} at $40°$C
e	Base for natural logarithm	2.71828 ($1/e = 0.368$)
	Electronic charge	1.602×10^{-19} coulomb
F	Faraday's constant	9.649×10^4 coulomb mol^{-1}
		9.649×10^4 J mol^{-1} V^{-1}
		2.306×10^4 cal mol^{-1} V^{-1}
		23.06 kcal mol^{-1} V^{-1}

(*continues*)

(*continued*)

Symbol	Description	Magnitude
g	Gravitational acceleration	9.780 m s^{-2} (sea level,^{a} 0° latitude)
		9.807 m s^{-2} (sea level,^{a} 45° latitude)
		9.832 m s^{-2} (sea level,^{a} 90° latitude)
		978.0 cm s^{-2} (sea level,^{a} 0° latitude)
		980.7 cm s^{-2} (sea level,^{a} 45° latitude)
		983.2 cm s^{-2} (sea level,^{a} 90° latitude)
h	Planck's constant	6.626 × 10^{-34} J s
		6.626 × 10^{-27} erg s
		0.4136 × 10^{-14} eV s
		1.584 × 10^{-37} kcal s
hc		1.986 × 10^{-25} J m
		1240 eV nm
H_{sub}	Heat of sublimation of water	51.37 kJ mol^{-1} (2.847 MJ kg^{-1}) at −10°C
		51.17 kJ mol^{-1} (2.835 MJ kg^{-1}) at −5°C
		51.00 kJ mol^{-1} (2.826 MJ kg^{-1}) at 0°C
		12.27 kcal mol^{-1} (680 cal g^{-1}) at −10°C
		12.22 kcal mol^{-1} (677 cal g^{-1}) at −5°C
		12.18 kcal mol^{-1} (675 cal g^{-1}) at 0°C
H_{vap}	Heat of vaporization of water	45.06 kJ mol^{-1} (2.501 MJ kg^{-1}) at 0°C
		44.63 kJ mol^{-1} (2.477 MJ kg^{-1}) at 10°C
		44.21 kJ mol^{-1} (2.454 MJ kg^{-1}) at 20°C
		44.00 kJ mol^{-1} (2.442 MJ kg^{-1}) at 25°C
		43.78 kJ mol^{-1} (2.430 MJ kg^{-1}) at 30°C
		43.35 kJ mol^{-1} (2.406 MJ kg^{-1}) at 40°C
		42.91 kJ mol^{-1} (2.382 MJ kg^{-1}) at 50°C
		40.68 kJ mol^{-1} (2.258 MJ kg^{-1}) at 100°C
k	Boltzmann's constant	1.381 × 10^{-23} J molecule^{-1} K^{-1}
		1.381 × 10^{-16} erg molecule^{-1} K^{-1}
		8.617 × 10^{-5} eV molecule^{-1} K^{-1}
kT		0.02354 eV molecule^{-1} at 0°C
		0.02526 eV molecule^{-1} at 20°C
		0.02569 eV molecule^{-1} at 25°C
		0.02699 eV molecule^{-1} at 40°C
K^{air}	Thermal conductivity coefficient of dry air (1 atm)^b	0.0237 W m^{-1} °C^{-1} at −10°C
		0.0243 W m^{-1} °C^{-1} at 0°C
		0.0250 W m^{-1} °C^{-1} at 10°C
		0.0257 W m^{-1} °C^{-1} at 20°C
		0.0264 W m^{-1} °C^{-1} at 30°C
		0.0270 W m^{-1} °C^{-1} at 40°C
		0.0277 W m^{-1} °C^{-1} at 50°C
	Thermal conductivity coefficient of moist air (100% relative humidity, 1 atm)	0.0242 W m^{-1} °C^{-1} at 0°C
		0.0255 W m^{-1} °C^{-1} at 20°C
		0.0264 W m^{-1} °C^{-1} at 40°C
K^{water}	Thermal conductivity coefficient of water	0.565 W m^{-1} °C^{-1} at 0°C
		0.599 W m^{-1} °C^{-1} at 20°C
		0.627 W m^{-1} °C^{-1} at 40°C
ln 2		0.6931
N	Avogadro's number	6.0220 × 10^{23} entities mol^{-1}
Nhc		0.1196 J mol^{-1} m
		119,600 kJ mol^{-1} nm
		28.60 kcal mol^{-1} μm
		28,600 kcal mol^{-1} nm
N^*_{wv}	Saturation mole fraction of water vapor (i.e., at 100% relative humidity) at 1 atm (0.1013 MPa)	See pp. 443–444 for values from −30 to 60°C

(*continues*)

(*continued*)

Symbol	Description	Magnitude
P_{wv}^*	Saturation vapor pressure of water	See pp. 443–444 for values from -30 to $60°C$
	Protonic charge	1.602×10^{-19} coulomb
R	Gas constant	8.314 J mol^{-1} K^{-1}
		1.987 cal mol^{-1} K^{-1}
		8.314 m^3 Pa mol^{-1} K^{-1}
		8.314×10^{-6} m^3 MPa mol^{-1} K^{-1}
		0.08205 liter atm mol^{-1} K^{-1}
		0.08314 liter bar mol^{-1} K^{-1}
		83.14 cm^3 bar mol^{-1} K^{-1}
RT		2.271×10^3 J mol^{-1} (m^3 Pa mol^{-1}) at $0°C$
		2.437×10^3 J mol^{-1} (m^3 Pa mol^{-1}) at $20°C$
		2.479×10^3 J mol^{-1} (m^3 Pa mol^{-1}) at $25°C$
		2.271×10^{-3} m^3 MPa mol^{-1} at $0°C$
		2.437×10^{-3} m^3 MPa mol^{-1} at $20°C$
		2.479×10^{-3} m^3 MPa mol^{-1} at $25°C$
		542.4 cal mol^{-1} at $0°C$
		582.2 cal mol^{-1} at $20°C$
		2.271 liter MPa mol^{-1} at $0°C$
		2.437 liter MPa mol^{-1} at $20°C$
		22.71 liter bar mol^{-1} at $0°C$
		24.37 liter bar mol^{-1} at $20°C$
		$22,710$ cm^3 bar mol^{-1} at $0°C$
		$24,370$ cm^3 bar mol^{-1} at $20°C$
		22.41 liter atm mol^{-1} at $0°C$
		24.05 liter atm mol^{-1} at $20°C$
$2.303\,RT$		5.612 kJ mol^{-1} at $20°C$
		5.708 kJ mol^{-1} at $25°C$
		1.342 kcal mol^{-1} at $20°C$
		1.364 kcal mol^{-1} at $25°C$
		$56,120$ cm^3 bar mol^{-1} at $20°C$
RT/F		25.3 mV at $20°C$
		25.7 mV at $25°C$
$2.303\,RT/F$		58.2 mV at $20°C$
		59.2 mV at $25°C$
		60.2 mV at $30°C$
$RT/\bar{V}_w$		135.0 MPa at $20°C$
		137.3 MPa at $25°C$
		32.31 cal cm^{-3} at $20°C$
		135.0 J cm^{-3} at $20°C$
		1350 bar at $20°C$
		1330 atm at $20°C$
$2.303\,RT/\bar{V}_w$		310.9 MPa at $20°C$
		316.2 MPa at $25°C$
		3063 atm at $20°C$
		3109 bar at $20°C$
S_c	Solar constant	1367 W m^{-2}
		1.959 cal cm^{-2} min^{-1}
		1.367×10^5 erg cm^{-2} s^{-1}
		0.1367 W cm^{-2}
	Thermal capacity of water (mass basis)	4218 J kg^{-1} °C^{-1} at $0°C$
		4182 J kg^{-1} °C^{-1} at $20°C$
		4179 J kg^{-1} °C^{-1} at $40°C$
		1.0074 cal g^{-1} °C^{-1} at $0°C$

(*continues*)

(continued)

Symbol	Description	Magnitude
		0.9988 cal g^{-1} °C^{-1} at 20°C
		0.9980 cal g^{-1} °C^{-1} at 40°C
	Thermal capacity of water (mole basis)	75.99 J mol^{-1} °C^{-1} at 0°C
		75.34 J mol^{-1} °C^{-1} at 20°C
		75.28 J mol^{-1} °C^{-1} at 40°C
		18.14 cal mol^{-1} °C^{-1} at 0°C
		17.99 cal mol^{-1} °C^{-1} at 20°C
		17.98 cal mol^{-1} °C^{-1} at 40°C
$\bar{V}_w$	Partial molal volume of water	1.805×10^{-5} m^3 mol^{-1} at 20°C
		18.05 cm^3 mol^{-1} at 20°C
ε_o	Permittivity of a vacuum	8.854×10^{-12} coulomb^2 m^{-2} N^{-1}
		8.854×10^{-12} coulomb m^{-1} V^{-1}
η_{air}	Viscosity of air (dry, 1 atm)	1.716×10^{-5} Pa s at 0°C
		1.813×10^{-5} Pa s at 20°C
		1.907×10^{-5} Pa s at 40°C
η_w	Viscosity of water	1.787×10^{-3} Pa s at 0°C
		1.307×10^{-3} Pa s at 10°C
		1.002×10^{-3} Pa s at 20°C
		0.798×10^{-3} Pa s at 30°C
		0.653×10^{-3} Pa s at 40°C
		0.547×10^{-3} Pa s at 50°C
		0.01002 dyn s cm^{-2} at 20°C
		0.01002 poise at 20°C
ν_{air}	Kinematic viscosity of air (dry, 1 atm)	1.327×10^{-5} m^2 s^{-1} at 0°C
		1.505×10^{-5} m^2 s^{-1} at 20°C
		1.691×10^{-5} m^2 s^{-1} at 40°C
ν_w	Kinematic viscosity of water	1.787×10^{-6} m^2 s^{-1} at 0°C
		1.004×10^{-6} m^2 s^{-1} at 20°C
		0.658×10^{-6} m^2 s^{-1} at 40°C
π	Circumference/diameter of circle	3.14159
ρ_{air}	Density of dry air (1 atm, 0.1013 MPa)	1.293 kg m^{-3} at 0°C
		1.205 kg m^{-3} at 20°C
		1.128 kg m^{-3} at 40°C
	Density of saturated air (1 atm)^c	1.290 kg m^{-3} at 0°C
		1.194 kg m^{-3} at 20°C
		1.097 kg m^{-3} at 40°C
ρ_w	Density of water	999.8 kg m^{-3} (0.9998 g cm^{-3}) at 0°C
		1000.0 kg m^{-3} (1.0000 g cm^{-3}) at 4°C
		999.7 kg m^{-3} (0.9997 g cm^{-3}) at 10°C
		998.2 kg m^{-3} (0.9982 g cm^{-3}) at 20°C
		995.6 kg m^{-3} (0.9956 g cm^{-3}) at 30°C
		992.2 kg m^{-3} (0.9922 g cm^{-3}) at 40°C
$\rho_w g$		0.00979 MPa m^{-1} (20°C, sea level, 45° latitude)
		0.0979 bar m^{-1} (20°C, sea level, 45° latitude)
		979 dyn cm^{-3} (20°C, sea level, 45° latitude)
		0.0966 atm m^{-1} (20°C, sea level, 45° latitude)
σ	Stefan–Boltzmann constant	5.670×10^{-8} W m^{-2} K^{-4}
		5.670×10^{-12} W cm^{-2} K^{-4}
		8.130×10^{-11} cal cm^{-2} min^{-1} K^{-4}
		5.670×10^{-5} erg cm^{-2} s^{-1} K^{-4}
σ_w	Surface tension of water	0.0756 N m^{-1} (Pa m) at 0°C
		0.0742 N m^{-1} (Pa m) at 10°C
		0.0728 N m^{-1} (Pa m) at 20°C
		0.0712 N m^{-1} (Pa m) at 30°C

(continues)

(*continued*)

Symbol	Description	Magnitude
		0.0696 N m^{-1} (Pa m) at 40°C
		7.28 × 10^{-8} MPa m at 20°C
		72.8 dyn cm^{-1} at 20°C
		7.18 × 10^{-5} atm cm at 20°C
		7.28 × 10^{-5} bar cm at 20°C

[a] The correction for height above sea level is -3.09×10^{-6} m s^{-2} per meter of altitude.
[b] The pressure sensitivity is very slight, with K^{air} increasing only about 0.0001 W m^{-1} °C^{-1} per atmosphere (0.1013 MPa) increase in pressure.
[c] Moist air is less dense than dry air at the same temperature and pressure because the molecular weight of water (18.0) is less than the average molecular weight for air (29.0).

Temperature (°C)	c_{wv}^* (g m^{-3})	c_{wv}^* (mol m^{-3})	N_{wv}^* (at 1 atm)	P_{wv}^* (kPa)
−30	0.34	0.019	0.00037	0.038
−25	0.55	0.031	0.00062	0.063
−20	0.88	0.049	0.00102	0.103
−19	0.97	0.054	0.00112	0.114
−18	1.06	0.059	0.00123	0.125
−17	1.16	0.064	0.00135	0.137
−16	1.27	0.070	0.00149	0.151
−15	1.39	0.077	0.00163	0.165
−14	1.52	0.084	0.00179	0.181
−13	1.65	0.092	0.00196	0.198
−12	1.80	0.100	0.00214	0.217
−11	1.96	0.109	0.00234	0.238
−10	2.14	0.119	0.00256	0.260
−9	2.33	0.129	0.00280	0.284
−8	2.53	0.141	0.00306	0.310
−7	2.75	0.153	0.00333	0.338
−6	2.99	0.166	0.00364	0.369
−5	3.25	0.180	0.00396	0.402
−4	3.52	0.195	0.00431	0.437
−3	3.82	0.212	0.00469	0.476
−2	4.14	0.230	0.00511	0.517
−1	4.48	0.249	0.00555	0.562
0	4.85	0.269	0.00604	0.611
1	5.20	0.288	0.00649	0.657
2	5.56	0.309	0.00697	0.706
3	5.95	0.330	0.00748	0.758
4	6.36	0.353	0.00803	0.816
5	6.80	0.378	0.00862	0.873
6	7.27	0.403	0.00923	0.935
7	7.76	0.431	0.00989	1.002
8	8.28	0.459	0.01059	1.073
9	8.82	0.490	0.01133	1.148
10	9.41	0.522	0.01212	1.228
11	10.02	0.556	0.01296	1.313
12	10.67	0.592	0.01384	1.403
13	11.35	0.630	0.01478	1.498
14	12.08	0.670	0.01578	1.599
15	12.84	0.713	0.01683	1.706

(*continues*)

(*continued*)

Temperature (°C)	c_{wv}^* (g m^{-3})	c_{wv}^* (mol m^{-3})	N_{wv}^* (at 1 atm)	P_{wv}^* (kPa)
16	13.64	0.757	0.01795	1.819
17	14.49	0.804	0.01913	1.938
18	15.38	0.854	0.02037	2.064
19	16.32	0.906	0.02169	2.198
20	17.31	0.961	0.02308	2.339
21	18.35	1.018	0.02455	2.488
22	19.44	1.079	0.02610	2.645
23	20.59	1.143	0.02774	2.810
24	21.80	1.210	0.02946	2.985
25	23.07	1.280	0.03128	3.169
26	24.40	1.354	0.03319	3.363
27	25.79	1.432	0.03520	3.567
28	27.26	1.513	0.03723	3.782
29	28.79	1.598	0.03955	4.008
30	30.40	1.687	0.04190	4.246
31	32.08	1.781	0.04437	4.495
32	33.85	1.879	0.04696	4.758
33	35.70	1.981	0.04968	5.034
34	37.63	2.089	0.05253	5.323
35	39.65	2.201	0.05553	5.627
36	41.76	2.318	0.05868	5.945
37	43.97	2.441	0.06197	6.280
38	46.28	2.569	0.06543	6.630
39	48.69	2.703	0.06905	6.997
40	51.21	2.842	0.07285	7.381
41	53.83	2.988	0.07682	7.784
42	56.57	3.140	0.08098	8.205
43	59.43	3.299	0.08533	8.646
44	62.41	3.464	0.08989	9.108
45	65.52	3.637	0.09464	9.590
46	68.75	3.816	0.09962	10.09
47	72.12	4.003	0.1048	10.62
48	75.63	4.198	0.1102	11.17
49	79.28	4.401	0.1159	11.74
50	83.08	4.611	0.1218	12.34
55	104.5	5.798	0.1555	15.75
60	130.3	7.217	0.1967	19.93

APPENDIX II

Conversion Factors and Definitions

Quantity	Equals	Quantity	Equals
acre	43,560 ft^2		10^5 J m^{-3}
	4047 m^2		10^6 dyn cm^{-2}
	0.4047 ha		0.9869 atm
	4.047×10^{-3} km^2		1.020 kg cm^{-2} (at sea level,
ampere (A)	1 coulomb s^{-1}		45° latitude)^a
	1 V ohm^{-1}		750 mm Hg (at sea level,
Å	10^{-10} m		45° latitude)^a
	0.1 nm		10.22 m water (at sea level,
	10^{-8} cm		45° latitude, 20°C)^a
atm	0.1013 MPa		14.50 lb inch^{-2} (psi)
	1.013×10^5 Pa	becquerel (Bq)	1 disintegration s^{-1}
	1.013×10^5 N m^{-2}		2.703×10^{-11} curie
	1.013×10^5 J m^{-3}	British thermal	1055 J
	1.013 bar	unit (Btu)	252.0 cal
	1.013×10^6 dyn cm^{-2}	Btu h^{-1}	0.2931 W
	1.033×10^4 kg m^{-2} (at sea	calorie (cal)	4.184 J
	level, 45° latitude)^a		4.184×10^7 erg
	1.033 kg cm^{-2} (at sea level,	cal cm^{-2}	1 langley
	45° latitude)^a		4.184×10^4 J m^{-2}
	760 mm Hg (at sea level,	cal cm^{-2} min^{-1}	697.8 W m^{-2}
	45° latitude)^a		6.978×10^5 erg cm^{-2} s^{-1}
	10.35 m water (at sea level,		1 langley min^{-1}
	45° latitude, 20°C)^a	cal cm^{-1} °C^{-1} min^{-1}	6.978 W m^{-1} °C^{-1}
	14.70 lb inch^{-2} (psi)	cal cm^{-1} °C^{-1} s^{-1}	418.4 W m^{-1} °C^{-1}
bar	0.1 MPa	cal cm^{-3}	41.84 bar
	100 kPa	cal g^{-1}	4.184 J kg^{-1}
	10^5 Pa	cal m^{-2} s^{-1}	4.184 W m^{-2}
	10^5 N m^{-2}	cal min^{-1}	0.06978 J s^{-1}

(continues)

(*continued*)

Quantity	Equals	Quantity	Equals
	0.06978 W	foot (ft)	30.48 cm
cal s^{-1}	4.184 J s^{-1}		0.3048 m
	4.184 W	foot^2	9.290×10^{-2} m^2
candela (cd)	1 lumen steradian^{-1}	foot^3	2.832×10^{-2} m^3
° Celsius (°C)	(°C + 273.15) K		28.32 liters
	[(9/5)(°C) + 32] °F		6.229 gallons (U.S.)
cm	0.3937 inch	footcandle (fc)	1 lumen ft^{-2}
cm bar^{-1} s^{-1}	10^{-7} m Pa^{-1} s^{-1}		10.76 lux
cm^2 bar^{-1} s^{-1}	10^{-9} m^2 Pa^{-1} s^{-1}	foot-pound	1.356 J
	10^{-3} m^2 MPa^{-1} s^{-1}	foot s^{-1}	1.097 km h^{-1}
cm s^{-1}	0.01 m s^{-1}		0.6818 mile h^{-1}
	10 mm s^{-1}	g	1 dyn s^2 cm^{-1}
	0.03600 km h^{-1}		10^{-3} N s^2 m^{-1}
	0.02237 mile h^{-1}	g cm^{-2}	10 kg m^{-2}
cm^3	10^{-6} m^3	g cm^{-3}	1000 kg m^{-3}
	1 ml	g dm^{-2} h^{-1}	27.78 mg m^{-2} s^{-1}
CO$_2$ concentration	See Table 8.2	gallon (British)	4.546 liter
coulomb (C)	1 J V^{-1}		4.546×10^{-3} m^3
	1 ampere s	gallon (U.S.)	3.785 liter
coulomb V	1 J		3.785×10^{-3} m^3
curie (Ci)	3.7×10^{10} Bq		0.1337 ft^3
	3.7×10^{10} disintegrations s^{-1}	gallon (U.S.) acre^{-1}	9.354 liter ha^{-1}
dalton (Da)	1.661×10^{-24} g (1/12 mass of ^{12}C)	hectare (ha)	10^4 m^2
			2.471 acres
day	86,400 s	hertz (H)	1 cycle s^{-1}
	1,440 min	horsepower (hp)	745.7 W
degree (angle) (°)	0.01745 radian	hour (h)	3600 s
dyn	1 erg cm^{-1}	inch (in.)	25.40 mm
	1 g cm s^{-2}	inch^2	6.452×10^{-4} m^2
	10^{-5} N	inch^3	1.639×10^{-5} m^3
dyn cm	1 erg		0.01639 liter
	10^{-7} J	joule (J)	1 N m
dyn cm^{-1}	10^{-3} N m^{-1}		1 W s
dyn cm^{-2}	0.1 N m^{-2}		1 m^3 Pa
	10^{-6} bar		1 kg m^2 s^{-2}
einstein	1 mol (6.022×10^{23}) photons		1 coulomb V
erg	1 dyn cm		10^7 erg
	1 g cm^2 s^{-2}		0.2388 cal
	10^{-7} J		10 cm^3 bar
	6.242×10^{11} eV	J kg^{-1}	2.388×10^{-4} cal g^{-1}
	2.390×10^{-8} cal	J mol^{-1}	0.2388 cal mol^{-1}
	2.390×10^{-11} kcal	J s^{-1}	W
erg cm^{-2} s^{-1}	10^{-3} J m^{-2} s^{-1}	kcal	4.187 kJ
	10^{-3} W m^{-2}		4.187×10^{10} erg
	10^{-7} W cm^{-2}	kcal h^{-1}	1.163 W
	1.433×10^{-6} cal cm^{-2} min^{-1}	kcal mol^{-1}	4.187 kJ mol^{-1}
eV	1.602×10^{-19} J		0.04339 eV molecule^{-1}
	1.602×10^{-12} erg	kDa	1.661×10^{-21} g (1000 × 1/12 mass of ^{12}C)
eV molecule^{-1}	96.49 kJ mol^{-1}	kelvin (K)	(K −273.15) °C
	23.06 kcal mol^{-1}	kg	2.2046 lb[a]
farad (F)	1 coulomb V^{-1}		0.001 tonne (metric ton)
°Farenheit (°F)	(°F −32)(5/9) °C	kg ha^{-1}	0.893 lb acre^{-1}
		kg m^{-2}	0.1 g cm^{-2}

(*continues*)

(*continued*)

Quantity	Equals	Quantity	Equals
	1.422×10^{-3} lb in.$^{-2}$ (psi)		264.2 gallons (U.S.)
	9.807 Pa (9.807 N m^{-2}) (at sea level,	M	mol liter^{-1}
	45° latitude; see g, Appendix I)		1000 mol m^{-3}
kg m^{-3}	10^{-3} g cm^{-3}	mbar	0.1 kPa
kg H$_2$O	1 liter at 4°C		100 Pa
kJ mol^{-1}	0.01036 eV molecule^{-1}		10^{-3} bar
km	1000 m		10^3 dyne cm^{-2}
	0.6214 mile		0.9869×10^{-3} atm
km^2	100 ha	mho	1 ohm^{-1}
	247.1 acres		1 siemens
	0.3861 mile^2		1 ampere V^{-1}
km h^{-1}	0.2778 m s^{-1}	mg cm^{-3}	1 kg m^{-3}
	27.78 cm s^{-1}	mg dm^{-2} h^{-1}	2.778×10^{-2} mg m^{-2} s^{-1}
	0.6214 mile h^{-1}	Mg ha^{-1}	0.4461 ton (U.S.) ha^{-1}
knot	1.852 km h^{-1}	m s^{-1}	100 cm s^{-1}
	1.151 mile h^{-1}		3.600 km h^{-1}
kW	1000 J s^{-1}		2.237 mile h^{-1}
	3.600×10^6 J h^{-1}	micron	1 μm
	859.8 kcal h^{-1}	mile	1609 m
kW h	3.600×10^6 J		1.609 km
	8.598×10^5 cal		5280 ft
lambert	0.3183 candela cm^{-2}	mile^2	640 acres
	3183 candela m^{-2}		2.590×10^6 m^2
langley	1 cal cm^{-2}		2.590 km^2
	4.187×10^4 J m^{-2}	mile h^{-1}	0.4470 m s^{-1}
langley day^{-1}	0.4846 W m^{-2}		44.70 cm s^{-1}
langley min^{-1}	697.8 W m^{-2}		1.609 km h^{-1}
liter	0.001 m^3	MJ	0.2778 kW h
	0.03531 ft^3	ml	1 cm^3
	0.2200 gallon (British)		10^{-6} m^3
	0.2642 gallon (U.S.)	mm Hg	133.3 Pa
	1.057 quarts	mM	1 mol m^{-3}
liter atm	24.20 cal	month (mean)	2.630×10^6 s
	101.3 J	MPa	10^6 N m^{-2}
liter bar	100 J		10^6 J m^{-3}
	23.88 cal		10 bar
liter ha^{-1}	0.1069 gallon (U.S.) acre^{-1}		1.020×10^5 kg m^{-2}
liter H$_2$O	1 kg at 4°C		(at sea level, 45° latitude;
ln	2.303 log		see g, Appendix I)
log	0.4343 ln		9.872 atm
lumen (L) ft^{-2}	1 footcandle	mol m^{-3}	1 mM
	10.76 lux		1 μmol cm^{-3}
lumen m^{-2}	1 lux	nautical mile	1.151 mile
	0.09290 fc	ng cm^{-2} s^{-1}	10 μg m^{-2} s^{-1}
lux (lx)	1 lumen m^{-2}	newton (N)	1 kg m s^{-2}
	0.09290 footcandle		10^5 dyn
m	100 cm	N m	1 J
	3.280 ft	N m^{-1}	10^3 dyn cm^{-1}
	39.37 inch	N m^{-2}	1 Pa
m^2	10.76 ft^2		10^{-6} MPa
	10^{-4} ha		10 dyn cm^{-2}
m^3	35.32 ft^3		10^{-2} mbar
	1.308 yard^3	ounce (avoirdupois)	28.35 g
	220.0 gallons (British)	ounce (troy)	31.10 g

(*continues*)

(*continued*)

Quantity	Equals	Quantity	Equals
pascal (Pa)	1 N m^{-2}		1.333 mbar (at sea
	1 J m^{-3}		level, 45° latitude; see g,
	$1 \text{ kg m}^{-1} \text{ s}^{-2}$		Appendix I)
	10^{-5} bar		1333 dyn cm^{-2} (at sea level,
	9.869×10^{-6} atm		45° latitude; see g,
Photosynthesis	See Table 8.2		Appendix I)
poise (P)	1 dyn s cm^{-2}	Transpiration	See Table 8.2
	0.1 N s m^{-2}	V	1 ampere ohm
	0.1 Pa s		1 J coulomb^{-1}
pound (lb)[a]	0.4536 kg	watt (W)	1 J s^{-1}
pound acre^{-1}	1.121 kg ha^{-1}		$1 \text{ kg m}^2 \text{ s}^{-3}$
	703.1 kg m^{-2}		10^7 erg s^{-1}
	$0.07031 \text{ kg cm}^{-2}$		14.33 cal min^{-1}
pound inch^{-2} (psi)	6.895 kPa	W cm^{-2}	10^4 W m^{-2}
	0.06895 bar	W m^{-2}	$1 \text{ J m}^{-2} \text{ s}^{-1}$
	0.06805 atm		$10^3 \text{ erg cm}^{-2} \text{ s}^{-1}$
radian (rad)	57.30°		$1.433 \times 10^{-3} \text{ cal cm}^{-2} \text{ min}^{-1}$
revolution min^{-1} (rpm)	$6° \text{ s}^{-1}$		$1.433 \times 10^{-3} \text{ langley min}^{-1}$
	$0.1047 \text{ rad s}^{-1}$		$0.2388 \text{ cal m}^{-2} \text{ s}^{-1}$
s cm^{-1}	100 s m^{-1}		$2.388 \times 10^{-5} \text{ cal cm}^{-2} \text{ s}^{-1}$
s m^{-1}	$10^{-2} \text{ s cm}^{-1}$	W m^{-1} °C^{-1}	$2.388 \times 10^{-3} \text{ cal cm}^{-1} \text{ °C}^{-1} \text{ s}^{-1}$
siemens (S)	1 mho (ohm^{-1})		$0.1433 \text{ cal cm}^{-1} \text{ °C}^{-1} \text{ min}^{-1}$
	1 ampere V^{-1}	W s	1 J
therm	1×10^5 Btu	week	6.048×10^5 s
	1.055×10^8 J	yard	0.9144 m
ton (U.S.)	2000 lb	yard^2	0.8361 m^2
	907.2 kg	yard^3	0.7646 m^3
	1 ton (short)	year (normal	3.154×10^7 s
	0.8929 ton (long)	calendar)	5.256×10^5 min
ton (U.S.) acre^{-1}	$2.242 \text{ tonne ha}^{-1}$		8760 h
	2.242 Mg ha^{-1}		365 days
tonne (metric ton)	1000 kg	year (sidereal)	365.256 days
	1 Mg		3.156×10^7 s
	1.102 ton (U.S.)	year (solar)	365.242 days
tonne ha^{-1}	0.4461 ton (U.S.) acre^{-1}		3.156×10^7 s
tonne m^{-3}	1 g cm^{-3}	μg cm^{-2} s^{-1}	$10 \text{ mg m}^{-2} \text{ s}^{-1}$
torr	1 mm Hg		
	133.3 Pa (at sea level,		
	45° latitude; see g,		
	Appendix I)		

[a]Sometimes it proves convenient to express force in units of mass. To see why this is possible, consider the force F exerted by gravity on a body of mass m. This force is equal to mg, g being the gravitational acceleration (see Appendix I). Thus F/g can be used to represent a force but the units are those of mass. One atmosphere is quite often defined as 760 mm Hg (or 1.033 kg cm^{-2}), but the elevational and latitudinal effects on g should also be considered (see Appendix I).

APPENDIX III

Mathematical Relations

A. Prefixes (for units of measure)

a	atto	10^{-18}	da[a]	deka	10	
f	femto	10^{-15}	h[a]	hecto	10^2	
p	pico	10^{-12}	k	kilo	10^3	
n	nano	10^{-9}	M	mega	10^6	
μ	micro	10^{-6}	G	giga	10^9	
m	milli	10^{-3}	T	tera	10^{12}	
c[a]	centi	10^{-2}	P	peta	10^{15}	
d[a]	deci	10^{-1}	E	exa	10^{18}	

[a] Not recommended by SI (Système International, the internationally accepted system for units).

B. Areas and Volumes

The following relations pertain to a cube of length s on a side, a cylinder of radius r and length along the axis l, and a sphere of radius r.

Shape	Area	Volume	V/A
Cube	$6s^2$	s^3	$s/6$
Cylinder	$2\pi r l + 2\pi r^2$	$\pi r^2 l$	$rl/(2l + 2r)$
Sphere	$4\pi r^2$	$\frac{4}{3}\pi r^3$	$r/3$

C. Logarithms

The following relations are presented to facilitate the use of natural and common logarithms, their antilogarithms, and exponential functions. For those readers who are completely unfamiliar with such quantities, a textbook or handbook should be consulted.

$$\ln(xy) = \ln x + \ln y \qquad\qquad \ln x^a = a \ln x$$
$$\ln(x/y) = \ln x - \ln y \qquad\qquad \ln(1/x^a) = -a \ln x$$
$$\ln(1/x) = -\ln x \qquad\qquad\quad \ln x = 2.303 \log x$$
$$\ln 1 = 0 \qquad\qquad\qquad\qquad \log 1 = 0$$
$$\ln e = 1, \quad \ln 10 = 2.303 \qquad \log 10 = 1, \quad \log e = \frac{1}{2.303}$$
$$\ln e^y = y \qquad\qquad\qquad\qquad \log 10^y = y$$
$$e^{\ln y} = y \qquad\qquad\qquad\qquad 10^{\log y} = y$$

$$\ln(1 + x) = x - \frac{x^2}{2} + \frac{x^3}{3} - \frac{x^4}{4} + \cdots \qquad -1 < x \le 1$$

$$e^x = 1 + x + \frac{x^2}{2!} + \frac{x^3}{3!} + \cdots$$

D. Quadratic Equation

The quadratic equation has the following form:

$$ax^2 + bx + c = 0 \qquad (a \neq 0)$$

Its two solutions (two roots) are

$$x = \frac{-b \pm \sqrt{b^2 - 4ac}}{2a}$$

E. Trignometric Functions

Consider a right triangle of hypotenuse r:

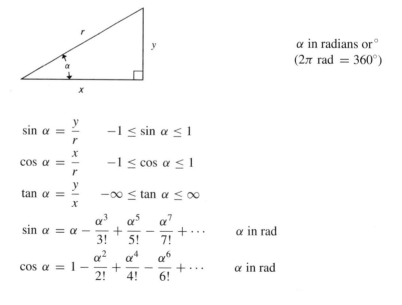

α in radians or $^\circ$
$(2\pi \text{ rad} = 360^\circ)$

$$\sin \alpha = \frac{y}{r} \qquad -1 \le \sin \alpha \le 1$$

$$\cos \alpha = \frac{x}{r} \qquad -1 \le \cos \alpha \le 1$$

$$\tan \alpha = \frac{y}{x} \qquad -\infty \le \tan \alpha \le \infty$$

$$\sin \alpha = \alpha - \frac{\alpha^3}{3!} + \frac{\alpha^5}{5!} - \frac{\alpha^7}{7!} + \cdots \qquad \alpha \text{ in rad}$$

$$\cos \alpha = 1 - \frac{\alpha^2}{2!} + \frac{\alpha^4}{4!} - \frac{\alpha^6}{6!} + \cdots \qquad \alpha \text{ in rad}$$

F. Differential Equations

As a final topic in this appendix, we will consider the application of the integral calculus to the solution of differential equations. A *differential equation* expresses the relationship between derivatives (first order as well as higher order) and various variables or functions. The procedure in solving differential equations is first to put the relation into a form that can be integrated and then to carry out suitable integrations so that the derivatives are eliminated. To complete the solution of a differential equation, we must incorporate the known values of the functions at particular values of the variables, the so-called *boundary conditions*. We will illustrate the handling of differential equations by a simple but extremely useful example.

One of the most important differential equations in biology has the following general form:

$$\frac{dy}{dt} = -ky \tag{III.1}$$

where k is a positive constant and t represents time. We encountered this equation in Chapter 4 (Eq. 4.9) in discussing the various competing pathways for the deexcitation of an excited singlet state. Equation (III.1) also describes the process of radioactive decay, where y is the amount of radioisotope present at any time t. The equation indicates that the rate of change in time of the amount of radioactive substance (dy/dt) is linearly proportional to the amount present at that time (y). Because the radioisotope decays in time, dy/dt is negative, and hence there is a minus sign in Eq. (III.1). Any process that can be described by Eq. (III.1), such as a chemical reaction, is called a first-order rate process, and k is known as the first-order rate constant.

To put Eq. (III.1) into a form suitable for integration, we must separate the variables (y and t) so that each one appears on only one side of the equation:

$$\frac{dy}{y} = -k\,dt \tag{III.2}$$

which follows from Eq. (III.2) upon multiplying each side by dt/y. (Note that the same initial process of separation of variables applies to the integration of a more complicated example in Chapter 3, namely, Eq. 3.11.) When the variables are separated so that a possible integrand, e.g., $-k\,dt$, is expressed in terms of only one variable, we can integrate that integrand. On the other hand, the integrand $-ky\,dt$ cannot be integrated as it stands— i.e, we cannot perform $\int -ky\,dt$—because we do not know how y depends on t. In fact, the very purpose of solving Eq. (III.1) is to determine the functional relationship between y and t.

Next, we will insert integral signs into Eq. (III.2) and perform the integration:

$$\int \frac{dy}{y} = \ln y = -\int k\,dt = -kt + C \tag{III.3}$$

When we take exponentials of quantities in Eq. (III.3), we obtain the following expression for y:

$$e^{\ln y} = y(t) = e^{-kt+C} = e^C e^{-kt} \tag{III.4}$$

where e^C is a constant. In Eq. (III.4) we have replaced y with $y(t)$ to emphasize that y depends on the independent variable t. When $t = 0$, Eq. (III.4) indicates that $y(0) = e^C e^{-k \times 0}$, which is simply e^C. Thus the constant e^C is the value taken on by the dependent variable y when t is 0. This latter relationship, $e^C = y(0)$, is referred to as a boundary condition (or initial condition because we are dealing with time). Incorporating this boundary condition into Eq. (III.4), we obtain the following solution to the differential equation represented by Eq. (III.1):

$$y(t) = y(0)e^{-kt} \tag{III.5}$$

Because of the factor e^{-kt}, Eq. (III.5) indicates that y decays exponentially with time for a first-order rate process. Moreover, $y(t)$ decreases to $1/e$ of its initial value $[y(0)]$ when t satisfies the following relation:

$$y(\tau) = \frac{1}{e}y(0) = y(0)e^{-k\tau} \tag{III.6}$$

where the value of time, τ, that satisfies Eq. (III.6) is known as the lifetime of the process whose decay or disappearance is being considered. Equation (III.6) indicates that $e^{-1} = e^{-k\tau}$, so the first-order rate constant is equal to the reciprocal of the lifetime τ (see Eq. 4.13). Therefore, the solution (Eq. III.5) of the partial differential equation (Eq. III.1) describing a first-order rate process becomes

$$y(t) = y(0)e^{-t/\tau} \tag{III.7}$$

APPENDIX IV

Gibbs Free Energy and Chemical Potential

THE CONCEPT of chemical potential is introduced in Chapter 2 and used throughout the rest of the book. In order not to overburden the text with mathematical details, certain points are stated without proof. Here we will justify the form of the pressure term in the chemical potential and also provide insight into how the expression for the Gibbs free energy arises.

A. Entropy and Equilibrium

A suitable point of departure is to reconsider the condition for equilibrium. The most general statement we can make concerning the attainment of equilibrium by a system is that it occurs when the entropy of the system plus its surroundings is at a maximum. Unfortunately, entropy has proved to be an elusive concept to master and a difficult quantity to measure. Moreover, reference to the surroundings—the "rest of the universe" in the somewhat grandiloquent language of physics—is a nuisance. Consequently, thermodynamicists sought a function that would help describe equilibrium but would depend only on readily measurable parameters of the system under consideration. As we will see, the Gibbs free energy is such a function for most applications in biology.

The concept of entropy (S) is really part of our day-to-day observations. We know that an isolated system will spontaneously change in certain ways—a system proceeds toward a state that is more random or less ordered than the initial one. For instance, neutral solutes will diffuse toward regions where they are less concentrated. In so doing, the system lowers its capacity for further spontaneous change. For all such processes ΔS is positive, whereas ΔS becomes zero and S achieves a maximum at equilibrium. Equilibrium means that no more spontaneous changes will take place; entropy is therefore an index for the capacity for spontaneous change. It would be more convenient in some ways if entropy had been originally defined with the opposite sign. In fact, some authors introduce the quantity *negentropy*, which equals $-S$ and reaches a minimum at equilibrium. In any case, we must ultimately use a precise mathematical definition for entropy, such as $dS = dQ/T$, where dQ refers to the heat gain or loss in some reversible reaction taking place at temperature T.

We can represent the total entropy of the universe, S_u, as the entropy of the system under consideration, S_s, plus the entropy of the rest of the universe, S_r. We can express this in symbols as follows:

$$S_u = S_s + S_r \quad \text{or} \quad dS_u = dS_s + dS_r \tag{IV.1}$$

An increase in S_u accompanies all real processes—this is the most succinct way of stating the second law of thermodynamics. S_u is maximum at equilibrium.

The heat absorbed by a system during some process is equal to the heat given up by the rest of the universe. Let us represent the infinitesimal heat exchange of the system by dQ_s. For an isothermal reaction or change, dQ_s is simply $-dQ_r$ because the heat must come from the rest of the universe. From the definition of entropy,[1] $dS = dQ/T$, we can obtain the following relationship:

$$dS_r = \frac{dQ_r}{T} = -\frac{dQ_s}{T} = -\frac{dU_s + P\,dV_s}{T} \tag{IV.2}$$

The last step in Eq. (IV.2) derives from the principle of the conservation of energy for the case when the only form of work involved is mechanical—a common assumption in stating the first law of thermodynamics. It is thus possible to express dQ_s as the sum of the change in internal energy (dU_s) plus a work term ($P\,dV_s$). The internal energy (U_s) is a function of the state of a system, i.e., its magnitude depends on the characteristics of the system but is independent of how the system got to that state. PV_s is also a well-defined variable. However, heat (Q_s) is not a function of the state of a system.

As we indicated previously, equilibrium occurs when the entropy of the universe is maximum. This means that dS_u then equals zero. By substituting Eq. (IV.2) into the differential form of Eq. (IV.1), we can express this equilibrium condition solely in terms of system parameters:

$$0 = dS_s + \left(-\frac{dU_s + P\,dV_s}{T} \right) \quad \text{or}$$

$$-T\,dS_s + dU_s + P\,dV_s = 0 \tag{IV.3}$$

Equation (IV.3) suggests that there is some function of the system that has an extremum at equilibrium. In other words, we might be able to find some expression determined by the parameters describing the system whose derivative is zero at equilibrium. If so, the abstract statement that the entropy of the universe is a maximum at equilibrium could then be replaced by a statement referring only to measurable attributes of the system—easily measurable ones, we hope.

In the 1870s Josiah Willard Gibbs—perhaps the most brilliant thermodynamicist to date—chose a simple set of terms that turned out to have the very properties for which we are searching. This function is now referred to as the *Gibbs free energy* and has the

---

[1] This definition really applies only to reversible reactions, which we can in principle use to approximate a given change; otherwise, dQ is not uniquely related to dS.

symbol G:

$$G = U + PV - TS \qquad \text{(IV.4a)}$$

which, upon differentiating, yields

$$dG = dU + P\,dV + V\,dP - T\,dS - S\,dT \qquad \text{(IV.4b)}$$

Equation (IV.4b) implies that, at constant temperature $(dT = 0)$ and constant pressure $(dP = 0)$, dG is

$$dG = dU + P\,dV - T\,dS \quad \text{at constant } T \text{ and } P \qquad \text{(IV.5)}$$

By comparing Eq. (IV.5) with the equilibrium condition expressed by Eq. (IV.3), we see that dG for a system equals zero at equilibrium at constant temperature and pressure. Moreover, G depends only on U, P, V, T, and S of the system. The extremum condition, $dG = 0$, actually occurs when G reaches a minimum at equilibrium. This useful attribute of the Gibbs free energy is strictly valid only when the overall system is at constant temperature and pressure, conditions that closely approximate those encountered in many biological situations. Thus our criterion for equilibrium shifts from a maximum of the entropy of the universe to a minimum in the Gibbs free energy of the system.

B. Gibbs Free Energy

We will now consider how the internal energy, U, changes when material enters or leaves a system. This will help us derive an expression for the Gibbs free energy that is quite useful for biological applications.

The internal energy of a system changes when substances enter or leave it. For convenience, we will consider a system of fixed volume and at the same temperature as the surroundings so that there are no heat exchanges. If dn_j moles of species j enter such a system, U increases by $\mu_j dn_j$, where μ_j is an intensive variable representing the free energy contribution to the system per mole of species j entering or leaving. Work is often expressed as the product of an intensive quantity (such as μ_j, P, T, E, and h) times an extensive one (dn_j, dV, dS, dQ, and dm, respectively), i.e., the amount of any kind of work depends on both some thermodynamic parameter characterizing the internal state of the system and the extent or amount of change for the system. In our current example the extensive variable describing the amount of change is dn_j, and μ_j represents the contribution to the internal energy of the system per mole of species j. When more than one species crosses the boundary of our system, which is at constant volume and the same temperature as the surroundings, the term $\sum_j \mu_j\,dn_j$ is added to dU, where dn_j is positive if the species enters the system and negative if it leaves. In the general case, when we consider all the ways that the internal energy of a system can change, we can represent dU as follows:

$$dU = dQ - P\,dV + \sum_j \mu_j\,dn_j \qquad \text{(IV.6)}$$

We now return to the development of a useful relation for the Gibbs free energy of a system. When dU as expressed by Eq. (IV.6) is substituted into dG as given by Eq. (IV.5), we obtain

$$dG = T\,dS - P\,dV + \sum_j \mu_j dn_j + P\,dV - T\,dS$$
$$= \sum_j \mu_j dn_j \tag{IV.7}$$

where dQ has been replaced by $T\,dS$. Equation (IV.7) indicates that the particular form chosen for the Gibbs free energy leads to a very simple expression for dG at constant T and P—namely, dG then depends only on μ_j and dn_j.

To obtain an expression for G, we must integrate Eq. (IV.7). To facilitate the integration we will define a new variable, α, such that $dn_j = n_j d\alpha$, where n_j is the total number of moles of species j present in the final system, i.e., n_j is a constant describing the final system. The subsequent integration from $\alpha = 0$ to $\alpha = 1$ corresponds to building up the system by a simultaneous addition of all the components in the same proportions that are present in the final system. (The intensive variable μ_j is also held constant for this integration pathway, i.e., the chemical potential of species j does not depend on the size of the system.) Using Eq. (IV.7) and this easy integration pathway, we obtain the following expression for the Gibbs free energy:

$$G = \int dG = \int \sum_j \mu_j dn_j = \int_0^1 \sum_j \mu_j n_j\,d\alpha$$
$$= \sum_j \mu_j n_j \int_0^1 d\alpha = \sum_j \mu_j n_j \tag{IV.8}$$

The well-known relation between G and the μ_j's in Eq. (IV.8) can also be obtained by a method that is more elegant mathematically but somewhat involved.

In Chapter 6 we presented without proof an expression for the Gibbs free energy (Eq. 6.1 is essentially Eq. IV.8) and also noted some of the properties of G. For instance, at constant temperature and pressure, the direction for a spontaneous change is toward a lower Gibbs free energy; minimum G is achieved at equilibrium. Hence, ΔG is negative for such spontaneous processes. Spontaneous processes can in principle be harnessed to do useful work, where the maximum amount of work possible at constant temperature and pressure is equal to the absolute value of ΔG (some of the energy is dissipated by inevitable inefficiencies such as frictional losses, so $-\Delta G$ represents the maximum work possible). To drive a reaction in the direction opposite to that in which it proceeds spontaneously requires a free energy input of at least ΔG.

C. Chemical Potential

We now examine the properties of the intensive variable μ_j. Equation (IV.8) $(G = \sum_j \mu_j n_j)$ suggests a very useful way of defining μ_j. In particular, if we keep μ_i and n_i constant, we obtain the following expression:

$$\mu_j = \left(\frac{\partial G}{\partial n_j}\right)_{\mu_i, n_i} = \left(\frac{\partial G}{\partial n_j}\right)_{T, P, E, h, n_i} \tag{IV.9}$$

where n_i and μ_i refer to all species other than species j. Because μ_i can depend on T, P, E (the electrical potential), h (the height in a gravitational field), and n_i, the act of keeping

μ_i constant during partial differentiation is the same as that of keeping T, P, E, h, and n_i constant, as indicated in Eq. (IV.9). Equation (IV.9) indicates that the chemical potential of species j is the partial molal Gibbs free energy of a system with respect to that species, and that it is obtained when T, P, E, h, and the amount of all other species are held constant. Thus, μ_j corresponds to the intensive contribution of species j to the extensive quantity G, the Gibbs free energy of the system.

In Chapters 2 and 3 we argued that μ_j depends on T, $a_j (a_j = \gamma_j c_j$; Eq. 2.5), P, E, and h in a solution, and that the partial pressure of species j, P_j, is also involved for the chemical potential in a vapor phase. We can summarize the two relations as follows:

$$\mu_j^{\text{liquid}} = \mu_j^* + RT \ln a_j + \bar{V}_j P + z_j FE + m_j g h \tag{IV.10a}$$

$$\mu_j^{\text{vapor}} = \mu_j^* + RT \ln \frac{P_j}{P_j^*} + m_j g h \tag{IV.10b}$$

The forms for the gravitational contribution $(m_j g h)$ and the electrical one $(z_j FE)$ can be easily understood. We showed in Chapter 3 that $RT \ln a_j$ is the correct form for the concentration term in μ_j. The reasons for the forms of the pressure terms in a liquid $(\bar{V}_j P)$ and in a gas $[RT \ln (P_j / P_j^*)]$ are not so obvious. Therefore, we will examine the pressure dependence of the chemical potential of species j in some detail.

D. Pressure Dependence of μ_j

To derive the pressure terms in the chemical potentials of solvents, solutes, and gases, we must rely on certain properties of partial derivatives as well as on commonly observed effects of pressure. To begin with, we will differentiate the chemical potential in Eq. (IV.9) with respect to P:

$$\left(\frac{\partial \mu_j}{\partial P} \right)_{T,E,h,n_i,n_j} = \left[\frac{\partial}{\partial P} \left(\frac{\partial G}{\partial n_j} \right)_{T,P,E,h,n_i} \right]_{T,E,h,n_i,n_j}$$
$$= \left[\frac{\partial}{\partial n_j} \left(\frac{\partial G}{\partial P} \right)_{T,E,h,n_i,n_j} \right]_{T,P,E,h,n_i} \tag{IV.11}$$

where we have reversed the order for partial differentiation with respect to P and n_j (this is permissible for functions such as G, which have well-defined and continuous first-order partial derivatives). The differential form of Eq. (IV.4) $(dG = dU + P\,dV + V\,dP - T\,dS - S\,dT)$ gives us a suitable form for dG. If we substitute dU given by Eq. (IV.6) $(dU = T\,dS - P\,dV + \sum_j \mu_j dn_j$, where dQ is replaced by $T\,dS$) into this expression for the derivative of the Gibbs free energy, we can express dG in the following useful form:

$$dG = V\,dP - S\,dT + \sum_j \mu_j \, dn_j \tag{IV.12}$$

Using Eq. (IV.12) we can readily determine the pressure dependence of the Gibbs free energy as needed in the last bracket of Eq. (IV.11)—namely, $(\partial G/\partial P)_{T,E,h,n_i,n_j} = V$ by Eq. (IV.12). Next, we have to consider the partial derivative of this V with respect to n_j (see the last equality of Eq. IV.11). Equation (2.6) indicates that $(\partial V/\partial n_j)_{T,P,E,h,n_i}$ is $\bar{V}_j$, the partial molal volume of species j. Substituting these partial derivatives into

Eq. (IV.11) leads to the following useful expression:

$$\left(\frac{\partial \mu_j}{\partial P}\right)_{T,E,h,n_i,n_j} = \bar{V}_j \tag{IV.13}$$

Equation (IV.13) is of pivotal importance in deriving the form of the pressure term in the chemical potentials of both liquid and vapor phases.

Let us first consider an integration of Eq. (IV.13) appropriate for a liquid. We will make use of the observation that the partial molal volume of a species in a solution does not depend on the pressure to any significant extent. For a solvent this means that the liquid generally is essentially incompressible. If we integrate Eq. (IV.13) with respect to P at constant T, E, h, n_i, and n_j with $\bar{V}_j$ independent of P, we obtain the following relations:

$$\int \frac{\partial \mu_j}{\partial P} dP = \int_{\mu_j^*}^{\mu_j^{\text{liquid}}} d\mu_j = \mu_j^{\text{liquid}} - \mu_j^*$$

$$= \int \bar{V}_j\, dP = \bar{V}_j \int dP = \bar{V}_j P + \text{``constant''} \tag{IV.14}$$

where the definite integral in the top line is taken from the chemical potential of species j in a standard state as the lower limit up to the general μ_j in a liquid as the upper limit. The integration of $\int \bar{V}_j\, dP$ leads to our pressure term $\bar{V}_j P$ plus a constant. Because the integration was performed while holding T, E, h, n_i, and n_j fixed, the "constant" can depend on all of these variables but not on P.

Equation (IV.14) indicates that the chemical potential of a liquid contains a pressure term of the form $\bar{V}_j P$. The other terms (μ_j^*, $RT \ln a_j$, $z_j FE$, and $m_j gh$; see Eq. IV.10a) do not depend on pressure, a condition used throughout this text. The experimental observation that gives us this very useful form for μ_j is that $\bar{V}_j$ is generally not influenced very much by pressure, e.g., a liquid is often essentially incompressible. If this should prove invalid under certain situations, $\bar{V}_j P$ would then not be a suitable term in the chemical potential of species j for expressing the pressure dependence in a solution.

Next we discuss the form of the pressure term in the chemical potential of a gas, where the assumption of incompressibility that we used for a liquid is not valid. Our point of departure is the perfect or ideal gas law:

$$P_j V = n_j RT \tag{IV.15}$$

where P_j is the partial pressure of species j and n_j is the number of moles of species j in volume V. Thus, we will assume that real gases behave like ideal gases, which appears to be justified for biological applications. Based on Eqs. (IV.15) and (2.6) [$\bar{V}_j = (\frac{\partial V}{\partial n_j})_{n_i,T,P,E,h}$], the partial molal volume $\bar{V}_j$ for gaseous species j is RT/P_j. We also note that the total pressure $P = \sum_j P_j$, where the summation is over all gases present (Dalton's law of partial pressures); hence, $dP = dP_j$ when n_i, T, and V are constant. When we integrate Eq. (IV.13), we thus find that the chemical potential of gaseous species j depends on the logarithm of its partial pressure:

$$\int \frac{\partial \mu_j}{\partial P} dP = \int_{\mu_j^*}^{\mu_j^{\text{vapor}}} d\mu_j = \mu_j^{\text{vapor}} - \mu_j^* = \int \bar{V}_j\, dP = \int \frac{RT}{P_j} dP_j$$

$$= RT \ln P_j + \text{``constant''} \tag{IV.16}$$

where the "constant" can depend on T, E, h, and n_i but not on n_j (or P_j). In particular, the "constant" $= -RT \ln P_j^* + m_j gh$, where P_j^* is the saturation partial pressure for species j at atmospheric pressure and some particular temperature. Hence, the chemical potential for species j in the vapor phase (μ_j^{vapor}) is

$$\mu_j^{\text{vapor}} = \mu_j^* + RT \ln P_j - RT \ln P_j^* + m_j gh \tag{IV.17}$$

$$= \mu_j^* + RT \ln \frac{P_j}{P_j^*} + m_j gh$$

which is essentially the same as Eqs. (2.20) and (IV.10b).

We have defined the standard state for gaseous species j, μ_j^*, as the chemical potential when the gas phase has a partial pressure for species j (P_j) equal to the saturation partial pressure (P_j^*), when we are at atmospheric pressure ($P = 0$) and the zero level for the gravitational term ($h = 0$), and for some specified temperature. Many physical chemistry texts ignore the gravitational term (we calculated that it has only a small effect for water vapor; see Chapter 2) and define the standard state for the condition when $P_j = 1$ atm and species j is the only species present ($P = P_j$). The chemical potential of such a standard state equals $\mu_j^* - RT \ln P_j^*$ in our symbols.

The partial pressure of some species in a vapor phase in equilibrium with a liquid depends slightly on the total pressure in the system—loosely speaking, when the pressure on the liquid is increased, more molecules are squeezed out of it into the vapor phase. The exact relationship between the pressures involved, which is known as the Gibbs equation, is as follows for water:

$$\frac{\partial (\ln P_{wv})}{\partial P} = \frac{\bar{V}_w}{RT} \quad \text{or} \quad \frac{\partial P_{wv}}{\partial P} = \frac{\bar{V}_w}{\bar{V}_{wv}} \tag{IV.18}$$

where the second equality follows from the derivative of a logarithm [$\partial \ln u / \partial x = (1/u)(\partial u / \partial x)$] and the ideal or perfect gas law [$P_{wv} V = n_{wv} RT$ (Eq. IV.15), so $\partial V / \partial n_{wv} = \bar{V}_{wv} = RT / P_{wv}$]. Because $\bar{V}_w$ is much less than $\bar{V}_{wv}$, the effect is quite small (e.g., at 20°C and atmospheric pressure, $\bar{V}_w = 1.8 \times 10^{-5}$ m^3 mol^{-1} and $\bar{V}_{wv} = 2.4 \times 10^{-2}$ m^3 mol^{-1}). From the first equality in Eq. (IV.18), we see that $\bar{V}_w \, dP = RT d \ln P_{wv}$. Hence, if the chemical potential of the liquid phase (μ_w) increases by $\bar{V}_w \, dP$ as an infinitesimal pressure is applied, then an equal increase, $RT \, d \ln P_{wv}$, occurs in μ_{wv} (see Eq. IV.10b for a definition of μ_{wv}), and hence we will still be in equilibrium ($\mu_w = \mu_{wv}$). This relation can be integrated to give $RT \ln P_{wv} = \bar{V}_w P + \text{constant}$, where the constant is $RT \ln P_{wv}^0$ and P_{wv}^0 is the partial pressure of water vapor at standard atmospheric pressure; hence, $RT \ln P_{wv}/P_{wv}^0 = \bar{V}_w P$, a relation used in Chapter 2 (see Section IVC). We note that effects of external pressure on P_{wv} can be of the same order of magnitude as deviations from the ideal gas law for water vapor, both of which are usually neglected in plant physiology.

E. Concentration Dependence of μ_j

We will complete our discussion of chemical potential by using Eq. (IV.17) to obtain the logarithmic term in concentration that is found for μ_j in a liquid phase. First, it should be pointed out that Eq. (IV.17) has no concentration term per se for the chemical potential of species j in a gas phase. However, the partial pressure of a species in a gas phase is really analogous to the concentration of a species in a liquid; e.g., $P_j V = n_j RT$ for gaseous species j (Eq. IV.15), and concentration means number/volume $= n_j / V = P_j / RT$.

Raoult's law states that at equilibrium the partial pressure of a particular gas above its volatile liquid is proportional to the mole fraction of that solvent in the liquid phase. A similar relation more appropriate for solutes is Henry's law, which states that P_j in the vapor phase is proportional to the N_j of that solute in the liquid phase. Although the proportionality coefficients in the two relations are different, they both indicate that P_j^{vapor} depends linearly on $N_j^{solution}$. For dilute solutions the concentration of species j, c_j, is proportional to its mole fraction, N_j (this is true for both solute and solvent). Thus, when P_j^{vapor} changes from one equilibrium condition to another, we expect a similar change in $c_j^{solution}$ because $\mu_j^{liquid} = \mu_j^{vapor}$ at equilibrium. In particular, Eq. (IV.17) indicates that μ_j^{vapor} depends on $RT \ln (P_j/P_j^*)$, and hence the chemical potential of a solvent or solute should contain a term of the form $RT \ln c_j$, as in fact it does (see Eqs. 2.4 and IV.10). As we discussed in Chapter 2, we should be concerned about the concentration that is thermodynamically active, $a_j (a_j = \gamma_j c_j$; Eq. 2.5), so the actual term in the chemical potential for a solute or solvent is $RT \ln a_j$, not $RT \ln c_j$. In Chapter 3, instead of the present argument based on Raoult's and Henry's laws, we used a comparison with Fick's first law to justify the $RT \ln a_j$ term. Moreover, the Boyle–Van't Hoff relation, which was derived assuming the $RT \ln a_j$ term, has been amply demonstrated experimentally. Consequently, the $RT \ln a_j$ term in the chemical potential for a solute or solvent can be justified or derived in a number of different ways, all of which depend on agreement with experimental observations.

INDEX

Page numbers in **boldface** refer to figures or structural formulas (entry also usually mentioned on that text page).

A^{mes}/A, 317–319, **318**, 334–336
Abscisic acid, 300, 341, 342
Abscission, leaf, 364, 380, 386
Absolute humidity, 311
Absorbance, 172
Absorptance, 262, 266
  leaf, 265–268, 283
Absorption band, 171, 187, 196
Absorption coefficient, 172–173, 175, 187,
      196, 266
  foliar, 362–364
  molar, 173
Absorption spectrum, 168, 171, 176
  carotenoids, **192**
  Chl a, **187**–188
  phycobilins, **195**
  phytochrome, **177**–178
Absorptivity, 262
Accessory pigments, 191, 192, 194
Acclimation, photosynthesis, 340, 343
Action spectrum, 168, 175–176
  O_2 evolution, **205**, **206**
  phytochrome, seed germination, **178**
Activation energy, 109–110
Active transport, 104, 111
  carriers, 120–122
  energy required, 115
  membrane resistance, 105–106
  Michaelis–Menten formalism, 119–121
  Na–K pump, 115, 119
  Nitella, 113–115
  phloem, 382–383
Activity, thermodynamic, 49, 86
Activity coefficient, 49
  ions, 86–87
  water, 52, 54, 56
Activity (concentration) term, chemical potential, 49,
      86–87, 235, 459–460

Adenine/Adenosine, 231, 236
Adenosine diphosphate (ADP), *see* ATP
Adenosine triphosphate (ATP), *see* ATP
Adhesion, 40–42, 73
Adiabatic lapse rate, 339
ADP, *see* ATP; Oxidative phosphorylation;
          Photophosphorylation
Advection, 356
Agave deserti, 337–339
Air boundary layer, *see* Boundary layer
Air density, 42, 442
Air gap, root–soil, 389–390
Air packets, *see* Eddy; Eddy diffusion coefficient
Albedo, 262
Alcohols, reflection coefficients, 129, 135–136
Algae, *see also Chara*; *Chlorella*; *Nitella*
  excitation transfer, 199–200
  pigments, 187, 192, 194–196, 206
Altitude, *see* Elevation
Anion, mobility, 94–95
Anode, 48
Antennae, pigments, 196, 202, 207, 209
Antiporter, 117, 242, 248–249, 300
Aphid stylet, 379–380
Apoplast, 7, 68, 374, 378, 385
Aquaporin, 117, 118
Arrhenius equation, 109
Arrhenius plot, **109**–110
Atmosphere, transmittance, 262
Atmospheric CO_2 level, 150, 184, 326, 331, 337, 345,
          394–395
Atomic orbitals, 155–156
ATP, 15, 18, 37, 47, 216, 230–**231**
  active transport, 119
  bonds, 231
  coupling to H^+, 243–244, 247–249
  energy currency, 147, 230, 234–235
  formation reaction, 231–233, 242

461

ATP (*continued*)
  proton ratio
    chloroplasts, 241, 243–244
    mitochondria, 248–249
  synthase, 243, 247, 248
  turnover, 223
ATP synthase (ATPase), 119, 243, 247, 248
Avogadro's number, 83, 107, 146, 440

Bacteria, 187, 207, 235
Bacteriochlorophyll, 187, 194, 201, 205, 207
Bandwidth, absorption band, 190
Basidiomycetes, transpiration, 314
Beer's law, 172–173, 362
Bicarbonate, 320–321, 322, 324
Biochemical reaction, 224, 232
Biosphere, energy flow, 143, 249–252
Blackbody, 148, 152, 250–251, 263
  absorptance/emittance, 263–264, 266
Bluff body, 271
  boundary layers, 273–274
  heat flux density, 275
Boiling point, water, 38
Boltzmann energy distribution, 106–**107**, 147, 189,
    250, 312
  energy levels, 167, 169, 170
Boltzmann factor, 106, 108, 188
Bouguer–Lambert–Beer law, 173
Boundary layer, **21**, 269–270
  air, 271–274, 276
    conductance/resistance, 296–298, 307–308, 338
    cylinder, 272–273
    flat plate, 272
    leaf, 272–273, 278, 282, 352
    mixing in, 272, 316
    sphere, 272–273
Bowen ratio, 358
Boyle–Van't Hoff relation, 60–62
  chloroplasts, 62–63
  irreversible thermodynamics, 134–136
Brownian movement, 8, 147
Bundle sheath cells, **328**–329, 379

C_3, 244
  enzymes, 326–327
  photosynthesis, 325, 329, **331**–332, 335, 379
  PPFD, 325, 335, 362
  WUE, 338, 343–345, 356
C_4, 244, 343–344
  anatomy, **328**–329
  enzymes, 329
  photosynthesis, 329, **331**–332, 335, 379
  PPFD, 325, 335
  WUE, 338, 343–345, 356
Cactus, 284–285, 286, 393
Calcium, 27, 102, 119, 300
Calculus, 9, 11, 24, 94, 451–452
Callose, 379

Calvin cycle, 327, 329
CAM, 60, 338–339, 343
Cambium, vascular, 59
Candela (candle), 148
Canopy, 283, 351–358
Capacitance
  membrane, electrical, 84–85, 105
  water storage, 391–393
    leaf, 391
    tree trunk, 391–393
Capacitor, **84**–85
Capillary rise, 41
  contact angle, 41–42
  height, 42–43
  xylem, 43
Carbohydrate, 184–185
Carbon dioxide, 8, 19
  atmospheric level, 150, 184, 326, 331, 337, 345,
    394–395
  cellular conductance, 334
  compensation point, 330–331
  concentration
    above canopy, 357
    plant community, 360, **361**, 364–366
    units, 315, 326
  conductance/resistance, 315–**317**, 321–324,
    329–334
  diffusion coefficient, 14, 316, 321, 439
  elevated, 345, 394–396
  fixation, 184–185; *see also* Photosynthesis
  flux density
    above canopy, 351–358
    leaf, 324–326, 329–334
    plant community, 359–366
  Michaelis constant, 325, 329
  partition coefficient, 320–321
  permeability coefficient, 297, 319, 322, 324
  photosynthesis, 184–185
    processing time, 203–204
  solubility, 320–321
  units, 315, 326
Carbonic anhydrase, 321, 329
α-Carotene, 191
β-Carotene, 191, 192, **193**
  absorption spectrum, **192**
Carotenes, 191, 192
Carotenoids, 191–193, 196
  absorption bands, **192**
  photooxidations, 194
Carrier, 19, 116–118, 119–121
Casparian strip, **6**, 7, 374
Cathode, 48
Cation, mobility, 94–95
Cavitation, 44, 376, 389
Cell sap, osmotic pressure, 55, 62–63
Cellular conductance, 322, 334
Cellulose, 1, 25–26, 31
  Young's modulus, 30

Cell growth, 76
Cell wall, 1, 25, **26**
  composition, 25–27
  diffusion across, 27–29, 321–322
  Donnan potential, 103–104
  elasticity, 30, 32, 65
  hydrostatic pressure in, 73–74, 368
  interstices, **26**, 27
    water relations, 43, 57, 73–74, 312, 378, 386
  mesophyll area, 317–319, **318**
  microfibrils, **26**
  middle lamella, 25, **26**
  permeability, 27–29, 321–322
  pits, 28
  plastic extension, 32
  Poiseuille flow, 378, 386
  primary, 25, **26**, 28
  resistance, CO_2, 317, 321–322
  secondary, 25, **26**, 28, 374
  stress–strain relations, 30–32
  water, 27, 43, 57, 73–74
  water potential, 72, 74
  yield threshold, 76
  Young's modulus, 30, 31
Celsius, 14
CF_o/CF_1, 243
Channel, membrane, 117–119, 248, 300
Chara, 3, 31
  growth, 76
  membranes, 89, 129
Charge compensation, 85
Charge number, 83
Chemical energy/electrical energy, 227–228
Chemical potential, 45–49, 82–83, 90, 93, 224–225,
    235, 455–460; *see also* Activity,
    thermodynamic; Concentration term;
    Electrical term; Gravitational term;
    Pressure term; Standard state
  protons, 240–241
  water, 57–58
  water vapor, 68–70
Chemical reaction, 224
  conventions, 225, 232–233
  equilibrium constant, 226
  Gibbs free energy, 224–225
Chemiosmotic hypothesis
  chloroplasts, 240–242
  mitochondria, 247–249
Chilling-sensitive plants, 110
Chl *a*, 185–190, **186**
  absorption spectrum, **187**–188
  excitation transfers, 190, 197, 199–200
  fluorescence, **187**–189, 190
  photosystems, 196, 206–207
  resonance transfer, 198–199
  vibrational sublevels, 188–**189**
Chl *b*, 186, 190, 196, 206
Chl *c*, 187

Chlorella, 3, 203, 205
Chloride, 99–103, 105, 113, 116, 210, **299**–300
Chlorobium chlorophyll, 187
Chlorophyll, 19, 185–190; *see also* Chl *a*; Chl *b*;
    Chl *c*; P_{680}; P_{700}
  absorption, 152, **187**–188, 265
  concentration, chloroplasts/leaves, 173, 196,
    203–204
  deexcitations, 159, 163–165
  electronic states, 159–161
  excitation frequency, 203–204
  fluorescence, 159, 166, **187**–190
  leaves, 173, 196, 203–204, 336
  polarized light, 189–190, 198
  redox properties, 238–239
  resonance transfer, 198–199
Chloroplast, 2, 17, **19**, 320; *see also* Electron transfer;
    Photophosphorylation; Photosynthesis
  bioenergetics, 236–244
  chlorophyll, 173, 203–204
  grana, 19
  ions, 63
  lamellar membranes, 19, 63
  light absorption, 173–174, 203–204
  membranes, 19
    resistance, CO_2, 317, 324
  number per cell, 4
  osmotic responses, 62–63, 135–136, 327–328
  photosynthesis, 325–326
  reflection coefficients, 135–136
  resistance, CO_2, 317, 319
  stroma, **19**, 317
  thylakoids, **19**, 208, **217**–218, 240–**241**, 243
  volume, 59, 62–63
Chromatophore, 19, 217, 242
Chromophore, 176, 193
Chromoplast, 191
Cis-trans isomerization, 158
Citric acid cycle, 18, 245, 249
Clay, 286, 366, 369, 376
Cloudlight, 260–261, 262
Clouds, 260, 269
CO_2, *see* Carbon dioxide
Cohesion, 40–42, 73
Cohesion theory, 384, 385
Colligative property, 53–54, 69
Colloid, 56, 57, 60, 61
Color, 145
Companion cell, **5**, 379, 383
Compensation point
  CO_2, 330–331
  light, 331, 356, 363–364
Competitive inhibition, **120**–122
Concentration (activity) term, chemical potential, 48,
    49, 86, 235, 459–460
Condensation (steaming), **358**
Conductance, 294
  boundary layer, 296–298, 307

Conductance (*continued*)
  cellular, CO_2, 322, 334
  CO_2, 316–317, 319, 334
  cuticle, 298, 303, 309
  Fick's first law, 296, 304–305
  intercellular air spaces, 298, 303
  leaf, 298, 307–308
  liquid phase, CO_2, 316–**317**, 334
  parallel, 308
  pressure dependence, 305–306, 339–340
  series, 308
  stomata, 298, 301–302, 344
  temperature dependence, 305–306, 339–340
  transpiration, 308–309
  water vapor, 307–309, 313
Conduction (heat), 258–259, 269
  across boundary layer, 274–275, 278
  cylinder, 275
  flat plate, 275
  soil, 287
  sphere, 275
Conductivity, 294
Conductivity coefficient, *see* Hydraulic conductivity
      coefficient; Water conductivity coefficient
Conjugate forces and fluxes, 124, 127
Conjugation, 174–175, 176
  carotenoids, 191, **193**
  chlorophyll, **186**
  phycobilins, **193**, 195
  phytochrome, 176–**177**
Constant field equation, 100
Contact angle, 41–42
  cell wall, 72–74
Continuity equation, 10–11
Convection (heat), 270, 274–275, 276
  forced vs free, 270, 277–278
  leaf, 278
Copper, 211, 214, 247
Cornfield, 354, 357, 360–361, 364–365
Cortex, root, **6**, 7, 374
Cotton, 30, 396
Coulomb's law, 44
Coupling factor, **241**, 243, **247**
Coupling sites, *see* Oxidative phosphorylation;
      Photophosphorylation
Crassulacean acid metabolism, 60, 338–339, 343
Cristae, mitochondrial, **18**
Crust, soil, 370
Cuticle, leaf, 3, 301
  conductance/resistance, 298, 300, 303, 307–308
Cuticular transpiration, 298, 303, 307, 309, 313, 314
Cutin, 3; *see also* Cuticle
Cuvette, leaf, 295
Cycles, biosphere, 253
Cyclic electron flow, **215**–216, **238**, 245–246
Cylinder
  boundary layer, 274

Fick's first law, 371
  heat flux density, 275
  volume flux density, 371
Cyt *a*, 212–213, 246
Cyt *b*, 211, 212–213, 238, 245
Cyt *c*, 18, 212–214, **213**, 245–247
Cyt *f*, 211, 213, 217, 238
Cytochrome b_6–*f* complex, 211, 212, 214, 216,
      217, 246
Cytochrome oxidase, 245, 247, 248
Cytochromes, 196, 212–213
  chloroplasts, 213–214, 238
  mitochondria, 245–247
Cytoplasm, 2; *see also* Cytosol
Cytoplasmic streaming, 15, 20, 21, 110
Cytosol, 2
  resistance, CO_2, 317, 323
  water potential components, 57, 61, 65

Daily changes, water, 385
  soil–plant–atmosphere, 393–394
  tree water storage, 391–392
Dalton, 14
Dalton's law, partial pressures, 305, 458
Damping depth, soil, 288
Darcy's law, 90, 368–369, 376
DCMU, 216
Debye–Hückel equation, 86–87
Deexcitations, 161–164, 451
Delayed fluorescence, 162, 165–166
Deuterium, 37
Dew formation, 279–282
Dew point (dew temperature), 280
Dielectric, 84
Dielectric constant, 44–45, 84
Differential equations, 451–452
Diffuse/direct radiation, 260–262
Diffusion, 9–10, 15, 48; *see also* Fick's first law;
      Fick's second law
  effective length, 303, 309
  energy barrier, 107–108, 110
  facilitated, 121–122
  into cell, 23–25
  time–distance relationship, 12–14, 23–25
Diffusional flux density, 127
Diffusion coefficient, 9, 15
  cell wall, 27–29
  eddy, 353–355, **359**–360
  gases, 14, 15, 439
  in membranes, 20
  pressure dependence, 15, 305–306, 339
  proteins, 14, 18
  small solutes, 14
  temperature dependence, 93, 305–306, 339
  viscosity, 15, 93
Diffusion potential, 93
  Donnan phase, 102–104

membrane, 98–101, 102
micropipettes, 95–96
solution, 94–96
Dilute solution, 54
Dimensionless numbers, 276–277
Dipole, electric, 154, 190, 198
Donnan phase, 102–**103**, 116, 366
Donnan potential, 102–104
Double bond, 158, 174–175
cis/trans, 158–159
Doublet, 155, 164
Drag (form), 270–271, 352, 360

Eddy, 270, 272, 353–354, 358, 359
Eddy diffusion coefficient
above canopy, 353–355
plant community, **359**–360
Effective length, diffusion, 303, 309
Einstein relation, 143
Elastic modulus
volumetric, 32, 68, 77–78
Young's, 30
Electrical circuit, 306–**307**, **317**, 329–**330**, 392
Electrical energy (work), 83, 228
conversion to chemical energy, 227–228
Electrical force, 44, 86, 144, 145
Electrical potential, 48, 50, 83
capacitor, 84–85
inside sphere, 85
membrane, 88, 95, 98, 100
Chara, 89
Nitella, 101–102
Nernst, 87–89, 104
Electrical resistance/resistivity, 90, 295–296
Electrical term, chemical potential, 48, 50, 82, 88, 235
Electric dipole, 154, 190, 198
Electrochemical potential, 83
Electrodes, **227**–229; *see also* Half-cells;
Micropipettes
Electrogenicity, 102, 104–106, 119
Electromagnetic spectrum, 145, 149
Electromagnetic wave (field), 144, 153–154
Electron
charge, 83, 439
energy/redox reactions, 227–228
orbitals, 155–157, 160
role in light absorption, 153–155
spin, 154–156
Electroneutrality, 83, 85, 99
Electron transfer (flow)
chloroplasts, 209–210, 217–218, 237–241
types, **215**–216, **238**
mitochondria, 245–247
Elevation
productivity, 337
water-use efficiency, 339–340
water vapor, 71, 443

Embolism, 44, 376, 389
Emerson enhancement effect, **205**–206, 216
Emissivity/emittance, 149, 251, 264
Endodermis, root, **6**–7, 374
Energy barrier, 108–109
Energy budget/balance
leaf, 258–259, 260
heat storage terms, 259
stem, 284–286
Energy currencies, 223, 230, 239; *see also* ATP;
NADPH
Energy flow, biosphere, 143, 249–252
Energy level diagram, **167**, 169
chlorophyll, **159**
Enhancement effect, Emerson, **205**–206, 216
Enthalpy, 185
Entropy, 8, 48, 185, 453–454
Environmental chamber, 150, 271
Environmental productivity index, 336–337
Enzyme, 108–110
EPI, 336–337, 395
Epidermis
leaf, **3**, 301
pubescence, 269, 309–310
root, **6**, 374
Equilibrium, 47, 89, 131, 224
chemical potential, 88, 94, 224
entropy, 453–455
Gibbs free energy, 47–48, 223–224, 250, 455–459
Equilibrium constant, 226, 233
Erythrocyte, 16, 25, 324
Evapotranspiration, 351, 357, 391, 396
Exchange diffusion, 122
Excitation transfer, 196–197, 199–200
efficiency, 199–200
resonance, 198–199
trapping, 201–202
Exodermis, root, 7
Extinction coefficient, 172–173

Facilitated diffusion, 121–122
FAD–FADH$_2$, **232**, 245
Farad, 84
Faraday's constant, 50, 83, 228, 439
Fatty acids, 16, 40
Feedback/feedforward, 340–341, 344
Ferredoxin, 211, 215–216, 217–218, 238–239
Ferredoxin–NADP$^+$ oxidoreductase, 211, 215, 216,
217–218
Ferrocytochrome *c*, 214
Fick's first law, 9–10, 22, 90, 93, 279, 353
conductance/resistance, 296, 301, 304–305
cylindrical geometry, 371
solvent, 75
Fick's second law, 11–13
Field capacity, 367
Fire ecology, 286

First-order process, 163, 164, 165, 451–452
Flavin adenine dinucleotide (FAD), 232, 246
Flavoprotein, 218, 246
Fluence, 149
Fluorescence, 161, 163, 170
    chlorophyll, **159**, 167, **187**–189; *see also* Chl *a*
    delayed, 162, 165–166
    depolarization, 190, 198
    lifetime, 162, 163, 170, 201
    resonance transfer, 198–199
Fluorescent lamp, 150
Flux/Flux density, 9, 48, 112, 294–295; *see also*
            Carbon dioxide; Photosynthesis;
            Transpiration; Water vapor
    cylindrical symmetry, 275, 371
    irreversible thermodynamics, 124–128
    solute, 90–92, 96–97
        irreversible thermodynamics, 136–137
    spherical symmetry, 275, 372
    velocity, 91–92
    volume, 74, 126–128
    water, 74–75
Flux ratio equation, *see* Ussing–Teorell equation
FMN, 245, 246, 249
Foliar absorption coefficient, 362–364
Food chain, 253
Force–flux relationship, 9, 90, 294–295
    irreversible thermodynamics, 124
Form drag, 270–271, 352, 360
Fourier's heat-transfer law, 274
Franck–Condon principle, 167, 169–171
Free energy, 45–48; *see also* Chemical potential;
            Gibbs free energy
Free radical, 212
Freezing, leaves, 269
Freezing point depression, 54, 55
Frequency, light, 145
Frost formation, 280, 281
Fucoxanthin, 192, **193**, 195, 199, 206
Fungus, 314, 374

Gas law, ideal (perfect), 69, 304–305, 326, 458
Gas-phase conductance, 344; *see also* Carbon
            dioxide; Conductance; Water vapor
Gas solubility, 320
Gate, membrane, 118–119
Germination, seed, 178–179
    action spectrum, **178**
    water uptake, 373
Gibbs equation, 70, 459
Gibbs free energy, 46–48, 50, 92, 223–224, 454–456
    ATP formation, 223–225
    biosphere, flow, 249–250, 252–253
    chemical reaction, 223–226
    electrical energy, 227–228
    equilibrium, 223–**224**
    equilibrium constant, 226

    glucose to ATP, 185, 249
    photosynthesis, 185
    reaction progress, 224
Girdling, 379
Global climate change, 345, 394–397
Global irradiation, 261–262, 265
Glucose, oxidation energy, 184, 249
Glycolate, 327
Glyoxysomes, 20
Goldman equation, 100–102
    Nitella, 101–102
Goldman–Hodgkin–Katz equation, 100
Grana, chloroplast, **19**
Grashof number, 277
Gravitational acceleration, 439, 448
Gravitational term, chemical potential, 48, 50, 90, 235
    water vapor, 71
Greenhouse gases, 395
Grotthuss–Draper law, 153, 175
Growth equations, cell, 76
Guard cell, **3**, 4, 119, 298–300, **299**; *see also* Stomata
Guttation, 285

H^+, *see* Proton
Hagen–Poiseuille law, 375; *see also* Poiseuille's law
Half-cells, 95, **227**–229; *see also* Redox potential
    hydrogen, 229
    $NADP^+$–NADPH, 235–236
    water–oxygen, 237–238
Half-time, 77, 161
Heat, latent/sensible, 259, 279–280
Heat capacity
    soil, 286–287
    volumetric, 284, 287, 439
Heat conduction/convection, *see* Conduction (heat);
            Convection (heat)
Heat convection coefficient, 276
Heat flux, density, 258–259
    conduction/convection, 269, 274–275
        cylinder, 275
        flat plate, 275
        soil, 287–288
        sphere, 275
    dew or frost formation, 279–281
    transpiration, 279–280
Heat of fusion, ice, 38–39, 440
Heat of sublimation, 281, 440
Heat of vaporization, water, 39, 280, 440
Heat storage, 284–285
    inflorescence (spadix), 284
    leaf, 259, 286
    stem, 284–286
    time constants, 285–286
Heme, **213**
Hemicellulose, 27
Henry's law, 460
Hill reaction, 208

Höfler diagram, **66**–67
Hormones, 271, 300, 341, 342
Humidity, *see* Absolute humidity; Relative
    humidity
Hund's rule, 156
Hydathodes, 385
Hydraulic conductivity coefficient, 128, 373
    soil, 369–371
Hydraulic conductivity/resistivity (phloem/xylem),
    387–389
Hydrogen bond, 38–40, **39**, 44, 45, 160, 384, 385
Hydrogen ion, **299**; *see also* pH; Proton
Hydrogen half-cell, 229
Hydrophylic/hydrophobic, 16–17, 40, 45
Hydrostatic pressure, 30–31, 52, 57, 59, 126, 235,
    367–368; *see also* Pressure term, chemical
    potential
    Höfler diagram, 66–67
    pressure–volume curve, 67–68
    yield threshold, 76

Ice
    frost, 279–281
    heat of fusion, 38–39
Ideal (perfect) gas law, 69, 304–305, 326, 458
Ideal solute/solution/solvent, 49, 50, 55
Illuminance, 149
Impermeability, reflection coefficients, 128–130
Incandescent lamp, 150, 152
Incipient plasmolysis, 64–65, 66–68
    irreversible thermodynamics, 132–133
Infrared, 145, 151, 171, 263
    absorption, leaf, 265–266
    emission, 153, 263–264
        leaf, 263–264
        sky, 263
Intercellular air spaces, **3**, 4, 8, 15, 72, 300, 312
    conductance/resistance, 298, 303–304, 307
Interface, *see* Cell wall; Matric pressure
Interstices, cell wall, 26, 27
    water, 43, 57, 73–74, 312, 378, 386
Invertase, 109
Ionic strength, 86, 233
Ions, chemical potential, 83, 87–88
IR, *see* Infrared
Iron, cytochrome, 213–214
Irradiance, 149
Irradiation, 149; *see also* Global irradiation; Solar
    irradiation; Sun
Irreversible thermodynamics, 123; *see also* Reflection
    coefficient
    Boyle–Van't Hoff law, 134–136
    conjugate forces and fluxes, 124, 127
    incipient plasmolysis, 132–134
    solute flux density, 136–137
    volume flux density, 126–128
Isomerization, 158–159

Isoprene, 191
Isoprenoid, 191, 212

Kelvin, 14
Kinematic viscosity, 277, 375, 442
Kinetic energy, 106, 123
Kirchhoff's electrical circuit laws, 330
Kirchhoff's radiation law, 266
Kranz anatomy, **328**
Krebs cycle, 18, 245, 249

Lambert–Beer law, 173
Lamellar membranes, chloroplast, 19, 63
Laminar sublayer, 272
Lamps, 150, 152
Lapse rate, 339
Latent heat, 259, 279–280, 357–358
Leaf, **3**–4; *see also* A^{mes}/A
    abscission, 364, 380, 386
    absorptance, 265–266, 267–269, 283
    air pressure, 312
    angle/orientation, 267, 283, 363
    area, plant community, 361–363
    boundary layers, 272–273, 278, 282, 352
        conductance/resistance, 296–298, 307–308
    cells, **3**–4
    conductance/resistance, 298, 307–309, 322,
        386–387
    cuticle, **3**
    energy budget, 258–259
    energy (heat) storage, 259, 286
    freezing, 269
    gas fluxes, measurement, 295
    IR absorption and emission, 263–266
    mesophyll, **3**–4
    net radiation, 266–269, 340
    PPFD and morphology, 318–319
    reflectance, 265, 268–269
    shaded, 283–284, 332
    shade/sun, 282, 318–319, 335, 379
    shape vs convection, 282
    silvery, 268–269
    solar tracking, 283
    temperature vs size, 268–269, 282, 313, 339–340
    water potential, 384–385
    water vapor, 312–313
    wilting, 283, 314, 341, 367, 385, 394
Leaf area index, 204, 362
Lecithin, **16**
Lenticels, 372
Lichen, dew, 281–282
Lifetime, 161, 164–165, 166, 452
Light, 144, 148–149
    absorption, 154, 161, 167; *see also* Absorption
        spectrum; Photosynthesis
        redox effects, 237–239
    color, 145

Light (*continued*)
   compensation point, 331–332, 363–364
   electric field, 144–145, 153
   energy, 146–147
   meter, 148–**149**
   speed (velocity), 145
   units, 149
   wave number, 180
   wave–particle duality, 143, 146
   wavelength, 144–146
Light-harvesting antennae, 196, 202, 207, 209, **217**
Light intensity, 149
Lignin, 26–27, 374
Lipid bilayer, membrane, 16–17
Liquid-phase conductance, CO_2, 316–**317**, 334
Loam, 286, 366
Lodging, crops, 271
Logarithm, 54, 450–451
Longwave radiation, 263; *see also* Infrared
Loop theorem, 330, 332
Lumen/lux, 148
Lutein, 192, **193**
   absorption spectrum, **192**

Manganese, 210
Magnesium, 63, 186, 233, 242
Magnetic force, 144, 145
Matric pressure/potential, 56–57, 60, 73, 368
Matrix, mitochondrial, **18**, 247–248
Mehler reaction, 216
Membrane, 16–19
   capacitance, 84–85, 105
   composition, 17
   permeability, 20, 22–23, 378
   phase changes, 110
   potential, 89, 114
      Chara, 89
      measurement, 95
      Nitella, 101–102
   resistance, 105–106
   Q_{10} for diffusion, 107–108
   tensile strength, 59
   water flow, 74–75, 378
Membrane channel, 117–119, 248, 299–300
Mesophyll cells, **2**, 3, 318
   area, 317–319, **318**, 334–336
Mesophyll resistance, CO_2, 317, 319, 323–324
Mesophyte, 72, 303
Methane, 395
Micelle, 45
Michaelis–Menten formalism, 119–121, 122
   constant, 120, 325, 329
   photosynthesis, 325
Microbodies, 2, 20
Microfibril, cell wall, 26, 30
Micropipettes, **95**–96, 118
Middle lamella, cell wall, 25, **26**

Midpoint redox potential, *see* Redox potential
Mie scattering, 261
Mitochondria, 2, **17**
   bioenergetics, 244–247
   cristae, **18**
   CO_2 flux, 329, 332
   electron transfer components, 245–247
   membranes, 17–19
   oxidative phosphorylation, 244, 247–249
   photorespiration, 327–328
   protein complexes, 246–248
Mobility, 91–92
   ions, 94–95
Modulus of elasticity, 32
Molal volume, 50
Molality, 51, 55–56
Molar absorption coefficient, 173
Molarity, 4, 55–56
Molecular orbitals, 156–158, **157**
   porphyrin ring, 213
Molecular weight, 14, 15
Mole fraction, 49, 311, 326
Münch hypothesis, 381, 383
Mycorrhizae, 374

NAD^+–NADH, 235, 245–246, 249
$NADP^+$–NADPH, 184, 211, 215
   energy currency, 223, 230, 239
   redox couple, 211, 230, 235–236
n electron, 157–158, 160
Nernst equation, 88–89, 112
Nernst–Planck equation, 92–93
Nernst potential, 87–90, 104
   K^+, *Chara*, 89–90
   Nitella, 101–102, 113–115
Net radiation, 266–269
Newton's law of cooling, 276
Newton's second law, 42
Newton's third law, 124
Nicotinamide adenine dinucleotide (NAD), 235, 244, 245
Nicotinamide adenine dinucleotide phosphate, *see* NADP
Nigericin, 242, 248
Nitella, 3, **114**
   active transport, 113–115
   cell wall elastic modulus, 30, 32
   growth, 76
   membranes, 101–102, 113–114, 129
Nobel prizes, 88, 118, 124, 147, 185, 202, 240, 243
Node, energy budget, 284–285
Noncyclic electron flow, **215**–216, **238**, 239, 244
Nonequilibrium, 123; *see also* Irreversible thermodynamics
Nonosmotic (nonwater) volume, 60, 62–63
Nonselectivity, reflection coefficients, 128–130
Nuclear vibration, 168–170

Nusselt number, 276
Nutrient, 336, 366, 386, 396

Ohm's law, 90, 105, 124, 387
Olive oil, 22, 51–52
Onsager coefficient, 124
Onsager reciprocity relation, 124, 125, 127
Optical density, 172
Optical path length, 173
Orbital, 155–157
Osmolal, 55
Osmometer, **53**
Osmosis, 119
Osmotic coefficient, 60, 135
Osmotic potential, 53, 58; *see also* Osmotic pressure
Osmotic pressure, 52–56, 126, 131
    Boyle–Van't Hoff relation, 60–62, 134–135
    cell sap, 55, 62–63
    cellular compartments, 61, 65, 102
    chloroplasts, 62–63
    Höfler diagram, 66–67
    incipient plasmolysis, 64–65, 133–134
    pressure–volume curves, 67–68
    Van't Hoff relation, 54–56
    wilting, 385
Oxidation–reduction potential, *see* Redox potential
Oxidative phosphorylation, 247–249
Oxidize, 197, 209, 229; *see also* Redox potential
Oxygen
    permeability coefficient, 23
    Rubisco, 326, 327
Oxygen evolution, photosynthesis, 37, 184, 208–209,
        217–218, 237–238, 243
    quantum yield, 205
Ozone, 151, 395

P_{680}, 190, 196, 209, 211, 237–238
P_{700}, 190, 196, 206, 209, 211, 214, 237–238, 239
Palisade mesophyll, **3**–4, 318–319
PAR, *see* PPFD
Parallel conductances/resistances, 307–308
Partial molal volume, 50
Partial pressures, Dalton's law, 305, 458
Partition coefficient, 22, 51–52, 129–130, 135
    CO_2, 320–321
    membrane, 22, 129–130
Patch-clamp technique, 118
Pauli exclusion principle, 155–156
Pectic acids/pectin, 26–27, 102
Perfect (ideal) gas law, 69, 304–305, 326, 458
Perforation plate, **5**, 375, 377, 388
Pericycle, root, 7, 374
Periderm, 372
Permeability coefficient, 22–23, 24, 99, 297
    cell wall, 27–28
    CO_2, 319, 322, 324
    irreversible thermodynamics, 130, 135, 136

lipid solubility, 22, 130, 135
Nitella, 101
plasma membrane, 23
reflection coefficient, 130, 135
series barriers, 33
solutes, 23, 136
water, 23, 75
Permittivity of vacuum, 44
Peroxisomes, 2, 20, 327–328
Petioles, 270, 379
pH
    ATP formation, 232, 241–242
    cellular, 119
    chemiosmotic hypothesis, 240–242
    CO_2 solubility, 322
    Donnan phase, 103
    equilibrium constant, 231–232
Phase transition, membrane, 110
Phenomenological coefficient, 124, 125
Phenomenological equation, 124
Pheophytin, 210, 237, 239
Phloem, 4–5, 7–8
    callose, 379
    cells, 4–5, 378–379
    contents, 380, 382–383
    flow, 380–381
    hydraulic conductivity, 388
    hydrostatic pressure, 381–383
    loading/sink/source/unloading, 383
    P protein, 379
    sieve cell/plate/tube, **5**–6, 378–379
    solute velocity, 380
    water potential, 381–383
Phosphoenolpyruvate carboxylase, 329, 338
Phospholipid, **16**, 17
Phosphorescence, **159**, 162, 166
Phosphorylation, *see* Oxidative phosphorylation;
        Photophosphorylation
Photochemistry
    laws, 153
    photosynthesis, 184, 197, 202
    reaction, 165, 197
    vision, 159
Photoelectric effect, 143, 147
Photoisomerization, 158–159, 176
Photometer, 148–**149**, 150
Photon, 146; *see also* Light
    absorption, redox potentials, 237–239
    meter, 148–**149**
    processing time, 204
Photophosphorylation, 208, 216–217, 218, 228, 242
    Gibbs free energy, 239–240
    lag, 242
    mechanisms, 240–242
Photorespiration, 326–329, 343
Photostationary state, 179
Photosynthate, 8, 379

Photosynthesis, 184–185; *see also* Carbon dioxide;
        Electron transfer; Photosphorylation
    acclimation, 340, 343
    action spectrum, 205, 206
    biochemistry, 184–185, 209
    canopy, 356
    chloroplasts, 324–326
    conductance/resistance, 315–317
    dark reactions, 184, 240
    efficiency, 185, 240, 252
    electrical circuit, 317
    electron flow, 208–218
    energy stored, 185, 259
    enhancement, 205–206, 216
    fluxes, 324–326, 332–334
    Michaelis–Menten, 324–326
    net, 329–334
    O_2 evolution, 37, 184, 205, 209, 217
    photochemistry, 184, 197, 202, 204
    PPFD, 325–326, 331–332, 335
    primary events, 143, 184, 190, 197
    processing time, 204, 325–326
    productivity, 183, 252
    quantum requirements, 184, 209–210, 252, 343
    rates, 333, 335–336
    red drop, 205
    temperature, 325, 328, 340, 343
    units, 315, 336
Photosynthetically active radiation (PAR), 148;
        see PPFD
Photosynthetic photon flux density, 148; *see* PPFD
Photosynthetic unit, 202, 203
    excitation processing, 203–204, 209, 326
Photosystem I, 196, 206–208, 209–211, 214, 238
Photosystem II, 196, 206–208, 209–211, 214,
        237–238
Phycobilins, 191, 193, 194–196
Phycobilisomes, 194, 207
Phycocyanin/phycoerythrin, 193, 194–196, 199, 200
    absorption spectra, 195
    enhancement of O_2 evolution, 206
Phytochrome, 176–179
    absorption spectrum, 177–178
    action spectrum, 178
Phytol, 186, 187, 189
π electron, 158, 160–161
Pits, cell wall, 28, 377
Planck's constant, 146, 154
Planck's radiation distribution formula, 152–153,
        250, 266
Plant community, gas fluxes, 359–366
Plant resistance, water flow, 387–388
Plasma membrane, 2, 16
    permeability, 16, 20, 25, 322, 324, 378
    resistance, CO_2, 317, 322–323
Plasmodesmata, 28–29, 374
Plasmolysis, 65, 132–133

Plastic extension, cell wall, 32
Plastocyanin, 211, 214, 217–218, 238
Plastoquinone, 212, 214, 217, 239
Plastoquinone A, 211, 212, 237, 238
Poiseuille's law, 90, 130, 375–378
    phloem, 381, 383
    soil pores, 376
    xylem, 376–378, 388
Polarized light, 189–190, 198
Polyhydroxy alcohols, 135–136
Porphyrin, 186, 190, 213
Porter, 117
Potassium, 89, 96, 99–103, 113–116, 299–300
PPF, *see* PPFD
PPFD, 148, 266
    A^{mes}/A, 318–319
    leaf angle/morphology, 318–319, 363
    photosynthetic rates, 325–326, 331–332, 335
    plant community, 361–364
    processing time, 203–204
PPFD index, 336
Pressure, 30; *see also* Hydrostatic pressure
    diffusion coefficients, 15, 305–306, 339
Pressure bomb (chamber), 58, 67–68
Pressure potential, 58
Pressure probe, 77
Pressure term, chemical potential, 48, 52, 57, 83, 225,
        457–459
    flow, 90, 235, 369, 371–372, 375
Pressure–volume curve, 67–68
Primary cell wall, 25, 26, 28
Primary productivity, 183
Proteins
    diffusion coefficients, 14, 18
    membrane, 18
Proton, 102
    chemiosmotic hypothesis, 241–242, 248–249
    fluxes, 102, 115, 117, 218, 299
    gradients, 218, 240–241, 248
    membrane potentials, 102, 106
    per ATP
        chloroplasts, 243–244
        mitochondria, 248–249
    pump, 106, 119
    transporter, 117, 243, 248
Protoplasmic streaming, 15, 20, 21, 110
Protoplast, 1, 25
Pseudocyclic electron flow, 215–216, 238
Pubescence, 269, 310
Pulvinus, 118, 283
Pyrophosphate, 236
Pyrrole, 176

Q_{10}, 107–108, 110
Quadratic equation, 450
Quantum, light, 146, 148
Quantum mechanics, 143–144, 153, 157, 175

Quantum meter, 148–149
Quantum yield (efficiency), 166, 201
    photosynthesis, 166, 201, 205, 244, 343
Quencher, 166
Quinol, 212
Quinone, 210, 212, 218, 237, 239

Radiance, 149
Radiant flux, 149
Radiation, *see also* Global irradiation; Solar
        irradiation; Sun
    balance, net, 266–269
    lakes, oceans, 151, 195
    terminology, 149, 262
Radiation distribution formula, Planck, 152–153,
        250, 266
Radiationless transition, **159**, 162, 163
Radioactivity, 111, 451
Radiometer, 148–**149**
Raoult's law, 69, 460
Rate constant, reaction, 108–109, 166
    first-order, 163, 166, 451
    second-order, 164
Rayleigh scattering, 261
Reaction
    first-order, 163, 166, 451–452
    photochemical, 165
Reaction center, 202, 204, 205, 208, 209
Reciprocity relation, Onsager, 124, 125, 127
Red drop, photosynthesis, **205**
Redox couple, **227**, 229; *see also* Redox potential
Redox potential, 228–230; *see also specific*
        molecules
    chloroplast components, 211, 237–239
    light absorption, 237–239
    midpoint, 230, 239
    mitochondrial components, 245–246
Reduce, 197, 209, 237; *see also* Redox potential
Reflectance, 262
    leaf, 265, 268–269
Reflection coefficient, 128, 132–133
    alcohols, 129, 135–136
    Boyle–Van't Hoff relation, 134–136
    chloroplast, 135–136
    impermeability, 128–130
    incipient plasmolysis, 132–134
    nonselectivity, 128–130
    partition/permeability coefficient, 128–129, 135
Reflectivity, 262
Relative humidity, 68, 70, 71–72, 311; *see also* Water
        vapor
    in leaf, 72, 73, 300, 312
Relative molecular mass, 14
Resistance, 90, 294–295
    above canopy, 355
    boundary layer, 296–298
    cell wall, 317, 321–322

    chloroplast, 317, 324
    CO_2, 316–**317**
    cuticle, 298, 303
    cytosol, 317, 323
    effective lengths, 309
    Fick's first law, 296, 301
    intercellular air spaces, 298, 303–304, 307
    leaf, 298, 307–308
    mesophyll, 317, 323, 324
    parallel, 307
    photosynthesis, 332–334
    plant, 387–388
    plant community, 360, **366**
    plasma membrane, 317, 322–323
    series, 75, 307
    soil, 387–390
    soil–plant–atmosphere, 387–390
    stomata, 298, 301–302, 307
    storage, 392–393
    transpiration, 308
    water flow, 386–391
    water vapor, 306–308
    xylem, 386–388
Resistivity, 90, 295–296
Resonance transfer, 198–**199**
Respiration, 244–245, 249
    leaf, 332
    plant, 365
    soil, 364
Reynolds number, 277
    xylem, 375, 377
Rhodopseudomonas spheroides, 202, 242
Riboflavin, 246
Ribulose-1,5-bisphosphate carboxylase/oxygenase,
        326–328, 343
Rice, 270, 396
Rieske-Fe S center, 214
Root, **6**–7, 289, 374
    area, 387–388
    casparian strip, **6**–7, 374
    cells, **6**–7, 374
    hairs, **6**, 7, 371
    hydraulic conductivity, 389–390
    pressure, 43, 385
    water traversal, 386–387
    water uptake, 371, 389–390
Root–soil air gap, 389–390
Rotational sub-sublevels, 171, 188
Roughness length, 353
Rubisco, 326–328, 343, 396

Saltbridge, 95, 227
Salt gland, 386
Sand, 286, 366, 369
Saturation vapor pressure, 68–69, 311, 443–444; *see*
        also Water vapor
Scaling, 395

Secondary cell wall, 25, **26**, 28
Seed coat, 373
Seed germination, 178–179
    action spectrum, **178**
    water uptake, 373
Semipermeable membrane, 53, 59
Sensible heat, 259, 280, 357–358
Series conductances/resistances, 33, 75, 307, 308
Series expansion, 54, 98, 450
Shaded leaf, 283
Shade leaf, 282, 318–319, 332, 335, 379
Shortwave irradiation, 263; *see also* Solar irradiation
SI system, 4–5, 9, 14, 30, 51, 146, 449
Sieve cell/plate/tube, **5**, 6, 378–379
Similarity principle, 296, 316, 355
Singlet, **155**
Sky, effective temperature, 263, 267–269
Skylight, **261**
Sodium, 99–103, 113–116
Sodium–potassium pump, 115, 116
Soil, 286, 366–367
    air/water interfaces, 57, **367**
    damping depth, 288
    dry/drying, 370, 385, 389–390, 394
    energy balance, 287–288
    field capacity, 367
    flux/flux density, 368–369
        cylinder, 371
        sphere, 372–373
    heat capacity, 286–287
    hydraulic conductivity, 369–370, 376, 389–390
    hydrostatic pressure, 57, 368–370
    matric pressure (potential), 57, 368
    nutrients, 336, 366
    osmotic pressure, 367
    pores, **367**–368, 376
    resistance, 389–390
    respiration, 364
    temperature/thermal properties, 286–289
    water potential, 367–368, 384–385
        wilting, 367, 385, 394
    water vapor flux, 288, 371
Soil–plant–atmosphere continuum, 384–387,
        393–394
Solar constant, 150–151, 249, 259, 262, 441
Solar irradiation, **150**–151, **260**–262; *see also* Sun
Solar tracking, leaf, 283
Solid angle, 148
Solute, 49, 82, *see also specific substances*
Solute flux density, 91–92, 96–98, 136–137
Solvent, 49, 51 173
    electronic energy levels, **160**–161, 190
Soret band, 187, 188, 194, 210
Specific activity, 111
Specific heat, 259
Sphere
    boundary layer, 274

    heat flux density, 275
    volume flux density, 372
Spin, electron, 154–155
Spin multiplicity, 155–156
Spongy mesophyll, **3**–4, 318–319
Standard state, chemical potential, 49, 51
Starch, 379
Stark–Einstein law, 153
Stationary state, 129, 131–132
Steady rate, 372
Steady state, 94, 97 131, 372
Steaming (condensation), **358**
Stefan–Boltzmann law, 250–251, 263–**264**
Stem, water flow, 387
Steradian, 148
Stokes shift, 188
Stomata, 3–4, 119, 298–300
    area, 300, 302
    conductance/resistance, 301–302, 307, 344
        end correction, 301–302
    control, 299–300, 307, 314, 334, 344
        photosynthesis, 334, 343–344
        water-use efficiency, 341–343
    opening, 118–119, 299–300, 338
        hormones, 300, 341, 343
        oscillation/closing, 299
    sunken, 309
Strain, 30
Stroma, chloroplast, **19**
Suberin, 7, 374
Sublevels, electronic state, 167–171
Sublimation, 281
Subsidiary cells, 299, 304
Succinate, 245, 246
Succinate oxidase, 109
Succulents, 284, 338
Sucrose, phloem, 6, 380, 382–383
Sun, 143
    altitude, 262
    chlorophyll excitation, 203–204
    energy radiated, 143, 251–252
    irradiation (light), **150**–151, **260**–262
    surface temperature, 152, 251
Sunlight, **150**–151, 260–262
Sun leaf, 282, 318–319, 335, 379
Surface free energy, 40
Surface tension, water, 40, 42, 442
    cell wall, 72–74
Surfactant, 40
Symplasm, 29, 68, 374, 378, 385
Symporter, **117**, 249, 300, 383
Système International, 4–5, 9, 14, 30, 51, 146, 449

Temperature, 258, 386
    absolute zero, 14
    change, time constant, 285–286
    dew point, 280

diffusion coefficients, 93
  kinetic energy, 106–108
  leaf angle/orientation, 283
  leaf size, 278, 282
  radiation, 250–251
  sky, 263, 268–269
  soil, 288–289
Temperature coefficient (Q_{10}), 107–108, 110
Temperature index, 336–337
Tensile strength, 43
  membranes, 59
  water, 43–44
Terpenoid, 191
Tetrapyrrole, 176, 186, 187, 194, 195, 213
Thermal conductivity coefficient, 274
  air, 275, 278, 440
  soil, 287
  water, 263, 440
Thermal radiation, 223–224; *see also* Infrared
Thermodynamics, 46, 123, 223–224
  first law, 258, 454
  second law, 454
Thylakoid, **19**, 208, **217**–218, 240–**241**, 243
Time, light absorption, 161; *see also* Lifetime
Time constant
  temperature change, 285–286
  volume change, 77–78
  water storage, 392–393
Time–distance relationship, diffusion, 12–14, 24
Tonoplast, 2, 59, 379
Tracheid, 5, 374, 377, 387
Translational energy, 171
Transmittance, 262, 265
Transpiration, 15, 295; *see also* Water; Water vapor
  canopy, 356, 357
  conductance/resistance network, 306–308
  cuticular, 298, 303, 307, 309, 313, 314
  fungi, 314
  heat flux density, 279
  leaf, 279, 313–315
  units, 315
Transpiration ratio, 357; *see also* Water-use efficiency
Transporter, 117, 119, 121, 248
Trap chl, 197, 201–202, 203, 248; *see also* P_{680}; P_{700}
Tree, 4, 271, 388
  capacitance, 391–393
  capillary rise, 43
  diffuse-porous/ring-porous, 377
Tricarboxylic acid (TCA) cycle, 18, 245, 249
Trichome, 269, 310
Trignometric function, 450
Triplet, **155**, 163
Tungsten lamp, 150, 152
Turbulence, 270, 272, 274
Turbulence intensity, 270, 272–273, 274
Turgor loss point, 64–65, 66–67
Turgor pressure, 30; *see also* Hydrostatic pressure

Ubiquinone, 245–246, 249
Ultraviolet, 145, 151, 395
Uncoupler, phosphorylation, 219, 242, 248, 419
Unstirred layers, 21, 28, 270; *see also* Boundary layer
Ussing–Teorell equation, 110, 113–114, 122
UV, 145, 151, 395

Vacuole, 2, 59–60
Valinomycin, 248
van den Honert relation, 386
van der Waals forces, 39, 44
Van't Hoff relation, 54–56
Variable, extensive/intensive, 226, 455
Vascular cambium, 4, 7, 391
Vascular tissue, *see* Phloem; Xylem
Velocity, mean, 127
  solute, 91–92
  water, 74
Vessel, xylem, **5**, 374–375
Vibrational sublevels, 167–169, 170
  chlorophyll, 188–189
Viscoelastic, cell wall, 32
Viscosity, 44, 276–277, 375–376, 381, 442
  diffusion coefficients, 15, 93
  kinematic, 277, 375, 442
  membrane, 18
  temperature, 93
  water, 44, 375
Vision, photochemistry, 159
Volume
  changes, time constant, 77–78
  chemical reactions, 225
  nonosmotic (nonwater), 60, 62–63
  partial molal, 50
    water, 52
Volume flux density, 74, 126–127, 131
  cylinder, 371
  sphere, 372
  water, 74
Volumetric elastic modulus, 32, 68, 77–78
Volumetric heat capacity, 284, 439
von Karman constant, 353

Water, 37; *see also* Water potential; Water vapor
  activity, 52, 54, 56
  activity coefficient, 56–57
  capacitance, 391–393
  conductivity coefficient, 74–75, 128
  dielectric constant, 37, 44–45
  flux density, 74–75, 125–127
    cell wall, 378, 386
    membrane, 378
    resistances and areas, 386–389
    soil–plant–atmosphere, 384–391
    xylem, 375–377
  heat of fusion, 38–39, 440
  heat of vaporization, 39, 440

Water (*continued*)
hydrogen bonding, 38–40, **39**, 44, 45
light absorption, 150, 151, 195
metastable, 44
osmotic pressure, 52–54, 82
oxidation, photosynthesis, 37, 184, 209, 217–218, 238
partial molal volume, 52
permeability coefficient, 23, 75
solvent, 37, 44–45
specific heat, 259
surface tension, 40, 42
tensile strength, 43–44
thermal capacity, 39, 441–442
velocity, 74, 276
viscosity, 44, 375, 442
volume flux density, 74, 386–387
Water conductivity coefficient, 74–75, 128
Water index, 336–337
Water–oxygen half-cell, 237
Water potential, 57–58, 64, 66–67
air, 70–71, 384
daily changes, 391–392, 394
leaf, 72, 385
measurement, 58
phloem, 381–383
soil, 367–369, 384–385
wilting, 367, 385, 394
water vapor, 68–70
xylem, 58, 382, 384–385
Water-use efficiency, 336–345, 395
C_3 vs C_4 plants, 337–339
elevation, 339–340
plant community, 356
stomatal control, 340–343
Water vapor, 68
altitude, 71, 340
chemical potential, 68–70, 151, 304–305
concentration/mole fraction
above canopy, 337
leaf, 310–313
plant community, 360–**361**
conductance/resistance
boundary layer, 297–298, 307
cuticle, 298, 303, 307
intercellular air spaces, 298, 303–304
leaf (total), 307–309
stomata, 298, 302, 307
diffusion coefficient, 14, 439
effective length, diffusion, 309
flux density
above canopy, 353
leaf, 310–315
plant community, 360–361
soil, 288, 370
light absorption, 150, 151
partial pressure, 69, 72

saturation, 68, 280, 311
numerical values, 311, 443–444
soil, 288, 370–371, 389–390
temperature, 443–444
water potential, 70
Wave number, 180, 261, 263
Wavelength, light, 144–146
Weed, 343, 344
Wettable Walls, 40, 72–74
Wheat, 270, 352, 396
Wien's displacement law, 152–153, 263
Wilting, leaf, 283, 314, 341, 367
soil water potential, 367, 385, 394
Wind, 70, 270–271
form drag, 270–271, 352–353, 368
speed, 270, 273, 278, 281
above canopy, 270, 352–353, 354
boundary layers, 272–274
eddy diffusion coefficient, 354
free/forced convection, 270, 277–278
plant community, 359
within plants, 70, 312
Wood, 4, 271, 391
Work, 46–47, 223
electrical, 83
WUE, *see* Water-use efficiency

Xanthophyll, 192; *see also* Fucoxanthin
Xerophyte, 72, 302, 303, 318, 385; *see also* Cactus; CAM
Xylem, 4–6, 8, 374–375
capillary rise, 43
cavitation, 44, 376, 389
cells, 4–6, **5**, 25, 374–375
diffuse-porous/ring-porous, 377
embolism, 44, 376, 389
hydraulic conductivity/resistivity, 387–389
perforation plate, **5**, 375
pressure gradients, 376–**377**
resistance, 387–389
Reynolds number, 375, 377
root, **6**–7
sap, 4–5, 44, 385
tracheids, **5**, 374, 377, 387
velocity, 376
vessel, **5**, 374–375
members, **5**
water flow, 376–378
water potential, 58, 382, 384–385
daily changes, 391–392, 394

Yield threshold, cell wall, 76
Young–Laplace equation, 368
Young's (Young and Dupré) equation, 41
Young's modulus, 30, 31

Zero plane displacement, 353